T0136021

# BABOQUIVARI
## MOUNTAIN PLANTS

# BABOQUIVARI MOUNTAIN PLANTS

## Identification, Ecology, and Ethnobotany

DANIEL F. AUSTIN

with linguistic consultant, David L. Shaul

The University of Arizona Press Tucson

The University of Arizona Press


www.uapress.arizona.edu

Library of Congress Cataloging-in-Publication Data
Austin, Daniel F.
Baboquivari mountain plants : identification, ecology, and ethnobotany / Daniel F. Austin ; with linguistic consultant, David L. Shaul.
p. cm.
Includes bibliographical references and index.
ISBN 978-0-8165-2837-0
1. Mountain plants—Arizona—Baboquivari Peak—Identification. 2. Mountain plants—Ecology—Arizona. 3. Indians of North America—Ethnobotany—Arizona. I. Title.
QK147.A92 2010
581.7′5309791—dc22
2009046444

Publication of this book is made possible in part by a grant from the Arizona Native Plant Society and the proceeds of a permanent endowment created with the assistance of a Challenge Grant from the National Endowment for the Humanities, a federal agency.

♻
Manufactured in the United States of America on acid-free, archival-quality paper containing a minimum of 30% post-consumer waste and processed chlorine free.

15 14 13 12 11 10 6 5 4 3 2 1

## DEDICATION

John and Mary W. Austin, my paternal grand-parents, who never saw the desert or the mountains;

Uncle Hugh Austin, who taught me all he could about the Tennessee River area and organisms he loved;

William Franklin Austin, my father, who taught me not only how to fish and hunt, but to observe and see;

Iva Lorene Austin, my mother, who fueled my interest in living things around me and encouraged me to learn—she even allowed me to go late to grade school one morning, delayed by watching a monarch butterfly hatch from its chrysalis;

and

Sandra K. Austin, my ever-patient and helpful wife, who taught me English, my "second language"

# CONTENTS

# PREFACE

The Baboquivari Mountains lie southwest of the city of Tucson along the eastern border of the Tohono O'odham Reservation (formerly the Papago Reservation). This compilation gives a glimpse of some common plants that grow in that range. I like to think of the information as an introduction to *Si'ihe e'es* (Elder Brother's garden). *Si'ihe* is an alternate name for *I'itoi*, one of the principal deities of the Tohono O'odham—who believe that he lives in a cave on the western side of Baboquivari Peak.

How a person with a history and reputation for working in wetlands and the wet tropics came to compose such a discussion deserves some explanation.

I grew up a "River Person," living near the banks of the confluence of the Ohio, Tennessee, and Clark's Rivers in the land of the Chickasaws, the Jackson Purchase of western Kentucky. After a beginning as a swamper, I went to graduate school at Washington University in St. Louis beside the Missouri and Mississippi Rivers. There I enjoyed a wider breadth of cultures and foods than I had known, and eventually I learned to cope with crowded humanity. Part of that learning was in St. Louis itself, where I met people from French, Germanic, and Italian heritage and more African-English dialects than in Kentucky. The next period was when I did research along the Amazon River in Belém, Brazil. Living and studying beside the world's largest river, I learned to speak Portuguese with a mixture of people whose ancestors included Africans, Europeans, and indigenous Tupían groups.

Once I obtained a Ph.D., my wife Sandra and I moved to another wet area, beside the Atlantic Ocean in southeastern Florida. The Everglades and Big Cypress Swamp were just west of the ridge where we lived, so I really was still on the banks of a river. This time it was the "River of Grass," as it was named by Marjorie Stoneman Douglas (1947). The cultures in Florida were different from those I had previously experienced, and they broadened not only my views, but also my linguistic contacts. The English-speakers were mostly from various suburbs in New York. Additionally, I associated with Hispanics from most parts of Latin America, Seminoles (both Creek-Muskogee and Mikasuki speakers), and Creole-French-speaking Haitians (Austin 2004).

After living 31 years in the Florida wetlands, Sandra and I moved to a settlement southwest of Tucson in 2001. This time, the closest things to rivers were a drained channel called the Santa Cruz River some 30 miles to the east of our home, and the Altar River over 40 miles to the south in Sonora. That move also led to learning a new set of cultures, languages, and foods. My Caribbean and southern Mexican Spanish is not the same as that spoken in southern Arizona but it is close enough to communicate. Tohono O'odham and the other nearby languages are quite alien to the Muskogean languages I sampled in Florida.

Near the middle of 2002, I learned from Steve McLaughlin (University of Arizona) that the Buenos Aires National Wildlife Refuge was looking for help with a vegetation survey. The Refuge started about 20 miles south of my house, and I contacted them. Dan Cohan, who was preparing GIS maps of the Refuge with a vegetation component, invited me for an interview and offered me a two-month job to make field surveys. The salary from the government was less than I had paid graduate students in Florida, but I had already sampled the Refuge—particularly the Baboquivari Mountains along its western border—so I took the opportunity. Instead of working in Florida temperatures of high 80s °F and middle 90s °F with the humidity at 90+%, I began the Arizona field studies in the monsoon season at thermometer readings of 100+ °F with humidity that ranged upward between 40% and 50%.

Doing the surveys for the Buenos Aires National Wildlife Refuge was challenging for a person who had studied the southeastern United States and Caribbean region for so many years. Surprisingly, a number of genera were shared between the northern Caribbean and southern Arizona. These two regions of diametrically opposed climates separated by over 3,000 miles even share a few species. Still, most of the plants were as new to me as were those in Florida to the students in my undergraduate field classes. Steve McLaughlin, Phillip Jenkins, Wendy Hodgson, and others accompanied me in the field several times, and Phil and Steve identified plant samples that I brought to the University of Arizona herbarium as documentation.

My favorite area on the Refuge was Brown Canyon, a region that traverses the mountains from about 3,400 feet to 5,942 feet at the head of the valley. Near the upper end is Baboquivari Peak itself at 7,734 feet. Refuge personnel told me that wildlife artist Ray Harm was one of the former landowners in the canyon. I had led Harm into a swamp in western Kentucky about 36 years earlier, so it felt like kismet to be in the basin. Although it was not a river, the intermittent stream also helped.

My enchantment with the Baboquivari Mountains and their plants continues to grow. I continue visiting the area during different seasons with an array of colleagues who, with their different experiences and emphases, show me more each trip.

To aid my own memory of the species and how they fit into the regional ecology, I began writing down some of the aspects of their lives. I added to this some of how the species fit into the cultures of the region, beginning with the Tohono O'odham and their Piman relatives in Arizona and Mexico. Finally, I added in names from a large sample of the people living between coastal California and middle Texas. The associations that emerged I find fascinating.

While this text focuses on plants in Brown Canyon, the species were chosen because they are characteristic of the Baboquivari Mountains. Moreover, the species are found in the surrounding mountain ranges—several as far west as the Ajos and most as far east as the Chiricahuas. Only a few plants have more restricted ranges in the area. This coverage makes this the first guide to the Baboquivari mountain plants.

The material presented here contains a small sample of what I have learned between August 2001 and now. There is much more to be discovered, as the Baboquivari mountain chain and their flora are among the least studied in this part of Arizona. Most of the times I go into the mountains, I find plants to add to their flora. One of the plants I found had never before been discovered in the United States, although it is frequent in Sonora. This "River Person" is still learning.

*Dan Austin*
Robles Junction
August 2008

# ACKNOWLEDGMENTS

I am grateful for information, help, and suggestions from Jonathan Amith, M. Kat Anderson, Rick and Wendy Brusca, Stephen L. Buchmann, Dan Cohan, Allen Dart, Mike and Jean Duever, William W. Dunmire, Eckart Eich, Richard S. Felger, Sally Gall, Roseann Hanson, Ray Harm, Mary Hunnicutt, Nancy P. Janson, Philip D. Jenkins, Allan Kirkpatrick, John E. Koontz, Scott Kozma, Jack B. Martin, Steven P. McLaughlin, Art A. Needleman, Margaret "Meg" Quinn, Nathan F. Sayre, Justin O. Schmidt, John C. Semple, Kathie Senter, David Shaul, Thomas E. Sheridan, David Siegel, Beryl Simpson, Richard Spellenberg, Bonnie Swarbrick, José Luis Tapia-Muñoz, Jan Timbrook, Fran and Gary E. Tuell, Caroline Wilson, and George Yatskievych.

I am especially indebted to people who read parts or all of the manuscript, offered constructive criticisms, and corrected errors. These individuals include Allan Dart, Mark Dimmitt, Richard S. Felger, Ray Harm, and Gary E. Tuell. Ildiko Palyka created the map in the introduction. David Shaul spent hours with me going over the sometimes curious transcriptions from old publications and transcribing them into proper modern phonetics. In addition, he examined every line of the "Linguistic Conventions and Pronunciation" segment to help me eliminate my errors.

Illustrations are used by permission from a variety of sources. Prominent among these are the original drawings by Lucretia B. Hamilton at the University of Arizona Herbarium. Some of these appeared in state publications like Parker (1958), and elsewhere. Other artists and sources include the following: Bobbi Angell (Henrickson and Johnston n.d.), Margaret B. Austin (Dayton 1937), Nancy Bartels (Hess and Henrickson 1987, courtesy of James Henrickson), Felicia Bond (courtesy of James Henrickson), Michael Chamberland (permission of artist), Hermione Dreja (Dayton 1937), F. Emil (Britton and Brown 1896–1898), Charles E. Faxon (Sargent 1890–1902), Vivian Frasier (Correll and Correll 1972), Meredith Gregg (Huisinga and Ayers 1999, courtesy of Tina Ayers), Lucretia B. Hamilton (originals @ ARIZ), Wendy C. Hodgson (permission of artist), Penelope N. Honychurch-Billingham (permission of artist), A. C. Hoyle (Dayton 1937), Regina O. Hughes (Reed 1971), Leta Hughey (Dayton 1937), Leta Hughey and Agnes Chase (Hitchcock and Chase 1950), Robert DeWitt Ivey (Ivey 2003), Edmund C. Jaeger (Jaeger 1941), Janice R. Janish (Abrams 1951), Matthew B. Johnson and N. L. Nicholson (permission of artist), Elnor L. Keplinger (Dayton 1937), Jerome D. Laudermilk (Benson 1957), Marjorie C. Leggitt (Carter 1988), Bruno Manara (Missouri Botanical Garden), William C. Martin (Martin and Hutchins 1986), Mark Mohlenbrock (Daniel 1984), Molly Ogorzaly (Simpson 1989), Kittie F. Parker (Parker 1958), E. M. Paulton (Mickel 1979), Raul Puente (Puente and Daniel 2001), M. Sharp (Arizona Rare Plant Field Guide 2001), Bonnie Swarbrick (permission of artist), Gene A. Walker (Dayton 1937), Edna May Whitehorn (Hitchcock and Chase 1950), Brian Wignall (Phillips and Comus 2000, courtesy of Brian Wignall), and Truman G. Yuncker (Yuncker 1921).

Individual artists and authors named above gave permission to use their work. In addition, permission to use drawings has been provided by *Arizona Rare Plant Field Guide* (2001); Arizona-Sonora Desert Museum Press, Tucson, AZ; Margaret Norem, editor of *Desert Plants*, 2120 E. Allen Road, Tucson, AZ; L. B. Hamilton drawings used by permission of the University of Arizona Herbarium; Henrickson, J. and M. C. Johnston. (n.d.), *A Flora of the Chihuahuan Desert Region*. Unpublished manuscript. Permission from Henrickson given to use illustrations; Houghton-Mifflin and Company; John Mickel, New York Botanical Garden, Bronx, NY; Amy McPherson, Missouri Botanical Garden Press, Steyermark et al. 1998. *Flora of the Venezuelan Guayana* (Vol. 4:391); The Board of Trustees of the Leland Stanford Jr. University, Stanford University Press gave permission to use illustrations from Jaeger, E. C. *Desert Wild Flowers*, © 1940, 1941; renewed 1967, 1969; Abrams and Ferris, *Illustrated Flora of the Pacific States*, Volumes 1–4. © 1960; Wiggins and Porter, *Flora of the Galapagos Islands*, © 1971; and Wiggins, Ira, *Flora of Baja California*, © 1980.

# BABOQUIVARI
## MOUNTAIN PLANTS

# Introduction

The Baboquivari Mountains lie southwest of Tucson in south-central Arizona, just west of the Altar Valley in Pima County. These are among the few mountains in southeastern Arizona that retain an indigenous name. "Baboquivari" is based on the Tohono O'odham name *waw giwulk* <*waw kiwulik, vav kivolik, vav giwulk*>, meaning "a neck between two heads" or "rock drawn in at the middle" (Granger 1960, Wright 1979). Saxton and Saxton (1973) translate it as "constricted peak"; Shaul (2007) gives *waw* (rock, cliff), and *giwulk* (constricted).

This sacred mountain is the center of the Tohono O'odham universe, in part because *I'itoi* (The Drinker, from *ih'e*, to drink) lives in a cave on its slopes. Their version is that the peak formerly was shaped like an hourglass. O'odham farmers nearby wanted more land and visited I'itoi to ask him to move the mountain. Their greed resulted in the top of the mountain breaking off, *Cewagĭ O'odham* (Cloud Man), who also lives in the mountains, becoming angry, and their never getting enough rain to cultivate the new lands (Wright 1979, Nabhan 1982).

The western side of the mountains is part of the Tohono O'odham Reservation, and much of the eastern slopes run down to the Buenos Aires National Wildlife Refuge. Sasabe is a border town on the southern end of this mountain chain that has both Arizonan and Sonoran villages. One Tohono O'odham name for Sasabe is *kui tatk*, mesquite root, although Lumholtz (1912) applied the name to a different locality farther west and north; he said they called modern Sasabe *ṣaṣawk* <*sháshovuk*>, echo.

The valley on the eastern side of the Baboquivari Mountains is called the Altar in the southern end and the Avra in the north. *El Altar* is the name of a place in northern Sonora where the first cleric to visit the area, Father Kino, stopped in 1692. This locality is where the Altar River has its source near the Mexican border to flow south, and the Altar Valley there extends north. The downstream northern end of the valley, called the Avra Valley (from *La Abra*, the opening), is actually the northern part of the Altar Valley. The Buenos Aires National Wildlife Refuge spans the Altar Valley and goes up the associated foothills of the Baboquivari, Cerro Colorado, Las Guijas, Pozo Verde, and San Luis Mountains (map 1).

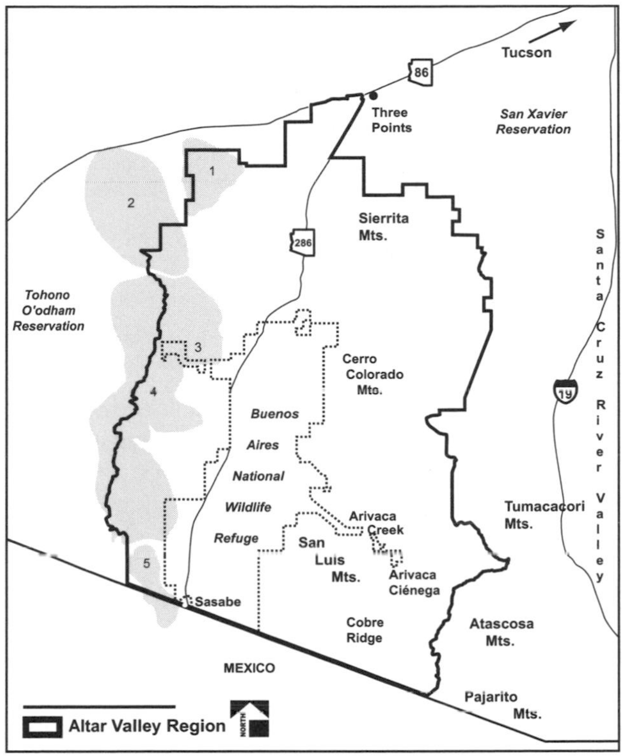


Location map showing the major features mentioned in the text. Shaded areas are mountains. 1, Coyote Mountains; 2, Quinlan Mountains; 3, Brown Canyon, the primary study area; 4, Baboquivari Mountains; 5, Pozo Verde Mountains.

The Baboquivari Mountains are pivotal in the ecology of the region, partly because the chain forms part of the eastern limit of the Sonoran Desert as defined by Shreve (1942), McLaughlin (1989, 1992b), Turner et al. (1995), and McLaughlin and Bowers (1999), among others.

The area's floristic regions have been analyzed recently in two manners. The study by Brown (1982) compares only dominant species and dubs the communities in the Altar Valley as the Semidesert Grasslands, part of the Chihuahuan Province. In that view, the Baboquivari Mountains and the Altar Valley to their east are part of a transition zone between the actual Chihuahuan and Sonoran Deserts, and that is where Shreve had considered them in 1917. Indeed, the Baboquivari Mountains form the boundary between the Chihuahuan and Sonoran provinces.

**TABLE 1. Comparison of Size and Species Richness of Local Floras**

| Size | # Species | Region | Source |
|---|---|---|---|
| 942,270 ha (land and sea) [2,328,399.9 acres] | 165 | Reserva de la Biósfera Alto Golfo de California y Delta del Río Colorado, Sonora, Mexico | Felger and Broyles 2007 |
| 348,033.7 ha [860,010 acres] | 414 | Cabeza Prieta National Wildlife Refuge, Pima County, Arizona | Felger and Broyles 2007 |
| 714,656 ha [1,765,953.4 acres] | 425 | Reserva de la Biósfera El Pinacate y Gran Desierto de Altar, Sonora, Mexico | Felger and Broyles 2007 |
| 200,860.5 ha [496,337 acres] | 439 | Sonoran Desert National Monument | Felger and Broyles 2007 |
| 814.6 ha [2,013 acres] (ca. 75% of canyon; Refuge only) | 482 | Brown Canyon, Buenos Aires National Wildlife Refuge, Pima County, Arizona | Hanson 1997, Austin unpubl. 2007 |
| 109,265,123,405 ha [2.7 million acres] | 497 | Barry M. Goldwater Air Force Range, Pima County, Arizona | Felger and Broyles 2007 |
| 1,499,999.9 ha [3,706,580.7 acres] | 589 | Gran Desierto in northwestern Sonora, Mexico | Felger 2000 |
| 40,000 ha [98,842.1 acres] | 610 | Tucson Mountains, Pima County, Arizona | Rondeau et al. 1996 |
| 930.8 ha [2,300 acres] | 624 | Sycamore Canyon, Pajaritos Mountains, Santa Cruz County, Arizona | Toolin et al. 1979 |
| 449,134.9 ha [1,109,836.4 acres] | 634 | Organ Pipe Cactus National Park, Pima County, Arizona | Felger and Broyles 2007 |
| ca. 99.99 ha [247.1 acres] | 695 | Cañón del Nacapule, Sonora, Mexico | Felger 1999 |
| 133,825.1 ha [330,689 acres] | 695 | Flora of southwestern Arizona (Organ Pipe Cactus National Monument, Cabeza Prieta National Wildlife Refuge, and the Tinajas Altas Region, Pima County) | Felger et al., in prep. |
| ~54,413 ha [~134,499 acres] (ca. 210 square mi.) | 785 | Baboquivari Mountains | Appendix 1 |
| 47,752.9 ha [118,000 acres] (including Brown Canyon) | 796 | Buenos Aires National Wildlife Refuge, Pima County, Arizona | McLaughlin 1992a, Hanson 1997, Austin unpubl. 2003 |
| 51,799.96 ha [128,000.5 acres] (200 square miles) | 986 | Rincon Mountains, Pima County, Arizona | Bowers and McLaughlin 1987 |
| 31,600 ha [122 square miles] | 994 | Huachuca Mountains, Cochise County, Arizona | Bowers and McLaughlin 1996 |

Higher in the mountains, some species are from the Petran Montane Coniferous Forest Province, others belong to the Madrean Montane Coniferous Forest Province, while the lower eastern slope and Altar Valley are within the Semidesert Grasslands (Brown 1982, Paredes-A. et al. 2000). The lower western slope of the mountains is in the Arizona Upland subdivision of Sonoran Desertscrub. Thus, the mountains have a mixture of plants from four floristic sources.

McLaughlin (1986, 1989, 1992b) and Bowers and McLaughlin (1987) take another approach. Instead of comparing dominant species, they statistically analyze a series of floras within the western United States. Their results indicate a classification of units with somewhat different compositions and with different borders. McLaughlin (1986, 1989, 1992b) and Bowers and McLaughlin (1987) identified the Apachian Floristic area east of the Baboquivari Mountains and the Sonoran Floristic area to their west. The Apachian Floristic area along the Baboquivari Mountains has the same delimitation as the Semidesert Grasslands of Brown (1982), although the limits are different in other parts of its range. The main difference is that the Apachian Floristic area is part of a larger unit called the Madrean Floristic area. Even in Bowers and McLaughlin's (1987) classification, the Baboquivari Mountains still prove to be an overlap of four floristic areas.

This mixture of four distinct floras results in a higher species diversity than each would otherwise support. For example, 785 species are documented in the Baboquivari Mountains (table 1). This figure is high given the richness of nearby floras. The much larger Organ Pipe Cactus National Park has 634 species; similarly, the large Buenos Aires National Wildlife Refuge in the valley supports 796 species (table 1). For the comparatively small area, the Baboquivari Mountains contain an elevated number of species over those two nearby regions. However, the number is near that predicted by the elevation gradient (Bowers and McLaughlin 1982, McLaughlin 1993).

Relatively little study has been made of the Baboquivari Mountains compared to nearby areas. There has been no previous inventory of plants growing in the Baboquivari Mountains, and Bowers (1981) found an unpublished partial list of only Thomas Canyon. Sporadic unpublished lists and studies have been made in various parts of the mountains since the early 1900s, but to date no one has compiled all the data.

# History and Human Influences

## Early History

There is disagreement about who occupied the Altar Valley and Arivaca Creek (map 1, see page 2) before the arrival of the Spanish. There are archaeological sites scattered throughout the Altar Valley from the Mexican border far into Arizona that belonged to the people called the Hohokam (from O'odham *huhugkam*, "those who have finished," Shaul and Hill 1998). These people preceded the O'odham in the region, and their habitation sites span the period from about CE 400 to 1450 (Reid and Whittlesey 1997).

Archaeological sites from 1450 onward are usually ascribed to the O'odham (Fish and Fish 1992). There are two well-known O'odham groups in the area, the Akimel O'odham (river people, also called the "Pima") and the Tohono O'odham (desert people, formerly called the "Papago").

When the Europeans arrived in the southwestern United States, there were three groups of O'odham living in the region, the Hiá Ceḍ O'odham (sand people) near the Ajo Mountains, the Tohono O'odham in the lands between there and the Baboquivari Mountains, and the Akimel O'odham, who farmed along the Santa Cruz and Gila Rivers (Crosswhite 1981). People along the San Pedro River have been called part of the Akimel O'odham by some and separated as the Sobaipuri by others, but these terms are not mutually exclusive.

The land in and west of the Baboquivari Mountains was part of that claimed and occupied by the Tohono O'odham before and at the time of arrival of the Spanish in 1692 when Padre Eusebio Francisco Kino visited the village in Pimería Alta that he called San Xavier (Pfefferkorn 1794–1795, Rea 1997, Hodgson 2001, Dunmire 2004). The Tohono O'odham were "Two Village" people (Fontana 1983a). Tohono O'odham lived in scattered permanent villages of a few hundred individuals in the lowlands, but they also had seasonal villages in the mountain canyons and foothills where they went during the winter and spring. Their mountain villages had permanent springs of water called "wells" (*wawhia <wahía, vaya>*, e.g., Ka:w Wawhia <Ke Vaya, Kahw Kowhia>, Badger Well). During the summer, these O'odham lived in their lowland villages with only temporary wells (Oidag, O'oidag; e.g., Ge Oidag, Big Field), and they relied on reservoirs called *charcos* for drinking water. These people lived by farming, gathering wild resources, and hunting, and they were skilled at all of these occupations. The Tohono O'odham were especially good at growing short-season plants using *akĭ ciñ <ak-ciñ>* (arroyo mouth) agriculture while in the lowlands (Castetter and Bell 1942, Nabhan 1983b).

It is certain that the Sobaipuri occupied the San Pedro River valley, but sources disagree about who lived farther west. Those people were apparently first called "*los sobas y jípuris*" by the Spanish, a phrase that may have morphed into "Sobaipuri" (Bolton 1936). Some insist that the Sobaipuri lived only on the San Pedro River (Farish 1915–1918, Center for Desert Archaeology 2005), while others maintain that Sobaipuri lived there and on the Santa Cruz (Di Peso 1953, Ezell 1983, Sheridan 1995). Still others argue that the Sobaipuri extended their territory to the Baboquivari Mountains and dominated the Arivaca Creek and Altar Valley region. Both Santamaría (1959) and Sobarzo (1991) pointed out that the Soba lived from the Altar Valley of Sonora to the Gulf of California. Just where the Jípuris lived is not clear. The Sobaipuri probably lived from the San Pedro River to the gulf. Little seems to be recorded about these O'odham, probably because they were decimated by disease and Apache raids so quickly, and because of their intermarriage with the Apache, Akimel O'odham, and Tohono O'odham (Bourke 1890, Castetter and Underhill 1935, Russell 1908, Ezell 1983). However, their riverine occupation of the San Pedro and Santa Cruz regions is more like the Akimel O'odham than the desert-farming Tohono O'odham, and both Fontana (1983a,b) and Ezell (1983) were convinced that the Sobaipuri were more like the former (Pima) than the latter (Papago).

The O'odham were the sole occupants of the region except for occasional raids by Apaches that probably began in the 1400s or 1500s when they migrated into the region from the north (Ezell 1983, Sheridan 1995, Sheridan and Parezo 1996, Wilcox 1981). However, Castetter and Underhill (1935) suggest that the Apaches did not become a serious problem to the O'odham until after the late 1600s.

On the other hand, the Apache depredation of the Sobaipuri along the Santa Cruz and San Pedro Rivers was worse earlier than in places farther west. Along those rivers, the Sobaipuri were strongly influenced by the missions, and the Apaches took advantage of the added Spanish wealth, especially in cattle and horses (Ezell 1983). The San Pedro River region was vacated by the Sobaipuri when they were forcefully removed closer to the missions farther west in 1762 by the Spanish military (Sheridan 2006). Apache raids continued across southern Arizona throughout Spanish, Mexican, and subsequent Anglo times until attacks were finally terminated with subjugation of the various Athabascan groups in the late 1800s.

With the arrival of the Spanish missionaries, various pressures were placed on the O'odham, and their lands began to be usurped by the outsiders. Padre Kino established a cattle ranch in the Santa Cruz River valley in 1697 (Sheridan 2006), and on April 28, 1700, he began building the foundation for the church that would later be rebuilt and is now so famous, Mission San Xavier del Bac (Granger 1960).

There were still Sobaipuri O'odham living on the Santa Cruz River when Padre Kino established Mission San Xavier del Bac. By the early 1800s, European diseases and Apache pressure had pushed these people into a single village. When the Americans arrived in the 1850s, this native group was living in the Sobaipuri village of Wa:k <Wahk> (Bac of Kino; the modern community of San Xavier del Bac). The Sobaipuri were replaced about 1865 after the Tohono O'odham began moving into the Santa Cruz Valley from farther west. The remaining Sobaipuri intermarried with the Tohono O'odham (Castetter and Underhill 1935).

Arivaca Creek, the Ciénega, and Altar Valley are between the Santa Cruz River and the Baboquivari Mountains (map 1). The creek and valley were occupied by the O'odham, whom historic records from about 1751 call "Piatos." (Ezell [1983] thought the word "Piatos" to be a contraction of "*Pimas Altos*"; Sobarzo [1991] believed that it derived from the Latin *pius, paitus*, pious, or *expiatus, expiare*, to make amends, atone.) These people revolted against the overbearing priests at the missions in 1751. They were led by Luis Oacpicapigua of Sáric, who had been appointed by Ortiz Parilla of the Spanish military as "governor and captain general" of the northern O'odham (Sheridan 2006). The Jesuits resented Oacpicapigua's elevation in status, believing that it was their right to appoint O'odham officials.

After Oacpicapigua's death, his sons and other followers allied with O'odham from the Gila River area, and these "apostate Pimas" continued harassing the Spanish for several years (Sheridan 2006). Eventually a tenuous peace was established, and the rebellion died along with many O'odham because of recurrent epidemics. By the time that Eva Antonia Wilbur-Cruce (1987) was growing up in the early 1900s, the people in the Arivaca region considered themselves "Papago" (Tohono O'odham), and she saw them leave the creek and valley for the reservation that was established in 1916 west of the Baboquivari Mountains.

Padre Kino and the other clerics did their best to congregate the people near Mission San Xavier del Bac in the Santa Cruz Valley. Still, there were numerous individuals and villages in the Altar Valley into the early 1900s (Wilbur-Cruce 1987). These people had a large village on the south side of Las Guijas Mountains, along with smaller encampments elsewhere in the vicinity of the present town of Arivaca. These villages thrived in spite of Las Guijas Mine, which was active there in the 1860s and 1870s (Granger 1960).

The O'odham particularly gravitated to Arivaca Creek and the Ciénega because of the year-round abundance of water. The fish in the creek were also an important factor in feeding these wandering groups. Wilbur-Cruce (1987) provided a first-hand view of the people and their attitudes when the Tohono O'odham were forced from this area onto the reservation by the American government. By 1874, land had been set aside for the Tohono O'odham at San Xavier, and in 1916, a second and larger reservation area was established west of the Baboquivari Mountains with its headquarters at Sells (Granger 1960). That reservation, originally the Papago Reservation and now the Tohono O'odham Reservation, was established by President Woodrow Wilson. It has undergone several adjustments, and it now contains 2,774,370 acres, thus giving the Tohono O'odham the second-largest reservation in the United States (that of the Navajos is the largest).

Land ownership in the Baboquivari Mountains has followed a mosaic pattern for more than 100 years. Currently, the western side of the mountains is part of the Tohono O'odham Reservation, and biological inventories by outsiders have been prohibited there since the 1980s. That restriction was begun when the Tohono O'odham closed the Baboquivari District to campers and hikers in May 1976 (McCool 1981). Although the Tohono O'odham Nation is conducting its own inventory, it is reluctant to share that information with outsiders.

The eastern side of the Baboquivari Mountains has an even more checkered history and a spotty documentation of organisms living there. As on the western side, there are numerous canyons running from the lowlands in the Altar Valley up into the highlands. These canyons have served as focal points for ranches and access.

The town of Arivaca, with its *ciénega* (marsh) and creek lying between the Las Guijas and San Luis Mountains on the east side of the Altar Valley, played an important role in the history of the region. This area was claimed by the O'odham until Augustin Ortiz purchased the site of Arivaca at public auction in 1812 (Sheridan 1995). The region soon became an important mining and ranching center, and the O'odham became minor elements on the land.

## Mining History

Pfefferkorn (1794–1795) recorded that the Spanish mines in Sonora and what is now Arizona were *placeres*. Those that yielded gold were *placeres de oro*, and those that produced silver were *placeres de plata*. Typically, these mines were in streams and rivers where the miners took a pan of water with sediments and washed out the ores. There were other mines, too, and Pfefferkorn mentions a lead mine in Sonora, but those that received the most attention provided the precious metals.

Hamilton (1881) thought that the Spanish were mining in what is now Pima County by the late seventeenth century. By the middle 1800s, placer mines had been replaced by shaft mines in the Baboquivari, Cerro Colorado, Las Guijas, and Santa Rita Mountains. At the time Hamilton (1881) was writing, the Black Hawk, Oro Fino, and Silver Chief mines were still active in the Baboquivari Mountains. Subsequently, Ralph and Chau (1993–2005) found records that located more than 20 named mines in that range. Early mining focused on gold and silver, while later 22 additional minerals were removed, including tungsten in the 1950s and early 1960s.

There were already 29 silver mines in the Cerro Colorado Mountains north of Arivaca by 1854, when speculators Charles Debrille Poston and Herman Ehrenberg visited Arizona after the Gadsden Purchase to learn about investment possibilities (Granger 1960). Poston bought the 17,000-acre Ortiz hacienda in 1856, and he and Samuel Heintzelman established it as part of the Sonora Mining and Exploration Company, formed that same year (Sheridan 1995). The most famous mine in the Cerro Colorado Mountains was the Heintzelman Mine (also known as the Cerro Colorado Mine and the Silver Queen Mine), the superintendent of which was John Poston, brother of Charles. The company's mining operation in the Cerro Colorado Mountains and in Arivaca was touted as "the most important Mining Company on this Continent" (Sheridan 1995).

Because of the financial "Panic of 1857," Apache raids, and the absence of a railroad leading from the processing plant at Arivaca, the Cerro Colorado mines were closed in 1861 (Granger 1960, Sheridan 1995). However, by 1870, soon after the Civil War, there were 85 people working at the Cerro Colorado mines, and in 1880, improvements were made at Arivaca. The mines remained active until about 1937 (Ralph and Chau 1993–2005).

Mining and ranching both consumed large amounts of wood, with mining using more. Not only was wood the fuel everyone used, but on the ranches, it was important for fences and corrals (Bahre 1991, Sayre 2002). Juniper (*Juniperus*), oak (*Quercus*), and Mesquite (*Prosopis*) were the preferred kinds, but madroño (*Arctostaphylos*), sumac (*Rhus*), mountain mahogany (*Cercocarpus*), silk-tassel (*Garrya*), and deer-brush (*Ceanothus*) were used as the supplies of other species dwindled (Bahre 1991).

Mining during the period was made possible by an abundance of trees that were used for shoring up the shafts to keep the roofs from collapsing, in building homes, and as fuel for the miners. Vast distances around mines were stripped of their timber (Lanner 1981a; Bahre 1984, 1991; Bahre and Hutchinson 1985; Turner et al. 2003). A local example is the "Tombstone Woodshed" that once covered about 31% of Cochise County and parts of both Pima and Santa Cruz Counties (Bahre 1991). This "woodshed" covered some 2,000 square miles and was temporarily deforested during mining.

Bahre (1991), who studied the area from the Santa Cruz River east, found that the first mountains "extensively" logged were the Santa Ritas, and that the forests were cut from the 1850s through the 1880s for mine timbers and lumber for Tubac, Tucson, and "nearby settlements." Bahre found that woodcutting camps were turning out 7,000–10,000 feet of timber per day in the Huachucas in 1880.

Although no historical documentation has been found, mining interests in the Cerro Colorado, Las Guijas, and elsewhere on the eastern side of the Altar Valley probably sent crews into the Baboquivari Mountains to cut timber (Tuell, personal communication, Dec. 2004). We do know that the Heintzelman Mine in the Cerro Colorado Mountains used charcoal from "local wood" from the 1870s to 1880 (Bahre 1991). Earlier mines there also used local wood.

Remains of roadbeds found in both Brown and Jaguar Canyons may date from that period. Wood used by the ranches in the valleys and the mines on the slopes of regional mountains surely contributed to the unusual and spotty distribution of some tree species today. Bahre (1991) found that elsewhere in southeastern Arizona, removal of wood resulted in structural changes to forest and woodland, such as differential frequencies of species, but not areal extent.

There was a woodcutters' camp in Thomas Canyon from the 1860s to at least the 1890s called Baboquivari Camp (Roskruge 1888, Tuell, personal communication, 2005). The Roskruge field notes of 1886 show a road leading from the Secundino (spelled "Segundino" on the 1893 official Pima County map) ranch toward the west-northwest. Sayre (2002:38–39) says that this road is labeled "Wood Road," and he concludes that this track led to the Baboquivari Mountains where wood was being cut. From the direction of the road and the presence of the Baboquivari Camp, that canyon was probably the wood-cutting headquarters for the mountains at the time.

There are few stumps in either Brown or Jaguar Canyons, and this lack has led some to speculate that there was no logging. However, as Bahre (1991) and others point out, wood was so scarce by the 1890s that stumps and roots were dug up and burned. Moreover, woodcutters had to go 20–30 miles to get fuel. There is some pollarding (stump sprouting) in the Baboquivari Mountains, thus suggesting that oaks and perhaps other species were harvested above ground, yet remained alive. Pollarding is a condition that Bahre (1991) frequently found in nearby mountain ranges.

## Ranching and Recent History

Although the Rancho de la Osa of 1812 was one of the first ranches in the area (reputedly occupying an adobe trading post built by the Franciscan Father Kino, the first European-built structure in Arizona), the boom in ranching in the Altar Valley and the Baboquivari Mountains did not begin until after the Civil War (1861–1865) and the Homestead Act of 1863 (USFWS 2003, Sayre 2002). Soon after those dates, settlers began moving into the region, and ranches spread along the eastern face of the Baboquivari Mountains and adjacent grasslands in the Altar Valley.

Early major ranches in the valley included the Aros, Buenos Aires, La Osa, Palo Alto, Pozo Nuevo, Redondo, and Segundo (Campa 1970, USFWS 2003). Large ranches in the 1860s and 1870s were owned by Pedro Aguirre, Esteban Aros, Jesus Robles, and the Redondo family. The ranch with the most published history was the Buenos Aires (originally spelled "Buenos Ayres"). This enterprise began with Pedro Aguirre, Jr., who first ran a stagecoach and freight line between Tucson and the mining towns of Arivaca in Arizona and Altar in Sonora, Mexico. He added a homestead in 1864, but by the tenth United States census of 1880 he was listed as living in Arivaca (Sayre 2002), probably because it was the closest settlement. Between 1891 and 1904, he acquired the lands around the artificial Aguirre Lake that he had built earlier to water his cattle. This ranch eventually ushered in a cattle boom and overgrazing in the Altar Valley that crashed with the droughts of 1891–1893 and 1898–1904. Although the boom passed, cattle ranching in the valley did not die.

Other ranches that formed during and after the cattle boom include the Poso Bueno (now Poso [Pozo] Nuevo in the northern end of the Buenos Aires National Wildlife Refuge), Ronstadt, Santa Margarita, and Warren (now King-Anvil). Some of the more important of these ranches nestled against the Baboquivari Mountains still remain today and include the Aros, Elkhorn, King-Anvil, Las Delicias, Palo Alto, and Santa Margarita. These ranches are based in different canyons or canyon drainages. Sayre (2002) has given the most complete history of the region in a single source.

There is limited access to the Baboquivari Mountains because most of the 30-mile chain is on Tohono O'odham land or private ranches. The Nature Conservancy has a right-of-way to land in the head of Thomas Canyon that is owned by a private ranch. This canyon provides access to the Baboquivari Wilderness, which is maintained by the Bureau of Land Management (BLM 2005).

Brown Canyon, now largely owned by the Buenos Aires National Wildlife Refuge, is also available to visitors. That canyon has been studied more than any other part of the mountains, and details from that canyon are used to provide an introduction to the flora and fauna elsewhere in the Baboquivari chain.

## Brown Canyon

We do not know what the prehistoric Hohokam or historic O'odham called the canyon. Remains of occupation sites in the lower part indicate that the Hohokam and O'odham used the area to harvest the abundant plants and animals. Also, there are pictographs and bedrock mortars. Presumably, the historic O'odham also used water and food resources as they did elsewhere in the valley, but there seem to be no records; it is possible that all of the occupation sites are prehistoric.

The canyon was identified as "Sycamore Canyon" (or "Gulch") in the field notes of surveyor George Roskruge in 1886. On Roskruge's 1893 map of Pima County, a camp labeled "Sycamore" is shown at the canyon mouth; the wash running from it is called "Sycamore Wash." That would have been a more descriptive name, although certainly not unique, as there are still dozens of Sycamore Canyons in the state, but Anglo bias and politics entered the picture.

The canyon finally was named for Rollin Carr Brown (b. Indiana, October 24, 1844, d. May 5, 1937), who arrived in Tucson on March 6, 1873, after serving for two years in the U.S. Army. He bought a small parcel in the canyon on the eastern side of the Baboquivari Mountains in 1879. The land had already been a ranch held by two individuals known as Sopher and Rainey (Kitt 1926–1929). Steere (n.d.) was unable to learn more about those men and speculated that the two were probably squatters.

In addition to running a cattle ranch at the mouth of the canyon, Rollin Brown was a newspaper editor and publisher, served in the territorial legislature, and acted as a Tucson school trustee in 1880–1881 (Kitt 1926–1929). Subsequently, the canyon and wash below it were named for him (Granger 1960), although Brown Canyon does not appear until after the General Land Office Survey Maps of 1910–1920.

A man named Perkins, who owned the Rancho de la Osa northwest of Sasabe, homesteaded part of the canyon from 1910 to 1930 (Pima County Map 1917, Tuell, personal communication, 2005). Then, Lee Mast owned the middle part of the canyon from 1951 to the 1970s (Jones 1975). Mast modified the two buildings of the Perkins homestead into a two-story configuration. He called it the Rancho del Baboquivari Lodge, and it served as a part-time hunting lodge for his friends.

Mast sold the land to a man named Blankenship, who, in turn, sold it to a consortium of five. The former Rancho del Baboquivari hunting lodge was obtained from this consortium and became the Environmental Education Center after the Buenos Aires National Wildlife Refuge purchased their land in the 1990s.

A ranch in the lower part of the canyon was owned by Roy Edwards (Jones 1975), a brand inspector, until it was purchased by Ray and Cathy Harm in 1983 (Harm, personal communication, 2006). The Harms ran a stocker cattle ranch on their land. Ray Harm's heart attack in 1989 made the Harms realize that they needed to be nearer to health and communication facilities. His health also prompted a discussion with Wayne Shifflit, then manager of the Buenos Aires National Wildlife Refuge, who told Harm that the ecological value of the canyon would be perpetuated if the U.S. Fish and Wildlife Service owned the complete area. That view appealed to Harm, who had worked with conservation groups for many years. The Harms sold their land to the Refuge in 1994, and they were subsequently instrumental in getting other landowners in the canyon to sell to the Refuge.

Donald E. and Nancy P. Janson purchased the area above the Blankenship lands in 1967 and sold it to the Refuge in 1997. The Jansons dubbed their holdings El Rancho del Zopilote Feliz (Happy Vulture Ranch), and they used it as a working cattle ranch (although Sayre [2002] writes that the Jansons used the ranch as a "weekend getaway and tax shelter"). During most of the week, the couple lived in Tucson, and cowboys cared for the ranch the rest of the time. Windmills, roads, barns, the house, and the corrals remaining in the canyon above the Education Center were built during this period. This family hired Mexican nationals to work as cowhands on the ranch. It was those cowboys who created the many rock walls, over a mile in total, that remain on the property (Sayre 2002).

Karen F. Allison and Maria Connie Lackey owned the last 80-acre parcel obtained by the Refuge in 2001. These women designed and built La Casita, which is now used by researchers and other guests in the canyon. The house also serves as a model for efficient design at comparatively low cost. The women subsequently bought land in Three Points (Robles Junction) and built a house exactly like the one that took them 10 years to finish in Brown Canyon (Senter, personal communication, 2004).

The Buenos Aires National Wildlife Refuge now manages most of the canyon (USFWS 2003). Their holdings reach from the Baboquivari Wilderness to the canyon mouth. Both the north and south sides of the Refuge holdings in the canyon are owned by the state of Arizona (Cohan in USFWS 2003). There is a single, small, private strip of land between the canyon and valley segments of the Refuge. That gap is held by the Arizona State Trust and by the Santa Margarita Ranch, which has its headquarters in adjacent Thomas Canyon.

The Refuge maintains three buildings and two trails near the bottom of the canyon. The Harm House is the first building in the canyon at 3,855 feet (1,175 m) elevation. In 2009, this structure was the residence of volunteers, who are often the caretakers of the Environmental Education Center. Next up the canyon is La Casita, which allows researchers access to the area. The Environmental Education Center at 4,160 feet (1,268 m) elevation is the former hunting lodge. From the Center, educational and environmental groups have access to the canyon for study and research.

In addition to the buildings, the Refuge maintains two low-impact trails that lead from the Environmental Education Center up the canyon almost three miles. About halfway up from the Center, there is a branch canyon beside the metal "catch corrals" at 4,260 feet (1,298 m) elevation. The branch canyon has been known since 1994 as Jaguar Canyon because that elusive Mexican animal has been seen near there. The Arizona Fish and Wildlife Service conducts surveys there for these rare animals, and the trail is typically used only by researchers. The other branch of the trail continues up Brown Canyon to a bridge in a volcanic dike known as The Arch at 4,774 feet (1,455 m) elevation. That rock formation, the fifth-largest natural bridge in Arizona, is the termination point for groups so that the remainder of the canyon is undisturbed. Not far above The Arch, the Baboquivari Wilderness lands begin.

## Selection of Species

There are 187 species discussed and illustrated in the body of the text. These taxa were selected because they are among the most common and obvious in the Baboquivari Mountains and associated ranges, at least seasonally. Many kinds are perennial and are present year-round. A few are ephemeral and will be visible for only short periods, typically during the spring or summer–fall seasons.

To provide the most information in compact format, related data are grouped under headings. Names, both scientific and common, are essential for exchange of information about organisms. Therefore, the species are listed alphabetically by family and then by their scientific names. Common names are given in two places; a heading of the locally used names precedes the technical name. Once the scientific name is given, other names with ethnologically pertinent data are added. This scenario is followed by other information that provides data on the ecological and ethnobotanical relevance of the species in the region.

# Format of Presentation of Species

## *Scientific name* (Family)

Recent studies suggest that some genera should be subdivided, but the data are not always convincing. Typically, however, there is enough support by specialists to incorporate the newer interpretations. Within the text, many of these genera are mentioned, and selected pertinent literature is cited.

Family names and limits are in flux as molecular genetics and other studies continue providing new data. Where morphological and molecular data reach consensus, family names proposed by the Angiosperm Phylogeny Group have been adopted. Names of nonflowering plant families follow the Editorial Committee of *Flora of North America North of Mexico* (1993).

## Common Names

The most frequently used vernacular names are listed separately for each species. Where available, these names are given in the three dominant languages in the area, English, Spanish, and Tohono O'odham.

**Other names:** Some plants have several names within single languages. Multiple names are applied for numerous reasons, including uneven knowledge. People who have not learned a name for a plant may make one up to help themselves remember, or to have something to tell a person questioning them.

However, there are other reasons for applying various appellations to the same plant. Sometimes the greatest variety occurs among lesser known species. Among other reasons are those given to Vestal (1952) by the Navajo. He was told that most plants have three names: the real name, the way-in-which-it-is-used name, and a descriptive name. His collaborator said that "many people do not know the first two names, as the descriptive name is given when you send someone after a plant who doesn't know the plant. All descriptive names are the ones used when you tell a person to get the plant and he doesn't know it well." The "real name" is used when addressing the plant, and it is usually concealed during interviews. There are even taboo names that should not be revealed to outsiders; the "real name" is often among those.

Perhaps more remarkable than finding a great diversity of names within a genus or species of plants is discovering that some have unique names across related languages (cognates). These folk-generic or folk-specific names are found for organisms that were of extreme importance as food, medicine, material goods, or in a religious context. At least in this sample, some of these names may be traced across single and related languages that span thousands of miles in extent. Some can be tracked back in historical records at least 300–400 years. Linguists even suggest that some words can be traced back thousands of years (cf. Fowler 1972, Shaul and Hill 1998, Hill 2001, 2002).

A list is given of some names applied in various languages within and outside the region (table 2). For each name, the meaning (when available) and the etymology (where possible) are provided. For names with currently accepted orthography available, that spelling is given first and those used in other sources are in angle brackets. I believe it is more instructive to provide the alternatives that will be found in the literature than to homogenize the variation into a single "correct" spelling.

Names have been taken mostly from the literature. English, Spanish, and scattered words from other languages are largely from Austin (2004), Johnson-G. and Carillo-M. (1977), Kearney and Peebles (1951), Lehr (1978), Martínez (1969, 1979), Martínez et al. (1995), SEINet (2009), and Standley (1920–1926). These names were supplemented and compared with the English and Spanish names given in all the other references cited. Mayan names are from Alcorn (1984), Anderson et al. (2003), and Austin (2004). Chumash comes from Timbrook (1990, 2007); Ohlone from Bocek (1984); and Yuki from Curtin (1957).

Uto-Aztecan words come from a large number of sources. Akimel O'odham names are from Curtin (1947, 1949), Rea (1997), and Russell (1908); Hiá Ceḍ O'odham names are from Felger (2000), Felger and Broyles (2007), Felger et al. (in prep.), and Nabhan et al. (1989); and Tohono O'odham names are from Castetter (1935), Castetter and Underhill (1935), Lumholtz (1912), Mathiot (1973), Nabhan (1983b), Saxton and Saxton (1969), Saxton et al. (1983), and

**TABLE 2. Linguistic relationships of the indigenous people mentioned in the text (based on Ortiz 1983 and Gordon 2005). The names used are those in the literature, followed, when available, by the name the people call themselves in their own language, if different.**

| Language Family, Subfamily | Language/Their name for themselves, if different | Geographic Region |
|---|---|---|
| Algic, Algonquian | Malecite-Passamaquoddy | Canada |
| Algic, Algonquian | Menominee/*Mamaceqtaw* | Wisconsin |
| Algic, Algonquian | Mesquakie | Iowa |
| Algic, Algonquian | Ojibwa | Canada |
| Arawakan, Caribbean | Taino | Bahamas, Cuba, Hispaniola, Puerto Rico |
| Caddoan, Northern | Pawnee | N Oklahoma |
| Carib, Northern | Carib | Surinam |
| Chumash [extinct] | Chumash (Barbareño, Ineseño, Ventureño) [Mahuna seems to be a subgroup; originally mentioned by Romero 1954 and Mahr 1955] | SW California |
| Gulf | Natchez/*Nah'-Chee* | Oklahoma |
| Hokan, Northern | Atsugewi | NE California |
| Hokan, Northern | Karok | NW California |
| Hokan, Seri | Seri/*Comcáac* | Sonora |
| Hokan, Washo | Washo/*Washiu* | California-Nevada border |
| Hokan, Yuman | Cocopa/*K$^w$apá* | Lower Colorado River |
| Hokan, Yuman | Kiliwa | Baja California |
| Hokan, Yuman | Kumiai [formerly Diegueño, Tipais] | San Diego, Imperial Valley, Baja California |
| Hokan, Yuman | Maricopa/*Pipatsje* | Colorado River |

**TABLE 2. *(continued)***

| Language Family, Subfamily | Language/Their name for themselves, if different | Geographic Region |
|---|---|---|
| Hokan, Yuman | Mohave/*Hàmakhá:v* | Colorado River |
| Hokan, Yuman | Pai/*Havasupai, Yavapai, Walapai* | Arizona |
| Hokan, Yuman | Paipai/*Akwa'ala* | Baja California |
| Hokan, Yuman | Quechan | SE California |
| Hokan, Yuman | Yuma/*K-wichhna* | Colorado River |
| Iroquoian, Southern Iroquoian | Cherokee/*Tsálăgĭ, Aníyûñ'wiya'i* | North Carolina |
| Keres, Eastern | Keres [San Felipe] | New Mexico |
| Keres, Western [Keresan] | Keres [Acoma, Laguna dialects] | New Mexico |
| Kiowa Tanoan, Tewa-Tiwa | Tewa | Arizona (Hano on the Hopi Reservation), New Mexico |
| Kiowa-Tanoan, Tewa-Tiwa | Isleta [Western Tiwa] | New Mexico |
| Language isolate | Kutenai | SE British Columbia |
| Language Isolate | Zuni/*Šiwi, ˀAšiwi* | New Mexico |
| Mayan, Chontal-Tzeltalan | Chontal | Mexico, Oaxaca |
| Mayan, Yucatecan | Yucatec Maya | Mexico, Yucatán |
| Mixe-Zoque, Mixe | Mixe | Mexico, NE Oaxaca |
| Mixe-Zoque, Zoque | Zoque | Mexico, Chiapas, Oaxaca, Veracruz |
| Muskogean, Eastern | Koasati | Louisiana |
| Muskogean, Eastern | Mikasuki | Florida |
| Muskogean, Eastern | Muskogee (Creek) | Florida, Georgia, Oklahoma |

(*continued*)

**TABLE 2. (*continued*)**

| Language Family, Subfamily | Language/Their name for themselves, if different | Geographic Region |
|---|---|---|
| Na-Dene, Eastern Apache | Chiricahua-Mescalero/*Aiaha* | SE Arizona, SE New Mexico |
| Na-Dene, Eastern Apache | Jicarilla | NW New Mexico |
| Na-Dene, Eastern Apache | Lipan/*Náizhan* | W Texas |
| Na-Dene, Tlingit | Tlingit | SE Alaska and nearby Canada |
| Na-Dene, Western Apache-Navajo | Navajo/*Diné* | Four Corners area (Arizona, Colorado, New Mexico, Utah) |
| Na-Dene, Western Apache-Navajo | Western Apache/*Indé* | East-central Arizona |
| Oto-Manguean, Mixtecan | Mixtec | Mexico, Oaxaca |
| Oto-Manguean, Otomian | Mazahua | Michoacán |
| Oto-Manguean, Otomian | Otomí | Mexico, Veracruz |
| Oto-Manguean, Zapotecan | Zapotec | Mexico, Oaxaca |
| Penutian, Maiduan | Nisenan | Central California |
| Penutian, Yok-Utian | Miwok | Coastal California |
| Penutian, Yok-Utian | Ohlone [formerly Costanoan] | California |
| Quechuan | Quechua | Andes of Ecuador, Peru |
| Salishan, Northern | Shuswap | E central British Columbia |
| Salishan, Northern | Thompson/*Nlaka'pamux* | SW British Columbia |
| Salishan, Southern | Okanagon [= Okanagon-Coville] | S British Columbia |
| Salishan, Southern | Sanpoil [a dialect of Okanagon] | British Columbia |
| Salishan, Straits | Saanich [Salish, Straits] | Vancouver Island, British Columbia |

**TABLE 2. (*continued*)**

| Language Family, Subfamily | Language/Their name for themselves, if different | Geographic Region |
|---|---|---|
| Salishan, Twana | Salish | Vancouver Island, British Columbia; Puget Sound, Washington |
| Salishan, Twana | Skagit | Puget Sound, Washington |
| Siouan, Siouan Proper | Ho-Chunk [Winnebago] | Nebraska, Wisconsin |
| Siouan, Siouan Proper | Lakota-Dakota | Nebraska to South Dakota, Montana |
| Siouan, Siouan Proper | Omaha-Ponca | Nebraska, Oklahoma |
| Tarascan | Purepecha [formerly Tarascan] | Mexico, Michoacán, Tamaulipas |
| Totonacan | Totonac | Mexico, Veracruz |
| Uto-Aztecan, Aztecan | Náhuatl [Aztec] | Mexico |
| Uto-Aztecan, Cahitan | Mayo/*Yoremem* | Mexico, Sonora |
| Uto-Aztecan, Cahitan | Ópata | Mexico, Sonora |
| Uto-Aztecan, Cahitan | Yaqui/*Yoeme* | Arizona; Mexico, Sonora |
| Uto-Aztecan, Central Numic | Comanche | W Oklahoma |
| Uto-Aztecan, Central Numic | Panamint/*Tümpisa* | SW Nevada, SE California |
| Uto-Aztecan, Central Numic | Shoshoni/*Nünü* | Nevada to Montana and central California |
| Uto-Aztecan, Hopi | Hopi/*Hópitu* or *Hopitushínumu* | Arizona |
| Uto-Aztecan, Northern Numic | Northern Paiute | N Nevada |
| Uto-Aztecan, Opatan | Eudeve | Mexico, Sonora |
| Uto-Aztecan, Southern Numic | Kawaiisu | SE California |
| Uto-Aztecan, Southern Numic | Southern Paiute/*Nɨngwɨ* | N Arizona, California |

(*continued*)

**TABLE 2. (*continued*)**

| Language Family, Subfamily | Language/Their name for themselves, if different | Geographic Region |
|---|---|---|
| Uto-Aztecan, Southern Numic | Ute/*Núu* | Four Corners area, Arizona, Colorado, New Mexico, Utah |
| Uto-Aztecan, Takic | Cahuilla/*Iviatim* | California |
| Uto-Aztecan, Takic | Cupeño | California |
| Uto-Aztecan, Takic | Luiseño [including Juaneño or *Acjachemen*]/*Chamteela* | California |
| Uto-Aztecan, Tarahumaran | Guarijío | Mexico, Chihuahua |
| Uto-Aztecan, Tarahumaran | Tarahumara/*Rarámuri* | Mexico, Chihuahua, Sonora |
| Uto-Aztecan, Tepiman | Mountain Pima/*O:b No'ok* | Mexico, Chihuahua, Sonora |
| Uto-Aztecan, Tepiman | Névome | Mexico, Sonora |
| Uto-Aztecan, Tepiman | Northern Tepehuan/*Ó'dami* | Mexico, Chihuahua |
| Uto-Aztecan, Tepiman | O'odham (*Akimel O'odham, Hiá Cẹḍ O'odham, Tohono O'odham*) | Arizona; Mexico, Sonora |
| Uto-Aztecan, Tübatulabal | Tübatulabal | S central California |
| Uto-Aztecan, Western Numic | Mono | E central California |
| Yuki | Yuki | Northern California |

Shaul (2007). Mountain Pima (*O:b No'ok*) words are from Escalante-H. and Estrada-F. (1993), Reina-G. (1993), and Shaul (1994). Tübatulabal words come from Voegelin (1938, 1958).

While Gordon (2005), Pennington (1980), and others recognize the "Pima Bajo," that term subsumes several distinct linguistic groups, including Mountain Pima and Névome (Nevome, Nébome). Words from these languages are distinguished where possible, following Santamaría (1959), Pennington (1979), Shaul (1983), Lionnet (1985), and Sobarzo (1991).

Other Tepiman and Tarahumaran names are from the publications by Bye (1979b), Pennington (1963, 1969, 1980), Hilton (1993), Miller (1996), Thord-Gray (1955), and Yetman and Felger (2002). Mayo is from Yetman and Van Devender (2001); Yaqui is from Molina et al. (1999). Words from the extinct Cahitan language Ópata (Eudeve, Dohema, Heve) follow Santamaría (1959), Pennington (1979), Shaul (1983), Lionnet (1985), and Sobarzo (1991).

Other Uto-Aztecan names are from the following sources: Hopi (Fewkes 1896, Whiting 1939, Hill et al.

1998); Takic is from Bean and Saubel (1972, Cahuilla), Romero (1954, Cahuilla), Curtis (1907–1930, several languages), Harrington (1933, Juaneño dialect of Luiseño), Fowler (1972, several languages), and Munro (1990, Cahuilla and Luiseño). Numic names are complicated because there are numerous dialects and word variants. Those for Ute and Shoshoni are mostly from Chamberlin (1909, 1911), although they were checked against entries in Train et al. (1957) and the Southern Ute Tribe (1979). Panamint is from Dayley (1989); Northern Paiute is from Fowler and Leland (1967) and Fowler (1989). Curtis (1907–1930) recorded names from several Numic and Takic languages. Other words came from Bye (1972, Southern Paiute), Kelly (1939, Southern Paiute), and Smith (1972, Shoshoni). Kawaiisu follows Zigmond (1981). Data for various tribes in Merriam (1979) are problematic because of the unorthodox word transcriptions. A few names in his records are included, although a linguist examining his lists may discover more that have not been published elsewhere.

Hokan names include both Seri and Yuman languages. Seri words follow Felger and Moser (1985). Yuman words come from Castetter and Bell (1951), Crawford (1989), Jöel (1976), Kay (1996), Kelly (1977), Mekeel (1935), Owen (1963), Stewart (1965), and Watahomigie et al. (1982).

Athapascan words are from Basehart (1974), Bray (1998), Castetter and Opler (1936), Elmore (1944), Mayes and Lacy (1989), Mayes and Roeminger (1994), Vestal (1952), Wyman and Harris (1941), Wyman and Harris (1951), and Young and Morgan (1980).

Names for *Phaseolus acutifolius* are partly from Nabhan and Felger (1978); names for *Proboscidea parviflora* are from Bretting (1986). These lists were supplemented with other sources that cite additional languages for those plants.

Current Tepiman orthography is from Shaul (2007, personal communication, 2007), except for Akimel O'odham, which is from Rea (1997). Names with no alternative spelling(s) are transcribed as they were in the original sources. Some of those words in Tepiman languages were unintelligible to Shaul (personal communication, 2007) and remain as found.

Orthography for other languages is from Hill et al. (1998) for Hopi, Young and Morgan (1980) for Navajo, Bray (1998) for Western Apache, and Castetter and Opler (1936) for Chiricahua and Mescalero Apache.

Synonyms as written in the sources listed, either dialectic variants or alternate transcriptions, are in angle brackets: for example, <*uhmug, umu'k, 'umug*>. Variations in names, that is, words dropped, substituted, or added, are in brackets: for example, *palma* [*palmilla*].

**Botanical description:** A brief description of the dominant features of each plant is given in mostly nontechnical language. This information is intended to augment that visible in the accompanying drawing.

**Habitat:** Some of the places are listed where the plants are most often found. These localities are followed by the range of elevations known within Arizona. Outside of Arizona, plants may grow higher or lower than those listed.

**Range:** Known occurrence is listed alphabetically by state within the United States. This listing is followed by areas within Mexico, and then by other parts of the known range outside these two political entities. These data are based on literature reports, and on herbarium records in SEINet (2009) and TROPICOS (2007).

**Seasonality:** Flowering season within Arizona is given by months. Depending on a number of factors, such as rainfall and cold weather, plants may be in flower before or after the months listed. Those "unusual" occurrences are given in parentheses.

**Status:** Information is given on whether or not the species is native to the region or introduced from elsewhere (alien, exotic). Sometimes it is not easy to determine the place of origin, and additional comments are given in those circumstances.

**Ecological significance:** From the information available about the species, selected items are presented to provide the user with locally pertinent data. For some plants, there is an abundance of ecological data available; for others, virtually none is published. Some of those gaps are pointed out. Unless a source is given, these comments are personal observations.

**Human uses:** Many of the plants discussed are or were used by one or more cultures in the region. Applications given emphasize indigenous peoples because those purposes are usually simply repeated or reflected in later cultures. Hispanics were flexible and adopted many of the medicinal and other indigenous uses of local plants. Anglos and other cultures have been more reluctant to accept these uses.

The species being discussed are often compared with others in the same genus, or even others in the same family, because nonscientists do not always consider plants in the same manner as do scientists. People typically "map" species in a broader sense than that covered by the Linnaean binomials. Because of that perspective, people frequently experimented, and still experiment, with similar plants around them regardless of the taxonomic views of scientists. This divergence of views may be noted by comparison of the common names.

**Derivation of the name:** The etymology of the scientific name is given. As will be seen, some are firmly established, while others are dubious. Linnaeus in the 1750s and later biologists have sometimes been capricious in naming plants, and they used epithets that have multiple meanings or possible origins. Most of the derivations are from Quattrocchi (1999), but they have been checked in other sources. His derivations are occasionally at odds with original stated meanings or historical interpretations.

**Miscellaneous:** Any additional information that seems pertinent or interesting to put the species in context is listed in this category. Some species have so much data in the other headings that it has not been possible to include this information. Where possible, other species are noted and traits are given to distinguish them from the one illustrated.

## Linguistic Conventions and Pronunciation

Names in English, Spanish, and other European languages are pronounced according to the rules of each. A number of phonetic conventions have been used for indigenous languages. Explanations here are taken from several linguistic sources cited elsewhere in this work, along with information from David L. Shaul.

There are tones of particular significance in the Apachean, Keresan, and Tanoan languages. All of these languages have both high and low tones. High tones are marked on all vowels with an acute accent (as *á, é, í, ó, ú*); vowels without an accent (as *a, e, i, o, u*) have low tones.

In addition, long vowels (see *aa* below for explanation) and diphthongs may have falling and rising tones. Falling tones occur when there is a high pitch on the first vowel of a sequence (as in *áa, ée, íi, óo, úu*), and rising tones occur when there is a high tone on the second vowel (as in *aá, eé, ií, oó, uú*). Falling and rising tones may occur on combinations of two different vowels; for example, *ái aí* and *áu aú*.

In the Tanoan languages, there is an additional mid tone, written with a circumflex accent, as in: *â, ê, î, ô, û*. Mid tones do not occur in Tanoan languages with long vowels or diphthongs.

**TABLE 3. Pronunciation Guide**

| Symbol | Example | Explanation |
|---|---|---|
| . | *ho.gisi'* | The period marks a syllable boundary. |
| :,<br>aa | *ka:w,*<br>*kaaw* | The colon (:) or repeated vowels represent a long sound. Compare the *o* in English wrote with the *o* in *code*; the *o* in *code* takes longer to say and is a long vowel. |
| ', ˀ | *'a'ud* | Glottal stop, the catch in "*oh*-oh!" The apostrophe (') is used in most southwestern languages to indicate a glottal stop. In other languages, the phonetic symbol (ˀ) is applied; this symbol looks like a question mark without the dot at the bottom. Old reports have written glottal stops with a question mark (?). Because the question mark is confusing, the symbol has been replaced with the current phonetic marker (ˊ). |
| ᵃ | *ta'ᵃñaeŋ* | A raised vowel indicates whispered, made without vibrating the vocal cords. |
| a | *o'odham* | Pronounced much like the *a* of English f*a*ther, Spanish *vaca*. In Apachean, Keresan, and Tanoan languages, this vowel is a low tone. |
| á | *áłtsíní iilt'áá'í* | Vowels with an acute accent in Apachean, Keresan, and Tanoan languages have a high tone. |
| ą | *ii't'ąą'* | Nasalized, much the same quality as *an* in French d*an*s. Other vowels may also be nasalized (ę, į, ǫ, ų). |
| áa | *náałshoih* | Long vowel with a falling tone. May occur in other vowels and diphthongs. |
| aá | *dích''ií'* | Long vowel with a rising tone. May occur in other long vowels (eé, ií, aí, etc.). |

*(continued)*

**TABLE 3. *(continued)***

| Symbol | Example | Explanation |
|---|---|---|
| ąą | *ch'iląągo* | Nasalized long vowel. Other vowels may also be nasalized (*ęę, įį, ǫǫ, ųų*). |
| à, è, ì, ò, ù | *abè* | Falling tones in Tanoan. |
| b | *basorí, chi'ilib* | Comparable to *p* of English *sp*ot, in contrast to *p* in *p*ot. A voiceless, unaspirated bilabial stop. |
| c | *casol* | In Seri, as *c* in *c*up. |
| c,<br>č,<br>ch,<br>tch,<br>tx | *cu:wĭ,*<br>*čoa,*<br>*chá'oł,*<br>*tchohokia,*<br>*txatitįiootl'ij* | In most spellings of native languages, there is no *c* as in English *c*at, but the letter is pronounced as *ch*, like *ch*at or *ch*ase. That same sound is indicated by both c and č. Older reports transcribed the sound as *ch* or *tch*; Navajo sometimes had *tx* (or even *tq*). |
| d | *'a'ud* | Comparable to *t* of English s*t*ay. A voiceless, unaspirated dental or alveolar stop. Sometimes described as a soft dental stop with the tip of the tongue. |
| ḍ,<br>D,<br>th | *jeweḍ,*<br>*jeweD,*<br>*jeweth* | ḍ (pronounced as "dh," much like the *d* in English wor*d*. Also written "D" or "th"). A retroflex *d* made with the tip of the tongue on the alveolar palatal area. Used in O'odham with the alternate spelling examples given. |
| dl | *yishdloh* | Comparable to English *gl* and *bl*, as in *gl*ow and *bl*ow. |
| dz | *dził* | Roughly equivalent to English *dz* in a*dze*. |
| e | *akimel, je:j* | Pronounced like the *e* of English m*e*t, or Spanish *pepino* in several languages; in O'odham, e is pronounced as *u* in English p*u*t or p*u*ll. |
| e | *hebe* | In Seri, as *a* in c*a*t. |
| ée | *hogéesh* | Long vowel with a falling tone in Apachean languages. |
| g | *sitagwiv, hiósig* | Comparable to *k* of English s*k*in, in contrast to *k* of *k*in. An unvoiced, unaspirated back palatal stop. Elder Ute speakers sometimes use the fricative (i.e., /ɣ/). When terminal, the *g* is often confused with the *k*, as the sounds are intermediate. |
| ĝ,<br>$g^{w}$,<br>gh | *ˀǫ́ǫ́rųnáaĝtį,*<br>*narag$^{w}$anɨ(m)bɨ,*<br>*bi'ghą'* | No equivalent in English or Spanish. A uvular fricative. In Apachean languages, a uvular fricative, now written *gh*, although formerly both <ĝ> and <$g^{w}$> were used. Produced by raising the back portion of the tongue to a position near the palate so that, when a stream of air is forced through the narrow passage, accompanied by a vibration of the vocal cords, there results a "growling sound." |

**TABLE 3. (*continued*)**

| Symbol | Example | Explanation |
|---|---|---|
| gw | *pa'gwanûp* | A labialized *g*, as in *gu* in the name *Gu*adalupe. |
| h | *hosh*,<br>*hehe* | Comparable to *h* of English *h*ot. In Seri, a glottal stop, *hehe* as *ˀeˀe*. |
| h/j | *moh <moj>* | The word before the <> is an English transcription for that written <inside angle brackets>. |
| hw | *ihwagi*,<br>*hʷos* | Roughly comparable to the *wh* of English *wh*ee or *wh*irl in dialects whose speakers pronounce the phoneme as hw-. |
| i | *iivdhat, s-totoñik* | Pronounced like English *i* in b*i*t when in last syllables preceding an ending consonant, or as *i* in pol*i*ce when elsewhere in a word. |
| į | *k'įį'* | Nasalized vowel. |
| ɨ | *tsɨkɨnɨn, tivaˀnɨbɨ* | Called the "barred-i." Pronounced as *u* in p*u*t or p*u*ll. |
| íi | *níil'įį'* | Long vowel with a falling tone in Apachean languages. |
| j | *bįįhjaa'* | In Navajo and Apachean, like the *j* of English *j*oke or *j*am. |
| j | *caaöj* | Voiceless velar fricative in Seri, like *ch* in Scottish lo*ch* or *ch* in German i*ch*. |
| k | *s-onk* | Comparable to the *k* of English s*k*ill. Gila O'odham has a distinct terminal *k*, as in *s-onk* (salty) (Rea 1997). |
| k' | *ndeelk'id* | In Apachean, Keresan, Tanoan, and Zuni, pronounced like the t', except that the back part of the tongue is raised against the back palatal area in a k-position. Made by running the English combination "kic*k*'up!" together. |
| kw,<br>kʷ | *kwàntsoki*,<br>*kʷa:p* | Comparable to the *qu* of English *qu*ick or *qu*ill. Aspirated, labialized back palatal stop. |
| l | *pala* | Comparable to English *l* in ye*ll*ow or *l*id. In O'odham <l>, is nearest the single flap *r* in Spanish, as in *entrada*. |
| ł | *ma'ał*,<br>*siml* | Called "barred-l," this is the ll of Welsh, as in the man's name Llewlyn; similar to English we*ll* (without the vocal cords vibrating on the *ll*). Pronounced approximately as *thl*, has no equivalent in English but approaches we*ll*. A voiceless fricative. In Seri, this sound is written as a final *l*. |

(*continued*)

**TABLE 3. *(continued)***

| Symbol | Example | Explanation |
|---|---|---|
| l/r | *melhog,*<br>*mïrok,*<br>*mïro'k* | The l/r sound in O'odham is a single tap of the tongue against the roof of the mouth, like the *tt* in English be*tt*er or *r* in Spanish *pero.* |
| ll | *collálle,*<br>*coyaje,*<br>*coyaye* | The double ll in Spanish or Hispanized words is sometimes transcribed in two ways, as <y> and <j>. |
| ly, $l_y$ | *ˀí-ly, uvaanál*$_y$*a* | Like the *lli* in mi*lli*on. |
| m, (m) | *sipu(m)bivɨ* | Equivalent to English *m* in *m*an. A nasal stop. |
| n | *tł'oh nástasí* | Equivalent to English *n* in *n*o or *n*ot. A nasal stop. |
| ñ | *hahaiñig* | The *ñ* has the sound of the *ny* in English ca*ny*on. A nasal consonant. |
| ŋ | *ŋwaejoka* | The velar nasal consonant is indicated by the *ŋ*. Pronounced as *ng* in English si*ng*. |
| o | *not* | The *o* in Tepiman languages is pronounced as *aw*, as in *au* in c*au*ght, *a* in *a*ll, or *o* in d*o*g (in dialects where this vowel is not the same as the one in c*o*t). Thus, the O'odham word *not* would be pronounced "nawt." Elsewhere, pronounced like the *o* in English wr*o*te, except in Ute, where it is like the English h*o*rse or p*o*rch. |
| ö | *haamxö caacöl* | In Seri, as *wh* in *wh*ich. |
| óo | *'nóosh'ní* | Vowel with a falling tone in Apachean languages. |
| p' | *'okup'e* | Made by running the English combination "kee*p'*up!" together. |
| q | *hulaqal* | Similar to the the *q* in Ira*q*. |
| qu | *coquée* | In Seri, as *c* in *c*up. |
| r | *teepar* | In Seri, a single tap. Like English *tt* in be*tt*er or *r* in Spanish *pero.* |
| ṣ,<br>š,<br>sh | *taṣ,*<br>*šušida,*<br>*utush* | Similar to English *sh*e or *sh*ip. |
| t | *tó* | Equivalent to *t* in English s*t*op, but made by pressing the tip of the tongue against the upper teeth (instead of against the roof of the mouth, as in English). |

**TABLE 3. (*continued*)**

| Symbol | Example | Explanation |
|---|---|---|
| t' | *nát'oh* | In Athapascan languages, made by running the English combination "mee*t'*up!" together. |
| ts | *kwàntsoki* | About equivalent to the *ts* of English Pa*ts*y or ca*ts*. |
| u | *a'uḍ* | For O'odham, pronounced like the *u* in br*u*te or *oo* in f*oo*d; sometimes like the *a* in *a*ll or the *o* in d*o*g or *o* in g*o*ld. |
| u̲ | *sųųvų̲́* | Voiceless or whispered vowels in Ute are underlined, instead of being indicated by a raised vowel. |
| ü | *wiyattampü* | Pronounced as *u* in p*u*t or p*u*ll, like the ɨ (see above). |
| v | *vi'ibgam* | In Ute, this is tone usually like the English equivalent, but some older speakers use the bilabial sound (i.e., /β/), with both lips vibrating against each other. |
| w | *wéˀe, wi:bkam* | Comparable to the *w* of English *w*ad or *w*ill. |
| x | *xitomat, xiu* | The *x* in Náhuatl and Mayan words pronounced as *sh* in English *sh*ip. |
| x | *hexe* | In Seri, a voiceless uvular fricative as *re* in French lett*re*. |
| y | *mayi, yolotl* | Comparable to the *y* of English *y*oke or *y*es. |
| z | *ootizx* | A voiced spirant similar to the *z* of English bu*zz*. |
| z | *cocazn-ootizx* | In Seri, as *sh* in *sh*ip. |
| zh | *łizhin, chínk'ózhé* | A voiced blade-palatal spirant similar to the sound of *s* in English plea*s*ure or *z* in a*z*ure. |
| θ | *kaθódn$_y$iúva* | The Greek theta (θ) represents the English *th* in *th*in (not the *th* in *th*y). |
| ω | *mωRc* | The *ω* is pronounced as omega is in Greek. |

# Common Plants

### *Anisacanthus thurberi* (Acanthaceae)<br>Desert Honeysuckle, *cola de gallo*

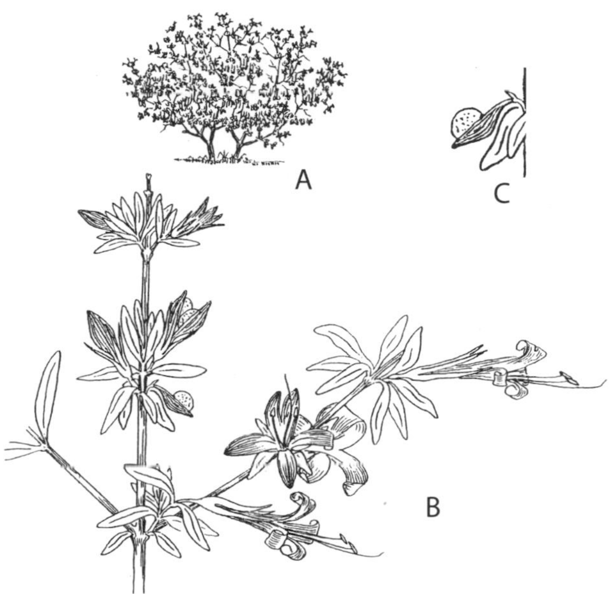


*Anisacanathus thurberi*. A. Growth form. B. Branch of flowering plant. C. Lower half of fruit with seed. Artist: R. D. Ivey (Ivey 2003).

**Other names:** ENGLISH: buckbrush, [Thurber's] desert honeysuckle (from Middle English *hunisuccle, -soukil*, apparently extended from *hunisuce, honysouke*; in use by ca. 1265; akin to "honey-suck," Old English *hunigsúge, -súce*, from *hunig*, honey + *súgan, súcan*, to suck; in use by ca. 725), hummingbird bush; SPANISH: *chuparrosa* (literally, rose sucker, alluding to visits by hummingbirds, Sonora), *cola de gallo* (rooster tail, Sonora), *colegallo* <*colegaiyo, colegayo*> (probably a rural pronunciation of *cola de gallo*, Chihuahua, Sonora), *hierba de cáncer* (cancer herb, Mexico; a name elsewhere given to *Acalypha*, *Cuphea*, and *Salvia*), *taparosa* (perhaps a misunderstanding of *chuparrosa*); UTO-AZTECAN: *lustich* <*lustiej*> (Guarijío), *muicle* (maybe from Náhuatl *mo huitli*, blue, but the allusion is obscure; a name given farther south to *Jacobinia*)

**Botanical description:** Shrub 2–2.5 m tall, with exfoliating bark and two vertical lines of sparse pubescence on the stems. Leaves lanceolate, 4–6 cm long, 1–2 cm wide, puberulent or glabrous. Flowers tubular, 2–3.5 cm long, red, yellow, or orange. Fruits capsular, 12–14 mm long.

**Habitat:** In canyons, along washes. 762–1,676 m (2,500–5,500 ft)

**Range:** Arizona (Apache to Yavapai, south to Cochise, Pima, and Santa Cruz counties), southwestern New Mexico; Mexico (Chihuahua, Durango, and Sonora).

**Seasonality:** With adequate moisture, the shrub may be in flower much of the year, but most often from March to April and sometimes again in October to November.

**Status:** Native.

**Ecological significance:** These are prime birdflowers and are visited by all of the resident and migratory hummingbirds in the range of Desert Honeysuckle. The timing of the peak flowering is more or less synchronized to hummingbird migrations. Daniel (2004) says that the flowers are also visited by bees (Apidae). Plants in my yard on the slopes of the Sierrita Mountains are visited regularly by: Anna's (*Calypte anna*), Black-Chinned (*Archilochus alexandri*), Broad-Billed (*Cynanthus latirostris*), Costa's (*Calypte costae*), and Rufous Hummingbirds (*Selasphorus rufus*). When the birds are visiting the flowers, the two stamens brush their throats to deposit pollen. Females especially have a bright band of yellow pollen across their throats that resembles jeweled necklaces against their gray neck feathers.

Verdins (*Auriparus flaviceps*) are fond of nectar, and they recognize the flowers as a potential source. While they cannot reach the nectar the "proper" way by reaching down thc tube, sometimes they bite holes in the bases of the corollas to "steal" a drink.

This shrub is a thornscrub species (Brown 1982). Plants grow along watercourses in areas with dependable summer rain (Turner et al. 1995). Dayton (1937) considered the shrubs "fairly good to very good" for sheep and cattle, but they are not their favorites, and animals eat it only when other forage is scarce (Kearney and Peebles 1951); probably deer do the same.

**Human uses:** Plants are being used in xeriscape and in plantings to attract native birds.

**Derivation of the name:** *Anisacanthus* comes from Greek *aniso*, unequal, and *akantha*, thorn or prickle. The name apparently alludes to the uneven corolla lobes. *Thurberi* commemorates George Thurber, 1821–1890, quartermaster of the Mexican

Boundary Survey (1850–1853) and editor of the *American Agriculturist.*

**Miscellaneous:** The genus *Anisacanthus* contains eight species that grow in the southwestern United States and Mexico. Other members of the family in the canyon are as follows:

*Carlowrightia arizonica* (*lemilla*) is recognized by being shrubby, having interrupted spikes, open white to cream-colored or purple corollas, and a yellow spot on the face of the posterior lobes. *Lemilla* grows in the Trans-Pecos region of Texas, west to Arizona, and from northern Mexico (Baja California, Chihuahua, and Sonora) south to Costa Rica. *Carlowrightia* flowers from April to May or even into September, and it is pollinated by flies (Bombyliidae) and bees (Halictidae) (Daniel 2004).

*Dicliptera resupinata* (*alfalfilla*) does not have a terminal spike-like flower cluster, and it is distinguished by deeply two-lipped rose-purple corollas and a pair of thin, valve-like, closely appressed, nearly orbicular bractlets. Plants grow in Arizona, southwestern New Mexico, and adjacent Mexico (Baja California Sur, Sonora, south to Jalisco, Michoacán). The species is pollinated by flies (Bombyliidae) and butterflies (Daniel 2004).

*Elytraria imbricata* (purple scaly-stem, *cordoncillo*) is herbaceous, has terminal spicate flower clusters, and is recognized by its lack of retinacula (levers that eject seeds). Corollas are purple. The species grows from the Trans-Pecos region of Texas, west to Arizona, and south to South America. It flowers from August to May, with peaks in March and September. Onavas Pima made a hot drink of the leaves (Pennington 1980).

*Justicia longii* (tube-tongue) is recognized by its long, slender, cylindrical corolla tube. Flowers are white, open in the evening, and stay open well into the following day, but they are fragrant only at night when blossoms are visited by hawkmoths (Daniel 2004). Kearney and Peebles (1951) knew this manifestation only from southern Arizona, but the type collection was made in 1855 in Sonora (Daniel 2004).

## *Tetramerium nervosum* (Acanthaceae) Hairy Fournwort, *cola de víbora*

**Other names:** ENGLISH: hairy fournwort ("fourn" unexplained, unless it means in 4s as does the genus, "wort," plant, Arizona, New Mexico, Texas); SPANISH: *calza de puerco* (pig's breeches, Ecuador), *chuparrosa* (rose sucker, Sonora; also applied to *Anisacanthus*, q.v.), *cola de víbora* (snake's tail, Guarijío), *olotillo* (little heart, from Náhuatl *yolotl* <*yollotl*>, heart, interior, Michoacán, Sonora), *rama de toro* (bull's branch, Sonora; for *T. tenuissimum*); MAYAN: *sak-chi'ilib* <*sak-ch'ilih, sak-ch-ilib*> (*sak*, white, *ch'iilib*, shrub, Maya), *xhuayunhak* <*xwayon-k'aak*> (*xhuayun*, usually refers to *Talisia olivaeformis*, Sapindaceae, *ak*, vine, Maya; also used for *Dodonaea*; could this be a confusion of names?); UTO-AZTECAN: *saya huehuásira* (Guarijío); (= *T. hispidum*)

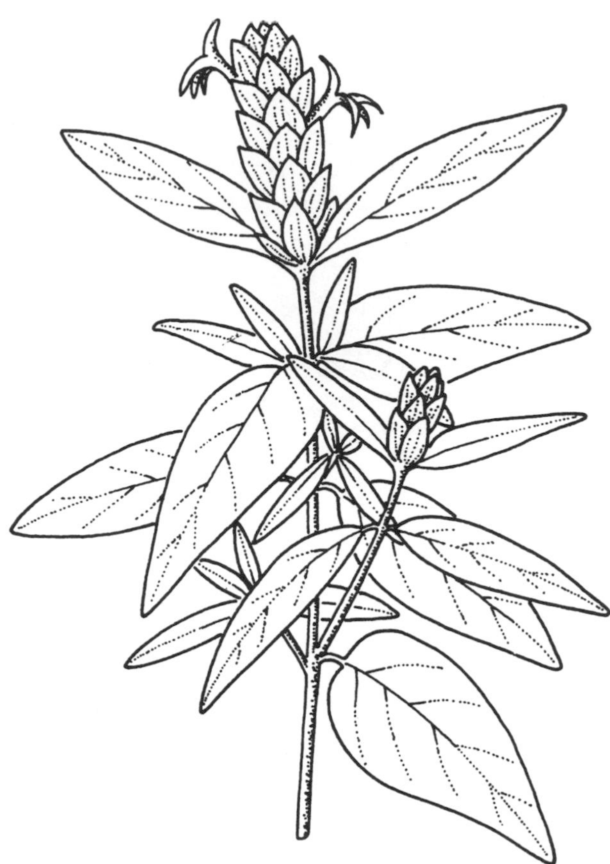

*Tetramerium nervosum*. Flowering branch. Artist: M. Mohlenbrock (Daniel 1984).

**Botanical description:** Herbs to 30 cm tall, the stems round, branched, brittle, softly pubescent or glabrous. Leaf blades lanceolate to ovate-lanceolate, 1–7 cm long, 0.5–2.5 cm wide, obtuse apically, rounded to cuneate basally, with soft pubescence, on petioles to 8 mm long. Inflorescences in terminal and lateral spikes to 9 cm long, 7–8 mm wide, with obvious lanceolate to ovate-lanceolate bracts in 4 ranks. Calices are similar to bractlets but slightly longer, narrowly lanceolate-aristate, 2.5 mm long, 0.5 mm wide.

Corollas are 1 cm long, white to deep yellow with scattered purple markings, the tube slender, the lips 5 mm long, the upper lip entire, the lower 3-lobed, the lobes elliptic, 3 mm long. Capsules are 4.5 mm long. Seeds are papillose.

**Habitat:** Canyons, along washes, streamside, grasslands. 900–1,500 m (3,000–5,000 ft).

**Range:** Arizona (Graham, Pima, and Santa Cruz counties); Mexico (Baja California, Chihuahua, and Sonora), south to Peru.

**Seasonality:** Flowering April to October.

**Status:** Native.

**Ecological significance:** These herbs are most common along washes and canyon margins where they are protected by other plants from excessive heat and cold; they also receive more moisture in these areas. The species is not listed by Felger (2000) or Felger et al. (in prep.) for the drier parts of southwestern Arizona and northwestern Sonora. The species barely makes it into the United States in Arizona and is not found in any of the surrounding states. Daniel (2004) considers this a thornscrub species, although he notes that it also grows in other habitats.

Although Daniel (2004) had data showing that *Tetramerium* flowers were visited by bombyliid flies, Daniel et al. (2008) suggested that at least part of the genus is butterfly pollinated. Moreover, there is a mechanism in the family that assures pollination. The stigma is erect, separate from the stamens, and receptive in the morning, but by afternoon it curves downward whether pollinated or not. That downward curve brings the stigma into contact with the pollen in the anthers and ensures pollination.

**Human uses:** People in Yucatán use a decoction of the flowers and leaves as a diuretic (Martínez 1969), and at childbirth to aid expulsion of the placenta (Hocking 1997). Guarijío livestock often eat *Tetramerium* (Yetman and Felger 2002). The Maya in Yucatán use *T. nervosum* "*para la supresión de los loquios*" (calmant) (Martínez 1969), as animal food, and to make brooms (Anderson et al. 2003). The name *olotillo* suggests that perhaps it has been used to treat heart problems, but no other reference has been found. Among the Mayo, *T. abditum* is called *rama del toro* (branch of the bull) and is eaten by livestock (Yetman and Van Devender 2001).

For decades, the special structures inside the fruits that fling the seeds away when the capsule dehisces were called "ejaculators" (from Latin, "I throw forth") by practicing botanists. The proper name for those is "retinacula" (plural of Latin *retinaculum*, a tether, from *retinere*, to retain), as used by Lawrence (1951) and others. This structure is actually the funiculus (the stalk by which the ovule is attached to the ovary wall or placenta), which is curved like a hook, and which retains the seed until mature (Jackson 1928).

**Derivation of the name:** *Tetramerium* means four-parted, in reference to the 4-ranked flowers of the inflorescence. *Nervosum*, with nerves or veins, probably refers to the obvious veins of the bracts.

**Miscellaneous:** This plant is the northernmost representative of a genus that encompasses 29 tropical American species (Daniel 2004).

## *Agave palmeri* (Agavaceae) Palmer's Agave, *lechuguilla, a'uḍ*

*Agave palmeri.* A. Leaf, with inset of variation. B. Flower cluster. C. Flowers, longitudinal section. D. Fruits. E. Seeds. F. Life-form. Artists: W. C. Hodgson (Hodgson 2001), A–E; R. D. Ivey, F (Ivey 2003).

**Other names:** ENGLISH: century plant (from the mistaken view that individuals took 100 years to blossom; by 1843 when J. L. Stephens used the term in his book *Incidents of Travel in Yucatan*, but he implied that the name had already been in use a long time), Palmer's agave [century plant] (a book name, based on the scientific names, Arizona, New Mexico); SPANISH: *lechuguilla* (little lettuce), *maguey* (see *A. parviflora* for derivation), *zapalote* (from Náhuatl *zapalotl*, the name for *A. tequilana* farther south, Sinaloa); ATHAPASCAN: *ikaz* (Western Apache; *nadah* <*nada*>, the roasted heart), *inada* (Chiricahua and Mescalero Apache), *noodah* (Navajo); UTO-AZTECAN: *amul* (Cahuilla, Cupeño, Luiseño; for *A. deserti*), *a’uḍ* <*a’ud*, *‘a’udh*, *a’ut*, *a’o’t*> (Tohono O’odham), *‘a’ud* (Hiá Ceḍ O’odham), *awé* (Guarijío), *čawé* <*chawi-ki*, *chgawe-ke*, *chawi*> (also called *sóko* or *méke*, Tarahumara), *ku’u* (Yaqui), *kwàntsoki* <*kwa ni*, *kwá:ni*> (*kwàn*, agave fruit, *tsoki*, upright plant, Hopi), *‘me* (*čoa*, the heart, Tarahumara), *nanta* [*nántA*] (Southern Paiute; perhaps a loan from Western Apache *nadah*); YUMAN: *m’ałʸ* <*ma’ał*> (Cocopa), *mavil* (Maricopa), *ūmúhl* (Kumiai), *vathi’l* (Mohave), *viyál* (Walapai)

**Botanical description:** Rosettes usually simple, rarely late suckering, 5–12 dm tall. Leaves narrowed above the base, lanceolate, long-acuminate, rigid, somewhat guttered, pale to light glaucous green or reddish-tinged, the margins nearly straight or undulate with or without small bases under closely set, regular, slender, flexed teeth. Flowers are 45–55 mm long, narrow, pale greenish yellow to waxy white. Capsules are oblong to pear-shaped, 3.5–6 cm long, short to long apiculate. Seeds are sooty black, 5–7 mm along the straight edge, 4–5 mm broad, thin, flat.

**Habitat:** Oak woodlands and grama grasslands. 1,066–2,286 m (3,500–7,500 ft).

**Range:** Arizona (Cochise, Gila, Graham, Pima, and Santa Cruz counties), southwestern New Mexico; Mexico (Chihuahua and Sonora).

**Seasonality:** Flowering June to August.

**Status:** Native.

**Ecological significance:** When *Agave* is in flower, the mountainsides are highlighted with splashes of yellow of both the candelabra clusters of *A. palmeri* and the similarly colored spikes of *A. schottii*. During the day, the flowers are visited by both Hooded and Scott’s Orioles that drink deeply of the abundant nectar provided by the blossoms. Hummingbirds also stop for nectar. In the night, the endangered Lesser Long-nosed Bat (*Leptonycteris curasoae*) is a primary pollinator of *A. palmeri* where the two occur together (Nabhan et al. 2004).

**Human uses:** Akimel O’odham, Apache, Comanche, Hopi, Mohave, Paiute, Pueblos, Ute, Tohono O’odham, and Yuma use the species for food, beverage, and fibers (Castetter 1935, Castetter and Opler 1936, Gallagher 1976, Hodgson 2001, Reagan 1929). Gentry (1982) reported that it was among the sweeter agaves, with little or no sapogenin. Leaves are cut from the rosettes, leaving the “hearts,” which were buried in a stone-lined pit and baked. This preparation method was used for *Agave* species among people from Baja California to the Grand Canyon, east among the Apache, and south through mainland Mexico. *Agave palmeri* is probably the species in remains of Mogollon people in New Mexico recorded by Kaplan (1963).

The resultant baked heart of agave is sweet because of alteration during baking of the abundant stored starches into sugars. The smoky flavor created during baking is not appreciated by all. Indeed, the product is an acquired taste that is widespread among people throughout most of Mexico. Although *A. palmeri* was eaten, it was commonly used as a source of fiber. Others make a *mezcal bacanora* called *lechuguilla* from *A. palmeri*. *Mezcal bacanora* now refers to an alcoholic drink; originally *mezcal* (from Náhuatl *mexcalli*, cooked heart of agave) and *bacanora* (Cahita *vaki*, cooked food, *onore*, cactus [*bisnaga*]).

Roasted stalks and young leaves of some species are eaten by some people, like the western Tarahumara (Pennington 1969). The Seri have no folk generic for agaves, but they give each a unique species name like *haamxö caacöl* (large agave, *A. colorata*).

Although not part of the native Baboquivari flora, *A. murpheyi* (Hohokam Agave) is cultivated by the Tohono O’odham in the canyons (Hodgson 2001). Because the “native” range of *A. murpheyi* is much farther north, the Hohokam Agave on the Tohono O’Odham Reservation probably was obtained from people living within its range. Hodgson (2001) recorded that the Apache, Akimel, and Tohono O’odham harvest and cultivate the species. Laura Kermen, a Tohono O’odham, told Hodgson in 1988 that her grandfather and others harvested Hohokam Agave in Sycamore Canyon on the western side of the Baboquivari Mountains. She was still cultivating *A. murpheyi* in her yard at that time.

**Derivation of the name:** *Agave* is from Greek *agavos*, meaning admirable, noble, splendid. *Palmeri* commemorates Edward Palmer, 1831–1911, a physician and avid collector of plants and animals in the American Southwest. He wrote about indigenous plant uses (Palmer 1871, 1878).

## *Agave parviflora* (Agavaceae) Santa Cruz Striped Agave, *a'uḍ*

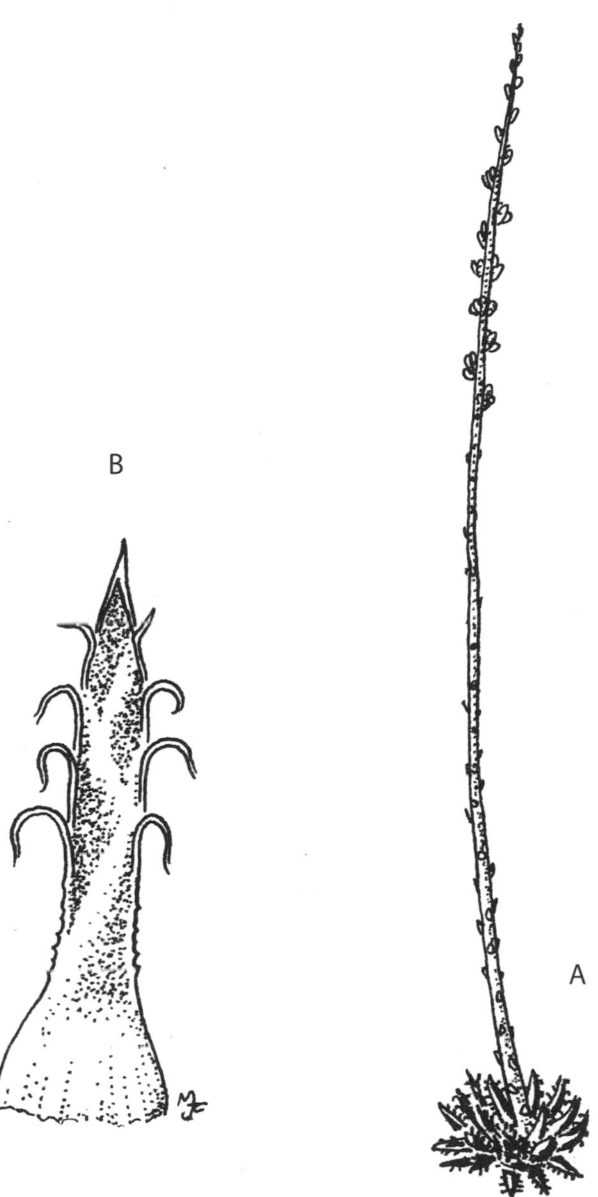


*Agave parviflora*. A. Life-form. B. Detail of leaf. Artist: M. Chamberland (courtesy Michael Chamberland).

**Other names:** ENGLISH: century plant (see *A. palmeri* for derivation); SPANISH: *maguey* (from the Caribbean Taino language for the genus; used by Oviedo in the 1500s and later widely adopted into Spanish); UTO-AZTECAN: *a'uḍ* <*'a'udh, a'o't*> (Tohono O'odham)

**Botanical description:** Succulent rosettes, single or growing in clusters (cespitose), 10–15 cm tall and 15–20 cm broad. Leaves 6–10 cm long, 0.8–1 cm wide, oblong-linear, widest at or above the middle, plane above, convex below, green, white bud-printed above and below, the margin conspicuously white-filiferous with long thread-like fibers, also minutely toothed near the base, the spine weak-subulate, 5–8 mm long, brown to grayish white. Spikes 10–18 dm tall, laxly flowered through the upper half of shaft, which is often reddish. Bracts of lowest flowers 1–3 cm long, much smaller above, scarious. Flowers pale yellow, in clusters of 2s, 3s, or 4s, 13–15 mm long; tepals 2–3 mm long, the outer slightly longer than the inner, the latter much broader than long, all erect to incurved. Filaments erect 10–12 mm long, inserted near base of the tube; anthers 5–6 mm long. Capsules are orbicular to oblong, 6–10 mm wide. Seeds are half-round, black, wedge-shaped, 3 mm long.

**Habitat:** Mountainous regions, on open slopes of desert grasslands and oak woodlands. 1,100–1,400 m (3,600–4,600 ft).

**Range:** Arizona (Pima and Santa Cruz counties); Mexico (Chihuahua and Sonora).

**Seasonality:** Flowering May to August.

**Status:** Native. The type collection was made by Arthur Schott (1813–1875) in the Pajarito Mountains in 1855. He was survey artist for the Emory Expedition of the Mexican Boundary Survey, 1849–1857, and also the first assistant surveyor under Emory. Schott was one of three artists enlisted to depict the border country landscape and its people from Texas west to Yuma, in what is now Arizona. The plants in the Baboquivari Mountains are subspecies *parviflora*; the similar ssp. *flexiflora* is endemic to Sonora.

**Ecological significance:** Gentry (1982) noted that the species was well named by John Torrey in 1859 because it has the smallest flowers in the genus. The flower lobes (tepals) are little more than broad lobes around the apex of the small cylindrical tube. Gentry found that flowers are pollinated by bumblebees (*Bombus*) and carpenter bees (*Xylocopa*).

This plant is one of the endangered species in Arizona (Arizona Rare Plant Committee 2001). Natural rarity, collector pressure, and habitat loss are among the factors that cause these diminutive succulents to appear on the rare plant list. These small plants are infrequent and occur in small colonies in the Baboquivari Mountains and across the Altar Valley in Las Guijas Mountains. Both sites are on the Buenos Aires National Wildlife Refuge.

**Human uses:** *Agave* fascinated European explorers when plants were discovered, and they immediately took samples back to the Old World. In many arid regions, such as parts of southern Europe, the Middle East, and northern Africa, some species have become naturalized.

Hearts of *A. parviflora* are sweet and edible but were not harvested because they are too small to use (Gentry 1982). The compact life-form has made *A. parviflora* popular with hobbyists and horticulturalists because they make attractive potted plants. People put potted plants on window sills and patios, where they appreciate the white bud-printed leaves with the threads separating from the margins. Sometimes owners over-water the *Agave*, and it grows larger than predicted from the size of wild plants. Gentry (1982) warned that "water must be rationed to maintain their natural compact habit."

**Derivation of the name:** *Agave* is from Greek *agavos*, meaning admirable, noble, splendid. *Parviflora* means small-flowered.

**Miscellaneous:** *Agave* was divided into two subgenera by Gentry (1982). He put the *Agave* with spicate flower-clusters in subgenus *Littaea* and those with paniculate, umbrella-like inflorescences in subgenus *Agave*. *Agave parviflora* belongs to subgenus *Littaea*. The genus is native to the New World, and it has 70 species in subgenus *Littaea* and 116 in subgenus *Agave* in continental North America (Gentry 1982). Gentry's treatment does not include species from elsewhere in the Americas, and Hodgson (1999) suggested that the genus contains about 200 species.

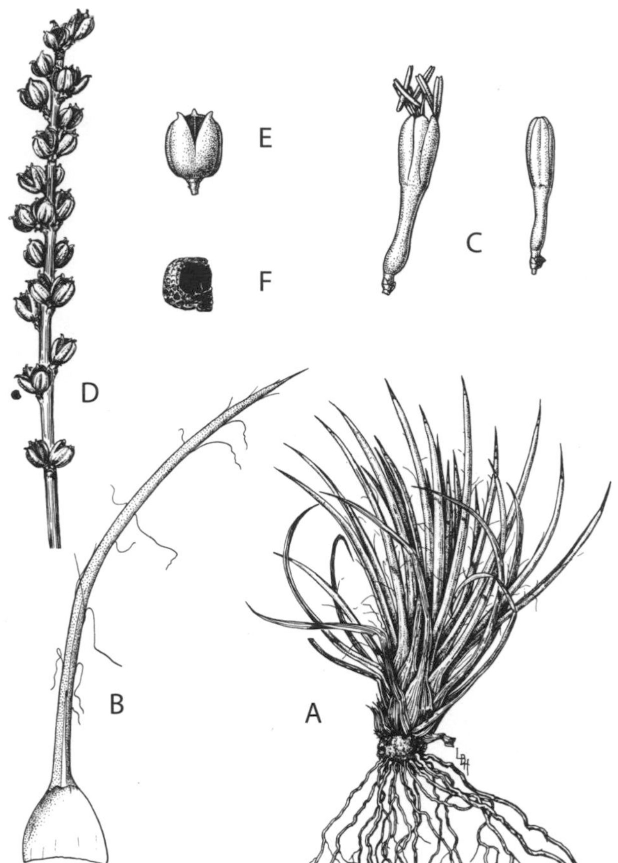


*Agave schottii*. A. Detail of basal rosette of leaves. B. Detail of individual leaves. C. Detail of bud and flower. D. Apex of fruiting stalk. E. Capsule. F. Seed. Artists: L. B. Hamilton, B–F (Benson and Darrow 1945); R. D. Ivey, A (Ivey 2003).

## *Agave schottii* (Agavaceae) Shin-Dagger, *amole, a'uḍ*

**Other names:** ENGLISH: century plant (see *A. palmeri* for etymology), shin dagger (from the way the sharp-pointed leaves stab the front part of the legs between the ankle and knee; "shin" in use since about CE 1000 for that region of the leg; from Old English *scinu*, cognate with West Frisian *skine*, Northern Frisian *skenn*, Middle Low German and Middle Dutch *schêne* [Dutch *scheen*], Old High German *scina*, *scena*, *sciena*, shin, needle; related to German *schiene*, thin wooden or metal plate, Swedish *skena*, splint, tire, rail; the fundamental meaning appears to be "thin or narrow piece"); SPANISH: *amole* (soap, Sonora), *amolillo* <amoliyo> (little soap, Sonora), *chugilla* [*churiqui*] (from Spanish *lechuguilla*, Mountain Pima), *maguey* (see *A. parviflora*), Schott's century plant (a book name, New Mexico); UTO-AZTECAN: *a'uḍ* <*'a'udh*, *a'o't*> (Tohono O'odham), *mayi* (Mountain Pima), *'utko je:j* (mothers' stalks, Hiá Ceḍ O'odham)

**Botanical description:** Succulent rosettes, densely clustered (cespitose), yellowish green to green. Leaves narrowly linear, 0.7–1.2 cm wide, 25–40(–50) cm long, widest at the base, straight, incurved, or falcate, pliant, flat or somewhat convex above, deeply convex below, smooth above and below, margins with a narrow brown border and sparse brittle threads, the spine 8–12 mm long, grayish, fine, weak and brittle. Spikes are 1.8–2.5 m tall, slender, frequently crooked, flowering in the upper 1/3 to 1/4 of the shaft. Bracts are straw-colored, filiform, needle-shaped (acicular), 2–4 cm long. Flowers yellow, 3–4 cm long, 1, 2, or 3 on stout pedicels that are 3–5 mm long, with setaceous bracteoles 10–15 mm long on lower spike but shorter above; tepals 10–16 mm long, yellow, unequally spreading at anthesis, the outer without a keel. Filaments erect, 15–

22 mm long, inserted high in the flower tube at 6–9 mm above the base; anthers 10–15 mm long. Capsules rounded to apiculate, 10–20 mm long. Seeds are 3–3.5 mm long.

**Habitat:** Exposed mountainsides and ridge crests, particularly grama [*Bouteloua*] grasslands and oak woodlands. 900–2,130 m (3,000–7,000 ft).

**Range:** Arizona (Cochise, Gila, Pima, Pinal, and Santa Cruz counties), southwestern New Mexico; Mexico (northwestern Chihuahua and Sonora).

**Seasonality:** May to October.

**Status:** Native. The type collection was made by Arthur Schott in the Pajarito Mountains (Santa Cruz County) in 1855. Born in Stuttgart, Württemberg, in 1814, this scientist, artist, and musician died in Washington, D.C., in 1875. He made significant contributions to Texas as a "special scientific collector" for the U.S. Boundary Commission beginning in late 1851. He made notes regarding the animals, plants, and geology, and he collected botanical, zoological, and geological specimens. In addition to his skills as a naturalist, geologist, and engineer, Schott was also a talented musician, poet, and artist. His drawings published in Emory's report included illustrations of the Seminoles, Lipan Apache, and Kiowa peoples.

**Ecological significance:** Flowers are apparently pollinated by both bats and bees. Gentry (1982) suggested that there were structural changes in the flowers to accommodate the bats, and he implied that there was less alteration for the bees. When in flower, the slender spikes accent the more showy stalks of *A. palmeri*.

Although most cattle ranchers and hikers view these armed plants with distaste or alarm, Shin-Dagger grows in exposed spots, where it contributes greatly to soil building and holds the thin soils that exist on the steep, rocky slopes. Ridge tops may be so thick with colonies of these *Agave* that their sharp-pointed leaves prevent passage of humans and most other animals.

**Human uses:** Leaves are mashed, mixed with water, and used locally in Mexico to wash clothes (Gentry 1982). Indeed, plants are widely known in Mexico and southern Arizona as *amole* or *amolillo*. Their hearts are not eaten by people because of the bitter constituents, including up to 1 percent sapogenins such as chlorogenin, manogenin, and tigogenin (Gentry 1982).

Although Gentry (1982) commented that "[t]he plants have little if any value as ornamentals," they are nonetheless planted in some areas. Their armament is particularly useful for discouraging traffic across sensitive places. Certainly planting near windows would prevent approach and entry as well as the iron bars more often used. Besides, their flower-spikes provide wonderful displays in the late spring and early summer.

**Derivation of the name:** *Agave* is from Greek *agavos*, meaning admirable, noble, splendid. *Schottii* commemorates its collector Arthur Schott (Gentry 1982; see Status above).

**Miscellaneous:** J. W. Toumey described the Arizona endemic *A. schottii* var. *treleasei* in 1901. Gentry (1982) accepted that variety with reluctance. Subsequently, Wendy Hodgson (Arizona Rare Plant Committee 2001) has found that var. *treleasei* is known from less than a dozen clones in northeastern Pima County. This variety differs from var. *schottii* by having wider leaves that lack marginal fibers, shorter lateral branches on the inflorescence, and larger, deeper-yellow flowers. She speculated that *A. schottii* var. *treleasei* may be hybrids of *A. chrysantha* and/or *A. palmeri* and *A. schottii* var. *schottii*.

## *Dasylirion wheeleri* (Ruscaceae) Desert-Spoon, *sotol, 'umug*

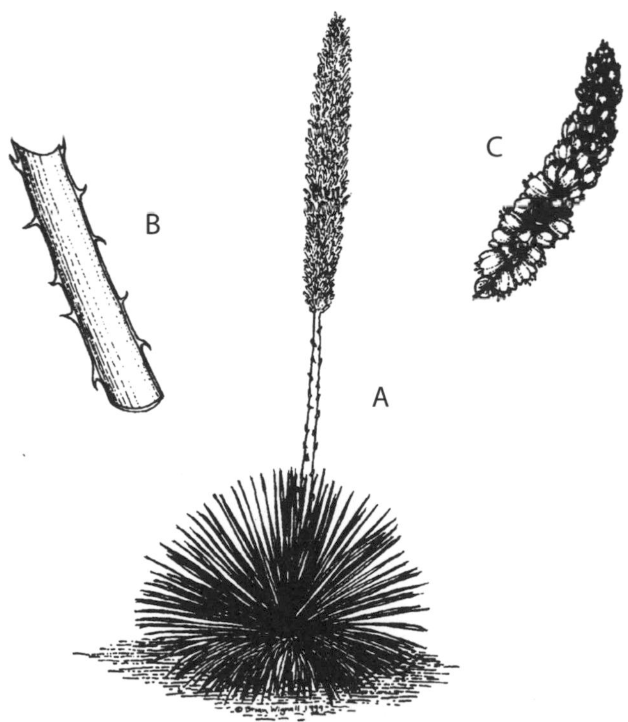


*Dasylirion wheeleri*. A. Flowering plant. B. Enlarged segment of leaf. C. Enlarged segment of flower cluster. Artist: B. Wignall (Phillips and Comus 2000).

**Other names:** ENGLISH: desert-spoon (possibly from Apache use of leaf bases, see below); SPANISH: *cucharilla* (little spoon; for *D. acrotriche* in San Luis Potosí), *sanó* (it heals, but possibly a loan-word from Guarajío), *palma* [*palmilla*] ([little] palm, Chihuahua), *sotol* (maybe from Náhuatl *tzontli*, hair), *tehuizote* (from Náhuatl *tehuizotl*, based on *te*, many, *itzo*, spines); ATHAPASCAN: *igabaané* <*ekibaane, k'ashbaané*> (Western Apache), *kokiše* <*kogice*> (fire stick? Chiricahua and Mescalero Apache); UTO-AZTECAN: *húumug* (Onavas Pima), *seré* <*selé*> (Guarijío), *seréke* <*sere-ke*> (Tarahumara), *seriki* <*shereki*> (straight, Mountain Pima), *umoga* (Mountain Pima), *umug* <*uhmug, umu'k, 'umug*> (Akimel and Tohono O'odham), *šušida kúrui* (Northern Tepehuan)

**Botanical description:** Bushy perennials, with stocky stems 0.5–1.5 m tall. Leaves are 1.5–3 cm broad above, flaring and clasping at bases, 6–10 dm long, armed with sharp, slender, straight to antrorse teeth that are yellow to brownish, 2–5 mm long, 5–15 mm apart. Panicles are slender, 3–4 m long. Flowers on numerous, ascending, short lateral branches, subtended by broad, scarious, fimbriate bractlets. Fruits are obovate, 5–6 mm wide, 6–7 mm long, the apical notch about 1 mm deep, apices of the wings acute to obtuse. Seeds are 3.5–4 mm long, brownish.

**Habitat:** Rocky slopes, grasslands. 1,220–1,830 m (4,000–6,000 ft).

**Range:** Arizona (Cochise, Gila, Graham, Greenlee, Pima, Pinal, and Santa Cruz counties), New Mexico, western Texas; Mexico (Chihuahua and Sonora to Nuevo León).

**Seasonality:** Flowering May to August.

**Status:** Native.

**Ecological significance:** Bighorn Sheep are reported to have grazed on *sotol* (Kearney and Peebles 1951).

**Human uses:** Desert-Spoon alludes to a former practice of making spoons from the leaf bases by Apaches and later in southwestern novelty shops (Castetter and Opler 1936, Gentry 1972). After the teeth are removed, leaves are used to make mats and baskets, for thatching, and the fiber from leaf buds is made into cordage (Gentry 1972, Rea 1997). Trunks have been used for posts in houses and corrals. Apache, Tarahumara, and Tohono O'odham roasted and ate the heart and emerging flower stalks like *Agave* (Castetter and Opler 1936, Hodgson 2001).

People from the Apache and Akimel O'odham to the Onavas Pima made many products from the leaves (Gallagher 1976, Pennington 1980, Moerman 1998, Hodgson 2001). Onavas Pima and Tohono O'odham found the fibers too coarse to make clothing, but they wove them into sleeping mats that could be rolled up and carried about (Castetter and Underhill 1935, Pennington 1980). Those mats (*main* <*ma'in*> to all Pimans) were similar to the Mexican *petate*, about six by three feet, and had rounded corners. Similar mats were made by the Hopi from Cattail (*Typha domingensis*) leaves, but they had wicker edges (Castetter and Underhill 1935). Tohono O'odham also used the leaf fibers to make cradle mats, back mats for the carrying frame, headbands, head rings, and two kinds of baskets. Mescalero Apache use the leaf stalks in a headdress of Mountain Spirit dancers (Moerman 1998). The soft meristematic tissues are also fermented and distilled by Onavas Pima, Northern Tepehuan, Hispanics, and others into a potent alcoholic drink called *sotol*, *tesgüino*, or *vino bacanora* (Pennington 1969, 1980, Gentry 1972, Hodgson 2001).

Tarahumara use scrapings of the emerging flower stalk to treat headache (Pennington 1963). They roll the material into small balls and put them on the head, sometimes holding them in place with cords. This plant is ceremonial among the Hopi; kachinas give fruiting stalks and leaves to children, especially young women (Whiting 1939).

**Derivation of the name:** *Dasylirion* is from Greek *dasys*, thick, and *leiron*, a lily, referring to the arrangement of the many small flowers. *Wheeleri* is named for Louis Cutter Wheeler, 1910–1980, a professor at the University of Southern California, Los Angeles, who specialized in the Euphorbiaceae.

**Miscellaneous:** The traditional placement of this genus is in the Agavaceae, along with several other southwestern monocotyledons. However, comparatively recent studies of molecular genetics and other aspects of *Dasylirion* suggest that the genus should be put in the Ruscaceae (Judd et al. 2008). *Dasylirion* is closely related to American *Nolina* and the Old World genera *Dracaena* and *Sansevieria*. *Dasylirion* comprises 15 species endemic to southwestern North America.

## *Amaranthus fimbriatus* (Amaranthaceae) Fringed Pigweed, *bledo, cuhugia*

**Other names:** ENGLISH: amaranth, [fringed] pigweed (eaten only by pigs), red root; SPANISH: *bledo* (Sinaloa), *quelite* (from Náhuatl *quilitl*, edible greens),

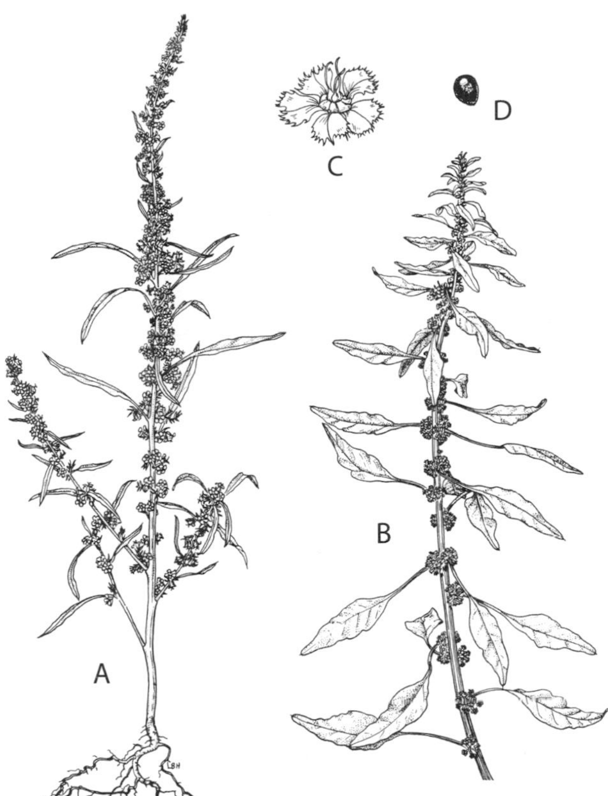


*Amaranthus fimbriatus*. A. Flowering plant with typical leaves. B. Flowering plant with more robust leaves. C. Mature female flower. D. Seed. Artist: L. B. Hamilton (Parker 1958).

*quelite de las aguas* (watery greens, Arizona, Sonora); ATHAPASCAN: *góchi bichan* (*góchi*, pig, *bichan*, from it, Western Apache), *ndaji* (black eye, Chiricahua and Mescalero Apache), *tł'ohdeeí'idí* (*tł'ohdeeí*, grass whose seeds fall, *'idí*, heaped, Navajo); HOKAN: *ziim caitic* (*ziim*, soft, Seri; *ziim* is used as the name of *Amaranthus fimbriatus*, *A. watsonii*, *Chenopodium*, and *Salsola*); KIOWA TANOAN: *su* (Tewa); UTO-AZTECAN: *ats* (Shoshoni), *basorí* <*wasorí*, *waso-ri*> (Tarahumara), *chuuhuggia* <*chu-hy-ki-ia*, *tchohokia*> (night carrying, Akimel O'odham), *cuhugia* <*cuhk-kia*, *chuhugia*, *teuhukia*> (Tohono O'odham), *cuhk-kia* <*cuhugia*> (Hiá Ceḍ O'odham), *hué* (Mayo), *hue-hué* (Guarijío), *poosiw* <*pó:siowu*> (based on *possi*, grain, seed, kernel, Hopi), *tucugusa* (Nevome), *tukya* (*tuk*, black, *ya*, derivative verb ending, Mountain Pima), *tukya* <*tungi'ia*> (Onavas Pima), *wéʾe* <*wéeʾe*> (Yaqui); YUMAN: *agwáva* <*agwávic*> (Maricopa), *agwáve* (Havasupai), *akwa'av* <*akwavdh*> (Mohave), *akwav* (Yuma), *kʷa:p* <*ko.p*> (mature plants, Cocopa; the immature plants or greens are *xpši*: <*hdhpši*>)

**Botanical description:** Monecious annuals. Stems usually less than 1 m tall, upright with mostly straight branches, branched throughout, pale green or often red, especially late in the season, glabrous or villous. Leaves alternate, the petioles slender, grading into the blade, the blade lanceolate to narrowly lanceolate, 2.5–5 cm long, 1–2 cm wide, infrequently wider, cuneate basally, acuminate apically. Inflorescences are terminal and axillary; bracts are herbaceous and not prickly. Pistillate flowers urn-shaped, their sepals 5, fringed (rarely nearly entire), white with green veins. Stamens are 3. Fruits are circumscissile. Seeds are 0.85–1 mm long, 0.75–0.9 mm wide, obovoid-lenticular, red-brown to blackish when fully ripe.

**Habitat:** Sandy washes, hillsides in open sites. 30–1,200 m (100–4,000 ft).

**Range:** Arizona (western Coconino, La Paz, Maricopa, and Mohave south to Cochise, Graham, Greenlee, Gila, Pima, Santa Cruz, and Yuma counties), southern California, Nevada, and southern Utah; Mexico (Baja California, Sinaloa, and Sonora).

**Seasonality:** Flowering (July) August to October.

**Status:** Native.

**Ecological significance:** The Fringed Amaranth is also known in the Coyote and Sierrita Mountains and in numerous sites in the Altar Valley. Plants are most abundant in washes and arroyos, but they are sometimes common in open desert, desert grasslands, and rocky slopes. Fringed Pigweeds are never as abundant as *A. palmeri*, but grow instead as scattered individuals, sometimes widely separated.

Seeds of this herb fall in a "seed rain" (about 16% of those produced), and few are retained in the "seed bank" (about 1% of those falling) (Price and Joyner 1997). One of the differences between seed fall and the seed bank is attributable to seed predation by various grain-eating animals, such as Merriam's Kangaroo Rat (*Dipodomys merriami*) (Soholt 1973).

*Amaranthus* is one of the major allergy-causing plants in the region because of its wind-borne pollen. When pigweeds are in flower, pollen causes "hay fever," although *A. fimbriatus* is rarely abundant enough to contribute greatly to the problem. The more common *A. palmeri* (see below) is more often the culprit.

**Human uses:** Cahuilla in southern California gather the seeds in the late summer and eat them (Bean and Saubel 1972). *Amaranthus fimbriatus* is one of the Seri favorites, with the seeds and the herbage being eaten (Felger and Moser 1985). Because the species is less common than *A. watsonii*, it is used less often, but seeds are stored for future use in pottery containers. Seeds are remarkable in *A. fimbriatus*, in that they feel somewhat oily and appear shiny. A small container of them actually looks and feels oily, although each seed individually does not seem to be like that. It takes a long time to gather a cupful.

All people who use the seeds probably gather them, as do the Seri (Felger and Moser 1985). Branches are held over a deer skin or cloth, and rolled gently in the hands, thereby causing the seeds and flowers to fall. Afterwards, the material gathered is winnowed to separate the seeds from the chaff. Seeds are toasted, ground into flour, and prepared as gruel. Seri often mix the flour or gruel with turtle oil.

Seeds are an important resource and are stored in pottery *ollas*. Seri compare eating the seeds of *Amaranthus* with those of cacti, saying: "[I]t is like eating the seeds of the columnar cacti" (Felger and Moser 1985). In addition, Seri consider the leafy young and tender green shoots an important source of cooked greens. The shoots and leaves are cooked in water, and the water is squeezed out by hand, a handful at a time. These greens are sometimes cooked "in a bit of sea turtle oil" or mixed with honey.

**Derivation of the name:** *Amaranthus* is derived from Greek *amarantos*, unfading, in allusion to the lasting quality of the flowers, persistent bracts, and sepals. *Fimbriatus* refers to the margins of the fringed (i.e., fimbrate) sepals.

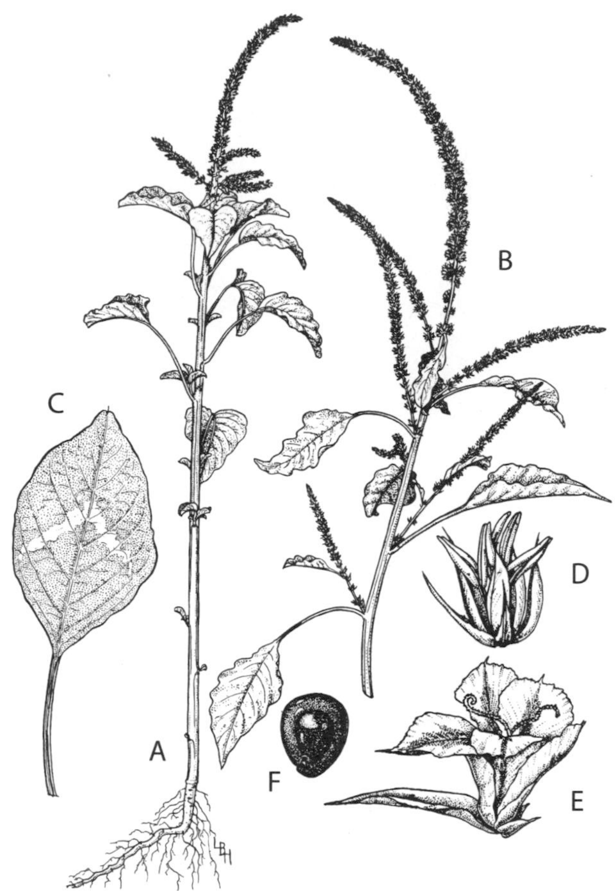

*Amaranthus palmeri*. A. Flowering plant. B. Flowering branch, enlarged. C. Basal leaf. D. Male flower showing stamens. E. Female flower. F. Seed. Artist: L. B. Hamilton (Parker 1958).

## *Amaranthus palmeri* (Amaranthaceae) Pigweed, *bledo, cuhugia*

**Other names:** ENGLISH: [Palmer's] careless [-weed] (not selective about where it grows), Palmer amaranth, pigweed (eaten only by pigs), red root; SPANISH: *bledo* (Sinaloa), *quelite* (from Náhuatl *quilitl*, edible greens), *quelite de las aguas* (watery greens); ATHAPASCAN: *góchi bichan*, *it'ąą dit'ógé* <*it'ą ditote*> (*góchi*, pig, *bichan*, from it, *it'ąą*, spinach, *dit'ógé*, crumbling, Western Apache), *ndaji* (black eye, Chiricahua and Mescalero Apache), *tł'ohdeeí'idí* <*y'oh de.sk'idi*> (*tł'oh*, grass, *ndeelk'id*, ridged, Navajo), *tł'ohdeeí hoshí* (*tł'ohdeeí*, grass whose seeds fall, '*idí*, heaped, *hoshí*, spiny, Navajo); KERES: *shiipa* (Acoma), *tsetayi* (Laguna); KIOWA TANOAN: *su* (Tewa); OTO-MANGUEAN: *ajna* (Mazahua; for *A. hybridus*); UTO-AZTECAN: *ats* (Shoshoni), *basorí* <*wasorí, waso-ri*> (Tarahumara), *chuuhuggia* <*chu-hy-ki-ia, tchohokia*> (night carrying, Akimel O'odham), *cuhugia* <*cuhkkia, chuhugia, teuhukia*> (Tohono O'odham), *cuhkkia* <*cuhugia*> (Hiá Ceḍ O'odham), *hué* (Mayo), *huehué* (Guarijío fide Yetman and Felger 2002), *keríba* (Guarijío fide Miller 1996), *poosiw* <*pó:siowu*> (based on *possi*, grain, seed, kernel, Hopi), *qo'u* [*qó'*$^{u}$] (Southern Paiute), *tucugusa* (Nevome), *tukya* (Mountain Pima), *tungi'ia* (Onavas Pima), *wéˀe* <*wéeˀe*> (Yaqui); YUMAN: *agwáva* <*agwávic*> (Maricopa), *agwáve* (Havasupai), *akwavdh* (Mohave), *akwav* (Yuma), *k*$^{w}$*a:p* <*ko.p*> (Cocopa; the greens were *xpši*: <*hdhpši*>)

**Botanical description:** Dioecious annuals, stems 6–10 dm tall (or taller), branched throughout, glabrous or villous. Leaves alternate, the petioles slender, the blades 1–6 cm long, ovate to lanceolate, basally cuneate to rounded, acute or shortly acuminate apically. Flowers are in terminal and axillary clusters or panicles. Bracts are 4–6 mm long, much longer than sepals. Staminate flowers with 5 stamens and 5 sepals, the pistillate flowers with 5 recurved sepals. Fruits are circumscissile. Seeds are 1–1.3 mm wide, dark brown.

**Habitat:** Washes, roadsides, disturbed grasslands. Up to 1,678 m (5,500 ft).

**Range:** Arizona (Coconino, Yavapai, and Greenlee south to Cochise, Pima, Santa Cruz, and Yuma counties), California, and Texas to Kansas; south to central Mexico.

**Seasonality:** Flowering July to September.

**Status:** Native.

**Ecological significance:** These are plants of disturbed sites, and may dominate floodplains to the near exclusion of other herbs. As such, pigweed provides seasonally abundant seeds for many granivorous birds and small mammals. Martin et al. (1951) recorded over 40 bird species that ate seeds from *Amaranthus*. Among the birds that eat the seeds are both ducks and Killdeer (*Charadrius vociferus*) (DeVlaming and Proctor 1968); moreover, these two authors found that seeds were 61% viable from the Killdeer and 75% from the ducks. Among mammals, both rabbits and kangaroo rats eat the seeds.

During the summer rains of 2002, the southern Altar Valley was dominated for miles by these herbs. Some of the patches were so thick that few other plants grew among them. Plots that measured 10 × 10 m in these areas usually had fewer than 20 species total, and ca. 90% of the biomass was the *Amaranthus*. Until the grasshopper season began, the sites harbored few insects. This dominance of *A. palmeri* in the area was caused at least in part because plants are allelopathic (Menges 1987). Experimental studies found that plants give off chemicals that inhibit the growth of other species, thus reducing their competition (Menges 1987).

**Human uses:** Acoma, Apache, Cahuilla, Cochimí, Cochiti, Cocopa, Guarijío, Hohokam, Hopi, Jemez, Laguna, Mayo, Mohave, Navajo, Akimel O'odham, Tohono O'odham, Quechan, and Tarahumara eat the young leaves or young plants of *Amaranthus* as greens (Castetter 1935, Gallagher 1976, Gasser 1981, Hodgson 2001, Marsh 1969, Reagan 1929, Whiting 1939, Yetman and Felger 2002). Wild and cultivated amaranths are used for grain by several groups of people in Baja California, as well as by Cahuilla, Cocopa, Guarijío, Onavas Pima, Tarahumara, Akimel O'odham, and Tohono O'odham, but not by Mayo (Pennington 1963, 1980, Hodgson 2001, Yetman and Van Devender 2001). Castetter and Underhill (1935) wrote without comment that seeds of amaranth or pigweed were "important among the Aztecs." These two imply that the Aztec plant was *A. palmeri*, but it was actually *A. leucocarpus* (Austin 2004).

Aztecs made dough from *A. leucocarpus* seeds (*tzoale* <*tzouali, zoale, soale*> from Náhuatl *tzoalli*). The *tzoalli* seeds were used to make a statue of *Uitzilopochtli* (*uitzilin*, hummingbird, *opochtli*, left-handed) to celebrate his festival during the month of *toxcatl*, which occurred during the last part of April and the first part of May. The first day of the celebration was dedicated to *Tezcatlipoca* (*tezcatl*, mirror, *poca*, brilliant) and culminated in the sacrifice of a human to this deity. Eight days later, Aztecs began the celebration of *Uitzilopochtli*. *Tzoale* seeds were also made into sweets and candies that were offered to their deities (Austin 2004).

**Derivation of the name:** *Amaranthus* is derived from Greek *amarantos*, unfading, in allusion to the lasting quality of the flowers, bracts, and sepals. *Palmeri* commemorates Edward Palmer, 1831–1911, a physician and avid collector of plants and animals in the American Southwest. He wrote about indigenous uses (Palmer 1871, 1878).

## ***Rhus aromatica* (Anacardiaceae) Skunk-Bush, *agritas***

**Other names:** ENGLISH: three-leaf sumac (New Mexico), ill-scented sumac ("sumac" is from Arabic *summaq*, their name for *Rhus coriaria*), lemon[ade]-berry, skunk-brush [-bush], squaw-[berry] bush (from Narragansett *squaws*, Massachusetts *squa*, woman, with related forms in many other Algonquian languages; in English by 1634; because *Rhus* stems were often used by women to make baskets); SPANISH: *acedillo* (little sour one, Mexico), *agrillo* (sour one, Durango), *agritas* <*agrito*> (sour one, Coahuila, Sonora), *chiquihuite* (California, from Náhuatl *chiquihuitl*, basket, originally made from *mimbres*, which may have been *Chilopsis* or *Salix*), *lambrisco* (wormy, probably from laxative use; from *lombriz*, Tamaulipas), *lemita* (little lemon, New Mexico), *lima* (lime, Mountain Pima), *saladillo* (little salty one, Mexico), *sidra* (cider, Mexico); ATHAPASCAN: *ch'ił lichíí'* (*ch'ił*, berry [lit. plant], *lichíí'*, red, Navajo), *chiiłchin* <*čilčin, cil cin, tchiiłtchin, tsiiłchin*> (*chiił*, plant, *chin*, fragrant, Navajo), *chínk'ózhé* <*tsínk'ózhé, nk'oze*> (Western Apache), *k'į́į'* (to peel, Navajo), *tciłtci*

*Rhus aromatica* var. *trilobata*. A. Fruiting branch. Artist: L. B. Hamilton (Humphrey 1960).

(smelly wood, Chiricahua and Mescalero Apache); CHUMASH: *shu'nay* (Barbareño Chumash), *shuna'y* (Ineseño Chumash); KERES: *yaa* (Acoma, Laguna); KIOWA TANOAN: *sapi'in* (*sa*, tobacco, *pi*, red, Tewa); UTO-AZTECAN: *divi'uka* (Mountain Pima), *iičivɨ* (Kawaiisu), *i'iši* <*'ici, c'i'ci, cïï, si'ibi*> (Southern Paiute), *motambiäts* (Ute), *selet* <*selit, silit*> (Cahuilla), *si-l* (Tübatulabal), *suuvi* <*sú:vi*> (Hopi), *sųųvų̱́*, *ˀisívų* (Ute), *tsiibi* <*cübi, si:vi*> (from *tsiivo* <*cükü*>, pungent, spicy, Hopi), *ai'tcĭb* [*dit(b), i'tcĭb*] (Shoshoni); YUMAN: *keθe'é* <*gith'e*> (Walapai); LANGUAGE ISOLATE: *ko'se o'tsi* (*ko'tse*, biting, *o'tsi*, man; because of taste, Zuni); (= *Rhus trilobata*)

**Botanical description:** Shrub, upright to 2 m tall, stems ascending, pungently fragrant when bruised. Leaves deciduous, with petioles to 25 mm long, the leaflets 3, thin or somewhat thickened, sessile or almost sessile, at first puberulent, later glabrate, the terminal leaflet broadly ovate to rhombic-ovate, 2–9 cm long, 1.5–9 cm wide, crenate-dentate to serrate near the apex, entire and abruptly cuneate basally, the lateral leaflets oval, 3.5–4.5 cm long, 1.7–3.5 cm wide, crenate-dentate to serrate near the apex, entire and obtuse basally. Inflorescences a terminal compound spike, to 6 cm long and 3 cm wide. Flowers are pale yellow. Fruits are red, almost globose, 5–6 mm wide, pubescent with simple and glandular hairs.

**Habitat:** Canyons, slopes. 760–2,290 m (2,500–7,500 ft).

**Range:** Arizona (almost throughout state), California, New Mexico, Texas, north to western Colorado, and Iowa; Alberta; Mexico (Chihuahua, Coahuila, Sonora, and south to Oaxaca).

**Seasonality:** Flowering (February) March to June (August).

**Status:** Native. This species is var. *trilobata*.

**Ecological significance:** Fall coloring of the leaves is spectacularly red and yellow, and Squaw-Bush stands out among the browns and greens of other plants. The colors make this shrub resemble Poison Ivy (*Toxidodendron radicans*, q.v.), which is often abundant nearby. Red fruits may persist into November, although they are often already eaten by birds and other animals. At least in California, the regeneration is through stump-sprouts after fires (Youngblood n.d.).

**Human uses:** People in the area have been using Squaw-Bush since at least CE 1200, the date of seeds in coprolites from the Four Corners region (Dunmire and Tierney 1997). The Navajos believe that this plant was given to them by the Bat (Wyman and Harris 1951). Castetter (1935) recorded Apache, Acoma, Laguna, and Navajo as eating fruits. As suggested by the Zuni name, the fruits have a spicy taste. I cannot help wondering if their taste paved the way for adoption of the Chile (*Capsicum annuum*) when it was introduced by the Spanish.

Palmer (1878) commented on the distinctive odor of *R. aromatica*; use of the species was so common that its odor was "always recognizable about Indian camps, and never leaves articles made from it." Plant parts have been used in baskets, snowshoes, loom rods and anchors, arrow points and shafts, digging sticks, scrapers, awls, and a number of ceremonial objects. Navajo use stems to make hoops for the Chiricahua Wind Way (Wyman and Harris 1941, 1951). Basket-making with this sumac extends from Ancestral Puebloans to modern Hopi, and others who use them in baskets include Apaches, Cahuilla, Chumash, Havasupai, Jemez, Kawaiisu, Keres, Luiseño, Navajo, Paiute, Panamint, Pomo, Tewa, Utes, Wintun, and Zuni (Castetter and Opler 1936, Curtin 1947, Elmore 1944, Gallagher 1976, Mayes and Lacy 1989, Moerman 1998, Romero 1954, Stevenson 1915, Timbrook 1990, Whiting 1939, Wyman and Harris 1951, Zigmond 1981).

Dyes from *R. aromatica* are used by Hopi, Navajo, and Puebloans, and many people make a tart drink from the fruits (Deschinny 1984, Hodgson 2001, Whiting 1939, Wyman and Harris 1951). At least Blackfoot, Cahuilla, Comanche, Hopi, Keres, Kickapoo, Kiowa, Navajo, Paiute, and Yokia use the fruits in medicines. Probably all tribes ate fruits, too, as did Apaches and Hopi (Fewkes 1896, Reagan 1929, Romero 1954). Kickapoo, for example, made a decoction of leaves to bathe women on the fourth day after childbirth, and the liquid was drunk for 40 days during the "involution" period (Latorre and Latorre 1977). Kickapoo also use *Rhus* to treat diarrhea, colic, bad breath, stomach ulcers, and to heal sores. Wood is one of the four kiva fuels (Hough 1898, Whiting 1939).

**Derivation of the name:** *Rhus* is the ancient name for sumac used by the Greek philosopher Theophrastus, 372–287 BCE, probably based on Greek *rhodos*, red, from the tannin in the stems and the edible red fruits. *Aromatica* means fragrant; *trilobata* refers to the 3-lobed leaflets and/or the 3 leaflets of the compound leaves.

**Miscellaneous:** *Rhus virens* is uncommon in Jaguar Canyon and flowers July to December; it is recognized by 5–7 leaflets on pinnate leaves.

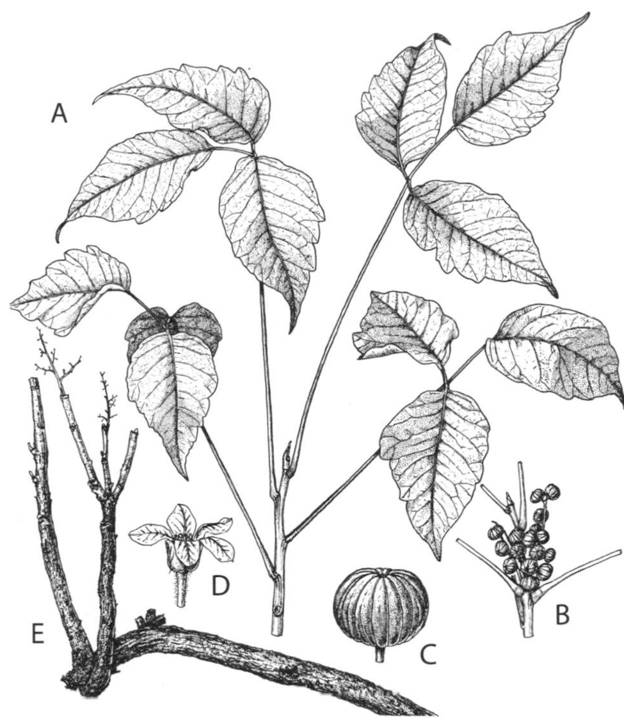


*Toxicodendron radicans.* A. Leafy branch with characteristic 3-foliolate leaves. B. Fruiting branch. C. Waxy berry. D. Flower. E. Underground rootstock with emergent branches. Artist: L. B. Hamilton (from original artwork).

## *Toxicodendron radicans* (Anacardiaceae) Poison-Ivy, *hiedra*

**Other names:** ENGLISH: poison ivy [oak] ("ivy" and "oak" refer to the resemblance of the leaves to *Hedera* and *Quercus*; "ivy" came from Old English *ifig*, and dates to about CE 800, originally applied to *Hedera helix*, and later to other evergreens; see also *Quercus*); SPANISH: *guardalagua* (perhaps from *guau del agua*, Jalisco), *guau* (originally a Taino word applied to the Caribbean poisonous anacardiaceous shrub *Comocladia*), *hiedra* <*yedra*> (ivy, New Mexico, Texas, Chihuahua, Durango, Sinaloa, Sonora, Nuevo León, Tamaulipas), *hiedra mala* (evil ivy, Arizona, New Mexico, Michoacán, Sonora), *hincha huevos* (egg [testes] lifters, Veracruz), *mala mujer* (bad woman, Jalisco, San Luis Potosí, Veracruz), *sumaque* <*zumaque*> (Mexico; taken into English as "sumac" by the 1400s and earlier into Spanish from Arabic, *summaq*); ATHAPASCAN: *k'ishíshjiį́įzh* <*k'isisjijz, k'íshíshjį́įzh*> (smashed down sumac, Navajo); TARASCAN: *bemberecua* (Purépecha); MAYAN: *chechén* (*cheché*, raw or unseasoned, meaning the rash, Maya); OTO-MANGUEAN: *mexie* <*mexi, mexye, meye*> (Otomí); UTO-AZTECAN: *ayáal* (Luiseño, Juaneño dialect), *ˀíya-l* (Cahuilla), *que cáguare* (Guarijío; apparently Spanish for "that which defecates," but surely from some other source), *ta'dabi* (Shoshoni), *tumorag* <*tumuraga*> (Mountain Pima), *t'ycob* (Mountain Pima); YUKI: *kiń-macho* (*kin*, to cry, Yuki)

**Botanical description:** Small shrub or vine, stems slender, glabrous or densely pubescent, sometimes with aerial roots and/or subterranean stolons. Leaves 3-foliolate, the leaflets variable, ovate to elliptic or rhombic to obovate, entire or irregularly serrate or dentate to regularly lobed-dentate with 3–7 rounded blunt or almost acute lobes, apically rounded to acute or acuminate, base rounded to cuneate, terminal leaflet 3–20 cm long, 1.5–13 cm wide, lateral leaflets somewhat smaller. Inflorescences composed of 3 lateral panicles. Sepals are usually 5, persistent. Corollas are white, imbricate in the bud, spreading at anthesis. Fruits are mostly whitish or cream-colored, globose,

5–7 mm wide, glabrous or pubescent. Seeds are about 4 mm wide.

**Habitat:** Canyons, ravines, washes, most common near streams. 900–2,440 m (3,000–8,000 ft).

**Range:** Arizona (Apache to Coconino, south to Cochise, Pima, and Santa Cruz counties), throughout most of North America; Mexico (Chihuahua, Sonora south to Chiapas, Guerrero, Jalisco, Oaxaca, and Veracruz).

**Seasonality:** Flowering April to September.

**Status:** Native.

**Ecological significance:** *Toxicodendron* requires considerable moisture and is most frequent in the shaded canyons. There, Poison-Ivy may spread and cover dozens of square meters of ground, thus supressing other plants.

West of the Baboquivari Mountains in southwestern Arizona, Poison-Ivy is recorded only from Alamo Canyon in the Ajo Mountains at about 3,200 feet (Felger et al. in prep.). The species has not been found in the Gran Desierto region of northwestern Sonora (Felger 2000).

As with most of the family, the flowers are small but produce considerable nectar. Flowers are pollinated by various insects; however, Diptera and Hymenoptera seem to be most important. Fruits are eaten and dispersed by many birds and mammals. Flickers (*Colaptes* spp.) are especially fond of the drupes, and Martin et al. (1951) listed 59 species of game-birds and song-birds that had been found eating the fruits. In addition, six mammals were listed.

**Human uses:** According to Bartolomé de las Casas in the 1500s: "*la leche de este árbol es ponzoñosa e della e de otras cosas hacen los indios la yerba que ponen en las flechas con que matan*" (the sap of this tree is poisonous, and from it and other things the Indians make a mixture that they put on their arrows with which they kill). This sap was also one of the ingredients in a Navajo arrow poison (Wyman and Harris 1941, Vestal 1952). In the eastern United States, several tribes have used Poison-Ivy in medicines, and it was listed in the United States Pharmacopoeia from 1820 to 1905 (Austin 2004).

Guarijío had no uses for Poison-Ivy, but they provided Yetman and Felger (2002), and, earlier, Gentry (1942, 1963), with advice about the poisonous nature of the herb. Guarijío grind *huehué* (*Amaranthus*) into a powder and rub it on body parts affected by Poison-Ivy.

**Derivation of the name:** *Toxicodendron* was named from the Greek *toxikon*, poison for smearing on arrows, and *dendron*, tree. *Radicans* means with roots, thereby alluding to the aerial roots produced by climbing plants.

**Miscellaneous:** There is disagreement on the delimitation of *Rhus* and *Toxicodendron*. Regarding this topic, Judd et al. (2008) wrote that the fruits of *Rhus* are glandular-pubescent, while those of *Toxicodendron* are glabrous, and that "if combined, the resulting group would not be monophyletic."

## *Bowlesia incana* (Apiaceae) Hairy Bowlesia

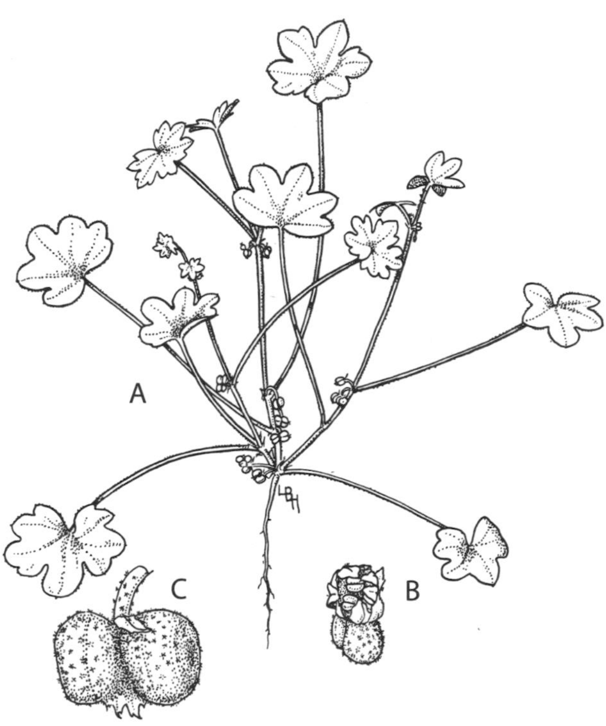


*Bowlesia incana.* A. Habit of plant with inconspicuous flowers and fruits at leaf bases. B. Flower. C. Fruit. Artist: L. B. Hamilton (Parker 1958).

**Other names:** ENGLISH: hairy bowlesia (a book name), miner's lettuce (Arizona; usually applied to *Claytonia perfoliata*, q.v.)

**Botanical description:** Annuals from a slender taproot, stellate-pubescent to glabrate, prostrate to almost erect, the stems 1–5 dm long. Leaves are opposite, the petioles to 7 cm long, the blade almost orbicular in outline, to 3 cm long and 4.5 cm wide, palmately 5- to 7-lobed, the lobes entire to dentate. Stipules are thin, dry, membranous (scarious), margins appearing torn (lacerate). Peduncles extend to 2 cm long, axillary. Umbels are 2-flowered to 6-flowered. Flowers are tiny, white or purplish, on short branches or sessile, with the calyx teeth ciliate. Fruits are sessile or almost sessile, 1–1.5 mm long, and 2–3 mm wide.

**Habitat:** Below bushes in habitats ranging from desertscrub to oak woodlands. 304–1,370 m (1,000–4,500 ft).

**Range:** Arizona (Coconino, Graham, Mohave, Pima, and Yuma counties), central California to southern Texas; Mexico (Baja California, and Sonora).

**Seasonality:** Flowering (January) February to April.

**Status:** Native.

**Ecological significance:** These delicate annuals appear year after year in the same places, often where there is shade from either other plants or on north-facing rocky slopes. *Bowlesia* "hides" in the protection of these areas to take advantage of the extra moisture. Soon after the heat begins, the herbs dry up and disappear. The flowers are so small and inconspicuous below the foliage that *Bowlesia* may be self-pollinating.

A study conducted in California found that these small herbs grew more commonly on the north and west side of trees in grasslands and less often on the south and east sides (Parker and Muller 1982). That preference was for the cooler and wetter sides of the tree. In these grasslands, the herb had less than 4% cover, but it occurred below 60% of the oak trees.

The only animals known to disperse *Bowlesia* are packrats and Collared Peccary (Anderson and Van Devender 1991, Corn and Warren 1985). The packrats carry pieces of the plants and the peccary eat the seeds. Both are probably effective in dispersal.

Mildred Mathias and Lincoln Constance (1965) originally suggested that the species had been introduced from South America. Correll and Johnston (1970) followed them and considered *Bowlesia* adventive in Texas, even though J. L. Berlandier, the first plant collector in that state, found it in 1828. Both Kearney and Peebles (1951) and Munz and Keck (1973) considered the species native to North America, and later Constance in Hickman (1993) decided that the species was native to Arizona and California. Thus, opinions on nativity have been divided for about 40 years.

Now, a combined study of herbivores and molecular clocks indicates that *Bowlesia* and insects were in North America long before their proposed introduction in the 1740s. Pellmyr et al. (1998) checked the interactions between *Bowlesia* and the specialized moth herbivore *Greya powelli* (Lepidoptera: Prodoxidae) in California and determined that *B. incana* and the animal had been in North America for about 2.3–3.8 million years. Moreover, the moth that feeds only on *Bowlesia* does not occur in South America. Therefore, *Bowlesia* has been in North America much longer than the proposed 250 or so years—it is native here.

**Human uses:** At least some individuals in Arizona call these plants "Miner's Lettuce" and eat them. That name historically is applied to *Claytonia* (q.v.) and is the "wrong" application for *Bowlesia*. I find the flavor of the leaves too strong, much like potent Parsley (*Petroselinum crispum*) or Cilantro (*Coriandrum sativum*) to be palatable. Perhaps if *Bowlesia* was used as seasoning and not eaten straight, it might be more acceptable.

However, indiscriminate "testing" of plants in this family is ill-advised. As Foster and Duke (1990) wrote, "Amateurs fooling with plants in the parsley family are playing herbal roulette." Their comment was generated by several violently poisonous members of the family, but especially Hemlock (*Conium verosa*), with which the classical Greek philosopher Socrates committed suicide in the spring of 399 BCE.

**Derivation of the name:** The genus *Bowlesia* is dedicated to William Bowles, 1705–1780, an Irish writer on Spanish natural history. *Incana* means white pubescent.

**Miscellaneous:** There are 15 species in the genus, with 12 of those being confined to South America and two growing in North America. *Bowlesia* is related to *Eremocharis* (Chile, Peru), *Hydrocotyle* (world), and *Trachymene* (SE Asia to Australia), and all of these plants belong to the subfamily Hydrocotyloideae (Pickering and Fairbrothers 1970, Plunkett et al. 1996).

## *Daucus pusillus* (Apiaceae) American Carrot, *zanahoria silvestre*

**Other names:** ENGLISH: rattlesnake weed (California, New Mexico), seed ticks (referring to the way the fruit segments adhere to clothes and skin, like ticks), American [wild] carrot (from French *carotte*, based on Latin *carota*, an adaptation of Greek *karoton*; in use in English by 1533; Dioscorides, CE 40–80, called *carota* the *pastinaca silvestris*); SPANISH: *hierba de la víbora* <*yerba de la víbora*> (rattlesnake herb, a name applied to at least 12 species in

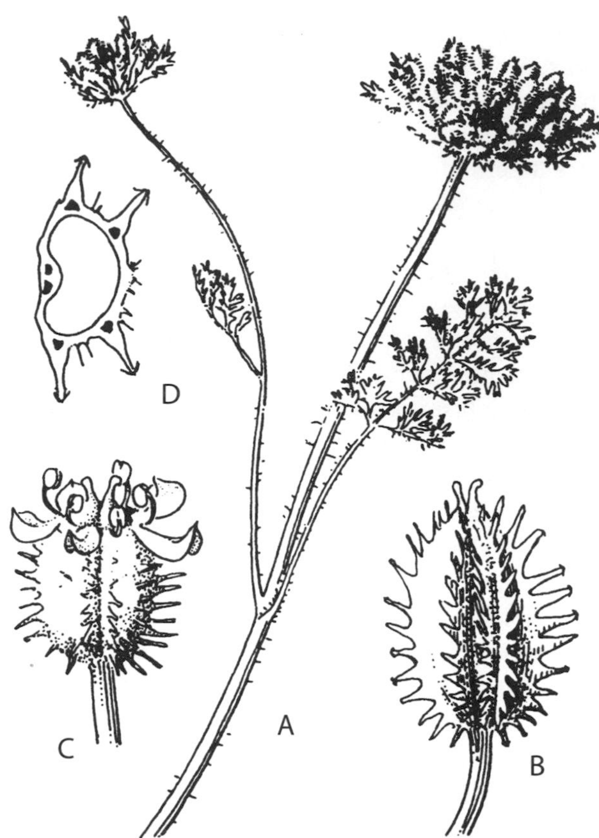


*Daucus pusillus*. A. Flowering branch. B. Fruit. C. Flower. D. Cross-section of flower. Artist: J. R. Janish (Abrams 1951).

7 families, Mexico, New Mexico), *zanahoria silvestre* (wild carrot; *zanahoria*, spelled *çanahoria* in 1557, is from Arabic *isfanaria* or *sannaria*; the word became *cenoura* in Portuguese); ATHAPASCAN: *bikéghad łitsogí* (*bikéghad*, root, *łitsogí*, yellow, Western Apache), *chąąsht'ezhiitsoh* (carrot, also used for *D. carota*, Navajo); UTO-AZTECAN: *ˀǫ́ą̨rų̨náaĝtį̨*, *tónąci* (*ˀǫ́ą̨qarį̨*, yellow, *tunáąvi*, root, Ute), *sanooria* (loan from Spanish, Yaqui)

**Botanical description:** Annuals that are winter-spring ephemerals, with stiff white (hispid) trichomes that are sometimes papilla-based on stems and inflorescence branches. Stems are slender, 7–50 cm tall. Leaves are highly dissected. Umbels are densely flowered, on stout peduncles 3.5–27 cm long, the bracts leafy. Sepals are absent. Petals are 0.6 mm long, white to pale yellow. Fruits are bur-like, the body dark-colored, 3 mm long, intricately sculptured with yellow barb-tipped spines.

**Habitat:** Grasslands, canyons. Up to 1,300 m (4,300 ft).

**Range:** Arizona (Greenlee to Mohave and Yavapai, south to Cochise, Pima, Santa Cruz, and Yuma counties), California, north to Washington and British Columbia, and east to Florida and South Carolina; Mexico (Baja California, Sonora, and east).

**Seasonality:** Flowering March to May (June).

**Status:** Native.

**Ecological significance:** This plant is a permanent component of the seed bank in desert grasslands in the Altar Valley and surrounding areas. Although *D. pusillus* does not seem to be present in dry years, carrots appear after abundant rain, particularly in sites that received extra water in runoff or ponding. The bristly fruits are involved with animal dispersal (Judd et al. 2008). Fruits certainly stick to human clothing and skin.

**Human uses:** Clallam (Olympic Peninsula, Washington), Cowichan (Vancouver Island, British Columbia), Navajo (see Table 2), Saanich, and Salish eat the roots raw, dried, and cooked (Castetter 1935, Elmore 1944, Kearney and Peebles 1951, Moerman 1998). Roots are small and hard, with the intense flavor of raw carrots. When winter rains have been sufficient, roots are large enough to add flavor to other foods. By themselves, it would take many of the roots to fill a hungry stomach.

California's Miwok and Ohlone consider the herbs medicinal, and the Mendocino people think them a good-luck talisman in gambling (Bocek 1984, Moerman 1998). Hocking (1997) reported that the plants contain compounds that promote wound healing. Unidentified people in California chewed the roots to relieve the effects of rattlesnake bites (Mathias 1994).

Classical Mediterranean taxonomy was not the same as that today. The Greeks said *staphylinos* for the root of what we call *D. carota*, the wild carrot, and people cooked and ate them, although each was rarely larger than a finger. Dioscorides (fl. CE 40–80) wrote that people mashed the leaves with honey to clean ulcers. The Romans called the same plants either *carota* or *pastinaca*, apparently reserving the latter name for those cultivated (Austin 2004).

As late as the 1500s, the Carrot was known in French as *carota* (by 1536) and *pastenade* (by 1561), and in Italian as *carota* (by 1561) and *pastinaca* (by 1597). By contrast, the Parsnip (in English by 1398, from *pastinaca*) was called *siser* or *pastinaca* (Latin), *cheruy* (French), *rape gialle* (French turnip, Italian), or *chiriuia* (Spanish). Leonard Fuchs called the Parsnip *sisaro* in 1542.

That array of names resulted in those used today in the regions formerly under the rule of the Roman Empire. For the Carrot, Charles de l'Ecluse in 1576 and Joseph Pitton de Tournefort in 1700 preferred *Daucus*, while Gaspar Bauhin used *Pastinaca* in 1623. Finally, in 1753, Linnaeus used *Daucus* for the technical name of the Carrot, and the Parsnip became *Pastinaca*.

**Derivation of the name:** *Daucus* is from Greek *daukos*, sweet or wild carrot. *Pusillus* means small.

**Miscellaneous:** This herb is easily confused with *Yabea microcarpa* (q.v.), another small herbaceous umbel that grows in the canyons. Both may be most abundant along streamsides and can be distinguished by examining the fruits, which are barbed in *Daucus* and hooked in *Yabea*.

## *Spermolepis echinata* (Apiaceae) Scale-Seed

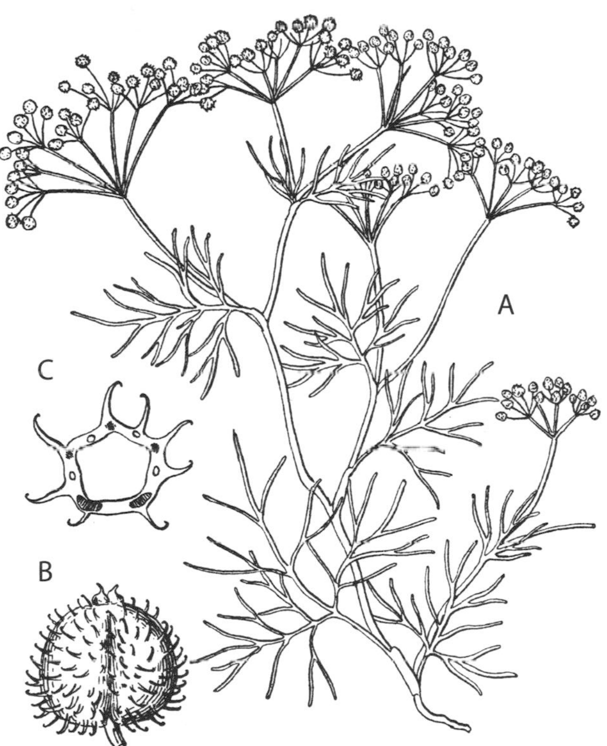


*Spermolepis echinata*. A. Flowering branch. B. Fruit. C. Section of fruit. Artist: F. Emil (Britton and Brown 1896–1898).

**Other names:** ENGLISH: beggar's lice [beggars'-lice] (name applied to *Galium*; also in the United States to certain boraginaceous plants, whose prickly fruit or seeds stick to the clothes; and to Spanish needles [*Bidens*] by 1880, later to *Spermolepis*, Texas), bristly scale-seed (translation of the scientific name, New Mexico, Texas)

**Botanical description:** Herbs, annual, plants low and often spreading, to 4 dm high. Leaves are ovate, to 25 mm long, 2 cm wide, 3 times compound (ternate), the ultimate divisions filiform, mucronulate. Peduncles extend to 65 mm long, the involucel of a few filiform toothed to glabrous bractlets shorter than the pedicels. Rays of umbels are 5–14, almost erect, unequal, 1–15 mm long, the pedicels 1–7 mm long, the central sessile umbellets 1-flowered. Flowers are white, the calyx teeth not well formed (obsolete). Fruits are ovoid, 1.5–2 mm long and broad, ribs not obvious, covered with short spiny bristles (echinate).

**Habitat:** Canyons, slopes, streamsides. 300–1,500 m (1,000–5,000 ft).

**Range:** Arizona (Cochise, Gila, Graham, Maricopa, Pima, and Pinal counties), southern California, New Mexico, Texas, east to Louisiana, and Missouri; Mexico (Chihuahua? Sonora, and Coahuila).

**Seasonality:** Flowering February to May.

**Status:** Native.

**Ecological significance:** These small herbs are confined to areas that are wet in the springtime. At that time, Scale-Seed flowers along with other wildflowers, although the blossoms on this one are so small that the species barely qualifies as a "wildflower." While the herbs are more common in moist lowlands, plants extend up into mountains from canyons. Scale-Seed is, for example, known in the Chiricahua and Galiuro Mountains (Cochise county), in the Santa Catalina, Tucson, and San Luís Mountains (Pima county), and in the Atascosa and Pajarito Mountains (Santa Cruz county).

The moth *Grammia genura* (Arctiidae) preferentially feeds on *Spermolepis*, but also on *Bowlesia incana* (q.v.) and on *Lomatium nevadense*, when available (Singer and Stireman 2001). Both of those other herbs grow in Brown Canyon, and the larvae of this moth with wide-ranging food tastes may be expected on them.

Scale-Seed may be locally common and typically grow uphill from stands of *Yabea microcarpa* (q.v.). Those two plants may be confused with each other and with *Daucus pusillus* (q.v.). Stems and leaves are rough-hairy (hispid) in *Yabea*, and the fruits are covered with short upwardly hooked bristles. Stems and leaves will be either glabrous or somewhat roughened, but not rough-hairy, in *Daucus* and *Spermolepis*. These two plants can be distinguished by examining the fruits, which are barbed in *Daucus* and hooked in *Spermolepis* and *Yabea*. Moreover, there are no calyx teeth or disk-like enlargements at the base of the style (stylopodium) in *Daucus*, while the calyx teeth are prominent, and the stylopodium is conic in *Yabea*.

Scale-Seed extends west of the Baboquivari Mountains into the Sonoran Desert. There, the herbs are most common in the Ajo Mountains, but they extend into washes in the eastern part of Cabeza Prieta National Wildlife Refuge (Felger et al. in prep.). The species extends into California only in the Borrego Valley (Hickman 1993).

*Spermolepis echinata* grows west of the related *S. divaricata* (fool's parsley, forked seed-scale, rough-fruited spermolepis).

**Human uses:** No uses found for this species. However, *S. divaricata* is an herb that grows from Virginia and Florida to Kansas and Texas (Correll and Johnston 1970), and it is considered a poisonous, nausea-causing plant (Hocking 1997). Although Hocking (1997) listed the species as poisonous, it is not considered as such by Hardin and Arena (1974) or by Turner and Szczawinski (1991), so *S. divaricata* must not be too bad. Still, all members of this family (Apiaceae) are to be regarded with caution. As Foster and Duke (1990) wrote, "Amateurs fooling with plants in the parsley family are playing herbal roulette." Their comment was written specifically for *Zizea aurea*, but it applies to the whole family.

**Derivation of the name:** *Spermolepis* is from Greek *sperma*, seed, and *lepis*, scale, thus alluding to the scruffy or bristly fruits of some species. *Echinata* means prickly, which is another reference to the fruits.

**Miscellaneous:** This genus comprises five species (Mabberley 1997). *Spermolepis* grows in North America, is disjunct to Argentina, and has one endemic species in Hawaii (*S. hawaiiensis*).

## ***Yabea microcarpa*** **(Apiaceae)** **California Hedge-Parsley**

**Other names:** ENGLISH: false carrot (Arizona, New Mexico; "carrot" is based on Latin *carota*, an adaptation of Greek *karoton*; in use by 1533, having come from French *carotte*; Dioscorides, CE 40–80, called *carota* the *pastinaca silvestris*), [California] hedge-parsley ("hedge" is a row of bushes or low trees; e.g., hawthorn, or privet, planted closely to form a boundary between pieces of land or at the sides of a road, in use by 785; derived from Old English **hecg*, *hegg*, corresponding to Early Frisian *hegge*, Middle Dutch *hegghe*, Dutch *hegge*, *heg*, Old High German *hegga*, *hecka*, and German *hecke*; "parsley" is *Petroselinum sativum*; from Latin *petroselinum*, akin to Greek *petroselinon*; in English by ca. 1000; "hedge-parsley" applied by 1830 to the genus *Torilis*, an umbelliferous plant with finely divided leaves; also applied to *Caucalis*), (= *Caucalis microcarpa*)

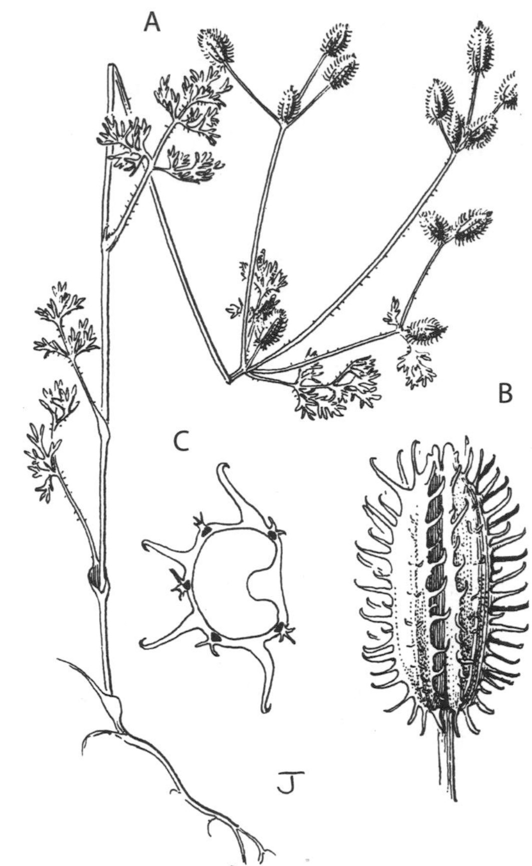

*Yabea microcarpa*. A. Fruiting plant. B. Fruit detail. C. Cross section of fruit. Artist: J. R. Janish. (Wiggins 1980)

**Botanical description:** Herbs, annual, stems 8–40 cm tall, pubescent with spreading hispid hairs, from a slender taproot. Leaves 2–3 (rarely 4) times pinnate or ternate-pinnate, with 3–4 opposite pairs of lateral primary leaflets, the blades 1–5 cm long, oblong or ovate in outline, on petioles 1–4.5 cm long or the upper leaves sessile, the lowest pair of primary leaflets almost half as long as the leaf blade, sessile or on short petioles, the ultimate segments 1–8 mm long, 0.5–2 mm wide. Umbels 1–4, on peduncles 3–10 cm long, the involucre resembling the upper leaves or a little smaller, rays of umbel usually 2–7, these are 1.5–

10 cm long, often about as long as the peduncles. Petals white, stamens white, styles short. Fruits 3–7 mm long, stalk supporting the 2 halves of the fruit (carpophores) bifid for about one-fifth its length.

**Habitat:** Streamsides, washes. Up to 1,370 m (4,500 ft).

**Range:** Arizona (Coconino, Gila, Mohave south to Pima, and Santa Cruz counties), Idaho, and New Mexico; north to British Columbia; Mexico (Baja California and Sonora).

**Seasonality:** Flowering April to May.

**Status:** Native.

**Ecological significance:** These are plants of wet places and are usually confined to the margins of the streams in the spring when there is water runoff from winter rain and snowmelt. The only place *Yabea* has been found west of the Baboquivari Mountains is in the Ajo and Bates Mountains of Organ Pipe Cactus National Monument (Felger et al. in prep.).

**Human uses:** Ebeling (1986) reported that unidentified people in California used the roots for food and medicine. Mathias (1994) added that people chewed the roots to relieve the effects of rattlesnake bites.

**Derivation of the name:** *Yabea* was dedicated to Yoshitaka (Yoshitada) Yabe, 1876–1931, professor of botany at Tokyo University. Among his writings is a revision of the Apiaceae of Japan and Korea. The genus was named in his honor in 1916 when Boris Mikhailovic Koso-Polansky realized that the American plants were different from the *Caucalis* in Europe. *Microcarpa* means small-fruited.

**Miscellaneous:** *Yabea* is a monotypic genus endemic to western North America (Mabberley 1997). *Caucalis*, the genus in which *Y. microcarpa* was formerly placed, is endemic to southern Europe, where *C. platycarpos* (bur parsley) is the only species known. Both *Daucus* and the Old World *Caucalis* are in the tribe Caucalidae, while *Spermolepis* is in the Apioideae with them, but in the tribe Apieae; *Yabea* is in the Torilidinae (Downie et al. 2001).

These herbaceous umbels are easily confused with both *Daucus* and *Spermolepis*. All three may be most abundant along streamsides. Stems and leaves are rough-hairy (hispid) in *Yabea*, and the fruits are covered with short upwardly hooked bristles. Stems and leaves will be either glabrous or somewhat roughened, but not rough-hairy in the other two plants. These two species can be distinguished by examining the fruits, which are barbed in *Daucus* and hooked in *Spermolepis* and *Yabea*. Moreover, there are no calyx teeth or disk-like enlargements at the base of the style (stylopodium) in *Daucus*, while the calyx teeth are prominent, and the stylopodium is conic in *Yabea*.

Another member of the family in the mountains is *Lomatium nevadense* (wild parsley). This plant is bigger than the others, and its fruits are flattened. Wild Parsley is another species in the flora that follows the high and moist Mogollon Rim down across the state into southeastern Arizona. Where the climate is xeric, even if still high, Wild Parsley is usually absent. The species does, however, occur in southwestern Arizona in Barry M. Goldwater Air Force Range and Organ Pipe Cactus National Wildlife Refuge (Felger et al. in prep.).

## *Amsonia kearneyana* (Apocynaceae) Kearney's Blue-Star

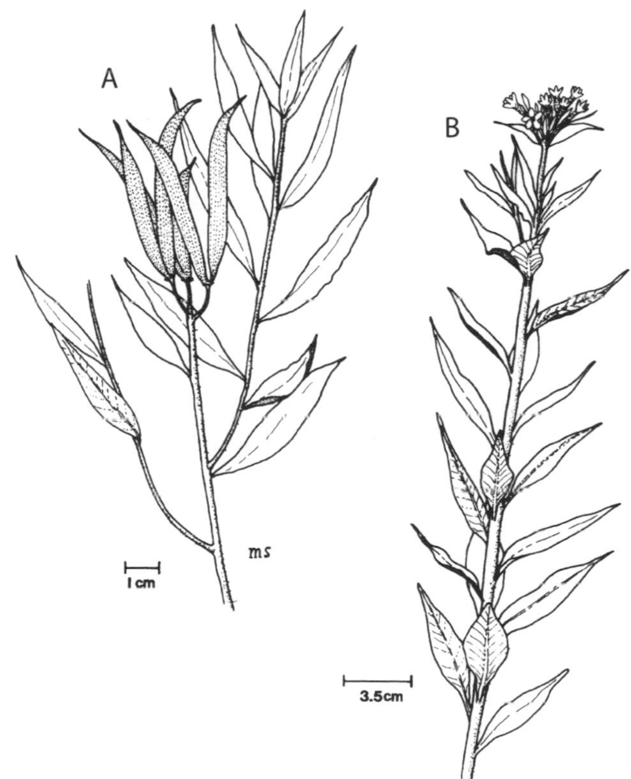


*Amsonia kearneyana*. A. Branch with fruit. B. Branch with flowers. Artist: M. Sharp (Arizona Rare Plant Committee 2001).

**Other names:** ENGLISH: [Kearney's] blue-star (applied to an Australian grass, *Chloris ventricosa*, by 1886; later expanded to *Amsonia*; "blue" was originally applied to the color of the sky, recorded in the literature about 1300), Kearney's amsonia (a book name)

**Botanical description:** Perennial herbs, multi-stemmed, up to 90 cm tall, with milky sap. Leaves alternate, lanceolate to linear-lanceolate, the lower 6–10 cm long and 11–17 mm wide, the upper 4–6 cm long, 3–8 mm wide, entire, soft spreading plant-hairs along the magins. Flowers are in terminal clusters. Corollas salverform, white to pale pinkish to bluish, 12–15 mm long, slightly constricted below the opening, the lobes 2–4 mm long. Fruits folicular, paired, 3–10 cm long. Seeds are cylindrical, corky, 8–11 mm long, 3–4 mm wide.

**Habitat:** Dry, open slopes, 20–30 degrees, in Madrean evergreen woodland and interior chaparral transition zones on stable, partly shaded, coarse alluvium along washes. 1,220–1,830 m (4,000–6,000 ft)

**Range:** Endemic to the slopes of the Baboquivari Mountains where it has a well-isolated range farther southwest than any other Arizona species except *Amsonia palmeri*.

**Seasonality:** Flowering March to May; fruiting June through August.

**Status:** Endemic native.

**Ecological significance:** Originally Kearney's Blue-Star was thought to be endemic to a few sites on the Tohono O'odham Reservation. Additional field studies revealed that it is more widespread. In 1988–1989, a population was established on the Buenos Aires National Wildlife Refuge because of concerns for the stability of other known sites. The floods of 1990 caused a marked reduction of the established plants, but several persist and reproduce. Later, a wild colony that contained about 300 plants was discovered in the canyon above those planted.

A variety of bees and butterflies visit the flowers. Donovan and Topinka (2004) found skippers (Hesperidae), swallowtails (Papilionidae), and bees (Anthophoridae and Halictidae) to be the major flower visitors. Seeds are probably wind- and water-dispersed. The Stinkbug (*Chlorochroa ligata*) consumes the seeds during wet years and may cause dramatic crop loss. When the species was first discovered, it was thought to be a sterile hybrid because the seeds were not viable, probably because of predations by this stinkbug (McLaughlin 1982).

**Human uses:** *Amsonia kearneyana* has no known uses, although several species contain alkaloids and some others are cultivated as ornamentals.

**Derivation of the name:** *Amsonia* is named for American physician Charles Amson, who resided in Gloucester, Virginia, in 1760, and was a friend of John Clayton (see *Claytonia*). *Kearneyana* is named for Thomas H. Kearney, 1874–1956, one of the authors of the *Arizona Flora* (1951).

**Miscellaneous:** The genus contains 20 species native to North America and Japan (Mabberley 1997). *Amsonia* is one of many bicentric genera that have American and Asian species, a phenomenon that was discovered in the 1800s and discussed by many people, including Asa Gray of Harvard University.

*Amsonia* is retained in the Apocynaceae because it has traditionally been placed there. However, there is a growing body of data showing that the Apocynaceae and Asclepiadaceae are artificial groups that should be combined (Judd et al. 2008).

This species is one of eight in Arizona, with the nearest being *A. grandiflora* from near Patagonia and Ruby (Santa Cruz county). *Amsonia peeblesii* from Apache and Coconino counties is also endangered (Arizona Rare Plant Committee 2001). Also growing in the Baboquivari Mountains is *A. palmeri*, although the species is recorded only in the western canyons. *Amsonia palmeri* was recorded by Kearney and Peebles (1951) from Coconino, Mohave, and Yavapai counties; it has subsequently been found in Cochise, Graham, Maricopa, Pima, Pinal, and Santa Cruz counties. It is also in New Mexico (Hidalgo County), where Charles Wright collected it in 1851. *Amsonia palmeri* is also known in Chihuahua, Mexico. Felger et al. (in prep.) do not include it in the flora west of the Baboquivari Mountains.

*Amsonia palmeri* is distinguished from *A. kearneyana* by having fruits (follicles) that are somewhat constricted between the seeds. Moreover, *A. palmeri* has corollas with lobes mostly over 4 mm long (2–7 mm), and a tube 10–18 mm long. McLaughlin (1994) found that the seeds were the best way to separate the two, as they are 3–4 mm wide in *A. kearneyana* and usually less than 2.5 mm wide in *A. palmeri*. Foliage is glabrous or sparsely pubescent in *A. palmeri* and pilose in *A. kearneyana*.

## *Asclepias linaria* (Apocynaceae) Pine-Needle Milkweed, *hierba de cuervo, ban wi:bam*

**Other names:** ENGLISH: pine-needle milkweed (pine-needle refers to the shape of the leaves; MILK + WEED, akin to Middle Dutch *melcwiet*, Old Saxon *melkwid*; in use since the 1590s); SPANISH: *algodoncillo* (little cotton, Durango; see *Gossypium*),

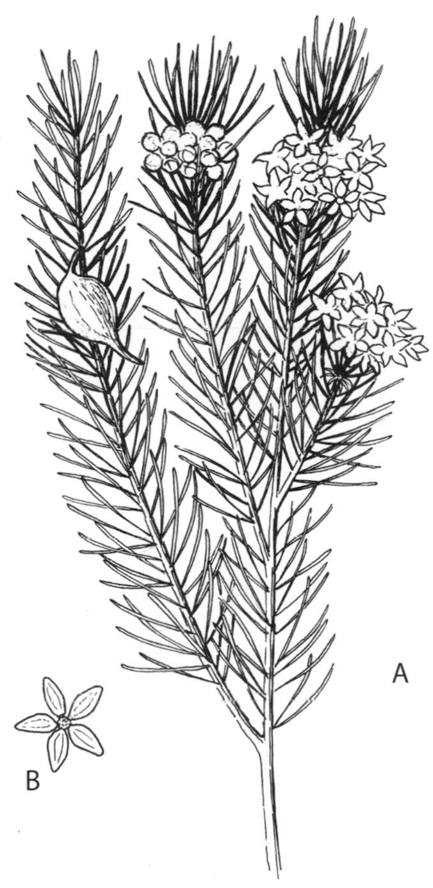


*Asclepias linaria.* A. Branch with buds, flowers, and fruit. B. Apical view of corona segment of corolla, with gynostegium. Artist: B. Swarbrick (courtesy Bonnie Swarbrick).

*hierba de la víbora* (rattlesnake herb, used by Guarijío; a name applied to at least 12 species in 7 families, Sonora), *hierba de la punzada* (puncture herb, Durango), *hierba del cuervo* (raven herb, Sonora), *inmortal* (immortal, Mexico; a name usually given to other plants), *lechestrenza* (milk braids, Mexico), *lechuguilla* (little lettuce, Mexico), *pinillo* (little pine, Edo. México, San Luis Potosí), *plumerillo* (little feathery one, Aguascalientes), *romerillo* (little rosemary, Edo. México), *solimán* (Edo. México), *teperromero* (Mexico), *tlalayote* <*tlalayotle*> (from Náhuatl *tlalli*, earth, *ayotli*, gourd, Mexico), *terbisco* <*torovisco*> (Durango, Hidalgo), *venenillo* (little poisonous one, Edo. México, San Luis Potosí); ATHAPASCAN: *ch'il 'abee'e* (*ch'il*, plant, *'abee'e*, milk, Navajo), *dé'iłchéhé izee'* (*dé'iłchéhé*, red ant, *izee*, medicine, Western Apache), *na'ashǫ'iidą́ą́'* (*na'ashǫ'ii*, reptile, *dą́ą́'*, food, Navajo), *tł'iish izee'* (*tł'iish*, snake, *izee'*, medicine, Western Apache), *tłoibee* (milk plant, Chiricahua and Mescalero Apache); UTO-AZTECAN: *alí okága* (little [plant] with leaves resembling those of pine, Tarahumara), *ban wi:bam* (*ban*, coyote, *wi:bam*, milkweed, Tohono O'odham), *kivat* <*kiyal*> (Cahuilla), *taˀámaˀávi* (*taˀávi*, milk, *maˀávi*, plant, Ute), *tezonpatli* (*tezon*, stone, *patli*, medicine, Náhuatl), *tlalnóchitl* (from Náhuatl *tlalli*, earth, *nochitl*, tuna, Náhuatl), *tlalacxoyatl* (from *tlalli*, earth, *acxoyatl*, spiny plant used by Aztec priests to draw blood, Náhuatl), *tlalochtli* (*tlalli*, earth, *ocitl*, pulque, Náhuatl), *wïi*$^{s}$ (Southern Paiute)

**Botanical description:** Herbs, mostly herbaceous, but sometimes woody at base, to 1.5 m tall. Leaves are spirally arranged, narrowly linear, 2–6 cm long, 0.5–1 mm wide, tightly revolute and almost rounded. Umbellate cymes axillary, 10–30 flowered. Calyx lobes are subulate, 2.5–3 mm long. Corolla lobes are lanceolate-ovate, 3–4 mm long, abruptly acute to acuminate, greenish to white. Staminal column elevated above corolla lobes. Hoods are spreading-ascending to nearly erect, 2.5–3 mm long, ovate in outline, greenish white. Follicles are ovoid-attenuate, 3.5–6 cm long, 1–1.2 cm wide, glabrous. Seeds are 4–5 mm long, granular, with a long tuft of plumose white fluff at apex.

**Habitat:** Dry rocky slopes and canyons, particularly in the oak woodlands. 457–1,828 m (1,500–6,000 ft).

**Range:** Arizona (Cochise, Gila, Graham, Maricopa, Pima, Pinal, and Santa Cruz counties); Mexico (Sonora and south throughout most of country).

**Seasonality:** Flowering February to October, but mostly September to October in the canyon.

**Status:** Native.

**Ecological significance:** These milkweeds, like others, have a specialized pollination mechanism. The flowers contain modified corolla lobes that comprise the hood. Between these lobes, the pollen sacks (pollinia) are almost hidden. The crevices between the hood lobes are so close that the feet of visiting insects slip between them and become hung. When the insect, typically bees or butterflies, struggles to remove its leg, it usually removes the pollen sacks attached to its foot or leg. When the insect lands on a new flower cluster, the foot and leg sometimes are again caught, this time leaving the pollen container from the other flower.

Sometimes this mechanism is devastating to the flower visitors. When insects, such as butterflies and moths with proboscides (long tongue-like mouthparts), visit, it may not be their legs that become stuck. When their proboscides are stuck in the flowers, these animals sometimes are unable to extract them, and they perish.

This mechanism is so finely tuned that the frequency of pollinations to those of flower visits is low. However, when there is a successful pollination, the individual fruits produce dozens or even hundreds of seeds. The success rate of pollination and of recruitment is clearly enough to keep the population of milkweeds at least constant, if not growing.

Queen Butterflies (*Danaus gilippus*) oviposit on the foliage and sometimes visit the flowers. However, even these "milkweed butterflies" prefer nectar from other plants.

**Human uses:** Some Guarijío boil the branches and apply them directly to snakebites (Yetman and Felger 2002). Tarahumara know that *Asclepias* is extremely poisonous, but they use it cautiously to treat headaches, stomach cramps, and as a laxative (Pennington 1969). According to López-E. and Hinojosa-G. (1988), this plant is used in Sonora to treat headaches and toothaches. Martínez (1969) found *A. linaria* being used as a laxative, but noted that "*su uso es infiel y peligroso*" (its use is unreliable and dangerous).

**Derivation of the name:** *Asclepias* was named for the legendary Greek physician and deity *Asklepios* or *Aesculapius.* According to Greek stories, this renowned individual was the son of Apollo. He was born in a forested area of *Pelion* (*Pilion*) in *Thessaly* and grew up there under the watchful eye of the wise centaur *Chiron*, who taught him the art of medicine. *Linaria* connotes resembling *Linum*, or flax.

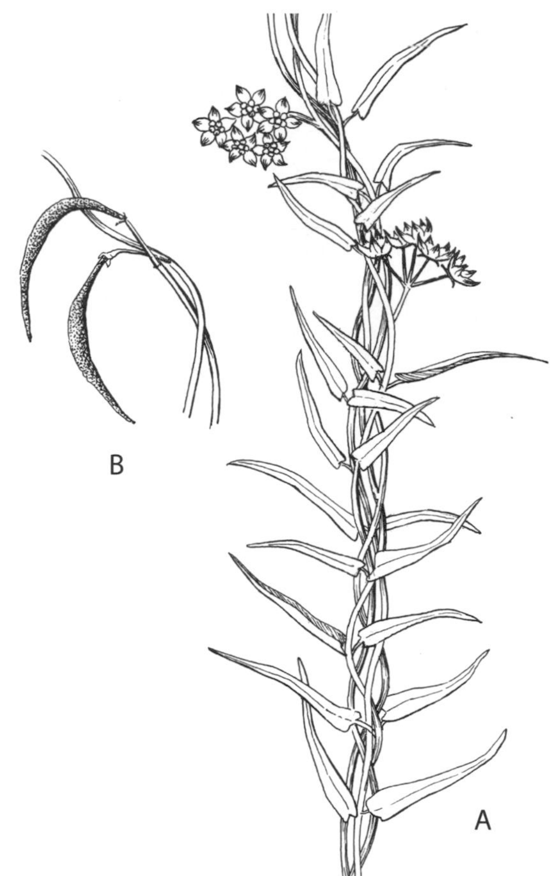


*Funastrum cynanchoides*. A. Flowering branches. B. Fruiting branches. Artist: B. Swarbrick (courtesy Bonnie Swarbrick).

## *Funastrum cynanchoides* (Apocynaceae) Climbing Milkweed, *güirote, wi:bam*

**Other names:** ENGLISH: Climbing milkweed (climbing refers to the twining stems; milkweed, see *Asclepias*), fringed twine-weed [vine] (Arizona, New Mexico); SPANISH: *güichire* (from Cahita), *hierba lechosa* (milky herb, Sonora), *güirote lechoso* (milky "vine," *güirote* <*huirote*> is a Cahita or Ópata word, probably originally applied to *Cardiospermum*, Sinaloa, Sonora), *platanito* (little banana [literally "flat one"], Sonora), *mata nene* (baby killer, Sonora), *sandia de la pasion* (watermelon of the crucifixion, Sonora); HOKAN: *hexe* (Seri); UTO-AZTECAN: *huichuri* <*huichoori*> (Mayo), *vibam* (Mountain Pima), *viibam* (milk it has, Akimel O'odham), *vi:bam* <*vi'ibgam*> (Hiá Ceḍ O'odham), *wi:bam* <*wi'ibgam*> (Tohono O'odham); (= *Sarcostemma cynanchoides*)

**Botanical description:** Trailing or twining from a deep, tuberous root. Stems 1 m or more long, much-branched, glabrous or puberulent. Leaves to 6 cm long, to 3.5 cm wide, usually smaller, broadly to narrowly ovate-lanceolate, acute to acuminate apically, cordate to hastate basally, sparsely pubescent above and below, with one or more glands on the midrib near the base. Inflorescences are umbellate, to 20-flowered. Calices lobes are ovate to narrowly ovate, 2–3 mm long, pilosulose without. Corollas are rotate-campanulate, greenish white to purple or pinkish, the lobes 5–7 mm long. Follicles are fusiform, to 7 cm long, attenuate apically, puberulent.

**Habitat:** Canyons, streamsides, washes, 450–1,370 m (1,500–4,500 ft).

**Range:** Arizona (Coconino and Yavapai to Greenlee, south to Cochise, Pima, and Santa Cruz counties), California, New Mexico, Oklahoma, western Texas, and southern Utah; Mexico (Baja California, Sonora to Tamaulipas, south to Jalisco, and Querétero).

**Seasonality:** Flowering (April) May to September.

**Status:** Native. The only variant found in the canyon is ssp. *cynanchoides.*

**Ecological significance:** These plants thrive where there is a little extra moisture. Rea (1997) found them most abundant near the *akĭ chiñ* (arroyo mouth) agricultural fields of the Akimel O'odham. The vines are among the comparatively few of that life-form in the deserts, and are typically confined to lower canyons and below in the grasslands and desertscrub.

Flowers are visited by moths, bees, and a variety of other insects (Holm 1950), as are those of *Funastrum clausum* in Florida. When an insect lands on a flower cluster, the group tilts, and the insect grasps the projecting anther wings. That action results in pollen sacs (pollinia) sticking to it, and these sacs are withdrawn when the insect again moves. When the pods open in the fall, the seeds are borne away on parachutes of long plumes of white silk attached to the apex of the seeds.

**Human uses:** Tohono O'odham eat the fruits of this subspecies, raw or cooked (Kearney and Peebles 1951). All other reports are for ssp. *hartwegii*; possibly both subspecies are equally usable. Rea (1997) speculated that *ban viibam* (coyote's *viibam*) was ssp. *cynanchoides.* Seri eat the flowers as a snack while walking through the desert. Felger and Moser (1985) thought the flowers had a faintly onion-like taste. Sap or green pods are used by Akimel, Hiá Ceḍ, and Tohono O'odham as chewing gum (Curtin 1949, Rea 1997).

Mayo boil *Funastrum* plants to produce a wash to relieve scorpion stings, and the milky sap is applied to calluses and corns to remove them (Yetman and Van Devender 2001). Seri use a similar decoction to relieve headaches, and it is drunk to alleviate symptoms of Black Widow Spider bites (Felger and Moser 1985). Seri cook the roots in water and use the liquid as eye drops.

**Derivation of the name:** *Funastrum* is based on the Latin words *funis*, rope, cord, and the suffix *-astrum*, resembling, alluding to the twining stems. *Sarcostemma*, the old generic name, is from Greek *sarkos*, fleshy, and *stemma*, crown, because the floral corona is fleshy. *Cynanchoides* means that it resembles *Cynanchum*, another climbing milkweed genus.

**Miscellaneous:** *Funastrum cynanchoides* ssp. *hartwegii* is the subspecies with broader (2.5–4.8 cm), longer (11–15.5 cm) leaves and white flowers. This variant grows from southern California, southwestern Utah, and western Texas south to Baja California and central Mexico. Typically, it is found in dryer sites than ssp. *cynanchoides.*

Another plant in the canyons with a twining lifeform is *Gonolobus arizonicus* (= *Matelea arizonica*). This broadleaf climber is widespread in the Baboquivari Mountains, although it is uncommon in Brown and Jaguar Canyons. This milkweed is also known in Sycamore Canyon in Santa Cruz County (Pajarito Mountains), and in the Rincon and Santa Catalina Mountains. In Arizona, it is known only in Maricopa and Pima counties. Kearney and Peebles (1951) listed this climbing milkweed as endemic to southern Arizona; they were not quite right, because *Gonolobus arizonicus* also occurs in northern Mexico.

## *Aralia humilis* (Araliaceae) Spikenard, *palo santo*

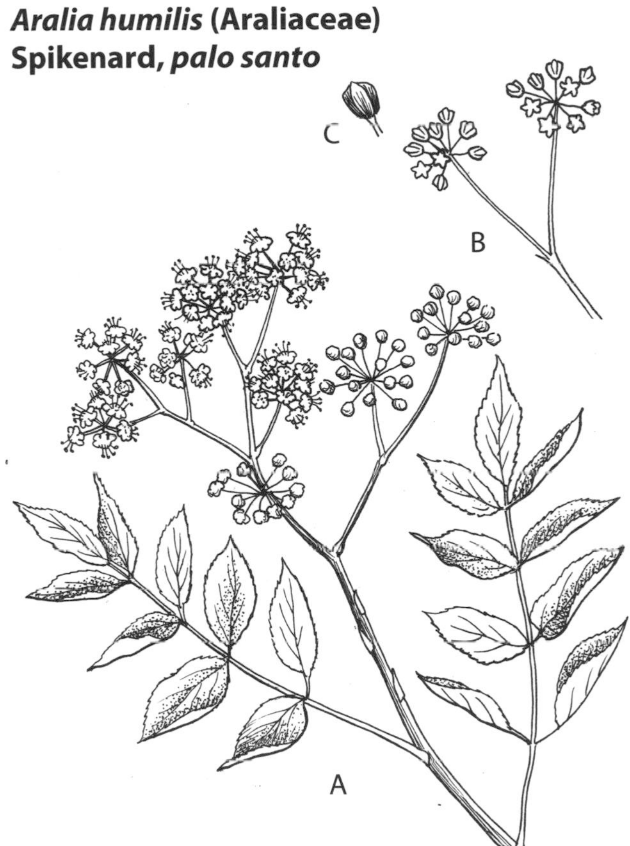


*Aralia humilis*. A. Branch with buds and flowers. B. Fruiting branch. C. Immature fruit. Artist: B. Swarbrick (courtesy Bonnie Swarbrick).

**Other names:** ENGLISH: [Arizona] spikenard (derived from Latin *spica*, spike, plus *nardi*, aromatic resin; an aromatic substance employed in Roman times in the preparation of a costly ointment or oil, originally made from *Nardostachys grandiflora*, Valerianaceae; used in English by CE 1350); SPANISH: *cuajilotillo* (a name usually given to *Aralia pubescens*, growing from central Sonora to Oaxaca; for some reason, these shrubs are compared to the *cuajilote*, from Náhuatl *cuahuitl*, tree, *xilotl*, *Parmentiera edulis*), *curguatón* (Chihuahua), *palo santo* (holy tree, Sonora), *petatillo* (little matting or mat, Mexico, from Náhuatl *petatl*, matting, mat), *tepatete* <*tepetate*> (Sonora; from *tepate* or *Datura*, reference obscure, or perhaps from *petatl*, mat, Náhuatl); UTO-AZTECAN: *teniparé* <*tanipari*> (Guarijío)

**Botanical description:** Shrubs or small trees, 2.5–5 m tall or rarely trees 6–10 m with a broad, open crown. Leaves winter deciduous (drought deciduous?), aromatic, odd pinnate, highly variable in shape and size, 15–38 cm long, petioled; leaflets 5–11, usually 7–9, on slender stalks, the blades thin, 5–13 cm long, 3–8 cm wide, ovate to ovate-triangular or oval, tips acuminate, the margins usually toothed. Flowers are monoecious or perhaps also bisexual, greenish white, on slender pedicels in umbellate-globose clusters on paniculate inflorescences. Berries are 5 mm in diameter, red when ripe, with 5 conspicuous grooves.

**Habitat:** Canyon bottoms, hill slopes, pine-oak woodlands. 1,066–1,524 m (3,500–5,000 ft), in Mexico 700–1,800 m

**Range:** Arizona (Cochise, Pima, and Santa Cruz counties) and New Mexico; Mexico (Chihuahua and Sonora).

**Seasonality:** Flowering July to October.

**Status:** Native.

**Ecological significance:** By October, or sometimes earlier, the butterflies have discovered the flowers of these shrubs. It is not unusual, in a good year, to see sulphurs of several species, Painted Ladies (*Vanessa cardui*), Queens (*Danaus gilippus*), and occasional Monarch (*Danaus plexippus*) butterflies visiting the flowers for nectar. These abundant visits, and probably those from other insects such as flies and bees, result in a heavy fruit set.

Immature fruits as early as September or as late as November make a contrasting purple and green cluster that catches the human eye; fruits are more arresting when fully red later in the season. If fruit colors do that with people, they surely do the same for fruit-eating birds that live in the vicinity. Felger et al. (2001) suggest that *A. scopulorum* of Baja California Sur may be the closest relative of *A. humilis*.

**Human uses:** Guarijíos gather and dry the bark of *Aralia* to make a tea to treat kidney problems and diarrhea (Yetman and Felger 2002). Miller (1996) said the wood of *Aralia* was used to make the *nuca del violín* (neck of the violin). Hocking (1997) gives information on 10 other *Aralia* species used by people. Most of those listed are in the Old World, but several are in the Americas.

Of those species in the New World, *A. hispida*, *A. nudicaulis*, *A. racemosa*, and *A. spinosa* have been used by people in the eastern United States and Canada. Several indigenous tribes use these plants, including the groups in the Southeast (Austin 2004). A common use among those species and tribes is as a diuretic, although *A. nudicaulis* has been added to root beer in place of "true salsaparilla" (*Smilax*). In the northeastern United States, Manassah Cutler wrote in 1785 that *A. nudicaulis* had roots that ". . . are aromatic and nutritious. They have been found beneficial in debilitated habits. It is said the Indians would subsist upon them, for a long time, in their war and hunting excursions." The same author wrote of *A. racemosa* that "[i]t is aromatic. The berries give spirits an agreeable flavour. The bark of the root and berries are a good stomachic." *Aralia racemosa*, too, was considered a good remedy for diarrhea, but it was mostly used to treat pulmonary ailments.

**Derivation of the name:** *Aralia* is from French-Canadian *aralie*, the original specimens having been sent by the Quebec physician Michel Sarrasin de l'Étang (1659–1734) to J. P. Tournefort (1656–1708) under that name. The origin of the word is unknown, although Coffey (1993) said that it was "[t]he Indian name." He does not name the tribe, but in that region of Quebec it may have been Algonquian or Iroquoian. *Humilis* means low-growing, small, in reference to the stems.

**Miscellaneous:** There are 36 or more species of *Aralia* in North America, eastern Asia, and Malesia (Mabberley 1997). Some kinds are cultivated as ornamentals, while others are medicinal and even edible (*A. chinensis*).

## *Aristolochia watsonii* (Aristolochiaceae) Indian-Root, *hierba de indio*

**Other names:** ENGLISH: Indian-root (Arizona), Watson's Dutchman's pipe (from similarity of the flower shape to the curved pipe used by the fictional Sherlock Holmes; in use by 1857), [Arizona] snake-

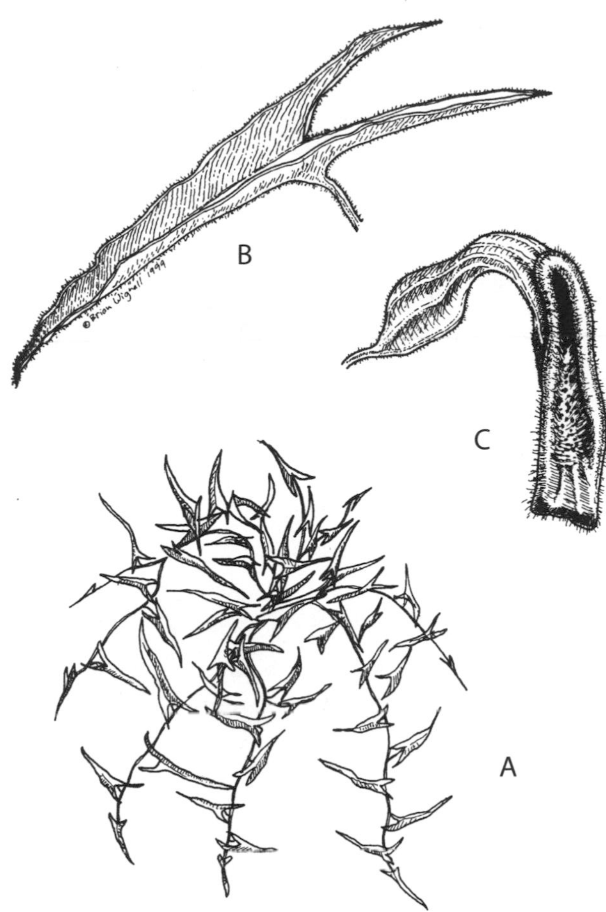


*Aristolochia watsonii.* A. Growth form. B. Leaf, enlarged. C. Apex of flower, enlarged. Artist: B. Wignall (Phillips and Comus 2000).

root (because of use to treat snakebite, in use by 1635); SPANISH: *huaco* <*guaco*> (from Taino, originally applied to *Mikania*, then to other genera used to treat snakebite), *hierba* <*yerba*> *de[l] indio* (Indian's herb, Arizona, Baja California, Sonora); HOKAN: *hatáast an ihiit* (what gets between the teeth, Seri); UTO-AZTECAN: *guasena jubiaria* (Mayo), *yerbalind* (probably from Spanish, *yerba de india*, Indian's herb, Mountain Pima)

**Botanical description:** Perennials from a single thickened, carrot-shaped root; dying back to root during drought and with freezing weather. Stems slender, herbaceous, trailing or twining, often less than 30 cm, although typically 1–1.5 m long in shaded, moist habitats. Larger leaves are 2.5–12 cm long, the blades narrowly triangular-hastate, or the lower leaves broadly triangular to triangular-hastate, the basal lobes as long as or longer than the petioles. Flowers are 3.5–5 cm long, solitary in leaf axils. Calyx tube slightly inflated surrounding style and stamens (just above ovary) and narrowed at the throat, the limb more or less "tooth-shaped" (1-lobed), yellow-green with brown-purple spots mostly along 5 prominent veins, the margin and tip dark maroon. Unopened capsules ovoid, 1.6–2.5 cm long, with a narrow ridge or wing along the midrib of each of 5 valves. Seeds flattened, blackish.

**Habitat:** Slopes and ridges, often below bushes and shrubs. 600–1,370 m (2,000–4,500 ft).

**Range:** Arizona (Gila, Greenlee, and southern Mohave to Cochise, Pima, Santa Cruz, and Yuma counties), southwestern New Mexico; Mexico (Baja California Sur, Chihuahua, and Sonora).

**Seasonality:** Flowering April to October.

**Status:** Native.

**Ecological significance:** This plant is one of the smallest of the 300 species of *Aristolochia* (Felger 2000). These vines are more common in the Altar Valley and lower in the canyon, but they grow well above the Environmental Education Center in Brown Canyon. Where individuals occur, *A. watsonii* "hides" below bushes and shrubs where they presumably take advantage of microclimates—and probably pollinators.

Pollination in *Aristolochia* is by flies, and those visiting *A. watsonii* are in the Ceratopogonidae (biting midges, black flies, "no-see-ums," punkies). The unusual aspect of these flies is that they are small, blood-sucking insects that pester humans and other mammals in the humid summer (Crosswhite and Crosswhite 1984, Dimmitt 2000). Instead of these flowers smelling like carrion or dung to attract flies, as do many members of the genus, *A. watsonii* gives off a musty odor and physically resembles a mouse's ear. The odor and visual cues fool the flies into thinking they will find a blood meal. These flies are trapped in the flowers long enough to accidentally pollinate them.

Leaves are often consumed by the black caterpillars with red "spines," of the Pipevine Swallowtail Butterflies (*Battus philenor*). These animals sequester the aristolochic acid from Indian-Roots, which renders the lepidopterans unpalatable to birds. This species is the only butterfly able to feed on *Aristolochia* because of toxic aristolochine, aristolochic acid, and volatile oils.

**Human uses:** The Old World species have long been important to people who used them medicinally (Austin 2004). Their names for *Aristolochia* point to a long utilization—*aristoloche* (French), *aristolochia* (Italian), *aristoloquia* (Portuguese), *culurain* (Gaelic), *Osterluzei* (derived from *Aristolochia*; thus, it has nothing to do with East [*Ost*] or Easter [*Ostern*], and *luzei* has no meaning), Desert Birthwort (from use to aid childbirth), *legeholurt* (*lege*, doctor, *hol*, hollow, *urt*, herb, Norwegian).

Indian-Root was used by indigenous people and settlers as a snakebite remedy (Kearney and Peebles 1951, Felger, personal communication, June 2005). The freshly dug roots are considered edible by some. Hicks (1966) said roots were "good to eat," although Michael Moore (2005) compared the taste to "a combination of mothballs and aspirin." Tea brewed from the roots is widely used as a tonic to relieve upset stomachs and aid digestion. Through much of its range, the leaves are boiled and the tea drunk to expel worms and treat stomachaches (Martin et al. 1998).

The Seri name refers to Indian-Root's being placed on the tooth for use as medicine (Felger and Moser 1985). A decoction of *Aristolochia* is held in the mouth to relieve toothaches or the dry root is held in the fire and then placed over the tooth cavity.

**Derivation of the name:** *Aristolochia* is based on Greek *aristo*, best, and *lochia*, delivery, from the use among European people as an aid in childbirth. The curved flower, with the summit and base together, suggested to them the human fetus in the womb, at least according to the Doctrine of Signatures (Bennett 2007). *Watsonii* is dedicated to Sereno Watson, 1826–1892, curator of the Gray Herbarium at Harvard University from 1888–1892 and critical student of western American plants.

## *Ageratina herbacea* (Asteraceae) Snakeroot

**Other names:** ENGLISH: ageratina (a book name based on the genus), [fragrant] snakeroot (one of many plants used by indigenous people to treat snakebites, New Mexico), white thoroughwort (because the stem appears to grow "thorough" [through] the leaves in some species; "wort" is an old English word meaning plant); SPANISH: *mata* (plant), *tabardillo* (an old name for several different plants, although originally applied to *Sanicula*, Apiaceae; now usually applied to *Calliandra* in Baja California and Sinaloa); ATHAPASCAN: *bił háách'i* [*bikąˀí*] <*bilha.zef'n*> ([male] wind odor, its scent is carried on the breeze, Navajo); (= *Eupatorium herbaceum*)

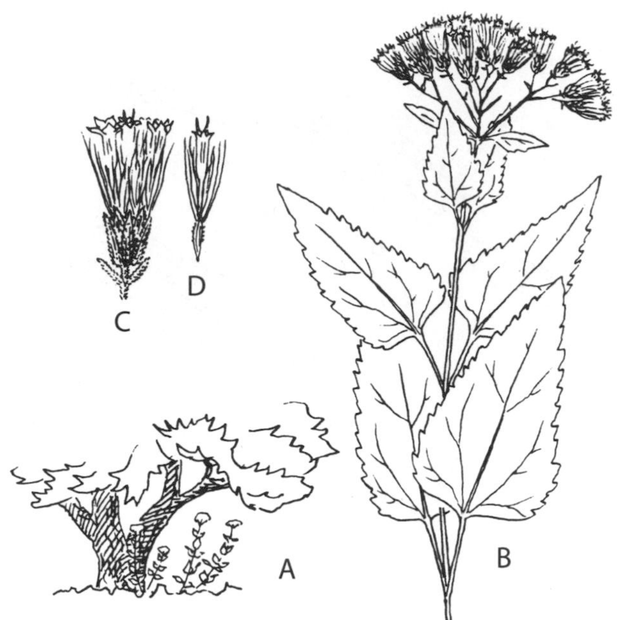


*Ageratina herbacea.* A. Typical habitat for the herbs below other plants. B. Flowering branch. C. Flower head. D. Individual flower. Artist: R. D. Ivey (Ivey 2003).

**Botanical description:** Perennial herb from a woody caudex. Stems 4–7 dm tall, branched above. Leaves mostly opposite, the blades 1.5–6 cm long, 0.5–4 cm wide, ovate, the bases cordate or truncate, coarsely crenulate-serrate, acute. Heads are numerous, in densely corymbose clusters. Involucres are 3.5–5 mm wide, 3–4 mm high; bracts are green, puberulent, almost equal in length. Corollas are white. Achenes are black, 1.5–2 mm long.

**Habitat:** Farther north, Snakeroot is associated with Ponderosa Pine forests and spruce-fir communities; here, along streambeds. 1,524–2,743 m (5,000–9,000 ft).

**Range:** Arizona (Cochise, Coconino, Greenlee, Mohave, Navajo, Pima, and Santa Cruz counties), California, Colorado, New Mexico, and Utah; Mexico (Sonora).

**Seasonality:** Flowering June to October.

**Status:** Native.

**Ecological significance:** In Brown and Jaguar Canyons, *A. herbacea* is confined to narrow banks of intermittent streams where it grows in the shade of sycamores (*Platanus wrightii*) and oaks (*Quercus*). The flowers are visited by a variety of insects.

*Ageratina herbacea* has also been found in Baboquivari and Fresnal Canyons. This species is one of two in *Ageratina* known from the Baboquivari

Mountains (cf. Appendix). Snakeroot is also known in the Quinlan, Santa Catalina, and Sierrita Mountains; *A. herbacea* is not known in southwestern Arizona or northwestern Sonora (Felger 2000, Felger et al. in prep.).

**Human uses:** Navajo use *A. herbacea* for a "chant lotion" (Wyman and Harris 1941). Although the name implies simply using the mixture externally, the "lotions" are used to relieve a variety of ailments, including headaches, fever, lameness, and general body aches and pains, as well as coughs, colds, and chills. Zuni use the related *A. occidentalis* to treat rheumatism and swelling (Stevenson 1915). Several tribes in the eastern United States used *A. altissima* similarly, and that species was subseqently considered diuretic, diaphoretic, antispasmodic, aromatic, and vulnerary by Anglo physicians (Austin 2004).

The genus has a reputation as a poisonous plant, with *A. altissima* the best known. White Snakeroot (*A. altissima*), called *noota ikheesh* (*noti'*, tooth, *ikhsnsh*, medicine) by the Choctaw and Chickasaw, became known as the "Milk-Sickness Plant" (from the technical physician's name *morbeo lacteo*). *Morbeo lacteo*, also called the "Trembles," was a mysterious and feared disease that attacked farm families and their livestock in the 1800s (Austin 2004). The malady was more common in the South and Midwest, particularly in North Carolina, Ohio, Indiana, and Illinois. In 1818, Abraham Lincoln's mother, Nancy Hanks Lincoln, died of this disease in Little Pigeon Creek, Indiana. At least partly because of that, Thomas Lincoln and his new wife Sarah and young Abraham, moved to Illinois. Although the cause of the problem was determined in the 1830s, county officials in Kentucky were still complaining in 1852 that the disease was so prevalent that it deterred settlers.

A Shawnee woman showed Illinois physician Anna Pierce the poisonous plant that caused the disease. That fugitive from the forced relocations of local tribes was known only as "Aunt Shawnee." In spite of that Anglo prejudice, she explained the relationship between White Snakeroot and both Milk Sickness and Trembles. However, it was not until 1917 that physicians convinced themselves that Snakeroot caused the problems.

Tremetol, a toxic alcohol, is the chemical causing "milk-sick," or milk sickness in humans or trembling in other animals. *Ageratina* causes ketosis (excess ketone bodies; overly acidic body chemistry) and is transmitted through cow milk to people. The breath takes on a fruity odor as acetone diffuses from the lungs. The symptoms are progressive and include lassitude, nausea, vomiting, stomach pains, intense thirst, prostration, coma, and death.

**Derivation of the name:** *Ageratina* is a diminutive form of the genus *Ageratum*, from Greek *ageraton*, not growing old, an allusion to the long-lasting flowers. Perhaps the term was originally applied by Pliny, CE 23–79, to what is now *Achillea* or Yarrow. *Herbacea* means "not woody," a comparison with similar species having woody stems.

**Miscellaneous:** There are 250–290 species in the genus, which ranges from the eastern United States (3 spp.) to South America (Mabberley 1997, Nesom 2006). The studied species, including *A. herbacea*, are diploids with 2n = 34 chromosomes. According to Watanabe et al. (1995), that number gave rise to plants with 2n = 20, in an unexpected decrease by loss of individual chromosomes through fusion of fragments of them being attached to others within the cells and the loss of their centromers. Previously, it was thought that the higher numbers had been derived from the lower.

### ***Ambrosia confertiflora* (Asteraceae)**<br>**Slim-leaf Bursage, *estafiate, mo'o taḍ***

**Other names:** ENGLISH: bursage [field, weakleaf burr] ragweed ("ragweed" from "ragwort," where "rag" denotes the incised margins of the leaves, in use by 1658; first applied to *Senecio*, then by 1866 to *Ambrosia*, New Mexico), slim-leaf [weak-leaf] bursage (an allusion to the prickly fruit and aromatic herbage; "bur" any rough or prickly seed-vessel or flower-head of a plant, in use by ca. 1330; akin to Danish *borre*, bur, burdock, and Swedish *borre*, sea-urchin; "sage" from Latin *salvia*; akin to Middle English *sauge*, French *sauge*, by the thirteenth century); SPANISH: *altamisa* [*del campo*] ([wild] corruption of *ambrosia*, Mexico), *estafiate* (from Náhuatl, see *Artemisia*, Mountain Pima), *yerba del sapo* (toad herb, New Mexico); ATHAPASCAN: *ch'ił diwozh* <*c'il dahwosi* [*dohwosi*]> (*diwozh*, spiny, *ch'ił*, plant, Navajo), *dziłk'ijí jatáál bilátah łitso* <*ʒiłk'iʒví zan'ił bilátah łoci*> (Mountaintop Way lotion with yellow flowers, Navajo; for *A. acanthicarpa*); HOKAN: *paxáaza* (Seri); KIOWA TANOAN: *waejoka* (Tewa); UTO-AZTECAN: *chíchibo* (Mayo), *mo'ostadk* (Hiá Ceḍ O'odham), *mo'o taḍ* <*mo'otaḍk, mo'otadk, mo'ostalk, mo'otari*> (to stick

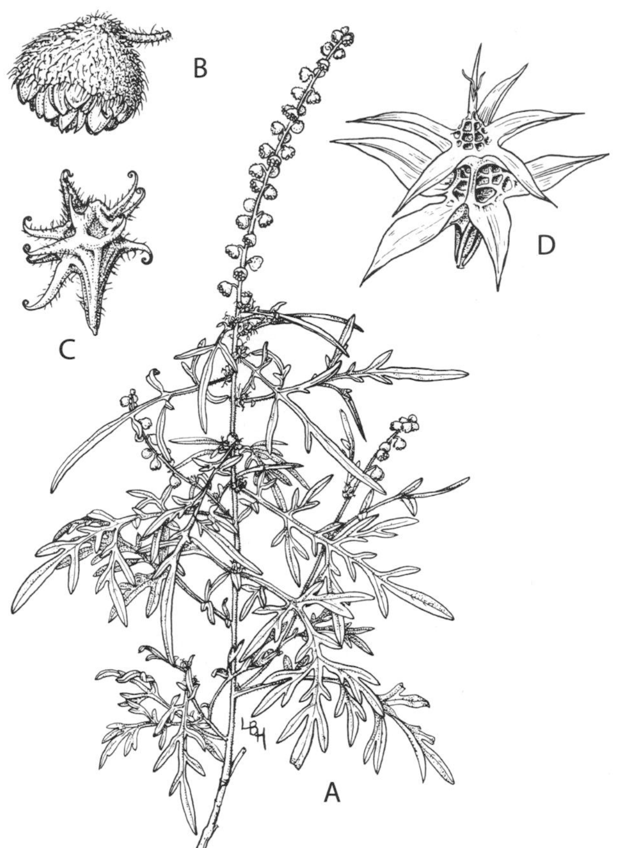


*Ambrosia confertiflora*. A. Flowering branch. B. Enlarged male flower cluster. C. Mature female flower (bur), with curved, hooked spines. D. Mature female flower of *A. acanthicarpa* for comparison. Artist: L. B. Hamilton (Parker 1958).

its head out, an allusion to broom rape [*Orobanche cooperi*], Tohono O'odham), *mo'o taḍk je:j* (mother of broom rape, Akimel O'odham), *mo'o taḍ* <*mo'otur*> (Onavas Pima; used for *A. acanthicarpa*), *musha* (Mountain Pima), *ñuñuwĭ je:j* (mother of vultures, Tohono O'odham), *pawya* <*pawíya*> (Hopi), *tatṣagi* <*taḍshagi, tatshagi*> (possibly a confusion with *totṣagi*, foamy, Tohono O'odham), *tu'rosip* (*turovi*, black, *sip*, juice, Shoshoni), *wahsapári* (Guarijío; for *A. acanthocarpa*)

**Botanical description:** Perennial, forming colonies, stems 0.3–0.6(–1.8) m tall. Leaves are alternate, 1–4 times pinnatifid, often with 1 to several pairs of small lobes on the petiolar bases below the main blade. Staminate heads are short-stalked. Fruits (achene) are ca. 5 mm long, the spines usually 10–20, rarely none, 1–2 mm long, with hooked tips.

**Habitat:** Washes, roadsides, disturbed grasslands. 304–1,981 m (1,000–6,500 ft).

**Range:** Arizona (Cochise, Coconino, Graham, Mohave, Pima, and Santa Cruz counties), California, to Colorado, and Texas; Mexico (Baja California and Sonora to Tamaulipas).

**Seasonality:** Flowering (April) September to October.

**Status:** Native.

**Ecological significance:** These are plants of disturbed sites, although the perennial habit may allow them to persist for decades. Presumably, the achenes of *A. confertiflora*, like many in the genus, are important foods for many birds (Martin et al. 1951). The hooked spines on the fruits are adaptations to animal dispersal. These fruits are probably cracked open and eaten, while *A. artemisiifolia* (farther east) and *A. trifida* (in the lowlands of the Arivaca Ciénega) are eaten whole.

Pollen of this species and other members of the genus are important allergenic sources for many people with hay fever. A person allergic to one species is likely to be cross-sensitive to all the others.

**Human uses:** The only record of Sonora people eating the species is an old one among the Tohono O'odham by Castetter and Underhill (1935); they reported that the young herbage is gathered and cooked as greens in the summer. Given the presence of poisonous chemicals in most members of the family, edibility is suspect (Mabberley 1997). More than likely, there was confusion with the edible *Orobanche* often growing on *Ambrosia* (cf. Rea 1997).

Mayo make a tea to expel persistent intestinal parasites (Yetman and Van Devender 2001). The same tea is used to treat diarrhea. Some Akimel O'odham make a bitter tea from the roots to treat stomachaches (Rea 1997). Similarly, the related *A. acanthicarpa* is used as a medicine by the Zuni (Stevenson 1915). Navajo use ashes from *Ambrosia* in the Evil Way blackening (Vestal 1952) and in the Mountaintop Way to dust Meal Sprinkler's equipment (Wyman and Harris 1951).

Mayo make a layer of *Ambrosia* on top of the *latas* (roof slats) and cover it with dirt to make a leak-free roof (Yetman and Van Devender 2001).

**Derivation of the name:** *Ambrosia* is from Greek *ambrosios*, delicious, divine, immortal, derived from *a*, not, *mbros*, mortal; in Greek mythology, the food and drink of the gods. *Confertiflora* means "with clustered-flowers," thereby alluding to the crowded inflorescences.

**Miscellaneous:** The original Eurasian "ambrosia" was considered part of the food of the gods. Davidson (1999) summarized the term by noting that, in the narrow sense of eatables, it was the counterpart of "nectar," the "drink of the gods." In a broader sense, "ambrosia" may mean food and drink. Regardless, no one has any idea what the gods were supposed to be eating when applying this term. Certainly, it is unlikely to have been anything like the ragweeds that Linnaeus dubbed with this Greek name. Perhaps it was Linnaeus's perverse sense of humor that led him to call these allergy-causing plants that are poisonous to livestock the "nectar of the gods."

### *Ambrosia monogyra* (Asteraceae) Burro-Brush, *jécota, 'i:wadhoḍ*

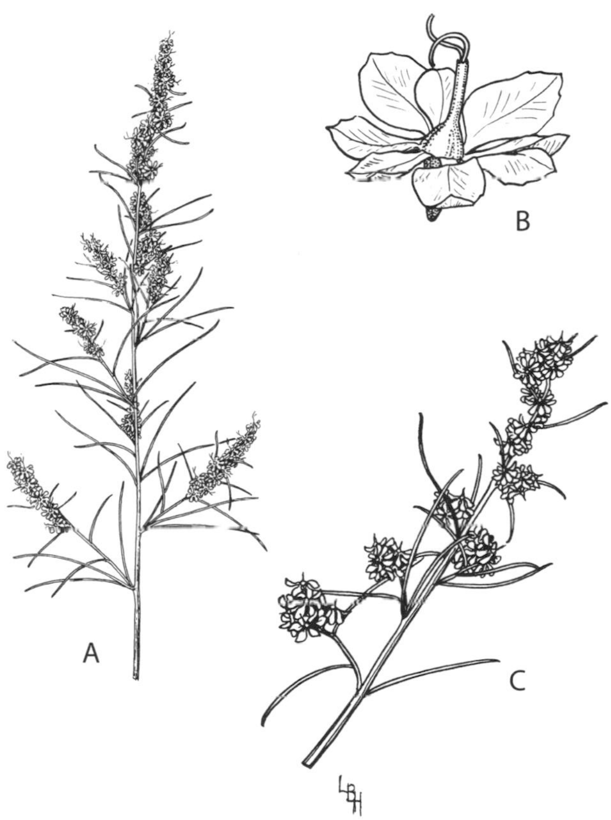


*Ambrosia monogyra.* A. Branch with both male and female flowers. B. Fruiting head, enlarged. C. Fruiting branch. Artist: L. B. Hamilton (from original artwork).

**Other names:** ENGLISH: arrow-wood (long, straight stems formerly used for arrow shafts), [single-whorl] burro-brush [bush] (nothing but a burro will eat it; the word "burro" came into English from Spanish about 1800), cheese-bush (used or because of leaf odor? Sonora), winged ragweed ("ragweed" from "ragwort," where "rag" denotes the incised margins of the leaves); SPANISH: *hécota* <*jécota, jejego*> (among Mayo, Guarijío, Onavas Pima; possibly from Náhuatl *xiotl*, mange, and medicinal use), *hierba del pasmo* (herb for treating *pasmo*, from Latin *spasmus*, Sonora), *romerillo* [*dulce*] ([sweet] little rosemary, Baja California, Sinaloa, Sonora); ATHAPASCAN: <*tłeł*> (Western Apache); UTO-AZTECAN: *'i:wadhoḍ* <*'i:watoḍ, i:watodh, iivadhoḍ*> (Tohono O'odham), *'i:vadhod* (Hiá Ceḍ O'odham), *iivdhat* (the word signifies something that is leafy and green, a "non-descript bright green cloud out in the desert," Akimel O'odham), *i'ivdag* <*i'ivdad*> (Onavas Pima), *jeco* (possibly cognate with Náhuatl *xiotl*, mange, Mayo, Guarijío), *païab* (Southern Paiute); YUMAN: *o'gach* (Walapai); (= *Hymenoclea monogyra*)

**Botanical description:** Perennial broom-like shrubs 5–20 dm tall, densely branched. Leaves are alternate, linear or filiform. Heads are in small globose axillary groups, ca. 5 mm high, of two kinds, the more numerous of staminate flowers and the less numerous of pistillate, both kinds present in the same cluster. Pistillate heads are 1-flowered, the involucre papery, whitish, with a basal obconical portion ca. 1 mm high and 9–12 spreading obovate outer phyllaries ca. 2 mm long, forming a saucer-shaped wing-like structure. Pappus is absent.

**Habitat:** Grasslands, washes. 300–1,200 m (1,000–4,000 ft).

**Range:** Arizona (Mohave to southern Yavapai, south to Cochise, Gila, Graham, Maricopa, Pima, Pinal, and Santa Cruz counties), southern California, New Mexico, and western Texas; Mexico (Baja California Sur, Chihuahua, Sinaloa, and Sonora).

**Seasonality:** Flowering September to October (November).

**Status:** Native.

**Ecological significance:** These plants are shrubs of arroyos, washes, and other temporally wet sites. Wherever the *jécota* grows, one may be assured that it will be seasonally wet or perhaps even flooded. Even areas with Burro-Brush that seem to be always dry will have water during rainstorms. Where these plants grow, particularly in dense stands, roots are efficient at erosion control (Dayton 1937).

Fruits are modified for water and probably wind dispersal. The northern limit closely follows the 1,300 m contour (Turner et al. 1995). Dayton (1937) found that the buds and sprouts of the shrub were an important source of food and water for desert rodents, especially Merriam's Kangaroo Rat (*Dipodomys merriami*).

Wet plants smell like a wet dog, but have to be really wet for the full impact.

**Human uses:** Akimel O'odham use the stems to make brooms (Rea 1997). Tohono O'odham use it for constructing the *u:kṣan <uksha>* (outdoor kitchen). Onavas Pima use the plant for cross-thatching roofs and side walls (Pennington 1980).

Yetman and Van Devender (2001) found the Mayo making a mat called a *zarzo* from the branches. Mayo hung this *zarzo* near the ceiling where cottage cheese (*panelas*) is put to cure safely away from animals. In addition, Mayo make a medicine of *A. monogyra* to treat mange in animals, itch in humans, and as a remedy to relieve colds, sores, fever, and hot flashes in women. Onavas Pima also make a poultice of the leaves to kill fleas and to reduce inflammation (Pennington 1980).

Hispanics in Sonora use Burro-Brush to treat abdominal pains and rheumatism, skin problems, sore throat, and swelling (López-E. and Hinojosa-G. 1988). Farther south in Mexico, *A. monogyra* is used as a remedy to treat abdominal pain (Standley 1920–1926).

The Seri use the related *A. salsola* and have a number of terms for it (Felger and Moser 1985). To them it is *casol cacat* (bitter aromatic plant), *casol coozlil* (sticky aromatic plant), and *casol ziix ic cöhíipe* (medicinal aromatic plant). Stems are used for firewood, for building brush houses, and for medicine

**Derivation of the name:** *Ambrosia* is from Greek *ambrosios*, see *A. confertiflora*. *Monogyra* means one circle, apparently alluding to the circular wing on the fruit. Formerly put in *Hymenoclea*, from Greek *hymen*, a membrane, and *kleis*, lock, key, referring to the bur.

**Miscellaneous:** Formerly, *Hymenoclea* was considered a genus that contained two species, *H. monogyra* and *H. salsola* (= *Ambrosia salsola*), both confined to southwestern North America and Mexico (Mabberley 1997). Both grow in the Gran Desierto of northwestern Sonora (Felger 2000).

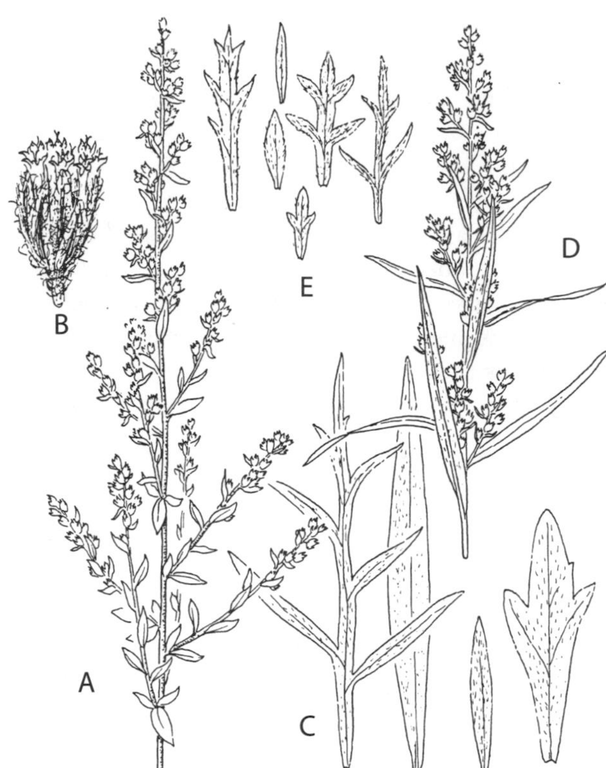


*Artemisia ludoviciana*. A–C. *A. ludoviciana* ssp. *albula*. A. Tip of flowering branch. B. Flower cluster. C. Leaf variation. D–E. *A. ludoviciana* ssp. *mexicana*. D. Tip of flowering branch. E. Leaf variation. Artist: B. Angell (courtesy James Henrickson).

## *Artemisia ludoviciana* (Asteraceae) Worm-Seed, *estafiate*

**Other names:** ENGLISH: [black, prairie, white] sage [brush] ("sage" from Latin *salvia* through *saluie* in Old English, *save* or *sauge* in Middle English, about fourth century; akin to *salbeia* or *salveia* in Old High German, *salvie* or *selve* in Middle Low German, *selie* in Dutch, *Salbie* in German, *salvia* in Italian and Spanish, *sauge* in French, *salva* in Portuguese), man-sage (Montana; translated from Cheyenne name), [Mexican] western mugwort (from Old English *mucgwyrt*, *mucg* meaning "midge"; cognate with Low German *muggart*, Old Saxon *muggia*, probably related to Greek *muia*, classical Latin *musca*, Albanian *mizë*, all in the sense of "fly"; "wort" is an old English word meaning plant), [Mexican] silver sage-brush (Coahuila), silver [Mexican] worm-wood (in use by 1400–1450, altered from Old English *wermod*, German *wermut*, *-muth*, whence French *vermout* and English vermouth), wormseed (seeds used as a vermifuge); SPANISH: *ajenjo [del pais]* ([country] absinth, New Mexico, Mexico; *ajenjo*, from Latin *absinthum*; akin to French *absinthe* and Italian *assezio*), *altamisa de la casa* (house ambrosia, Mexico), *chamiso cenizo* (ash-colored *chamiso*, from *chamiza*, brushwood, kindling, akin to Portuguese or Galician *chamiça*, from *chama*, flame, from Latin *flamma*, Mexico), *estafiate <astafiate, estafeate, istafiate>* (from Náhuatl *iztauhyatl*

<*iztahuatl*>, bitter salt, Chihuahua, Coahuila, San Luis Potosí, Sonora), *romerillo* (little rosemary, Mexico); ATHAPASCAN: *ch'ilzhóó'* <*céˀéžíh, ceˀezíh*> (rock sage, Navajo), *tsejintci* (strong-smelling sage, Chiricahua and Mescalero Apache); OTO-MANGUEAN: *j'mipzi* (Mazahua); UTO-AZTECAN: *chíchibo* (Mayo), *kȭsidab* [*koósiddúp, kosedap, kusedáp*] (Mono), *kȭsidava* (Northern Paiute), *musa, sanankdam* (*mu:ṣ*, clitoris; *sanuka*, plant used for abortions, *dam*, one that is, Mountain Pima), *ŕosáberi* (Tarahumara), *páakušh* (Luiseño), *páakwsech* (Luiseño, Juaneño dialect), *popohoppeh* (Shoshoni), *tavotqa* <*tavótka*> (Hopi), *tugusoovɨ* (Kawaiisu; for *A. douglasiana*)

**Botanical description:** Perennial herb, rarely slightly subshrubby near base, often almost rhizomatous, several- to many-stemmed from the base, 2–16 dm tall, all the herbage grayish- to whitish-pubescent (in some plants the upper surface of leaves is nearly glabrous). Leaves are entire or dissected, with the main axis and lobes usually 3–10 mm broad and exceedingly variable in shape. Involucre is usually grayish-woolly or whitish-woolly. Ray flowers present, occasionally fertile, usually infertile.

**Habitat:** Along watercourses. 1,500–2,000 m (4,900–6,562 ft).

**Range:** The species, as broadly defined with dozens of subspecies and varieties (Keck 1946), ranges from Alaska and Canada through much of the United States into northern Mexico (Baja California, Chihuahua, Coahuila, and Sonora south to Edo. México). Shultz (2006) recognized five varieties in North America outside of Mexico, with three in Arizona.

**Seasonality:** Flowering September to October.

**Status:** Native.

**Ecological significance:** Although it may grow higher up the slopes, *A. ludoviciana* is more common close to waterways and streams. This plant is a temperate and boreal species that reaches near its southern limit in Mexico. Indeed, most of the 350 or so species in the genus are in the north temperate parts of the world. Europe has 55 species; China has 170 (Mabberley 1997).

**Human uses:** The first records of Aztec *A. ludoviciana* uses were the Badiano Codex (within 30 years after Spanish arrival), the Florentino Codex (1529 and 1590), and Hernández's *Historia Natural* (compiled 1571–1577). All three note *iztauhyatl* being used for debility of the hands, rectal problems, urine retention, and other maladies. Later accounts added additional uses (Heinrich 2002). Over the next few centuries, uses shifted toward resembling historical Old World Hispanic applications of *A. absinthium*.

Europeans have long histories of using *A. absinthium* (Wormwood), *A. dracunculus* (Tarragon), and *A. vulgaris* (Mugwort). Modern use of *Artemisia* in Mexico is a mixture of Spanish Old World uses with those of indigenous Americans (Kay 1996). Today's Mexicans use the flowers and leaves as an aperitif, to expel intestinal worms, and as an emmenagogue (Martínez 1969, Heinrich 2002).

*Estafiate* does not appear to be part of the western border pharmacopoeia (Moerman 1998). Nearby, it was used by the Kickapoo, Tarahumara, Northern Tepehuan, and Navajo. Tarahumara use leaves to make a tea for women during menstruation (Pennington 1963). Northern Tepehuan also use Worm-Seed medicinally (Pennington 1969). Kickapoo use it to treat sores, stomach aches, wounds, and poison ivy rashes (Latorre and Latorre 1977). Havasupai use it in sweat baths (Moerman 1998). Navajo use *A. ludoviciana* in the Evil Way, Hand-Trembling Way, and as a Life Medicine (Vestal 1952, Wyman and Harris 1951).

Farther north, the Man-Sage is important medicine to the Blackfoot, Cheyenne, Comanche, Crow, Gros Ventre, Keres, Kiowa, Kutenai, Lakota, Mesquakie, Mewuk, Navajo, Okanagan, Paiute, Polikiah, Shoshoni, Thompson, Yokut, and Yurok (Train et al. 1957, Moerman 1998). Apache, Blackfoot, and Paiute use the seeds as food (Palmer 1878). This species is perhaps the most important ceremonial plant of the Cheyennes (Hart 1976).

**Derivation of the name:** The genus *Artemisia* comes from Greek *Artemis*, the Greek Diana, goddess of chastity, as the plant was thought to bring on early puberty. Pliny, CE 23–79, used the name for the Mugwort, *Artemisia vulgaris*. He said that the name was in honor of *Artemisia*, wife of *Mausolus*, king of *Caria*. *Ludoviciana* means from Louisiana, where it was first found.

## *Baccharis salicifolia* (Asteraceae) Seep-Willow, *batamote, ñehol*

**Other names:** ENGLISH: groundsel tree (maybe derived ca. CE 700 from Anglo-Saxon *grundeswelgiae*, pus-absorber; perhaps, but less convincingly, from *grundeswylige*, ground-absorber, from its rapid spread; originally used for *Senecio*, see *Packera*), mule's fat (Arizona, New Mexico), seep-willow (a willow-like plant that grows in wet places), [false, Gila] water wil-

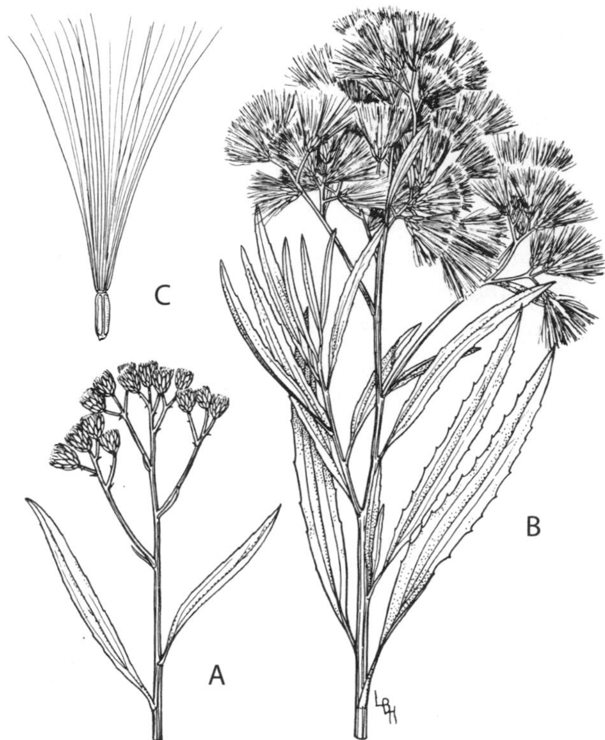


*Baccharis salicifolia.* A. Male flower clusters. B. Branch from female plant in fruit. C. Achene with tuft for dispersals. Artist: L. B. Hamilton (Parker 1958).

low [water-motie, water-wally]; SPANISH: *batamote* [*guatamote*] (from Cahita, Baja California, California, Sinaloa, Sonora), *hierba del carbonero* (charcoal maker's herb, Valley of Mexico), *jara* (arrow, from Hebrew *khara*, to cast, Guanajuato, Texas), *jaral* (Guanajuato, Tamaulipas), *jarilla* [*jarillo del río*] (little [river] arrow, Chihuahua, Durango, Sinaloa, Sonora), *vara dulce* (sweet bush, Chihuahua), *yerba del pasmo* (herb for *pasmo*, from Latin *spasmus*, Chihuahua); ATHAPASCAN: *k'ídzítso bi'tsiin łigai <k'íłcoi bicin łagai>* (*k'ídzítso*, yellow sprouts, *bi*, with, *'atsiin*, stems, *łigai*, white, Navajo), *tóeejí béé'ditó <tóˀiʒ$^{v}$i kéƛ'o>* (*tóeejí*, Water Way chant, *béé'di*, against it, *tó*, liquid, Navajo), *<tłeł>* (Western Apache); CHUMASH: *shu'* (Barbareño and Ineseño Chumash), *wita'* (Ventureño Chumash); HOKAN: *caaöj* (Seri); OTO-MANGUEAN: *mb'axu* (Mazahua); UTO-AZTECAN: *ba'asham <baashoma>* (*wakck*, soaked, *ham*, near/at, Mountain Pima), *bacho'ma <bachomo>* (Mayo), *bašam* (Onavas Pima), *čaguši <čagu'ši>* (Tarahumara), *guachomó <uachama>* (Guarijío; fide Yetman and Felger 2002), *guagualuasi* (mountain Guarijío; fide Yetman and Felger 2002), *ñehol* (servant, Tohono O'odham), *oágam* (brains or marrow, alluding to the stem pith, Akimel O'odham), *paq'ily <paki>* (Cahuilla), *pogosɨvɨ* (Kawaiisu), *ṣuṣk ku'agi <šu:šk kuagsig>* (*ku'agi*, wet, *ṣuṣk*, sandal/shoe, Tohono O'odham), *ṣu:ṣk kuasĭ <šu:šk, susk, kuagsig>* (Hiá Ceḍ, Sonora), *uachamo* (Mayo, Sonora), *waˀlurúbisi* [*waˀerúgesi*] (Guarijío; fide Miller 1996); YUMAN: *hamaséiva* (Havasupai), *hamḍavil* (Walapai), *hanta véel* (Mohave and Yuma), *xantavaíl*$_{y}$ (Maricopa), *xa'tam mual* (Paipai); (= *B. glutinosa*)

**Botanical description:** Shrubs, dioecious, 10–35 dm tall, the branchlets striate-angled, glabrous, sticky glandular (glutinous). Leaves are punctate, sessile to short-petioled, lanceolate to narrowly elliptic, tapering from middle to apex and to base, nearly entirely to prominently serrate, 3–10 cm long, 1–2 cm wide, distinctly 3-nerved. Heads are in a terminal corymb, often with several branches. Pistillate heads with 50 or more flowers, corollas 2–2.3 mm long. Staminate heads with 10–20 flowers, corollas 3–4 mm long. Achenes less than 1 mm long, glabrous, 5-nerved.

**Habitat:** Washes and other places where there are wet, sandy, gravelly, or poorly drained clay soils. Up to 1,525 m (5,000 ft), but usually lower.

**Range:** Arizona (Cochise, Coconino, Graham, Greenlee, Mohave, Pima, and Santa Cruz counties), California, and Colorado to Texas; Mexico (Baja California, Chihuahua, and Sonora to the Isthmus of Tehuantepec); Central and South America.

**Seasonality:** Flowering (March) July to August (December).

**Status:** Native.

**Ecological significance:** These shrubs are always in washes or streams that are at least seasonally wet. When the waterway is or has been moist for long periods, Seep Willows sometimes form large clumps. These clumps may form barriers along the margins that partly impede access to the waterway. Clumps also may form aisles in the streambeds.

**Human uses:** Akimel O'odham use the stems for children's arrows (Rea 1997). Branches also make sides for wattle-and-daub structures (*kosin* or *kii*), and to roof the *vatto* (arbor, ramada). Historically the final painting stage of pottery was to apply the pigment with a branch of *oágam*.

Seri use sections of stem, sometimes dyed black, as beads in necklaces (Felger and Moser 1985). Guarijío weave stems into wiers for trapping fish (Yetman and Felger 2002). Mayo say a layer of branches above the *latas* (ceiling cross-hatching) is cooling (Yetman and Van Devender 2001). Onavas Pima formerly used the stems to weave carrying baskets (Pennington 1980).

These Pimas apply the same name to *Baccharis* as to a carrying sack used to transport clay to make ceramics. Onavas Pima also use the branches to make sleeping mats (*main* <*ma'in*>, in Spanish *petate*), dove cages, fire drills, fish traps, and walls of houses.

Tea made from leaves is used to help Seri lose weight, as a contraceptive, and to stop blood loss after childbirth. Leaves heated over coals are placed on the head to relieve headaches or on body sores (Felger and Moser 1985). Some *Baccharis* may be the species Onavas Pima use to help wounds heal, but Pennington (1980) could not identify it.

Roots are a remedy for *mal de orín* (difficulty urinating). Leaves bound to the skin relieve headaches and fever among Guarijío (Yetman and Felger 2002). Mayos use leaves to reduce sweaty feet and body odors. Stems formerly were used to make arrows for hunting (Yetman and Van Devender 2001). Tarahumara crush branches and put them into pools to catch fish (Pennington 1958, 1963). In addition, Tarahumara make poultices from the branches to cure skin infections, and they also make a drink to relieve colds and coughs. Navajo also use *B. salicifolia* in medicines (Wyman and Harris 1951).

Lewis and Elvin-Lewis (2003) noted that almost all members of the Asteraceae contain sesquiterpene lactones, and many of these are cytotoxic. Additionally, *B. salicifolia* is known to contain several potent essential oils, including verboccidentafuran, chromolaenin, and germacrone (Loayza et al. 1995).

**Derivation of the name:** *Baccharis* may be from *Bacchus*, the Greek god of revelry. Or, the name may be from Greek *bakkaris*, an unguent made from 'asaron'; *bakcharis*, a name used by Dioscorides (fl. CE 40–80) for *Cyclamen hederaefolium* (Primulaceae). *Salicifolia* means having leaves like willow (*Salix*).

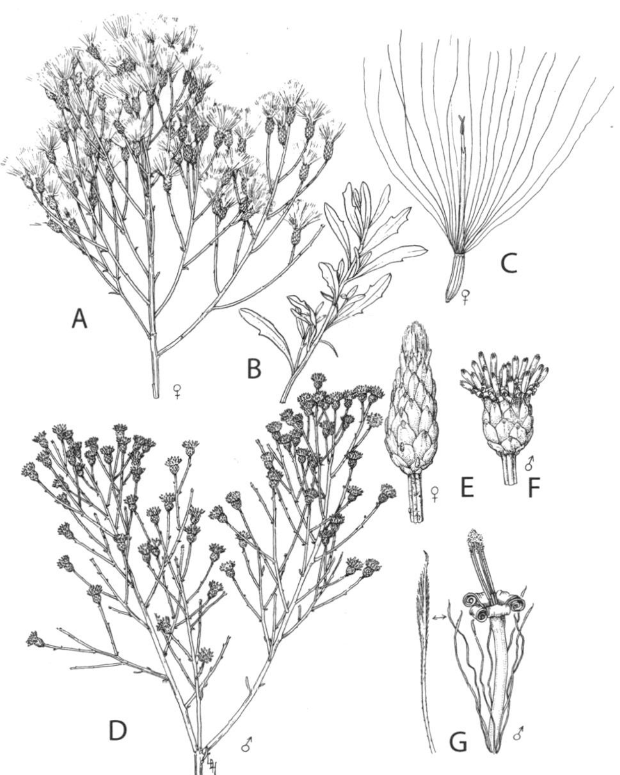


*Baccharis sarothroides*. A. Fruiting branch from female plant. B. Leafy branch from young plant. C. Female flower with elongate pappus, the corolla minute and the ovary large. D. Flowering branch from male plant. E. Head of female flowers. F. Head of male flowers. G. Male flower with enlargement of a pappus bristle. Artist: L. B. Hamilton (from original artwork).

## *Baccharis sarothroides* (Asteraceae) Desert-Broom, *romerillo, a:n*

**Other names:** ENGLISH: desert-broom (Arizona, New Mexico), grease-wood (a name originally applied to prickly chenopodiaceous shrubs, of genera such as *Sarcobatus* and *Atriplex* by 1851; later extended to *Baccharis* and *Larrea*; "grease" is from Old French *graisse*, fat, in English by ca. 1340), rosin bush ("rosin" is a variant of "resin," the residue from distilling turpentine from pine sap; in English by ca. 1350); SPANISH: *batamote* <*guatamote, huatemote*> (from Cahita, see *B. salicifolia*, Mexico), *escoba amarga* (bitter broom, Baja California), *hierba del pasmo* (herb for *pasmo*, from Latin *spasmus*, Baja California), *romerillo* (little rosemary, Sonora); HOKAN: *casol caacöl* (large *casol*, Seri); UTO-AZTECAN: *a:n* <*'a:ñ*> (Tohono O'odham), *shuushk vakchk* (wet [fide Rea 1997] sandals/shoes, Akimel O'odham), *ṣuṣk kuagǐ* <*su:sk, ṣuṣk kuagig*> (*ṣuṣk*, sandals/shoes, *kuagǐ*, firewood, Hiá Ceḍ, Sonora), *ṣuṣk wakc* <*ṣuuṣk wakchk, šu:šk uwakita*> (*ṣuṣk*, shoe/sandal, *wakc*, wet, Tohono O'odham); YUMAN: *ˀi:xʷír* (Cocopa)

**Botanical description:** Shrub, dioecious, 2–4 m tall, erect, glabrous, sticky glandular (glutinous), green, nearly leafless, with broom-like sharply angular-grooved branches. Leaves all nearly linear, entire, rigid, up to 2 cm long. Heads mostly solitary at the tips of the numerous branchlets, the involucre of female heads 6–8 mm high, cream-colored; involucre of male heads 3–4 mm high. Achenes 1.7–2.2 mm long, glabrous, 10-nerved.

**Habitat:** Along washes, roadsides, and disturbed sites. 304–1,676 m (1,000–5,500 ft).

**Range:** Arizona (Gila, Graham, Greenlee, Mohave, and Yavapai south to Cochise, Pima, Santa Cruz, and Yuma counties), California, New Mexico, and Nevada; Mexico (Baja California, Chihuahua, Sonora, and Sinaloa).

**Seasonality:** Flowering September to October.

**Status:** Native.

**Ecological significance:** These are premier heliophiles (sun-lovers) and pioneers. In the fall and early winter (November to December), their fruits are developed and spread in the wind by the hundreds of thousands. Sometimes, the propagules are so common that they resemble windrows of snow on cold days. If there is enough moisture in crevices and nooks below "nurse" plants, the seeds germinate to produce new brooms in places that humans often do not want them.

Desert-Broom is remarkably tolerant of abuse. I have seen a branch that was notched so deeply from rubbing against a stone that over three-quarters of the stem was missing. There was no obvious stress caused in the shrub, and that stem flowered as successfully as the others, ultimately producing abundant seeds.

**Human uses:** Akimel O'odham use these shrubs to make brooms and as a quick fuel, especially for starting fires (Rea 1997). Tohono O'odham use *Baccharis* for building the *u:kṣan* <*uksha*> (outside kitchen or windbreak) and Akimel O'odham use it for the roof of their *vatto* (arbor or ramada).

Tohono O'odham use the branches to make a tea-like drink for refreshment (Castetter and Underhill 1935). Seri make a tea of branches as a remedy for a cold and rub it on sore muscles for relief (Felger and Moser 1985). Mayo use the branches and leaves to make a tea to treat sores or itches (Yetman and Van Devender 2001).

The name *hierba del pasmo* (from Latin *spasmus*) refers to a malady considered endemic to the Americas. This disease is a general depression of the nervous system that affects both humans and domestic animals.

**Derivation of the name:** *Baccharis*, see *B. salicifolia*. *Sarothroides* means that the plant resembles a broom.

**Miscellaneous:** *Baccharis* is a genus of perhaps 350–450 species (Sundberg and Bogler 2006). Several other species grow in the Baboquivari Mountains, including *B. brachyphylla* (Short-Leaved Broom), *B. salicina* [= *B. emoryi*] (Emory Broom), *B. pteronioides* (*yerba del pasmo*), and *B. thesioides* (broom; *batamote de monte, hierba del pasmo*). *Baccharis brachyphylla* is a low plant (to 60 cm tall), woody only toward the base, that has untoothed leaves and is densely rough pubescent. The others are taller and woody. *Baccharis pteronioides* has solitary heads on short, leafy branches that are arranged like a raceme along other stems. Its leaves are obovate to oblanceolate and sharply toothed. *Baccharis salicina* has cuneate-oblanceolate leaves 3–8 mm wide that are distinctly 3-nerved. *Baccharis thesioides* has narrowly linear to linear-lanceolate leaves 1.5–8 mm wide that are usually closely, evenly, and sharply spinulose-serrulate.

## ***Bidens leptocephala*** **(Asteraceae)** **Few-Flower Beggar-Ticks, *acahual***

*Bidens leptocephala*. A. Flowering and fruiting branch. B. *Bidens bigelovii* flowering branch. C. Head. D. Inner and outer involucre bracts. E. Disc flower. F. Corollas spread open. G. Stamens spread open. H. Style branches. I & J. Outer achenes (sterile). K. Inner achene. Artist: V. Frasier (Correll and Correll 1972).

**Other names:** ENGLISH: bur marigold ("burr," meaning the sticky fruits; "marigold" is from the proper name Mary, presumably alluding to the Virgin Mary, plus gold, their color; in use by ca. CE 1300; the name of several yellow-flowered Asteraceae, originally *Calendula*), few-flower beggar-ticks [fewflower beggarticks] ("beggar" has meant one who habitually asks alms since ca. 1225; Thoreau apparently linked it with "ticks" in *Walden* in 1854 when he wrote, "It was over-run with Roman worm-wood and beggar-ticks, which last stuck to my clothes"; also used for *Desmodium*, q.v.), Spanish-needles (used by about 1845 for *B. bipinnata*); SPANISH: *acahual* [*acuahualillo*] (from *atl*, water, *cahualli*, abandoned; various Asteraceae, some used to ignite the *horno* or oven, Mexico), *aceitilla* (little oily one, Edo. México, San Luis Potosí), *mozote* (from Náhuatl *mozotl*, a name used for several Asteraceae farther south in Mexico); ATHAPASCAN: *ch'il hosh* (*ch'il*, plant, *hosh*, cactus, Navajo)

**Botanical description:** Annual herbs, stems 1–5 dm tall, petioles connate basally. Leaf blades 1.5–5.5 cm wide, 2–10 cm long, 1–2-pinnately divided into linear to ovate segments, rough-hairy to glabrous. Heads are obscurely radiate or apparently discoid, 4–8 mm wide, 3–5 mm tall at anthesis, on slender peduncles 2–8 cm long. Outer involucral bracts are 4–6, linear, the margins ciliate, 1–2.5 mm long, the inner about 1.5–5 mm long. Ray flowers are usually 2–3, the ligules ca. 2.5 mm long, pale yellow to nearly white. Achenes are 5–13, linear, somewhat 4-angled, 6–14 mm long, the 2 awns 1–3 mm long.

**Habitat:** Mostly along streams, preferring shaded, sandy soils. 900–1,830 m (3,000–6,000 ft).

**Range:** Arizona (Apache, Yavapai, south to Cochise, Graham, Gila, Santa Cruz, Pima, and Pinal counties), New Mexico, and western Texas; Mexico (Baja California, Chihuahua, and Sonora).

**Seasonality:** Flowering August to October.

**Status:** Native. The original (type) collection was made by J. C. Blumer in the Chiricahua Mountains in 1907.

**Ecological significance:** These herbs are locally frequent near intermittent streams in the canyons. If *B. leptocephala* is anything like its more tropical relative *B. pilosa*, every flying insect in the area will visit for a drink of nectar.

Like other members of the genus, the achenes are adapted for animal dispersal by sticking to their fur. The two "teeth" (pappus) on the achenes also are adept at becoming attached to the clothes of humans who unwittingly come too close. In the fall, *Bidens* fruits are produced in abundance and stick like "beggars" to any mammal that passes. Humans, too, are acceptable dispersers, and the needle-like fruits are sometimes like fur on their skin and clothes after passing an unnoticed herb.

**Human uses:** No uses found for *B. leptocephala*; however, several others are used. *Aceitilla*, an extract of several *Bidens* species, is prepared as a diuretic (Santamaría 1959).

The name *té de milpas* is used for *B. aurea* from Sonora south, and these herbs are made into a diuretic tea (Martínez 1969). López-E. and Hinojosa-G. (1988) record that species as being used to treat diabetes, chest problems, and as a diuretic.

The more tropical *B. pilosa* contains so many potent compounds that only the Dainty Sulphur Butterfly (*Nathalis iole*) is able to eat the leaves. Hsu (1986) and Foster and Duke (1990) say that this plant has phytosterin B, a compound that acts as a central nervous system depressant that lowers blood sugar. No wonder a single butterfly can eat the leaves. Probably other species in the genus also contain these toxic chemicals.

**Derivation of the name:** *Bidens* is from Latin *bis*, twice, *dens*, tooth, the fruit (cypsela) is 2-toothed. *Leptocephala* means small-headed, referring to the flower clusters that are tiny in comparison to other species in the genus.

**Miscellaneous:** *Bidens bigelovii* (also *acahual*, *acetilla*) is in Brown Canyon and on the western side of the mountains. The species mostly ranges to the east of the Babouqivari Mountains, also being known in New Mexico, Texas, Colorado, and Sonora. These two are similar, but *B. leptocephala* has 5–9 (rarely to 13) achenes, while *B. bigelovii* has 14–50 achenes.

*Bidens aurea* flowers from September to October along major washes, streams, and moist areas; 914–1,828 m (3,000–6,000 ft). That species is known as [Arizona] Beggar-Tick, Bur[Burr]-Marigold, *té de corral* (corral tea, Mexico), *té de milpa(s)* [*té de milpa amarillo*] ([yellow] cornfield tea, Sonora to at least the Edo. México), *xiuh-elo-quílitl* (from *xiuitl*, herb, *elotl*, young corn, *quílitl*, cooked herbs, Náhuatl). This herb grows from Arizona (Cochise, Pima, and Santa Cruz counties) and Mexico (Chihuahua and Sonora) to Guatemala.

## ***Brickellia californica*** **(Asteraceae)** **California Bricklebush, *pachaba***

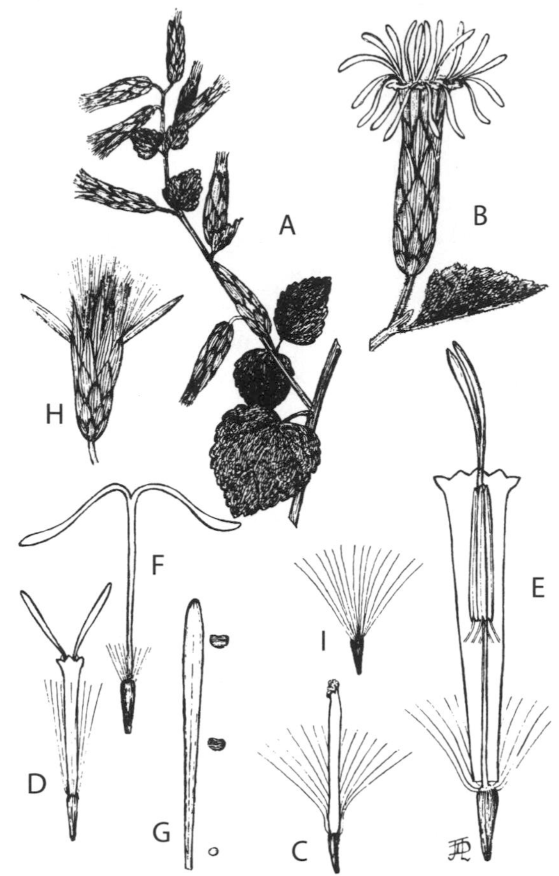


*Brickellia californica.* A. Flowering branch. B. Flower cluster. C. Flower bud. D. Disc flower. E. Dissected flower showing internal details. F. Gynoecium, with corolla and stamens removed. G. Stigma lobe. H. Fruiting cluster. I. Fruit with pappus bristles. Artist: J. D. Laudermilk (Benson 1957).

**Other names:** ENGLISH: [California] bricklebush [brickellbush] (a book name based on the genus), false boneset ("boneset," a name applied by 1670 to comfrey, *Symphytum officinale*; later applied to these American plants); SPANISH: *hamula* (hooked, from Latin *hamulus*, Mexico), *hierba <yerba> de la vaca* (cow herb in Baja California, New Mexico, and Mexico), *pachaba* (from Hopi *patcavu*, now widely used in Arizona), *prodigiosa* (marvelous, Mexico; applied to several *Brickellia*); ATHAPASCAN: *bił háách'i <bilha.zef'n>* (its scent is carried on the breeze, Navajo), *tséghą́ą́' 'adisxas <cek'i.nˀalcizi>* (chuckwalla [lit. scrapes on rock] *'adisxas*, scrapes, claws, *tséghą́ą́'*, on the rock, Navajo), *'azee' dich'íízh <ˀazeˀ dičíž>* (*'azee'*, medicine, *dich'íízh*, rough, Navajo); UTO-AZTECAN: *patcavu* (Hopi); YUMAN: *kwaq impal* (Paipai)

**Botanical description:** Shrub, usually more than 1 m tall, branched, the stems with glandular-puberulent deciduous papery outer bark. Leaves alternate, the petiole about 1/3 or 1/2 as long as the blade, the blade punctate and scabrous, broadly ovate to deltoid, basally truncate to cordate, margins dentate, crenate, somewhat lobed or rarely entire, apically acute to obtuse. Inflorescences are a leafy panicle or a series of terminal ascending branches, each with a raceme. Involucres are ca. 3 mm across, 9 mm high, often reddish-tinged. Achenes are tawny-pubescent.

**Habitat:** Along canyons, washes. 914–2,133 m (3,000–7,000 ft)

**Range:** Arizona (throughout), California, to Colorado, the Oklahoma panhandle, and western Texas; Mexico (Baja California, Chihuahua, and Sonora).

**Seasonality:** Flowering July to November.

**Status:** Native.

**Ecological significance:** This plant is an indicator of relict chaparral vegetation (Rondeau et al. 1996). In Brown Canyon, another species associated with this community is *Vauquelinia arizonica*, q.v. The shrub live oak (*Quercus turbinella*) so characteristic of chaparral has not been found in Brown or Thomas Canyons, but it is known from Spring Canyon near Fresnal Canyon on the western slopes. The oak is also known from Las Guijas Mountains across the Altar Valley to the east (McLaughlin 1992a).

**Human uses:** Hopi rub a related species on the head to relieve headaches (Whiting 1939). Kumiai (San Diego County, CA) use an infusion of the leaves to treat fevers. Navajo use *Brickellia* as a ceremonial emetic, to cure sores in infants, and to treat colds and fevers (Vestal 1952, Wyman and Harris 1951). They also use it to treat clan incest and in the Coyote Way. Sanel (Mendocino County, CA) use the leaves as a substitute for tea (Moerman 1998). Hocking (1997) wrote that seeds of *B. californica* are added to flour meal by the Shoshoni of Utah and Nevada; that seems to be an error for *B. grandiflora* (see below).

Another species in this region that has numerous recorded uses is *B. grandiflora* (Moerman 1998). That plant has seeds that the Shoshoni near the Great Salt Lake in Utah and nearby Nevada consider poisonous, but people still use them as medicine. The Shoshoni also grind the seeds into meal and use them as "baking powder" to improve cakes, apparently diluting the poisonous effects (Chamberlin 1911).

Keres grind the dried leaves of *B. grandiflora* with water and use the mixture as a salve to treat rheumatism (Moerman 1998). Keres also take an infusion of the leaves to relieve flatulence and overeating. This infusion is considered a liver medicine. These same Pueblo people use bunches of the stems to make brooms.

Navajo take a cold infusion of the leaves to relieve headaches, tuberculosis, and flu. Because the concoction is emetic, it is also used in ceremonies (Wyman and Harris 1941). The mixture is put with other plants to use as a ceremonial liniment for the Female Shooting Life Chant.

Farther south in Mexico, several species of *Brickellia* are famous as medicines (Martínez 1969). One of the best known is *B. cavanillesii* (*atanasia amarga*, *hierba del becerro*, and *gobernadora de Puebla*). This plant and other species are considered prime bitter tonics and stimulants for the stomach.

**Derivation of the name:** Named for John Brickell, 1749–1809, of Savanna, Georgia, an amateur botanist and helpful correspondent to G. H. Muhlenberg (see *Muhlenbergia*), John Fraser, and others. *Californica* from California, because the original specimen was found there.

**Miscellaneous:** There are seven other species of *Brickellia* known from Brown and Jaguar Canyons: *B. amplexicaulis*, *B. betonicifolia*, *B. coulteri*, *B. eupatorioides* var. *chlorolepis* (= *B. chlorolepis*, *Kuhnia rosmarinifolia*), *B. floribunda*, *B. grandiflora*, and *B. venosa*. These species vary considerably in aspect.

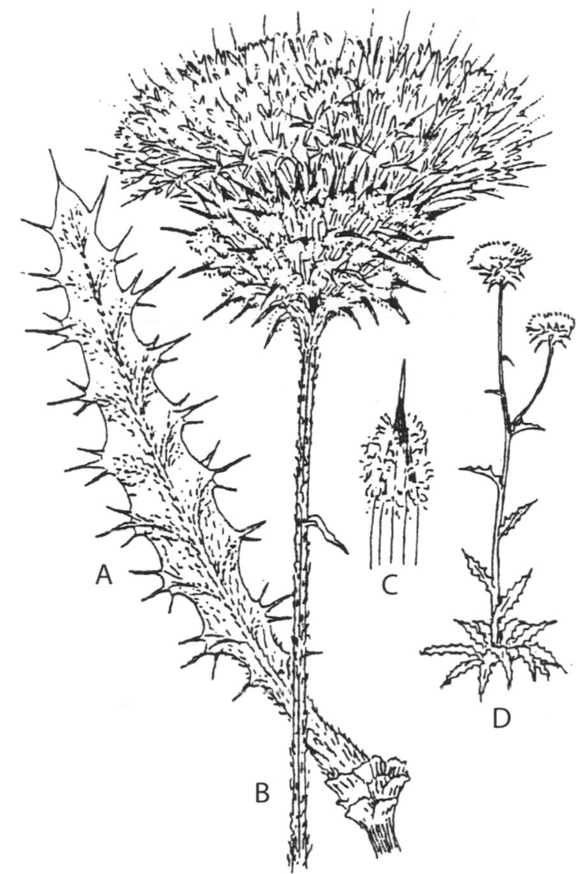


*Cirsium neomexicanum*. A. Basal leaf. B. Flower cluster. C. Involucre tip. D. Habit of plant. Artist: R. D. Ivey (Ivey 2003).

## *Cirsium neomexicanum* (Asteraceae) New Mexico Thistle, *cardo*, *gewul*

**Other names:** ENGLISH: [New Mexico, yellow] thistle (origin uncertain; the best that can be said is that the word is Germanic; formerly spelled *thistil*, *thistel*, *thystel* in English; there are variants in German and Dutch [*distel*], Old Norse [*thistell* or *thistill*], Swedish [*distel*], and Danish [*tidsel*]); SPANISH: *cardo* (Spanish; from Latin *carduus*, this was the name used by classical authors Virgil, 70–19 BCE, Pliny, CE 23–79, and by Linnaeus in 1753 for an array of bristly plants related to *Cirsium*), *cardillo* (little thistle, New Mexico); ATHAPASCAN: *'azee' hókánii* <*ˀazéˀ hukani*> (round medicine, Navajo), *'azee' yishdloh* (*'azee'*, medicine, *yishdloh*, laughing [medicinal plant that looks like it is laughing (when it goes into flower)], Navajo), *'azee' ditł'ooí* <*'aze'titl'oih*> (*'azee'*, medicine, *ditł'ooí*, fuzzy, Navajo), *hosh ikaz*, *kǫ' dahosh* <*goda hosh*> (*hosh*, cactus, *ikaz*, agave, *kǫ'*, fire, *dahosh*, on a cactus, Western Apache), *tłobindadatłidje* (plant with blue seeds, Chiricahua and Mescalero Apache); UTO-AZTECAN: *čiiyavɨ* (Kawaiisu; for all *Cirsium*), *čiyiyal* [*čiŋiya-l*] (Tübatulabal; for *C. occidentale*), *cuna* (Cupeño), *cunala* (Luiseño), *gewel* <*gewihol*> (Hiá Ceḍ O'odham), *gewul* (Tohono O'odham), *pa'bogo* [*pa'bogwo*], *tsĭñ'ga* (Shoshoni), *tsininga* <*tcíninga*, *ciniŋa*> (Hopi)

**Botanical description:** Herbaceous biennial, stout, 1.5–2 m tall, white-woolly. Basal leaves narrow to more or less elliptic, deeply and regularly sinuate-pinnate, strongly spiny, to ca. 4 dm long, the cauline leaves gradually reduced up the stem, strongly decurrent. Heads 4–5 cm broad, the involucre 2.5–4 cm high, almost globose, sparsely covered with pubescence resembling cobwebs (arachnoid), becoming glabrous with age; outer phyllaries reflexed, the middle rough or scurfy (squarrose), strongly spiny, innermost attenuate, scarcely spine-tipped. Flowers are pink or whitish, exserted.

**Habitat:** Canyons, grasslands, along roadsides. 304–1,981 m (1,000–6,500 ft).

**Range:** Arizona (Coconino, Yavapai and northern Mohave to Cochise, Pima, Santa Cruz, and Yuma counties), southern California, north to Colorado, Nevada, and east to New Mexico; Mexico (Sonora).

**Seasonality:** Flowering March to May (September).

**Status:** Native.

**Ecological significance:** Although these plants are now associated with disturbances caused by humans or livestock, the species was probably originally confined to washes, canyons, and other places where there was seasonal rearrangement by water. Hummingbirds and bees visit and pollinate the flowers. Although *C. arizonicum* has red flowers that are more adapted to bird pollination, hummingbirds recognize the flower clusters of *C. neomexicanum* as potential sources of food and take advantage of them.

*Cirsium* is biennial, and the basal rosettes of leaves appear in late January. Leaves are characteristically spiny-margined, making the the genus comparatively easy to recognize. It is only in the second year of life that plants produce flowers, and there may be many more sterile basal rosettes than flowering plants in the summer and fall blooming period.

**Human uses:** Navajo use the herb in a medicine to treat chills and fever, to wash sore eyes, as a treatment when one simply "feels bad all over," and as a Life Medicine (Vestal 1952, Wyman and Harris 1951). At least Navajo and Yavapai eat raw, peeled *Cirsium* stems (Moerman 1998, Wyman and Harris 1951). Similarly, the Cahuilla of southern California eat the base of the flowers (Bean and Saubel 1972). Curtin (1947) found people in New Mexico using *C. undulatum*, but not *C. neomexicanum*. Paiute eat the peeled stems of *C. pastoralis* raw and call them *izá'·ᵃkwasi* (coyote tail) (Kelly 1932). Shoshoni eat the stems of several species (Chamberlin 1911).

North of Mexico, some species of *Cirsium* have been used by the Abenaki, Apache (Chiricahua and Mescalero), Atsugewi, Blackfoot, Cahuilla, Cherokee, Cheyenne, Comanche, Cowichan, Cree, Delaware, Flathead, Shoshoni, Hopi, Iroquois, Kawaiisu, Kiowa, Kwakiutl, Mesquakie, Montagnais, Navajo, Nitinaht, Ojibwa, Okanagon, Ohlone, Paiute, Potawatomi, Quileute, Spokan, Sushwap, Thompson, Tübatulabal, and Western Keres (Bocek 1984, Castetter and Opler 1936, Elmore 1944, Gallagher 1976, Moerman 1998, Smith 1933, Voegelin 1938, Whiting 1939, Wyman and Harris 1951, Zigmond 1981). In Mexico, *C. mexicanum* is *cardo santo* (holy thistle), and it is used for pulmonary diseases in Guerrero (Martínez 1969). The Northern Tepehuan eat the leaves with *esquiate* (a corn dish similar to cornmeal mush) or the stalks raw (Pennington 1969); they call the herb *šiñáka*. Northern Tepehuan also make the roots into a medicine for chest pain.

Given the use of *C. tuberosum* as a food (salad and roots) in the Old World, I have to wonder if some of the uses were not transferred from there to the Americas. However, indigenous Americans independently discovered edibility of their species because the genus was in pre-European material in the Cordova Cave of New Mexico (Kaplan 1963).

**Derivation of the name:** *Cirsium* is from Greek *kirsion*, some kind of thistle. *Neomexicanum*, of New Mexico, indicates that the first specimens were collected in New Mexico.

**Miscellaneous:** A second species, *C. arizonicum*, occurs at higher elevations in the canyons, usually coming down to about 4,500 feet. While the flowers of *C. neomexicanum* are lavender, they are red in *C. arizonicum*. The heads are also narrower, almost tubular, while they are broader and essentially bell-shaped in *C. neomexicanum*.

## *Conyza canadensis* (Asteraceae) Horseweed, *cola de caballo, vopoksha*

**Other names:** ENGLISH: bitterweed (New Mexico), butterweed (corruption of "bitterweed"?), colt's tail (New Mexico), horsetail, [Canadian, smooth] horseweed (when "horse" is combined in names of plants, it often denotes a large, strong, or coarse kind, akin to use of *Rosz-* in German; in use by 1790 for *Collinsonia canadensis*; by 1870s, for *Conyza*), fleabane (repels fleas, New Mexico), fox tail (Dutch Antilles), pride weed (by ca. CE 1330 "pride" referred to internal organs, by middle 1700s to male genitals, New Mexico); SPANISH: *cola de caballo* (horse tail, Arizona, Sonora), *hierba del caballo* (horse's herb, Sonora), *hierba de burro* (donkey herb, Sinaloa), *pazotillo* (little skunk feces or little *Chenopodium ambrosioides*, from Náhuatl *epatl*, skunk, *tzotl*, filth, New Mexico), *yerba del aire* [*aigre*] (air [bitter] herb, California); ALGIC: *gababi'kwuna'tig* (knotted tree, Ojibwa), *no'sowini* (sweat, because it is used in the sweat bath,

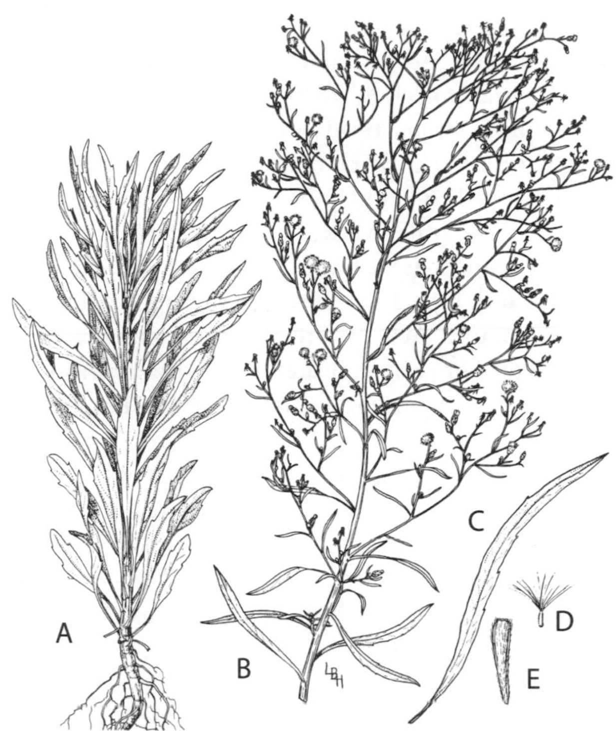


*Conyza canadensis*. A. Lower portion of plant. B. Upper portion of flowering plant. C. Enlarged leaf. D. Achene, with plume of hairs. E. Achene, enlarged. Artist: L. B. Hamilton (Parker 1958).

Mesquakie); ATHAPASCAN: *'azee' dilkǫǫh* <*ʔazeʔ dilkǫkí*> (*'azee'*, medicine, *dilkǫǫh*, smooth, Navajo), *dlǫǫdą́ą́'* <*ƛǫ́dą́ʔ*> (*dlǫǫ*, prairie dog, *dą́ą́'*, food, Navajo), *ne'etsah 'azee'* <*neʔecah ʔazéʔ*> (*ne'etsah*, pimple, *'azee'*, medicine, Navajo), *ne'etsah béé'ditó* <*neʔecah beƛoh*> (*né'étsah*, pimple, [lotion] *béé'di*, against it, *tó*, liquid, Navajo); CHUMASH: *wili'lik* (Barbareño Chumash), *wililik'* (Ineseño Chumash); MUSKOGEAN: *atakło:lasti* (*takłî*, bush, *lvste*, black, Creek), *takłô:cî* (black bush, Mikasuki), *vtakłv lvste* (*vtakłv*, plant, *lvste*, black, Muskogee); IROQUOIAN: *atsil-sun'ti* (fire making material, *ajila*, fire, Cherokee); SIOUAN: *canhlo'gan was'te'mna iye'cece* (resembling sweet-smelling weed, Lakota); UTO-AZTECAN: *monáhaña* (Hopi fide Hough 1898), *on'tĭmpiwai* [*on'tĭmpiwatsĭp*] (Shoshoni, also used for *Chenopodium*), *vopoksha* <*vopoghakam*> (quiver or stepchild, Akimel O'odham); LANGUAGE ISOLATE: *ha'mo uvteawe* (*ha* from *ha'li*, leaf, *mo* from *mo'li*, ball-shaped, *u'teuwe*, flowers, Zuni); (= *Erigeron canadensis*)

**Botanical description:** Mostly summer-fall ephemerals (annuals elsewhere), slender, erect, to 1–2 m tall, unbranched except the terminal, much-branched flowering portion, herbage sparsely pubescent. Leaves linear, reduced upwards, the larger 3–6 cm long, 2–6 mm wide, sessile, but narrowed below, margins entire or with a few shallow teeth, ciliate and strigose, sparsely pubescent with coarse white hairs, the midveins prominent, orange, resinous. Ray florets minute and numerous, white or white with pink tips. Achenes are 1 mm long.

**Habitat:** Grasslands, along trails, roadsides. 300–2,130 m (1,000–7,000 ft).

**Range:** Arizona (throughout the state), widely distributed in North America; Mexico (Chihuahua and Sonora); south to South America.

**Seasonality:** Flowering (March) August to October (November).

**Status:** Native.

**Ecological significance:** These are herbs of disturbed sites, often found along washes, streams, and other sunny places with "natural" and frequent changes. Human disturbance enhances their abundance.

Indigenous people of the eastern United States introduced the Europeans to Horseweeds, and *Conyza* was taken to England by 1640 as medicine. There, John Parkinson, herbalist to King Charles I, touted their marvels and uses (Dobelis 1986). Thirteen years later, the plants were in Paris's *Jardin des Plantes*, apparently as a weed.

**Human uses:** *Conyza* is used as medicine by at least the Blackfoot, Chumash, Cree, Hopi, Houma, Iroquois, Keres, Mesquakie, Navajo, Ojibwa, Potawatomi, Seminoles, and Zuni (Curtin 1947, Hough 1898, Moerman 1998, Wyman and Harris 1941, 1951). Indeed, people throughout the range of the species use it for medicine to treat a variety of ailments (Austin 2004). Curtin (1947) recorded uses for the herbs in New Mexico. Zuni, for example, crush the flowers and inhale them to cause sneezing to relieve rhinitis (Stevenson 1915). Navajo simply crush or moisten the leaves, or make them into a lotion, to treat skin problems and in the Eagle Way (Wyman and Harris 1941). Chumash grind *C. canadensis* to relieve pain, and they make a tea of it to help kidney problems (Timbrook 1990).

Rea (1997) had the Akimel O'odham word *vopoksha* explained to him as: "Well, someone's stepchild is an obligation, like a quiver that is strapped to your back. You can't get away from it. You just have to carry it along." That view is much like the way Horseweed grows. Once *C. canadensis* gets started in a yard or a garden, it is there almost forever. At least, these are medicinal weeds. Felger et al. (in prep.) do not consider the species native in southwestern Arizona. Perhaps that is why the local people have no uses for it except to feed it to livestock.

**Derivation of the name:** Latin *conyza*, and Greek *konyza*, a flea, a name originally used by Aristotle, 372–287 BCE, Theophrastus, 372–287 BCE, Nicanor, fl. 200 BCE, and Dioscorides, fl. CE 40–80, for two kinds of European composites. Christian Friedrich Lessing, 1809–1862, reapplied the name. *Canadensis*, from Canada, because Linnaeus used a specimen collected in Canada when he described the species in 1753.

## *Coreocarpus arizonicus* (Asteraceae) Little Lemonhead

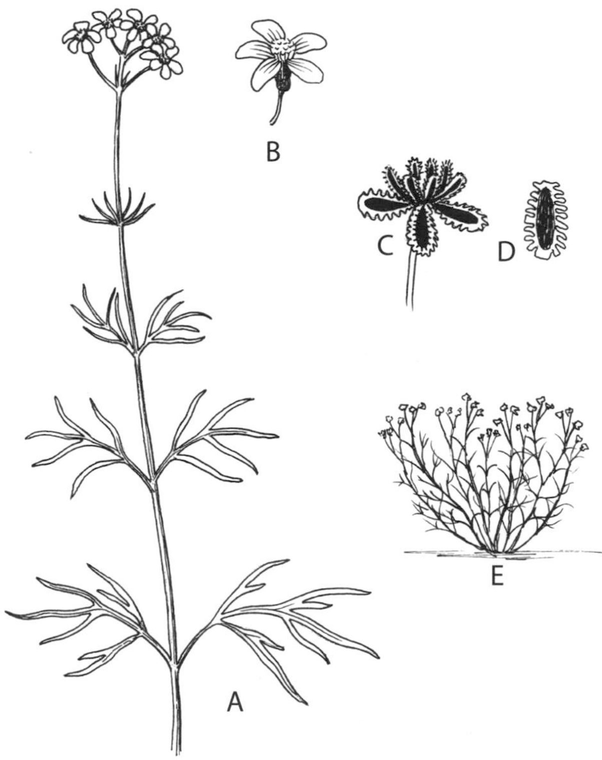


*Coreocarpus arizonicus*. A. Flowering branch. B. Individual head showing ray and disc florets. C. Fruit cluster. D. Individual fruit. E. Habit. Artist: B. Swarbrick (courtesy Bonnie Swarbrick).

**Other names:** ENGLISH: lemonhead (apparently a local name in Arizona, although probably not widely used; alluding to the yellow flower clusters)

**Botanical description:** Herbs, perennial, somewhat woody below, herbaceous above. Stems are glabrous, 2–6 dm tall. Leaves pinnately divided into 3–5 ovate divisions and then divided again into several linear lobes 1.5–3 mm wide, or the uppermost leaves only pinnately 3–5 parted into linear lobes, in outline broadly ovate to deltoid, to 10 cm long, with petioles 1–3 cm long, glabrous. Heads are in few-flowered cymes. Involucres are bell-shaped, the bracts broadly ovate to oblong, mostly 5–6 mm long, obtuse to abruptly acuminate or the inner acute, strongly purple-lined, the margins papery (scarious). Ray flowers are pale yellow, 3–5 mm long. Disc flowers are abruptly widened, 2–2.5 mm long. Achenes are ca. 3 mm long, the wing incised into broadly cuneate lobes, with 2 retrorsely spinulose awns present on some achenes, but absent from others.

**Habitat:** Shady canyon bottoms, upper Sonoran Zone and Oak Woodland, otherwise extending upward into Pine-Oak Woodland. Found at 400–1,700 m (1,300–5,600 ft) in Sonora, 1,200–1,400 m (4,000–4,600 ft) in the Rincon Mountains, and 914–1,676 m (3,000–5,500 ft) in Arizona.

**Range:** Arizona (western Cochise, eastern Pima, and Santa Cruz counties); Mexico (western Chihuahua, northern Sinaloa, and Sonora).

**Seasonality:** Flowering January to November, following rains.

**Status:** Native.

**Ecological significance:** In late October and early November, the flowers are visited by an array of butterflies, including members of at least the families Pieridae (sulphurs), Riodinidae (metalmarks), and Hesperiidae (skippers). In one short morning session, the Southern Dogface (*Colias cesonia*), Mexican Yellow (*Eurema mexicana*), Fatal Metalmark (*Calephelis nemesis*), Mormon Metalmark (*Apodemia mormo*), and Funereal Duskywing (*Ernnis funeralis*) were identified on the flowers. Several others species were seen but not identified.

Studies comparing molecular genetics, crossing experiments, and morphological traits have found that the genus is probably 1 million years old (Kimball et al. 2003). This conclusion was not drawn using a "molecular clock," but was estimated from the fact that the basal and apparently oldest species grows only on a volcanic island that appeared above the ocean about 1 million years ago.

As now understood, the genus *Coreocarpus* contains nine species that are confined to southwestern North America, especially to Mexico (Smith 1989, Turner 1996). Shreve and Wiggins (1964) recognized five species in the Sonoran Desert. Wiggins (1980) found four of those in Baja California. None grows in southwestern Arizona west of the Baboquivari Mountains (Felger et al. in prep.) or the Gran Desierto of northwestern Sonora (Felger 2000).

Smith (1989) was of the opinion that the genus was closely related to both *Bidens* and *Coreopsis*. The winged achenes are similar to *Coreopsis*, and their sometimes awned nature is more like that of *Bidens*. Indeed, *C. arizonicus* is one of the species that sometimes has awns and at other times lacks them.

Although historically floral and fruit characteristics have been considered to be conservative and most useful in reconstructing phylogenetic relationships, a study by Kimball and Crawford (2004) found that was not the case within *Coreocarpus* or other genera included within the tribe Coreopsideae. Their study suggests that vegetative traits are more conservative and likely to provide traits of relationships than those of reproductive structures.

**Human uses:** No uses found.

**Derivation of the name:** *Coreocarpus* is from Greek *koris*, bug, insect, and *carpus*, fruit. The genus was created in 1844 by George Bentham. *Arizonicus*, of Arizona, means that the first specimens were found in Arizona. Those specimens were collected by C. G. Pringle in the Santa Catalina Mountains in 1881, and were named by Asa Gray in 1882.

**Miscellaneous:** This plant is another of the DYCs (damned yellow composites) or *pinche compuestas amarillas* that are so abundant in much of the region. This genus resembles *Bidens aurea*, *Heliomeris*, *Tagetes lemmonii*, *Verbesina rothrockii*, and *Viguiera*, which also grow in the canyons. The seeds (achenes) of *Coreocarpus*, with their toothed or lobed margins (sometimes appearing scalloped), are distinctive, and at least a few fruits are typically present in most months. Pinnately compound leaves with filiform segments and no pellucid (translucent) dots are also unique.

## *Encelia farinosa* (Asteraceae) Brittle-Bush, *incienso, tohawes*

**Other names:** ENGLISH: [white] brittle-bush (stems break easily, Arizona, Sonora), goldenhills (doubtless referring to the change in the color of hillsides when in flower, Arizona); SPANISH: *hierba de las ánimas* (soul herb, Sonora), *hierba del bazo* <*vaso*> (enlarged spleen herb, Sonora), *hierba ceniza* (ashy herb, Sonora), *incienso* (incense, Baja California, Arizona, California, and New Mexico), *palo blanco* (white

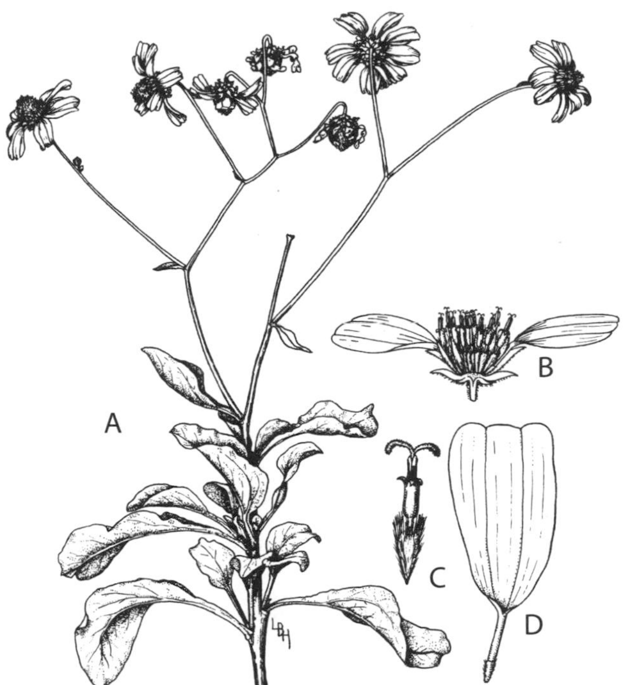


*Encelia farinosa*. A. Flowering branch. B. Cross section of flower-cluster. C. Fertile disc flower. D. Sterile ray flower. Artist: L. B. Hamilton (Benson and Darrow 1945).

bush, Sonora), *rama blanca* (white branch, Sonora), *yerba de la vaca* (cow herb, Paipai); ATHAPASCAN: *wóláchíí’ bitsiįį’ bił nát’oh* <*wóláčíˀbiciˀičíˀ bił nát’oh*> (*wóláchíí’ bitsiįį’ bił*, red-headed ant, *nát’oh*, tobacco, Navajo); HOKAN: *cotx* (acrid smell, Seri); UTO-AZTECAN: *choyoguo* (tar bush, Mayo, Sonora), *pa’akal* (Cahuilla), *tahavis* (Mountain Pima), *tohaves* (Hiá Ceḍ O’odham), *tohavs* (probably from *toha*, white, Akimel O’odham), *tohawes* (Tohono O’odham)

**Botanical description:** Small shrubs, not long-lived, aromatic, 0.5–1.6 m tall. Older stems are scarcely woody, with a usually rounded or hemispherical, often dense crown. Vegetative stems relatively thick, brittle, white woolly, glabrate with age. Leaves are highly variable with soil moisture, semipersistent, 3–10 cm long, including petiole, 1.4–3.6 cm wide, the blades mostly ovate, entire or nearly so, often white woolly or greener and thinner when grown during wet periods, whiter and thicker during dry times. Flowering branches with few-branched panicles, 8–30 cm long, are usually raised above the foliage. Phyllaries are graduated, green, lanceolate to ovate, sparsely hairy, the longer 3–5 mm long. Flowers showy, the rays bright yellow, 12–18 mm long, the disk florets yellow or maroon-brown. Achenes are 3.5–5 mm long, gray to blackish, obovate, shallowly notched apically.

**Habitat:** Dry, rocky slopes. Up to ca. 1,100 m (3,500 ft).

**Range:** Arizona (Coconino, Greenlee, Gila, Maricopa, Mohave, Pima, Pinal, Santa Cruz? and Yuma counties), southern California, southern Nevada, and southwestern Utah; northwestern Mexico (Baja California and Sonora).

**Seasonality:** November to May.

**Status:** Native.

**Ecological significance:** Leaves are allelopathic and release chemicals that are toxic to other plants (Kearney and Peebles 1951). Herbage is browsed by Bighorn Sheep.

During most seasons, a visitor to Brown Canyon would conclude that this species was absent entirely, or confined to the wash below the canyon. However, a visit during the flowering season produces an entirely different experience. At that time, the slopes above the entrance road turn bright yellow with flowering Brittle-Bush. Indeed, *Encelia* is common on those high, dry slopes, particularly on the south-facing side. Brittle-Bush is, however, also present on the north-facing slopes in fewer numbers.

Although the *Encelia* in Brown Canyon flowered in profusion during the spring of 2005, not a single flower appeared from 2006 to 2009. The prolonged drought of those years left the slopes devoid of the golden blanket of flowers produced by Brittle-Bush.

**Human uses:** In the O'odham Creation Story, *Jeweḍ Ma:kai* (Earth Shaman) created the *Taṣ* (Sun) and *Maṣad* (Moon). These two begat *Ban* (Coyote). The narrative says, "Out of the west beneath the tohafs [*tohavs*] bush the moon gave birth to Coyote. . . ." (Russell 1908). The version from the Thin Leather account noted that the Moon had duties and left *Ban*, who "was nourished on the earth" (Rea 1997). Thus, he became known by *Ñui* (Vulture) as *Tohavs* because of the fragrant herbs that he was laid upon.

Stems exude a yellowish resin that oozes from wounds. When heated, this resin becomes plastic and is used by Sonoran peoples as a glue or sealant, for medicine, and as incense (Felger and Moser 1985, Curtin 1949, Rea 1997). The gum was chewed by indigenous people, and it later was used as incense in the churches of Baja California and probably elsewhere (Castetter 1935, Kearney and Peebles 1951, Russell 1908).

Seri use *cotx icsipx* (brittle-bush its-resin) as glue when dry, and for hafting points on sea turtle harpoons (Felger and Moser 1985). Seri distinguish resin from the lower stems near the roots, and that from the upper branches (*cotx itáje*, brittle-bush its-saliva). The more gummy form from the upper stems is used to seal pottery and as a violin resin.

Seri heat a green twig without the bark in ashes and bite it to "harden a loose tooth" (Felger and Moser 1985). Roots mixed with *Asclepias* and *Euphorbia* are taken for toothaches and heart pain. The resin, *csipx*, is ground and sprinkled on sores. Tea of the leaves is used to treat rheumatism by the Mayo (Yetman and Van Devender 2001).

**Derivation of the name:** *Encelia* is of obscure origin, as with many of the names proposed by Michel Adanson (1727–1806). Possibly it is named for Christopher Encel [Christophorus Encelius, Christoph Entzel], d. 1583, author of *De Re Metallica. Farinosa* means "mealy" or "powdery" in reference to the drought leaves and stems that are white-pubescent.

**Miscellaneous:** The genus contains 15 species and reaches its greatest diversity in northwestern Mexico (Felger 2000). In wild regions, those species studied are self-incompatible. Cultivated plants hybridize and produce fertile offspring.

## *Ericameria laricifolia* (Asteraceae) Turpentine-Bush, *hierba del pasmo*

**Other names:** ENGLISH: gold-brush (because of flower color), larch-leaf [narrow-leaved] golden-weed (because of yellow flowers), turpentine-brush [bush] (from the odor); SPANISH: *hierba del pasmo* (herb for *pasmo*, see also *Ambrosia monogyra*), YUMAN: *xal shaB u* (Paipai); (= *Haplopappus lauricifolius*)

**Botanical description:** Rounded and much-branched, resinous-aromatic shrub, 3–8 dm tall. Leaves are 4–20 mm long, 1–2 mm wide, conspicuously dotted with "glands," the upper with smaller leafy twigs in the axils. Heads are 9–10 mm long, broadly top-shaped (turbinate) to bell-shaped, the phyllaries subulate, not strongly graduated and not in distinct vertical rows (shortest about half as long as the longest), relatively thin and chaffy, brown-stramineous, each with a thin, darker median strip running essentially the full length and unexpanded toward the apex. Disk corolla is 5.5–5.6 mm long, the ray flowers 0–6.

**Habitat:** Rocky slopes in canyons. 900–1,800 m (3,000–6,000 ft).

**Range:** Arizona (Greenlee to Mohave, south to Cochise, Pima, and Yuma counties), New Mexico,

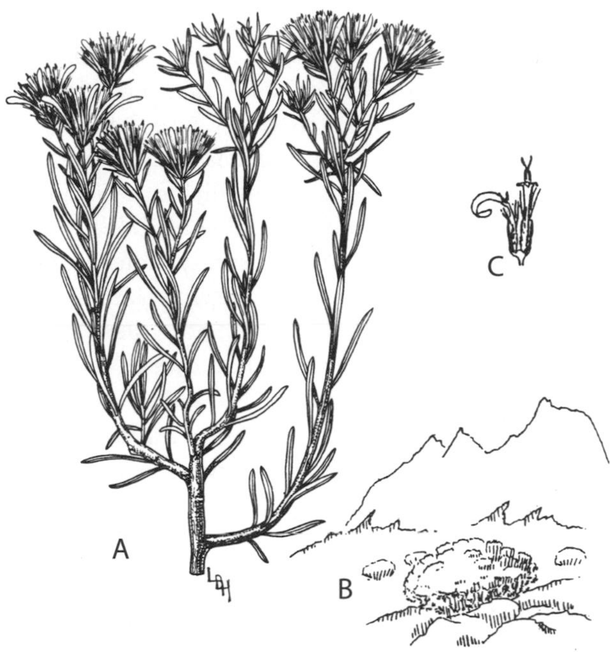


*Ericameria laricifolia*. A. Flowering branch. B. Growth form. C. Head with ray and disc flowers. Artists. L. B. Hamilton (Benson and Darrow 1945), A; R. D. Ivey (Ivey 2003), B–C.

western Texas, and southwestern Utah; Mexico (Sonora).

**Seasonality:** Flowering (March) August to November.

**Status:** Native.

**Ecological significance:** These plants are shrubs of desert mountains. The most common place to find Turpentine-Bush is in exposed spots on ridges, particularly those with crevices in the rocky outcrops. The shrubs are pioneers with lives long enough to partly hide that fact. Oddly enough, Turpentine-Bush occurs in the Madrean oak forests in the Baboquivari Mountains, but is not there in the Peloncillo Mountains of New Mexico (Moir 1979). In the New Mexico mountains studied by Moir (1979) the species is confined to desert grasslands and chaparral.

These are common Arizona plants that reach the southern limit of their range in northern Sonora. This species has only recently been documented near the Arizona border in Sonora (Van Devender and Reina 2005).

Apparently, the fungus *Aspergillus flavipes* is restricted to the root system (rhizosphere) of Turpentine-Bush. Zhou et al. (2004) extracted cytochalsan chemicals from this fungus that showed weak to moderate anti-cancer activity against several strains.

**Human uses:** Paipai make a tea for problems owing to *frio* (cold). That is an allusion to the hot-cold theory of medical treatment, a concept purportedly introduced by Europeans (Kay 1996). Turpentine-Bush is considered to be "hot"; the logic is that it will be effective against a "cold" problem. Other species similarly considered "hot," at least after Spanish arrival, are used by Akimel O'odham, Hopi, Navajo, Mexicans in Baja California, and Hispanics in New Mexico (Kay 1996). Regardless of their medical philosophies, *Ericameria* is used among the Cahuilla, Kawaiisu, Klamath, Kumiai, Luiseño, Miwok, Paiute, Shoshoni, and Tübatulabal of northern Arizona, California, Oregon, Utah, and Montana (Bean and Saubel 1972, Moerman 1998, Train et al. 1957, Voegelin 1938, Zigmond 1981). *Hierba del pasmo* makes the same reference, as that malady involves fever and aching bones, as well as blood stagnation. This *Ericameria* is said to contain rubber, especially in the roots (Munz and Keck 1973).

**Derivation of the name:** *Ericameria* may be composed of Greek *erike*, heather, broom, and *meris*, part, alluding to its resemblance to *Erica* (Ericaceae). *Laricifolia* means that the leaves resemble those of *Larix* (larch).

**Miscellaneous:** The genus *Ericameria* contains 36 species in western North America (Urbatsch et al. 2006). Although Thomas Nuttall segregated the genus from *Haplopappus* in 1840, other biologists, like Kearney and Peebles (1951), were reluctant to recognize the group. It was Lloyd Shinners (1918–1971) of Southern Methodist University who realized that Nuttall was correct and placed the species in *Ericameria* in 1950 to create the modern binomial *E. lauricifolia*.

*Ericameria cuneata* (Desert Rock-Goldenbush) grows in Brown Canyon and in several other canyons in the Baboquivari Mountains, as well as in the nearby Coyote and Quinlan Mountains. *Ericameria cuneata* differs from *E. laricifolia* by having leaves that are 3–5 mm wide and widest toward the tip. By contrast, *E. laricifolia* has linear leaves that are usually less than 2 mm wide. While Turpentine-Bush typically grows on ridges and slopes in the open sun, *E. cuneata* is more commonly found in rock crevices and cliff and rock faces in the shade. West of the Baboquivari Mountains, Desert Rock-Goldenbush reaches its westward limit in the Ajo Mountains. These perennial herbs or subshrubs were more widespread during the late Wisconsin, and ranged across the region more widely from about 9,000 to more than 37,000 years ago (Felger et al. in prep.). The species is known in southern Nevada, southeastern California, and southwestern Arizona, as well as in Mexico (Baja California and Sonora).

## ***Erigeron divergens*** **(Asteraceae)**<br>**Desert Fleabane, *hierba pulguera***

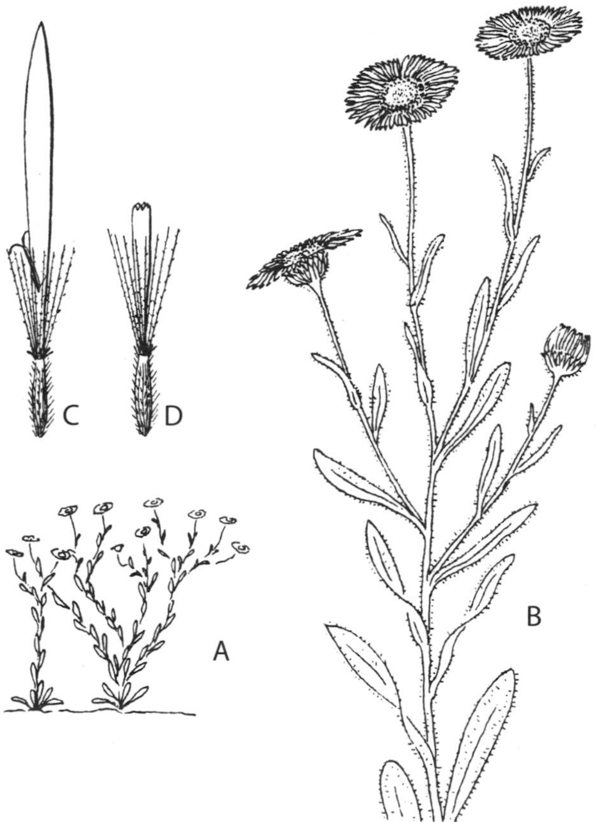


*Erigeron divergens*. A. Growth form. B. Flowering branch. C. Ray flower. D. Disc flower. Artist: R. D. Ivey, A, B (Ivey 2003); A. Hollick, C, D (Britton and Brown 1896–1898).

**Other names:** ENGLISH: [desert, spreading] fleabane ("fleabane" appeared in print in English when Wm. Turner published his book *Names of Herbes* in 1548; for hundreds of years the genus was thought to repel fleas), spreading daisy (Utah; *daeyes-eye* [daisy] used by ca. CE 1000; however, by Chaucer's time, ca. 1385, it was more recognizable as "daysyes," now called *Bellis perennis*; subsequently, applied to a number of Asteraceae); SPANISH: *hierba pulguera* (herb for fleas, Mexico); ATHAPASCAN: *ats'os níí'iinit* <*ˀacose níˀin'ił*> (*ats'os*, Plume Way, *níí'iinit*, snuff, Navajo), *'azee' ná'oołtádii* <*'azee'na'ołtxátiih*> (*azee*, medicine [which is used for], *ná'oołtádii*, untying [ceremonial knots], Navajo), *'azee'* [*ch'il*] *łibá* <*'aze'* [*c'il*] *labahi*, *labaˀigi*> (*'azee*, medicine, *ch'il*, plant, *łibá*, gray, Navajo), <*c'os be'yi'c'ol*, *béyi.c'ol*> ("vein spurter," *'ats'oos*, vein, maybe contraction of *biyah*, supporting it, and *choo'į*, used or useful, Navajo), *chį́į́h 'azee'* <*c'ís ˀazéˀ*> ([running] *chį́į́h*, nose, *'azee'*, medicine, Navajo), *dibetsétah ch'il* <*dibecetah ch'il*> (*dibetsétah*, bighorn's, *ch'il*, plant, Navajo), *k'aalógiidą́ą́* (*k'aalógii*, butterfly, *dą́ą́'*, food, Navajo), *na'ashjé'iidą́ą́'* (*na'ashjé'ii*, spider, *dą́ą́'*, food, Navajo), *wóláchíí' dą́ą́'* <*wolaciˀ da*> (*wóláchíí'*, red-ant, *dą́ą́'*, food, Navajo); UTO-AZTECAN: *tĭ'sas* [*dĭ'sas*, *toiyadĭsas*, *toidĭsas toi'yadatigora*] (Shoshoni; for *E. glabellus*)

**Botanical description:** Annual, biennial, or short-lived perennial herbs. Stems branched from the base and above, pubescent with spreading hairs, 0.5–7 cm tall. Basal leaves oblanceolate and spatulate, entire or sometimes lobed, usually 1–7 cm long, 2–10 mm wide, spreading-pubescent, petiolate, cauline leaves reduced toward apex. Heads are several to many, the involucres 7–11 mm wide, 4–5 mm high, finely glandular and hirsute with long spreading hairs. Ray flowers are blue, pink, or white, ca. 75–150, usually 5–10 mm long, sometimes hardly developing. Achenes are 2–4 nerved, sparsely hairy.

**Habitat:** Rocky slopes, canyons, grasslands. 300–2,750 m (1,000–9,000 ft).

**Range:** Arizona (throughout the state), California, New Mexico, Nevada, Texas, Utah, to South Dakota, and British Columbia; Mexico (Sonora).

**Seasonality:** Flowering February to October.

**Status:** Native.

**Ecological significance:** This variable species thrives in places that have been disturbed, either by humans, animals, or environmental events.

There are two forms of this species that appear within individual colonies. One has the ray flowers fully developed and "normal" (i.e., heads ca. 2 cm across). The other has reduced ray flowers (heads ca. 13 mm across). At first the colonies are uniquely the large-flowered plants but, in later years, they are replaced by more of the small-headed types. Perhaps part of this phenomenon may be attributable to different ploidy levels within the species (Nesom, personal communication, 2005).

The flowers of both sizes are favored by a variety of butterflies. Early in the spring, insects are hunted avidly in *E. divergens* by Lucy's Warblers. In the fall, several kinds of sparrows hunt through them for food.

**Human uses:** Navajo use the herb in a number of ways (Elmore 1944, Moerman 1998, Vestal 1952, Wyman and Harris 1941, 1951). Elmore (1944) recorded that their *'azee' ná'oołtádii* refers to the first of the five-night-long sings; *E. divergens* is chewed and blown on the ceremonial knots of the strings that are tied around the bundles of plants used.

An infusion is taken by women to aid childbirth, Fleabane was snuffed to relieve headache, and an infusion is used as an eyewash and to treat snakebites. In addition, Navajo use the herb ceremonially, make a medicine to treat infection caused by lightning, and use the species as a Life Medicine. Farther east, the Kiowa consider this Fleabane an omen of good fortune and bring it into the home (Vestal and Schultes 1939).

**Derivation of the name:** *Erigeron* is from Greek *eri*, early or the spring, or *erio*, woolly, and *geron*, old man, an allusion to an early-flowering hairy plant. *Divergens* means spreading, thus alluding to spreading trichomes on the stems, and perhaps also to the facility with which new plants volunteer.

**Miscellaneous:** *Erigeron divergens* has upper leaves that are unlobed. The other frequent species in Brown Canyon, *E. oreophilus* (Gorge Fleabane), has pinnate leaves throughout. While *E. divergens* is frequent in the lower parts of the canyons, *E. oreophilus* grows only in the upper reaches, coming down to about 1,200 m (4,000 ft). Within Arizona, that species grows up to 2,740 m (9,000 ft). *Erigeron oreophilus* has pinnate leaves like *E. neomexicanus*, but it also has glandular hairs as well as rough hairs.

Other species in this mountain chain include *E. arisolius* (Dry-Sun Fleabane) and *E. tracyi* (Running Fleabane). *Erigeron arisolius* has long been confused with *E. divergens*, but it differs by having erect buds; ray flowers reflexing, stems with coarse, thick-based hairs arising mostly from the cauline ribs; and minute, but prominently stalked, glands. By contrast, *E. divergens* has nodding buds, ray flowers straight or closing upwards, stems with relatively thin-based hairs arising evenly from the ribs and spaces between them, and sometimes minutely glandular and rarely stalked glands near the heads. *Erigeron tracyi* [= *E. colomexicanus*] has runners that connect plants.

## *Pseudognaphalium stramineum* (Asteraceae) Cotton-Batting, *gordolobo*

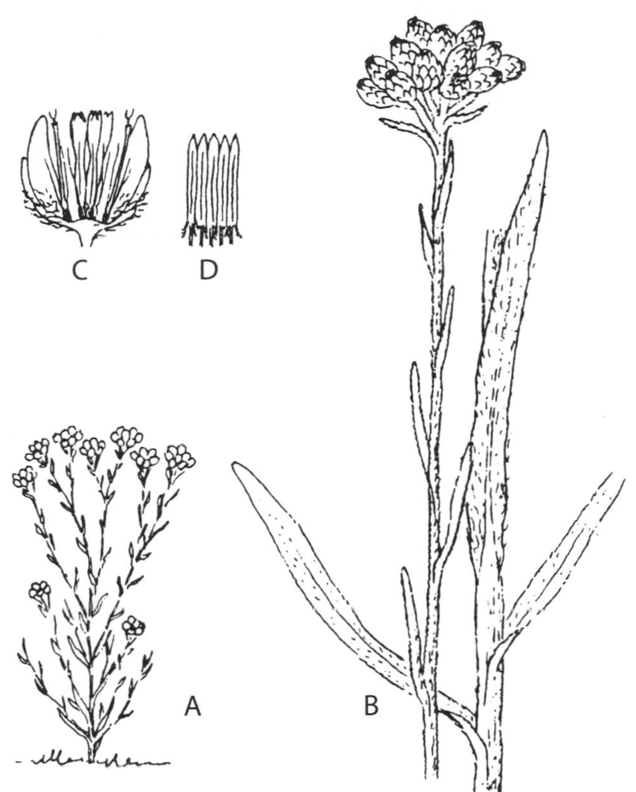


*Pseudognaphalium stramineum*. A. Habit. B. Flowering branch with detail of leaves. C. Detail of disc flowers. D. Staminal collar. Artist: R. D. Ivey (Ivey 2003).

**Other names:** ENGLISH: cotton batting (a name given to anything made of cotton; hence, an allusion to the pubescence), cotton weed, cudweed (chewed as a cud, like tobacco), [pearly] everlasting (for its long-lasting dried flowers); SPANISH: *gordolobo* (used in Mexico for the genus; from vulgar Latin *coda lupi*, wolf tail, originally applied in Spain to *Verbascum*, Scrophulariaceae; later brought to the New World and expanded to other plants); ATHAPASCAN: *'azee'disǫs* (*'azee'*, medicine [plant whose roots are], *disǫs*, glitter/glossy, Navajo); UTO-AZTECAN: *narag$^w$anɨ(m)bɨ* (Kawaiisu), *ŕosáberi* (Tarahumara), *toi'yadatibuda* (Shoshoni); (= *Gnaphalium chilense*)

**Botanical description:** Annual or biennial, several-stemmed, 2–6 dm tall, usually leafy with green-yellowish tomentum. Leaves are lanceolate to narrowly spatulate, 2–5 cm long, gradually reduced upwards. Heads are usually in a single close cluster (glomerule) at end of each stem, sometimes paniculate. Involucre is 5–6 mm high, green-yellow, the phyllaries all obtuse.

**Habitat:** Slopes and ridges above streams. 30–2,130 m (100–7,000 ft).

**Range:** Arizona (Apache, Coconino, Mohave, south to Cochise, Pima, Santa Cruz, and Yuma counties), California, New Mexico, Texas, and north to Montana and Washington; British Columbia; Mexico (Baja California, Chihuahua, and Sonora).

**Seasonality:** Flowering May to October.

**Status:** Native.

**Ecological significance:** These herbs grow among the rocks and crevices on slopes, particularly in shady spots below the canopy in oak woodlands. This *Pseudognaphalium* does not seem to be a sun-lover (heliophile) like many of the other species in the genus.

Flower heads first open narrowly, but become broad with age (Jaeger 1941). This plant is a cismontane species found along the edges of the deserts.

**Human uses:** Cotton-Batting is used by the Kawaiisu as a moxa to relieve pain (Zigmond 1981). A smoldering piece mixed with *Phragmites* and *Artemisia* is put on the painful spot. In addition, as the common names suggest, the genus is notable for being part of the ethnoflora in many parts of the world.

*Pseudognaphalium leucocephalum* (= *Gnaphalium leucocephalum*) is also present in the canyons, and it is known in many places as Pearly Everlasting, White Cudweed, or *gordo lobo*. *Pseudognaphalium leucocephalum* is known as *talcampacte* among the Mayo, who use it to make a tea to treat aches and fever (Yetman and Van Devender 2001). The same species is called *cusiteri* among the Guarijío, who use it for the same maladies as the Mayo and to promote digestion (Yetman and Felger 2002).

The Spanish Fray Francisco Ximénez wrote in 1615 of a cudweed called *tzonpoton* <*teçompotonic*> (*tzontli*, hair, *poton*, bad odor, stink) that it "cured the chest," and of one called *tlacochichic* (*tlacotl*, branch, *chichic*, bitter) having the same uses. Francisco Hernández wrote in 1651 that the Aztecs used the genus to treat ulcers (Kay 1996).

The Northern Tepehuan use at least two species, known in Spanish as *manzanilla del río* (river chamomile), *gordolobo*, and *avo yoošígai* (light [weight] flower, Northern Tepehuan) to relieve heart pains, stomach disorders, and coughs (Pennington 1969). Although the Northern Tepehuan name literally translates as "light," the word *liviano* used by Pennington is an old one in Spanish for plants used to treat lung problems. The Tarahumara use *P. wrightii*, known also as *manzanilla del río* or by their indigenous names *čiyowi*, *řasorá*, and *řasóko*, to relieve diarrhea and coughs (Pennington 1963).

**Derivation of the name:** *Pseudognaphalium* is based on Greek *pseudos*, false, and *glaphallion*, *knaphallon*, soft down, woolly, thereby referring to the pubescence. *Stramineum* is the Latin word for straw-colored, thus alluding to the pubescence.

**Miscellaneous:** The other cudweeds recorded for the canyons are *P. leucocephalum*, *Gamochaeta stagnale* (Purple Cudweed), and *G. wrightii* (Wright's Cudweed). *Pseudognaphalium leucocephalum* has decurrent leaves, white phyllary tips, and densely white-tomentose leaves. White Cudweed is typical of sunny streambeds and other wetter sites. *Pseudognaphalium canescens* leaves are not decurrent, the phyllary tips are white or straw-colored, and it grows on dry rocky ridges.

*Gamochaeta stagnale* (long misidentified as *Gnaphalium purpureum*) is unique in having spicate flower clusters, purple or brownish phyllary tips, and pappus bristles that are united basally and deciduous in a ring. Nesom (2004) explains why these small plants should be called *Gamochaeta stagnale*. He points out that *Gamochaeta stagnale* differs from *Gnaphalium purpureum* in being annual, having oblanceolate leaves equally tomentose on upper and lower surfaces, interrupted flower clusters, small basally tomentose heads, and phyllaries conspicuously purple at the base and along their basal margins.

## *Gutierrezia sarothrae* (Asteraceae) Snakeweed, *escoba de la víbora*, *siwĭ tatṣagĭ*

**Other names:** ENGLISH: broom [broom-weed, brown-weed, threadleaf] snakeweed ("snakeweed" seems to be a translation of the Spanish name), match [sheep, turpentine-bush, yellow-top]-weed; SPANISH: *collálle* <*coyaje, coyaye*> (probably from Náhuatl *coyolli*, rattle, an allusion to rattlesnakes, New Mexico), *escoba de la víbora* (snake's brush, New Mexico), *hierba de San Nicolás* (St. Nicholas's herb, Nuevo León), *rosita* (little rose, San Luis Potosí), *yerba de la víbora* (snake's herb, throughout Mexico); ATHAPASCAN: *ch'il diilyésii* [*yázhí*] <*č'il dilyési*, *c'il dihesi'* [*yazi*], *tc'iltiiyéesiih*> (*ch'il*, plant, *diilyésii*, refuge [plant (which is used as a) refuge (by small animals)], *yázhí*, little, Navajo), *ch'iląągo* (Western Apache), *tł'iish bichagosh'oh* (*tł'iish*, snake, *bichagosh'oh*, its

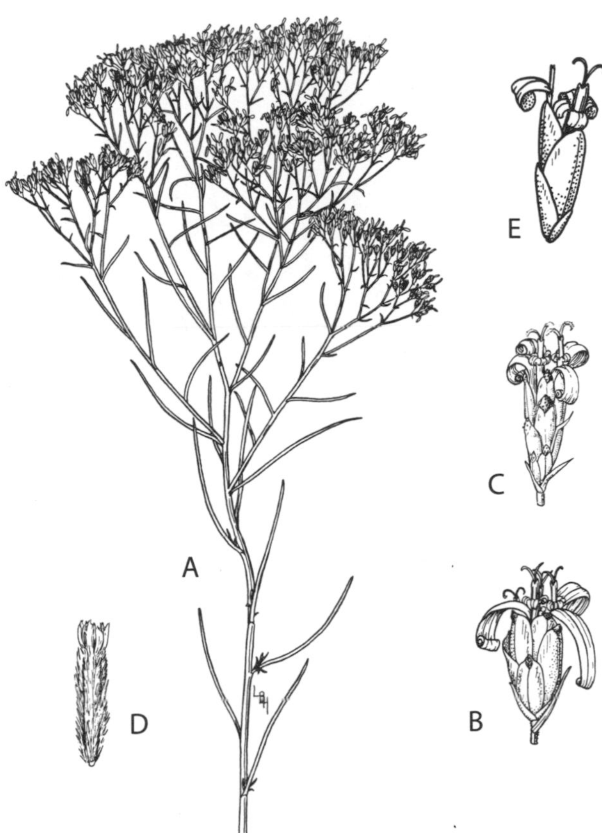


*Gutierrezia*. A–D. *G. sarothrae*. A. Branch tip. B. Flower cluster, with multiple ray and disc flowers. C. Flower cluster, with 4 ray flowers and 2 disc flowers. D. Achene. *G. microcephala*. E. Flower cluster, with single ray and disc flowers. Artist: L. B. Hamilton (Benson and Darrow 1945).

house/refuge, Western Apache); KIOWA TANOAN: *kojaji* (Tewa); UTO-AZTECAN: *jaribomenáguat* (Ópata fide Sobarzo 1991, Sonora), *kwitaweyampeh* (Shoshoni), *pohniyavɨ* (*pohniya*, skunk, *vɨ*, plant, Kawaiisu), *tsatsakwma'övi* <*tcatcákmá:'ɜvi*> (*tsatsakw*, small, *ma'övi*, snakeweed, Hopi), [*pas*] *maa'övi* <*pamnavi*> (*pas*, true, *maa'övi*, snakeweed, Hopi), *shpûmp* (Ute), *sqúmpï* (squirrel plant, Southern Paiute), *kû'kikoinûmp* (Shoshoni), *siw taḍsagi* (Hiá Ceḍ O'odham), *siw u'us* (*siw*, bitter, *u'us*, plant, bush, Tohono O'odham), *siwĭ tatṣagĭ* <*siwstaḍ*, *tatṣṣagi*, *tad.xxagĪ*> (*siwĭ*, bitter, *tatṣagĭ*, *Gutierrezia*, Tohono O'odham); YUMAN: *gohwa:yo* (Walapai); LANGUAGE ISOLATE: *kia'hapoko* (*kia* from *kia'we*, water, *ha'poko*, gathered together; so named because it drank abundant water, Zuni)

**Botanical description:** Shrubs, 30–100 cm tall, stems grayish, straw-colored or green above. Leaves are dimorphic, the cauline are 2–5 cm long and 2–4 mm wide, linear to linear-lanceolate, and with shorter, narrower, fasciculate axillary ones, often both lacking when flowering. Heads are clustered at branch ends, sessile. Ray flowers are 3 to 8 and 3–5.5 mm long. Disk flowers are 2–9 and 2–3 mm long. Achenes of disk flowers are abortive, those of ray flowers are fertile, 2–3 mm long, hairy.

**Habitat:** Grasslands to oak woodlands, especially overgrazed sites. 970–1,980 m (3,200–6,500 ft).

**Range:** Arizona (Apache to Mohave, south to Cochise, Pima, and Santa Cruz counties), California, Colorado, Utah, east to Kansas, and north to Saskatchewan; Mexico (Baja California, Chihuahua, Coahuila, and Sonora).

**Seasonality:** Flowering September to October.

**Status:** Native.

**Ecological significance:** *Gutierrezia* plants probably evolved to grow as pioneers in open, unvegetated sites. Snakeweed thrives in places where most other perennials cannot become established. Disturbance by livestock and humans caused a general increase in Snakeweeds over the past few hundred years since Europeans arrived in the New World. *Gutierrezia sarothrae* is known to have replaced thousands of acres of grasslands since Europeans arrived (Brown 1982).

**Human uses:** In old California, *G. sarothrae* was used to treat snakebite, thus giving rise to the Spanish names (Curtin 1947). Presumably, the practice of applying a chicken to the snakebite, as an adjunct to the tea from Snakeweed, evolved from indigenous use of some native bird.

Cahuilla, Hopi, Navajo, Tewa, and Zuni use this plant to relieve toothaches, digestive problems, and as a poultice (Bean and Saubel 1972, Kay 1996, Moerman 1998, Robbins et al. 1916, Stevenson 1915, Whiting 1939, Wyman and Harris 1951). Zuni make a decoction of snakeweed to relieve retention of urine (Stevenson 1915). Snakeweed is one of the Life Medicines of the Navajo, being used, for example, in the Enemy Way and Evil Way (Wyman and Harris 1941, 1951). Moore (2003) concluded that *G. microcephala* and *G. sarothrae* may be used interchangeably for soothing pain from aching legs, arthritis, headaches, sore legs, aching body, and rheumatism. Hopi and Tewa add *Gutierrezia* to roasting corn, and they attach a sprig of *Gutierrezia* to a *paho* (prayer stick) (Fewkes 1896, Robbins et al. 1916, Whiting 1939). Walapai use the plant in rain ceremonies and make brushes of the stems to remove spines (glochids) from *Opuntia* fruits (Watahomigie et al. 1982).

Hispanics in the region, particularly in New Mexico, consider Snakeweed a prime medicine for female complaints, including childbirth and excessive menstruation (Kay 1996). Caution is indicated during pregnancy because *Gutierrezia* has uterine stimulant and hemolytic activity. Babies with coughs also benefit from Snakeweed by drinking a tea of it with sugar.

**Derivation of the name:** *Gutierrezia* was named in 1816 to commemorate Spanish nobleman Pedro Gutierrez, a correspondent of the Jardin Botanico, Madrid. *Sarothrae*, from Greek *sarothron*, means that this plant resembles or was used to make a broom.

**Miscellaneous:** In some places, the most common species is *G. microcephala*. Typically, the difference between the two species is not the size of the flower-heads but the number of flowers per head. There are usually 1–2 ray flowers and 1–2 disc flowers in *G. microcephala*, while there are (2–)3–5(–8) ray flowers and (2–)3–8(–9) disc flowers in *G. sarothrae*. There are often 8–14 or more phyllaries in *G. sarothrae* and 4–6(8) phyllaries in *G. microcephala*.

The other species of *Gutierrezia* in Arizona are typically plants of lowlands. West of the Baboquivari Mountains, there is the annual *G. arizonica* (Felger 2000, Felger et al. in prep.).

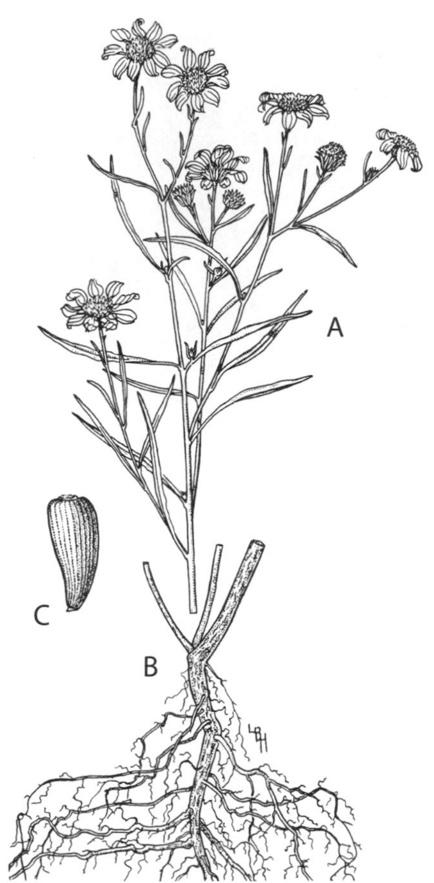


*Heliomeris longifolia*. A. Upper branch with flowers. B. Taproot of plant. C. Achene. Artist: L. B. Hamilton (Parker 1958).

## ***Heliomeris longifolia*** **(Asteraceae)** **Golden-Eye, *tacote***

**Other names:** ENGLISH: golden-eye (the OED gives this name for ducks, *Bucephala* spp., applied by John Ray in *Willughby's Ornithology* of 1678 as: "The Golden-eye. . . . The Irides of the Eyes are of a lovely yellow or gold-colour"; later applied to a fish, undated, and an insect by 1753; no plants are discussed); SPANISH: *tacote* <*tecote*> (used for several yellow flowering plants; from Náhuatl *tlacotl*, stem, Mexico), *tlalpopote* (from *tlal*, earth, *popotl*, broom, Mexico); ATHAPASCAN: *ch'il 'at'ąą' 'ałts'óózí* <*c'il bit'aˀ ˀa'lc'ozigi*> (*ch'il*, plant, *'at'ąą'*, forehead? *ałts'óózí*, slender, leaf, Navajo), *ndíyílii ts'ósí* <*ildf'ilí c'o's*> (*ndíyílii*, sunflower, *ts'ósí*, slender, Navajo); (= *Viguiera annua*)

**Botanical description:** Herbs, annual, erect, branched or almost simple, about 7 dm tall, from a taproot, the stem tuberculate-strigillose to strigose. Leaves opposite below, alternate above, linear or narrowly linear-lanceolate, attenuate at both ends, tuberculate-strigillose on both sides and gland-dotted below, 1-nerved, strongly revolute, 3–7 cm long, 1.5–3 cm wide, the upper smaller, on petioles to 4 mm long. Heads are with disk 5–7 mm high, 6–8 mm thick, the involucre 3–5 mm high, in 2 series, phyllaries lanceolate, acuminate, blunt-tipped, mucronulate, herbaceous, pubescent. Ray flowers are often 12, ovate, 5–15 mm long. Disk flowers are somewhat glandular below, 2.5–2.8 mm long. Achenes are lucid, black, glabrous, 1.8 mm long, with pappus absent.

**Habitat:** Canyons, washes, streamsides. 760–2,130 m (2,500–7,000 ft).

**Range:** Arizona (Apache to eastern Mohave, south to Cochise, Pima, and Santa Cruz counties), New Mexico, and western Texas; Mexico (Chihuahua and Sonora).

**Seasonality:** Flowering (July) August to October.

**Status:** Native. Plants in the canyons are var. *annua*.

**Ecological significance:** This species and similar plants at different times have been considered part of *Heliomeris* and *Viguiera*. Although the latest treatment in *Flora of North America* (2006, Vol. 21) separates the two genera, there is nothing in that publication that allows a nonspecialist to distinguish them. The two genera are confused and confusing. Therefore, both are discussed here, and the historic delimitation of species is followed.

Packrat middens have been found that contain seeds that belong to either *Helianthus* or *Viguiera* (Rea 1997). Because the achenes are so similar in these two genera, identification is problematical. However, Felger et al. (in prep.) record a *Viguiera* from Alamo Canyon, twigs and achenes, 1,150 ybp. Other samples from Montezuma's Head (Sierra Estrella, Pinal County), twigs and achenes, were 20,490 and 21,840 ybp, and five from Puerto Blanco Mountains were 3,400 to 9,860 ybp. Perhaps both genera came from the southwest and spread from there into the Great Plains. Recent identification of old *Helianthus* seeds in Mexico (Lentz et al. 2001) may support that view, although the identification has been contested (Heiser 2008).

**Human uses:** The only records found of *H. longifolia* var. *annua* being used were among the Navajo (Vestal 1952). These people use the herb in their Life Medicine, and their sheep and goats are given the plants as fodder.

Other species used include *H. multiflora* in the United States, and some in Mexico. *Heliomeris multiflora* has seeds (achenes) that are eaten by the Shoshoni (Chamberlin 1911). The Navajo use *Heliomeris* to feed sheep and deer, and they also make them into a witchcraft medicine (Vestal 1952). Hocking (1997) found that the Mexican *Viguiera buddleiaeformis* contains alkaloids that show promise in treating cancer. Bundles of the dried stems of *V. dentata* have been used to make torches. Farther down in southern Mexico, both *Viguiera excelsa* and *V. sphaerocephala* are used to treat wounds and internal problems.

**Derivation of the name:** *Heliomeris*, from Greek *helios*, sun, and *meris*, parts, refers to the flower heads and their resemblance to both the sunflower (*Helianthus*) and the sun. The other genus into which these species were previously placed, *Viguiera*, is named after French physician L. G. Alexandre Viguier, 1790–1867, author in 1814 of the natural history and medical aspects of *Argemone*. *Longifolia* means that it has long leaves. *Annua* indicates that the individual plants live but a single year.

**Miscellaneous:** Other species in Brown Canyon are *Viguiera dentata* var. *lancifolia* and *Heliomeris multiflora*. Both *V. dentata* and *H. multiflora* are perennial. Moreover, these two species have wider leaves (typically over 5 mm to 30 or more) than *H. annua* (1.5–3 mm).

## *Heterotheca subaxillaris* (Asteraceae) Camphor-Weed, *gordolobo*

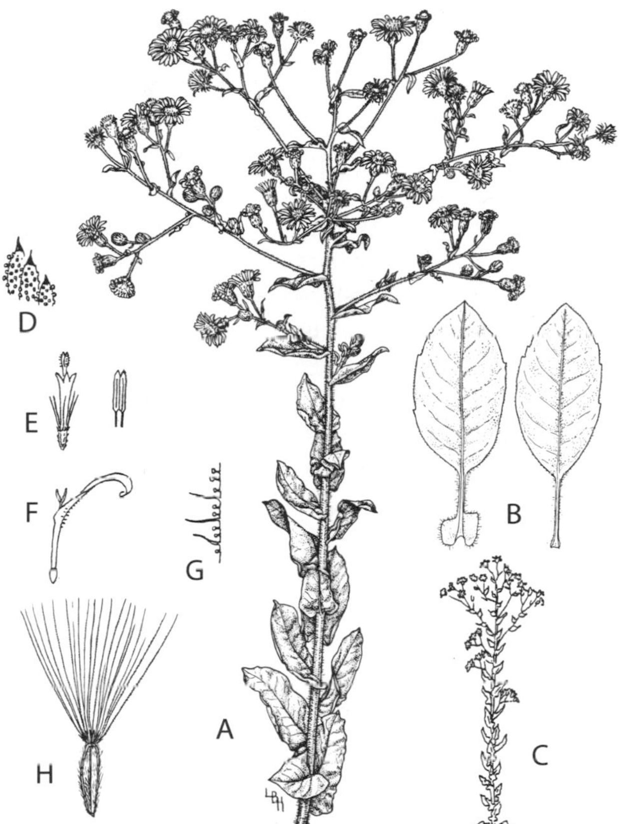


*Heterotheca subaxillaris*. A. Upper part of plant with sessile, clasping leaves. B. Two basal leaves showing variation. C. Growth form of plant. D. Involucral bracts of flowers with glandular hairs. E. Disc flower. F. Ray flower. G. Glandular and nonglandular hairs of stem. H. Achene from central disc flower. Artist: A–B, L. B. Hamilton (Parker 1958); C–H, R. D. Ivey (Ivey 2003).

**Other names:** ENGLISH: camphor-daisy [-weed] ("camphor" taken into English from French *camfre* or *camphre*, ultimately from Greek *caphura* and Arabic *kafur*, a vegetable oil with a bitter taste and characteristic smell; by at least 1570 it was combined with other words to indicate distinctive species, Arizona, New Mexico), false arnica, golden [gold]-aster ("golden," in reference to color, has been combined with various plants since at least 1570; the OED 2007 lists 33 combinations, and "aster" is not among them), telegraph plant; SPANISH: *árnica* (name used by Kickapoo in Cahuilla, others in Durango, Sonora; origin disputed, perhaps from Greek *arnakis*, lamb's-skin, from the leaf texture; or from Greek *ptarmiké*, a plant whose odor provokes sneezing, through Latin *ptarmica*, to Medieval Latin *armica*), *gordo lobo* (fat wolf, Chihuahua, Sonora; see *Pseudognaphalium* for etymology), *malamujer* ("bad woman," because the herbs have adhesive propagules, Mountain Pima); ATHAPASCAN: *wóláchíí' bi'gh ą́* *<wolaciˀ be.ga>* (*wóláchíí'*, red ant, *bi'ghą́*, [killer] lit. it kills them, Navajo); UTO-AZTECAN: *arniko* (from Spanish *árnica*, Mountain Pima), *haramkulyi* (*haram*, sticky, *kulyi*, old man, Mountain Pima); (= *H. psammophila*)

**Botanical description:** Annual aromatic herbs. Stems are occasionally over 1 m tall, rarely to almost 2 m, scabrous or short-hirsute. Basal and lower stem leaves petiolate, ovate to elliptical, 3–7 cm long, entire to dentate, upper leaves sessile, ovate to lanceolate, smaller, entire to serrate, rough on both surfaces (scabrous). Inflorescences are a loose flat-topped panicle. Flowers of both disks and rays, yellow. Disk achene is obovate, densely sericeous, ray achene nearly glabrous.

**Habitat:** Washes, roadsides, trails. 300–1,670 m (1,000–5,500 ft).

**Range:** Arizona (southern Navajo to Yavapai, south to Cochise, Graham, Pima, and Santa Cruz counties), New Mexico, Texas, east to Florida, and north to Delaware and Kansas; Mexico (Chihuahua and Sonora).

**Seasonality:** Flowering (March) August to November.

**Status:** Native.

**Ecological significance:** This plant thrives on disturbance. Before humans populated the New World, this plant was probably growing beside waterways and animal trails. Now, the seeds more often germinate and grow in areas where the soils are disturbed by humans.

**Human uses:** No record was found of the indigenous people using Camphor-Weed within the United States. Pennington (1973) noted that a lotion of the entire plant of *H. subaxillaris* was used on rheumatic areas by Mountain Pima. "Mexicans" (Kickapoo from southern Wisconsin who moved there in 1852 to avoid persecution in the United States) use *gordo lobo*. Kickapoo use *H. subaxillaris* to treat heat rash, burns, and itches, and to ease menstrual pains (Austin 2004). López-E. and Hinojosa-G. (1988) say that people in Sonora use it to treat asthma, emphysema, fever, hemorrhoids, and phlegm. The name *gordo lobo* suggests that the herbs are used like *Gnaphalium* and *Pseudognaphalium*. Moerman (1998), Whiting (1939), and Vestal (1952) found Cheyenne, Hopi, Isleta, Luiseño, and Navajo using *H. grandiflora* and *H. villosa*. Hopi use *patcávu* for *H. villosa*, the same name they give *Brickellia* (Whiting 1939). Mountain Pima have two names for another species and use it medicinally (Kay 1996, Laferrière 1991).

The discussion in Kay (1996) indicates that "False Arnica" is a widespread remedy in Mexico and in Tucson *yerbarias*. Given the paucity of other literary references to *Heterotheca*, her identification of the species may be incorrect—or nontaxonomists may have a broader definition of "arnica" than scientists. People define "arnica" differently, as the name originally was used for *Arnica*. Martínez (1969) also recorded that *H. inuloides* (*acahual*, *falsa árnica*, *cuauteteoc*) has long been used in Mexico to treat the same maladies.

Ajilvsgi (1984) wrote that Camphor-Weed is avoided by livestock. That comes as no surprise to anyone who has smelled the herbs, although John Semple (personal communication, Apr. 2003) found herbarium specimens with notes that it "might be used as a forb locally."

**Derivation of the name:** *Heterotheca* is from Greek *heteros*, different, and *theca*, case, thereby alluding to the dissimilar ray and disk achenes. *Subaxillaris* means having flower clusters near the angle (axil) where the leaves join the stem.

**Miscellaneous:** There was taxonomic confusion among *Heterotheca*, *Chrysopsis*, and *Pityopsis* until the species were studied by Semple (1977, 1981, 1996, Semple et al. 1980). The misunderstanding among the three genera was caused by convergence in traits, and a failure of previous students to appreciate the importance of what appeared to be trivial characteristics. After the studies by Semple and his students, it became apparent that there were differences in cytology, morphology, anatomy, habit, and habitat among these genera.

## *Hymenothrix wislizeni* (Asteraceae) Burro-Brush

**Other names:** ENGLISH: [Trans-Pecos] thimblehead ("thimble" came from Old English *thýmel*,

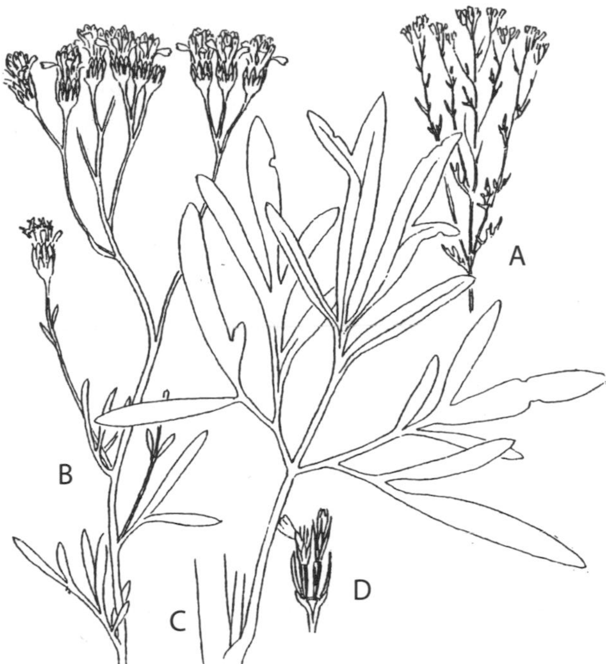


*Hymenothrix wislizenii.* A. Upper part of flowering plant. B. Flowering branch. C. Leaf and node. D. Detail of a flower cluster. Artist: R. D. Ivey (Ivey 2003).

meaning "thumb instrument," ca. 1000; akin to *handle*; the later English developed a *b* after *m*, as in *humble, nimble*; akin to Old Norse where *thumall* meant the thumb of a glove; perhaps a leather thumbstall was the earliest form of thimble; metal thimbles were introduced in the seventeenth century; combined with "head" recently, Arizona, California, Texas), Wislizen's burro-brush ("burro brush" is a name applied to at least *Ambrosia* and *Hymenothrix*; presumably only a burro will eat them)

**Botanical description:** Herbs, annual, erect, to 12 dm tall, branching above, the stems striate, herbage puberulent throughout. Leaves with 2–3 divisions again divided into 3s (biternate to triternate), 5–12 cm long, the divisions linear to oblong, acute to apiculate apically. Heads are in corymbose-paniculate clusters, the peduncles slender, to 1 cm long. Involucres are top-shaped, 3–5 mm high and wide, the bracts narrowly lanceolate-oblong, 1 mm wide or less. Ray flowers are 3–4 mm long, yellow, shallowly 3-toothed. Disc corollas are yellow, 4–5 mm long, lobes triangular, acute. Achenes are black, 3–4 mm long, sparsely hirsutelous, truncate apically, narrowing sharply to base, the pappus of 10–15 narrowly lanceolate scales with a strong midnerve turning into a barbed awn.

**Habitat:** Washes, less often higher sites in grasslands and in canyons. 760–1,650 m (2,500–5,500 ft).

**Range:** Arizona (Cochise, Gila, Graham, Greenlee, Pima, Pinal, and Santa Cruz counties) and southwestern New Mexico; Mexico (Chihuahua and Sonora).

**Seasonality:** Flowering June to December.

**Status:** Native.

**Ecological significance:** Seeds of these herbs are eagerly sought by Lesser Goldfinches (*Carduelis psaltria*), House Finches (*Carpodacus mexicanus*), and other granivorous birds in the fall when the tops turn white with fruits. Even the alien House Sparrow (*Passer domesticus*) recognizes the potential food source and attempts to pick off the fruits of low-hanging clusters by jumping at them. House Sparrows apparently are too heavy to land directly on *Hymenothrix*, as do the native species.

Not only birds but also Leaf-Cutter Ants (*Acromyrmex*) gather the fruits. In October and November, the ants gather fruits from several Asteraceae, including *Bebbia*, *Hymenothrix*, *Isocoma*, and *Machaeranthera*. These achenes appear in rings around the ant nests, apparently because the colony foragers brought them and the garden-tending ants rejected this "food" (Justin O. Schmidt, personal communication, July 2007).

When food is scarce in the early spring, Javelina (*Pecari tajacu*) sometimes graze the young leaves on the basal rosettes. These animals sometimes come to my front door to nibble these herbs.

The plants themselves are most often found along washes and other waterways where they thrive. Occasionally, Burro-Brush is found on upland areas, but the plants are often dwarfed there and never have as many flowers or fruits. Shreve and Wiggins (1964) indicate that *H. wislizeni* is most common in the lower mountains and foothills in the Sonoran Desert.

*Hymenothrix* has triploid members capable of producing reduced and unreduced gametes that pass on genes to diploid parent taxa, contrary to old dogma. An interspecific tetraploid taxon is found to be outcompeting one of its parental species for a significant part of its range (Pinkava 2003).

**Human uses:** None found.

**Derivation of the name:** *Hymenothrix* is from Greek *hymen*, a membrane, and *thrix*, hair, referring to the membranous bristles of the pappus. *Wislizeni* is named for German-born Friedrich (Frederick)

Adolph(us) Wislizenus, 1810–1889, who collected plants in the southwestern United States.

**Miscellaneous:** *Hymenothrix* is a genus of five species that was described by Asa Gray in 1849; all are found in southwestern North America (Felger 2000). The genus has been considered a member of tribe Helenieae and subtribe Gaillardiinae (Mabberley 1997). That tribe was proposed by Bentham in 1873, and it has long been thought to be unnatural. Within the group, Keil et al. (1988) were among the students of the genus who studied chromosome numbers and suggested the close relationship of *Hymenothrix* to *Bahia*. Subsequently, studies of internal relationships within the Helenioid Heliantheae, including the subtribe Gaillardiinae, based on molecular genetics show that *Hymenothrix* is part of the lineage now called the tribe Bahieae, subtribe Bahiinae, including the genera *Bahia*, *Florestina*, *Palafoxia*, and other southwestern genera (Baldwin et al. 2002).

## *Isocoma tenuisecta* (Asteraceae) Burro-Weed, *tatșagĭ*

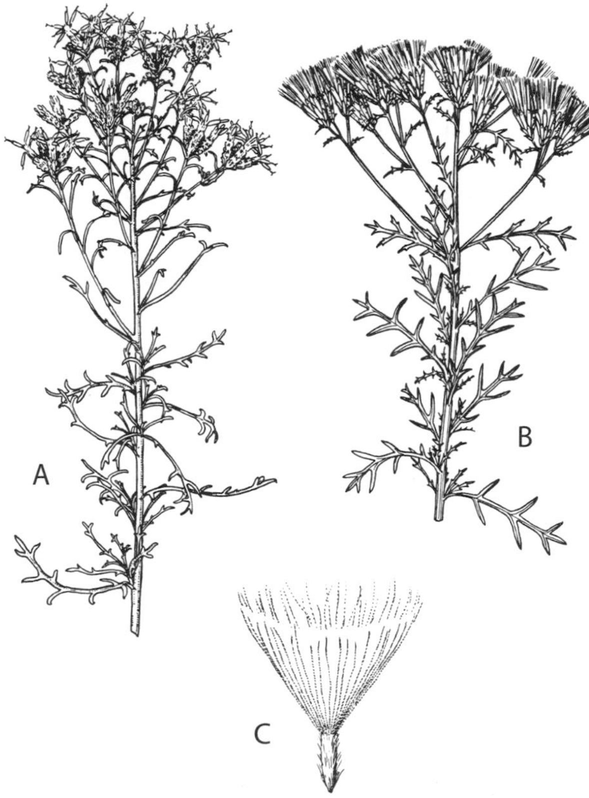


*Isocoma tenuisecta*. A. Leafy branch with flower clusters. B. Branch with fruits. C. Achene with turf of hairs. Artist: L. B. Hamilton (from original artwork).

**Other names:** ENGLISH: bitter-weed (in English by 1878), burro-weed (like *Hymenothrix*, only a burro will eat them; the word "burro" came into English from Spanish about 1800), golden-bush ("golden" in use since ca. 1300 to indicate something resembling gold, at least in color), shrine golden-weed ("weed" used since ca. 888 for a plant that was not attractive, or useful, and occupying space for desired species; from Old English *wéod*, cognate with Old Saxon *wiod*, Low German *wed*, Flemish *wied*, East Frisian *wiud*), shrine jimmy-weed ("jimmy" applied to a trembling disease caused by livestock eating poisonous plants, cf. Curtin 1949), turpentine-bush (comparing the odor of the leaves to turpentine, an oily resinous fluid originally taken from terebinth, later from pines); UTO-AZTECAN: *tatșagĭ* <*taḍshagi, tatshagi*> (probably from *siw tatșagi*; confusion with *totșagi*, foamy? Tohono O'odham, Arizona); (= *Haplopappus tenuisectus*)

**Botanical description:** Shrubs, erect, with numerous branches that form a compact rounded clump 3–8 dm tall, branchlets slender, yellowish, puberulent to glabrate. Leaves broadly oblong in outline, 1–4.2 cm long, 0.5–1.5 cm wide, pinnately parted into 2–4 pairs of linear acute lobes 1–1.5 mm wide, to 9 mm long, dotted with resinous glands. Heads in compact clusters (glomerules) cymosely grouped at branch tips. Involucres are top-shaped to bell-shaped, 4–5.5 mm long, 3–3.5 mm wide, the bracts narrowly lanceolate-ovate to linear in 3–5 series, resin-dotted. Ray flowers absent. Disc flowers are 6–12 or up to 15, yellow, 4–5 mm long, lobes linear, 1–1.3 mm long, glabrous. Achenes are mostly 2.5–3 mm long, rough-silky, the pappus bristles pale brownish, longer ones about equaling corollas.

**Habitat:** Grasslands, ridges. 600–2,600 m (2,000–5,500 ft).

**Range:** Arizona (Cochise, Gila, Graham, Greenlee, Pima, Pinal, and Santa Cruz counties); Mexico (Sonora). The species was formerly thought to range into Chihuahua, southeastern New Mexico, and western Texas (e.g., Correll and Johnston 1970). Those plants are now placed in *I. plurifolia* (Nesom 1991).

**Seasonality:** Flowering August to October (November).

**Status:** Native.

**Ecological significance:** Leaf-Cutter Ants (*Acromyrmex*) gather the fruits. In October and November, the ants gather fruits from several Asteraceae, including *Bebbia*, *Hymenothrix*, *Isocoma*, and *Machaeran-*

*thera*. These discarded "food" items appear in rings around the ant nests.

This shrubby herb has replaced thousands of acres of grasslands since the arrival of Europeans. The species has been considered an indicator of formerly healthy grasslands (Brown 1982).

Study by Bean et al. (2004) indicates that these shrubs are among the second group of plants that colonized old agricultural lands in the Santa Cruz River valley of southern Arizona. The first species to become established were annuals such as *Salsola* and *Sisymbrium*. These annuals were then joined by short-lived perennials, including *Isocoma* and *Baccharis sarothroides* (q.v.).

Although *I. tenuisecta* may live 20 years, one study indicates that most individuals do not survive for more than 7 (Goldberg and Turner 1986). While the shrubs have a root-system that may be 3–6 m long, *Isocoma* are often killed by drought. This loss of adults is offset by an increase of seedlings during and after dry periods (Ladyman 2005).

**Human uses:** Saxton and Saxton (1969) say that this species is a medicinal plant, but they give no further details. No other source has been found for O'odham or other people using *I. tenuisecta* medicinally, but both Curtin (1949) and Rea (1997) discuss other species under the name *sai u'us <sai oos>* (bad-smelling sticks or plant). The two species they discussed were used externally to treat local inflammation or other problems. There is either confusion or faulty elicitation because *taḍshagi* is also applied to *Ambrosia* and *Gutierrezia* (q.v.), or, more probably, the O'odham recognize three folk species within a folk genus.

This species, and presumably others in the genus, are poisonous to livestock. Horses appear to be the most sensitive and sheep the least (Bradley et al. 1998). Burro-Weed shares the pyrrolizidine alkaloid trematol with White Snakeroot (*Ageratina altissima*), hence the name Jimmy-Weed. For more on trematol poisoning, see *Ageratina herbacea*.

**Derivation of the name:** *Isocoma* is based on Greek *isos*, equal, and *kome*, hair of the head, "so called from its equal flowers" (from original description by Thomas Nuttall in 1841). *Tenuisecta* means that the leaves are divided into small segments.

**Miscellaneous:** *Isocoma* is one of several genera that were previously put in the "waste-can" genus *Haplopappus*. As now defined, *Icocoma* comprises 16 species, all confined to the southwestern United States and adjacent Mexico (Nesom 1991). The genus is in tribe Asterae. The shrubs have been compared with "goldenrods" (*Solidago*) by some who do not know their scientific name. That common name is at least putting them in the right tribe.

## *Lactuca serriola* (Asteraceae) Prickly Lettuce, *lechuga silvestre*

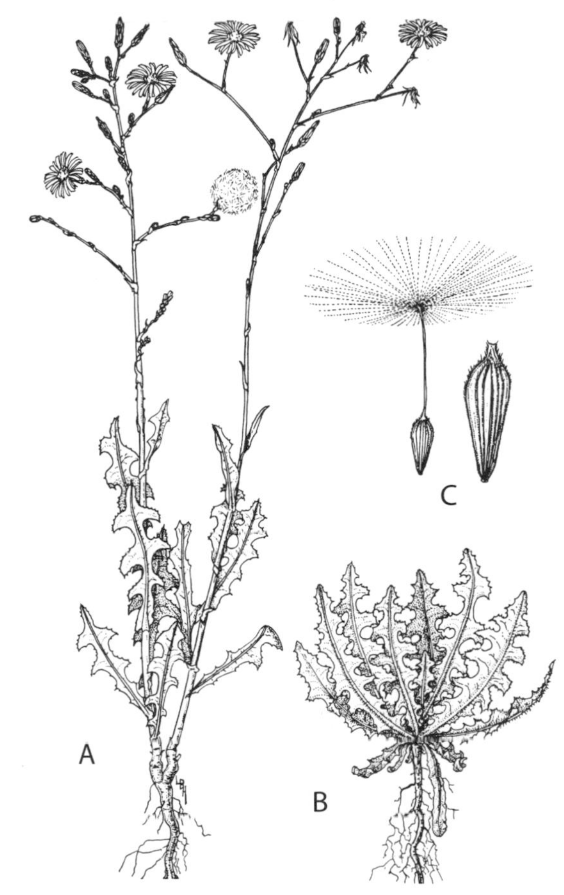


*Lactuca serriola*. A. Plant with flowers and fruits. B. Basal rosette of leaves on seedling. C. Detail of fruits, with and without pappus. Artist: L. B. Hamilton (Parker 1958).

**Other names:** ENGLISH: compass plant (so called because the leaves turn to follow the sun during the day), horse thistle, prickly [opium, acrid, wild] lettuce ("lettuce" was in English by ca. CE 1290; the origin is debated, with some deriving it from Latin *lactuca* or its adjectival derivative *lactucea*; akin to Middle English *letuse*, Old and Modern French *laitue*; those words were derived from *lac*, milk, referring to the latex; cognates are *laitue*, French; *lattich*, German; *lattuga*, Italian; *lechuga*, Spanish; *liatas*, Gaelic), wild opium (formerly considered narcotic); SPANISH:

*lecheras* (one who sells milk, Spain), *lechuga espinaca* [*silvestre*] ([wild] spinach lettuce, Sonora), *lechuguilla* (little lettuce, also used for *Agave* and *Sonchus* in Mexico); ATHAPASCAN: *azee' hókánii łibáhígíí* <*'azee'xokhánii'łipáhikíih*> (*'azee'*, medicine, *hókánii*, rising in tiers, *łibáhígíí*, gray [gray medicine (plant whose stems) rise in tiers], Navajo), *ch'il 'abe'* <*c'il ˀabeˀ*, *coh*, *nca.gi'*> (*ch'il*, plant, *'abe'*, milk, Navajo), *it'ąą'dotł'izhí* (*it'ąą*, spinach, *dotł'izhí*, turquoise, Western Apache); UTO-AZTECAN: *lechuuwa* (from Spanish *lechuga*, Yaqui), *mu'tcigĭp* [*mu'tcigi*], *paot'qa* (Shoshoni; for *L. ludoviciana*), *saĝwátųkápi* (*saĝwáĝarį*, green, *tųkáy*, food, Ute); YUMAN: *licú* [*ricú*] (Cocopa); sometimes misspelled "*scariola*"

**Botanical description:** Annual herbs from taproot, the stems 3–20 dm tall, erect, simple except near the top. Leaves are mostly pinnatifid, usually clasping, with the margins finely spinulose-dentate. Involucres are 8–11 mm long at anthesis, 9–13 mm in fruit. Corollas are yellow, often bluish or purplish at the tip. Achene body is 2.5–3 mm long, compressed but still lens-shaped (lenticular) in cross section, about a third as thick as broad, brown, oblanceolate to linear-obovate, usually 7 longitudinal nerves on both faces, rarely slightly winged on the upper margins, the beak filiform, 3–4 mm long.

**Habitat:** Disturbed grasslands up to oak woodlands, canyons, washes. 30–2,440 m (100–8,000 ft).

**Range:** Arizona (Coconino to Navajo, south to Cochise, Pima, Santa Cruz, and Yuma counties), through most of the United States and southern Canada; Mexico (Sonora).

**Seasonality:** Flowering May to September.

**Status:** Exotic. Introduced from Europe.

**Ecological significance:** Although these herbs are mostly weedy plants of disturbed areas, Prickly Lettuce sometimes occurs in wild, relatively unaltered places. In the Baboquivari Mountains, this lettuce is usually confined to the moist areas near permanent and temporary streams.

**Human uses:** People have used lettuce in Europe since at least the time of the first Roman emperor Augustus (63 BCE–CE 14), who had a statue built of the physician who prescribed lettuce for his illness (Dobelis 1986). It was long believed that the latex in *Lactuca* (called *lactucarium*, lettuce opium) was a narcotic, but it is only "slightly soporific" (Millspaugh 1892). Lettuce was greatly valued as a sedative and analgesic well into the nineteenth century. Chemical studies into the middle 1980s had identified no chemicals that would act in that way, and many have questioned its efficacy.

Other uses included being part of Roman banquet menus to prevent intoxication, and new mothers drinking a brew of leaves to increase milk flow. Navajo use the herbs in a compound mixture as a ceremonial emetic to treat snakebite in the *Hoozhónee* or Beauty Way (Wyman and Harris 1941). Hispanics in Sonora use wild lettuce to treat wounds, sores, and fevers, and they also use it as a diuretic, antiscorbutic, and against anemia (López-E. and Hinojosa-G. 1988).

People in Spain still consider the herbs edible (Tardío et al. 2005). Cultivated Lettuce (*L. sativa*) lacks the bitter sesquiterpene lactones in the wild species (Austin 2004). Some people dislike the bitter taste of the wild plants, but others eat the leaves.

**Derivation of the name:** *Lactuca* is Latin, based on *lacteus*, milky or full of milk. Both *lactuca* and *lacteus* are allusions to the milky sap. *Serriola* may be based on Latin *serriola*, a small saw, thereby referring to the teeth on the leaf blades, or perhaps to *escariola*, an old name for wild lettuce, from *escarius*, pertaining to eating, food, *escaria*, fit for eating.

**Miscellaneous:** Another species in Brown Canyon at about 4,000 feet is *L. graminifolia* (Blue Lettuce). The basal rosette on that species has leaves remarkably like *Taraxacum* (Dandelion) and may easily be mistaken for that genus. When *L. graminifolia* flowers in September, the blue corollas are unmistakable and the flowering stalk bears no resemblance to Dandelion. Blue Lettuce grows from North Carolina to Florida, and west to Texas, New Mexico, and Arizona. In Texas, the herbs grow in moist canyons in the Davis and Guadalupe Mountains (Correll and Johnston 1970). In Arizona, this species is absent from the Pajarito, Rincon, and Tucson Mountains (Bowers and McLaughlin 1987, Rondeau et al. 1996, Toolin et al. 1979). The species has not been found in either southwestern Arizona (Felger et al. in prep.) or the Gran Desierto of Sonora (Felger 2000).

## *Packera neomexicana* (Asteraceae) New Mexican Groundsel

**Other names:** ENGLISH: groundsel (maybe derived ca. CE 700 from Anglo-Saxon *grundeswelgiae*, pus-absorber; perhaps, but less convincingly, from *grundeswylige*, ground-absorber, from its rapid spread); ATHAPASCAN: *'azee' t'ááłah* <*ˀazéˀ lahdilt'ei*> (*azee'*, medicine, *t'ááłah*, alone, Navajo),

*Packera neomexicana*. A. Basal part of plant. B. Flowering apex. Artist: R. D. Ivey (Ivey 2003).

*chooyin 'azee'* <*co'in ˀazéˀ*> (*chooyin*, arthritis, *'azee'*, medicine, Navajo), *shash bi'iiłkóóh* <*šaš beˀiłkó*> (*shash*, bear, *bi*, its, *iiłkóóh*, emetic, Navajo), *tóbájíshchíní binát'oh* <*tóbáʒvíščíní binát'oh*> (*tóbájíshchíní*, Born for Water, *bi*, his, *nát'oh*, tobacco, Navajo); UTO-AZTECAN: *koatsĕmsĭtagwïv* (*koats* = ?, *ŭm*, possessive, *sitagwiv*, medicine, Ute), *tĭm'pidzanakwo* (Shoshoni), *muyitqa* <*muyítka*> (Hopi); (= *Senecio neomexicana*)

**Botanical description:** Perennial herbs from a taproot. Stems 14–40 cm tall, erect, the herbage tomentose. Leaves basal and cauline, the blades 1–5 cm long, 0.6–2 cm wide, oblanceolate to obovate or oval, dentate, serrate or almost entire, toothed to obtuse apically, cauline leaves reduced upward. Heads are few to many, corymbose or almost umbellate. Involucres are 4–7 mm long, 5–12 mm wide, lanceolate-attenuate, green or brown, with scarious margins. Ray flowers are 8–13, yellow, 4–10 mm long, the pappus white. Achenes are pubescent.

**Habitat:** Below trees in oak woodlands; elsewhere in chaparral, pine forests. 914–2,743 m (3,000–9,000 ft).

**Range:** Arizona (Apache, Coconino, Greenlee, Navajo, south to Cochise, Pima, and Santa Cruz counties), Colorado, New Mexico, and southern Utah; Mexico (Chihuahua and Sonora).

**Seasonality:** Flowering April to August.

**Status:** Native.

**Ecological significance:** These herbs grow below the oak woodland canopy at higher elevations in the canyons. Although the species reaches down to 3,000 feet within Arizona, it has not been found that low in the Baboquivari Mountains. In Brown Canyon, New Mexico Groundsel is not known below about 4,500 feet. The yellow flowers are visited by an array of insects, particularly butterflies.

New Mexico Groundsel is known nearby in the Pajarito, Rincon, Santa Catalina, and Santa Rita Mountains (Toolin et al. 1979, Bowers and McLaughlin 1987). The herbs have not been recorded for the Coyote, Quinlan, or Tucson Mountains (Rondeau et al. 1996), or west of the Baboquivari Mountains (Felger et al. in prep.).

**Human uses:** Navajo put a poultice of this *Packera* on burns, use it to treat "bear infections," in the Mountaintop Way, and consider it an antidote to narcotics (Wyman and Harris 1951). Navajo also use a cold infusion of this species as a lotion to bring luck while hunting (Vestal 1952).

Plants in both *Packera* and *Senecio* have been used by people for thousands of years in both hemispheres. Famous species in Europe are *Senecio cineraria*, *S. jacobaea*, and *S. vulgaris* (Austin 2004). Vickery (1995) found people still applying *S. vulgaris* to cuts, treating ague with it, and using it as a laxative. *Senecio jabocaea* had more association with witches and fairies in the British Isles, and it is known in Gaelic as *buadhghallan buidhe* <*buaghallan, boholàun, boholàun buidhe*> (*buadh*, virtuous, *ghallan*, branch, *buidhe*, yellow).

All species of *Senecio* yet studied contain potent pyrrolizidine alkaloids (e.g., senecionine), which cause liver damage if ingested (Lampe and McCann 1985, Lewis and Elvin-Lewis 2003). That damage occurs through acute venous occlusions in the liver (Budd-Chiari syndrome), and it can lead to cirrhosis and, in some cases, death (Lampe and McCann 1985).

In spite of their toxic properties, some *Packera* and *Senecio* were still being used externally in Europe in the 1970s as poultices for wounds and abscesses. Bown (1995) found that *S. aureus* still is grown in

Belorussia, central Russia, and the Ukraine for thc pharmaceutical industry.

**Derivation of the name:** *Packera* is named for John G. Packer, a Canadian botanist born in 1929. *Neomexicana* means from New Mexico, where it was first found. The old generic name, *Senecio*, is from Latin *senex*, an old man, alluding to the white pubescence of many species, or the white pappus.

**Miscellaneous:** Three other species are known from the Baboquivari Mountains. *Senecio carlomasonii* is a perennial like *S. neomexicanus*. This herb is larger, growing to 1–2 m tall, and is aromatic.

The other two species are annuals or weakly perennial. *Senecio lemmoni* has linear to lanceolate leaves that are irregularly dentate, at least toward the base, and radiate heads. This herb is spread across southern Arizona from near La Paz to Yavapai and from Yuma County east to Graham. *Senecio flaccidus* var. *monoensis* differs by having pinnately lobed leaves or entire blades, which are linear to linear-filiform, and broadly campanulate heads. This herb grows in Cochise and Graham counties west to Barry Goldwater Air Force Range and Organ Pipe Cactus National Monument (Felger et al. in prep.).

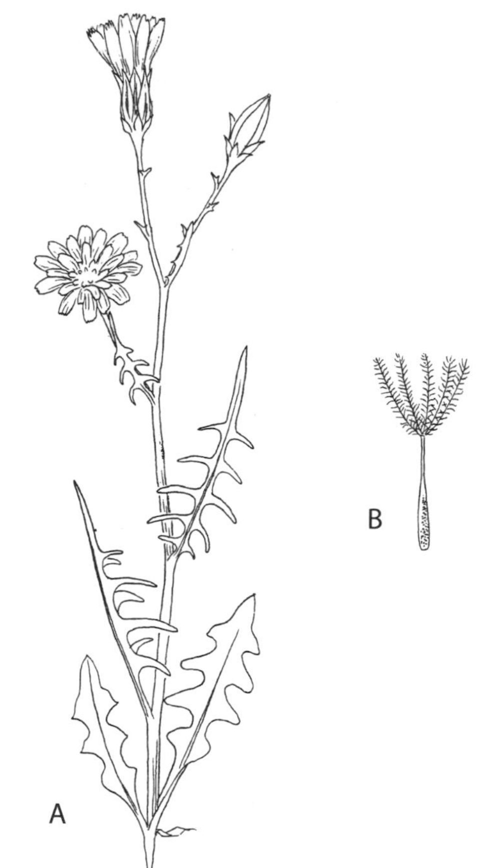


*Rafinesquia neomexicana*. A. Flowering branch. B. Achene with pappus. Artist: R. D. Ivey (Ivey 2003).

## *Rafinesquia neomexicana* (Asteraceae) Desert Chickory

**Other names:** ENGLISH: desert chicory (from French *cichorée*, now *chicorée*, cognate with Italian *cicórea*; both derived from Latin *cichorium*, *cichorum*, and Greek *kichora*, *kichoreia*, in turn taken from Egyptian *tybi*, January, the month in which they grow; also called succory or endive, *Cichorium intybus*; chicory in use by 1393 for *Cichorium*, later applied to *Rafinesquia*), [New Mexico] plume-seed (because of the parachute-like tuft on the fruits, Arizona, New Mexico, Texas); UTO-AZTECAN: *sí*[a] (Southern Paiute)

**Botanical description:** Annual, from taproot, the stems glabrous, branched at or near the base, ascending 2–5 dm tall, each branch ending in a head. Lower leaves pinnatifid, upper reduced and only about 5–10 mm long, clasping the stem, and with a few lateral lobes near the base. Heads are usually 20–25 mm high, the involucre narrowly campanulate or top-shaped, with the phyllaries membranous. Corollas are white, purplish white, or pale pink, 5-toothed apically. Achenes are ca. 11 mm long, with a top-shaped body but with the apex narrowing into a beak, brown, rough-hairy, the pappus plumose.

**Habitat:** Grasslands, foothills, washes. 60–900 m (200–3,000 ft).

**Range:** Arizona (Apache to Mohave, south to Gila, Graham, Maricopa, Pima, Pinal, and Yuma counties), southern California, New Mexico, Nevada, western Texas, and southern Utah; Mexico (Baja California and Sonora).

**Seasonality:** Flowering (February) March to May.

**Status:** Native.

**Ecological significance:** These annuals appear in open sites in the spring, along with *Phacelia* and other sun-loving (heliophilic) plants. This plant is one of the few composites in the region that obviously closes its "flowers" (heads) at night and reopens them the next morning. There are others that "sleep" but their heads are smaller and less conspicuous.

The herbs are most common along washes where the soils are often rearranged by floodwaters, but they also appear along roadsides. During years of drought, *R. neomexicana* is most frequent along roads where the water has been concentrated as runoff from the impenetrable surface of the blacktop.

**Human uses:** The Southern Paiute name is included tentatively. Sapir (1930) recorded some plant, perhaps *R. californica*, as being eaten by these people. They ate the stems raw. His botany is dreadful, but this possibility is worth exploring in the future.

**Derivation of the name:** *Rafinesquia* was dedicated, by his contemporary colleague Thomas Nuttall, to Constantine Samuel Rafinesque-Schmaltz, 1783–1840, a brilliant but eccentric and erratic naturalist who collected widely in eastern North America. *Neomexicana* means "New Mexico," where the species was first found.

Rafinesque was born in Constantinople of French parents and, according to him, started collecting plants at age 11. He got a job as a merchant's clerk when he arrived in Philadelphia in 1802, and he spent his free time identifying and ordering the bird collection of a local museum and botanizing. After spending 1805–1815 in Europe, he returned to eastern North America keen to write the first book about its flora. Unfortunately, he was shipwrecked off Block Island, Rhode Island, and he lost his books, manuscripts, drawings, and plant collections. The next decades were spent traveling and exploring the wilderness, mostly on foot, and studying plants, languages, and customs of indigenous people. Lamentably, yet true to his erratic behavior, he rarely referred to tribes or even regions when noting plant uses in his publications.

For a period, he taught at Transylvania University in Lexington, Kentucky. His teaching reputation was great, and his classes were given with "such enthusiasm and originality, using living botanical specimens to illustrate his lectures, that no one who sat in any of his classes ever forgot him" (Leighton 1987, quoted in Coffey 1993). Rafinesque was so eccentric in manners and appearance that many considered him crazy during his lifetime; he sometimes dressed like a pirate with a bandanna around his head instead of a hat. He claimed: "I like the free range of woods and glades. I hate the sight of fences like an Indian." He named species without regard for prior names and was a devout "splitter." It was the 1980s before most of the organismal names he proposed were seriously considered; before then, they usually were simply ignored. Rafinesque died a pauper in Philadelphia, and, tragically, his herbarium at the Academy of Natural Sciences in Philadelphia was thrown in the trash by Elisa Durand, who was the curator there in 1840 (Smith 1954).

**Miscellaneous:** *Rafinesquia* is a genus of two species restricted to the southwestern United States and adjacent Mexico. *Rafinesquia californica* grows in California, Utah, Arizona, and in Baja California.

This genus is in tribe Cichorieae (formerly Lactuceae) and is closely related to *Stephanomeria* (q.v.) in this region. Both *Cichorium* and *Lactuca* (q.v.) are in the same tribe (Gottlieb 2006).

## *Sonchus aspera* (Asteraceae) Sow-Thistle, *chinita, s-ho'idag ṣaipuk*

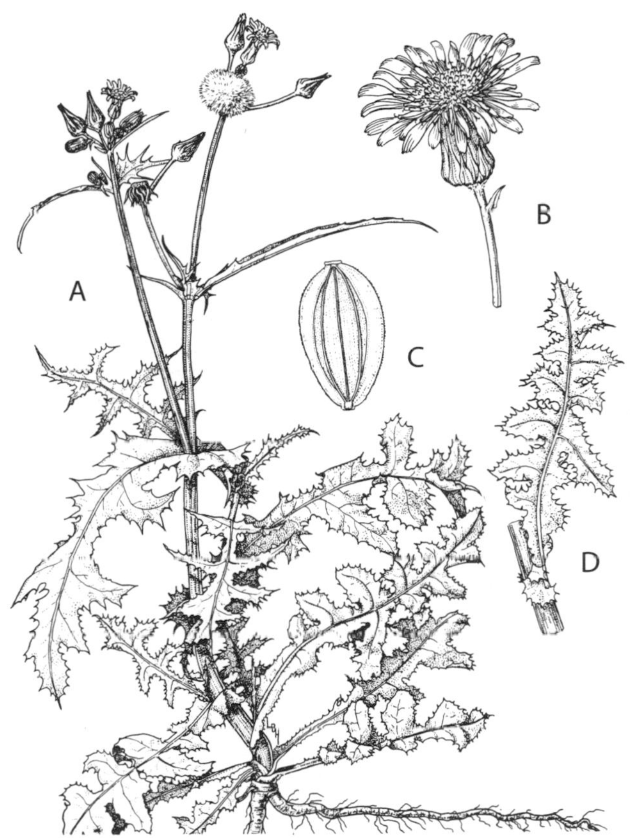


*Sonchus aspera*. A. Plant with flowers and fruits. B. Flower head. C. Achene, enlarged (minus pappus) to show ribbing. D. Detail of leaf showing basal lobes. Artist: L. B. Hamilton (Parker 1958).

**Other names:** ENGLISH: prickly [spiny] sow thistle (etymology obscure, first written ca. CE 1250 as *sugeisthstel*, then by CE 1387 equated with *rostrum porcinum*, a pig's snout; "thistle," spiny); SPANISH: *achicoria [chicoria] dulce* (sweet chickory, Arizona, Texas, Sonora; from Latin *cichorium, cichorum*, comparing the herbs with the succory or endive, *Cichorium inty-*

*bus*), *cardo lechero* (milky thistle, Spain), *cerraja* (a saw, Chihuahua, Durango), *chinita* (Arizona, Sonora; origin unclear; in Mexico *chinito* is the common name of the Cedar Waxwing, *Bombycilla cedrorum*), *lechiterna* (soft and milky, Spain); ATHAPASCAN: *'azee' hókánii łibáhígíí <azee'xokhánii'łipáhikíih>* (medicine, rising in tiers, gray [gray medicine (plant whose stems) rise in tiers], Navajo), *shá'inałał <šáˀinałał>* (revolving with the, *shá*, sun, Navajo); UTO-AZTECAN: *činaka <china-ri>* (Tarahumara), *ho'idkam 'i:vaki* (spiny greens, Hiá Ceḍ O'odham), *mu'tcigĭp* [*mo'tcigĭp*, *mu'tcigi*, *mo'tcigi*] (Shoshoni), *ho'idkam 'i:vakĭ* (*ho'idkam*, spiny, *'i:vakĭ*, eaten greens, Hiá Ceḍ O'odham), *s-ho'idag shaipag <shaipuk>* (*s-ho'idag*, spiny, thorny, *sa'i'*, bushy, *pag*, growing in many places, Tohono O'odham), *s-ho'idkam iivagi* (spiny eaten greens, Akimel O'odham), *si'imel iivagi* (lactating eaten greens, Akimel O'odham); YUMAN: *kee tá ha* (Mohave), *ma:xškáłʸ* [*ma:škáłʸ*] (Cocopa)

**Botanical description:** Annuals, nearly glabrous, the stems usually simple below and branched near the top. Leaves alternate, usually with roughly and harshly spinose-dentate margins, the lowest leaves usually pinnatifid, those of the midstem with a few lateral lobes or nearly lobeless and clasping, the upper reduced and clasping, with rounded auricles. Involucre is obconic at anthesis, 1 cm high, the phyllaries often glandular near the midnerve, green, scarious-margined. Flowers are numerous, perfect, fertile. Corollas are yellow, the ray about as long as the tube. Achenes are ca. 2–2.5 times as long as broad, widest near middle, somewhat roughened but the central ones nearly smooth.

**Habitat:** Disturbed ground. 45–2,440 m (150–8,000 ft).

**Range:** Arizona (Apache, Coconino, south to Cochise, Pima, Santa Cruz, and Yuma counties), nearly throughout North America; Mexico (Chihuahua and Sonora).

**Seasonality:** Flowering (February) April to October.

**Status:** Exotic. Native of Europe.

**Ecological significance:** This European species and the similar *S. oleraceus* were introduced early by Old World settlers, probably as edible and medicinal herbs. *Sonchus* soon escaped and is now a troublesome weed throughout much of North America. Although Sow-Thistle is a plant of disturbed places, individuals are most frequent in gardens and lawns. Farther west, Felger et al. (in prep.) found the herbs mostly at waterholes, particularly artificial ones.

The bristly fruits cause skin infections in some people.

**Human uses:** Europeans ate Sow-Thistle much as they did Dandelion (*Taraxacum*). Dioscorides (fl. CE 40–80) and Pliny (CE 23–79) praised *Sonchus* as food. Culpepper (1653) extolled Sow-Thistle as medicine and food. He wrote that "[t]he young tops are good as salad with oil and vinegar, for a scalding of water." Linnaeus listed three kinds (*Fet-tistel*, *Mjolf-tistel*, *Tota*) in his *Flora Oeconomica* (Linnaeus 1749). Considering their early introduction, people may have had several purposes in importing *Sonchus*. Vickery (1995) found people in the British Isles still using the sap to remove warts.

Akimel O'odham eat the stems of plants found growing in their cultivated fields and yards. Rea (1997) was told that the young stems can be eaten directly, while older stems are better when peeled. Some O'odham think people peel them to remove the milky sap, the source of the second name given to them, *si'imel iivagi*. The related *S. oleraceus*, *huai hehevo* (mule deer's eyelashes to the Akimel O'odham), is cooked while this species is not. Formerly, *S. asper* leaves were cooked and eaten. *Soncus asper* is also eaten by Luiseño in California, Mohave farther north in Arizona (Castetter and Bell 1951), and Tarahumara in Chihuahua (Thord-Gray 1955, Pennington 1963).

Navajo make a medicine of *Sonchus* to treat heart palpitations, but some consider it poisonous (Wyman and Harris 1951). Farther east, the Iroquois also use Sow-Thistle in medicine (Moerman 1998).

**Derivation of the name:** *Sonchus* is based on the Greek name *sonchos <sogkos, sogchos, sonkos>* used by Theophrastus (372–287 BCE) for a prickly plant, perhaps the Sow-Thistle; the Latin *sonchus* was applied by Pliny. *Asper* means rough.

## *Stephanomeria pauciflora* (Asteraceae) Desert-Straw, *pionilla*

**Other names:** ENGLISH: brown-plume wire-lettuce (referring to the wiry stems and leaves that resemble lettuce, *Lactuca*), desert-straw ("straw" is from Old English *stréaw*, with a Teutonic origin; cognates are *stroo* in Dutch, *stroh* in German, *strå* in Swedish, and *straa* in Danish; by ca. CE 1000 meant specifically the stalks of grasses, but subsequently has been applied to dried stems of many plants, Arizona), skeleton-weed ("skeleton" from Greek *skeleton*, neu-

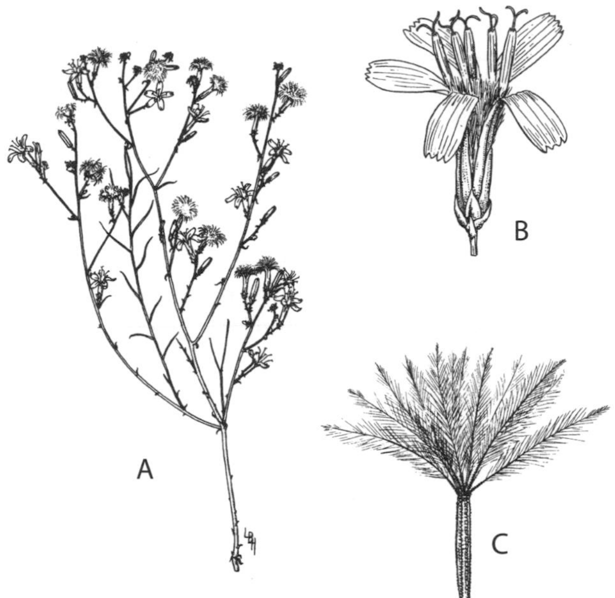


*Stephanomeria pauciflora*. A. Flowering branch. B. Flower cluster. C. Achene with plumed pappus. Artist: L. B. Hamilton (from original artwork).

ter of *skeletos*, dried up; the name "skeleton-weed" appeared by 1935, and later was defined as "a close relative of the dandelion and chicory, having a spindly habit of growth"); SPANISH: *pionilla* (little peonia, Mexico; reference obscure; a name usually applied to plants with red or red and black parts; see also *Erythrina*); ATHAPASCAN: *jeeh dootł'izh* [*ts'oh*, *ts'ósí*] <*jéˀdóy.is*, *ʒˇeˀ doƛ'iš* [*coh*, *c'o's*]> ([large, slender], *dootł'izh*, blue, *jeeh*, gum, resin, Navajo); HOKAN: *posapátx camoz* (what thinks it's a sweet-bush [*Bebbia juncea*], Seri); UTO-AZTECAN: *mo'agûp* (Shoshoni; for *S. exigua*), *piinga* <*pí:nga*> (Hopi), *sanako'ogadɨbɨ* (*sanako'ovɨbɨ*, chewing gum, *bɨ*, plant), Paiute); YUMAN: *hebe imixáa* (rootless plant, Seri)

**Botanical description:** Herbs, perennial, the stems woody at base, rigid, branched, forming rounded bushes 3–6 dm tall, pale, glabrous. Basal leaves 3–7 cm long, pinnatifid with narrow divisions, the upper leaves entire, often reduced to scales. Heads solitary, short-peduncled, 3–5 flowered. Involucre is 8–10 mm high, with 5 principal phyllaries. Achenes 5-angled, striate, often more or less rugulose, the pappus-bristles tinged brown, plumose, but these not extending down to the base.

**Habitat:** Canyons, slopes, streamsides, washes. 45–2,130 m (150–7,000 ft).

**Range:** Arizona (almost throughout state), California, New Mexico, Texas, Utah, and east to Kansas; Mexico (Baja California, Chihuahua, Coahuila, and Sonora).

**Seasonality:** Flowering throughout most of year, at least from May to October.

**Status:** Native.

**Ecological significance:** These perennials are sun-loving plants that occur mostly near washes and streambeds. The herbs thrive on disturbance, and, once established, persist as long as there is ample sunlight. Tops of *Stephanomeria* may be removed numerous times throughout the growing season, and the roots continue producing new growth during the next warm rainy period. Although the herbs seem to have no root to the Seri, and can be easily knocked over, they have energy enough to continue sending up new growth from the base. The small flower clusters are showy and must be pollinated by insects. Butterflies often visit for nectar.

**Human uses:** Hopi make a medicine to increase milk flow in new mothers, while Navajo consider the roots narcotic, and they use Desert-Straw to hasten delivery of the placenta, and as a Life Medicine (Moerman 1998, Vestal 1952, Whiting 1939, Wyman and Harris 1951). Navajo also employ *S. pauciflora* in paint ingredients for chant arrows used during the Evil Way and Night Way and use the root as chewing gum (Vestal 1952, Wyman and Harris 1951).

Kawaiisu and Navajo both use the thick latex as chewing gum (Vestal 1952, Zigmond 1981). Seri sometimes use Desert-Straw as shading material on the roofs of their frame houses (Felger and Moser 1985).

Moerman (1998) listed five other species used by different tribes in the West and Southwest. Hopi used *S. exigua* as a diuretic (Whiting 1939), and the Navajo used it as a treatment for measles (Vestal 1952). Apache and Zuni use *S. tenuifolia* to treat rattlesnake bites (Reagan 1929, Stevenson 1915), while the Shoshoni and Thompson use it for other problems (Train et al. 1957). Keres treat sore eyes with *S. runcinata* (Moerman 1998).

Farther north, the Cheyenne make a remedy to treat colds and tuberculosis from *S. spinosa*. The same species is also used to treat mumps or included with other medicines for all ailments. Paiute similarly consider some species as a panacea, and they apply *S. spinosa* to stop diarrhea, swellings, and toothaches, and also use it as an emetic (Train et al. 1957). Shoshoni make a decoction to stop vomiting, use it as an eyewash, and drink it as a tonic (Train et al. 1957).

In California, the Kawaiisu use *S. virgata* as an eye medicine (Zigmond 1981).

**Derivation of the name:** *Stephanomeria* is compounded from Greek *stephanos*, a crown or wreath, and *meris*, part or division. *Pauciflora* means few-flowered.

**Miscellaneous:** *Stephanomeria* is a genus of 16 species that grow in western North America (Gottlieb 2006). The genus is in the tribe Cichorieae (formerly Lactuceae). *Stephanomeria* is related to lettuce (*Lactuca*, q.v.) and Desert Chicory (*Rafinesquia*, q.v.). Although Old World Chicory (*Cichorium intybus*) is in the same tribe, researchers do not agree about which subtribe it should be put in.

## *Trixis californica* (Asteraceae) Trixis, *hierba de pasmo*

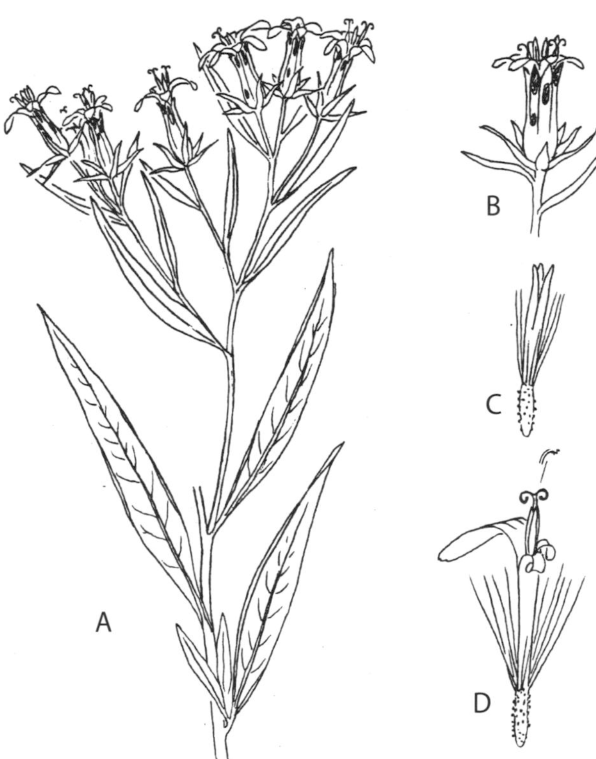


*Trixis californica*. A. Flowering branch. B. Flower cluster (head). C. Disc flower. D. Ray flower. Artist: R. D. Ivey (Ivey 2003).

**Other names:** ENGLISH: American [California] trixis (a book name); SPANISH: *cachano* (from an indigenous name of several Asteraceae, possibly originally a *Senecio*, New Mexico, Chihuahua, Coahuila; since the 1920s or so, this has been an allusion to the devil), *hierba de aire* (air herb, Sonora), *hierba de pasmo* (herb for *pasmo*, from Latin *spasmus*, Sonora), *ruina* (ruin, Sonora); HOKAN: *cocazn-ootizx* (rattlesnake's foreskin, Seri); UTO-AZTECAN: *hebai sa'igar* <*j'bai sa'igar*> (*hebai*, where, *sa'i*, plants, *gar*, are [*ga*, to have + *r*, at], Mountain Pima)

**Botanical description:** Shrub, branched, 0.3–1 m tall, the stems brittle, whitish. Leaves bright green, lanceolate or linear-lanceolate, 2–10 cm long, basally narrowed, margins entire to dentate, acute apically, almost sessile. Panicles terminal, dense, in flattened clusters. Heads 1.5–2 cm high, subtended by leafy bracts. Corollas are 13–17 mm long. Achenes are brown, glandular, 11–12 mm long, narrowed and beak-like at summit, with a tawny or whitish pappus of capillary bristles, 1 cm long.

**Habitat:** Canyons, rocky slopes. Up to 1,500 m (5,000 ft).

**Range:** Arizona (Coconino, Greenlee, Yavapai, south to Cochise, Pima, Santa Cruz, and Yuma counties), southern California, New Mexico, and western Texas; Mexico (Baja California, Chihuahua, Coahuila, Durango, Nuevo León, San Luis Potosí, Sinaloa, Sonora, Tamaulipas, and Zacatecas).

**Seasonality:** Flowering (February) April to October (following rains).

**Status:** Native.

**Ecological significance:** Leaves are cold and drought deciduous. Perhaps the seeds are dispersed by wind, but recruitment is infrequent. Turner et al. (1995) recorded that in four years of observation on a 557-square-meter plot at Tumamoc Hill in Tucson, mass germination occurred only once. Indeed, germination is episodic and occurs in some years with good rains but not in others (Bowers et al. 2004). Perhaps that uncommon recruitment partially explains the infrequency of the individual plants. There is often considerable distance between individuals. Once established, individual plants may live to be 40 years old (Bowers et al. 2004).

Jaeger (1941) thought that the "preferred" habitat in the Mohave Desert was crevices in rocks, beneath overhanging ledges, or in the shelter of bushes. In the Baboquivari Mountains, *hierba del pasmo* is more often in open sites.

These shrubs comprise a small percentage of the diets of mule deer in the winter, that is, less than one percent (Krausman et al. 1997). Given the potent compounds in *Trixis*, that might not seem surprising. However, deer are also known to eat Poison Ivy—a powerful plant (see *Toxicocendron*).

**Human uses:** Seri smoke the leaves like tobacco, and women use a root tea to hasten childbirth (Felger and Moser 1985). The name *hierba del pasmo* refers to a malady considered endemic to the Americas (see also *Ambrosia monogyra*, *Baccharis sarothroides*, *Bouvardia ternifolia*, *Ericameria laricifolia*). This disease is a general depression of the nervous system that affects both humans and domestic animals.

Hocking (1997) lists five other species that are used, including *T. divaricata* (Brazil, Colombia), *T. glandulifera* (Argentina), *T. inula* (Texas to Venezuela and Chile; *falsa árnica*, false arnica, *árnica del monte*, wild arnica), *T. longifolia* (Mexico; *lobo buase*), and *T. pringlei* (Mexico; *yerba de la víbora*, snake herb). The three species in Mexico are made into medicines to treat diabetes, earaches, rheumatism, snakebites, tumors and swelling, venereal disease, and wounds and ulcers. The others are remedies for amenorrhea, measles, urinary problems, and to aid conception. Pennington (1969) added *T.* cf. *radialis*, the *yerba del aire* or *ïvïli yoosigai* (air herb) among the Northern Tepehuan, which is used to help broken bones heal or to relieve an earache.

These reports are all the more intriguing because the genus *Trixis* is known to contain flavonoids (Ribeiro et al. 1997), sesquiterpene lactones (Bohlman and Zdero 1979), and trixanolides (Kotowicz et al. 2001). Among other activities, flavonoids from *Trixis* have been show to be trypanocidal against *Trypanosoma cruzi*, the protozoan that causes Chagas' disease.

**Derivation of the name:** *Trixis* is based on Greek *trixos*, threefold, akin to Latin *triplex*, referring to the outer corolla lip with three lobes. *Californica*, of California, means that the first specimens were collected there.

**Miscellaneous:** *Trixis* is a member of the tribe Mutisieae, which is recognized by having heads with bilabiate or ligulate flowers (Karis et al. 1992, Mabberley 1997). The flowers are also unusual in that they have anthers with "tails." There are 76–77 genera in the tribe, with ca. 1,000 species in the tropics, mostly in South America (Editorial Committee 2006).

## ***Xanthium strumarium* (Asteraceae) Cocklebur, *cadillo, waiwel***

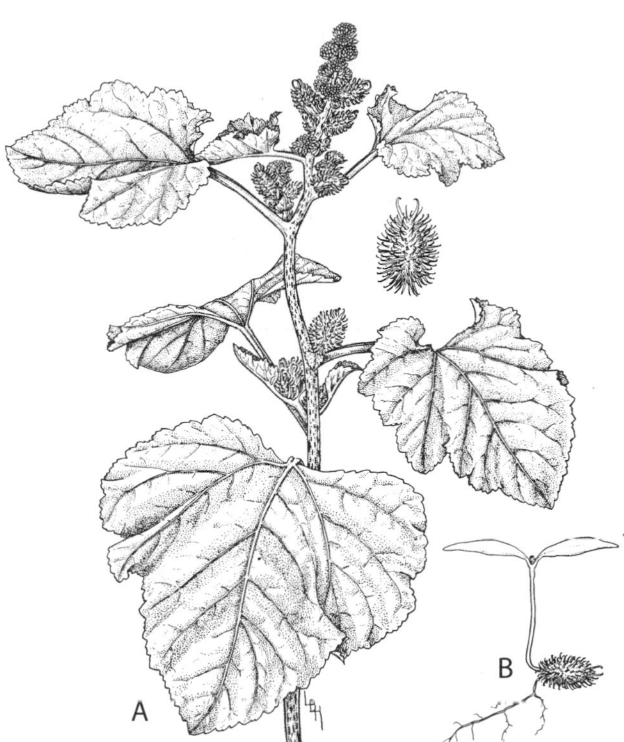


*Xanthium strumarium*. A. Branch with male flowers at apex and female flowers below. B. Female bur. C. Seedling emerging from bur. Artist: L. B. Hamilton (Parker 1958).

**Other names:** ENGLISH: buttonbur (in English by 1634; "button" is from Old French *boton*, modern French *bouton*, bud, knob, button; the ultimate origin is supposed to be Teutonic; akin to Spanish *boton*, Portuguese *botão*, Italian *bottone*; apparently connected with late Latin **bottare, buttare*, to thrust, put forth, whence French *bouter*, Spanish *botar*, Italian *bottare*), clotbur [clothbur] ("clotbur" used by Wm. Turner in 1548, meaning "ball-bur," England, Texas), common [spiny] cocklebur (from Old English *coccul*, *coccel*, probably originally applied to *Lychnis*, Caryophyllaceae, a weed in cultivated fields; used for *Xanthium* by 1866), ditchbur, sheepbur (notorious for getting tangled in sheep wool); SPANISH: *abrojo* (bur, Arizona to Texas, Tabasco), *cadillo* (bur, Arizona, New Mexico, Sonora), *huichapole <güichapol, güichapori, guachapore, guacaporo, huichaori, huachapore>* (*huicha*, spine from Cahita fide Sobarzo 1996; related to Náhuatl *huitztli*, California, Sonora to Puebla); ATHAPASCAN: *'ałta'neets'éhii <'alxa'niits'éhiih, ta'neets'éhii>* (interlocking [plant whose burs adhere (to one)], Navajo), *izee inlwozh <izee inkozee>* (*izee*, medicine, *inlwozh*, ditch, Western Apache), *ta'neets'éhii ntsxaaz <ˀdtani-c'ehí ŋ̀ca·gí>* (*ntsxaaz*, large cockle-bur, Navajo); CHUMASH: *sho'moy <shomoy>* (Barbareño Chumash), *mokoksh* (Ineseño Chumash); HOKAN: *cözazni caacöl* (large sandbur, Seri); KIOWA TANOAN: *ŋwaejoka* (*ŋwae*, thorn, *jo*, augmentative, *ka*, leaf, Tewa); UTO-AZTECAN: *atsiáŋwádova*, *atsiogopapa* (Northern Paiute), *bachapo'or* (from Náhuatl *güichapol*, Mountain Pima), *kámukníyų* (Ute), *kwĭ'tcĕmbogop* (bison fruit, Shoshoni), *paatso* <*pá:taco*, *pa:tcótco*> (Hopi), *vaiwa* <*vaiva*, *váiva*> (Akimel and Hiá Cẹḍ O'odham), *waiwel* <*vaivul*> (Tohono O'odham); YUMAN: *kmn*$^{y}$*a* (Cocopa), *qum nah* (Paipai), *wisapole* (Paipai); LANGUAGE ISOLATE: *mo'kĭachipa* (*mo* from *mo'li*, round [fruit], *kĭa'chipa*, prickly, Zuni)

**Botanical description:** Herbs, annual, monoecious, the stems mostly 0.4–2 m tall. Leaves alternate, the blades to 15 cm long, ovate to deltoid, almost orbicular or kidney-shaped, serrate or with shallow lobes on margins, rough-hairy (scabrous), the petiole elongate, to 15 cm long. Heads are small, axillary, nearly sessile, unisexual. Ray flowers are absent; staminate heads with 1–3 series of separate phyllaries. Disk flowers have minute corollas. Bur is 1–3 (sometimes more) cm long, 1–1.5 cm wide, terminated by 2 prominent spines, covered with stiff, hooked prickles ca. 5 mm long, bur and bases of spines densely pubescent. Achenes are 2 per bur.

**Habitat:** Washes, streamsides. 30–1,830 m (100–6,000 ft).

**Range:** Arizona (throughout state), widespread in the United States; Mexico (Baja California, Chihuahua, and Sonora).

**Seasonality:** Flowering August to September.

**Status:** Perhaps native in the United States.

**Ecological significance:** This species may be native to the Americas, but it is now widespread throughout the world as a weed. Plants are alien in the Baboquivari Mountains and probably were introduced as contaminants in animal food when Brown Canyon contained active ranchs. Fruits are clearly, and ably, adapted for animal dispersal. According to their web page (velcro.com), VELCRO® was created by Swiss inventor George de Mestral after walking through a patch of "cockleburrs."

Spencer (1957) says that every bur contains two seeds. One of these two germinates the first year, and the other ripens later and sprouts the second. Thus, each fruit may produce two plants. Because each plant has dozens of fruits, *Xanthium* provides a seedbank for numerous years.

Cocklebur is poisonous and potentially fatal to pigs and other livestock if ingested. The burs cause mechanical injury to humans and other animals. There is an old commentary that an irritable person acts like he "has a [cockle-]bur under his saddle."

**Human uses:** Navajo make a liniment of Cocklebur (Mayes and Lacy 1989). Akimel O'odham make medicines of the leaves and burs to treat several maladies in cattle, and historically *Xanthium* was used to treat sore eyes and diarrhea (Curtin 1949, Rea 1997). Seri make a tea of the bur to treat kidney pain (Felger and Moser 1985). Bocek (1984), Gallagher (1976), Moerman (1998), Robbins et al. (1916), Romero (1954), and Stevenson (1915) record that Cocklebur also is medicinal to the Apache, Cahuilla, Lakota, Navajo, Ohlone, Paiute, Tewa, and Zuni.

Apache, Ohlone, and Zuni have ground the seeds to make bread or *pinole*, largely as a famine food (Bocek 1984, Moerman 1998, Reagan 1929, Stevenson 1915). The seeds are used by the Keres in making paint for masked dancers. Leaves are burned for ceremonial blackening by the Navajo (Vestal 1952), and used in roasting pits as containers for roasting beans by the Akimel O'odham (Castetter 1935). John Gerarde in England wrote in 1633 that "it seemeth to be called *Xanthium* of the effect, for the Burre or fruite before it be fully withered, being stamped . . . maketh the haires of the head red."

**Derivation of the name:** *Xanthium* comes from the Greek word *xanthos* or *xanthion*, meaning yellow, which was some plant used for dyeing the hair. *Strumarium* is Latin "for swellings or tumors."

## *Zinnia acerosa* (Asteraceae)
## White Zinnia, *zinia del desierto*

**Other names:** ENGLISH: desert [white] zinnia; SPANISH: *zinia del desierto* (desert zinnia, Sonora); HOKAN: *saapom ipémt* (what purple prickly-pear is rubbed with, Seri), *cmajíic ihásaquim* (what women brush their hair with, Seri), *mojépe ihásaquim cmaam* (female saguaro hairbrush, Seri); (= *Zinnia pumila*)

**Botanical description:** Small shrubs to 16 cm tall, stem slender, branched, tomentose. Leaves opposite, linear, 1-nerved, with a sharp pointed apex, to 2 cm long and 2 mm wide, pubescent to almost glabrous. Heads are solitary and terminal on branches, with peduncles to 25 mm long, heads narrowly bell-shaped, about 6 mm high, the phyllaries oblong, herbaceous. Ray flowers are 4–6, white, oblong to almost orbicular, to 1 cm long, tubeless, green-nerved below. Ray achenes are oblanceolate, 3-angled, apically trun-

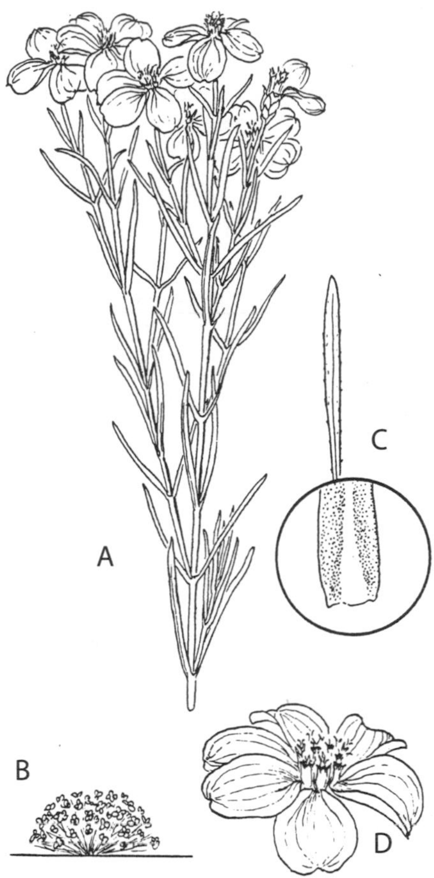


*Zinnia acerosa*. A. Flowering branch. B. Life-form. C. Leaf with inset of base. D. Flower cluster. Artist: unknown (courtesy James Henrickson).

cate, 2.6–4 mm long, the awns striped (striate), fused to the corolla nerves or 3 awns free above. Disk flowers are 8–13, whitish yellow but drying reddish, 6 mm long. Disk achenes are oblanceolate, 2.4–3.6 mm long, striate, the pappus usually of 2–3 unequal awns or these reduced.

**Habitat:** Grasslands, desertscrub, ridges. 760–1,500 m (2,500–5,000 ft).

**Range:** Arizona (central Yavapai to Graham, south to Cochise, Pima, Santa Cruz, and Yuma counties), New Mexico, and Trans-Pecos Texas; Mexico (Chihuahua, Coahuila, Sonora, Nuevo León, and south to Zacatecas).

**Seasonality:** Flowering (February) April to October.

**Status:** Native.

**Ecological significance:** These small plants are most often on caliche soils, where the calcium has formed almost impermeable layers below the surface. The shrubs are most common in desert grasslands, but they also occur in the lower reaches of Brown Canyon.

Mostly, the species grows in the lowlands in the Altar Valley and west of the Baboquivari Mountains. The bushes also go up into mountain ranges in other places, such as Las Guijas, Rincons, and Santa Ritas.

**Human uses:** Seri use *Zinnia* plants to remove the spines from fruits of *Opuntia violacea* before picking them to eat (Felger and Moser 1985). Although the Seri profess no medicinal use, they told Felger and Moser (1985) that the Tohono O'odham use *Z. acerosa* to make a medicine to treat diarrhea. Neither Castetter and Underhill (1935) nor Moerman (1998) found confirming evidence that this was correct. Rea (1997) did not find Akimel O'odham using them, and *Z. acerosa* has not been collected in their modern area. Perhaps the old knowledge has been lost among the O'odham. The Keres make a medicine of White Zinnia to treat swellings or aches (Moerman 1998). Keres also give *Zinnia* to children to help them learn to talk.

The Keres also use *Z. grandiflora*, a species growing in Arizona, Colorado, Kansas, New Mexico, Texas, and Mexico (Chihuahua, Coahuila, Durango, and Zacatecas) (Moerman 1998). This plant is given as a remedy for kidney problems and put in a bath to stop excessive sweating. Navajo and Zuni also make medicines from *Z. grandiflora*. Navajo called Great Plains Zinnia *'azee' 'alástsii'* <*'azee'táGaa'ii'*> (medicine, beard) or *zaaxwocíkíih* (which contracts the lips); they use it to treat throat trouble and stomach aches, and they take it as a cathartic and emetic (Vestal 1952). Zuni treat bruises, use it in the sweat bath to reduce fever, and apply it in an eyewash (Stevenson 1915). Zuni call it *tu'na ikĭapokĭa* (*tu'na*, eyes, *ikĭapokĭa*, to put into). In the sweatbath, Zuni powder *Z. grandiflora*, sprinkle it over hot stones, and have the patient sit beside it and inhale the fumes (Stevenson 1915).

**Derivation of the name:** *Zinnia* commemorates Johann Gottfried Zinn, 1727–1759, a German physician, professor, and director of the Botanical Gardens of Göttingen. He wrote the first book on the anatomy of the eye. *Acerosa* means sharp, an apparent allusion to the leaf tips.

**Miscellaneous:** The genus *Zinnia* comprises ca. 17 species that grow from the southwestern United States south to Argentina (Brickell and Zuk 1997, Mabberley 1997, Smith 2006). Among those species is the widespread and popular *Z. violacea* (= *Z. elegans*), an allopolyploid that was brought into cultivation by the

Aztecs before Europeans arrived in the New World. Now there are many cultivars with flowers of almost all colors except blue.

## *Tecoma stans* (Bignoniaceae) Trumpet-Bush, *lluvia de oro*

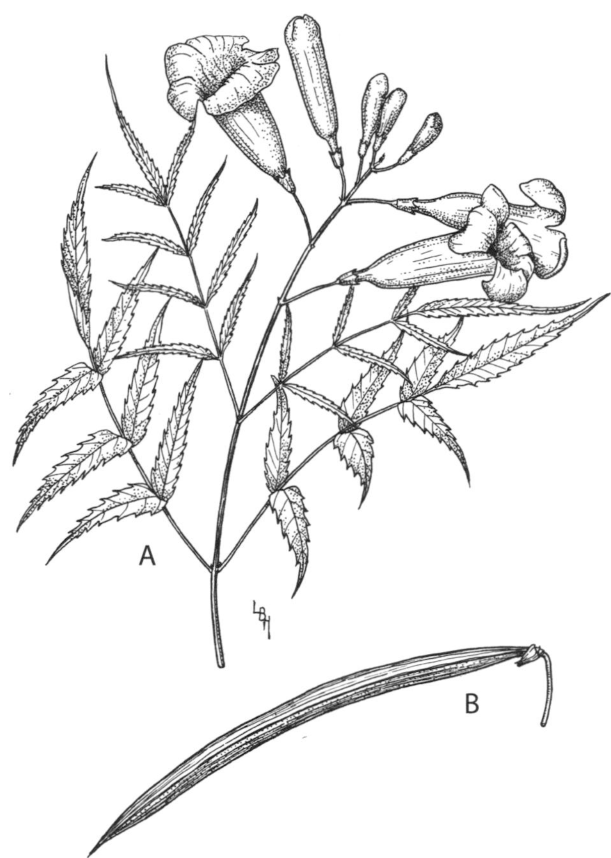


*Tecoma stans*. A. Flowering branch. B. Fruit. Artist: L. B. Hamilton (Benson and Darrow 1945).

**Other names:** ENGLISH: [yellow] trumpet-bush [-flower], yellow elder (allusion to similar leaves to *Sambucus*, q.v.); SPANISH: *caballito* (little horse, Sonora), *esperanza* (hope, Texas), *gloria* (Sinaloa, Sonora), *lluvia de oro* (golden shower, Arizona, Sinaloa, Sonora), *miñona* (Texas, Nuevo León), *palo amarillo* (yellow tree, Chihuahua), *palo de arco* (bow tree, Baja California, Chihuahua, Sonora, Oaxaca), *retamo* [*retama*] (Durango, Guerrero, Edo. México, Michoacán, Jalisco, San Luis Potosí), [*borla de, flor de, hierba de*] *San Pedro* [*San Nicolás*] ([St. Nicolas's] St. Peter's [tassel, flower, herb], Chiapas, Coahuila, Durango, Guanajuato, Edo. México, Michoacán, San Luis Potosí, Tamaulipas, Veracruz), *trompeta* (top, Durango), *trompetillo* (little top, Hidalgo), *tronador* [*tronadora*] (thunderstorm plant, Texas, Chihuahua, Guanajuato, Hidalgo, Edo. México, Sonora, Zacatecas); MAYAN: *kanlol* <*kanló, k'anlol, xkanlo*> (*kan*, yellow, *lol*, flower, Maya); OTO-MANGUEAN: *guie-bichi* (*guie*, flower, *bichi*, deer, Zapotec); UTO-AZTECAN: *kusí urámake* (Tarahumara), *nixtamaxochitl* <*nextamalxochitl*> (*nextamalli*, cornmeal dough, *xochitl*, flower, Náhuatl), *tulasúchil* (*tollin*, cattail, *xochitl*, flower, Náhuatl, Oaxaca), *wasáro* (Tarahumara)

**Botanical description:** Shrubs, erect, deciduous, to 6 m tall, usually shorter. Leaves opposite, 10–24 cm long, pinnately compound, the leaflets 7–9, less often 3–5, thin, 5–10 cm long, 0.8–2.3 cm wide, asymmetrical at the sessile base, lanceolate, the margins toothed, the terminal leaflet longer than the laterals. Inflorescences of terminal racemes or panicles are 5–20 cm long. Calices are 7.5–8 mm long, with 5 slender teeth. Corollas are yellow, 4–5 cm long, moderately bilateral, with a short basal tube and a swollen throat. Capsules are 10–15 cm long, linear, pendent.

**Habitat:** Canyons, rocky slopes. 900–1,600 m (3,000–5,500 ft).

**Range:** Arizona (Cochise, Pima, Pinal, and Santa Cruz counties), southern New Mexico, and western Texas; Mexico (Chihuahua and Sonora to Tamaulipas, south); Central America; Caribbean; South America.

**Seasonality:** Flowering May to August (October).

**Status:** Native. The Arizona and Sonora plants belong to var. *angustatum*. There are two varieties in the area (Felger et al. 2001). Variety *stans*, commonly cultivated, is a larger shrub or small tree with masses of brilliant yellow flowers and may be distinguished from var. *angustatum* by the larger stature and wider leaflets.

**Ecological significance:** *Tecoma* often grow on ridges among rocks and boulders, where plants receive some protection but are pruned by winter freezes. This tropical species is at the northern limits in Arizona and occupies the ecotone between Sonoran desertscrub and semidesert grasslands (Turner et al. 1995). There is both protection from frosts and adequate moisture in this ecotone. Both bumblebees (*Bombus*) and carpenter bees (*Xylocopa*) visit flowers, the former pollinating and the latter sometimes robbing by biting holes in the bases to get the nectar.

**Human uses:** Bernardino de Sahagún, writing between 1519 and 1540, recorded that the Aztecs used *Tecoma* in religious ceremonies. In Sonora, and elsewhere in Mexico, the flowers are brewed as tea to regulate blood pressure, and the leaves are made into a tea taken to treat respiratory and other illnesses (Felger 2000, Moore 2003, Standley 1920–1926). Tarahumara make medicines of the flowers to treat colds, chest pains, and sore eyes (Pennington 1963). According to Yetman and Van Devender (2001), the more tropical var. *stans* is cultivated in almost all Mayo villages as an ornamental, and for shade and medicine. Mayo pay little attention to the differences between varieties, using both to treat blood pressure problems and other maladies.

Wood is used by indigenous people to make bows (Lamb 1975), but plants in Arizona are usually too small to be used for normal bows. Stems might be large enough for the tiny bows made by the Northern Tepehuan (Pennington 1969). Tarahumara made arrows from the stems (Pennington 1963).

**Derivation of the name:** *Tecoma* is based on the Náhuatl name *tecomaxochitl* (*tecomatl*, jar, vase, *xochitl*, flower), given to plants with tubular flowers. *Stans* means in an erect position.

### ***Amsinckia menziesii* (Boraginaceae) Coast Fiddleneck, *cetkom***

**Other names:** ENGLISH: devil's lettuce, [fireweed] fiddleneck (comparing the flower cluster with the neck of the violin); UTO-AZTECAN: *cedkam* (Hiá Cẹd O'odham), *cetkom* *<chetkom>* (Tohono O'odham), *chedkoadag* *<tci-itkatak, djeh-t-ka-tak>* (Akimel O'odham), *tivaˀnɨbɨ* (Kawaiisu; for *A. tessellata*), *tso'hamp* [*tso'nap*] (Shoshoni; for *A. tessellata*, Chamberlin 1911), *tu'karûmp* (Ute; for *A. tessellata*), *kuniroûmp* (Shoshoni, Chamberlin 1909); YUMAN: *kacú:l nʸmpałʸ* (*kacú:l*, lizard, *nʸmpałʸ*, tongue, Cocopa); (= *A. intermedia*)

**Botanical description:** Annuals, usually with 1 to several main stems, 30–50 cm tall, plant-hairs bristly. Flower clusters conspicuously curved like a scorpion tail (helicoid). Flowers are self-fertile. Calices lobes are alike. Corollas are yellow, tube 10-nerved below insertion of the stamens. Nutlets 4, relatively deeply arched, the dorsal (back) side with a high ridge crest, sculptured with sharp, ragged ornamentations, the roughenings not crowded, 2.4–2.6 mm long.

**Habitat:** Along washes, under shrubs and trees in dry, rocky soil. To 914 m (3,000 ft)

**Range:** Arizona (Cochise, Coconino, Mohave, Pima, and Yuma counties), California, and New Mexico; Mexico (Sonora).

*Amsinckia intermedia*. A. Habit showing the "fiddleneck" cluster. B. Nutlet containing one seed. Artist: L. B. Hamilton (Parker 1958).

**Seasonality:** Flowering (February) March to May.

**Status:** Native. Our plants are var. *intermedia*.

**Ecological significance:** Usually these herbs grow as scattered plants, with one here and there. During unusually wet years, there may be patches of thousands of individuals, usually under trees. In those stands, the yellow flower clusters make them more obvious than when growing alone, and surely serve to attract more pollinators. More successful pollinations mean more seeds in the reserve soil bank for the next seasons.

Seeds of *Amsinckia* are an important food for quail. The genus represents more than 5% of the diet of California Quail (*Callipepla californica*) (Crispens et al. 1960), but its importance to the Gambel's Quail (*Callipepla gambelii*) and Masked Quail (*Colinus virginicus ridgewayi*) does not seem to be known. Indeed, although the seeds of this genus are highly ornamented, these structures do not seem to deter their being eaten by wildlife. Martin et al. (1951)

found that some Fiddleneck species were eaten by Mourning Doves (*Zenaida macroura*), several quail species, goldfinches, Horned Larks (*Eremophila alpestris*), meadowlarks, tohees, Cactus Wrens (*Campylorhynchus brunneicapillus*), and ground squirrels.

Although some species exhibit distyly, a mechanism for outcrossing, flowers of *A. menziesii* have a single style type and are self-fertilizing (Schoen et al. 1997). In distylous species, there are long- and short-styled plants. Pollen from long-styled plants may pollinate either long- or short-styled flowers, but that from short-styled plants cannot fertilize long-styled ones. Species with single-length styles are not all self-fertilizing, and the evolution of mating systems in *Amsinckia* is complex.

Plants contain alkaloids, typically the pyrrolizidine alkaloids. In general, these alkaloids are potent poisons, and children have been killed by overdoses of plant decoctions of other genera in the family. The pyrrolizidine alkaloids are notorious in both *Amsinckia* and the family (Austin 2004). These chemicals can cause cancer by interfering with DNA synthesis. Still, other studies suggest that some species have antitumor principles, insecticidal and antifungal properties, and wound-healing effects. In domestic animals and humans, the alkaloids may be passed to offspring in milk (Panter and James 1990).

*Amsinckia* is closely related to the yellow-flowered members of *Cryptantha* subgenus *Oreocarya*, and the genus is particularly allied with the *C. flava-confertiflora* complex (Felger 2000). *Amsinckia intermedia* and *A. tessellata* are thought to be evolutionarily derived within the genus.

**Human uses:** Although no modern O'odham profess to use *Amsinckia*, they retain names for them. Historical records indicate that O'odham once ate the leaves directly from *chedkoadag*, although they were aware that the dry plants caused dermatitis (Rea 1997). Maybe that itch is why the herbs are called "Devil's Lettuce."

In the late 1930s, Zigmond (1981) worked with the Kawaiisu. *Amsinckia* leaves were still an important source of spring greens. Only young leaves were eaten, and they were considered inedible after they got older and fruits matured. Because there are names given to these plants by the Shoshoni from Montana south to the O'odham of Arizona and Sonora, this genus must have been a more important food in the past.

Chumash of southwestern California eat the seeds in *pinole* (Timbrook 1990).

**Derivation of the name:** *Amsinckia* commemorates the German Wilhelm Ansinck, 1752–1831, burgomaster of Hamburg, Germany, and patron of botany. *Menziesii* is dedicated to Archibald Menzies, 1754–1842, surgeon and botanist. *Intermedia* means "intermediate," presumably between two other species in the genus.

## *Boechera perennans* (Brassicaceae) Rock Cress, *arábide*

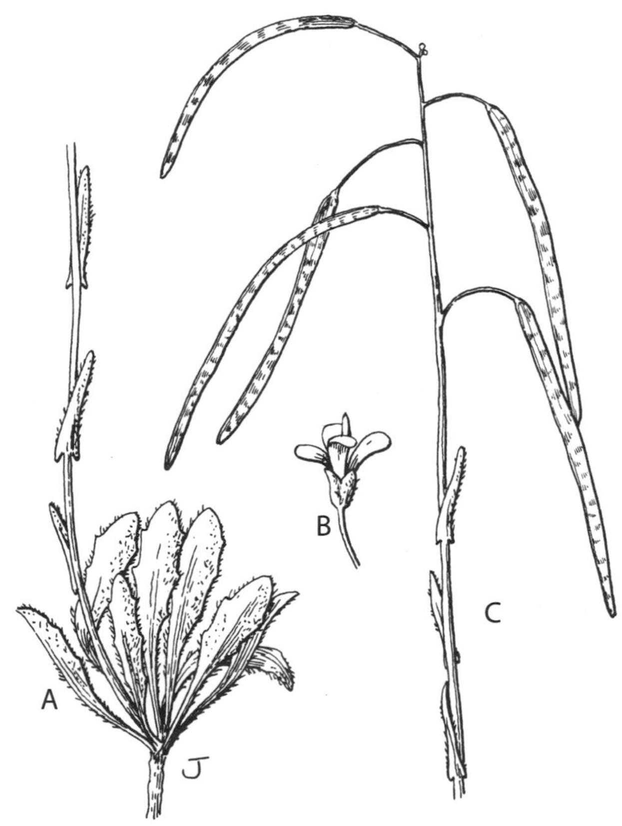


*Boechera perennans*. A. Base of plant. B. Flower. C. Fruiting branch apex. Artist: J. R. Janish (Wiggins 1980).

**Other names:** ENGLISH: [perennial] rock cress ("cress" is derived either from Old High German, *chresan*, to creep or creeper, or from Latin *crescere*, to grow; "rock" because the substrate where the herbs grow is often rocky); SPANISH: *arábide* (of Arabia, Mexico; there are cognate names in French and Italian for the genus); ATHAPASCAN: *'azee' naneeshtł'iizh* <*ˀazéˀ na'ne'sdizi*> (*'azee'*, medicine, *naneeshtł'iizh*, winding, Navajo), *'iiníziin ch'ił* <*'i'lyizin c'il*> (*'iiníziin*, witchcraft, *ch'ił*, plant, Navajo), *'atsé 'áłts'óózí*

<[*ˀosceˀ*] *y'osce ˀa.lc'ozgi*> (slender first one, Navajo); UTO-AZTECAN: *qta'komav* (Ute), *si'boiûp* (Shoshoni; for *A. holboellii*; also used for *Cleome*); (= *Arabis perennans*)

**Botanical description:** Perennial herbs. Stems 0.9–6 dm tall, several to many from a simple or branching herbaceous to woody base (caudex) that typically rises from between the basal rosette and a tuft of ascending leaves. Basal leaves pubescent with star-shaped hairs on margins and on surfaces, rarely with simple or forked hairs along petiole base, the stem (cauline) leaves longer than the internodes at least below, several, 0.7–4 cm long, 2–8 mm wide, oblong to lanceolate, entire to toothed, hairy or the upper glabrous. Flower cluster racemose. Pedicels of flowers 4–24 mm long, spreading to arched downward, glabrous or pubescent. Sepals are 3–4.5 mm long, often purplish, usually with star-shaped hairs. Petals are 5–7 mm long, pink to lavender, spatulate, erect to spreading. Fruits (siliques) are 20–60 mm long, 1.2–2 mm wide, spreading to pendulous, valves glabrous, nerveless or nerved at the base. Seeds are in a single line (uniseriate).

**Habitat:** Lower mountain slopes and canyons. 600–2,440 m (2,000–8,000 ft).

**Range:** Arizona (all counties), southern California, western Colorado, Nevada, and Utah; Mexico (Baja California and Sonora)

**Seasonality:** February to October.

**Status:** Native.

**Ecological significance:** In Baja California, these herbs grow from desert and chaparral-covered hills to the Yellow Pine belt (Wiggins 1980). By contrast, in Brown Canyon, *B. perennans* does not become frequent until an elevation of about 4,300 feet. Because these are predominantly herbs of the cooler seasons, *Boechera* is most obvious in the fall. At that time, the long, slender fruiting-stalks are often topped by the lavender flowers, thus making an eye-arresting display even though the parts are small.

**Human uses:** Navajo use Rock Cress to treat hiccups caused by a dry throat, to counteract bad dreams, and to make a lotion to relieve general body pains (Vestal 1952, Moerman 1998, Wyman and Harris 1951). Although not recorded elsewhere, Wyman and Harris (1941) note that *'iiníziin ch'ił* is the name of plants used to cure diseases caused by witchcraft. *Boechera perennans* is involved with a rain-making story that also includes the Turkey and the Beaver (Wyman and Harris 1951).

Several other *Boechera* species are used (Moerman 1998). The Keres use *B. fendleri* (as *Arabis fendleri*) as a stomach medicine. The Navajo consider it one of their Life Medicines (Vestal 1952). Navajo also use *B. holoellii* (as *A. holoellii*) in the *Keldzi Hatal* (Night Chant) Ceremony (Elmore 1944).

Farther away, the Shoshoni consider *B. puberula* (as *Arabis puberula*) good for making a liniment or mustard plaster (Train et al. 1957). *Boechera stricta* (as *A. drummondii*) is use by Okanagon, Salish, and Thompson people to treat pains in the back and kidney region, and to relieve venereal disease. The Okanagan chew roots of *B. sparsifolia* (as *A. sparsiflora*) to stop diarrhea and heartburn, as a contraceptive, and in an infusion for an eyewash (Moerman 1998).

**Derivation of the name:** Askel and Doris Löve named *Boechera* in 1976, a patronym for Tyge Böcher, a Danish botanist. Linnaeus created *Arabis*, "from Arabia," in his *Species Plantarum* of 1753. His plants were from the district *Petraea*, a stony desert. He continued the name used earlier by Bauhin in 1623, Royen in 1740, and Hallier in 1742 for Old World plants now called *Arabis alpina*. *Perennans* means living more than two years.

**Miscellaneous:** *Boechera* has 110 species, and all except *B. falcata* (of the Russian Far East) are North American. Of these, one species (*B. holboellii*) is endemic to Greenland. There are 10 species in Arizona that were called *Arabis* by Kearney and Peebles (1951) and Lehr (1978). After Askel and Doris Löve created *Boechera*, Bill Weber transferred the American species into the genus, stressing that it differs generically from the Old World *Arabis*. Rollins (1993) and Kartesz (1994) put *Boechera* in synonymy with *Arabis*, and Mabberley (1997) did not even mention *Boechera*. Judd et al. (2008) also recognize *Arabis* in the broad sense and do not mention *Boechera*. On the other hand, O'Kane and Al-Shehbaz (2003), Al-Shehbaz (2003), and Holmgren et al. (2005) recognize *Boechera*, and Felger and Broyles (2007) record the single species *B. perennans* from southwestern Arizona.

There is no consensus on the generic change, but the reputation that both Löve and Weber have for extreme splitting may have something to do with the reticence to accept their views. In spite of that, molecular, cytological, and morphological studies clearly show that *Arabis* and *Boechera* are distinct (cf. Al-Shehbaz 2003, O'Kane and Al-Shehbaz 2003, and papers cited therein).

## *Descurainia pinnata* (Brassicaceae) Tansy Mustard, *pamita, da:pk*

*Descurainia pinnata.* A. Flowering plant. B. Portion of fruit showing seed attachment. C. Fruit. D. Seed. Artist: L. B. Hamilton (Parker 1958).

**Other names:** ENGLISH: [pinnate, western, yellow] tansy mustard ("tansy" originally used for *Tanacetum vulgare*, Asteraceae; from Old French *tanesie* by thirteenth century, modern French *tanaisie*, shortened form of *athanasie*, from medieval Latin and Greek *athanasia*, immortality, also Italian *atanási* by 1611, Portuguese *atanasia* or *athanasia*; "mustard" is based on "must," juice of plants, originally from grapes, so called because the condiment was originally prepared from brassicaceous plants by making the ground seeds into a paste with grape juice or "must"; used in this sense in English by 1269–1278; cognate with French *moutarde*, Old Occitan *mostarda* by 1350, Italian *mostarda* by the fourteenth century, Spanish *mostaza* by ca. 1400, Portuguese *mostarda* by 1416, and Catalan *mostalla*); SPANISH: *pamita* [*palmita, pamitón*] (from Ópata *pamit*, Baja California, Sonora), *sinapismo* (from Latin *sinapis*, mustard); ATHAPASCAN: *'atsé <ˀosceˀ>* (first one, Navajo), *'atsé 'áłts'óózí <ˀosceˀ 'a.lc'ozigi>* (slender first one, Navajo), *'atsé ts'oh <ˀosceˀ coh>* (big first one, Navajo), *chooyin 'azee' <co' in ˀazéˀ>* (*chooyin*, arthritis, *azee'*, medicine, Navajo); KIOWA TANOAN: *'awae* (Hano Tewa); UTO-AZTECAN: *aasa <asa, á:sa, ˀasa>* (Hopi), *akavɨ* (*aka*, the seed, Kawaiisu) *ˀáṣi-la <asil, asily>* (Cahuilla), *atsa' <acá>* (red, Paiute), *da:pk* (smooth/slippery, Tohono O'odham), *ívagi* (Northern Tepehuan; cognate with Akimel O'odham *iivagĭ*, eaten greens, and Névome *hibagui*), *hahck* (Southern Paiute), *hasá <jasá>* (Guarijío), *ṣu:waḍ <shu'awat>* (Onavas Pima), *ṣu:'waḍ<cu'uwa't>* (Tohono O'odham), *shuu'uvad <rú-u-what, show-ou-wat>* (maybe *such'iavik*, recorded by Curtin 1949, is a faulty transcription, Akimel O'odham, Arizona), *su'uvad* (Hiá Ceḍ O'odham), *suavoli* (Northern Tepehuan); YUMAN: *akav* (Mohave), *ka siB* (Paipai), *kosen* (Cocopa), *kse.v ilokwak* (*ilokwak*, sour, Maricopa); LANGUAGE ISOLATE: *ai'yaho* (Zuni)

**Botanical description:** Annuals with erect stems 8–60 cm tall, mostly unbranched, leaves and lower stems whitish pubescent. Early leaves in a basal rosette, these and the larger stem leaves 3–7 cm long, 2–3 cm wide, 1–3 times pinnatifid, gradually reduced upwards. Inflorescences are glabrous or glandular, racemose, 3–45 cm long. Pedicels are spreading, 4–12 mm long. Flowers are pale yellow, ca. 1.5 mm wide. Fruits are 3.5–7.2 mm long, 1.2–1.6 mm wide, narrowly club-shaped. Seeds are 0.8 mm long, ovoid, orange-brown, surfaces minutely papillate in longitudinal lines.

**Habitat:** Desertscrub to oak woodlands. Up to 2,130 m (7,000 ft).

**Range:** Arizona (throughout the state), California, east to Arkansas and Florida; Mexico (Chihuahua, Sinaloa, and Sonora); the species throughout the United States and Canada.

**Seasonality:** Flowering (February) March to April.

**Status:** Native. Our plants are ssp. *halictorum*.

**Ecological significance:** These are herbs of disturbed sites where *Descurainia* may appear in great numbers in the early spring. During the spring of 2005, the landscape in the Altar Valley was yellow with their flowers. The fruits and dead plants persisted and were common enough to add a tan cast to many places in the landscape well into June. No plants were seen during the dry years of 2006 and 2007.

**Human uses:** Guarijío eat the leaves as greens and value the seeds as medicine (Gentry 1963, Yetman and Felger 2002). Guarijío call *hasá* the "*quelite de los indios*" (Indian greens). Tarahumara, Northern Tepehuan, and Tewa also use *Descurainia* as a potherb (Robbins et al. 1916, Pennington 1963, 1969). Formerly, plants were collected and sold to druggists in Alamos and Navajoa, Sonora. Akimel O'odham, Cahuilla, Mayo, Navajo, Paipai, and Utes also used the herb medicinally (Ford 1975, Moerman 1998, Hodgson 2001, Yetman and Van Devender 2001).

*Descurainia* seeds were eaten by the Hohokam of Arizona, and some think it was cultivated; seeds have been recovered from pottery in two archaeological sites (Bohrer et al. 1969, Hodgson 2001). Among the historic people, Tansy Mustard has been eaten by the Akimel O'odham, Atsugewi, Cahuilla, Cocopa, Shoshoni, Hopi, Kawaiisu, Keres, Maricopa, Mohave, Navajo, Paiute, Quechan, and Tohono O'odham (Bean and Saubel 1972, Castetter and Bell 1951, Hodgson 2001, Moerman 1998, Rea 1997, Russell 1908, Stewart 1965, Vestal and Schultes 1939, Whiting 1939).

Hopi and Tewa use *Descurainia* mixed with dark iron pigments as a dye (Fewkes 1896, Robbins et al. 1916, Whiting 1939). Both tribes use the pigment in decorating pottery.

**Derivation of the name:** *Descurainia* is named for the French pharmacist François Déscourain, 1658–1740, a friend of the French botanists Antoine and Bernard de Jussieu. *Pinnata* refers to the compound leaves.

**Miscellaneous:** The genus comprises 40 species that grow in temperate and cool parts of the world. Europe has a single species, *D. sophia*, that is called "Flixweed" (variation of "Flux-weed," an allusion to a use to control diarrhea in Britain), which is naturalized in North America. That alien grows near the Arivaca Ciénega.

## *Draba cuneifolia* (Brassicaceae) Spring Whitlow-Grass, *draba primaveral*

**Other names:** ENGLISH: wedge-leaf draba (a book name), [wedge-leaf] whitlow-grass ("whitlow" is a suppurative inflammatory sore or swelling in a finger or thumb, usually in the terminal joint); SPANISH: *draba primaveral* (spring draba, Mexico), *gasa* (from Guarijío *hasá*? Mexico), *sanguinaria menor* (little bloody one, a name usually given to plants that stop bleeding, Mexico); (= *Draba cuneifolia* var. *integrifolia*)

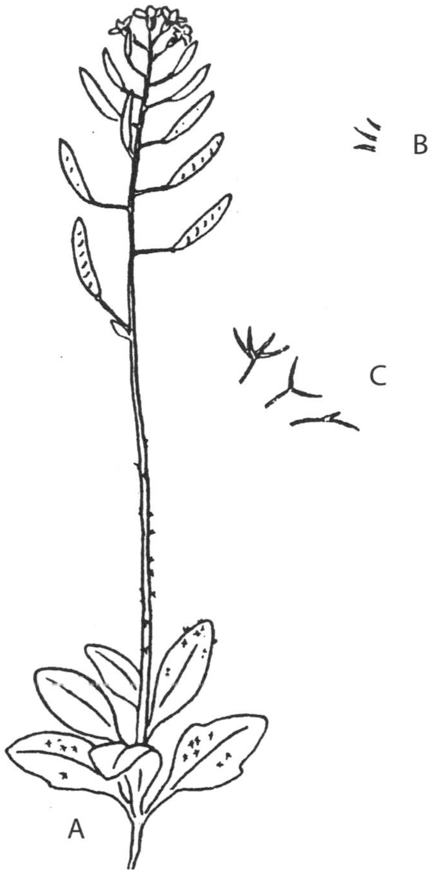

*Draba cuneifolia*. A. Flowering and fruiting plant. B. Simple hairs from fruits. C. Branched hairs from stems. Artist: R. D. Ivey (Ivey 2003).

**Botanical description:** Annuals, hirsute with stipitate branched trichomes. Stems to 3 dm tall, erect, usually branched from the base. Leaves near base to 5 cm long, usually obovate-cuneate, entire to more or less dentate, tapering into a short petiole or sessile, the cauline leaves few, on the lower half of the stems, to ca. 35 mm long, sessile, mostly obovate, entire to remotely dentate. Flower-clusters are dense and congested. Flowers with sepals are 1–2.5 mm long, oblong-ovate to linear. Petals are white, 3.5–5 mm long, spatulate and emarginate, sometimes smaller, linear, or absent. Fruits are silicles, 5–15 mm long, elliptic to oblong, hispid or glabrous. Seeds are 0.7 mm long, in 2 or several rows.

**Habitat:** Canyons, usually in the shade of trees. 300–2,130 m (1,000–7,000 ft).

**Range:** Arizona (throughout state), southern California, New Mexico, Texas, southwestern Utah, east to Florida, and north to Illinois and Washington; Mexico (Sonora to Zacatecas).

**Seasonality:** Flowering February to May.

**Status:** Native.

**Ecological significance:** This genus is circumboreal, with extensions into the tropical regions only at higher elevations. For example, Arizona has 11 species, Alaska has 32, and Europe has 44 (Kearney and Peebles 1951, Hultén 1968, Mabberley 1997). Spring Whitlow-Grass extends down into the upper parts of the canyons in the Baboquivari Mountains because *Draba* is intolerant of the hot weather in the lowlands.

**Human uses:** No uses were found for this species. The common names suggest that it has been used to treat sores, wounds, and other problems much like the species with documented records.

Other species are used by at least the Keres and Navajo. The Keres make *D. helleriana* into a drink used when one is not feeling well (Moerman 1998). Navajos use *Draba* in numerous ways, including in ceremonies as an emetic, to treat coughs, sore kidneys, gonorrhea, as an eyewash, and as a Life Medicine (Vestal 1952). The plant is considered potent, because Navajos also use it to protect themselves from witches. Navajos also use *D. rectifructa* and *D. reptans*, the former as a diuretic and the latter to heal sores (Elmore 1944, Vestal 1952).

**Derivation of the name:** *Draba* is from Greek *drabe*, acrid, applied by Dioscorides, fl. CE 40–80, to some cress. *Cuneifolia* means wedge-leaves; "integrifolia" means having unlobed leaves.

**Miscellaneous:** *Draba* is a genus of perhaps 300 species in northern temperate and boreal North America, Europe, Asia, and the mountains of South America.

Some recognize varieties in the entire range of the species. *Draba cuneifolia* var. *cuneifolia* grows from Texas through the southern and central United States, and into northern Mexico. *Draba cuneifolia* var. *cuneifolia* has both branched and simple trichomes on the vegetative parts and fruits, while var. *integrifolia* has only branched hairs or none. Felger et al. (in prep.) combine these, and also include var. *sonorae* with them.

Also in the Baboquivari Mountains are *D. helleriana* and *D. petrophila*. Perennial plants with yellow flowers are *D. petrophila*. Winter-annuals are either *D. cuneifolia* or *D. helleriana*. Plants with several pubescent cauline leaves are *D. helleriana*.

Another genus easily confused with *Draba* is *Dryopetalon*. While *Draba* has entire petals, *Dryopetalon runchatum* has them deeply lobed at the apex. These annual herbs appear in the understory of the oak woodlands in the canyons and also in the nearby Coyote and Quinlan Mountains. Instead of having entire or dentate leaves like *Draba*, these herbs have pinnatifid leaf-blades. The petals are deeply lobed to pinnately cleft, white, and curved (gibbous) at the base. *Dryopetalon* is known in Apache, Cochise, Gila, Graham, Greenlee, Maricopa, Pima, Pinal, Santa Cruz, and Yavapai counties; it is also known in New Mexico and northern Mexico (Chihuahua and Sonora). Flowers appear from February to May.

## *Lepidium virginicum* (Bassicaceae) Pepper-Grass, *pasote, ka:kowañi*

*Lepidium virginicum*. A. Flowering plant top. B. Young plant. C. Flowers. D. Fruit. E. Seed. Artist: R. O. Hughes (Reed 1971).

**Other names:** ENGLISH: [hairy] pepper-grass [weed] ("pepper-grass" was applied by 1475 to *Lepidium sativum*, because of the spicy taste of the fruits, subsequently to other species); SPANISH: *lentejilla* <*lentajilla*, *lentejuela*, *lentepilla*, *antejuela*, *antejuelilla*> (little lens, Texas, Chihuahua), *mostacilla* (little mustard, New Mexico), *pasote* (probably from Náhuatl *epatl*, skunk, *tzotl*, filth, comparing it to *Chenopodium ambrosioides*, Sonora); ATHAPASCAN: *'atsé* <*ˀosceˀ*, *y-'osce*> (first-one [*Descurainia*], Navajo), *'atsé 'ahiniidlin* <*ˀosceˀ ˀi'lt'I)'ˀi*> (resembling first-one, Navajo), *'azee' ch'il bilátah liga'iga* <*ˀazéˀ* [*c'il*] *bilatah halgai* [*liga'igi*]> (*halgai* [*liga'igi*], white, *ch'il bilátah*, flowers, *'azee'*, medicine, Navajo), *tsą́ą́halts'aa'* <*tshą́ axalts'aa'*, *tsáhalts'a'*> ([plant with pods] shaped [like the] interior stomach, Navajo); HOKAN: *coquée* (a term also used for chiles and black peppers, Seri), *isnáap ic is* (whose fruit is on one side, also their name for *L. lasiocarpum* and *Bouteloua*, Seri), *queeto oohit* (what Aldebaran eats, Seri; for *L. lasiocarpum*; "*queeto*" is the star Aldebaran, believed to cause plants to flower and fruit out of season); OTO-MANGUEAN: *ts'inrroí* (Mazahua); UTO-AZTECAN: *ătsăhámaˀa* (Northern Paiute), *avhaḍ* (Akimel O'odham), *ka:kowañi* <*ka: cowani*> (Tohono O'odham; for *L. lasiocarpum*), *ka:kowani* (Hiá Ceḍ O'odham; for *L. lasiocarpum*), *pasóre* (from Spanish, Guarijío), *ŕočíwari* <*dochi-wari*> (Tarahumara), *soowiidɨbɨ* (Kawaiisu; for *L. lasiocarpum*), *wa'tomasĭv* (Ute), *wu'buinûp* (Shoshoni); (= *Lepidium medium*, *Lepidium virginicum* var. *medium*)

**Botanical description:** Herbs, annual, 0.5–2.5 dm tall, with 1 to many erect to decumbent stems from base, herbage glabrous or puberulent to cinereous. Lower leaves spatulate to oblanceolate, 2–8 cm long, 0.5–1.4 cm wide, pinnatifid into toothed segments, the upper leaves reduced, merely toothed or slightly lobed, uppermost entire. Racemes erect, numerous, to 1.5–8 cm long. Sepals obovate to almost orbicular, about 1 mm long, with white margins. Petals white, 2–3 mm long. Pods almost orbicular, 2.5–3.2 mm long, shallowly and narrowly notched apically. Seeds reddish brown, about 1.2 mm long, narrowly winged.

**Habitat:** Grasslands, washes, canyons. Up to 2,290 m (7,500 ft).

**Range:** Arizona (most counties except the NE and SW corners), most of the United States; part of Canada; Mexico (Baja California, Chihuahua, and Sonora, south to at least the Valle de México).

**Seasonality:** Flowering February to August [September].

**Status:** Native.

**Ecological significance:** Regardless of the species, these are plants of disturbed sites. Dunmire and Tierney (1997) predict that any burned site will have *Lepidium* in warm weather after rains. Rea (1997) wrote, "They thrive in disturbed places on the floodplains, particularly fallow fields, yards, and waysides. One species or another may grow up on the bajadas between the black pebbles and desert pavement." Although he was talking specifically about the Gila River valley, his comments apply widely throughout the region.

*Lepidium virginicum* is usually confined to riparian areas and to washes and drainages in the desert grasslands. On those sites, this Pepper-Grass takes full advantage of the extra moisture that collects and funnels off after rains. Winter and summer rains may stimulate the species to germinate and flower in white patches.

**Human uses:** Although botanists recognize several species in the Southwest, differences are technical and not easily seen without magnification. Probably indigenous and other people utilize them all as one kind of plant. Doolittle (2000) said that *Lepidium* had been cultivated in "southern Arizona," but only the Tarahumara of Chihuahua are listed as cultivators by Hodgson (2001).

Tribes throughout the region were fond of *Lepidium* seeds (Hodgson 2001). More than likely, all the local peoples ate at least the seeds. Seeds or greens of one species or another were eaten by Kumiai, Havasupai, Kawaiisu, Luiseño, Paiute, Navajo, ancestral Puebloans, Seri, Tarahumara, Northern Tepehuan, and Tohono O'odham, plus several more northern and eastern tribes (Palmer 1878, Pennington 1969, Felger and Moser 1985, Dunmire and Tierney 1997, Hodgson 2001, Moerman 1998, Wyman and Harris 1951, Zigmond 1981). Although the Akimel O'odham have a name for *Lepidium* (Rea 1997), they no longer profess a use and simply classify them as *vashai* (grasses, greens). Tarahumara cultivate *Lepidium* and added the leaves as a spice to *esquiate*, which they call *keoíki* (a corn dish similar to cornmeal mush) (Pennington 1963).

*Lepidium* seeds are chewed by Isleta to relieve headache (Dunmire and Tierney 1997). Navajo use Pepper-Grass to relieve dizziness and intestinal problems (Wyman and Harris 1951). Navajo living around the Chaco Canyon now use the genus as a disinfectant (Hocking 1956). Historically, *Lepidium* was used medicinally at least as far south as the Maya of Yucatán (Austin 2004).

**Derivation of the name:** *Lepidium* is from the classical Greek name, *lepidion*, a little scale, perhaps an allusion to the flat fruit, but equally likely because of an Old World species' reputation for treating leprosy and other diseases that form scales on the skin. *Virginicum*, of Virginia, notes that the first specimens were from what is now that state.

## *Sisymbrium irio* (Brassicaceae)<br>London Rocket, *pamito, ban ciñiṣañ*

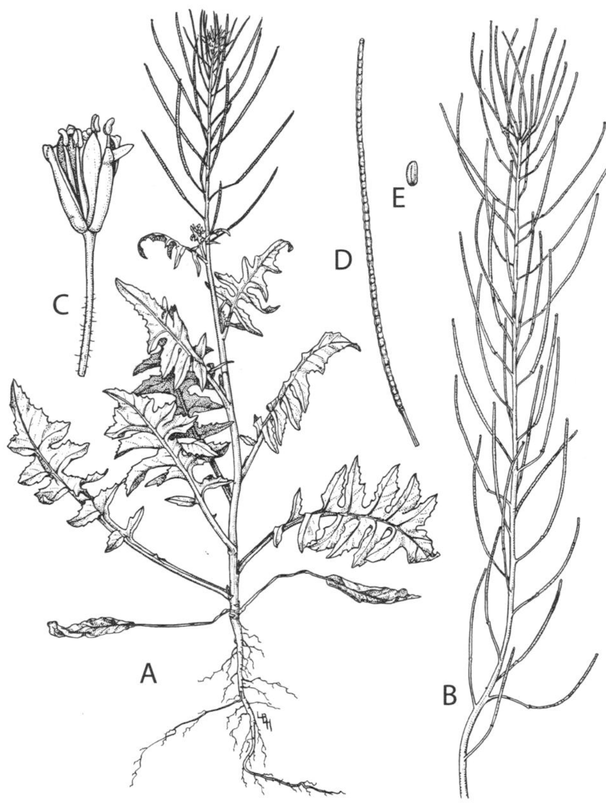


*Sisymbrium irio*. A. Plant with flowers and fruits. B. Fruiting branch. C. Flower. D. Fruit. E. Seed. Artist: L. B. Hamilton (Parker 1958).

**Other names:** ENGLISH: London rocket (presumably London is the place from which the genus was introduced; "rocket" is from Middle French *roquette*, which came from Old Italian *rochette*, a diminutive form of *ruca*, the garden rocket or arugula, *Eruca*; in English by ca. 1530), rocket mustard (see *Descurainia* for etymology); ATHAPASCAN: *bílátah dootł'izh* <*bilátah doƛiž*> (*dootł'izh*, blue, *bílátah*, flower, Navajo; for *S. elegans*, now *Thelypodiopsis elegans*); SPANISH: *mostaza* [*silvestre*] ([wild] mustard, Sonora), *pamita* (from Ópata *pamit*, Sonora; see also *Descurainia*); HOKAN: *cocóol* (name also given to *Descurainia*, Seri); UTO-AZTECAN: *ban ciñiṣañ* <*ban cinișani*> (coyote's mouth, Tohono O'odham), *ban cenṣañig* (coyote's mouth, Hiá Ceḍ, Sonora), *hiulit* (Tübatulabal; for *S. officinale*), *pamit* (Ópata), *pamitón* (Mayo), *poe'tcĕmĕn* (Ute; for *S. canescens*), *poi'ya* [*po'nak*] (Shoshoni; for *S. canescens*), *shuu'uvaḍ* (name also given to *Descurainia*, Akimel O'odham); YUMAN: *haskahl'* <*has káhl*> (Mohave)

**Botanical description:** Herbs, annual, stems branched near the base and above, to 6 dm tall, glabrous or sparsely hirsute above. Leaves petiolate, pinnately lobed or the upper nearly entire, glabrous or sparsely hirsute on petioles and lobe margins. Flowers are in terminal racemes. Calices are slightly shorter than petals. Petals are pale yellow, oblanceolate, to 4 mm long. Pedicels are slender, spreading-ascending, glabrous or sparsely hirsute. Siliques are slender, round in cross section (terete), glabrous, 3–5 cm long, ca. 1 mm wide. Seeds are oblong, wingless, less than 1 mm long.

**Habitat:** Canyons, streamsides, washes. Up to 1,370 m (4,500 ft).

**Range:** Arizona (western Coconino, Mohave, south to Cochise, Pima, and Santa Cruz, Yuma counties) and Texas; Mexico (Sonora).

**Seasonality:** Flowering March to July or October if there has been adequate rain.

**Status:** Exotic. Native of Europe.

**Ecological significance:** London Rocket was first found in Phoenix in 1909 (Rea 1997). This species was still a comparatively rare plant in 1951 when the *Flora of Arizona* was published (Kearney and Peebles 1951). That book recorded the herbs from only irrigated sites in southern Arizona. By the time Parker (1972) made her studies, she wrote that the species was "abundant throughout the irrigated lands of Arizona, in alfalfa, small grains, gardens, citrus, orchards, pastures, roadsides, and waste places." It was still confined to human-influenced places. Now the herbs grow in relatively undisturbed sites in the Baboquivari Mountains. *Ban ciñiṣañ* has not been found, however, in most of the arid region west of these mountains (Felger et al. in prep.).

These herbs are among the first green plants to appear in winter. However, London Rocket disappears except in shaded places when the weather becomes hot. *Sisymbrium* is a prolific seed producer, and that has doubtless contributed to its spread.

**Human uses:** Akimel O'odham, Cahuilla, Mohave, and Yuma eat the whole plants or leaves after they are boiled in several changes of water (Bean and Saubel 1972, Castetter and Bell 1951, Curtin 1949, Rea 1997, Stewart 1965). The Akimel O'odham also eat the seeds in water (*atole*), stored them as winter food, and made them into *pinole* (Curtin 1949, Rea 1997). Pennington (1980) recorded that Onavas Pima eat *S. auriculatum* as a potherb and called it *oibari* <*o'ibari*> (quick; cognate with Tohono O'odham *oiwi-d*, hurry).

Seri make an infusion of *S. irio* seeds to treat sore eyes (Felger and Moser 1985). Akimel O'odham also use the seeds to treat eye problems (Rea 1997). Seeds of *S. officinale* can be toxic if ingested in quantity, especially to people with heart problems (Dobelis 1986, Duke et al. 2002). Presumably, the same caveats apply to *S. irio*. In the Old World, *S. officinale* is mostly considered a medicinal herb; however, the foliage became a favorite of European settlers in the New World as a potherb and salad. According to Voegelin (1938), the Tübatulabal learned to eat leaves of *S. officinale* from Mexicans.

**Derivation of the name:** *Sisymbrium* is from Greek *sisymbrion*, a name given some sweet-smelling plant in the Brassicaceae. Pliny said it was an herb sacred to Venus. The name is akin to Hebrew *sis*, *sisa*, flower, *bor*, pit, well, Akkadian *burum*, well. *Irio* is a classical Latin name given to some kind of mustard.

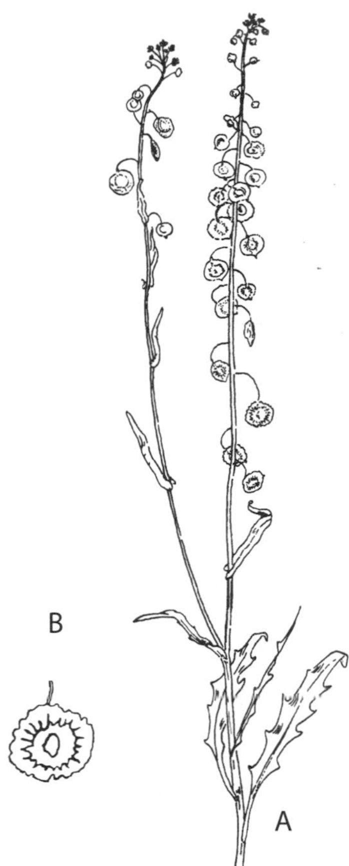


*Thysanocarpus curvipes*. A. Flowering and fruiting branch. B. Fruits, enlarged. Artist: E. C. Jaeger (Jaeger 1941).

## *Thysanocarpus curvipes* (Brassicaceae) Sand Fringe-Pod

**Other names:** ENGLISH: [common, sand] fringe [fringed]-pod ("sand" because that is the substrate where it typically grows; "fringe-pod" is an undated name given in California to *Thysanocarpus laciniatus*, later to other species in the genus; "fringe" from Middle English *frenge*, in turn from Old French *frenge* or *fringe* by 1316, modern French *frange* and Portuguese *fremja*, *fermja* all from popular Latin **frimbia*, altered from classical Latin *fimbria*, border, fringe), [hairy, sand] lace pod [lacepod, lacepod mustard] ("lace" in the thirteenth century was a net, noose, snare; also, a string or cord serving to draw together opposite edges [chiefly of articles of clothing, as bodices, stays, boots and shoes] by being passed in and out through eyelet-holes [or over hooks, studs, etc.] and pulled tight; by 1548, an ornamental braid used for trimming men's coats, etc.; a trimming of this; from Old French *laz*, *las*, modern French *lacs*, from popular Latin *laqueum*, a noose; cognates are Italian *laccio*, Spanish and Portuguese *lazo*)

**Botanical description:** Annual herbs, stems leafy, mostly erect, often 15–35 cm tall, single to few-branched, glabrous or sparsely pubescent with somewhat coarse, white hairs on leaves. Leaves linear-oblong to oblanceolate, with a few marginal teeth, the lower leaves petioled and usually in a rosette, the stem leaves clasping and sessile. Flowers minute, white to purplish in slender racemes, elongating as the fruit develops. Pedicels are slender. Fruits are nearly orbicular, slightly longer than wide, often 5.5–7.3 mm long, flat and thin, the wing usually perforated with small, evenly spaced holes. Seeds are 1 per fruit, not mucilaginous.

**Habitat:** Moist, sandy soils in canyons and washes. Rarely above 1,200 m (4,000 ft) in Arizona; higher in California.

**Range:** Arizona (Apache, Coconino, south to Cochise, Pima, and Santa Cruz counties), cismontane California, New Mexico, Nevada, Utah, and north to Montana; British Columbia; Mexico (Baja California and northern Sonora).

**Seasonality:** Flowering January to May.

**Status:** Native. Our plants have been called *T. elegans* and *T. curvipes* var. *elegans*, although Felger (2000) does not recognize either distinction.

**Ecological significance:** This species is the southeastern counterpart of *T. laciniatus* ([Common, Narrow-Leaf, Narrow-Leaved, Mountain] Fringe-Pod) that is known in northern Arizona from the Grand Canyon (Coconino County), Mohave County, and Prescott (Yavapai County) regions (Kearney and Peebles 1951). That species is mostly found through cismontane southern California from the western edge of the desert south to Santa Barbara and Inyo counties, south into Baja California, and east to Arizona. In spite of being known as "Mountain" Fringe-Pod, it grows only up to about 3,500 feet, while *T. curvipes* is known up to 5,000 feet in California.

West of the Baboquivari Mountains, *T. curvipes* is known in the Organ Pipe Cactus National Monument, and it extends into the Gran Desierto of northeastern Sonora on the Sierra Cipriano. In that Sonoran locality, it grows on a granitic bajada (Felger 2000).

**Human uses:** A decoction of the whole plant has been used to relieve stomachaches (Hocking 1997). Unidentified Mendocino people use the seeds in *pinole*. Ebeling (1986) noted that the herbage has a pungent and peppery taste, and that the seeds are parched, ground into fine flour, and mixed with other flours to improve the taste of *pinoles*.

**Derivation of the name:** *Thysanocarpus* is composed of Greek *thysanos*, a fringe or tassel, and *karpos*, fruit; the fruit is winged, and the margins ornamented in some. *Curvipes* is from Latin, *curvus*, curved, and *pes*, foot; "elegans" means elegant or graceful.

The genus and this species were named by William Jackson Hooker in 1830 in his *Flora Boreali-Americana* (Flora of North America). Then, in 1838, Thomas Nuttall, John Torrey, and Asa Gray named *T. laciniatus* in their *Flora of North America*.

**Miscellaneous:** *Thysanocarpus* is a genus of four species endemic to the western United States and nearby Mexico (Mabberley 1997, Felger 2000). The other two species in the United States are *T. conchuliferus* and *T. radians*. *Thysanocarpus conchuliferus* (Santa Cruz Island Fringe-Pod) is endemic to the Santa Cruz Islands in California. *Thysanocarpus radians* (Ribbed Fringe-Pod) occurs from Oregon south through the Sacramento Valley, bordering foothills, and west.

## ***Carnegiea gigantea*** **(Cactaceae)**
## **Saguaro, *sahuaro, ha:sañ***

**Other names:** ENGLISH: saguaro (from Spanish); SPANISH: *sahuaro* (Spanish from Cahita *sahuo*); ATHAPASCAN: *hosh 'aditsahii* <*xwoctítshahiih*> (*hosh*, cactus, *aditsahii*, awled, Navajo), *nanolzheegé* [*nanolzheegí*] (Western Apache), *xucntsai* (large cactus, Chiricahua and Mescalero Apache); HOKAN: *mojépe* <*moxéppe*> (Seri); UTO-AZTECAN: *bahidaj* (the fruit, Hiá Ceḍ and Tohono O'odham, Arizona), *haashañ* <*ha:canyi, hahshani*> (Akimel O'odham, Arizona), *ha:sañ* (Hiá Ceḍ, Sonora), *ha:sañ* (Tohono O'odham), *saguo* <*sauguo*> (Mayo), *sauwo* (Yaqui), *tudhua* (Ópata); YUMAN: *ʔa:ʔá* (Cocopa), *a'a'* (plant and fruit, Maricopa), *a'á'īl'íla* (Walapai, fruit *a'á'*)

**Botanical description:** Large columnar cactus, often 5–14 m, unbranched to several-branched, lower portion of stem thick, with 11–15 ribs, the longer central spines 3.7–11 cm long, stout, rigid. Upper parts of stems with 19–25 ribs, the spines 2.2–3 cm long, with closely set and nearly confluent, flower-bearing areoles. Flowers are 1 per areole, nocturnal, 8.5–13 cm long. Outer tepals are green, inner white. Fruits are 6–9.5 cm long just at dehiscence, ellipsoid to obovoid, mostly spineless.

**Habitat:** Rocky hillsides, desertscrub, grassland slopes; normally to 1,066 m (3,500 ft), but recorded at 1,417 m (4,649 ft) in upper Brown Canyon, where a single large individual stands atop a ridge.

**Range:** Arizona (Cochise, Gila, Graham, La Paz, Maricopa, Mohave, Pima, Pinal, Santa Cruz, Yavapai, and Yuma counties) and locally in southeastern California across from Yuma and Parker; Mexico (Sonora). Cf. Pinkava (1995).

**Seasonality:** Flowering May to June.

**Status:** Native.

**Ecological significance:** The endangered bat *Leptonycteris*, White-Winged Dove (*Zanaida asiatica*), and many other birds and bees drink nectar from the flowers. These cacti have a dual (night and day) pollination system—a "fail-safe" mechanism to assure a new crop of seeds. Bats visit at night and pollinate

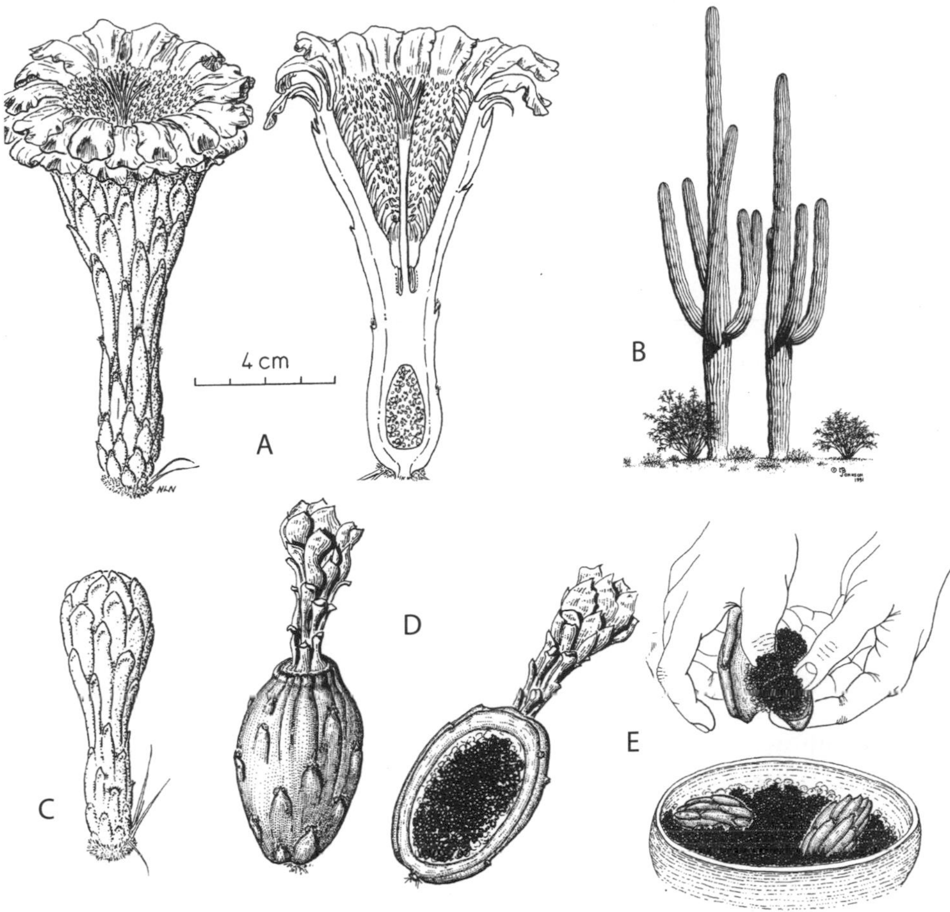


*Carnegiea gigantea*. A. Flower intact and longitudinal section. B. Growth form. C. Bud. D. Fruits intact and longitudinal section. E. Removing seeds and pulp for food. Artists: M. B. Johnson and N. L. Nicholson, A, B, C (courtesy M. B. Johnson); W. Hodgson D, E (courtesy Wendy Hodgson).

the flowers. Blossoms open about 1:00 a.m. but remain open well into the next day, when flowers are visited by birds and insects. An amazing variety of birds visit, including White-Winged Doves (*Zanaida asiatica*), woodpeckers of several kinds, and other nectar-feeding types with long bills.

Seeds are also important food for native bats, birds, and terrestrial mammals. This cactus is a keystone species in the upper Sonoran Desert.

**Human uses:** Although not described scientifically until 1848, from plants collected on the Gila River, the cacti were noted in California and Mexico by 1540 (Britton and Rose 1937). The first Anglo-Saxon observation of saguaro was made by James O. Pattie in 1825.

Mayo consider the fruit edible, but "hardly worth collecting" (Yetman and Van Devender 2001). Similarly, the Guarijío consider them inferior to other species, but eat them (Yetman and Felger 2002). Guarijío also consider the ribs useful, but weaker and shorter-lived than those of *etcho* (*Pachycereus pecten-aboriginum*) or *pitahaya* (*Stenocereus montanus*). Seri eat the fruits, make firedrills (*caaa*) and walking canes from the ribs, use the fruits in games, seal pots with the dry flowers, make shelters from the ribs, use the seeds in tanning, and the cacti figure in their supernatural beliefs (Felger and Moser 1974, 1985).

Among the northern Sonoran people, the cactus was markedly more significant (Bohrer et al. 1969, Gallagher 1976, Gasser 1981, Rea 1997, Hodgson 2001). The species is important food among the Hohokam, O'odham, Western Apache, and Yavapai. The names of months involving Saguaro among the O'odham seem to have changed with time, location, and individuals. Fontana (1980), Lumholtz (1912), Russell (1908), Underhill (1939, cited in Crosswhite 1980), and Saxton and Saxton (1969) call May the *kai cuckagid maṣad <kai chkukalig mashad>* ([saguaro]

seeds are turning black month, Akimel O'odham, Tohono O'odham) and June the *ha:ṣañ maṣad* <*hashani mashad*> (saguaro [harvest] month, Tohono O'odham). Mathiot (1973) recorded June as *s-toñ maṣad* <*stañ maxad*> (hot month, Tohono O'odham) and July as *ha:ṣañ-bahidaj maṣad* <*haashañ bahidag mashad, haaxañ-bahidag maxad*> (saguaro has fruit month, Akimel O'odham, Tohono O'odham). Rea (1997) agreed with *ha:ṣañ-bahidaj maṣad* but said it was "roughly June." Shaul (2007) names May as *ko'ok ha:ṣañ* (painful month, Tohono O'odham) and *pilkañ bahidag mašad* (wheat harvest month, Akimel O'odham); June as *ha:ṣañ bak mašad* (saguaro fruit month, Tohono O'odham) and *u'us wihogdag ha:ṣañ* (mesquite beans month, Tohono O'odham). Certainly, *pilkañ bahidag mašad* is post-European, because this Old World crop was introduced by the Spanish (Dunmire 2004). *Carnegiea* also served as material for tools and lodging among the O'odham.

**Derivation of the name:** *Carnegiea* was dedicated to Scottish-born American Andrew Carnegie, 1835–1919, steel industrialist, philanthropist, and patron of the sciences. *Gigantea*, giant, refers to the large size of the cacti.

**Miscellaneous:** This genus is monotypic, with this single species in the southwestern United States and northern Sonora. The species was first put in *Cereus* in 1848, and then *Carnegiea* was created by N. L. Britton and J. N. Rose in 1908, when they realized that it conformed to no other genus in the family. Although Mabberley (1997) says it is the largest of cacti, there are "a number of Mexican and South American species which are taller and would weigh more than *Carnegiea gigantea*" (Britton and Rose 1937).

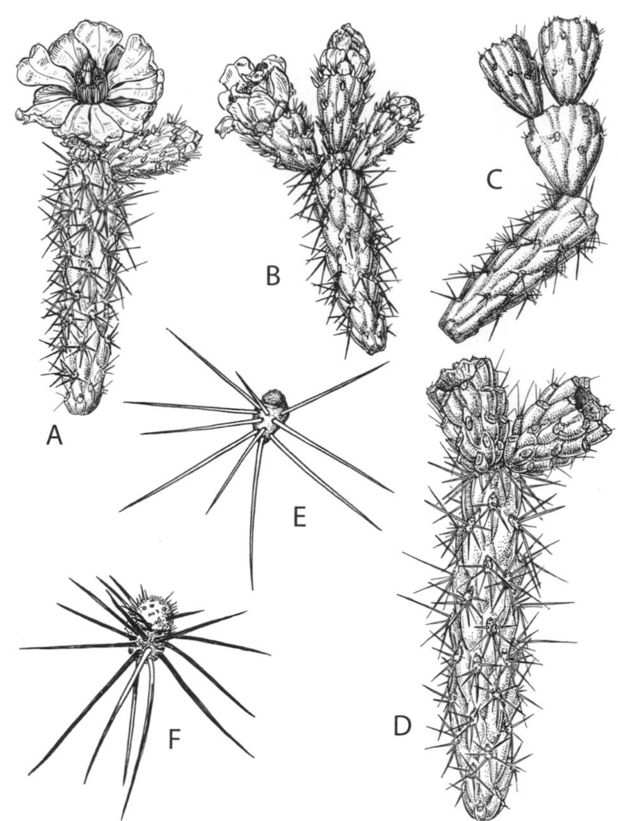


*Cylindropuntia*. A, D, F. *C. spinosior*. B, C, E. *C. versicolor*. A. Flowering branch, *C. spinosior*. B. Flowering branch, *C. versicolor*. C. Fruiting branch, *C. versicolor*. D. Fruiting branch, *C. spinosior*. E. Spines, *C. versicolor*. F. Spines, *C. spinosior*. Artist: L. B. Hamilton (from original artwork).

## *Cylindropuntia spinosior* (Cactaceae) Cane Cholla, *tasajo, ceolim*

**Other names:** ENGLISH: cane [handlegrip] cholla <*choya*> ("cholla" is Spanish for "skull" or "head" in allusion to the fruits; Sobarzo 1991 thought the word came from Latin *sciolo*, to know), walking-stick cactus (New Mexico); SPANISH: *tasajo* ([break into] pieces, Arizona, New Mexico, Chihuahua, Sonora); ATHAPASCAN: *hosh 'aditsahiitsoh* <*hosh 'aditsahii, xwoctítshahiih*> (*hosh*, cactus, *'aditsahii*, awl-like, *tsoh*, big, Navajo), *hosh ńchaagi* <*k'intsǫǫze*> (*hosh*, cactus, *ńchaagi*, big, Western Apache); UTO-AZTECAN: *ceolim* <*ciolim, cialim, tci'orim*> (Tohono O'odham), *'chi'odima'* (Hiá Ceḍ O'odham), *choa* (Yaqui; loan from Spanish), *hanam* <*hánam*> (Tohono O'odham), *ösö* <*ˀöso, ɜsɜ'*> (Hopi), *siviri* <*sivili*> (Cahita), *ušil* <*ˀusi-l*> (Tübatulabal), *úunvat* (Luiseño, Juaneño dialect), *wehcábori* [*wehcapó*] (Guarijío), *wiyaribi* (*wiya*, awl, Kawaiisu; for *C. acanthocarpa*), *wiyattampü* (Panamint); YUMAN: *atáta* (Havasupai), *atót* (Maricopa), *cac* $q^{wʔ}$*i:š* (*cac*, thorn, Cocopa), *ḍaqwi:s* (Walapai); (= *Opuntia spinosior*)

**Botanical description:** Shrubs or small trees, 0.4–2 m tall, with whorled branches. Stem segments green to purple, 1.3–3.5 cm in diameter, the tubercules obvious, oval to narrowly oval, 4.5–15 mm long. Areoles are yellow- to tan-felty, aging gray to black, broadly obdeltoid to elliptic, 4.5–7 mm long, 2–4 mm wide. Spines at most areoles, pale tan to (rarely) yellowish, pinkish, or red-brown, (4–)6–18(–24) spines per areole. Flowers rose to red-purple or bronze-purple, or yellow, perianth segments (tepals) spatulate, 18–35 mm long. Fruits yellow, sometimes tinged red or purple, broadly cylindric, pulpy-fleshy, 2.5 cm long, strongly tuberculate. Seeds are 4–5 mm long, pale yellow, almost orbicular to oval in outline, nearly flat.

**Habitat:** Desert and plains grasslands, extending up canyons. 300–2,000 m (1,000–6,600 ft).

**Range:** Arizona (Cochise, Gila, Graham, Greenlee, Maricopa, Pima, Pinal, Santa Cruz, and Yavapai counties) and New Mexico; Mexico (Chihuahua and Sonora).

**Seasonality:** Flowering May to June.

**Status:** Native.

**Ecological significance:** *Cylindropuntia spinosior* hybridizes with both *C. versicolor* and *C. fulgida* (Pinkava 1999). The hybrids typically have a morphology intermediate between the parents, often with changes in number of spines per areole, stem diameter, and fruit tubercules. Sometimes *C. acanthocarpa*, *C. arbuscula*, and *C. leptocaulis* hybridize with *C. spinosior* where they grow together. Near Tucson, hybrids have an intermediate stem and fruit size, and the flowers vary from purple to yellow. Some think that these variable-color plants are the result of hybridization between *C. acanthocarpa* and *C. spinosior* (Paredes-A. et al. 2000).

The species is widely scattered and rare in the northern part of Organ Pipe Cactus National Monument and Cabeza Prieta National Wildlife Refuge (Felger et al. in prep.). These authors suggest that most, if not all, of the specimens were planted there. Because this species was a popular food plant, perhaps the theory is correct.

**Human uses:** Tohono O'odham pit-bake the buds, fruits, and joints of some species of *Cylindropuntia*, formerly as a staple food. Cholla buds of various species are baked and eaten by at least the Akimel, Hiá Ceḍ and Tohono O'odham, Apache, Baja California people, Cahuilla, Cocopa, Hohokam, Seri, and Yavapai (Barrows 1900, Bean and Saubel 1972, Bohrer et al. 1969, Castetter and Bell 1951, Curtin 1949, Hodgson 2001, Felger and Moser 1985, Rea 1997, Russell 1908). Although the Kawaiisu have a name for cholla, they profess no use for them (Zigmond 1981). The cooked buds are moderately high in iron, but unusually high in calcium. Historically, these foods provided these desert dwellers with almost all of the needed daily requirements of calcium.

**Derivation of the name:** *Cylindropuntia* is from Greek *kylindros*, a cylinder, and *opuntia*, from Latin *herba opuntia*, from *Opus*, *Oputis*, a town of Locris in Greece; in Greek *Opous* or *Opountos*. *Spinosior* means very spiny.

**Miscellaneous:** Other species in the canyons are *Cylindropuntia arbuscula* (Pencil Cholla; *wipnoi*), *C. fulgida* var. *mammillata* (Chain[-Fruit] Cholla), *C. leptocaulis* (Desert Christmas Cactus; *tasajillo*; *wipnoi*), and *C. versicolor* (Staghorn Cholla).

Both *C. arbuscula* and *C. leptocaulis* are small versions of cholla. The fruits of *C. arbuscula* are 20–35 mm long, pale green but sometimes tinged with red to purple, with joints 7–10 mm wide. *Cylindropuntia leptocaulis* fruits are 9–15 mm long, yellow to scarlet, with joints 3–5 mm wide. All other species in the region have much wider stem joints, always over 1 cm wide.

*Cylindropuntia fulgida* has fruits suspended in chain-like clusters and is impenetrably armed with joints 3–5 cm thick, always greenish. Variety *mammillata*, in the lower part of the canyon, has short, clustered branches; it tends to be whitish from the spines. *Cylindropuntia versicolor* has mostly solitary fruits and is spiny, but the armament does not obscure the stems, and joints are 1–2 cm thick, often tinged red to purplish.

## ***Ferocactus wislizeni* (Cactaceae)**
## **Barrel Cactus, *biznaga, jiavuli***

**Other names:** ENGLISH: Arizona [fish-hook, candy] barrel cactus; SPANISH: *biznaga* [*de agua*, *gigantesca*, *hembra*] ([water, giant, female] barrel cactus; probably from Náhuatl *huitzli*, spine, and *nahuac*, around, or covered with spines; in Spanish the word *viznaga*, of Arabic origin, originally applied to the parsnip, *Pastinaca sativa*; the two words became confused in Mexico); ATHAPASCAN: *hosh tsał* <*hosh chaał*> (*hosh*, cactus, *tsał*, basketry awl, Western Apache), *hosh sidáhí* (*hosh*, spine, *sidáhí*, sitting, Navajo); HOKAN: *siml* <*simláa*> (true barrel cactus, Seri); UTO-AZTECAN: *chiávul* (Akimel O'odham), *hisil* <*hísely*> (cholla, Mountain Pima), *ibávoli* (Northern Tepehuan), *jiavul* (Hiá Ceḍ O'odham), *jiavuli* <*jiawul*, *tciaur*, *tjedvoli*> (devil, supposedly a loan word from Spanish *diablo*, Tohono O'odham), *kïče'apïl* (Tübatulabal), *nookwi'a*(*pi*) (Panamint, the fruit *tüükimpi*), *ono'e* (Yaqui), *táci* (Southern Paiute), *teˀíwe* (Guarijío); YUMAN: *miltát* <*milḍaḍ*> (Walapai), *miltót* (Maricopa), *multát* (Havasupai), *mułʸcác* (Cocopa)

**Botanical description:** Stems cylindroid or ovoid, 6–60 dm long, 3–6 dm wide, ribs 20–28. Central spines colored ashy gray, 4 per areole, strongly cross-

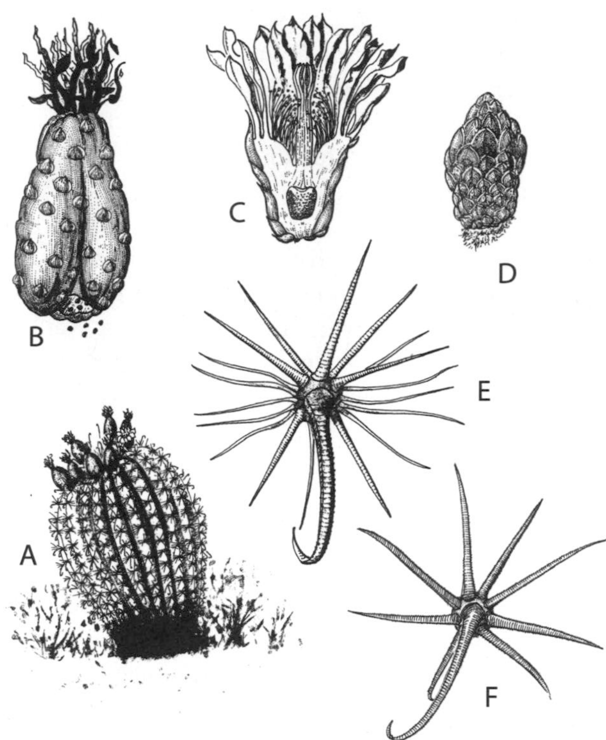


*Ferocactus wislizenii*. A. Growth form. B. Mature fruit. C. Flower (longitudinal section). D. Flower bud. E. Spine cluster. F. Spine cluster of *F. emoryi*. Artist: W. C. Hodgson (courtesy Wendy Hodgson).

ribbed, the lower ones hooked, to 4–5 cm long, 1.5–2 mm wide, flattened, radial spines 12–20 per areole, spreading, slender, curving irregularly, not cross-ribbed. Flowers 4.4–6.2 cm wide, petaloid perianth parts orange-yellow. Fruits are yellow, barrel-shaped, with numerous almost circular shallowly fringed scales, 3–4.4 cm long. Seeds are minutely reticulate, 2–2.5 mm long.

**Habitat:** Desertscrub, grasslands, canyons. Up to 1,450 m (4,750 ft).

**Range:** Arizona (Cochise, Greenlee, Maricopa, and Pima counties), southern New Mexico, and western Texas; Mexico (Chihuahua, Sinaloa, and Sonora).

**Seasonality:** Flowering July to September.

**Status:** Native.

**Ecological significance:** The Barrel Cactus grows in rocky, gravelly, or sandy soils of hills, flats, canyons, wash margins, and alluvial fans (Turner et al. 1995). Plants are limited to the north by cold, in part mediated by their self-shading ability with their spines. In Texas, the abrupt eastern limit is probably attributable to landscapes that are too xeric (Turner et al. 1995).

Flowers are pollinated by bees (genus *Lithurge*) that land on the stigmas and crawl onto the stamens (Turner et al. 1995, Dimmitt 2000). About four days is the life of the partly self-incompatible flower. Fruits and seeds are eaten by rodents (including Rock Squirrels, *Spermophilus variegatus*), birds, deer, Bighorn Sheep (*Ovis canadensis*), and Javelina (*Pecari tajacu*). Cactus Beetles (*Moneilema gigas*) eat Barrel Cactus itself.

At least three theories have been proposed regarding why this cactus tends to lean toward the south. Leaning is so great that plants sometimes fall. Plants grow slowly, about 4 cm tall in the first four years, but reaching 11 cm in the next four. For the first 10–15 years, growth in diameter nearly equals growth in height (Turner et al. 1995).

**Human uses:** The oldest records of these cacti being used by humans are among the Aztecs, who called the cacti *teocomitl* (divine vessel). The *biznaga* was sacred to the Aztec deity Mixcoatl. This god was their divinity of the home (Siméon 1885), who was also venerated by the Otomí and Chichimeca. The Mexicans had built two temples in honor of this deity, one of them with the name Teolalpan (from *teotl*, deity, *itztlí*, obsidian, *tlan*, near). In these temples, Aztecs made offerings of rabbits, hares, and other items, and celebrated *fiestas* for the divinity at the beginning of the month called *quecholli* (the fourteenth Aztec month of the year; also a bird, probably what is now called the Quetzal, *Pharomachrus moncinno*).

Pulp and/or seeds provided food for at least the Apache, Akimel, Hiá Ceḍ, and Tohono O'odham, Baja Californians, Guarijío, Tarahumara, Northern Tepehuan, and Seri (Felger and Moser 1985, Hrdlička 1908, Hodgson 2001, Moerman 1998, Pennington 1963, 1969, Rea 1997, Yetman and Felger 2002). As an example, the Akimel O'odham, Tohono O'odham, and Northern Tepehuan slice and boil the pulp of the stem for food during the spring (Castetter and Underhill 1935, Hrdlička 1908, Pennington 1969, Rea 1997).

Other species produce "water" in the stems that usually makes people ill if drunk. However, the Seri and Tohono O'odham consider water from *F. wislizeni* a substitute for fresh water (Castetter 1935, Felger and Moser 1985, Hodgson 2001). See Dimmitt (2000) for caveats to extracting moisture.

**Derivation of the name:** *Ferocactus* is based on Latin *ferox*, ferocious, and *cactus*, spiny. *Wislizeni* is named for German-born Friedrich (Frederick) Adolph(us) Wislizenus, 1810–1889, who collected plants in the southwestern United States and northern Mexico.

**Miscellaneous:** The related *F. emoryi* is known from Fresnal and Thomas Canyons, and in a few isolated places in Brown Canyon. *Ferocactus emoryi* differs from *F. wislizeni* in that it lacks bristles at the base of the larger spines in the clusters. Flowers tend to be redder in that species than in *F. wislizeni*, at least in Arizona. There is a single specimen cultivated by the Harm House.

### *Opuntia engelmannii* (Cactaceae) Desert Prickly-Pear, *nopal, i:bhai, navĭ*

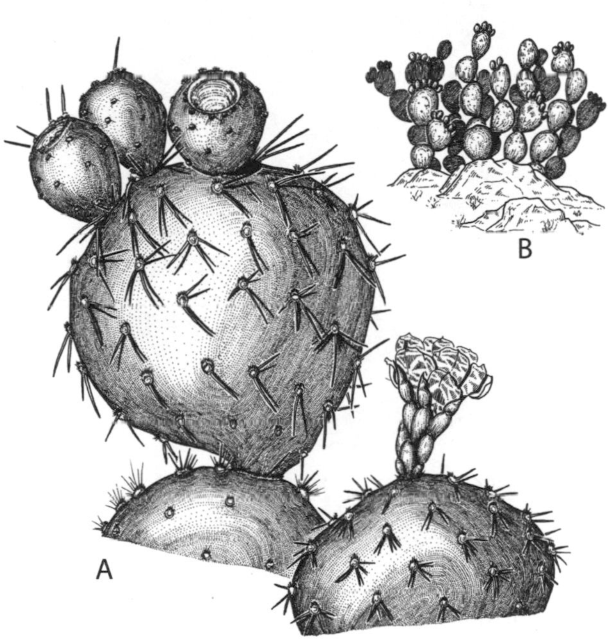


*Opuntia engelmannii.* A. Fruiting and flowering branches. B. Growth form. Artist: W. C. Hodgson (courtesy Wendy Hodgson).

**Other names:** ENGLISH: prickly-pear (called *prickle-pear* by Capt. John Smith in 1624, *prickly-pear* by 1760, because of spiny fruits shaped like a pear); SPANISH: *coyonoxtle <joconostle>* (from *coyotl*, the coyote, *nochtli*, tuna, Náhuatl), *cuija* (lizard, from Náhuatl *cuixa*), *huichacame <huichanabo>* (from Cahita *huicha*, spine, *ame*, spiny, Sonora), *nopal* [*cuixo*] ([lizard] prickly-pear, Sonora; the plant, from *nopalli*, Náhuatl), *tuna* [*cuija*] ([lizard] prickly-pear; *tuna* for cacti is from the Caribbean Taino language, Sonora), *vela de coyote* (coyote's candle, also used for *C. fulgida* in Sinaloa and Sonora); ATHAPASCAN: *gołtcide <gułtcide>* (Chiricahua and Mescalero Apache), *hosh nteelí <h$^w$os>* (*hosh*, cactus, *nteelí*, broad, Navajo), *hosh nteelí* [*ts'osé*] (*hosh*, cactus, *nteelí*, wide, *ts'osé*, slender, Western Apache); HOKAN: *heel hayéen ipáii* (prickly-pear used for face painting, Seri); KIOWA TANOAN: *sae* (Tewa); UTO-AZTECAN: *ai'gwobi* (Shoshoni), *ekupittsi* (Panamint), *i:bai <ibai>* (fruit, Onavas Pima), *'i:bhai <iibhai>* (fruit, Akimel O'odham, Hiá Ceḍ O'odham), *i:bhai* (*i'ibhai* [*ii'ibhai*], plural, fruit, Tohono O'odham), *ila'* (Guarijío), *irá* [*ira-ka, rihuirí*] (Tarahumara; individual species have their own names), *ïyal <i'yal>* (Tübatulabal), *naavo* (Yaqui), *náavut* (Luiseño), *nabo <nacoó>* (Cahita), *nabu* (Northern Paiute), *navet <náve-t, navit>* (Cahuilla), *napó* (Tarahumara), *nava* (Mountain Pima), *nav* (Hiá Ceḍ O'odham), *navĭ <naf, naw, nohwi>* (the plant, Akimel and Tohono O'odham), *navku(m)bɨ* (Kawaiisu; for *O. basilaris*), *návoi* (Northern Tepehuan), *návu* (the plant, Hopi), *navú-c* (Eudeve), *návūt* (Cupeño, Luiseño), *yöngö <yüñü, yɜ́:ngu>* (the fruit, Hopi); YUMAN: *ălâva* (Havasupai), *a'láv$^a$* <*alav*> (Walapai), *kal yap* (Maricopa), *tach pa* (Yuma), *xpa:* (Cocopa), *xté* (Paipai); (= *O. phaeacantha* var. *discata*)

**Botanical description:** Shrubby plant, 1.5–2 m tall, 1.5–5 m across, with ascending branches, joints obovate to almost orbicular, 1.5–3 dm long, 1.5–2 dm wide, 2–2.5 cm thick, smooth, glabrous, the areoles orbicular, with reddish glochids, spines 1–8, these 1.5–6 cm long. Flowers are 6–9.5 cm wide, yellow, inner petals broadly obovate, rounded apically. Fruits are obovoid, 2.5–4 cm wide, 4–7.5 cm long, red or purple, but with white to yellow flesh. Seeds are kidney-shaped, 3–4 mm wide, flattened, margin ridged.

**Habitat:** Grasslands, desertscrub. 300–1,980 m (1,000–6,500 ft).

**Range:** Arizona (through most of state), California, Nevada, New Mexico, and Texas; Mexico (Chihuahua, Coahuila, and Sonora).

**Seasonality:** Flowering April to June.

**Status:** Native

**Ecological significance:** These cacti appear to hybridize extensively with *O. phaeacantha*. Often the patches seem to be mixtures of various crosses and back-crosses of these two species. The "typical" *O. phaeacantha* is a lower plant with smaller pads, with fruits that are longer, narrower, and sweeter.

Fruits are food for birds, Coyotes (*Canis latrans*), and Javelinas (*Pecari tajacu*); surely, other mammals also. Indeed, Coyote (*Canis latrans*) scat in the fall is often blood-red from the fruits. Fruits of *O. phaeacantha* typically disappear from plants before those of *O. engelmannii*. Prickly-Pear plants are also primary sites for packrat nests.

**Human uses:** Juice from fruits is used as face paint, and fruits are eaten by Seri (Felger and Moser 1985). Fruits are used as food by Acoma, Akimel O'odham, Cahuilla, Cocopa, Kumiai, Hohokam, Hopi, Keres, Laguna, Maricopa, Mohave, San Felipe, Sia, and Tohono O'odham (Bean and Saubel 1972, Castetter and Bell 1951, Gasser 1981, Moerman 1998, Rea 1997, Stewart 1965, Whiting 1939). Castetter and Underhill (1935), Curtin (1949), and Rea (1997) were told that fruits of one kind were eaten by all with impunity (probably var. *discata*), while another (probably var. *major*) would cause chills in the unaccustomed. Hispanics dried the fruits into *quesos de tuna* (tuna cheeses) and stored or sold them for later use (Saunders 1920).

Although Curtin (1949) reported Akimel O'odham used Prickly-Pear pads as medicine, Rea (1997) found that only the Maricopa did that. The Akimel O'odham knew of the Maricopa use, but said they did not take advantage of it. Kumiai used the juice from pads to lubricate oxcart wheels.

Apache, Mountain Pima, and Ópata historically ate the cooked pads of native Prickly-Pears (Castetter and Opler 1936, Rea 1997). Apparently, some other O'odham did not eat the pads until Hispanics introduced *O. ficus-indica* (Rea 1997). Even though the Akimel O'odham knew that the Apache ate pads from native Prickly-Pear, they claimed not to have used it.

**Derivation of the name:** *Opuntia* was created by Philip Miller, who took it from Latin *herba opuntia*, based on *opus*, *oputis*, a town of Locris in Greece; in Greek *Opous* or *Opountos*. The Latin name was used by Theophrastus, 372–287 BCE, for some unknown plant, and Philip Miller applied it to the New World cacti. *Engelmannii* is devoted to George Engelmann, 1809–1884, a German-born physician in St. Louis, Missouri, and painstaking student of difficult groups like *Agave*, Cactaceae, and *Cuscuta*.

**Miscellaneous:** Also in canyons are *O. chlorotica* (Pancake Pear), *O. phaeacantha* var. *major* (Plains Prickly-Pear), and *O. violacea* var. *santa-rita* (Santa Rita Prickly-Pear). The *O. chlorotica* is found at higher elevations and has lighter-colored pads than either *O. engelmannii* or *O. phaeacantha*. Fruits on *O. phaeacantha* are longer and more slender than those on *O. engelmannii*. Santa Rita Prickly-Pear tends to have purplish pads, and the colors intensify with drought and cold.

### *Mammillaria grahamii* (Cactaceae) Fish-Hook Cactus, *cabeza de viejo, ba:ban ha-i:swigĭ*

*Mammillaria grahamii*. Flowering plant. Artist: unknown (Phillips and Comus 2000).

**Other names:** ENGLISH: fish-hook cactus [pincushion] (because of hooked spines, Arizona, Sonora), Graham's nipple cactus (a book name; this epithet did not appear until 1876, some 20 years after the genus was described), strawberry cactus (from fruit color, Arizona); SPANISH: *biznaguita* (little barrel cactus, Sonora), *cabeza de viejo* (old man's head, Sonora), *chilitos de viznaga* (little cactus chiles, San Luis Potosí), *choyita* (variant of "little *cholla*," Sonora), *churrito* (apparently "little diarrhea," but probably a variant of "little *cholla*," Sonora), *pitahayita* <*pitaiaya, pitajaya, pitahaya*> (little cactus fruit, from the Caribbean Taino language, originally *bitajaya*, recorded by the early 1500s; Sonora); ATHAPASCAN: *xuebi* (Chiricahua and Mescalero Apache); HOKAN: *hant iipzx* *itéja* (bladder of the arroyo, Seri); UTO-AZTECAN: *arimo'o* <*urimo'o*> (Onavas Pima), *ban cepla* (*ban*, coyote, *cepla*, be convex, Tohono O'odham), *ban cesani* (*cesani* = ? Hiá Ceḍ O'odham), *ba:ban ha-'is:vig* (coyotes' hedgehog-cactus [*Echinocereus*], Hiá Ceḍ O'odham), *ba:ban ha-i:swigĭ* <*bahban ha-ihswig, baaban ha-iiswikga*> (coyotes' hedgehog-cactus [*Echinocereus*], Tohono O'odham), *ban cekida* (coyote vaccination, Hiá Ceḍ and Tohono O'odham), *ban ha 'iswig* (coyote's hedgehog-cactus [*Echinocereus*], Tohono O'odham), *ban mauppa* <*baaban makuppa*> (coyote's paws, singular and plural, Akimel O'odham), *ban maupai* (like coyote paws, Akimel O'odham), *chikul hu'i* (*chikul*, mouse, *hu'i* = ? Yaqui), *chicul ónore* (Mayo, Sonora), *híkuri* (Tarahumara), *hue tchurí* <*we*$^{ʔ}$*cúri, wehcúri*> (Guarijío), *noog*$^{w}$*iyavɨ* (Kawaiisu), *mu'tsa* (Shoshoni; for genus), *tuur soigai* <*tu'i shogi*> (*tuur*, coal, *soigai*, owner, a reference to the fruits? Mountain Pima), *tori bichu* (Mayo, Sonora), *uvayu'u*$^{s}$ (Southern Paiute); YUMAN: $^{a}$*tat* (Walapai); (= *Mammillaria microcarpa*)

**Botanical description:** Stems simple or clustered, globose to cylindrical, to 16 cm tall, 5–6 cm wide. Tubercules on stem spirally arranged, conic, the apex rounded, to 7 mm tall and 6 mm wide at the base, dark green. Areoles are ovale where spines are attached. Spines are radially arranged, 15–35, needle-shaped, 6–12 mm long, smooth, rigid, whitish yellow, the central spines 1–4, hooked, 12–18 mm long. Flowers are near the apex, funnel-shaped, to 25 mm long, to 40 mm wide, pink to rose-purple, the outer segments with minutely fringed margins. Fruits are club-shaped, red when ripe, to 25 mm long and 8 mm wide. Seeds are rounded-obovoid, blackish, pitted.

**Habitat:** Desertscrub, grasslands, ridges, slopes. Up to 1,370 m (4,500 ft).

**Range:** Arizona (southeastern Mohave to Yavapai, southern Coconino, western Graham, south to Cochise, Pima, Santa Cruz, and Yuma counties), southeastern California, New Mexico, and Texas (El Paso County); Mexico (Chihuahua and Sonora).

**Seasonality:** Flowering July–August.

**Status:** Native.

**Ecological significance:** Fruits are adapted to prolong the time of dispersal by remaining green and inconspicuous for long periods before turning red. Although the fruits are mature and contain viable seeds, they remain green (Felger 2000). Because the contrast of colors of the fruits against the whitish stems is the Fishook Cactus's way of "advertising" to birds and mammals that the fruits are ripe, the green forms remain inconspicuous. Once the colored crop of fruits is removed from the cactus and dispersed, the green ones change to red.

There are several flushes of flowering during the summer. Each of these flowering periods is about five days after the first two or three soaking summer "monsoon" rains (Mark Dimmitt in Felger 2000). Seri told Felger and Moser (1985) that the fruits were eaten by Javelina (*Pecari tajacu*).

**Human uses:** Curtin (1949) records that the Akimel O'odham rubbed the red fruit on arrow shafts to color them, and they boil the cactus to make a medicine to treat earaches. By the time that Rea (1997) talked with these people, the only use he could find was that people pounded open the stems to get water in an emergency. That use for water seems to be the allusion in the Seri name.

Some other people still use *Mammillaria*. Castetter and Opler (1936) found that the Chiricahua and Mescalero Apache eat the fruits. Western Apaches formerly ate the pulp (Reagan 1929). Guarijío told both Gentry (1942) and Yetman and Felger (2002) that children "carefully pluck" the fruits and eat them like candy. Mountain Pima, Onavas Pima, and Seri children also eat the fresh fruits (Felger and Moser 1985, Laferrière 1991, Pennington 1980). Onavas Pima and Seri use the juice to treat earaches. Bye (1979a) found that Tarahumara shamans use the latex-filled species as a hallucinogen.

**Derivation of the name:** *Mammillaria* is based on Latin *mammilla*, a breast, nipple, referring to the small tubercules lining the stems of plants. *Grahamii* commemorates Colonel James Duncan Graham (1799–1868), a member of the Boundary Survey party of 1851. Mount Graham was also named after him (Powell and Weedin 2004).

**Miscellaneous:** The other species in the canyons is *M. macdougalii* ([Macdougal's] Pin-Cushion Cactus, *biznaga*). This plant is flat, often barely reaching above the ground and looking a bit like a squashed Barrel Cactus; it also has latex in the stems. Tarahumara mashed the fruits and added them to *atole* as food (Pennington 1963).

## *Sambucus nigra* (Caprifoliaceae) Mexican Elderberry, *tápiro, tahapidam*

**Other names:** ENGLISH: [blue-] desert elderberry ("elder" has been in English since at least

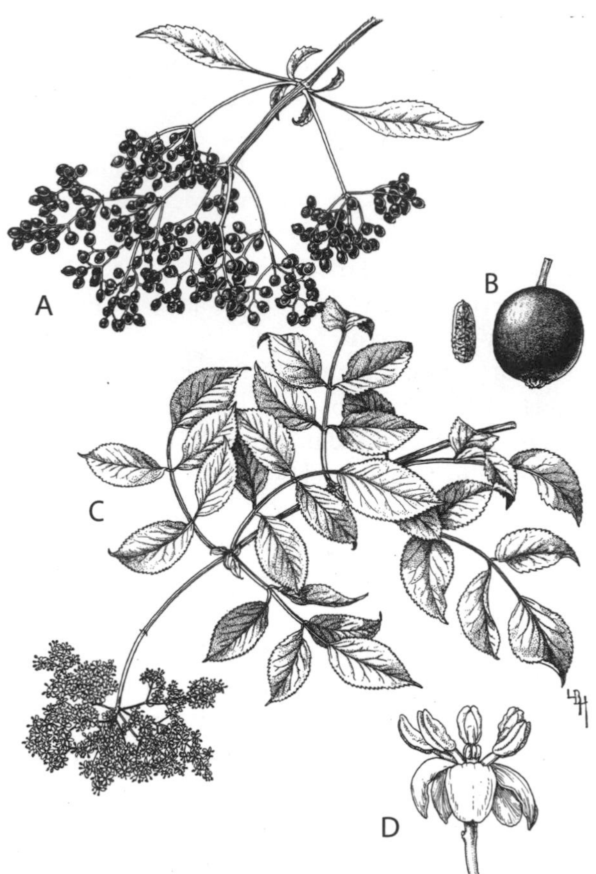


*Sambucus nigra* ssp. *cerulea*. A. Fruiting branch. B. Fruits and seed, enlarged. C. Flowering branch. D. Flower, enlarged. Artist: L. B. Hamilton (from original artwork).

CE 700, when it was spelled *ellaern*, *ellen* [Old English] or *ellern* [Anglo-Saxon]; related words in Middle Low German were spelled *elderne*, *alhorn*, or *elhorn*, *hyld* [Danish], and *hyll* [Swedish and Norwegian], and these are related to *Holle*, *Holunder*, *Holder* or *Elhorn* [German]; those names and the Danish *Hyldemör* or *Hyllefrao* allude to the Norse goddess *Freda*), Mexican elder; SPANISH: *capiro* (misspelled *tápiro*? q.v.), *capulín* [*silvestre*] ([wild] cherry, Mexico; see *Prunus* for etymology), *ocoquihui* (maybe from Náhuatl *ocotl*, torch, *noquia*, diarrhea), *saúco* [*azul*] <*sauco*> ([blue] elder, California, Chihuahua, Sonora south), *tápiro* (Arizona, Sonora); ATHAPASCAN: *'atsiniltł'ish 'ii't'ąą'* <*ˀacinlƛiš ˀilt'ąˀí*> (*atsiniltł'ish*, lightning, *ii't'ąą'*, with leaves [like zigzag], Navajo), *ch'ił bitsiin łizhin* <*č'il bicin łižin*> (*ch'ił*, plant, *bitsiin*, with stems, *łizhin*, black, Navajo), *ch'ilhazhé* <*suul*> (Western Apache), *jiłhazhí* (*ch'il*, plant, *hazhí*, biting, Navajo; see also *Celtis reticulata*), *tsizoł* (Chiricahua and Mescalero Apache); CHUMASH: *qayas* (Chumash); HOKAN: *baaduˀ* <*páru*> (Washo); MIXE-ZOQUE: *coyapa* (Zoque), *xiiksh* (Mixe); OTO-MANGUEAN: *bixhumí* (Zapotec), *nttzirza* (Otomí), *yutnucate* (Mixtec); TARASCAN: *cumdemba* <*cumdumba*, *cumtempa*, *condumbo*> (Purépecha); UTO-AZTECAN: *dahapdam* (Akimel O'odham), *hauk u'usi* <*hauk u'ushi*> (*hauk*, light [weight], *u'usi*, wood, Mountain Pima), *hubu'* <*hub·ú*> (Northern Paiute), *huvúhya* (Mono), *hungwat* <*hun-kwat*> (Cahuilla), *huvúi* (Western Paiute), *kuhupïl* <*kuhupɨ-l*> (Tübatulabal), *kunugívū* (Mono), *kunuguvɨ* (Kawaiisu; *kunuguvuˀivi*, the berry), *kunuki*(*ppüh*) (Panamint), *ku:ta* (Luiseño), *kúūt* (Cupeño), *kuuhuutɨ* (Serrano), *pa'gonogwĭp* [*pa'go-nogĭp*] (Shoshoni), *pɨgūbūxia*, *hūbū́xia*, *saíinoiyaˀa*, *sainṓwaiyuˀu* (Northern Paiute), *rotosí* (Tarahumara), *sau* (from Spanish *saúco*, Mountain Pima), *sauko* [*saokó*] (from Spanish, Guarijío), *tahapidam* (Hiá Cẹd O'odham, Tohono O'odham), *tóisavui* (Western Paiute), *xometl* <*azumiatl*, *azumiatl*, *xomét*> (Náhuatl, San Luis Potosí, Veracruz); YUKI: *kēwēmäm* <*kíwimám*, *kiwimöm*, *kiwi*> (Yuki); YUMAN: *kopáhl* (Kumiai), *tal tal* (Paipai), *xsa:wk* (Cocopa); (= *Sambucus mexicana*)

**Botanical description:** Small trees or large shrubs, relatively short-lived, trunks single or mostly multiple. Leaves partially to fully deciduous in the summer, new growth and leaves appearing with summer rains, opposite, odd-pinnate, 8.5–22 cm long or more, the leaflets 3–7, highly variable, 4.5–11.5 cm long, mostly lanceolate, the margins serrate-toothed. Inflorescences are dense, terminal, compound umbel-like cymes, 6–17 cm wide. Flowers are creamy white, 5–6.5 mm wide. Corollas are 5-lobed, ovary inferior. Fruits are rounded, berry-like drupes, ca. 5 mm wide.

**Habitat:** Canyons, washes, streamsides. 300–1,200 m (1,000–4,000 ft).

**Range:** Arizona (Navajo to Mohave, south to Cochise, Pima, and Santa Cruz counties) and Pacific North America to Montana; British Columbia, Alberta; south to Mexico (Baja California, Chihuahua, Durango, Nuevo León, Sinaloa, and Sonora).

**Seasonality:** Flowering March to June.

**Status:** Native. Arizona plants are ssp. *cerulea*.

**Ecological significance:** These shrubs or small trees are indicators of permanently moist or even wet sites. Many birds and other animals depend seasonally on the fruits. In the early 1900s, the trees sometimes were covered by birds eating the fruits (Wilbur-Cruce 1987).

**Human uses:** Fruits are eaten by Akimel O'odham, Apache, Cahuilla, Hiá Ceḍ O'odham, Kawaiisu, Kumiai, Luiseño, Paiute, Tohono O'odham, and Yavapai (Bean and Saubel 1972, Felger and Broyles 2007, Hodgson 2001, Kelly 1932, Moerman 1998, Nabhan et al. 1982, Rea 1997, Reagan 1929, Walker et al. 2004, Zigmond 1981). *Sambucus* is used as medicine by the Acjachemen, Akimel O'odham, Aztecs, Cahuilla, Chumash, Guarijío, Kumiai, and Paiute (Bean and Saubel 1972, Martínez 1969, Moerman 1998, Rea 1997, Timbrook 1990, Train et al. 1957, Walker et al. 2004, Yetman and Felger 2002). Teas from the flowers is considered either simply refreshing or medicinal by different groups (Gallagher 1976, Yetman and Felger 2002). The Navajo *jiłhazhí*, with cognates in Apache, refers to putting a piece of the branches of both *Celtis* and *Sambucus* in the mouth when thirsty and lacking water (Young and Morgan 1980). In spite of all these uses, people have been poisoned by the cyanides and have experienced severe diarrhea (Duke et al. 2002, Kay 1996, Lewis and Elvin-Lewis 2003).

Mountain Pima use the mashed leaves to treat headache (Pennington 1973). Onavas Pima make a lotion of the leaves to bathe women during childbirth (Pennington 1980). Navajo consider *Sambucus* protection from lightning (Wyman and Harris 1951).

Cahuilla use the berries as a black dye for baskets, and the stems to make an orange dye (Bean and Saubel 1972). Twigs are used to make whistles by Cahuilla, and in games by the Kumiai. The California Acjachemen (Juaneño Band of Mission people) call the plants the "tree of music" and make a "clapperstick" from the stems (Walker et al. 2004). The clapperstick is a percussion instrument used to keep time in the accompaniment of songs, and it is used by all tribes in California.

Farther east, *S. nigra* ssp. *canadensis* is used by many of the eastern tribes in most of the same ways as ssp. *cerulea* (Austin 2004). Europeans brought an ancient history of using Elderberry when they arrived in the New World. Some of the common names there allude to *Freda*, who is the Norse goddess of fertility and the most revered of all goddesses. She is called *Vrouw Arda*, *Vrouw Herta*, or *Vrouw Holle* in the Netherlands. This "Lady Elder" lives in a *Sambucus* and became the symbol of reincarnation. She and Elderberry represent the European "Tree of Life" (Austin 2004). The importance of the genus *Sambucus* in the Iberian Peninsula was discussed by Vallès et al. (2004).

**Derivation of the name:** *Sambucus* may be from Greek *sambuce*, an ancient flute-like musical instrument made from the readily removed tubes of bark or by reaming out the pith from young shoots. Alternatively, perhaps from *sambuca <sambuke, sambuc>* (Latin *sambúca*, Greek *sambúka*, cognate to Aramaic *sabbeka*), the name may connote "a triangular stringed instrument with a shrill tone." Could the word be related to Greek *sabakos*, rotten, because of the unpleasant smell of the foliage? *Nigra* means black; *cerulea* means blue; presumably, both names allude to the fruit color.

## ***Silene antirrhina*** **(Caryophyllaceae)** **Sleepy Catchfly, *silene***

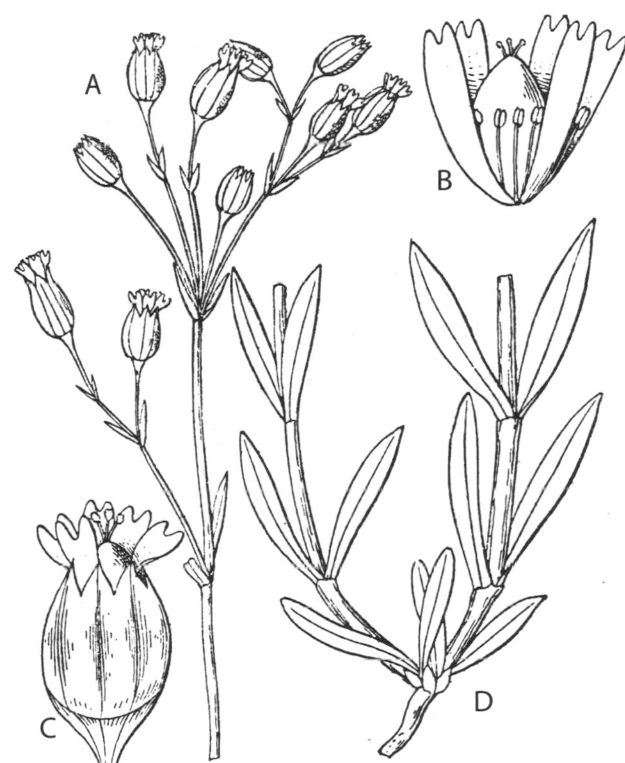


*Silene antirrhina*. A. Flowering tip of branch. B. Flower. C. Fruit. D. Basal portion of plant. Artist: F. Emil (Britton and Brown 1896–1898).

**Other names:** ENGLISH: campion (etymology obscure, compared with both "champion" and French *compagnon*, but neither certain), sleepy [silene] catchfly [silene] ("sleepy" is probably applied because the flowers close in the morning; "catchfly" is a name originally given to *Silene armeria*; in use by 1597 when John Gerarde wrote: "I have called it Catchflie, or Lime woort, The whole plant, as wel leaues as stalkes, and also the flowers, are couered ouer with a most thicke and clammie matter like vnto Birde lime"), gartner-pink (South), snapdragon catchfly (Massachusetts); SPANISH: *silene* (from Latin *sileni*, aged Satyr; cognates occur in French and Italian); UTO-AZTECAN: *oi'tcuyo* (*oi'tcu*, bird, Shoshoni)

**Botanical description:** Herbs, annual, stems erect or ascending, to 8 dm tall, branched, puberulent to glabrous, often with sticky bands below the nodes. Basal leaves oblanceolate to spatulate, those of the stem linear to lanceolate or oblanceolate, to 7 cm long, 15 mm wide. Inflorescences are numerous, slender-pedicelled. Flowering calyx is spindle-shaped (fusiform), in fruit becoming ovoid-campanulate, to 1 cm long, glabrous, with 10 straight nerves, the short triangular apical teeth often purplish. Petals are small, disappearing early, white to pink or purplish, often notched at apex (emarginate). Capsules are almost sessile within the enclosing calyx, ovoid, to 1 cm long.

**Habitat:** Moist banks and slopes near streams and washes. Up to 1,830 m (6,000 ft).

**Range:** Arizona (Coconino, Mohave, south to Cochise, Pima, and Santa Cruz counties), New Mexico, and Texas through most of temperate United States; Canada; Mexico (Sonora).

**Seasonality:** Flowering March to May.

**Status:** Native.

**Ecological significance:** Many of the species in *Silene* are pollinated by Lepidoptera. Those that flower in the day are visited by butterflies of various kinds, but sulphurs and whites (Pieridae) predominate because they "prefer" white flowers. The remaining species open their blossoms in the evening and are visited by moths, particularly Sphinx Moths. Although no observations have been made of *S. antirrhina* in the evening, the flowers open at dusk and close in the morning. Flowers are probably pollinated by moths. Thoreau wrote in 1852 on "the ordinarily curled-up petals scarcely noticeable at the end of the large oval calyx." He was noting that the flowers were closing, because he continued with comments from Asa Gray and Thomas Bigelow about the nocturnal flowers.

As Englishman John Gerarde noted in 1597, and as people had before him, the sticky bands of plant-hairs around the stem prevent crawling insects from having free access to the flowers. These bands are effectively "locks" that prevent the "wrong" (i.e., non-pollinating) insects from pilfering nectar. In many members of *Silene*, it is common to find ants and other insects stuck to those bands.

**Human uses:** None found for *S. antirrhina*, although Chamberlin (1911) gave the Shoshoni common name for it. In my experience, people have names only for plants they use or for those that are somehow culturally significant to them.

Hocking (1997) lists 12 species of *Silene* that have been used in various parts of the world. Only 2 of those (*S. menziesii*, *S. virginica*) are native to the Americas. Moerman (1998) includes 10 species that are used by indigenous peoples. Most of the 10 are native and are considered medicinal.

Among the native species in the Southwest that have uses are *S. acaulis*, *S. douglasii*, *S. drummondii*, *S. laciniata*, *S. noctiflora*, and *S. scouleri*. Shoshoni in Utah and Nevada use *S. acaulis* to treat children with colic (Chamberlin 1911). These same people use *S. douglasii* as an emetic, and the Navajo make a treatment of it for Coyote bites on humans and their animals (Vestal 1952). Keres rub crushed *S. laciniata* on ant bites (Moerman 1998), and Navajo treat burns and dog or Coyote bites with it (Vestal 1952); Navajo use *S. noctiflora* to relieve prairie dog bites (Vestal 1952). Navajo also use *S. drummondii* as a Life Medicine (Vestal 1952). Shoshoni take *S. scouleri* as an emetic (Chamberlin 1911). It would appear that the saponins in *Silene* support the use in traditional medicines (E. Eich, personal communication, July 2008).

**Derivation of the name:** *Silene* is based on a comparison with the sticky secretions of many species. There are two possible derivations: from Greek *sialon*, saliva, or from Greek *Seilenos*, Latin *Silenus*, the Asiatic woodland deity (satyr) who was the intoxicated foster father of Bacchus. *Seilenos* was described as being covered with foam, and he taught men to play reed instruments. The patronym is most likely, as aged satyrs were called *sileni*, and Linnaeus often took names from Greek mythology. *Antirrhina*, from Greek *anti*, against, opposed, *rhinos*, nose or snout; perhaps the name compares these herbs with *Antirrhinum*, the Snapdragon, whose flowers resemble a dragon.

**Miscellaneous:** Kartesz (1994) admits for the United States, Canada, and Greenland 76 species of *Silene*. That number is a small part of the ca. 700 that are known from the Northern Hemisphere (Mabberley 1997).

## ***Atriplex canescens*** **(Amaranthaceae)** **Wingscale, *chamiso cenizo, 'onk 'i:wagi***

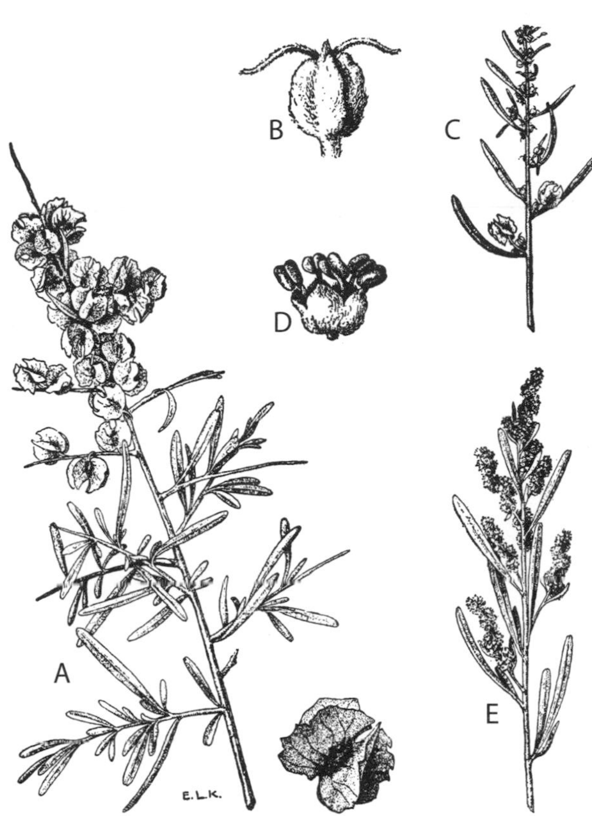


*Atriplex canescens*. A. Fruiting branch. B. Female flower. C. Female flowering branch. D. Male flower. E. Male flowering branch. Artist: E. L. Keplinger (Dayton 1937).

**Other names:** ENGLISH: buckwheat shrub ("buckwheat" or "beech-wheat" came from the Latin *fagus*, beech, and Greek *pyrum*, fruit or seed, comparing the introduced *Fagopyrum esculentum* from Asia with one that was native and familiar (*Fagus*); recorded by Wm. Turner's *Names of Herbs* in 1548; cognate with Dutch *boekweit* and German *Buchweize*), four-winged salt-bush ("salt-bush" alludes to the common occurrence in salty soils; in use for *Atriplex* by 1863), grease-wood (a name originally applied to prickly chenopodiaceous shrubs, of genera such as *Atriplex* and *Sarcobatus* by 1851; later extended to *Baccharis* and *Larrea*; "grease" is from Old French *graisse*, fat, in English by ca. 1340), orach (from French *arroche*; in use in English by ca. 1430), [salt, wafer]-sage (see *Artemisia*), shad-scale (winged fruits resemble fish [shad] scales), [wheel-] wing-scale (both "wheel" and "wing" refer to the shape of the bracts; "scale," as Old French *escale*, modern French *écale*, meant a husk, pod, or chip of stone by the twelfth century); SPANISH: *ceniso* <*cenizo*> (ashy one, Baja California, Chihuahua, Sonora), *chamiso* <*chamiza*> (see *Artemisia ludoviciana*, Baja California, Chihuahua, Sonora, New Mexico), *chamiso cenizo* [*blanco*] (ashy [white] chamiso, Mexico), *costilla de vaca* (cow's rib, Zacatecas), *saladillo* (little salty one, Baja California, Chihuahua); ATHAPASCAN: *díwózhiiłbeii* <*dóyʷóžiłbáˀí*, *tiwójiiłpáih*> (gray greasewood, Navajo); HOKAN: *hataj-isijc* (immature vulva, Seri), *hataj-ixp* (white vulva, Seri); KIOWA TANOAN: *ta'ᵃñaeŋ* (*ñaeŋ*, nest, Tewa); UTO-AZTECAN: *cïw'wïïbïl* (Tübatulabal), *dzi'cûp* (Shoshoni), *koksvul sha'i* (cocoon bush, Akimel O'odham), *murunavɨ* (Kawaiisu), *'onk 'i:vakĭ* (salty greens, Hiá Ceḍ), *'onk 'i:wagi* <*teu'ari*> (salty greens, Tohono O'odham), *sha'ashkadk iibadkam* (it has rough fruit, Akimel O'odham), *suwvi* <*cüovi*, *súovi*> (Hopi), *taˀibi* [*tónova*] (Northern Paiute); YUMAN: *ḍasilk* (Walapai), *mu'kwapt* (Paipai), *pˀay* <*pa'ai*> (Cocopa; for *A. lentiformis*); LANGUAGE ISOLATE: *ke'mwe* (*ke*, plant, *ma'we*, salt, Zuni)

**Botanical description:** Shrubs, usually over 4 dm (to 25 dm) tall, woody throughout, loosely to densely branched, the stems erect, stout, gray-scurfy. Leaves are 1–5 cm long, sessile or nearly so, linear-spatulate to narrowly oblong, cuneate at the base, usually obtuse at the apex, entire, thick, with a gray scurf, becoming glabrous. Male and female flowers are on separate plants. Staminate flowers in dense spikes of long terminal panicles. Bracts united to the summit, conspicuously 4-winged from the sides and back of the bracts, the whole bract 4–15 mm long. Seeds are brown, 1.5–2.5 mm broad.

**Habitat:** Gravelly and sandy soils of washes and other floodplains. To 1,980 m (6,500 ft).

**Range:** Western North America from South Dakota to southeastern Oregon and south; Mexico (Baja California, Chihuahua, Coahuila, San Luis Potosí, Sonora to Sinaloa, and Zacatecas).

**Seasonality:** Flowering June to August.

**Status:** Native.

**Ecological significance:** The genus *Atriplex* is composed of plants that are unusually salt tolerant: hence, their common name, "salt-bush." This genus has been promoted as a promising forage source (Dayton 1937). *Atriplex canescens* is as high in nutritive value as Alfalfa (*Medicago sativa*), and is eaten by sheep and cattle (Ayensu 1975). Pronghorn (*Antilocapra americana*), deer, rodents, and quail (Gambel's, *Callipepla gambelii*, and Masked Bobwhite, *Colinus virginianus ridgewayi*) are attracted and feed on its fruits (Martin et al. 1951).

*Atriplex* is one of the few known plants that is able to change from one sex to the other, particularly from female to male, in response to stress such as drought (Freeman et al. 1984).

Because these shrubs often grow in drainage areas where evaporation concentrates salts, this species is among the most succesful plants there. However, the shrubs are not confined to these conditions and grow on slopes in the lowland Buenos Aires National Wildlife Refuge (Austin 2003).

**Human uses:** Seri make an emetic tea from the leaves, and branches are used as roofing material (Felger and Moser 1985). The name *koksvul sha'i* (cocoon bush, Akimel O'odham) alludes to the cocoons that the Pascola festival dancers use on their legs (Rea 1997). When Akimel O'odham basket weavers cannot locate the *edam* (quail bush, *Atriplex lentiformis*), they substitute *A. canescens*.

Leaves are chewed for gastralgia and as an emetic by Navajo (Vestal 1952, Wyman and Harris 1951), as a physic by the Shoshoni (Train et al. 1957), and to treat diabetes in Mexico (Hocking 1997). Hopi, Isleta, Jemez, and Zuni have many medical and other uses (Dunmire and Tierney 1995, Moerman 1998, Stevenson 1915, Whiting 1939). Havasupai make a soapy lather of the leaves to wash the hair and treat rashes and itches (Moerman 1998). Apache use it similarly (Gallagher 1976).

Navajo use the shrub as an emetic in the Evil Way and Wind Way ceremonies; they also add leaves to roasting corn like salt, and seeds are parched and ground into meal for use alone or with other flour (Vestal 1952). *Atriplex* is a source of dyes (Elmore 1944, Mayes and Lacy 1989), including imparting a green hue to blue cornmeal used to make *piki* bread among the Hopi (Castetter 1935, Whiting 1939). In Sonora, Wingscales are used to expel intestinal parasites (López-E. and Hinojosa-G. 1988).

**Derivation of the name:** *Atriplex* (from Greek *a*, not, *triplex*, to nourish, because it was thought to rob the soil) was an ancient Latin name used by Pliny, CE 23–79, for the Orach, the kitchen vegetable, *A. hortensis*. *Canescens* means gray, hoary, in allusion to the surface pubescence.

## *Chenopodium berlandieri* (Amaranthaceae) Pit-Seed Goosefoot, *bledo, cual, iwagi, ṣu:'uwaḍ*

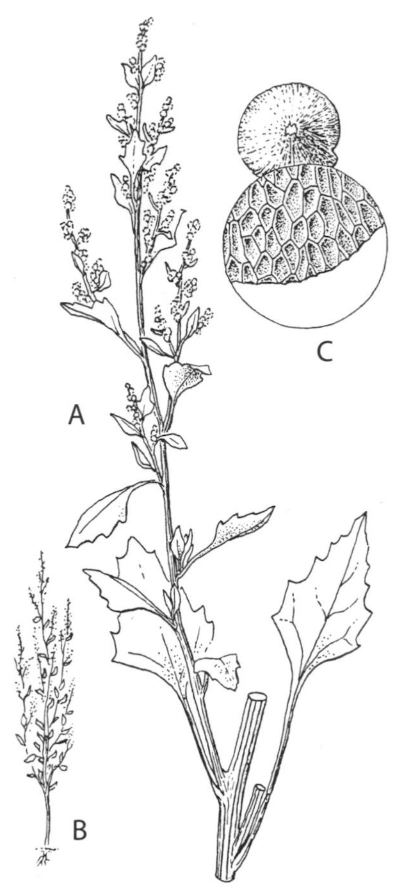


*Chenopodium berlandieri*. A. Flowering branch. B. Base of plant. C. Fruit. Artist: F. Emil (courtesy James Henrickson).

**Other names:** ENGLISH: pit-seed goosefoot (from Greek *chen*, a goose, and *pous*, foot, in allusion to the shape of the leaves; *Chenopodium* probably the herb called *chenopus* by Pliny, CE 23–79); FRENCH: *belle dame sauvage* (wild *Chenopodium*, Louisiana in 1758); SPANISH: *bledo* (akin to blite, in English by 1420), *chuale* <*chual*, *choale*, *guaute*> (from Náhuatl, *tzohualli*, Mexico), *quelite* [*cenizo*, *salado*] ([ashy, salty] greens, Mexico); ATHAPASCAN: *dích'íí' 'ik'eht'ąą'* <*dokǫ́z ˀikt'ąˀí*> (*'ik'eht'ąą'*, with leaves like, *dích''íí'*, bitter [*Atriplex*], Navajo), *it'ąą dit'ógé*, *it'ąą nch'ii'é* <*it'ą inkozee*> (*it'ąą*, spinach, *dit'ógé*, crumbling, *nch'ii'é*, bitter, Western Apache), *ita* (leaf, Chiricahua and Mescalero Apache), *łibá'ígí* <*łabáˀígí*> (*łibá*, gray, *ígí*, the one that is, Navajo), *łichíí'* <*łčíˀ*> (the one that is red, Navajo), *tł'ohdeeí* [*ts'oh, ts'yaa*] <*y'oh de'* [*c'o's, ci'yah*]> (*tł'ohdeeí*, grass whose seeds fall, *ts'oh*, big, *ts'yaa*, under trees [contraction of *tsin 'ayaa*], Navajo), *tł'oh łigai* <*tł'oh łigaii*> (*tl'oh*, grass, *łigai*, white, Navajo); CHUMASH: *we'lel* (Barbareño Chumash), *welel* (Ineseño Chumash); GULF: *choupichoul* (Natchez name given by Du Pratz in 1758); UTO-AZTECAN: *čuá* <*činaka, čuaka, chu-aka, chu-ya*> (Tarahumara), *cotasula* (Guarijío), *cuatztli* (Náhuatl), *cual* (Tohono O'odham), *kokoncher* <*kokeynchár*> (*kokon*, raven, *cher*, in, Mountain Pima), *sirwa, höhöla* (*höhöla*, rise up, tilt, lift, Hopi), *huaquilitl* (*quauhtli*, eagle, *quilitl*, edible greens, Náhuatl), *ho:ohal* (Tübatulabal), *huauzontle* <*huauzontl, huauzontli, huanzoncle, guausoncles, guauzoncles*> (*huatli*, goosefoot, *tzontli*, hair, Náhuatl), *i'ûpi* (Shoshoni; other species with different names), *iwagi* <*ihwagi*> (edible greens, Tohono O'odham), *kapa* (Yaqui), *ki'awet* <*kehawut, keit, kit*> (Cahuilla), *kobu* <*cobu*> (Nevome), *kovĭ* (Akimel O'odham), *koovi* (Kawaiisu), kö'yo (Northern Paiute), *michihuatli* (from *michin*, fish, Náhuatl), *'onk 'i:wagi* (salty edible greens, Tonoho O'odham), *uauhtli* (name of *Chenopodium* seed, Náhuatl), *síswa* (Hopi), *ṣu:'uwaḍ* <*šu'awat*> (Onavas Pima), *ṣu:'uwaḍ* (Tohono O'odham), *üapa* (Northern Paiute; for *C. nevadense*), *uha* (Onavas Pima), *uauhquilitl* (*uauhtli*, name of *Chenopodium* seed, *quilitl*, edible greens, Náhuatl), *waha* <*waja, wakk*> (*waka*, pulled out/extracted [as picked leaves or seeds], Mountain Pima), *wɨtā* <*wa·'ta'*> (Northern Paiute), *wiwida* <*guiguida*> (to plant goosefoot or to scatter salt on food, Mountain Pima); YUMAN: *hdhpši* (*quelite*, for greens, Cocopa), *kwa'thami* <*quoth ah me*> (Mohave); LANGUAGE ISOLATE: *kia'tsanna* (*'kia'we*, seeds, *tsan'na*, small, Zuni); (= *C. murale* of literature)

**Botanical description:** Winter-spring ephemerals, 0.5–1 m tall, coarse and erect, the stems often pink- or red-striped. Leaves 3–10 cm long, the blades mealy and pale on both surfaces, mostly lanceolate, ovate, or rhombic, the petioles well developed. Sepals keeled, mostly enclosing fruit. Seeds 1.2–1.4 mm wide, surfaces covered with a coarse but minute cellular pattern (honeycomb depressions), shiny black after removal of pericarp, the margins obtuse, without a distinct rim.

**Habitat:** Along washes, rocky slopes. To 1,980 m (6,500 ft).

**Range:** Arizona (Apache to Mohave, Pima, and Santa Cruz counties), California, north to Oregon, Washington, east to Florida, and Virginia; Mexico (Baja California, Chihuahua, Coahuila, Guanajuato, Nuevo León, Sonora, Tamaulipas, Tlaxcala, Veracruz, and Yucatán).

**Seasonality:** Flowering June to August.

**Status:** Native, sometimes cultivated. The wild form is var. *sinuatum*.

**Ecological significance:** This plant is the wild form of a species formerly cultivated by most of the indigenous people living in the southern United States (Austin 2004). The wild plants are most common near waterways.

**Human uses:** Seeds and/or leaves are used as food in the Southwest, at least among the Akimel and Tohono O'odham, Apache, Baja Californian tribes, Cahuilla, Chumash, Cocopa, Havasupai, Hopi, Isleta, Kawaiisu, Keres, Kumiai, Malecite, Mohave, Navajo, Paiute, Onavas Pima, Northern Tepehuan, Tarahumara, Yaqui, Yavapai, and Zuni (Austin 2004, Bean and Saubel 1972, Castetter 1935, Castetter and Bell 1942, 1951, Castetter and Opler 1936, Elmore 1944, Gallagher 1976, Moerman 1998, Pennington 1969, 1980, Rea 1997, Reagan 1929, Stevenson 1915, Thord-Gray 1955, Timbrook 1990, Vestal 1952, Whiting 1939, Wyman and Harris 1951, Zigmond 1981). Navajo also use the herbs, with others, in the Mountain Chant and as fly and mosquito repellent (Elmore 1944, Franciscan Fathers 1910).

The Hopewellian people brought *C. berlandieri* into cultivation before Maize was introduced into the Mississippi Valley between 170 BCE and CE 60 (Riley et al. 1994). By the time Europeans arrived in the New World, most eastern tribes had ceased growing Goosefoot (Austin 2004). *Chenopodium berlandieri* was cultivated longer in the southwestern United States, and it is still cultivated in Mexico. Bean and Saubel (1972) and Hodgson (2001) found that Cahuilla and Louiseño gathered *Chenopodium* seeds and ground them into flour. Akimel O'odham parched and made the seeds into flour for *pinole* or combined it with other meal before it was eaten (Rea 1997). Akimel O'odham stopped growing *kovĭ* for food in the 1870s when the drought years hit. Yavapais collect the fruit clusters in burden baskets, spread them on a flat surface, and beat them with a stick to separate the seeds. Winnowed seeds are then parched with coals in a basket, ground, boiled, and eaten (Moerman 1998). The same use pattern holds from California, Arizona, and New Mexico south into Mexico.

While Felger and Broyles (2007) list several of the O'odham names for *C. murale*, Rea (1997) had been unable to find a Piman name for that species. The Hiá Ceḍ O'odham that Felger and Broyles found calling *C. murale* by these names either transferred the old name for *C. berlandieri* to this introduced Old World weed or they include *C. murale* within their folk genus concept. *'Onk 'i:wagi* includes *Atriplex*, q.v.

**Derivation of the name:** *Chenopodium* is from Greek *chen*, a goose, and *pous*, foot, in allusion to the shape of the leaves. Probably this is the potherb called *chenopus* in Latin and in Greek by Pliny, CE 23–79. *Berlandieri* was named for its discoverer Jean Louis Berlandier, 1805–1851; *sinuatum* means that the leaf margins are wavy.

**Miscellaneous:** Probably the most frequent *Chenopodium* in Brown Canyon in the fall monsoon season is *C. neomexicana*. This species is characterized by the fishy odor of the herbage, among other traits. These plants sometimes form large patches fairly high in the canyons of the Baboquivari Mountains to about 1,219 m (4,000 ft). This herb grows in Arizona, New Mexico, and Texas. Given the abundance of this species and the confusion in the literature over the species utilized by indigenous people, I would be surprised if seeds were not eaten, particularly since the name *michihuatli* refers to the fishy smell.

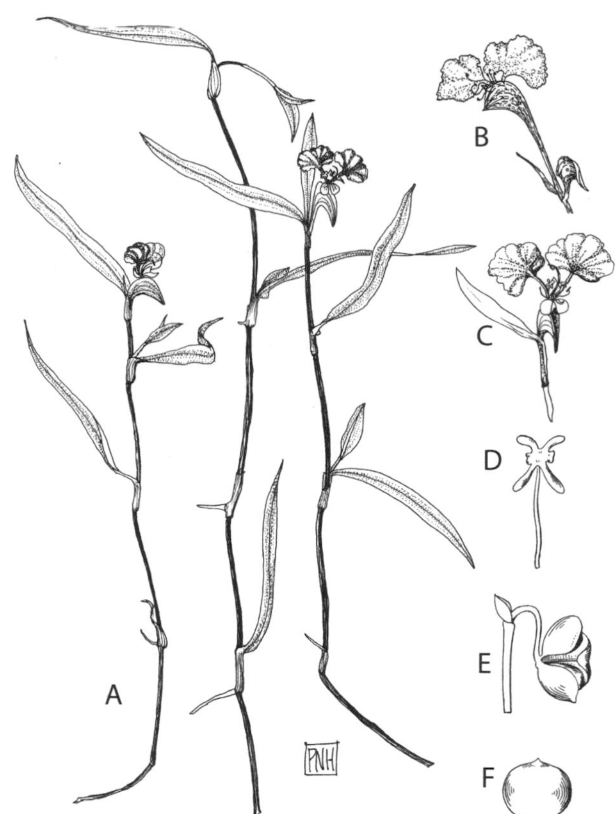


*Commelina erecta*. A. Flowering plants. B. Side view of flower and bract. C. Front view of flower. D. Anther, enlarged. E. Fruit, enlarged. F. Seed, enlarged. Artist: P. N. Honychurch, A–C (courtesy P. N. Honychurch-Billingham); A. Hollick, D–F (Britton and Brown 1896–1898).

## ***Commelina erecta* (Commelinaceae)**
## **Day Flower, *hierba de pollo***

**Other names:** ENGLISH: bluebird (a creole name, Nicaragua), [white-mouth] day flower [whitemouth dayflower] (alluding to the single-day duration of each flower), dew-flower, little bamboo (Belize), widow's tears; SPANISH: *atlic* (drinkable, Náhuatl, San Luis Potosí), *canutillo* (widespread Spanish name for the genus, maybe "little cane," from *caña*), *cuna de niño* (baby's cradle, Sonora), *espuelitas* (little spurs, Coahuila), *hierba de pollo* (chicken herb, Texas, Sonora), *perrito* (little dog, Mountain Pima), *sinvergüenza* (without shame, Sonora), *yerba de la borrego* (sheep herb, Northern Tepehuan); ATHAPASCAN: *áłtsíní iilt'áá'í* <*k.lciní'j.lt':j..'i*> (mariposa-like [*Calichortus*] leaf, Navajo); CARIB: *tamakusi* (Carib, Surinam); MAYAN: *pah-tsá* (*pah*, something bitter, *tsá*, to mix in, Maya), *utek'* (Huastec, San Luis Potosí), *x-habul-ha* (*ha*, water, *bul*, submerge, Maya), *ya'ax-ha-xiu* (*ya'ax*, green, *ha*, water, *xiw*, herb, Maya); MUSKOGEAN: *okí ahissí* <*oybá ahissí*> (water [*oybá*, literally "rain"] medicine, Koasati, Louisiana); UTO-AZTECAN: *kasalá* (Guarijío), *matlalitztic* <*mataliste, mataliz, matalís*> (from *matalin*, dark green, *itztic*, cold, Náhuatl, Tabasco, El Salvador), *osi* (Mayo, Sonora)

**Botanical description:** Perennial herbs, overwintering by tuberous roots, the stems to 3 m long (usually shorter in Arizona), at first erect, later decumbent. Leaves are linear to ovate-lanceolate, to 15 cm long, 14–35 mm wide, rough-hairy above, glabrous below. Flower clusters in specialized leaf (spathe) which is folded around the buds and fruits, 13–35 mm long, 8–23 mm high, glabrous to heavily pubescent. Corollas have 2 large blue petals and 1 small white one, 24–40 mm wide. Capsules are inconspicuous.

**Habitat:** Canyons, washes, below trees. 1,100–1,500 m (3,500–5,000 ft); going down to sea level outside Arizona.

**Range:** Arizona (Cochise, Graham, Pima, and Santa Cruz counties) to southeastern Colorado, east to Florida, New York, and Wisconsin; Mexico (Chihuahua and Sonora).

**Seasonality:** Flowering July to September.

**Status:** Native.

**Ecological significance:** Flowers are visited by bees, probably for both nectar and pollen. Primary visitors to flowers are flies (Syrphidae) and bees, including native *Agapostemon*, *Bombus*, and alien *Apis* (Faden 1992). One of the unusual aspects of the flowers is the hairs associated with the stamens. These hairs attract visitors to the flowers—either toward or away from the pollen. The hairs also affect how the insects move once within the flower (Faden 1992).

These are pioneer plants of unstable substrates, but they may persist for long periods so long as there is a period of ample sunlight. However, *Commelina* thrives in moist sites, and the sunlight and temperature are not always coincident. Because these herbs have ample water in their stems and other parts, they tolerate drought surprisingly well.

Seeds of Day Flower are important sources of protein in Bobwhite Quail diets (Peoples et al. 1994). These seeds also provide up to 93% of the nonprotein nitrogen in quail diets.

**Human uses:** The Mayan name *pah-tsá* perhaps is used because the *Commelina* is mixed with maize dough to aid fermentation in the preparation of *pozol* [*posole*] (Austin 2004).

Aztecs used several species to relieve pain, particularly during childbirth, and to treat bloody diarrhea (Martínez 1969). In Mexico, people from the Mountain Pima (Chihuahua) to the Huastecs (San Luis Potosí) use the water trapped in the inflorescence bracts for eye problems (Alcorn 1984, Pennington 1973); it is similarly used in Bolivia (von Reis-Altschul 1973) and Belize (Balick et al. 2000). Northern Tepehuan put seeds under the eyelids to reduce swelling (Pennington 1969). Mountain Pima also mash up the entire plant to treat injuries (Pennington 1973).

Widespread use of this genus and other members of the family (see also *Tradescantia*) is somewhat surprising because of the calcium oxylate crystals, plus other unidentified irritating chemicals, contained in the tissues (Nellis 1997). Some people are more sensitive to those compounds than others. If absentmindedly nibbled, the leaves may cause severe burning and irritation in the mucous membranes of the mouth and throat.

**Derivation of the name:** *Commelina* commemorates three Dutch botanists: Jan Commelin or Commelijn, 1629–1692; his nephew Kaspar, 1667–1732; and a third relative who died when young. The two blue petals reminded Linnaeus of Jan and Kaspar, and the small white one of their relative. *Erecta* means that the stems are upright.

**Miscellaneous:** *Commelina* contains ca. 170 species (Puente and Faden 2001). Although the genus is widespread, most of the species are tropical. There are only two species that reach Arizona, *C. dianthifolia* and *C. erecta*. The former species ranges much farther north in the state, reaching two-thirds of the way to the northern border, while *C. erecta* is confined to the southeastern counties.

## *Tradescantia occidentalis* (Commelinaceae) Prairie Spider-Wort, *hierba de pollo*

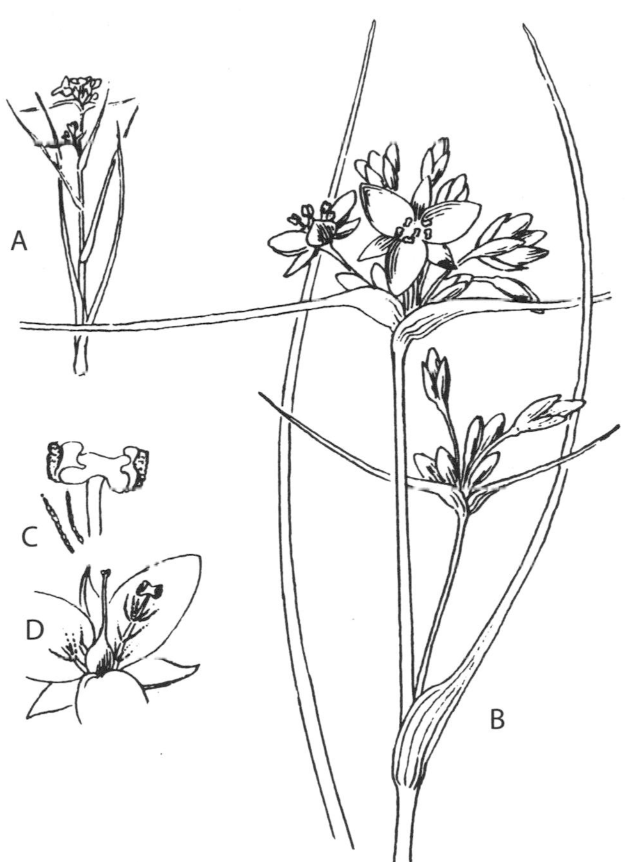


*Tradescantia occidentalis* A. Flowering branch. B. Flower clusters. C. Anther detail. D. Flower detail. Artist: R. D. Ivey (Ivey 2003).

**Other names:** ENGLISH: [western] spider-wort ("spider-plant," was used by 1629 by John Parkinson, see below for origin), wandering jew (compares the sprawling plants with a legendary person who was condemned to wander over the earth without rest until the Day of Judgment; according to a popular belief first mentioned in the thirteenth century, and widely current at least until the sixteenth century, this person insulted Jesus on his way to the Cross; often referred to as the proverbial type of restless and profitless traveling from place to place); SPANISH: *hierba de pollo* (chicken herb, Mexico); ATHAPASCAN: *áłtsíní iilt'áá'í <kilcíni ˀilt'ąí, k.lciní'j.lt':j..'i>* (mariposa-like [*Calichortus*] leaf, Navajo), *áłtsíní tsoh <ki-lcini' coh>* (*tsoh*, big, *áłtsíní*, child, Navajo); KERES: *bashu* (Acoma, Laguna); UTO-AZTECAN: *pasómi* (Hopi)

**Botanical description:** Herbs, perennial, from stout, fleshy roots, the stems erect to ascending, straight, branched, glabrous, glaucous and somewhat succulent. Leaves are linear-lanceolate, long-acuminate, to 5 dm long and 2 cm wide, stiff and widely spreading or sickle-shaped, the sheath inflated, to 3 cm long and 35 mm wide. Cymes are umbellate, few- to many-flowered, terminal, solitary, the enclosing bracts leaf-like, glabrous, glaucous, sharply reflexed, to 21 cm long and 2 cm wide. Sepals are elliptic, acute to acuminate, 4–10 mm long, glaucous or with some rose or purple, glandular-puberulent. Petals 3, broadly ovate, 7–16 mm long, blue to rose and magenta. Filaments are pilose with long hairs. Capsules are oblong-trigonal, 4–7 mm long. Seeds are roughly compressed-oblong, 2–4 mm long.

**Habitat:** Canyons, streamsides, washes. 760–2,130 m (2,500–7,000 ft).

**Range:** Arizona (Apache, Navajo, south to Coconino, Graham, and Pima counties), Colorado, New Mexico, Texas, through Great Plains east to Arkansas, and Wisconsin.

**Seasonality:** Flowering April to September.

**Status:** Native.

**Ecological significance:** This herb is a temperate eastern species that reaches into the Southwest in mountain canyons and other wet, cool spots. In the Baboquivari Mountains, the herbs are always close to temporary or permanent streams.

There are long staminal hairs that look much like spider legs. These hairs are eaten by pollinating insects (Mabberley 1997).

**Human uses:** Acoma, Hopi, Keres, Laguna, and Navajo eat its tender shoots raw or cooked (Bailey 1940, Castetter 1935, Hough 1898, Moerman 1998, Hodgson 2001); the Hopi until the 1930s (Whiting 1939). Raw consumption is remarkable because the members of the family are loaded with needle-like oxalic acid crystals that cause extreme irritation of mucous membranes in at least some individuals.

Navajo use *Tradescantia* to treat internal injuries and in medicine for the Night Way (Mayes and Lacy 1989); also used as an aphrodisiac (Moerman 1998). Northern Tepehuan put seeds under the eyelids to reduce swelling (Pennington 1969).

There are several versions of how *Tradescantia* got the name "spider-wort" (Coffey 1993, Mabberley 1997). In all, "wort" is an old English word for "plant." One version is that the name compares the stamens with spider legs. Another is that the sap from broken stems forms filaments like a spider's web. Some think that "spiderwort" was chosen because the angular leaf arrangement suggested a squatting spider. Finally, the story is told that leaves were formerly infused in wine as a treatment for spider bites.

**Derivation of the name:** *Tradescantia* was named for John Tradescant, ca. 1570/1575–1638, gardener to the king Charles I of England. He visited Archangel, Russia, in 1618 and sailed on Sir Robert Maxwell's 1620 expedition against the Barbary pirates. Finally, he settled in Lambeth, established a garden and museum, and became gardener to the king. His son John Tradescant the Younger (1608–1662) was collecting plants in North America in 1638 when his father died. The son then returned to succeed his father as royal gardener. *Occidentalis* means western.

**Miscellaneous:** Robert Brown first observed protoplasmic streaming in 1828 in *Tradescantia.* He noted that phenomenon in the staminal hairs where it is easily observed under a low-power microscope. There are now 70 American species in the genus because it has been combined with *Rhoeo*, *Setcreasea*, and *Zebrina* (Mabberley 1997).

## *Cuscuta erosa* (Convolvulaceae) Dodder, *yerba sin raiz, wegĭ washai*

**Other names:** ENGLISH: dodder (from *dother* or *dither*, trembling, shudder, or shiver, from 1658, now dialectic; applied to *Cuscuta* since about 1265; originally meaning yellow, as in the yolk of an egg); SPANISH: *hongo de juve* (Jove's fungus, Mountain Pima), *yerba sin raiz* (herb without roots, New Mexico); ATHAPASCAN: *tálkáá' 'ąą dilk'ooł <talkáˀ dahíkal>* (*tálkáá'*, on the surface of water, *'ąą dilk'ooł*, spreads, Navajo), *tsiighájíłchi* (based on *tsiighá*, hair, Navajo); UTO-AZTECAN: *čigɨpiwanavɨ* [*čigɨpižiya*

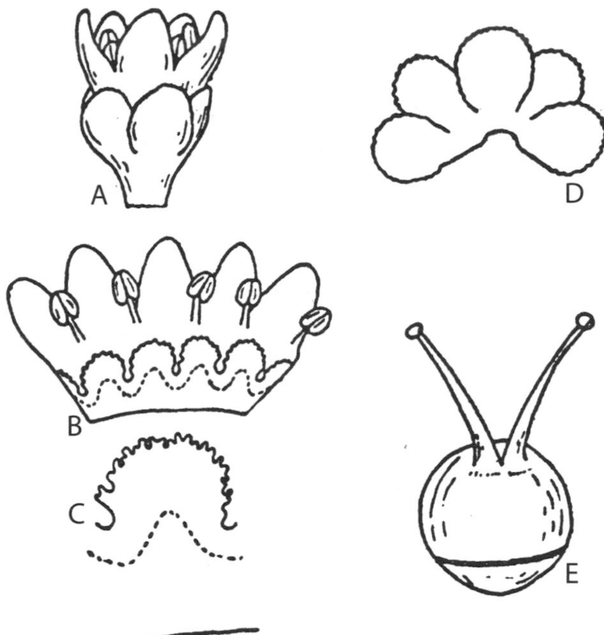


*Cuscuta erosa.* A. Flower. B. Opened corolla. C. Enlarged scale. D. Opened calyx. E. Fruit. Artist: T. G. Yuncker (Yuncker 1921).

*waneena*] (lizard's net, Kawaiisu; for *C. californica*), *vamaḍgiikoa* <*vammat geekwa*> (snake headdress or crown, Akimel O'odham), *vamaḍ givuḍ* (snake belt, Akimel O'odham), *vamaḍ vijin* (snake spider's web, Akimel O'odham), *vepegi vasai* (reddish grass, Hiá Ceḍ O'odham; for genus), *viva viikorgara* (*wiw*, tobacco, *viikorgara*, perhaps including *vi'ikon*, erode, left over, residual, Mountain Pima; for *C. campestris*), *wamaḍ gi:ko* (*wamaḍ*, snake, *gi:ko*, crown, Tohono O'odham), *wegĭ washai* <*wegei* or *wepegĭ waṣai*> (reddish grass, Tohono O'odham), *yurihuiri* (lit. *yori* = white, but probably from *iuco*, rain, *huiri*, twiner, Cahita)

**Botanical description:** Stems white. Inflorescences dense in cymose clusters, closely twisted about the host. Flowers are white, drying reddish brown, 3–4 mm long. Calices are campanulate, shorter than or equaling the corolla tube, the lobes orbicular, obtuse, denticulate, cupped, overlapping, membranous at the edges, fleshier in the median portions, sometimes with a dorsal ridge or projection. Corollas are campanulate, lobes ovate-oblong, obtuse, some with a small acumen projecting from the end of the thickened, vein-like median dorsal ridge, lobes upright or spreading, about as long as or slightly shorter than corolla tube. Fruits are circumscissile, globose, carrying the withered corolla, usually 1-seeded. Seeds are 1.5 mm long, globose, ovate, with a short, linear line or dot on the hilum.

**Habitat:** Mountain canyons, washes, below trees. Up to about 1,100 m (3,500 ft).

**Range:** Arizona (Pima County); Mexico (northern Sonora).

**Seasonality:** Flowering (July) September to October.

**Status:** Native.

**Ecological significance:** This parasite has been found on *Abutilon*, *Amaranthus*, *Ambrosia*, *Anisacanthus*, *Boerhavia*, *Gomphrena*, *Ipomoea*, *Justicia*, and *Kallstroemia*. As with all members of *Cuscuta*, the species begins from seeds in the ground that send up shoots that "feel" around for a host. When they locate a support, the stems produce haustoria (suckers) that tap into the vascular system of the chlorophyllous plant and feed the dodder. Because dodders do not have chlorophyll, they either find a host or die. Still, because these parasites usually grow near so many other plants, finding a host is not often a problem.

Although *Cuscuta* is virtually unstudied, the suspicion is that most species are self-fertilizing (autogamous). Seeds may simply fall below *Cuscuta*, as they certainly do in some species, to produce extremely local populations. Perhaps some seeds are spread during high water or even by birds, but no one is sure. The only species in the genus that have become widespread are those that have been accidentally carried by humans as contaminants in seeds of other plants.

**Human uses:** Akimel O'odham do not distinguish species but use folk generic terms for *Cuscuta* (Rea 1997). Historically, Akimel O'odham considered them poisonous, and they believed that, "if a snake sees them take the plant [on which it grows], the snake will get after them" (Curtin 1949). Now, Akimel O'odham regard *Cuscuta* more as an agricultural pest (Rea 1997).

Navajo include *Cuscuta* in a "form genus" (i.e., folk genus) that includes *Berula erecta*, *Ranunculus cymbalaria*, and *Nasturtium alpinum* (Wyman and Harris 1941). While these two authors included *C. umbellata* in their study of medical plants, no use was listed. Vestal (1952) noted that another person used the "spreads over water" name given by Wyman and Harris (1941) for *C. umbellata*, but gave it for *C. curta*. Vestal (1952) said that *C. curta* was a Plume Way emetic.

When I lived in Belém, Brazil, I was surprised to find people encouraging *Cuscuta* on their Croton (*Codiaeum*) hedges in their front yard gardens. When we asked them about the parasites, people told us they called them *cabelos de Jesus* (Jesus' hair). They considered them not only ornamental, but also medicinal. The name seemed to be associated with the Doctrine of Signatures. However, when I looked into the literature, I discovered that an older name was *cabellos de Venus* (Venus's hair). Because both Venus and Jesus are associated with love, albeit different kinds of love, both etymologies are equally appropriate.

While people derived from European temperate regions uniformly regard *Cuscuta* as a pest, people from other cultures do not. Among the indigenous people of North America, for example, there are records of Cahuilla, Cherokee, Kawaiisu, Kumiai, Navajo, Paiute, and Pawnee using *Cuscuta* for food, medicine, cleaning "pads," and dyes (Austin 2004, Bean and Saubel 1972, Castetter 1935, Moerman 1998, Train et al. 1957, Zigmond 1981).

**Derivation of the name:** *Cuscuta* is from Arabic *kusuta*, *kshuta*, *keskhut*, or *kusu*, "a tangled wisp of hair"; akin to Aramaic *kesatha* (Austin 1979). *Erosa* means "love" in Greek, "gnawed away" in Latin, a reference to the sepal margins.

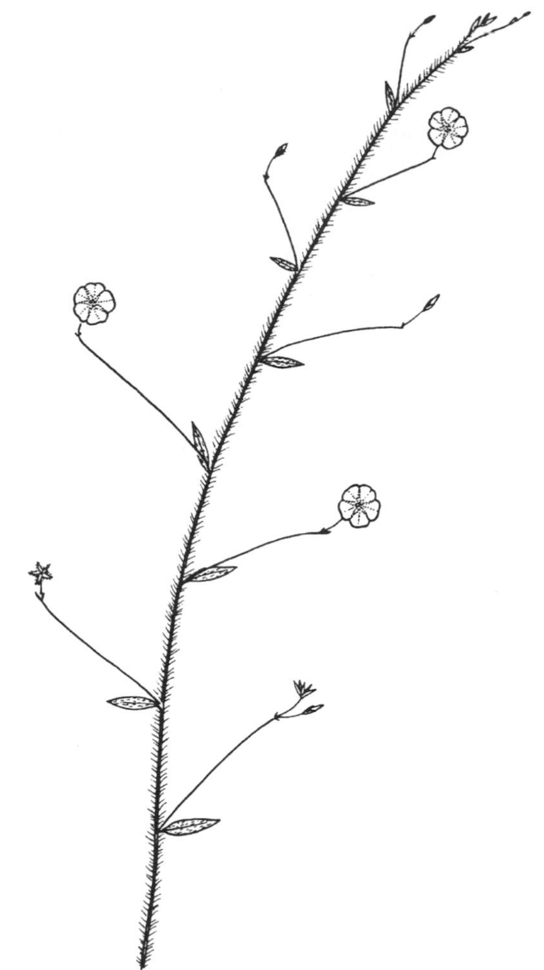

*Evolvulus alsinoides*. Flowering branch. Artist: B. Manara (Steyermark et al. 1998).

## *Evolvulus alsinoides* (Convolvulaceae) Blue-Eyes, *oreja de ratón*

**Other names:** ENGLISH: blue-eyes (from Spanish), [slender] dwarf morning-glory (Arizona, New Mexico); SPANISH: *cenicito* (little ashy one, El Salvador), *ojitos azules* (little blue-eyes, Yucatán), *ojo de víbora* (snake's eye, Mexico and adjacent Texas), *oreja de ratón* (mouse's ear, Sonora, El Salvador), *pata de paloma* [*pate paloma*] (dove's foot, Honduras), *quiebra-cajete* (box-breaker, Guatemala); MAYAN: *sian-xiw* <*sia-siu*, *xia-xiu*, *xiatiu*> (*sian*, enchantment, *xiw*, herb, Maya), *tsoots ts'ul* (Spaniard's hair, Maya), *x-havay* <*x-haway*> (means leprosy and other contagious skin diseases, Maya)

**Botanical description:** Herbs, the stems prostrate or ascending, 6–50 cm long, loosely appressed pilose and with some hairs spreading. Leaves ovate, oblong or elliptic to lanceolate, 8–22 mm long, 3.5–11 mm wide, the apex obtuse and mucronulate, the base acute to rounded, sparsely to densely pilose on both surfaces, with strongly and loosely appressed, soft, short, grayish trichomes. Inflorescences are 1 or 2 flowers on filiform peduncles, shorter than or longer than leaves. Flowers are on pedicels 2–4 mm long, short pilose; the bracteoles linear-subulate. Sepals are lanceolate, 2–2.5 mm long, acuminate, short pilose. Corollas are pale blue or white, rotate, (5–)7–10 mm wide. Fruits are globose, 4-valved, glabrous. Seeds are 1–4, ovoid, tan to brown, glabrous.

**Habitat:** Rocky sites, canyons, grasslands. 750–1,500 m (2,500–5,000 ft).

**Range:** Arizona (Cochise, Pima, Pinal, and Santa Cruz counties), New Mexico, Texas, and east to Georgia and Florida; Mexico (Baja California, Chihuahua, Coahuila, Sonora, and south to Chiapas; Central and South America.

**Seasonality:** Flowering April to September.

**Status:** Native.

**Ecological significance:** This plant is primarily a grassland species through most of its range in the Americas. While the species is more common in the foothills and lowlands in the Altar Valley grasslands, it does grow in Baboquivari, Brown, Sabino, and Thomas Canyons. Besides being in the Baboquivari Mountains, the species grows in foothills and mountains elsewhere in Pima County. Known sites for the species include the Ajo Mountains, Baboquivari Mountains, Comobabi Mountains, Coyote Mountains, Rincon Mountains, San Luís Mountains, Santa Catalina Mountains, Santa Rita Mountains, Sierrita Mountains, Tucson Mountains, and Waterman Mountains.

*Evolvulus alsinoides* is spread around the world, mostly in the tropics. The last revision of the group (Ooststroom 1934) recognized 15 varieties. That is all the more interesting because *E. alsinoides* is almost certainly native to the Americas, along with all its close relatives. Presumably, the species was carried early by the Spanish explorers as a medicine, and evolved in each locality much as the House Sparrow (*Passer domesticus*) supposedly has in its new homes.

**Human uses:** In Yucatán, Blue-Eyes are known as enchantment herbs (*sian-xiw*). This name implies that the herbs have an ancient and religious significance. The Maya also call them *x-havay* (leprosy and other contagious skin diseases) and *tsoots ts'ul* (Spaniard's hair). The fact that other countries that are occupied by descendants of the Maya have names for them further suggests a long history. In El Salvador, *Evolvulus* is *cenicito* (little ashy one) or *oreja de ratón* (mouse's ear). In Guatemala, *E. alsinoides* is *quiebra-cajete* (box-breaker), an allusion to medical treatment of bowel problems. In Honduras, people say *pata de paloma* [*pate paloma*] (dove's foot). Farther north in Mexico and adjacent Texas, plants are *ojo de víbora* (snake's eye), so maybe people in northern Mexico used them like the Maya.

At least the Tarahumara used the seeds as beads (Pennington 1963). Because the seeds are tiny (1.5–2 mm), the beadwork must have been extremely fine.

**Derivation of the name:** *Evolvulus* is from Latin words meaning without twining. This name was used by Linnaeus to distinguish the genus from the similar *Convolvulus* (with twining). *Alsinoides* means resembling *Alsine*, a genus of Caryophyllaceae.

**Miscellaneous:** *Evolvulus* is a genus of about 100 species. Most of the species are South American, and many of those extend into Central and North America. There are a few that are restricted to North America, with *E. nuttallianus* being mostly in the United States. Lower in the canyon, mostly below the Harm House, the common species is *E. arizonicus*. This plant has larger corollas than *E. alsinoides*; they are mostly 12–22 mm wide.

## *Ipomoea cristulata* (Convolvulaceae) Scarlet Creeper, *bi:bhiag*

*Ipomoea cristulata*. A. Flowering branch. B. Fruit. C. Seed. Artist: L. B. Hamilton (Parker 1958).

**Other names:** ENGLISH: scarlet creeper [morning-glory] ("creeper," written in 1626 by Francis Bacon for plants that crawl on the ground or climb a support; "morning glory," written by Frederick Pursh in 1814 for members of this family), Transpecos morning glory (New Mexico); SPANISH: *heguerilla* (a name usually given to *Ricinus communis*, Castor Bean; perhaps because seeds of both are laxative?), *redadera* (twiner, Mountain Pima); ATHAPASCAN: *tł'é' godigáhá* (*tł'é'*, evening, *godigáhá*, that is going to happen, Western Apache); UTO-AZTECAN: *bi:bhiag* (Tohono O'odham), *kusáˀrupų* (Ute), *situlyi* <shiitulyi> (maybe *sitol*, syrup, alluding to nectar or *sikori*, round, meaning the flower or fruit, *tulyi*, twiner, Mountain Pima); (= *Ipomoea coccinea* misapplied)

**Botanical description:** Annual herbs, the stems twining, glabrous or pilose on the nodes. Leaves with blades ovate, 1.5–10 cm long, 1–7 cm wide, typically the lower are entire and the upper 3–5-parted or all palmately parted or lobed, the margins irregularly dentate, the base cordate to ± truncate. Inflorescences are cymose or rarely solitary. Flowers are 3–7, on peduncles 3–6(–25) cm long. Sepals are unequal, the outer oblong, 3–3.5 mm long, with a ± terminal arista 3–5 mm long, glabrous, the inner oblong, 4–5.5 mm long, with a ± terminal arista 2.5–3.5 mm long. Corollas are salverform, 1.8–2.6 cm long, red or red-orange, glabrous, the limb 1–1.5 cm wide. Fruits are ± globose, 7–8 mm wide, with an apiculum 2 mm long. Seeds are 1–4, 3.5–5 mm long, ovoid, black to dark brown, finely tomentose.

**Habitat:** Canyons, washes; grasslands, chaparral, Madrean oak woodlands, Ponderosa pine zones. 730–2,800 m (2,400–9,190 ft).

**Range:** Arizona (southern Mohave to Apache, Yavapai, south to Cochise, Gila, Graham, Greenlee, Maricopa, Pima, Pinal, Santa Cruz, and Yuma counties), New Mexico, and Texas; Mexico (Baja California Sur, Chihuahua, Coahuila, Nuevo León, Sonora, Tamaulipas south to Distrito Federal, Edo. México, and Jalisco).

**Seasonality:** Flowering May to November

**Status:** Native.

**Ecological significance:** Flowers are pollinated by butterflies and hummingbirds. Although hummingbirds visit the flowers, the vines are not their favorite plants. These creepers are most frequent in canyons and washes where the creepers sometimes cover large expanses of other plants. When *I. cristulata* is in flower, the red blossoms are eye-catching. Given the frequency of occurrence in these areas, the seeds may be at least partly dispersed by water. However, more likely, seeds are eaten and spread by doves and other birds and simply have enough water to flourish in the low places.

**Human uses:** Grown for ornament. Although *I. cristulata* is not as showy as the more popular *I. quamoclit*, it produces similar flowers. Indeed, both of the species are in the same taxonomic group (*Ipomoea* series *Mina*) that has most of its species adapted for bird and butterfly pollination. Whereas, *I. quamoclit* was probably brought into cultivation by the Aztecs, *I. cristulata* is a temperate species that evolved in the wild from tropical ancestors.

**Derivation of the name:** *Ipomoea* is based on Greek *ips, ipos*, worm or Bindweed [*Convolvulus*], *homios, hoimios*, resembling, referring to the twining habit. *Cristulata* means crested, a notation of the ornamentation of the sepals.

**Miscellaneous:** Most of the literature on *Ipomoea* in the United States uses the incorrect names *I. coccinea* or *I. hederifolia*. Those are two distinct species with ranges not or barely overlapping that of *I. cristulata*. *Ipomoea coccinea* is a species of the southeastern United States, but it extends west to eastern Kansas, Oklahoma, and Texas. *Ipomoea hederifolia* is a tropical species naturalized in Florida and nearby states, and sporadically into west-central Texas. *Ipomoea cristulata* grows in the Chihuahuan Desert and extends north into the western Great Plains. All three have red, salver-shaped flowers, but sepals are different. *Ipomoea hederifolia* has sepals that are 4–4.5 mm long, while those of the other two are 6–8 mm long, rarely only 5. The easiest way to distinguish between these two species is that the leaf blades are 3–7-lobed or parted to entire on the same plant in *I. cristulata*. Blades are entire or simply angle-toothed in *I. coccinea*.

## *Ipomoea hederacea* (Convolvulaceae) Woolly Morning Glory, *trompillo, bi:bhiag*

**Other names:** ENGLISH: [ivy-leaf] woolly morning glory; SPANISH: *enredadera de campanilla* (bell twiner, Mexico), *flor de verano* (summer flower, Mexico), *manto* [*de la Virgen, mexicano*] ([Virgin's, Mexican] mantle, Mexico), *redadera* (twiner, Mountain Pima), *trompillo* [*morado*] ([purple] little top, Arizona, New Mexico, Sinaloa, Sonora; diminutive of *trompo*, top); ATHAPASCAN: *tł'é'godigáhá* (see *I. cristulata*, Western Apache); UTO-AZTECAN: *bi:bhiag* (Hiá Ceḍ O'odham and Tohono O'odham), *kusá'rupų* (Ute)

**Botanical description:** Annual herbs, the stems twining, densely to sparsely pubescent throughout. Leaves have blades ovate to ± orbicular, 5–12 cm wide and long, entire to 3–5-lobed, basally cordate, the lobes apically acute to acuminate, pubescent. Inflorescences are cymose. Flowers are 1–3(–6), on peduncles 5–10 cm long. Sepals ± equal, 12–24 mm long, 4–5 mm wide, herbaceous, lanceolate, abruptly narrowed from the ovate base into a narrow linear-lanceolate apex, usually curved, at least in fruit, apex sometimes strongly curved, densely long-hirsute

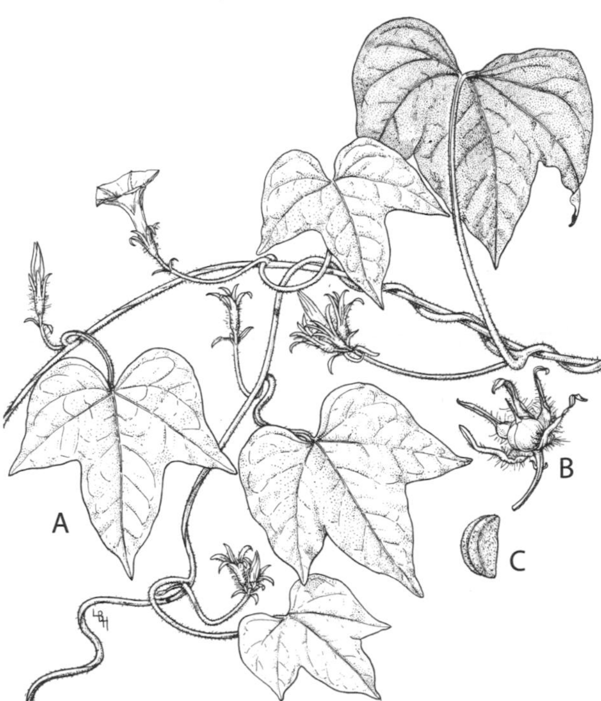


*Ipomoea hederacea*. A. Portion of plant with flowers. B. Fruits surrounded by calyx lobes. C. Seed. Artist: L. B. Hamilton (Parker 1958).

at least on the basal 1/3. Corollas are funnelform, 2–3.7(–4.5) cm long, light blue, with the inside of the tube white or pale yellow, the limb 1.7–3.5 cm wide. Fruits ± globose, somewhat depressed, 8–12 mm wide, enclosed within the sepals. Seeds 1–4, 4–4.5 mm long, pear-shaped, black to dark brown, densely hairy with short trichomes.

**Habitat:** Washes, canyons. 300–1,900 m (980–6,230 ft).

**Range:** Arizona (Cochise, Coconino, Gila, Graham, Pima, Pinal, Santa Cruz, Yavapai, and Yuma counties), New Mexico, Texas; Mexico (Baja California Sur, Chihuahua, Nuevo León, Sonora, Tamaulipas, south sporadically to Guerrero, Oaxaca, and Chiapas, where it has probably been introduced as a contaminant in rice); South America; Old World.

**Seasonality:** Flowering August to November.

**Status:** Native.

**Ecological significance:** Although formerly thought to be adventive in the Southwest (Austin 1990, 1998a), this twiner is an indigenous species of canyons and washes across the Southern United States from Arizona into the Southeast (Austin 2006b). From that region, the species has been widely introduced for ornament and medicine. Although the distribution in the southwestern states is mostly along canyons and washes, the seeds are more likely to be dispersed by birds than water. While seeds float, owing to air pockets within the folded embryo, they are too widespread to use that method exclusively.

**Human uses:** No records were found of people using this species in northwestern Mexico or the United States, but elsewhere the seeds have been used as substitutes for those of *I. nil*. Both are drastic purgatives (Williams 1970, Austin et al. 2001). Both species are among those at one time lumped under the name *jalapa* (from the town Xalapa in Veracruz). Roots and seeds sold under *jalapa* in the 1800s were an important medicine in Europe, where people were obsessed with the health of their colons (Lewis and Elvin-Lewis 2003). Considering that many people had virtually no vegetables in their diets, perhaps their concern was valid.

**Derivation of the name:** *Ipomoea* is based on Greek *ips*, *ipos*, worm or bindweed, *homios*, *hoimios*, resembling, referring to the twining habit. *Hederacea* means that it has leaves that resemble Ivy (*Hedera helix*).

**Miscellaneous:** From the time Europeans began learning about American morning glories, I. *hederacea* has been confused with *I. nil*. Linnaeus and all of his predecessors confused the two, as people still do. The differences have been discussed numerous times (e.g., Shinners 1965, Austin 1986), but the species are still confused and confusing to the novice. Sepals of *I. hederacea* are abruptly narrowed, and the long subacute apices are strongly spreading or curved, at least in fruit. By contrast, sepals of *I. nil* are gradually narrowed, and the long acute apices are more or less erect, straight, and scarcely spreading. Moreover, *I. hederacea* is a temperate species, occurring within the tropics only at higher altitudes. By contrast, *I. nil* is a tropical species found mostly in the lowlands and low latitudes. No specimens of *I. nil* have been examined from Arizona outside cultivation (Austin 1998a).

A complicating factor to recognizing these two species is that *I. nil* is widely cultivated. Plants that grow in gardens have been selected for larger and more diversely colored flowers since at least the 1600s, and they often bear little resemblance to wild plants. Still, except for a few of the bizarre cultigens developed by the Japanese (cf. Austin et al. 2001), sepals are still recognizable as *I. nil*.

Scattered among these species are other blue-flowered morning glories. Those plants are *I. barba-*

*tisepala* and *I. purpurea*. The former has smaller blue flowers and usually 5-lobed leaves; the latter has larger purple flowers with abruptly acuminate sepal apices.

## *Garrya wrightii* (Garryaceae) Silk-Tassel, *chichicahuile*

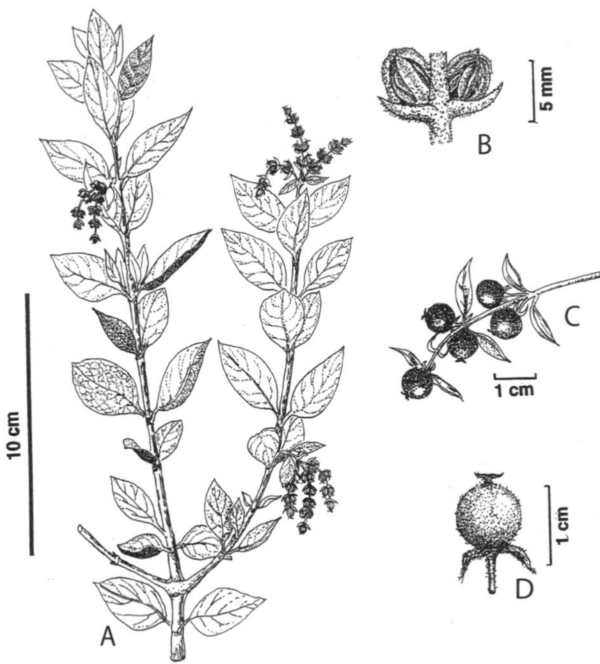


*Garrya wrightii*. A. Flowering branch. B. Staminate flowers. C. Branch with fruits. D. Mature fruit. Artist: R. Puente (Puente and Daniel 2001, courtesy Raul Puente).

**Other names:** ENGLISH: coffee-berry bush ("coffee" is from Turkish *quawah*, pronounced *kahveh*, coming into European languages about 1600; German physician Leonard Rauwolf, 1535–1596, was the first European who mentioned coffee after his trip to the Levant in 1573–1576; about 200 years after Europeans settled in the New World, the French introduced *Coffea arabica* into Martinique in 1717, and the Dutch took it to Surinam shortly before that; coffee reached Jamaica in 1728), fever-bush (used to treat fever), quinine-bush ("quinine" is from Quechua *quina*, bark, extracted from *Cinchona*, and was introduced into medical practice about 1820; later other bitter plants received a name comparing them with the original), silk-tassel (origin obscure, but comparing the flower cluster with a tassel made of silk); *chichicahuile* (from Náhuatl, *chichic*, bitter, *cuahuitl*, tree, Mexico); UTO-AZTECAN: *kánïnkwap* (Southern Paiute)

**Botanical description:** Shrubs or small trees are 1–4 m tall. Leaves are elliptic to broadly elliptic or oblong-elliptic, almost obtuse to acute and mucronulate at the apex, 3–5.5 cm long, 1.5–3 cm wide, when young sericeous, with age glabrous or nearly so, the margins roughened or somewhat roughened with tiny bumps (muriculate). Lowermost floral bracts are leaf-like (foliaceous). Fruits are dark blue, 4–7 mm in diameter.

**Habitat:** Canyons, dry slopes. 900–2,400 m (3,000–8,000 ft).

**Range:** Arizona (Coconino, Gila, Greenlee, Pima, and Santa Cruz counties), New Mexico, and western Texas; Mexico (Chihuahua and Sonora).

**Seasonality:** Flowering March to August.

**Status:** Native.

**Ecological significance:** Lamb (1975) states without qualification that Silk-Tassel "is found in the pinyon-juniper zone from about 5,000 to 8,000 feet elevation." Most others also would agree that the species is part of the higher-elevation communities. However, this conclusion is an example of the polyclimax theory in that the species is not confined to the "community" where it "belongs." This tree is one of several species in the Baboquivari Mountain canyons that extend down below the pinyon-juniper zone into the oak woodlands below 5,000 feet. Another species from that zone in the area is *P. discolor*, the pinyon itself (see *Pinus*). However, there is nothing of the dominance that may be seen in other, higher regions that support the typical array of species of the pinyon-juniper zone.

**Human uses:** Saunders (1920) thought the bitter chemicals in the bark of Silk-Tassel might be the same as those in Dogwood (*Cornus*) because the trees are related. *Garrya* contains an alkaloid called garryin that is used medicinally (Kearney and Peebles 1951, Masamune 1964), while that compound has not been found in Dogwood (Austin 2004). Most of the records of *Garrya* use are from outside Arizona. For example, in California and farther north, both *G. elliptica* and *G. fremontii* are used to treat fever (Hocking 1997). In Mexico, *G. laurifolia* (*palo verde*; *čivo bopotai*; *čivo*, goat from Spanish, *vopotai*, beard or fur, Northern Tepehuan) and other species are used to treat diarrhea. The bark of *G. elliptica* contains at least five alkaloids, including delphinine, which is otherwise known only from *Aconitum* and *Delphinium* in the Ranunculaceae (Mabberley 1997). Several of these alkaloids are highly toxic.

In Arizona, the Havasupai use *Garrya* to make whistles (Moerman 1998). Small quantities of rubber have been extracted from the species (Kearney and Peebles 1951). Palmer (1878) said that *G. flavescens* was used by indigenous people in Arizona to dye materials violet. The leaves were made into a tea to treat ague and colds, but Palmer does not specify by which tribes.

*Čivo bopotai* is used by the Northern Tepehuan to make arrows, violin bows, and the balls for their "kickball" game (Pennington 1969). *Garrya ovata*, called *šivátu úši* (goat tree, Northern Tepehuan) and *cuauhchichic* (*cuahuitl*, tree, *chichic*, bitter, Náhuatl, Sonora), is also used to make violin bows. People in Sonora use *G. ovata* to treat diarrhea (Martínez 1969, López-E. and Hinojosa-G. 1988).

**Derivation of the name:** *Garrya* is dedicated to Nicholas (or Michel) Garry, the first secretary of the Hudson Bay Company, who befriended David Douglas, an explorer of the late 1700s and early 1800s. The genus was proposed by Douglas but not published until John Lindley named it in 1834. *Wrightii* commemorates Charles Wright, 1811–1885, explorer and plant collector in Texas, Cuba, and his native Connecticut.

**Miscellaneous:** *Garrya* comprises 13 species ranging from Washington State to Panama. For many years, the genus was put in the Cornaceae. Now there are data indicating that it should be isolated into its own family, the Garryaceae, while still being considered related to the dogwoods.

## *Cucurbita digitata* (Cucurbitaceae) Coyote Gourd, *chichicoyote*, *aḍawĭ*

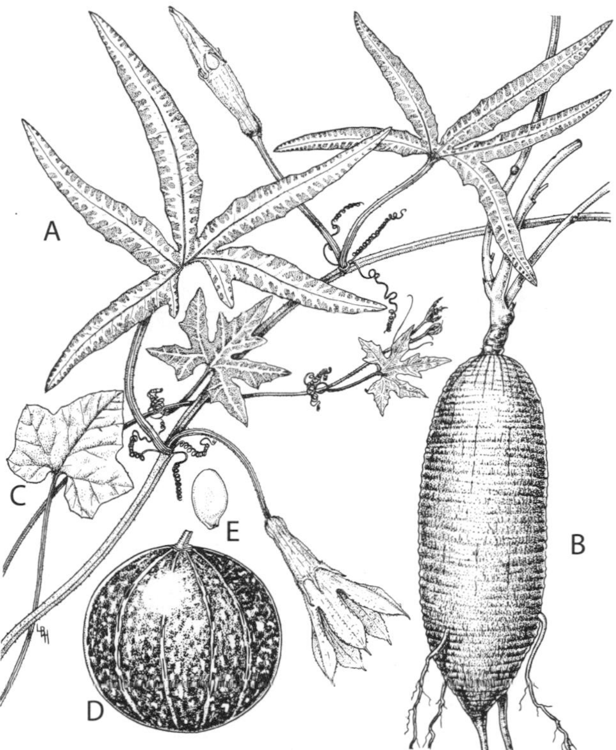


*Cucurbita digitata*. A. Branch of plant with flower. B. Fleshy root. C. Leaf of young plant. D. Fruit. E. Seed. Artist: L. B. Hamilton (Parker 1958).

**Other names:** ENGLISH: coyote gourd (Coyotes, *Canis latrans*, eat the fruit pulp and seeds; "gourd" is from Latin *cucurbita*, appearing in English by 1303), finger-leafed gourd (New Mexico); SPANISH: *calabacilla* (little gourd, Arizona, Sonora; derived from Arabic *qar'ah yábisah*, dry gourd, a name originally for *Lagenaria*), *calabaza amarga* (bitter gourd, Arizona, Sonora), *chichicoyote* <*chichicayote, chichi coyota*> (coyote's breasts, from Náhuatl, *chichi*, breast, literally "to nurse," *coyotli*, the coyote, Sonora), *meloncillo* (little melon, from Latin *melopepo*; "melon" appeared in English by CE 1387, Arizona), *melon de coyote* (coyote melon, Arizona, Sonora); ATHAPASCAN: *be'iłkan dee'é* [*joołé*], *naadołkal* <*nat dil kaali*> (*be'iłkan*, squash, *dee'é* = ? *joołé*, gourd/ball, *naadołkal*, gourd, Western Apache), *ndilkal* (Navajo; for *Cucurbita*); HOKAN: *ziix is cmasol* (yellow-fruited thing, Seri); UTO AZTECAN: *'ad* (Hiá Ced O'odham, Arizona, Sonora), *'adavĭ* (Akimel O'odham, Arizona), *'adawĭ* (Hiá Ced O'odham), *aḍawĭ* (Tohono O'odham), *a:ḍ* (Tohono O'odham), *aláwe* (Guarijío), *ara* (Mountain Pima), *mösipatnga* (*mösi*, food packet, *patanga*, gourd, Hopi), *nekhish* <*nekish*> (Cahuilla), *patnga* (Hopi; typically applied to *C. moschata*; in combination with other modifiers for other species), *teta'ahao* (Yaqui), *whsáraaĝanápų* (Ute); YUMAN: *xamach* (Paipai), *xa:más* (Cocopa)

**Botanical description:** Perennials from deep, thickened roots. Stems are relatively coarse, sprawling to several meters, or climbing shrubs and trees. Herbage is rough-hairy. Tendrils 3–5 branched from base. Leaves are digitately cleft, the lobes 6–13 cm long, the upper surface whitish near the midrib, the petioles 4–9 cm long; juvenile leaves are smaller. Corollas are yellow, 7–10 cm wide. Gourds are 8–9.4 cm wide, round, smooth, green with whitish stripes and mottling, yellow at maturity. Seeds are 9–11 mm long, 5.5–6.5 mm wide, smooth, compressed, white.

**Habitat:** Washes, canyons, grasslands. 30–1,524 m (100 to ca. 5,000 ft).

**Range:** Arizona (Graham to southern Yavapai, Cochise, Pima, Santa Cruz, and Yuma counties), southeastern California, and southwestern New Mexico; Mexico (Baja California and northern Sonora).

**Seasonality:** Flowering June to October.

**Status:** Native.

**Ecological significance:** Gourds are visited and pollinated by insects called "digger-bees" or "gourd-bees" (*Peponapis, Xenoglossa*) (Hurd and Linsley 1971, Buchmann, personal communication, Aug. 2005). Indeed, the females sleep below the vines in ground burrows and wake earlier than other bees to begin gathering pollen from the flowers. Males often sleep within the gourd flowers, making locating females easier. These bees arise long before the sun comes up and begin visiting flowers, usually when it is too cool for other bees to be moving. By the time the sun is a short way above the horizon, flowers are beginning to close. Between 8 and 10 a.m., flowers are closed, excluding late-rising visitors.

As the common name "coyote gourd" suggests, those canids eat the fruits. When fruits are not yet ripe in the fall, and after the Prickly-Pear fruits are gone, these mammals crack the gourds and eat the pulp and seeds. Coyote scat in some areas seems to consist of nothing but undigested gourd seeds. Many farmers and ranchers, however, find *Cucurbita* undesirable because it occupies ground where they prefer other plants to grow (Parker 1972).

**Human uses:** The Náhuatl name *chichicoyotli* started with an ancient practice in Mexico. That name originated because women smeared their nipples with juice from the fruits to give them a bitter taste. When infants tried to drink, they were "tricked into thinking that *chichis* are no good any more!" Gary Nabhan (1985) was told the story in Baja California, and versions have appeared in a variety of other places.

Seeds of wild gourds previously were used for food (Hodgson 2001), and they are rich in oils and some carbohydrates. Once the seeds are soaked in sodium-bicarbonate, they can be eaten.

People throughout the range of gourds of several species use the root and fruits for medicine and cleaning. Nabhan (1985) relates an amusing incident between a big city photographer and an O'odham woman regarding cleaning clothes with gourds. Some even told him that the water mixture with gourds was ". . . just like Clorox."

**Derivation of the name:** *Cucurbita* is the classical Latin name for a type of gourd. Heiser (1979) recorded that the word "gourd" itself originally meant "fool," "idiot," and "adulteress," but that is not confirmed by the Oxford English Dictionary. Indeed, in the OED (2007) that seems to be a recent meaning. The original *cucurbita* was what is now called *Lagenaria* (Austin 2004), but Linnaeus applied that Old World name to a New World species in 1753. *Digitata* refers to the leaves that resemble the fingers of a hand.

**Miscellaneous:** *Melon loco* (Crazy Melon, *Apodanthera undulata*) also grows in Brown Canyon. This plant is distinctive and easily separated from other gourds in the area by its toothed and wavy-margined 5-lobed kidney-shaped leaves and its longitudinally ribbed fruits. *Melon loco* flowers at the same time as Coyote Gourd, and it is known from western Texas to southeastern Arizona and Mexico (Chihuahua, Coahuila, Durango, Hidalgo, Nayarit, Puebla, Sonora, and Zacatecas). *Apodanthera*'s range stops on the eastern side of the Baboquivari Mountain chain.

## *Echinopepon wrightii* (Cucurbitaceae) Wild Balsam-Apple

**Other names:** ENGLISH: balsam-apple (applied by 1578 to *Mormodica*; later to this genus, Arizona, New Mexico; "balsam" came into English from Latin *balsamum*, about CE 1000, meaning an aromatic vegetable juice; "apple" from Old English *appel*, the fruit, appearing ca. CE 885; akin to *aepplas*, of the eye; it is not clear which is the earlier application, the fruit or the eye; suggestive is the Gaelic expression *clach-na-sula* [stone of the eye], which means the apple or the pupil of the eye; De Candolle suggested in 1886 that "apple" might have been derived from an Indo-European root of *ab*, *av*, *af*, or *op*, akin to Norwegian *elpe*, Russian *iabloko*, Lithuanian *obolys*), spiny cucumber (New Mexico)

**Botanical description:** Climbers, annual, with slender stems. Leaves are shallowly lobed, cordate basally. Staminate flowers are in long racemes, the corolla 7–8 mm wide, glandular-dotted, the lobes triangular. Pistillate white flowers are solitary. Fruits are ovoid, to 1.5 cm wide, with glandular-hirsute spines, opening by an apical lid, commonly 3-celled.

**Habitat:** Canyons, along streams. 900–1,280 m (3,000–4,200 ft).

*Echinopepon wrightii*. A. Branch with male flowers, female flowers, and fruit. B. Leaf on branch with tendrils. Artist: R. D. Ivey (Ivey 2003).

**Range:** Arizona (Cochise, Pima, and Santa Cruz counties) and southwestern New Mexico; Mexico (Chihuahua and Sonora); south to Nicaragua.

**Seasonality:** Flowering July to October.

**Status:** Native.

**Ecological significance:** This vining herb is a frequent part of the understory at margins where there is ample light reaching the ground. There is often a profusion of growth that shows up as a dark, rich green in comparison to many of the other plants flowering in this season. Although the species is recorded down to 900 m (3,000 ft), plants are uncommon in other parts of the Altar Valley at that elevation. In Brown Canyon near the Education Center, Wild Balsam-Apple is uncommon below about 1,280 m (4,200 ft).

*Echinopepon* reaches its northern limit in Arizona and New Mexico. The vines range south into Central America, with the center of diversity for the genus on the Pacific coast of Mexico, especially in the Sierra Madre Occidental and the Sierra Madre del Sur (Monro and Stafford 1998). In that area, the majority of species grow above 1,000 feet. Rodríguez (1995) discussed the distribution of species in the genus.

There are 7–18 species now recognized in *Echinopepon*, all American (Monro and Stafford 1998, Jeffrey 2001). Jeffrey (2001) gave no details on why he considered the group to comprise only 7 species when Monro and Stafford (1998) had recognized 18.

The genus is similar to *Echinocystis*, and some have suggested that the two may not be distinct (Mabberley 1997). Célestin Alfred Cogniaux considered the group as *Echinocystis* section *Echinopepon* in 1881, but that classification has not been followed in recent years. That monotypic genus, with the single species being *Echinocystis lobata*, is mostly a plant of eastern North America. *Echinocystis* was used medicinally by indigenous people, but there seem to be no records of *Echinopepon* having been used. Both genera are superficially similar to *Marah*, but that is a perennial climber.

**Human uses:** None found for this species.

In the Valley of Toluca, Mexico, *E. milliflorus* is a weed in Maize fields (Vieyra-O. and Vibrans 2001). There, the *mestizos* gather 74 weedy species from these fields as forage for their livestock, and *E. milleflorus* is among the most frequently used. While the foliage of cucurbits is notorious for causing gastric problems (Deena Decker-Walters, personal communication, 2005), and *Sicyos deppei* is not given to horses by those villagers, people did include the *Echinopepon* for other animals. *Echinopepon coulteri* of Mexico and Costa Rica has edible fruits (Williams 1981).

The related *Echinocystis lobata* is used by people in the Taos Pueblo to treat rheumatism (Curtin 1947). The Menomini brew a bitter tea from the roots and use it as an analgesic and love potion (Smith 1923).

**Derivation of the name:** *Echinopepon* is from Greek *echinos*, a hedgehog, and *pepon*, originally ripe, later for a melon. *Wrightii* commemorates Charles Wright, 1811–1885, explorer of Texas, Cuba, and his native Connecticut.

**Miscellaneous:** More obvious in the canyons in the springtime is *Marah gilensis* (big root, wild cucumber). That robust climber blankets the vegetation below the Education Center with dark green herbage that contrasts markedly with the bare trees between February and March. Big Root flowers as soon as the branches are long enough, extending into April. The species is known from Greenlee to Mohave and south to Cochise, Pima, and Pinal counties, all south of the Mogollon Rim, except one in Coconino County; also in New Mexico; Mexico (Sonora?). Flowers are not showy in either species, but *Marah* is a perennial from a large tuberous root, while *Echinopepon* is an annual.

### ***Juniperus deppeana*** **(Cupressaceae)** **Alligator Juniper, *huata***

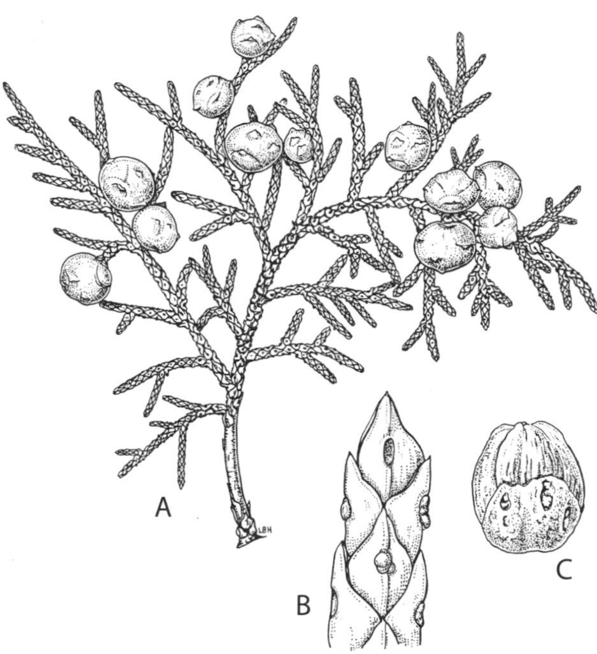


*Juniperus deppeana*. A. Branch with berry-like female cones. B. Mature scale-like leaves in alternating pairs. Note glandular pits on their backs. C. Seed. Artist: L. B. Hamilton (Parker 1958).

**Other names:** ENGLISH: cedar (from Latin *cedros*, in English by ca. CE 1000), juniper (from Latin *ieniperu*, *gigiperus*; in English by the 1400s; cognates include *enebro*, Spanish by 1557; *einer*, Norwegian; *geneure* <*ieneuve*>, French by 1550; *genever*, Dutch by 1549; *genève* <*genèvier*>, modern French; *ginepro*, Italian by 1551; *zimbro*, Portuguese), Mexican sandarac (from Greek *sandarake*, but probably a loan word there; used for red arsenic sulfide by 1398, transferred to a tree gum by 1655); SPANISH: *cedro chivo* (goat cedar, New Mexico; sometimes spelled *cedro chino*, curly cedar, Puebla), *huata* (Sonora; probably from Uto-Aztecan *waˀat* fide Timbrook 2007; however, Santamaría 1959 pointed out that it might be a variant of *guata*, originally a Spanish word referring to cotton branches used in processions, later and dialectic meaning the "belly"), *sabino* (from Latin *sabina*, juniper), *tascate* (Mountain Pima), *tlascal* <*tlaxcal*> (Hidalgo, from Náhuatl *tlascalli*, grill, comal); ATHAPASCAN: *diltáłé* <*tatle*>, <*diltáłétchí'*> (popping bark, Western Apache), *dilt'áłi* (popping bark, Navajo), *gad* <*kat*> (Navajo), *gad* [*izee*] (*gad*, juniper, *izee*, medicine, Western Apache), *gad ni'eli* (floating juniper, Navajo; for *J. scopulorum*), *kálhtē* (Jicarilla Apache), *tałehntsai* (large juniper, Chiricahua and Mescalero Apache); HOKAN: *pal* (Washo); UTO-AZTECAN: *aóri* (Guarijío), *awarí* <*awori-ki*, *aorí*> (Tarahumara), [*ban*] *ga'a* ([*ban*, coyote] (*ga'a*, *Cupressus arizonica*; based on reflex of proto-Uto-Aztecan **wa'ac*, proto-Tepiman **ga'a*, Mountain Pima), *gayi* (Northern Tepehuan), *hohu* (*ho*, juniper, *hu*, wood, Hopi), *kneumapee* (?Hopi), *nooahntup* (Southern Paiute; for *J. osteosperma*), *samapi* (Panamint), *tahkali* (Yaqui), *táscale* <*tásate*, *taxcate*> (Mountain Pima, maybe cognate with Náhuatl *tlascalli*, grill, comal, Chihuahua, Sonora, Durango), *waˀadabɨ* (Kawaiisu; for *J. californica*), *wadul* (tree, *wa'at* [*waˀa-t*], the fruit, Tübatulabal; for *J. californica*, *J. utahensis*), *wa'a*$^{g}$ (Southern Paiute), *waapin* (Shoshoni; for *J. scophulorum*), *wa'ápų* <*wap*> (Ute), *wap* (Comanche), *wa'p* (Mono), *wápi* (Northern and Western Paiute), *wa'pi* [*wap*, *wa'ap*] (to burn, Shoshoni; full name *wa'apopi*, meaning fire, match, or kindling, from *wa'ap*, juniper, and *o'pi*, wood), *waˀat* (Luiseño), *yuyily*, *iswat* (Cahuilla); YUMAN: *tc'auka* (Havasupai), *tcóq* <*joq*> (Walapai)

**Botanical description:** Usually trees, 7–15 m tall or sometimes larger, dioecious. Stems are gray, deeply fissured into rectangular plates on old trunks. Leaves are usually in 4 rows, alternating in pairs at right angles (decussate), closely appressed, scale-like, the gland obvious. Pollen cones are terminal, 3–4 mm long, oblong. Seed cones are terminal, 8–20 mm long, almost spherical to broadly ellipsoid, maturing bluish to red-tan or red-brown in second year, glaucous, dry, hard. Seeds are usually 4–5 per cone, 6–9 mm long, ovoid to oblong or irregular, brown.

**Habitat:** Canyons, ridges. 1,350–2,900 m (4,400–9,600 ft).

**Range:** Arizona (Apache, Cochise, Coconino, Gila, Graham, Greenlee, Navajo, Pima, Pinal, Santa Cruz, and Yavapai counties), New Mexico, and Texas; Mexico (Chihuahua, Coahuila, Puebla, and northeastern Sonora).

**Seasonality:** Reproductive February to March.

**Status:** Native.

**Ecological significance:** These trees are typically plants of higher elevations, and they are characteristic of the juniper-pinyon zone growing between 1,370 and 1,980 m (4,500–6,500 ft) elevation. This community is often considered the transition between lowlands and uplands.

One of the largest living trees found in Brown Canyon is just beside the southern end of the arch at about 4,800 ft. Another with a 32.5-inch d.b.h. grows on the stream margin just above the uppermost windmill. Below that point, most plants are young seedlings and not yet reproductive. One old individual with a trunk 25 inches in diameter lived for many years at the junction of Brown and Jaguar Canyons. It was living in 2002, but dead by 2004.

West of the Baboquivari Mountains, *J. deppeana* is replaced with *J. coahuilensis,* which grows in Barry M. Goldwater Air Force Range, Organ Pipe Cactus National Wildlife Refuge, and the Sonoran Desert National Monument (Felger et al. in prep.).

**Human uses:** Junipers have been used by all cultures that are familiar with them, including the classical Old World Arabs, Gaels, Greeks, Romans, and Germanic groups (Austin 2004). Seeds and "fruits" were found at all prehistoric levels in the Cordova Cave, New Mexico (Kaplan 1963). Fruits (technically seed cones) of *J. deppeana* provide food for at least Apache, Isleta, Navajo, San Felipe, and Yavapai (Castetter and Opler 1936, Moerman 1998, Reagan 1929, Vestal 1952, Wyman and Harris 1951). Wood ashes are used by Hispanics and other people in preparing tortillas (Hocking 1997). Yavapai, and many others, use the wood as lumber and fuel (Hocking 1997). For treating rheumatism, the Tarahumara make a wash of the branches and use it as a vapor (Pennington 1969). Tarahumara also have numerous religious uses for *Juniperus* (Thord-Gray 1955), and use it to prepare runners in the *rarahipa* <*rara-hi-pa*> (foot-race game).

The volatile oils in the leaves are used by a variety of people to treat rheumatism and neuralgia. Mountain Pima use the branches to make a drink to treat colds (Pennington 1973). However, Duke et al. (2002) admonish that extracts from *Juniperus* may be fatal.

**Derivation of the name:** *Juniperus* is the classical name; from Latin *ieniperum, gigiperus,* names used by Virgil, 70–19 BCE, and Pliny, CE 23–79. *Deppeana* commemorates Ferdinand Deppe, 1794–ca. 1860, a German botanist who traveled in Mexico from 1824 until 1827, and later with Wilhelm Scheide. Deppe first traveled under the sponsorship of Count Von Sack, the *Zweiter Ober-Jägermeister* and chamberlin to the king of Prussia.

## *Cyperus squarrosus* (Cyperaceae) Bearded Flat-Sedge, *tulillo, waṣai s-u:w*

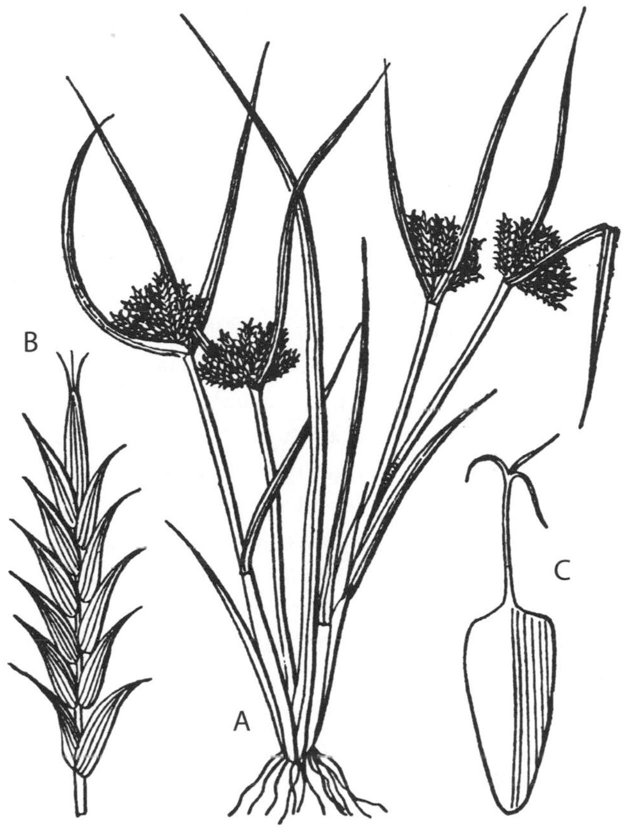


*Cyperus squarrosus*. A. Flowering plant. B. Spikelet. C. Achene. Artist: F. Emil (Britton and Brown 1896–1898).

**Other names:** ENGLISH: bearded flat-sedge ("bearded" alluding to the "fringed" spikelets; "sedge" from Old English *seíg*, related to Low German *segge*, both probably from the Latin *secare*, to cut; akin to Old Celtic *seska*, Irish *seisg*, Welch *hesg*, Breton *hesq*, Arizona, New Mexico), [dwarf] marsh sedge ("marsh" is fertile alluvial land on a river or coast; in use by 1180; cognate with Old Frisian *mersk*, West Frisian *marsk*, *mask*, *mersk*, Middle Dutch *marsch*, *meersch*, *mersch*, Dutch *mars*, *meers,* water meadow, pasture, Middle Low German *marsch*, *masch*, *mersch*, water meadow, German *Marsch*, Danish *marsk*), nut-sedge (a sedge with nut-like storage regions on the roots, in use by

1861 in North America for another species); SPANISH: *apoyamate* (from Náhuatl *apoyamatli*), <*grulla*> (the word looks like Spanish *grillo*, cricket, Mountain Pima), *tule* (usually used for *Typha* or *Scirpus*), *tulillo* (little sedge); ATHAPASCAN: *teeł níyiz* <*te.l ni'izi*> (round cattail, Navajo; also for *Eleocharis* and *Schoenoplectus*), *tłoliyesze* (plant that stands next to horses, Chiricahua and Mescalero Apache); UTO-AZTECAN: *to'ora* (from Spanish *tule* or *totora*? Mountain Pima), *vashai s-uuv* (scented grass, Akimel O'odham), *waṣai s-u:w* (Tohono O'odham); YUMAN: *ˀaráwp* <*kwarao*> (Cocopa; for *C. esculentus*); (= *C. aristatus*)

**Botanical description:** Nonseasonal ephemerals, tufted, 1.5–18 cm tall. Leaves are few, soft, basal or nearly so, usually less than 1 mm wide. Bracts below inflorescence are leafy, the largest bracts longer than the inflorescence. Spikelets are 5–15 mm long, flattened, in compact clusters, sessile or on short rays. Each spikelet scale has a prominent recurved awn-like tip, the awn tips giving an unusual "fringed" appearance to the spikelets; scales are often reddish bronze to yellowish with green margins. Stamens are 1(–2) or staminodes. Achene is 3-sided, the style 3-branched.

**Habitat:** Moist areas, oak woodlands. 260–2,280 m (850–7,500 ft).

**Range:** Arizona (Coconino and Navajo, south through Cochise, Pima, and Santa Cruz counties) and throughout most of North America; Mexico (Chihuahua and Sonora); south to South America.

**Seasonality:** Flowering August to September.

**Status:** Native.

**Ecological significance:** These sedges grow in moist areas in oak woodlands. Felger (2000) notes that this species is widespread in permanently or temporarily wet soils in the Sonoran Desert. However, the species is found around the world in temperate and tropical regions. This *Cyperus* is the smallest sedge in the Sonoran Desert.

Bowers (1980) reported *C. squarrosus* as abundant in moist soil near the pond at Quitobaquito. The sedge has not been found there recently, but it is common in comparable habitats in open places along the nearby Río Sonoyta in Sonora. More than likely, *C. squarrosus* has been extirpated from Quitobaquito since modification of the pond and/or since the livestock were removed and open ground in wetland habitats has filled with *Scirpus* (Felger et al. 1992).

**Human uses:** The O'odham name probably applies to one of the lowland nut-grasses, but may have been of more general use (Rea 1997). Curtin (1949) records use of the chewed tubers of *C. esculentus* to relieve coughs and colds, snakebites, and to give spirit to their horses when going on rabbit hunts (*kuushada*). Rea (1997) was told that people still chew the tubers for medicine.

Dry *Cyperus squarrosus* has the strong odor of Slippery Elm bark (*Ulmus rubra*) (Kearney and Peebles 1951).

**Derivation of the name:** *Cyperus* is from the ancient Greek *kyperos* or *kyperios*, the name of the galingale or sedge. *Squarrosus* means scaly or rough, an allusion to the "fringed" awn-tips.

**Miscellaneous:** The other sedges in Brown Canyon are *Cyperus flavicomus* and *Cyperus pringlei*. These two species are distinguished by not having the "fringed" awn-tips. *Cyperus flavicomus* has 3 styles and/or stigma lobes, lens-shaped seeds, and scales that have broad, white, hyaline margins. This species is also in the Coyote Mountains. *Cyperus pringlei* has 2 styles and/or stigma lobes, 3-angled seeds, and lacks the hyaline scale margins. Based on collections, *C. pringlei* is the most common species in the mountains, and it has been found in Baboquivari, Moristo, South, Thomas, and Toro Canyons, as well as in the Coyote Mountains. *Cyperus pringlei* has not been found west of the Baboquivari Mountains in Arizona or in the Gran Desierto of Sonora (Felger 2000, Felger et al. in prep.).

## *Woodsia phillipsii* (Dryopteridaceae) Mexican Cliff Fern, *helecho*

**Other names:** ENGLISH: [Mexican, Phillips'] cliff fern ("cliff" is from Old English *clif* neuter, plural *clifu*, originally *cleofu*; akin to Old Saxon *kli*, Low German *clif*, *clef*, Middle Dutch *clif*, *clef*, plural *clve*, Dutch *clif*; used in English by 854; "fern" is from Old English *fearn*, Old Teutonic **farno*, Old Aryan **porno*, whence Sanskrit *parna*, meaning wing, feather, leaf; the ancient meaning of the word is surely "feather"; for the transferred application cf. Greek *pterov*, feather, *pteris*, fern); ATHAPASCAN: *tułbáí* <*tułil bidą*> (gray water, Western Apache); SPANISH: *helecho* (fern, Mexico, Spain); UTO-AZTECAN: *wačigónuri* (Tarahumara); (= *W. mexicana* of authors)

**Botanical description:** Leaf stalks (stipes) are 1.5–9 cm long, with a single color, pale brown scales and 2-colored scales having pale brown margins and dark brown central portions at the base of the stipe. Leaf blade is lanceolate, 5–35 cm long, 1.5–6 cm wide,

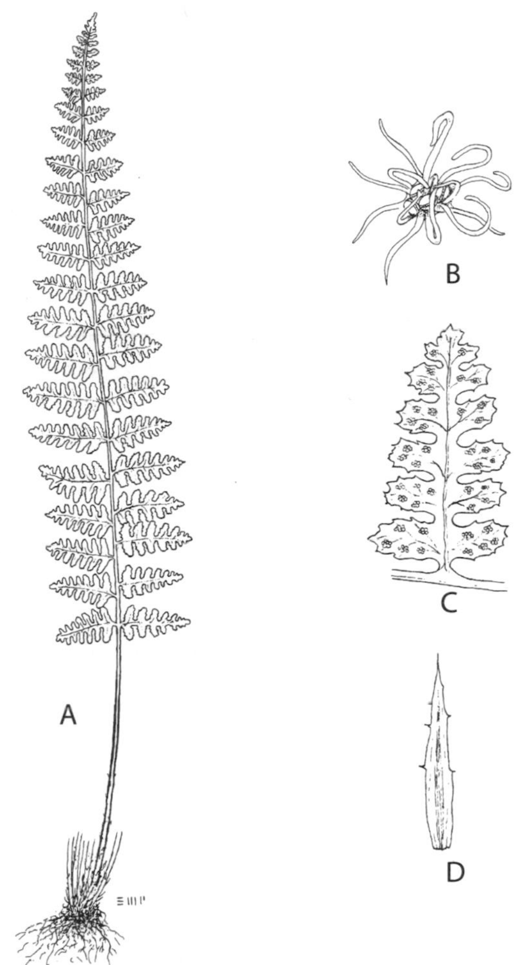


*Woodsia phillipsii*. A. Frond with basal roots. B. Indusium. C. Fertile lower side of pinnule. D. Scale. Artist: E. M. Paulton (Mickel 1979).

attenuate and sometimes truncate at the base, acute-acuminate apically, pinnate-pinnatifid in small specimens, 2-pinnate-pinnatifid at the base of the pinnae in large ones, the pinnae usually 7–18 pairs, only the lowest few basal pairs increasingly distant, elongate-deltoid to elliptic, with lobes often truncate at the apex, with few to many elongate glands on the blade and stipe and multicellular marginal hairs. Indusia are small, with conspicuous marginal hairs.

**Habitat:** Canyons, rocky slopes, washes. 1,100–2,900 m (3,500–9,500 ft).

**Range:** Arizona (Apache, Coconino, Greenlee, south to Cochise, Pima, and Santa Cruz, Yuma counties), New Mexico, and western Texas; Mexico (Chihuahua and Sonora).

**Seasonality:** Reproductive during rainy seasons.

**Status:** Native.

**Ecological significance:** These ferns that grow on rocks (epipetric) are delicate plants that often need more water than some of the other genera in the Baboquivari Mountains. *Woodsia* is mostly a temperate and cool-temperate genus of 25–30 species nearing the margins of the genus distribution in southern Arizona (Windham and Rabe 1993, Mabberley 1997). While *W. phillipsii* reaches its northern limit in the southwestern United States, the related *W. mexicana* extends deeper into Mexico. The type collection of *W. phillipsii* was made in the Chiricahua Mountains where the ferns grew in pine woods at 6,500 feet elevation.

Windham and Rabe (1993) consider diploid *W. phillipsii* probably to be one parent of hybrid polyploid *W. neomexicana*. Similarly, *W. phillipsii* was involved in the hybrid origin of *W. mexicana*. The remarkable thing about the diploid *W. phillipsii* is that they have 2n = 84 chromosomes. Its near relatives, and the species to which it is thought to have given rise, have 2n = 152 chromosomes. While this figure sounds like a highly derived number of chromosomes, Haufler and Soltis (1986) argued that the 2n = 84 number is diploid and not derived.

**Human uses:** None found for this species. Surely, sorting out which species have been used has been complicated by the past misunderstanding of "*W. mexicana*" (Windham and Rabe 1993). Indigenous people did not split the species as finely as botanists; they had no dissecting microscopes to examine almost invisible traits. Plus, ferns are infrequently culturally important to indigenous groups.

Tarahumara use a decoction of *W. mexicana* leaves to relieve aches and pains (Pennington 1963). *Woodsia neomexicana* is used by the Keres as a cleansing agent at childbirth (Moerman 1998). The Navajo include *W. neomexicana* in a compound mixture taken and applied as a lotion and as a Life Medicine (Vestal 1952). The Okanagan consider *W. scopulina* to be an indicator of water when one is traveling through the mountains (Moerman 1998). Western Apache use *W. plummerae* to make a tea as a beverage (Gallagher 1976). Surely the name *tułbáí* indicates use for more than a beverage since it is also used for *Datura* (q.v.) and *Ephedra* (q.v.).

**Derivation of the name:** *Woodsia* is named for Joseph Woods, 1776–1864, a British architect and Fellow of the Linnean Society, author of *The Tourist's Flora*, among other publications. *Phillipsii* commemorates the collector of the type specimen, Walter Sargent Phillips, 1905–1975. He collected that specimen in Rucker Canyon on the western side of the Chiricahua Mountains on 7 October 1945.

**Miscellaneous:** The other species in the Baboquivari Mountains is *W. plummerae*, and it is similar to *W. phillipsii*. *Woodsia plummerae* has indusia composed of relatively broad segments that are multiseriate for most of their length but may be branched or divided distally. *Woodsia phillipsii* has indusia that are composed of narrow, usually filamentous segments that are uniseriate for most of their length.

## *Ephedra trifurca* (Ephedraceae) Mormon Tea, *canutillo, ku:pag*

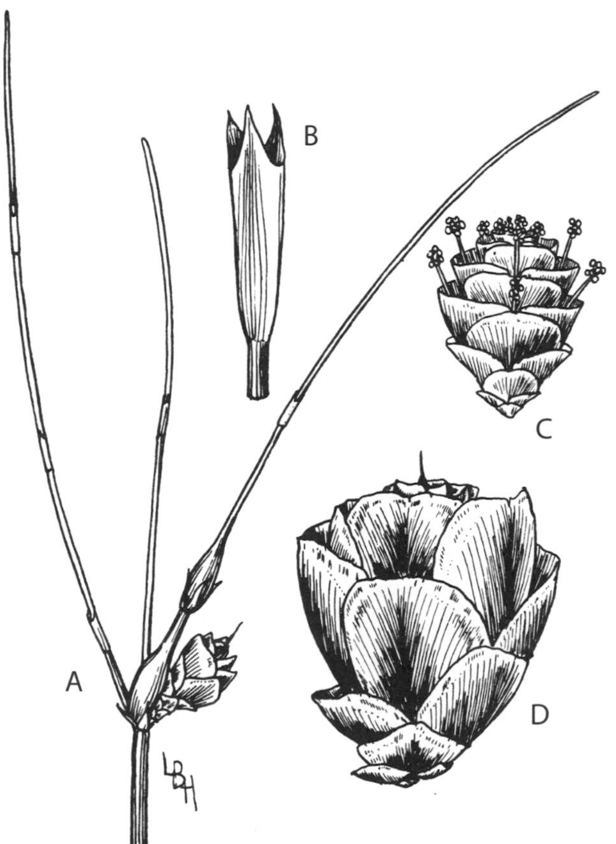


*Ephedra trifurca*. A. Branch with ovulate cone below swollen branch (gall) caused by insects. B. Node with 3 coalescent leaves, enlarged. C. Male cone, enlarged. D. Female cone, enlarged. Artist: L. B. Hamilton (Benson and Darrow 1945).

**Other names:** ENGLISH: joint fir (in use by 1866 for Gnetaceae, later for *Ephedra*), longleaf [ephedra, desert, Mexican, Mormon, teamster's] tea (because of wide use as beverage), mountain rush ("rush" is of uncertain etymology, but perhaps Germanic, from Old English *risc* <*rix, rise, rixe*>, Middle Dutch *risch*, and Middle Low German *risch(e)* <*rysse, risk, ryse*>, dating to ca. CE 725), three-forked ephedra (a book name); SPANISH: *cañatilla* [*canatilla*] (little pipe or cane, Arizona, Texas), *canutillo* [*del campo*] ([wild] little pipe or cane, New Mexico, Sonora), *hierba de la coyuntura* (jointed herb, Mexico), *itamo real* (royal spurge, Coahuila; this common name also is applied to *Aristolochia*, *Pedilanthus*, *Pellaea*, and *Turnera*), *popotillo* (from Náhuatl *pototl*, a slender, hollow stick or broom, Chihuahua, New Mexico, Texas), *tepopote* (from Náhuatl *tetl*, stone, plus *popotl*, northeastern Baja California, Chihuahua, Coahuila, Sonora, Texas); ATHAPASCAN: *tł'oh 'azihii* (*tł'oh*, grass, *'azihii*, voice, often referring to being hoarse, Navajo), *tułbáí* <*tułil bidą*> (graywater, Western Apache); UTO-AZTECAN: *ku:pag* (Tohono O'odham), *kuupag* (Akimel O'odham), *kuuvid nonovi* <*koovit nawnov*> (Pronghorn's foreleg, Akimel O'odham), *ku:pag* <*ku'upok*> (Hiá Ceḍ O'odham), *ösvi* <*'ɜ́sivi*> (Hopi), *u'us ti* <*oo-oosti*> (sticks tea, Akimel O'odham), *sudupi*, *tóyatukdúva* (*tōyɨvi*, mountain, *tudúva*, tea, Northern Paiute; for *E. viridis*), *tutúpï* (Southern Paiute; for *E. viridis*), *tutupivɨ* (Kawaiisu; for *E. viridis*), *tųtų́pųvų* (Ute), *tuttumpi* (Panamint), *tuttumpin* (Shoshoni), *tutut* (Cahuilla), *ṭúvūt* (Cupeño, Luiseño), *u'tuudul* (Tübatulabal; for *E. viridis*); YUMAN: *djimawaí* (for *E. viridis*, Havasupai), *ˀi:šíw* (Cocopa), *jumway* (Walapai)

**Botanical description:** Sprawling dioecious shrubs, 1–2.5 m, reaching 3–4 m across, usually with several trunks. Lower branches are often decumbent or spreading near the ground, thick and woody, the twigs bright green, long, slender, arching. Leaves are 3 per node, 5–10 mm long, fused to ca. 2/3 their length, narrowly acute, semipersistent, splitting and fraying with age. Cones are 1 to several per node, sessile, 1-seeded.

**Habitat:** Desertscrub. Sea level to 1,600 m (5,250 ft).

**Range:** Arizona (Gila to Greenlee, Mohave, Yavapai, south to Cochise, Pima, and Santa Cruz counties), southern California, New Mexico, and southwestern Texas; Mexico (Baja California, Chihuahua, Coahuila, and Sonora).

**Seasonality:** Cones appear (February) March to April.

**Status:** Native.

**Ecological significance:** Maximum observed longevity is 50 years (Turner et al. 1995). Seedling recruitment is uncommon, but may occur where moisture is favorable. Dispersal is wind-mediated by the keeled wings on the bracts of the fruiting structures (Judd et al. 2008). However, birds are also implicated as they are attracted to the brightly colored (yellow, orange, and red, in different species) outer bracts.

Honeybees collect the pollen, although the genus is generally considered wind-pollinated. Judd et al. (2008) also note that the insects are attracted by the nectar produced by the ovulate cones (strobili).

**Human uses:** Akimel O'odham, Apache, Cocopa, Navajo, and Zuni use *E. trifurca* medicinally (Gallagher 1976, Moerman 1998, Rea 1997, Reagan 1929). The Akimel O'odham with whom Rea (1997) works know only *E. aspera*, calling it by the folk genus *kuupag*, which Curtin (1947) rendered *koopat*. Her consultants called both *E. aspera* and *E. trifurca u'us ti* <*oo-oosti*>, but had other names for both species. Rea (1997) thought that the *u'us ti* was simply someone's descriptive name used in place of *kuupag*. His consultants thought that these names referred to other plants, not *Ephedra*.

This genus is a notorious diuretic and stimulant, because of the ephedrine and other compounds that it contains. Curtin found *Ephedra* being used as an "antileutic" (against venereal disease). Cocopa, like the Akimel O'odham, make a medicine for sores (Curtin 1947, Moerman 1998). Rollin C. Brown, for whom Brown Canyon is named, started a business in 1901 to make "Papago Tea" from *Ephedra*. He claimed that the Tohono O'odham used *ku:pag* for liver trouble (Kitt 1926–1929). Navajo use *E. trifurca* for gastric distress (Hocking 1956).

Hocking (1997) says that *E. trifurca* is used like *E. torreyana* and *E. antisyphilitica*. Many species are used similarly, and some current residents of Altar Valley say that as kids their mothers made them drink tea of *E. trifurca* to prevent having "a snotty nose." Presumably, it contains ephedrine, which acts as an antihistamine.

**Derivation of the name:** *Ephedra* is from Greek *epi*, upon, and *hedra*, seat, a name used by Pliny, CE 23–79, for what is now *Hippuris*. Linnaeus reapplied the word for these shrubs. *Trifurca* means three-lobed.

**Miscellaneous:** Formerly thought to be more closely related to gymnosperms, *Ephedra* is now placed near the tropical family Gnetaceae. That relationship makes *Ephedra* intermediate between the gymnosperms (pines and relatives) and angiosperms (flowering plants). Indicators of that intermediate position are the following properties: having seeds that are not enclosed in an ovary (typical of gymnosperms), having wood with vessels, exhibiting double fertilization, and displaying somewhat flower-like structures like angiosperms. Molecular genetics is ambivalent on the topic (Judd et al. 2008).

*Ephedra* is comprised of ca. 60 species (Stevenson 1993). The genus grows from the Mediterranean to China, in the western United States and Mexico, and in the Andes (Mabberley 1997).

## ***Acalypha neomexicana*** **(Euphorbiaceae) New Mexican Copper-Leaf,** ***hierba de cáncer***

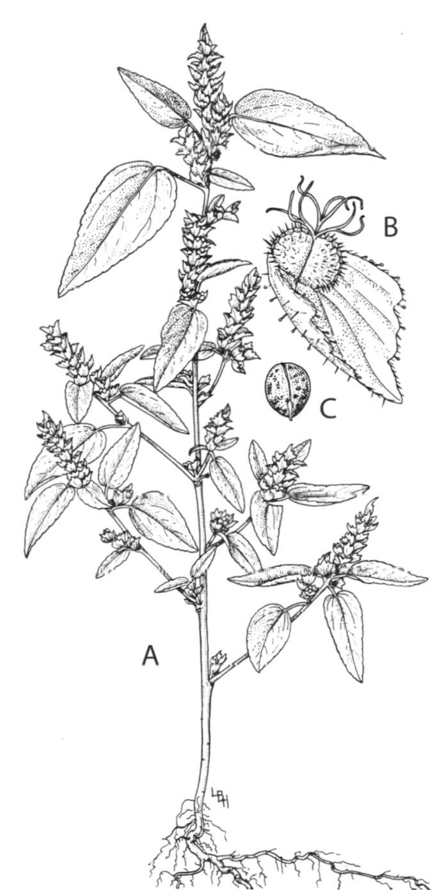


*Acalypha neomexicana*. A. Flowering plant. B. Young fruit showing three-branched styles and the associated bract. C. Seed. Artist: L. B. Hamilton (Parker 1958).

**Other names:** ENGLISH: New Mexican copperleaf (some of the species having leaves that turn coppery-reddish in the fall), three-seeded mercury

("mercury" was originally applied to a European member of this family, *Mercurialis annua*; formerly the herb was grown for medicinal qualities and used especially in enemas; in use for the Old World plant about 1398, later used for other genera; see below); SPANISH: *hierba de cáncer* (a term for the genus through much of Mexico)

**Botanical description:** Annual herb, 7–60 cm tall. Leaves alternate, lance-shaped, shallowly toothed, 2.5–13 cm long, truncate basally, acute apically. Flowers in terminal or axillary spikes, unisexual, with the male above and the female below near the base; male flowers with 6–8 stamens, the female with 3 styles divided into long thread-like parts. Bracts beneath the female flowers resemble leaves, conspicuously veined, the middle tooth elongated. Capsules are 3-lobed, breaking into single-seeded parts, ca. 2 mm long. Seeds are granular, reddish to dark brown, sometimes with brown spots, egg-shaped, 1.6 mm.

**Habitat:** Desert grasslands, washes, streams, canyons. 760–2,290 m (2,500–7,500 ft).

**Range:** Arizona (Greenlee to Yavapai, south to Cochise, Pima, and Santa Cruz counties) and New Mexico; Mexico (Chihuahua, Sonora, and south to San Luis Potosí, Aguascalientes).

**Seasonality:** Flowering August to October.

**Status:** Native.

**Ecological significance:** Although the flowers and fruits are green and inconspicuous, these small herbs are actually common in the canyons in the monsoon season. There, *Acalypha* is usually part of the green stratum that covers the ground below trees and shrubs. A careful examination of the herb layer often reveals *hierba de cáncer.*

**Human uses:** No uses have been found for *Acalypha neomexicana*, but the similar *A. ostryaefolia* in the lower parts of the Buenos Aires National Wildlife Refuge contains cyanogenic compounds, as do several other members of this genus (Nahrstedt et al. 2006). Several *Acalypha* are used elsewhere in the world. Related *Mercurialis annua* also has cyanogenic compounds (Seigler 1994).

The common name comparing *Acalypha* to "Mercury" suggests a recognition, if not a confirmation of use, that this plant has utility similar to that historic Old World medicine. The herbs called *Mercurialis annua* by Linnaeus in 1753 have a venerable history (Dobelis 1986). The story that was inherited by Europeans from the Greek physicians like Hippocrates (ca. 460–ca. 377 BCE) and Dioscorides (fl. CE 40–80) was that Hermes, equated with the Roman Mercury, the messenger of the gods, discovered the medicinal properties of the Old World plants. By the 1500s, the herbs were known as "Mercury" in Germanic and Romance languages to commemorate that Roman story. Early herbalists like Leonard Fuchs in 1542, Henry Lyte's 1578 translation of Rembert Dodoens's Dutch herbal of 1574, John Gerarde in 1597, and Nicholas Culpepper in 1653 wrote of the virtues of plants known in the pharmaceutical and medical trades as "Mercurialis" or "Herba Mercurialis."

When the New World was discovered, the Europeans were well acquainted with "Herba Mercurialis" and various "false mercuries," such as "Baron's mercury" (*Mercurialis annua*), "Dog's [French] mercury" (*Mercurialis perennis*), "Maiden mercury" (*Mercurialis tomentosa*), and "Scotch mercury" (*Digitalis purpurea*). Europeans even began using the name "Mercury" and "English mercury" about 1450 for the edible garden herb *Chenopodium bonus-henricus*, although that is usually known as "Good King Henry" or "Allgood." Dodoens in 1574 was perhaps the first to point out the confusion, and Gerarde confirmed the problem in 1597.

In the New World, Europeans found plants that strongly resembled the Old World species in either morphology or physical reactions with the human body. Among the New World species are those called "Three-Seeded Mercury" (*Acalypha*). In the Southeast, the genus *Stillingia* has been called "Marcory" (mispronunciation of "mercury"); even Poison-ivy (*Toxicodendron radicans*) was called a "Mercury" (Austin 2004).

**Derivation of the name:** *Acalypha* is taken from the Greek *acalephe*, nettle, as explained by Linnaeus. *Neomexicana*, from New Mexico, notes that the first specimens of this species were found in that state.

**Miscellaneous:** The species of *Acalypha* on the western Baboquivari slope canyons is *A. californica*. That differs from *A. neomexicana* in that it is a shrub and not an annual. *Acalypha californica* grows from the western side of the Baboquivari Mountains to Quijotoa on the Tohono O'odham Reservation, the Ajo Mountains, and Organ Pipe Cactus National Monument, south through Sonora into Sinaloa, in Baja California Sur, and in southern California (Felger 2000). Equally common in the Altar Valley is *A. ostryaefolia*. This species is distinguished from *A. neomexicana* when plants are reproductive by the bracts below the flowers that have comb-like (pectinate) lobes along the margins. The bracts are only notched or shallowly toothed in *A. neomexicana*.

### *Chamaeysce albomarginata* (Euphorbiaceae) Rattlesnake Weed, *golondrina*, *wi:bkam*

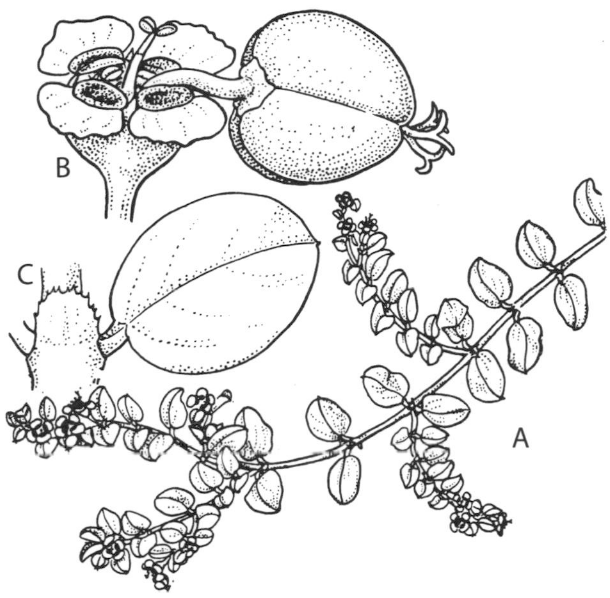


*Chamaeysce albomarginata*. A. Branch with flowers and fruits. B. Fruits attached to cyathium with four petal-like appendages and four glands of the involucre. C. Leaf with one white membranous scale at base of petiole. Artist: L. B. Hamilton (Parker 1958).

**Other names:** ENGLISH: rattlesnake weed (a translation of the Spanish name), white-margin sandmat (a reference to the prostrate habit and the ability to make a small "mat" on the ground surface, Arizona, New Mexico); SPANISH: [*yerba de la*] *golondrina* (swallow [herb]; used for the genus over much of Latin America), *hierba de la víbora* (rattlesnake herb, a name applied to at least 12 species in 7 families, Mexico); ATHAPASCAN: <*c'os be'i'c'oi*> ("vein spurter," *'ats'oos*, vein, maybe contraction of *biyah*, supporting it, and *choo'į*, used or useful, Navajo); UTO-AZTECAN: *corape* (Ópata, Sonora), *memeya* ("*Derramarse por todos lados*" [leak everywhere], Náhuatl, Mexico), *qénxamal* (Luiseño; for *E. polycarpa*), *qénxmal* (Luiseño, Juaneño dialect), *sïmïndïŋ tibohišn* (*sïmïndïŋ*, rattlesnake, *tibohišn*, medicine, Tübatulabal), *temal hepi'* (earth's milk, Cahuilla), *tɨ vikagivɨ* (earth collar, Kawaiisu), *tuvukpi* <*tuvúkpi*> (Hopi), *vii'ipkam* <*veeipkam*> (*viib*, to have milk, *kam*, attributive, Akimel O'odham, Arizona), *wi:bkam* (*wi:b*, milk, or *wipi*, breast, *kam*, attributive, Tohono O'odham); LANGUAGE ISOLATE: *i'kwikĭakĭa tsan'na* (*i'kwikĭakĭa*, make milk, *tsan'nu*, small, Zuni); (= *Euphorbia albomarginata*)

**Botanical description:** Perennial herbs with thickened roots, the stems dying back in drought, but new growth appearing any time of year with rains, prostrate, rooting at the nodes, herbage green, glabrous. Leaves short petioled, the blades 2–8 mm long, broadly ovate to orbicular, often with a red blotch in the middle, the margins entire. Stipules are relatively large, united into a triangular white scale with a fringed margin. Floral cups (cyathia) are solitary at nodes, the appendages white and showy. Capsules are ovoid, sharply triangular, 1.3–2 mm long. Seeds are 0.9–1.4 mm long, the upper surfaces with a rounded ridge crest, moderately flattened on either side of the ridge, and excavated on both sides of the septum on the ventral surface; mucilaginous when moistened.

**Habitat:** Roadsides, trails, mostly on clay and loam flats. 304–1,828 m (1,000–6,000 ft).

**Range:** Arizona (throughout most of state), California, southeastern Colorado, Kansas, New Mexico, Nevada, western Oklahoma, western Texas, and Utah; Mexico (Baja California, Chihuahua, Coahuila, Durango, and Sonora).

**Seasonality:** Flowering (February) August to October.

**Status:** Native.

**Ecological significance:** These are pioneer plants that grow in open sites. When the grasses and shrubs begin to get thick, these small herbs decline and die. Seeds of *Chamaeysce* are dispersed by ants (Yoshihiro et al. 2005). The seeds have special "food-bodies" (elaisomes), and those tiny structures attract these small foraging insects. Ants carry the seeds back to their nests, eat the food, and then discard the seeds outside. The seeds, having been put in a rich garbage-heap with plenty of open space and reduced competition, germinate to provide new colonies. At least, the species studied have this mechanism; no one has studied this one to see if it conforms.

These herbs are pollinated by a variety of insects and set few seeds unless the animals visit flowers. Ehrenfeld (1976) found 175 different species visiting the flowers on *C. albomarginata*, with individuals from ants and bees (Hymenoptera), flies (Diptera), beetles (Coleoptera), and bugs (Hemiptera). She noted that the pollination was a "mess and spoil" type and not specialized.

**Human uses:** Indigenous people, like almost everyone else, essentially consider all species of *Chamaesyce* the same. Rea (1997) speculated that probably all of the *Chamaesyce* species have the pan-Piman name similar to *vii'ipkam*. The noticeable trait of bleeding "milk" (white latex) prompted names from the Aztec *memeya* (Siméon 1885) to Akimel O'odham *vii'ipkam* (Rea 1997).

Ópata in northwestern Mexico use *corape* to treat wounds (Kay 1996). Northern Tepehuan use two other species to clean and treat sores, wounds, and inflammations (Pennington 1969). One of those species is also made into a tea to relieve colic. Akimel O'odham and their neighbors treat snakebites, scorpion stings, and stomach and bowel problems with several species (Curtin 1949, Rea 1997, Russell 1908). Apache, Keres, Zuni, and Navajo also use *Chamaesyce* and *Euphorbia* as remedies for several maladies (Elmore 1944, Moerman 1998, Reagan 1929, Stevenson 1915, Vestal 1952, Wyman and Harris 1951).

**Derivation of the name:** *Chamaesyce* is a name used by Dioscorides, fl. CE 40–80, from Greek *chamai*, on the ground, *sykon*, a fig; plants are prostrate and have fig-like fruits, or at least S. F. Gray, who named them, thought so. *Albomarginata* means having white margins, those along the margins of the cyathium or floral cup.

**Miscellaneous:** *Chamaesyce melanadenia* is at least as common in the canyons as *C. albomarginata*, or more so. *Chamaesyce albomarginata* has glabrous fruits and roots at the nodes while *C. melanadenia* fruits are pubescent, and their stems do not root at the nodes.

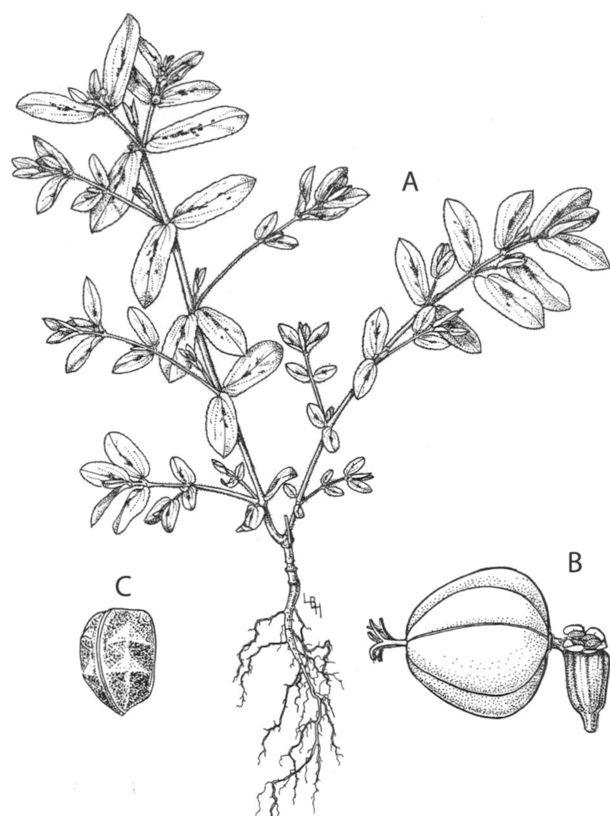


*Chamaeysce hyssopifolia*. A. Habit of plant. B. Fruits attached to cyathium. C. Seed. Artist: L. B. Hamilton (Parker 1958).

## *Chamaeysce hyssopifolia* (Euphorbiaceae) Hyssop Spurge, *golondrina, wi:bkam*

**Other names:** ENGLISH: hyssop spurge ("hyssop" is from Hebrew *esov*; loaned to Greeks about CE 1000, who transcribed it *ussupos*; later spelled in Old English as *ysopo, ysope, ysop, isop*, finally to hyssop in modern English; "spurge" from Old French *espurge*, modern French *épurge*, from *espurgier*, to free from or rid of impurity; in use for these kinds of plants by 1387), [hyssop-leaf] sandmat (Arizona, New Mexico, Florida); PORTUGUESE: *burra leitera* (donkey's milk, Brazil), *erva de leite* (milk herb, Brazil), *erva de santa luzia* (St. Lucia's herb, Brazil), *pau de leite* (milk tree, Brazil); SPANISH: *golondrina* (swallow, Mexico); UTO-AZTECAN: *vipgam* (it has much milk, Onavas Pima), *wi:bkam* (it has much milk, Tohono O'odham)

**Botanical description:** Annual herbs, to 6 dm tall, the main stems erect, mostly glabrous. Leaves opposite, the blades lanceolate to oblong, 5–30 mm long, apically rounded to acute, basally inequilateral and rounded to truncate, the margins serrate, sometimes pilose at least near base, the petioles 1–1.5 mm long, the stipules mostly united or partly distinct on distal nodes. Cyathia solitary but clustered, the involucre turbinate, 1.2–1.7 mm long, the glands circular to slightly elliptic, the appendages white to reddish or purplish, half-moon shaped (semilunate). Capsules are ovoid-triangular, 1.5–2.1 mm long. Seeds are ovoid-quadrangular, 1–1.4 mm long, the angles definite, larger or dorsal 2 facets with 2 or 3 transverse grooves (sulci) and low rounded ridges, pale brownish to whitish, without a protuberance (caruncle).

**Habitat:** Open rocky sites, grasslands, deserts. 304–1,828 m (1,000–6,000 ft).

**Range:** Arizona (Gila, Graham, Pinal, Maricopa, Yavapai south to Cochise, Pima, and Santa Cruz counties) east to southern Florida; Mexico (Chihuahua and Sonora); to South America.

**Seasonality:** Flowering July to September (November).

**Status:** Native, or perhaps introduced during the Spanish colonial period.

**Ecological significance:** This annual is extremely sensitive to moisture changes, and it may be one of the dominant species in the herb layer after summer rains. As the soil begins to dry out, it is among the first species to wither and die.

This spurge is locally abundant in moist canyons in western Texas (Brewster, El Paso, Jeff Davis, and Presidio counties). In the Altar Valley, it is less particular about where it grows, being found seasonally in many sites. All of the places it grows have been recently disturbed by some factor, either natural like stream margins, or "artificially" by humans and livestock. This species is an excellent indicator of disturbance.

The species does not require the visits of insects for successful pollination, and it abundantly sets seed (Ehrenfeld 1976).

**Human uses:** None found in this area. Perhaps its recent arrival accounts for no local human uses. In what is perhaps its homeland in South America, the species is a popular medicinal herb (Mors et al. 2000). Could the species be confused with *C. hypericifolia*? That species is used by almost all cultures of people who live where it grows. For example, the Onavas Pima crush that *Chamaesyce* and use it to relieve scorpion stings and ant bites (Pennington 1980). Those plants grow from Colombia and Venezuela north to Florida, Georgia, and Texas (Correll and Johnston 1970). Few people can distinguish the two, but differences are the inconspicuous stipules, seed size, and fruit size—not traits that the average person will easily observe. Moreover, these characteristics mostly need magnification to determine, and few "real" people carry lenses in their pockets.

Hyssop Spurges are used medicinally in Brazil, Cuba, and Jamaica (Hocking 1997, Mors et al. 2000). In Jamaica, the herbs are made into an infusion to treat colds; people in Cuba consider *Chamaesyce* a diuretic and an emmenagogue; and the plants are used to treat inflamed eyelids and in poultices for erysipelas in Brazil. The latex is applied to warts, calluses, and ringworms in an attempt to remove them.

**Derivation of the name:** *Chamaesyce*, see *C. albomarginata* for etymology. *Hyssopifolia* means that the herbs have leaves resembling hyssop (*Hyssopus*) of the mint family (Lamiaceae).

**Miscellaneous:** There is continuing debate on whether or not to segregate these species into *Chamaesyce* or submerge them into *Euphorbia*. For example, Felger (2000) and Felger et al. (in prep.) put the species in *Euphorbia*. People in other parts of the world keep them in *Chamaesyce* (e.g., Wunderlin and Hansen 2003). At least in Florida, the opposite leaves have oblique bases in *Chamaesyce*, while those of *Euphorbia* are alternate at least below the flower clusters, and the bases are equal. There are other technical differences, but those two are the most easily seen.

## ***Manihot angustiloba*** **(Euphorbiaceae)** **Narrow-Leaved Cassava,** ***yuca del cerro***

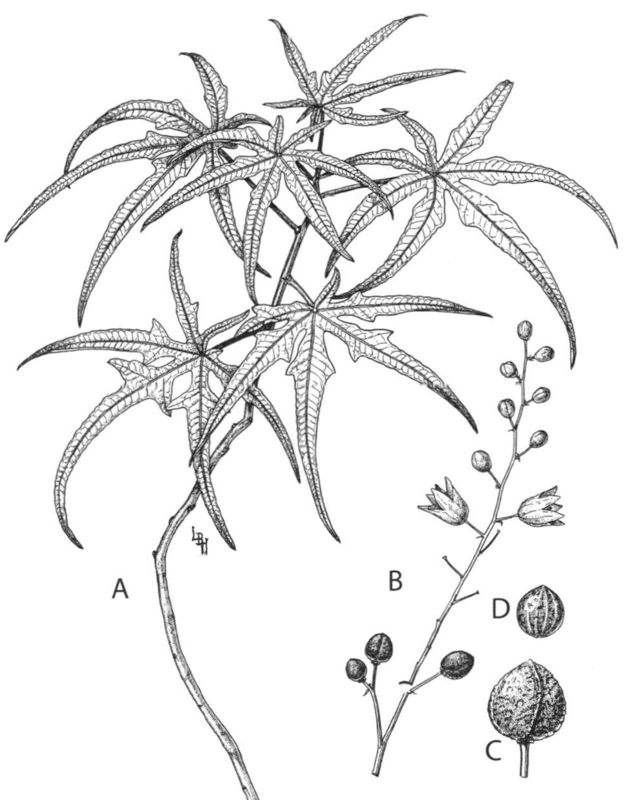


*Manihot angustiloba*. A. Leafy branch. B. Flower cluster, with male flowers above and fruits below. C. Fruit, enlarged. D. Seed, enlarged. Artist: L. B. Hamilton (from original artwork).

**Other names:** ENGLISH: desert mountain manihot (Arizona, New Mexico), narrow-leaved cassava (see derivation below); SPANISH: *pato de gallo* (rooster foot, Chihuahua), *yuca del cerro* (wild [mountain] manioc, Chihuahua, Sonora; *yuca* was the word the Caribbean Tainos used for *Manihot*; subsequently adopted into American Spanish and applied to many plants with tuberous roots regardless of their taxonomic relationships)

**Botanical description:** Shrubs or suffrutescent perennials, 0.3–2 m tall, herbage glabrous, from a woody rootstock, stipules 1–2 mm long, falling early. Leaf blades are divided nearly to the base into 5–7 lanceolate to linear-lanceolate lobes, 4–15 cm long, 3–12 mm wide, all or most of primary lobes sinuate to again laterally lobed, the blade dark green above, paler below, prominently nerved. Inflorescences are few-flowered. Staminate calices are green to yellow-green, 9–11 mm long, glabrous, on slender pedicels 6–8 mm long. Capsules are rugose to low-tuberculate, 12–15 mm high, to 2.5 cm wide, 3-lobed or fewer. Seeds resemble beetles (scaraboid), sometimes nearly orbicular, 8–10 mm wide, 10–12 mm long, gray to pale buff with dark brown to olive green mottling and streaks, caruncle orbicular to oblong.

**Habitat:** Rocky slopes, hillsides, outwash slopes. 900–1,500 m (3,000–5,000 ft).

**Range:** Arizona (Cochise, Pima, and Santa Cruz counties) and New Mexico; Mexico (Baja California, Chihuahua, Edo. México, and Sonora to Jalisco).

**Seasonality:** Flowering (June) July to August (September).

**Status:** Native.

**Ecological significance:** These shrubs are almost always found on rocky slopes and among the boulders and rubble on the sides of canyons. There, among the rocks, *Manihot* finds enough seasonal moisture to survive. Most members of the family are poisonous, containing cyanogenic glycosides in many organs (Mabberley 1997).

**Human uses:** None found for this species.

The most famous member of the genus is *M. esculenta*. One of the earliest records of manioc came from Fernández de Oviedo y Valdés's book of 1526 where he wrote: "[*La yuca es*] *alimento para sustenar la vida, licores dulces y agrio que sirven como miel y vinagre, potaje que comen y gustan los indios, ramas para leña cuando no hubiere otras, y veneno tan potente y malo. . . . Otra particularidad . . . en alguna parte del continente se hace muy buen vino del mismo*" (The yuca is food that sustains life; sweet and sour liquids that serve as honey and vinegar; a stew that is eaten and enjoyed by the Indians; firewood from the branches when there is no other; and a potent and deadly poison. . . . Another feature . . . in some parts of the continent they make a good wine from it).

Other names for *Manihot* and its products are cassava (from Taino *caçábi*, bread), manioc (see derivation of genus below), tapioca (from Tupí-Guaraní *tipioca*; based on *tipi*, residue, dregs, and *og*, *ók* to squeeze out; used by Georg Marcgrave in 1648), *yuca* (from Taino *yuca*, originally the root of *M. esculenta*, while the plant was called *yucubia*) or *mandioca* (Portuguese, see derivation of genus). *Manihot esculenta* is one of the most important food plants in the world (Austin 2002). Other species yield rubber (*M. dichotoma*, *M. glaziovii*).

Tarahumara use an unidentified species in Chihuahua to obtain a yellow dye (Pennington 1963), using the entire plant. Perhaps the one they use is the same species (*M. chlorosticta*; called *algodoncillo*) used by the Onavas Pima to reduce fevers (Pennington 1980). Northern Tepehuan consider a species in their area good forage for livestock (Pennington 1969).

**Derivation of the name:** *Manihot* is taken from the Brazilian Tupí name, *mãddi'og*, also giving rise to *mandioca* in Portuguese and manioc in English. *Angustiloba* means narrow-lobed.

**Miscellaneous:** Also found in the Baboquivari Mountains is *M. davisiae*, an endangered species in the region (Arizona Rare Plant Committee 2001). The differences between this taxon and *M. angustiloba* are minimal. The major distinction is that the ends of the leaf lobes are broadened in *M. davisiae*, while they remain narrow in *M. angustiloba*. Plants that conform with that taxon have been found in the Baboquivari, Santa Rita, and Santa Catalina Mountains.

## *Acacia greggii* (Fabaceae)
## Catclaw Acacia, *uña de gato*, *'u:paḍ*

**Other names:** ENGLISH: [long-flower] catclaw acacia (comparing the thorns with a cat's claws, and using the old Greek *akakia* for thorny), devil's claw, tear-blanket (California); SPANISH: *algarroba* (usually used for *Prosopis*, adapted from Arabic *al-kharrubah* [the bean-pod], originally applied to carob or *Ceratonia siliqua*, reapplied to other species in the New World), *gatuño* (cat claw, Chihuahua), *patitos* (little feet, New Mexico), *tepame* (Mexico), *tésoto* [*tesota*, *tésota*] (from Cahita *teso* + *ta*, Sonora), *uña de gato* (cat's claw, New Mexico, Chihuahua); ATHAPAS-

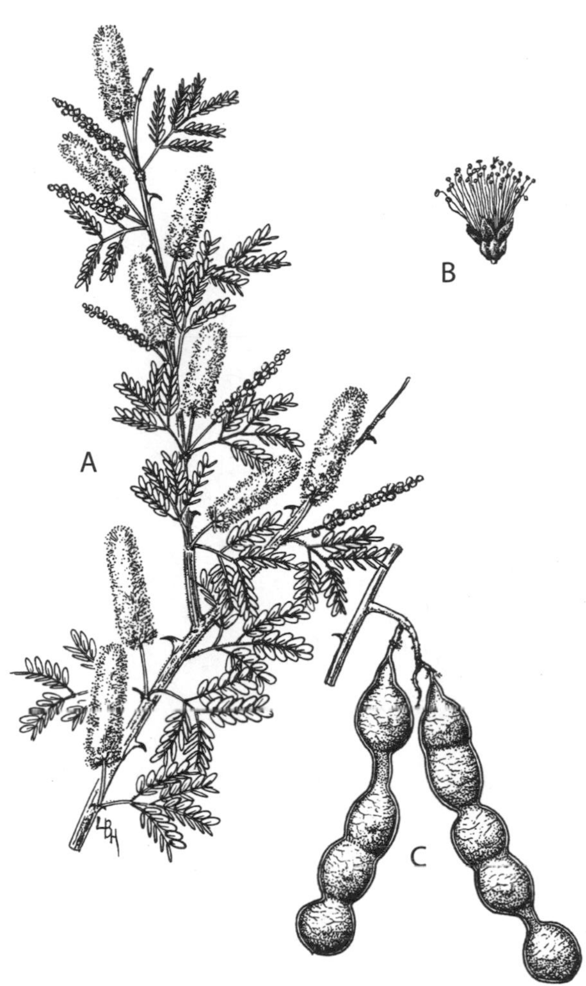


*Acacia greggii*. A. Flowering branch. B. Enlarged single flower; note 20 or more stamens. C. Fruiting branch. Artists: L. B. Hamilton, A, C (Humphrey 1960); E. L. Keplinger, B (Dayton 1937).

CAN: *ch'il yíjish <ch'il gotiza>* (*ch'il*, bush, *yíjish*, that scratches, Western Apache); HOKAN: *tis* (name for harpoon or its point, Seri); UTO-AZTECAN: *huʾupa kekʾala* (*huʾupa*, mesquite, *kekʾala*, mangy, Yaqui), *sichingily <sichingal, sichingil>* (Cahuilla), *teso* (Cahita), *tümippüh* (Panamint), *'u:paḍ <'u:padh, uupat>* (Hiá Ceḍ and Tohono O'odham), *uupaḍ* (Akimel O'odham); YUMAN: *ka'djása* (Havasupai), *kitcás*[a] *<gijes>* (Walapai)

**Botanical description:** Shrubs or small trees 2–3(–8) m tall. Stems have recurved prickles, usually along the internodes or occasionally at a node. Pinnae 1–3 pairs per leaf. Flowers are sweet-scented, disposed in cylindrical, spike-like racemes, the flowering portion often 1.8–3 cm long, the filaments creamy white. Pods are 5–15 cm long, 1–2 cm wide.

**Habitat:** Washes, slopes, canyons. Sea level to 1,525 m (5,000 ft).

**Range:** Arizona (Coconino and Mohave to Greenlee, Cochise, Pima, and Yuma counties), southeastern California, southern Nevada, and Texas to southwestern Utah; Mexico (Baja California Sur, across the border Mexican states, and Durango).

**Seasonality:** Flowering (April) May to (June) September.

**Status:** Native.

**Ecological significance:** Foliage has a high cyanide content, but pods often are chewed by rodents (Felger et al. 2001). Egg cases of Praying Mantis are commonly found on the branches. This plant is a favored nesting site for the Verdin (*Auriparus flaviceps*), especially the highest branches (Jaeger 1941). At least in California, the lower branches are used as a "haven" for rabbits.

The species has been recorded from all around the Baboquivari Mountains, including in the Altar Valley on the Buenos Aires National Wildlife Refuge, and in several mountain ranges (Ajo, Rincon, Roskruge, Santa Catalina, and Santa Rita) and surrounding areas. The shrubs extend west across the Tohono O'odham Reservation to the Reserva de la Biósfera El Pinacate y Gran Desierto de Altar in Sonora (Felger 2000).

**Human uses:** Fruits ripen at the beginning of the monsoon rains. Few groups seem to use the pods or seeds, probably because they are bitter (Hodgson 2001). Still, Akimel O'odham, Apaches, Cahuilla, Kumiai, and Seri formerly roasted and ground the fruits (Bean and Saubel 1972, Hodgson 2001, Rea 1997, Russell 1908). Seeds were pounded into a meal that was eaten in *atole* or cakes (Barrows 1900, Castetter 1935, Bean and Saubel 1972). People in New Mexico suck the nectar from the flowers (Curtin 1947).

Wood is strong and hard; Seri use it for tools, bows for violins, hunting bows, and carrying yokes (Felger 1977, Felger and Moser 1985). Yaqui also make bows from the wood (Rea 1997). As the Seri name *tis* implies, the wood was preferred for making harpoon points for fish and turtles before metal was introduced (Felger and Moser 1985).

The Tohono O'odham also made a scraper for working animal skins from the branches (Castetter and Underhill 1935). Women among the Tohono O'odham keep dried buds and blossoms as a perfume sachet among their possessions.

**Derivation of the name:** *Acacia* was taken from Greek, *akakia*, tip, thorn, sharp point, thus alluding to the armament of most species. *Greggii* commemorates Josiah Gregg, 1806–1850, who collected plants in Mexico and died in the wilderness in northern California, which he reached in 1849. He was a frontier trader and author who knew a little about medicine and surgery. Therefore, he became known as "doctor." His books are considered frontier classics.

**Miscellaneous:** There are three other *Acacia* in the Baboquivari Mountains. In Brown and Thomas Canyons and lower in Altar Valley is *Acacia angustissima* (White Ball Acacia; *timbe*), which grows in grasslands, on rocky slopes, and in washes. The shrub's small white globes of flowers appear from June to September. This species is a favorite food of the Masked Bobwhite Quails (*Colinus virginianus ridgewayi*). The White-Thorn Acacia, *largoncillo*, or *gidag* (*Acacia constricta*), is more common on the slopes of the Sierrita Mountains than in the Baboquivaris, and it grows in grasslands and on rocky slopes. The yellow flower clusters appear from May to August.

*Acacia farnesiana* (Sweet Acacia; *huisache*, from Náhuatl *huitztli*, thorn or spine, and *ixachi*, a great amount) is rare and is represented by a few plants near the Environmental Education Center in central Brown Canyon. This plant is more common on the western slopes of the Baboquivari Mountains, and it is always on rocky slopes and along washes. This small tree or shrub also grows in the Pajarito Mountains (Santa Cruz County) (Toolin et al. 1979). The yellow flower heads appear on these trees or shrubs from April to June.

## *Calliandra eriophylla* (Fabaceae) Fairy Duster, *huajillo, cu:wĭ wuipo*

**Other names:** ENGLISH: fairy duster [fairy-duster] (a duster used by fairies; "fairy" in English by 1320, through Old French *faerie* and *fae*, derived from Latin *fata*, the Fates), false [bastard, mock] mesquite [catclaw]; SPANISH: *brasilillo* (little Brazil-wood, usually applied to spiny *Rhamnus*, New Mexico, Chihuahua), *cabellito* [*cabellos, pelo de ángel*] (little [angel] hair, Mexico), *cabeza* [*de*] *ángel* (angel head, Baja California), *charresquillo* (little thicket, San Luis Potosí), *cósahui* [*del norte*] (comparing it with the *cozahuico* or *Manilkara*, Sonora), *huajillo* <*guajillo*> (little *guaje*, from Náhuatl *huaxin* <*hoatzin, hoaxin, uaxim, waxim*>, ultimately from *huitzli*, spiny, a com-

*Calliandra eriophylla*. A. Flowering branch. B. Detail of flower. Artist: E. I. Keplinger (Dayton 1937).

parison with *Acacia farnesiana*, Mexico), *mezquitilla* (little mesquite, Mexico), *plumita* (little plume, interior Mexico); HOKAN: *haxz iztim* (dog's hipbone, Seri); UTO-AZTECAN: *cu:wĭ wuipo* <*cu:wi wu:pui*> (jack-rabbit eyes, Tohono O'odham), *taaseyueylalá* <*ta-a-sey-ueylalá*> (Guarijío)

**Botanical description:** Woody shrubs, much-branched, 0.5–0.8 m tall, with firm but flexible stems and grayish bark, the young twigs, peduncles, and leafstalks densely to moderately pubescent with short white hairs, the leaflets sparsely hairy. Leaves gradually drought deciduous and frost sensitive, the pinnae usually 3 pairs, sometimes 1–2; leaflets 2.5–6 mm long, 6–15 pairs per pinna. Flowers are sessile, in few-flowered, stalked heads. Stamens are 1.5–2 cm long, showy, white to pinkish, opening at night, drooping in the daytime; anthers small. Pods are 4–5.6 cm long.

**Habitat:** Rocky slopes, mostly grasslands and desertscrub. To 1,600 m (5,500 ft).

**Range:** Arizona (Greenlee to southern Mohave, south to Cochise, Pima, Santa Cruz, and Yuma counties), southeastern California, southwestern New Mexico, Nevada, Utah; Mexico (Baja California, Chihuahua, Sonora, and south to Chiapas). Another variety grows in southern Texas.

**Seasonality:** Flowering (February) March to April (September).

**Status:** Native.

**Ecological significance:** These small shrubs occur in almost every habitat on the Buenos Aires National Wildlife Refuge. Part of that abundance may be an artifact because Fairy Duster is able to survive where native grasses have been overgrazed. These shrubs are favored by deer and are among the most valuable browse for wild and domestic animals in southeastern Arizona (Dayton 1937, Kearney and Peebles 1951, Turner et al. 1995). When the flowers open in the spring, and sometimes again in the fall, they may be visited by resident and migrating hummingbirds. Pollen adheres to butterfly wings (Turner et al. 1995). Because the flowers open at night, moths are perhaps their primary visitors. White-Lined Sphinx Moths (*Hyles lineata*), sometimes a half-dozen at a time, spend considerable time visiting flowers. Seeds may be eaten by birds, and that may be the allusion in the Mexican name *huajillo*, as one theory is that the word originally referred to the quail.

**Human uses:** Yavapai take a decoction of leaves and stems after childbirth (Moerman 1998). Although nothing has been found to indicate that Fairy Duster contains tannins, a commonly used plant hemostat, those compounds are known in other species (Getachew et al. 2000). Tarahumara treat gonorrhea with *Calliandra* (Pennington 1963), first boiling the entire plant for several hours and then setting the mixture aside for several days. Each morning for three months, a half-gourdful of this decoction is drunk before eating.

Seri call *Calliandra*, *Hoffmanseggia*, and *Krameria* by the same name. The latter two are used to make a red dye, but Felger and Moser (1985) made no comment on *Calliandra*. Presumably, it too yields a dye, but Felger (personal communication, Dec. 2004) was doubtful. Perhaps the Seri, like the Tarahumara, use shavings from *Calliandra* stems as a fixing agent for dyes (Pennington 1963).

Farther south, other species are used, including *C. anomala* (*cabellos de ángel*, *cabellitos*, *tlacoxiloxochitl*), growing from Chihuahua to Oaxaca, and *C. houstoniana* (*tabardillo*) in Sinaloa. The former has been used to treat eye problems, inflammations, fevers, and other problems since at least the time of the Aztecs (Martínez 1969). *Tabardillo* is used to treat gum problems, malaria, and heart trouble.

**Derivation of the name:** *Calliandra* is from Greek *kalli*, beautiful, and *andros*, anther, stamen. *Eriophylla* is from Greek *erio*, wool, and *phylla*, leaf.

**Miscellaneous:** One other species, *C. humilis* var. *reticulata*, is known from the Baboquivari Mountains, but it has not been found in Brown Canyon. *Calliandra humilis* is herbaceous above ground, and it rarely reaches more than 20 cm tall. It has 10–18 pairs of leaflets. *Zapoteca formosa* is similar and has been considered a *Calliandra*.

## *Desmodium batocaulon* (Fabaceae) Tick-Clover, *trifolio*

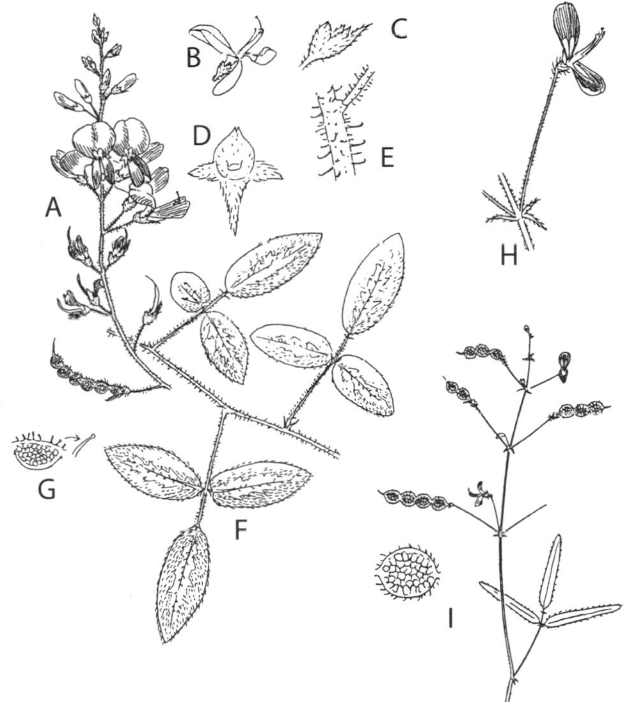


*Desmodium*. A–G. *D. batocaulon*. H–I. *D. rosei*. A. Flowering and fruiting branch. B. Flower. C. Calyx. D. Ovary. E. Detail of hairs on stems. F. Leafy branch. G. Detail of legume segment and hairs. H. Flowering and fruiting branch. I. Detail of flower and pubescent stem. J. Detail of legume. Artist: R. D. Ivey (Ivey 2003).

**Other names:** ENGLISH: beggar's-ticks ("beggar" has meant one who habitually asks alms since ca. 1225; Thoreau apparently linked it with "ticks" in *Walden* in 1854 when he wrote, "It was over-run with Roman worm-wood and beggar-ticks, which last stuck to my clothes"; also used for *Bidens*, q.v.), simple-leaf [Wright's] tick-clover ("clover," the form *clover* is rare before 1600 [one example of *clouere* ca. 1265], and did not prevail much before 1700; the usual Middle English and sixteenth-century form was *claver*; the earliest Old English glossaries have *clabre, clafre*; Dutch *klaver*, Danish *klever, klöver*, Norwegian *klöver, klyver*, Swedish *klöfwer*, originally applied to *Trifolium* by ca. CE 1000), [San Pedro] tick-trefoil [ticktrefoil] (named because the seeds adhere like ticks to the fur of animals; in use by the time of Thoreau in 1853, who wrote of "*Desmodium nudiflorum*, naked-flowered tick trefoil, some already with loments round-angled"; "trefoil," from Latin *trifolium, tri-* three + *folium* leaf, whence Italian *trifoglio*; Portuguese *trefueil*, French *trèfle* by 1450); SPANISH: *trifolio* (Mexico); ATHAPASCAN: *naa'oł'í yilt'ą́ą'í* (*naa'oł'í*, bean, *yilt'ą́ą'í*, leaves resembling, Navajo)

**Botanical description:** Perennial herbs from woody caudex. Stems are slender, climbing or prostrate, 3–15 dm long, densely pilose with hook-tipped hairs (uncinate) when young. Stipules are deltoid-ovate, 2–3 mm long, persistent. Leaflets are 3, lanceolate-ovate to lanceolate-oblong, 5–25 mm wide, 1.5–6 cm long, acute to obtuse and minutely apiculate apically, basally rounded, usually with a lighter-green region along the midrib. Flower cluster is racemose-paniculate, pilose. Calices are nearly glabrous, 4 mm long, teeth deltoid-lanceolate, 1–2 mm long. Corollas are purple to lavender, the banner 6–8 mm long. Fruits and loments, 4–7 jointed, deeply notched beneath, shallowly above.

**Habitat:** Rocky slopes, stream banks. 1,100–1,980 m (3,500–6,500 ft); in Mexico 1,250–2,300 m (Martin et al. 1998).

**Range:** Arizona (Cochise, Graham, Greenlee, Pima, and Santa Cruz counties) and the boot heel of southwestern New Mexico; Mexico (Chihuahua and Sonora).

**Seasonality:** Flowering June to September.

**Status:** Native.

**Ecological significance:** This species is on the western fringe of its range in the Baboquivari Mountains. The genus is rare in the arid lands west of these mountains in the southwestern part of the state, with only *D. procumbens* being known (Felger et al. in prep.). The genus is absent, for example, from California, Colorado, and Utah (Welsh et al. 1987, Weber 1990, Hickman 1993). Few species of *Desmodium* come west of the middle of the Great Plains, roughly the 100th parallel (Barkley 1977, 1986).

The six species known from Brown Canyon and the seven from the Baboquivari Mountains have their apparent affinities in the mountainous regions of Mexico and not in the eastern United States. All of the seven species are shared with Mexico and regions to the south, and none is found in the eastern half of Texas (Correll and Johnston 1970).

**Human uses:** No use found for this species, or for any others in the southwestern United States. However, the situation is totally different in the eastern United States and the Caribbean (Austin 2004). For example, Moerman (1998) found eight species being used as medicines in the eastern United States. Records for use exist among three major linguistic groups, the Algonquian, Iroquoian, and Muskogean people.

**Derivation of the name:** *Desmodium* is from Greek *desmos*, a bond, chain, or headband, plus Greek *-idium*, small. Fernald (1950), Diggs et al. (1999), and perhaps everyone else except Quattrocchi (1999) thought the name referred to the segmented fruits; Quattrocchi could not make up his mind and said it referred to the "stamens joined together or to the pods or to the long racemes of flower or the flexible branches." *Batocaulon* is from Greek *batos*, bramble, thorn-bush, and *kaulos*, stem.

**Miscellaneous:** *Desmodium* is a genus of about 450 species that grows in the warm parts of the world, being particularly rich in Eastern Asia and Brazil (Mabberley 1997).

In Brown Canyon *D. grahami* (3-foliolate; fruit less deeply notched above than below), *D. procumbens* (3-foliolate; fruit about as deeply notched above and below, segments contorted), *D. psilocarpum* (single leaflet), and *D. rosei* (3-foliolate, leaflets linear to linear-lanceolate; fruit less deeply notched above than below, segments orbicular to elliptic, margins flat) all have been recorded. All flower from August to September.

## *Erythrina flabelliformis* (Fabaceae) Southwest Coral Bean, *chilicote, ba:wui*

**Other names:** ENGLISH: Indian bean, Southwestern coral bean (an allusion to the coral color),

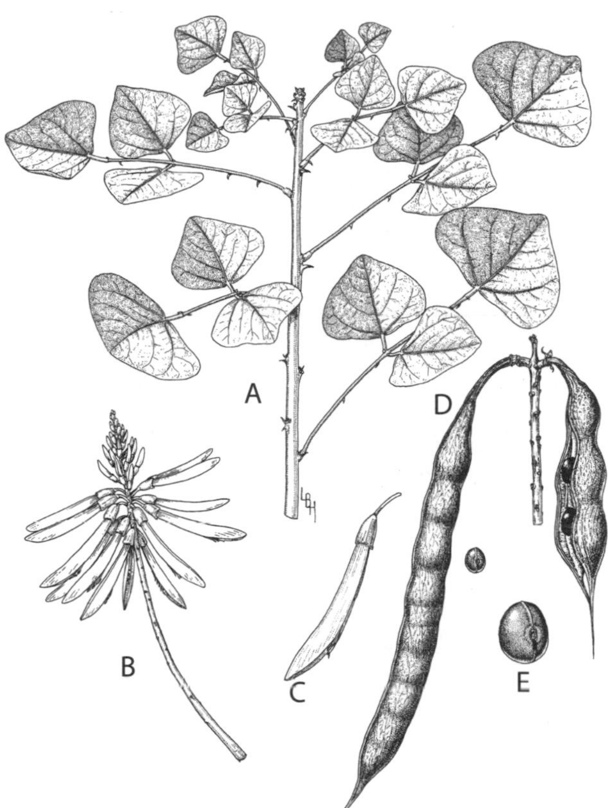


*Erythrina flabelliformis*. A. Branch with leaves. B. Flower cluster. C. Flower, enlarged. D. Fruiting branch, with seed. E. Seed, enlarged. Artist: L. B. Hamilton (from original artwork).

coral-tree; SPANISH: *chijol* (from Náhuatl *chixolli*, swollen pod), *chilicote* (based on the Zapotec word *chilla*, luck or oracle, alluding to a custom of predicting the future with the seeds, Chihuahua, Durango, Sonora), *colorín* (red one, Chihuahua, Durango, Sonora, south), *coralina* (little red one, Baja California), *corcho* (cork, Sonora), *frijolillo* (little bean, Chihuahua), *guaposi* (Mexico), *peonía* [*pieoneo*] (Chihuahua, Sonora; also used for *Cyperus esculentus* in the Valley of Mexico), *pionilla* (little peony, from Greek *paeonia*, based on the name Paeon or Paion, the physician of the immortal gods, originally applied to the peony, *Paeonia*, because of the medicinal red or viable, and black or inviable, seeds; later applied to numerous plants with red and/or black seeds), *tristesa* (sadness, Chihuahua), *zompantle* <*tzompantle, zumpanilla*> (from Náhuatl *tzon-pantli*, hair banner); HOKAN: *xloolcö* (Seri); UTO-AZTECAN: *baowi* <*bavul, bawi*> (bean, Onavas Pima), *bawui* (Akimel O'odham), *ba:wui* (bean, Tohono O'odham), *caposí* (Tarahumara), *chiloko'ot* (cognate with Náhuatl or from Spanish, Mountain Pima), *chirikote* (Yaqui), *jévero* (Mayo), *kaposí* <*apoši, apošiki*> (Tarahumara), *tzinacancuáhuitl* (*tzinacan*, bat, *cuáhuitli*, tree, Náhuatl), *xoloco* (from *xolotl*, servant, *co*, came; a place near Tenochtitlan on a small stream by the same name; this is where Montezuma II met the Spanish, Náhuatl), *waspósi* (Guarijío)

**Botanical description:** Shrubs to 1.5 m tall, the stems light tan, leafless much of the year, with scattered spines that develop into conical tubercules on large plants. Leaves are compound, with 3 leaflets, these triangular, each 4–7 cm long, 5–10 cm wide. Racemes are terminal on branches. Flowers are 2–5 cm long, bright red. Legumes are brown, knobby (torulose), 1.2–2.5 dm long. Seeds are 12–14 mm long, red, or occasionally yellowish.

**Habitat:** Slopes and ridges. 900–1,680 m (3,000–5,500 ft), farther south in Sonora, 100–1,950 m.

**Range:** Arizona (Cochise, Pima, and Santa Cruz counties) and southwestern New Mexico; Mexico (Baja California, Chihuahua, Sonora, south to Jalisco, Michoacán, and Morelos).

**Seasonality:** Flowering May to June (August).

**Status:** Native.

**Ecological significance:** The species is a good indicator of winter temperatures because it is cold-sensitive. When slopes are covered with Coral Bean, it is certain that they are in a warm belt. Stems freeze at −2.2 to −4.4°C (Turner et al. 1995). Plants often have thick stems a few inches above ground that produce small branches. This pattern results from repeated freezing in wintertime.

Flowers are preferred by resident and migrating hummingbirds, who extract nectar from the elongate folded banner that mimics a tubular corolla. Other birds may be partly responsible for dispersal of the seeds, as they also seek out red. However, seeds or whole pods also fall from *Erythrina* and are washed down hillsides in water. When seeds land in a favorable site, they germinate.

**Human uses:** When Padre Kino passed the village of *Topawa* (it is a bean) in 1699, he found Tohono O'odham boys playing a game with red seeds called *mawi* (probably incorrectly transcribed *ba:wui*) (Granger 1960). That sounds like a meaningless child's game, but it likely reflects the importance of *ba:wui* for divining the future, as it did among people farther south (Austin 2004). The Zapotec-derived name *chilicote* (based on "oracle") passed into Spanish, and the O'odham use continued until at least the recent past, if not now. Russell (1908) mentions the game, and Rea (1997) found that the village of Bapchule was one of the last places on the Gila Reservation where this gambling game persists.

Although the species is not native to the Akimel O'odham region, some still remembered the plants (Rea 1997). There is even the old song *Gidval Ñe' ñei* (Blue Swallow Song) that mentions *Bawuigam Do'ag* (Coral-Bean Mountain). Sylvester Matthias identified it as the place we now call Kitt Peak (Rea 1997).

Onavas Pima carve the wood into masks used during Pascola (Pennington 1980). Yaqui carve the soft wood into *santerias* and other items, and they and Onavas Pima make rust-colored or yellow dye from the bark (Pennington 1963, 1980). When the bark is used as a dye, either lime or urine is the mordant. As indicated by the name *corcho*, the wood is used to make corks (Gentry 1942, Pennington 1980).

Tarahumara made a drink of *Erythrina* seeds that is used to treat intestinal disorders (Pennington 1963). The beans are toasted and ground and then mixed with water. A small drink is emetic. Crushed beans are used to treat toothaches. Medical use is dangerous, since the seeds contain toxic alkaloids. Indeed, a few eastern Tarahumara told Pennington (1963) that seeds mashed and mixed with food were used to poison humans. Northern Tepehuan use small portions of seeds to concoct a laxative (Pennington 1969). The Seri make a tea from the cooked seeds to treat diarrhea (Felger and Moser 1985).

Northern Tepehuan use the seeds in making necklaces, which they call *baoáivukai* (Pennington 1969). Sometimes they use the seeds to polish the outside of pots.

**Derivation of the name:** *Erythrina* is from Greek *erythros*, red; flowers and seeds are red in most species. *Flabelliformis* is from Latin *flabella*, a fan, and *formis*, shaped, alluding to the leaflets.

*Eysenhardtia orthocarpa*. A. Branch tip with leaf and flower cluster. B. Fruiting branch. Artist: L. B. Hamilton (Benson and Darrow 1945).

## *Eysenhardtia orthocarpa* (Fabaceae) Kidney-Wood, *palo dulce*

**Other names:** ENGLISH: kidney-wood; SPANISH: *matariqui* (comparing it with *matarique*, an herbaceous Asteraceae; the word is based on Cahita spelled as *maturi* and *maturín* in Yaqui fide Sobarzo 1991, although Laferrière 1991 said from Tarahumara, Sonora), *palo dulce* (sweet wood, Sonora to Oaxaca); UTO-AZTECAN: *ba'iva* (bean, Mountain Pima), *bakuhuío* <*baiguo*> (breeze, Mayo), *koksigam* (mistranscription of *hiosgam*, flower? Mountain Pima), *pa'wió* <*pahuió*> (Guarijío), *tenipari* (Mountain Guarijío)

**Botanical description:** Shrubs or small trees 1.5–6 m tall. Leaves are odd-pinnate, 5–15 cm long, the leaflets 21–41, oblong, 8–20 mm long, finely puberulent to glabrous, punctate below with minute, dark glands. Racemes are 5–12 cm long. Flowers are white, 5–7 mm long. Legumes are 1-seeded, indehiscent.

**Habitat:** Canyons. 1,066–1,828 m (3,500–6,000 ft).

**Range:** Arizona (Cochise, Graham, Pima, Pinal, and Santa Cruz counties) and southwestern New Mexico; Mexico (Chihuahua, Sinaloa, Sonora, and Jalisco).

**Seasonality:** Flowering (April) May to August (September).

**Status:** Native.

**Ecological significance:** This plant is a thorn-scrub species (Brown 1982), and Turner et al. (1995) recognized two varieties. The variety that belongs to the Madrean evergreen woodlands is *orthocarpa*, while var. *tenuifolia* is part of the Sonoran desert-scrub and Sinaloan thorn-scrub. Thus, plants in Brown Canyon are var. *orthocarpa*, and they grow along waterways and on canyon slopes and hillsides, where they are disjunct from the main southern populations. *Eysenhardtia polystachya* replaces this species southward in Mexico. *Eysenhardtia orthocarpa* does not extend west of the Baboquivari Mountains into the drier lands of the Sonoran Desert (Felger 2000, Felger et al. 2001, Felger et al. in prep.).

*Eysenhardtia* are browsed by White-Tail and Mule Deer. Kidney-Wood is cultivated in southern Arizona, and it is hardy to −8° C (17°F), thus making it well suited as a patio tree (Felger et al. 2001). Kidney-Wood's glands on its leaves have a pungent smell when crushed, like tangerines and then mangos (Felger et al. 2001).

**Human uses:** The species is been widely used to treat urinary problems through Mexico (Standley 1920–1926); hence, the common name "Kidney-Wood." Mayo also consider the tree a remedy for fever (Yetman and Van Devender 2001). The name "*palo dulce*" seems to refer to livestock relishing the branches and growing fat. Onavas Pima crush the bark of *E. polystachya*, steep it in water, and use it as a body lotion to reduce fever (Pennington 1980). Mountain Pima use it to treat bruises, upset stomach, and coughs (Pennington 1973, Reina-G. 1993).

Wood continues being used for tool handles and dwellings (Gentry 1942, Yetman and Van Devender 2001, Yetman and Felger 2002). Guarijío also burn the wood for fuel; they say that fence posts made from *paˀwió* never rot.

Mayo use the wood alone to produce a blue to pinkish dye (Yetman and Van Devender 2001). Color varies, depending on the boiling time. When it is boiled with *tajimsi* (*Krameria erecta*), wool is permanently colored purple.

Northern Tepehuan call *E. amorphoides* the *áli sákoi* (*áli*, little, *sákoi*, *Lysiloma watsoni*), and they use its branches to build fences (Pennington 1969). They use branches of the *Lysiloma* in the same way and sometimes use the same common name for it. Northern Tepehuans profess no use for *E. polystachya*. The Tarahumara, on the other hand, use the bark to make a medicine to treat pain caused by internal injuries (Pennington 1963).

**Derivation of the name:** *Eysenhardtia* is dedicated to German Carl (Karl) Wilhelm Eysenhardt, 1794–1825, professor of botany at Königsberg. *Orthocarpa* means straight fruit.

**Miscellaneous:** The related *E. polystachya* is a Mexican tree that was erroneously reported for Arizona. That species was formerly imported into Europe as a medicine. That tree, too, is called Kidney-Wood, or by the pharmaceutical name with the same meaning, *lignum-nephriticum*. Wood chips of *E. polystachya* are said to produce a peacock-blue phosphorescence when placed in water against a black background (Mabberley 1997). This species is known as *coatli* (water serpent, Náhuatl), *palo* [*vara*, *varaduz*] *dulce* (sweet tree, Durango), *palo cuate* (snake [or twin] tree, from Náhuatl *coatli*, Sinaloa), *rosilla* (Sinaloa), *taray* [*del país*] (maybe originally the name of *Tamarix*; Durango, Nuevo León), and *urza* (Otomí). Most of the 10 species in the genus are largely confined to Central America.

## *Mimosa aculeaticarpa* (Fabaceae) Wait-a-Bit, *uña de gato*

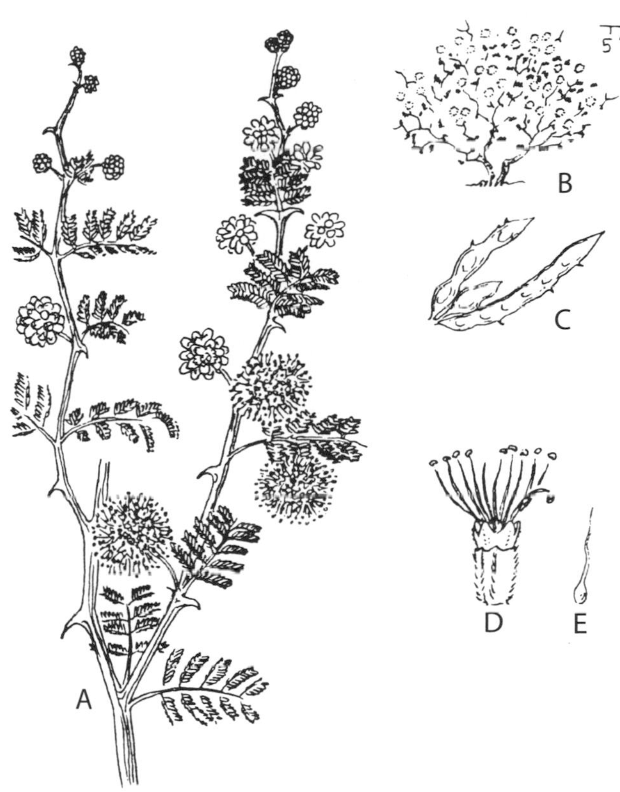


*Mimosa aculeaticarpa*. A. Flowering branch. B. Aspect of plant. C. Fruit. D. Flower; note only 10 stamens. E. Gynoecium. Artist: R. D. Ivey (Ivey 2003).

**Other names:** ENGLISH: cat's claw, mimosa (New Mexico), wait-a-bit [minute] (because it stops passersby); SPANISH: *brenales* (maybe from *breñal*, an area cleared of weeds, Mexico), *chaparro* (thicket, Oaxaca; for other species), *garabatillo* [*garavatillo*] (little hook, Aguascalientes; for other species), *gato* (cat, Sonora), *gatuño* [*garroño*] (cat claw, Chihuahua), *raspillas* (scratcher, Tamaulipas; for other species), *uña de gato* (cat's claw, Chihuahua, Arizona, New Mexico); ATHAPASCAN: *ch'il yíjish* <*ch'il gojiza*> (*ch'il*, plant, *yíjish*, scratching it, Western Apache); (= *M. biuncifera*)

**Botanical description:** Shrubs to ca. 2 m tall, rounded, branches numerous, the many prickles not conical but somewhat flattened basally and apically markedly recurved. Leaves are pinnately compound, the pinnae 4–8 pairs, less often 10; leaflets 5–9 pairs per pinna, 2 or more mm long, linear-oblong. Flowers are in globes, white to pink. Petals are united for half their length or more. Pods are linear, curved to almost straight, 20–35 mm long, 3–4 mm wide, the valves separating at maturity from the margins but remaining intact, not breaking into joints, the margins with short recurved prickles (absent in some plants).

**Habitat:** Desertscrub to oak woodlands, canyons, washes. 900–1,830 m (3,000–6,000 ft).

**Range:** Arizona (southern Apache, Mohave, south to Cochise, Pima, and Santa Cruz counties), New Mexico, and western Texas; Mexico (Chihuahua, Coahuila, and Sonora).

**Seasonality:** Flowering (May) June to August.

**Status:** Native. Our plants are var. *biuncifera*.

**Ecological significance:** This species does not occur west of the Baboquivari Mountains, and it is not in Cabeza Prieta National Wildlife Refuge, Organ Pipe Cactus National Monument, or the Tinajas Altas Mountains (Felger et al. in prep.). The shrubs do, however, follow the Mogollon Rim across Arizona from southwestern New Mexico to Kingman (Lamb 1975). The shrubs are more characteristic of the Chihuahuan Desert than the Sonoran. There is a disjunct population north of the Baboquivari Mountains in the Hualapai Mountains, Mohave County.

**Human uses:** No uses found for this species.

Guarijío and Mayo use *M. asperata* (Mayo: *yete ogua*, *rama dormilona*), *M. distachya* (Guarijío: *gatuña*; Mayo: *nésuquera*), and *M. palmeri* (Guarijío and Mayo: *chopo*, *cho'opó*) (Yetman and Van Devender 2001, Yetman and Felger 2002).

Mayo use *yete ogua* (sleepy bush) to induce sleep, particularly in children. A branch with the thorns removed is placed under the pillow. These people use *M. distachya* as firewood, and some make brooms of the branches. Guarijío also use the branches for brooms. *Mimosa palmeri* is used by Guarajío and Mayo to make fence posts, handles of implements, and charcoal. The bark is used by both peoples to clean the teeth. Guarijío also make medicine from it to relieve cough and diarrhea. Some use the bark to tan hides.

**Derivation of the name:** *Mimosa* is based on Greek *mimus*, mimic, referring to the sensitive leaves of some species. *Aculeaticarpa* means having spiny fruits; *biuncifera* means having two-hooks, a reference to the paired stipular thorns on the stems.

**Miscellaneous:** Two other species, *M. dysocarpa* (*gatuño*, *uña de gato*, Velvet Pod Mimosa) and *M. grahamii* (*gatuña*) are also present. *Mimosa dysocarpa* (*garároa*, Tarahumara) is recognized by the cylindrical pink spikes when in flower, and later by the unarmed, velvety fruits. This species also stops in the Baboquivari Mountains, but it ranges east into southwestern New Mexico and Texas, where it was first found; it also appears in Chihuahua, Durango, and Sonora. *Mimosa dysocarpa* is frequent at lower elevations in the Buenos Aires National Wildlife Refuge.

*Mimosa grahami*, which the Mountain Pima call <*chishpurahwa*>, *mihischo tara* <*miishuktara*> (cat's paw), is similar to *M. aculeaticarpa* in aspect and fruits, but it is distinguished by white-ball flower clusters and pods that do not break into one-seeded segments. *Mimosa grahami* has longer leaf rachises, longer and more numerous pinnae, and slightly larger leaflets. The prickles are neither consistently stipular nor recurved, occur at the internodes and on the leaf rachises, and are often longer than those of var. *biuncifera*. Sterile, *Mimosa grahami* looks more like *M. dysocarpa*. These shrubs are more common at higher elevations than the other two, being known from 4,000 to 6,000 ft. The species is also known in northern Sonora and Chihuahua.

## *Phaseolus acutifolius* (Fabaceae)
## Tepary, *frijoles*, *ban bawĭ*

**Other names:** ENGLISH: [Mexican] tepary [bean] ("tepary" appeared in print in English in 1912), Texas bean; SPANISH: *ejotillo* (little bean, from Náhuatl *exotl*, bean, Sinaloa), *escomite* (Sonora to Chiapas), *frijolillo* (little bean, Arizona), *frijol tepari* (tepary bean, Mexico; *frijol* is derived from Latin *phaseolus*), *garbancillo bolando* (false little garbanzo, Sonora; "garbanzo" came into English from Spanish where it perhaps is a loan-word from Basque *garau*, seed, *antzu*, dry; possibly related to Greek *erebinsos*, as Portuguese is *ervanço*), *hurimuni* (Spanish from Cahita

*Phaseolus acutifolius*. A. Flowering and fruiting branch. B. Dehisced legume. C. Seed. Artist: W. C. Hodgson (courtesy Wendy Hodgson).

*yorimuni*, white bean, Sonora), *patol* (Spanish name used by Onavas Pima; from Náhuatl *patolli*, bean), *tépari* [*tepary, del monte*] ([wild] tepary, Sonora); ATHAPASCAN: *bé'ísts'ǫ́z* *<bes ch'oz>* (Western Apache), *naa'ołí* *<naˀołí>* (bean, Navajo), *náohlētsostēt* (Jicarilla Apache); HOKAN: *marik* (Kumiai), *teepar* (Seri; from Spanish fide Felger and Moser 1985); UTO-AZTECAN: *bavĭ* *<bavī, pavi, paf(î), pawi>* (Akimel and Hiá Ceḍ O'odham; related to *bawui*, *Erythrina*, q.v.), *ban bawĭ* (coyote bean, Tohono O'odham), *baw, meena* (*baw* cognate with *bawĭ*; *meena* cognate with *muni*, Mountain Pima), *cepulina bawĭ* (tepary bean, Tohono O'odham), *mori* (Hopi; Whiting 1939 said they called this *tcatcaímori*, from *tcatcaí*, small, *mori*, bean), *mo:ri:* *<mori, móri, morí'>* (Southern Paiute), *mu:ñ* (Tohono O'odham), *muni* (Cahita), *muni* (Ópata), *múni* [*muniki*] (Eudeve, Guarijío, Tarahumara), *muní* [*muniki*] (Tarahumara), *pošol* (Onavas Pima; they recognize two kinds, *vi pošol,* or small bean, and *gugu pošol*, large bean), *seˀelaim* *<séˀelaim>* (white cultivar; *heseim* for brown, Yaqui; spelled *selaim* by Sobarzo 1991 for Cahita), *tepar* (*tepa*, bean, *ri*, genitive, Ópata, probably the word that gave rise to tepary in English; however, Castetter and Underhill 1935 said the origin was from the Tohono O'odham phrase *ṭ pawi*, "it is a bean"), *tepar* (Eudeve), *te:pari* (Ópata), *tevinymalen* [*tevasmalem*] (Cahuilla), *yori muni* (*yori*, white, Caucasian, *muni*, bean, Tarahumara), *yori muni* (*yori*, white, Caucasian, *muni*, bean, Mayo); YUMAN: *$^{a}$matíg$^{a}$* (Walapai), *madíga* (Havasupai), *marék* (Yuma), *marika* (Yavapai), *maRik* (Mohave), *maRík* (Maricopa), *mri:k* *<merik>* (Cocopa); LANGUAGE ISOLATE: *nókwína, pintu pa, tsikapú ult* (Zuni)

**Botanical description:** Annual herbs from slender taproots, stems twining or decumbent. Leaves are 3-foliolate, with leaflets linear to lanceolate to narrowly rhombic-ovate, acute, 2.5–8 times as long as broad, rarely lobed. Peduncles are 5–40 mm long, with a raceme of blossoms 5–30 mm long, flowers 1–4 per inflorescence. Flowers are 8–10 mm long at anthesis. Calices are 4 mm long, pubescent, about equally 5-lobed. Corollas are creamy blue to lavender. Legumes are curvilinear, 4–7 cm long. Seeds are multicolored or brown.

**Habitat:** Washes, streamsides. 900–1,830 m (3,000 –6,000 ft).

**Range:** Arizona (Cochise, Gila, Greenlee, Pima, Pinal, and Santa Cruz counties), New Mexico, and western Texas; Mexico (Sonora to San Luis Potosí).

**Seasonality:** Flowering August to October.

**Status:** Native.

**Ecological significance:** Regardless of the varietal names used for the plants (and different authorities have distinct interpretations), there are two basic kinds of wild plants—those that grow up on the slopes and those that are confined to the floodplain washes. Because the forms on the washes mature, fruit, and die quickly, it has been suggested that they were ancestral to the cultivated plants. At least the cultivated forms mature and fruit within two months, before the summer rains have evaporated (Nabhan 1983a, Nabhan 1985, Rea 1997). That short lifespan makes these twiners better adapted to desert habitats than many other plants.

**Human uses:** This species is the wild ancestor of the plants cultivated by the Tohono O'odham that they call *bawĭ*. Their old name, "Papago," is reputed to have been derived from the Akimel O'odham name for them, the *Papah* <*papawi*> (bean) *O'odham* <*o'otam*> (people) (Castetter and Underhill 1935). Fontana (1983a) gives more details on the origin, and noted that derivation from *bá·bawĭ o'odham* (bean people) was correct.

These beans are cultivated and eaten by at least the Akimel and Tohono O'odham, Cocopa, Havasupai, Hopi, Kamia, Seri, Sia, and Tarahumara (Castetter and Underhill 1935, Felger and Moser 1985, Moerman 1998, Pennington 1963, Rea 1997, Whiting 1939). Rea (1997) and Hodgson (2001) provide details on teparies among both the Akimel and Tohono O'odham. At one time, there were 47 different varieties of this desert cultigen in southern Arizona, all derived from wild ancestors. The derivation of the cultivars has been hypothesized to have been farther south (Rea 1997). Onavas Pima still used the name *bav*, but they applied it to *P. vulgaris* by the time Pennington (1980) worked with them.

Tohono O'odham bite the herbage and hold it between their teeth to relieve toothaches (Castetter and Underhill 1935). Keres make the seeds into flour for ceremonial use, but that was considered highly unusual (Moerman 1998).

**Derivation of the name:** *Phaseolus* is from Greek *phaseolos*, a little boat, light vessel, referring to the similarity between the pod and the craft. The name was used by Dioscorides, fl. CE 40–80, for what Agnes Arber (1986) identified as *Vigna unguiculata*, and was also known as *faseolus* or *phaseolus* to Romans. *Acutifolius* means that it has sharp-pointed leaf apices.

**Miscellaneous:** *Phaseolus* was domesticated in more than one place in the New World, with *P. coccineus* (Scarlet Runner) having been derived in Mexico (Kaplan and Kaplan 1992), and both *P. lunatus* (Butter [Lima] Bean) and *P. vulgaris* (Common [French, Haricot, Kidney, Pinto] Bean) independently derived in South America and in Mexico (Simpson and Conner-Ogorzaly 1995). Although the word "bean" was originally applied to what we now call Fava Beans (*Vicia faba*) in all the languages of Europe, it soon shifted to the New World plants when their fruits were brought back to the Old World. Because Fava Beans are mildly to fatally toxic in people genetically deficient in G-6-PD (the enzyme glucose-6-phosphate dehydrogenase), their consumption must be limited (Desowitz 1987). That problem does not occur with *Phaseolus*, even though it is hard to digest; it provides often-needed protein and fiber.

## *Prosopis velutina* (Fabaceae)
## Velvet Mesquite, *mezquite, kui*

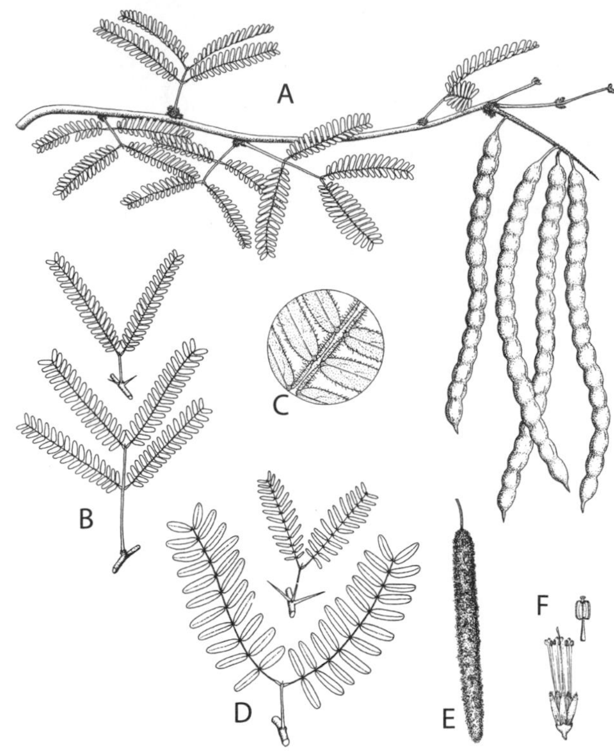


*Prosopis velutina.* A. Fruiting branch. B. Leaf variation. C. Detail of pubescence. D. Leaves of *P. glandulosa* var. *torreyana* for comparison. E. Inflorescence. F. Flower and stamen. Artist: M. B. Johnson (courtesy M. B. Johnson).

**Other names:** ENGLISH: mesquite (in English from Spanish by 1572); SPANISH: *algarroba* <*algoroba*> (Texas, Colima; adapted from Arabic *al-kharrubah* or *al-jarrúba*, originally applied to carob, *Ceratonia siliqua*, reapplied to other species in the New World), *chachaka* <*chúcata*> (Michoacán; from Maya *cha*, resin, *chac*, red), *mezquite* (Sonora, from Náhuatl *mizquitl*), *péchita* (from Ópata *péchit*, for the fruits; Arizona, Chihuahua, Sonora); ATHAPASCAN: *iyah* <*iiyáá*> (the pod, akin to *jeeh*, the resin, Western Apache), *nastane* <*natase*> (that which lies about, Chiricahua and Mescalero Apache); HOKAN: *haas* <*ˀaas*> (Seri); OTO-MANGUEAN: *tají* (Otomí); TARASCAN: *tziritzecua* (Purépecha); UTO-AZTECAN: *ˀé:-la* (Luiseño), *hu'upa* (Yaqui), *ˀíly* <*ily, il*> (Cahuilla; for *P. glandulosa*), *kui* (Akimel, Hiá Ceḍ, and Tohono O'odham, the bean *wihog* or *vihog*), *kui* <*k'ui*> (Onavas Pima), *meskít* (Mountain Pima), *ohpimpü* (Panamint), *opi(m)bɨ* (Kawaiisu), *quiot* (Ópata, Sonora), *sako* (Mountain Pima), *uhpalá* (Guarijío), *upárai* (Northern Tepehuan); YUMAN: *anáhl* (Kumiai), $^{a}$*nāl*$^{a}$ <*anāl*$^{e}$, *na:l*> (Walapai, the bean *iyá'*), *ana'l*$_{y}$ (Maricopa, the bean, *iya'* <*iyác*>), *ava* (Mohave, the bean, *avya'*), *eva*$^{c}$ (Yuma), *iyáa* (Havasupai), *kwayúły* <*anyal*> (Cocopa, the bean, *ˀa:*)

**Botanical description:** Large woody shrubs or trees, herbage and inflorescences pubescent. Leaves are compound, at least some and usually most with 2 pairs of pinnae (bijugate), the new growth, spring, and drought leaves tend have a single pair of pinnae (jugate), leaflets about 3 times longer than wide, often 6–13 mm long by 2–3.5 mm wide, 14–24 pairs per pinna. Flowers are mimosoid, small, yellow or cream, numerous in densely flowered cylindric racemes or spikes. Calices are shallowly toothed. Stamens are 10 in 2 series and separate. Pods are indehiscent, straight or curved, with a carbohydrate-rich mesocarp.

**Habitat:** Slopes, canyons, washes. Up to 1,370 m (4,500 ft).

**Range:** Arizona (Cochise, Coconino, Gila, Graham, Greenlee, La Paz, Maricopa, Mohave, Pima, Pinal, Santa Cruz, Yavapai, and Yuma counties) and southwestern New Mexico; Mexico (Chihuahua and Sonora).

**Seasonality:** Flowering March to June, sometimes again in late summer with abundant rains. Pods mostly ripen in early summer and sporadically through summer and fall.

**Status:** Native.

**Ecological significance:** Appearance of leaves on mesquite is a sure sign that winter is over. Although 220–240 flowers may be produced for each inflorescence, few develop into fruits. About 26/10,000 set fruit, and only 7 of those reach maturity (Turner et al. 1995). This small number is surprising, because several native bees and alien Honeybees (*Apis mellifera*) are effective pollinators (Keys et al. 1995). Although roots may extend down more than 50 m, most grow laterally within inches of the surface. The nitrogen-fixing bacteria on roots make the trees effective contributors to soil richness.

In late summer, often August, the Mesquite trees are festooned with Mesquite Bugs (*Thassus acutangulus*). These true bugs (family Coreidae, the leaf-footed bugs) feed on the sap of Mesquite trees during this season and reproduce. Insect-eating bats feast on these abundant and apparently easily caught but stinking bugs. Bat roosts often have their floors strewn with discarded wings each morning.

Before introduction of cattle, Mesquites were trees of waterways, where they formed large and impressive *bosques*. Introduction of livestock, with subsequent lowering of water tables and other associated changes induced by European-American cultures, resulted in drastic expansion into former grasslands (Turner et al. 1995). The changes in the water table caused by the 1887 earthquake also contributed to its spread.

**Human uses:** Mesquites are used by all cultures in their range, as contributors to food, shelter, weapons, material industries, medicine, cosmetics, fuel, and religious and ritual practices (Felger 1977). Fruits of all species were of extreme importance in the border region between Mexico and the United States from early pre-Columbian times to the present (Felger and Nabhan 1976). Many places still have boulders with cylindrical holes (*ce:po'o*, bedrock mortar, Tohono O'odham) in them where indigenous people pounded the pulp from the seeds of the fruits. In this area, *Prosopis* fruits are used by at least the Apache, Cahita, Cahuilla, Cocopa, Eudeve, Havasupai, Hohokam, Kawaiisu, Kumiai, Kiliwa, Maricopa, Mayo, Akimel O'odham, Hiá Ceḍ O'odham, Tohono O'odham, Ópata, Onavas Pima, Paiute, Quechan, Seri, Shoshoni, Yaqui, and Yavapai (Castetter 1935, Castetter and Opler 1936, Castetter and Bell 1951, Felger 1977, Felger and Broyles 2007, Felger and Moser 1985, Gallagher 1976, Hodgson 2001, Moerman 1998, Pennington 1980, Rea 1997, Zigmond 1981). Both Akimel O'odham and Tohono O'odham use *Prosopis* for medicine (Castetter and Underhill 1935, Curtin 1949, Rea 1997).

The Tohono O'odham called what is now the town of Sasabe on the Mexican border *kui tatk*, Mesquite root, but no records of their using roots as cord and binding have been found. It is known that the Seri and three Yuman tribes (Cocopa, Kamia, and Mohave) make the roots into twine for several products (Felger 1977). Among other things, the Seri bound their reed canoes with Mesquite root cords (Felger and Moser 1985). Kissell (1916) found that the O'odham use Mesquite fibers to sew their large grain-storage baskets.

**Derivation of the name:** *Prosopis* was originally the Greek name of some prickly plant; Jaeger (1941) thought it was the Butter-Bur, *Pedasites*, in the Asteraceae. Why Linnaeus reapplied it to these American legumes is unexplained, unless it is because they, too, are spiny. *Velutina* means velvety or pubescent, an allusion to leaf pubescence.

## ***Senna lindheimeriana*** **(Fabaceae)**
## **Hairy Senna, *hierba de piojo, ko'owĭ ta:tamĭ***

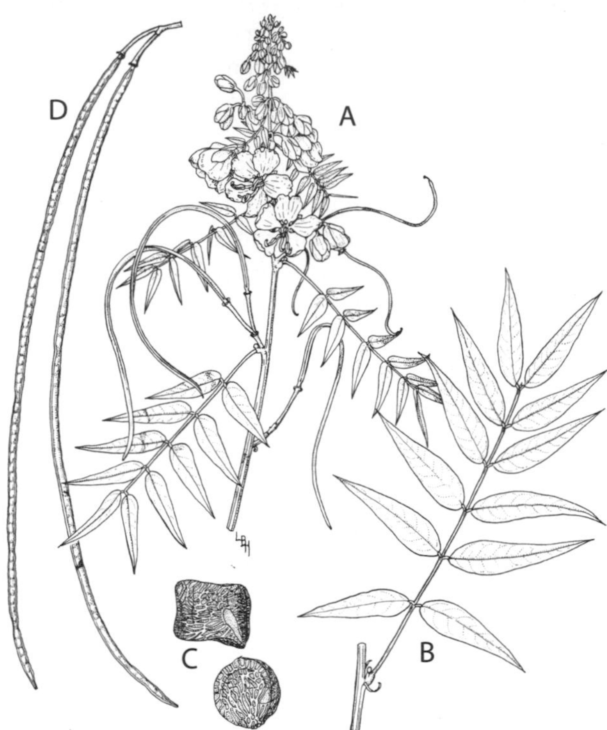


*Senna lindheimeriana*. A. Flowering branch with immature fruits. B. Compound leaf; note gland at the base of the petiole. C. Two views of seeds. D. Legumes. Artist: L. B. Hamilton (Parker 1958).

**Other names:** ENGLISH: [hairy] senna (from Arabic into English by 1543 when Teraheron wrote in Vigo's book on surgery: "Sena hath lytle braunches, and the leafe of fenugreke"), velvet leaf wild sensitive plant (New Mexico); SPANISH: *ejotillo* (little bean, from Náhuatl *exotl* or *ejotl*, Sonora; for "*Cassia occidentalis*" fide López-E. and Hinojosa-G. 1988), *hierba de piojo* (louse herb, Sonora), *nahuaposte* <*nahuapozte*, *nahuapoztli*> (maybe from *nahua*, the Aztecs, *poztectli*, broken, an allusion to the laxative properties, Náhuatl, Jalisco), *viche* (from Zapotec *bichi*, dry, Sinaloa); MAYAN: *salche* <*zalché*> (maybe incorrectly transcribed *ba'alche*, animal, Maya); UTO-AZTECAN: *ko'owĭ ta:tamĭ* (*ko'owĭ*, rattlesnake, *ta:tamĭ*, teeth, Tohono O'odham; probably an allusion to the long slender legumes); (= *Senna hirsuta*, *Senna leptocarpa*)

**Botanical description:** Erect perennial herbs, 1–2 m tall, with 1 to several grooved (sulcate) stems, velvety-pubescent, from deep woody root. Leaves are pinnately compound, with 5–8 pairs of leaflets that are oblong to elliptic, acute or obtuse, mucronate, asymmetrical basally, 2.5–5 cm long, 1–2 cm wide, sericeous above and below, the petiolar glands like bristles (setaceous), 1 between each pair of leaflets. Racemes are about as long as leaves, from upper axils. Sepals are ovate to elliptic, 6–8 mm long. Petals are golden yellow, 12–15 mm long. Legumes are linear, compressed, straight or curved, 3.5–6 cm long, 6–8 mm wide, apiculate. Seeds are dark brown, irregularly ovoid, margined or rugose.

**Habitat:** Grasslands, canyons, washes. 900–1,670 m (3,000–5,500 ft).

**Range:** Arizona (Cochise, Pima, and Santa Cruz counties), New Mexico, and western Texas; Mexico (Chihuahua, Coahuila, Nuevo León, Sonora, and Tamaulipas).

**Seasonality:** Flowering August to October.

**Status:** Native.

**Ecological significance:** *Senna* and *Solanum* are the only genera in the southern Arizona region that have buzz-pollinated flowers. Bees visiting the flowers remove pollen by "buzzing" or "drumming." The insects change the frequency of their wing vibrations when they settle on these flowers, and that causes the pollen to be ejected from the anthers. Insects then gather the pollen and remove it to their hives. Bees engaged in this activity are often heard before being seen. There is a decided change in the pitch of their wingbeats between flying and landing.

These weedy plants grow along roadsides and trails. However, Hairy Senna also is frequent in canyons, streambeds, and washes where the soils are similarly disturbed. It is likely that their adaptation to those habitats prepared them to utilize the collected moisure and disturbance of human-influenced places such as roads.

**Human uses:** Leaves are toxic but not fatal to grazing animals, and they are also a potent laxative (Diggs et al. 1999). Most, if not all, of the *Senna* species owe their action to anthraquinones, although lectins and other compounds are known (Lampe and McCann 1985, Lewis and Elvin-Lewis 2003, Perkins and Payne 1978). The exact action on the body seems to be dose-dependent.

No uses by people have been found for this species; however, many other *Senna* are utilized in medicines (Felger and Moser 1985, Hocking 1997, Moerman 1998, Yetman and Felger 2002, Yetman and Van Devender 2001). As examples, the Northern Tepehuan make "*Senna leptadenia*" (correctly *Chamaecrista nictitans*) into a febrifuge (Pennington 1969), and standardized *Senna* is the basis of of the commercial preparation Senokot®.

**Derivation of the name:** *Senna* was named by Philip Miller, 1691–1771, who adapted the Arabic *sana* or *sanna*. *Lindheimeriana* commemorates Ferdinand Jakob Lindheimer, 1801–1879, a German-born collector of Texas plants (1836–1879) and a resident of New Braunfels, Texas.

**Miscellaneous:** Although most literature considers *Cassia leptocarpa* distinct from *S. lindheimeriana*, P. D. Jenkins (personal communication, Jan. 2005) considers them synonyms. *Chamaecrista nictitans* (syn. *Cassia leptadenia*) is known in several mountain canyons. This herb is distinguished from *S. lindheimeriana* by shorter stature and the 12–16 pairs of leaflets. Also, *Chamaecrista nictitans* usually does not have glands on the petioles of the leaves, while *S. lindheimeriana* has them.

*Senna covesii* (desert senna, *hojasen*) may grow on hillsides in the mountains, but it is documented only as far south as the Sierrita Mountains. That plant has only 2–3 pairs of leaflets, and is found mostly at lower elevations in the valleys on both sides of the Baboquivari Mountains.

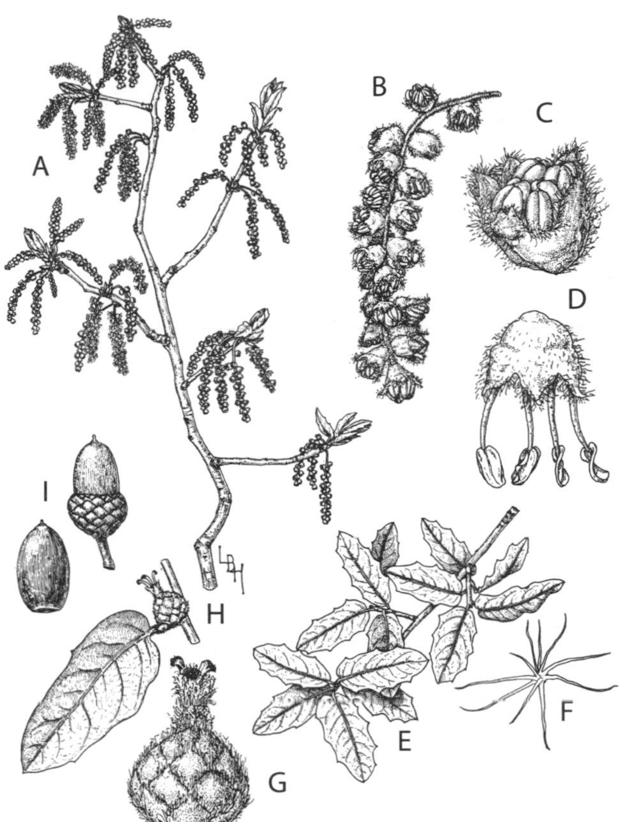


*Quercus emoryi*. A. Branch with young leaves and male flower clusters. B. Flower cluster (catkin), enlarged. C. Individual male flower cluster before anthesis, enlarged. D. Individual male flower cluster after anthesis, enlarged. E. Mature leaves. F. Star-shaped plant hair. G. Female flower, enlarged. H. Female flower in leaf axil. I. Acorns, with and without cup, enlarged. Artist: L. B. Hamilton (from original artwork).

## *Quercus emoryi* (Fagaceae)
## Emory Oak, *bellota, toa*

**Other names:** ENGLISH: black oak ("oak" was in English by CE 749; related to Old High German *eih*, and Middle High German *eich*; now *Eiche*, German), blackjack oak ("blackjack" was applied to *Q. nigra* in the eastern United States by John Bartram in 1765; he wrote, "Ye oaks black which is reconed ye best fire wood thay call them black Jacks seldom grow above A foot diameter"), Emory oak (named for topographical engineer Major W. H. Emory, 1811–1887); SPANISH: *bellota* (acorn; derived from Arabic *ballûta*, into Spanish by ca. 1212; before then, *lande* from Latin *glande*), *encino* [*bellotero, colorado, duraznillo, negro, prieto*] ([acorn, red-colored, little peach, black, black] evergreen oaks, Chihuahua, Sonora), *roble negro* (black [deciduous] oak); ATHAPASCAN: *chéch'il* <*chéchi'il, tséch'il*> (*ché*, rock, *ch'il*, plant, Navajo), *chích'il* [*nteelí, łichí'é*] (*tsee*, rock, *ch'il*, plant, *nteelí*, wide, *łichí'é*, red, Western Apache), *natókatsé* (Jicarilla Apache), *tcitcile* <*tcintcile*> (Chiricahua and Mescalero Apache); HOKAN: *mállūŋ* (Washo); KIOWA TANOAN: *kwae* (Tewa); OTO-MANGUEAN: *xilojo* (Mazahua; for *Quercus*); UTO-AZTECAN: *doha* <*roha, rojuá*> (Tarahumara), *héhat* (Luiseño), *híhat* (Cupeño), *koowi* (Yaqui), *k*$^{w}$*iiyavɨ* (tree, Kawaiisu; the acorn *k*$^{w}$*iiyara*; for *Q. kelloggii*), *kwingvi* <*kwí:ngvi*> (Hopi), *kwi'niûp* [*ku'niûp*] (Shoshoni; *ku'niroûmp*, acorn), *kwiávų* <*kwi'ûv*> (*tǫ*ˀ*mą́pų*, acorn, Ute), *qwiya* (Southern Paiute), *qwinyal* <*kwinyil*> (Cahuilla), *tcakicavü*$^{ʔü}$ (Northern Paiute; for *Q. kelloggii*), *toa* [*doa*] (Akimel and Tohono O'odham; altered to *toji* in Sonoran Spanish), *tohá* (Eudeve), *tohé* <*tohá, tohí*> (Guarijío), *tua* <*toha*> (Mountain Pima), *umíčari* (probably cognate with Cahita *cusi* [Sobarzo 1991], Tarahumara), *ṭúva* (Cupeño), *veyotam* (from Spanish *bellota*; used only for *Q. emoryi*, Yaqui), *viyóōdi* <*viyóīdi*> (from Spanish *bellota*, Akimel O'odham), *wia* <*wía*> (Northern Paiute; for *Q. kellogii*), *wíat* (Cupeño), *wiŋiyal* [*winiya*] (Tübatulabal; *wa'ant*, acorn; for *Q. californica*), *wiyampippüh* (*wiyan*(*pi*), acorn, Panamint), *wíya* (Mono), *wiyo:di* <*wi:yoda*> (from Spanish *bellota*, Tohono O'odham); YUKI: *has* (Yuki); YUMAN: *īsñó* (Kumiai), *ṣn*$^{y}$*a:* <*senya, sn*$^{y}$*a:*> (acorn or plant, Cocopa; for genus or *Q. turbinella*), *sñaiw* (Paipai), *tinyík* <*ḍinyikḍa*> (Walapai)

**Botanical description:** Trees, often 10–20 m (or more) tall, with a well-developed single trunk and spreading canopy. Bark is black, rough. Leaves are 5–9 cm long, the blade firm, the upper surface dark green, smooth and glossy, the lower surface dull green and smooth except for a conspicuous patch of fuzz at the base of the blade, the apex spine-tipped, margins entire or with a few short, spine-tipped teeth. Acorns are annual, nearly sessile.

**Habitat:** Foothills and canyons. 1,036–2,130 m (3,400–7,000 ft).

**Range:** Arizona (southern Coconino to Greenlee, south to Cochise, Pima, and Santa Cruz counties), southwestern New Mexico, and western Texas; Mexico (Chihuahua, Coahuila, Durango, Nuevo León, and Sonora).

**Seasonality:** Flowering April to May.

**Status:** Native.

**Ecological significance:** This tree is the most abundant low-elevation oak in northern Sonora and southern Arizona (Felger et al. 2001), although in Brown Canyon *Q. oblongifolia* is more common. The acorns from this tree are important food for many species of wildlife, and the branches harbor insects that are sought by numerous birds, resident and migratory. Trees also provide shelter for many animals.

**Human uses:** In Náhuatl, the name for oaks is *ahuatl* <*ahoatl, aoatl*> (Siméon 1885). Early Mexicans held the oaks in such high regard that the Aztecs named cities such as *Ahuatepec* (oak hill), *Ahuachichilpa* (in the red oaks), and *Ahuatlán* (near the oaks). Wood of *Q. emoryi* is harder than most other common oaks in the border region and is a preferred firewood. *Bellotas* are harvested in quantity and sold in Sonoran markets and in southern Arizona (Felger et al. 2001). Acorns are eaten fresh and sold in local *cantinas*, being unusually sweet for a black oak (Dayton 1937). Castetter and Opler (1936), Pennington (1963), Moerman (1998), and Hodgson (2001) found records of Apache, Tarahumara, Tohono O'odham, and Yavapai using acorns of *Q. emoryi* as food.

Palmer (1878) recorded a sample of sugar made from these oaks. No other source has been located describing sugar from oaks.

Historically, oaks have been so important for people that Old World languages have different words for them (Austin 2004). Germanic people base their name for oaks on Old Frisian "*k*." The Old Frisian name evolved into the German words given above plus Dutch *eik*, *eek*, Old Icelandic *eik*, Old Swedish *ek* (Swedish *ek*), and Danish *eg*. Romance language words are based on both Latin and earlier languages. In Spanish, the trees are *encinos* (evergreen or "live" species) or *robles* (deciduous species; from Latin *robore*, to be strong). Portuguese use the term *carvalho* (from the pre-Roman base word *carb* or *carv*, branched), and *roble* is applied to only European *Q. robur*. To French-speakers, oaks are *chêne*, related to *chesne* as written in 1550. In Hebrew, oaks are *allon* or *elon*, both derived from the base-word *el* (God). Italians say *quercia*, based on Latin *quercus*. Classical Greeks said *dryas*, and considered the oak the preferred tree of Zeus. Celtic people had at least two words for oak. One was *dair*, the fourth letter of their alphabet, and an ancient holy site is the modern Irish city of Kildare (church in the oaks). In Gaulois, oak was *cassanus*. Both *cassanus* and *quercus* are probably akin to Akkadian *kassu*, strong.

**Derivation of the name:** *Quercus* is the Latin name applied since the days of the Roman Empire. *Emoryi* commemorates Major William Hemsley Emory (1811–1887), army officer and director of the Mexican Boundary Survey.

**Miscellaneous:** *Quercus hypoleucoides* is known from a single mature tree in Brown Canyon. The species is also known in Baboquivari Canyon and in the Quinlan Mountains. Silver-Leaf Oak is distinctive in having a glabrous upper surface and dense white covering on the lower leaf blade. These are trees growing mostly between 1,100–2,700 m (3,600–9,000 ft) and barely come down into Brown Canyon to about 4,500 feet.

### *Quercus oblongifolia* (Fagaceae) Mexican Blue Oak, *encino azul, toa*

**Other names:** ENGLISH: [Mexican] blue oak ("blue" refers to the color of the leaves, especially when viewed from a distance), white live oak; SPANISH: *bellota* [*de cochi*] ([pig's] acorn, from Arabic *ballûta*), *encino azul* (blue evergreen or "live" oak,

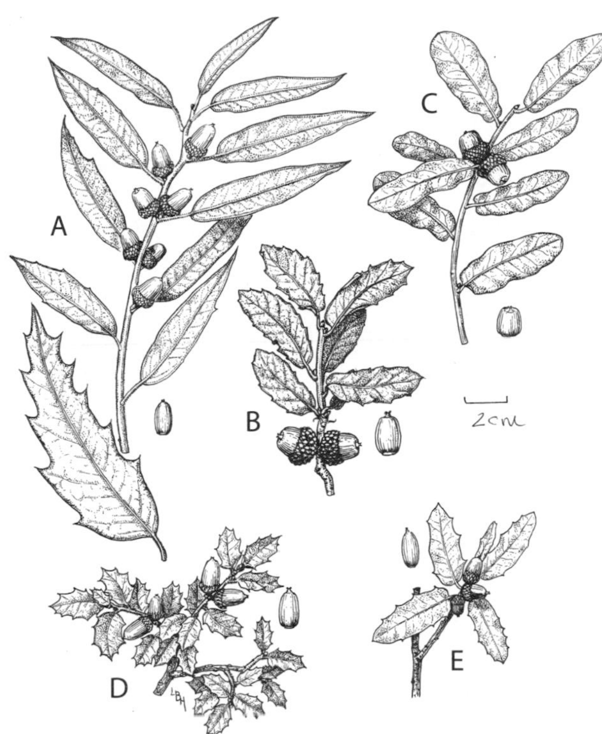


*Quercus oblongifolia*, compared with other local oaks. A. *Q. hypoleucoides*. B. *Q. arizonica*. C. *Q. oblongifolia*. D. *Q. turbinella*. E. *Q. emoryi*. Artist: L. B. Hamilton (Humphrey 1960).

Sonora); ATHAPASCAN: *chích'il łibayé* (*chích'il*, oak, *łibayé*, gray? Western Apache); KIOWA TANOAN: *kwae* (Tewa); UTO-AZTECAN: *héhat* (Luiseño), *híhat* (Cupeño), *kwiávų* <*kwi'ûv*> (Ute; *tǫˀmǫ́pų*, acorn), *kwingvi* <*kwí:ngvi*> (Hopi), *qwinyal* <*kwinyil*> (Cahuilla), *toa* (Akimel and Tohono O'odham), *tua* <*tu'a*, *toha*> [~*to'e*, cognate with *toa*] Mountain Pima), *wiyampippüh* (*wiyan*(*pi*), acorn, Panamint), *wiyo:di* <*wi:yoda*> (from Spanish *bellota*, Tohono O'odham)

**Botanical description:** Small to medium-sized trees, crown spreading. Bark is pale gray, fissured and checkered. Leaves are glabrous, 3–5 cm long, 1–4.5 cm wide, almost sessile or petiolate, the blades firm, mostly oblong, varying to elliptic or narrowly obovate, surfaces glaucous blue-green, the tip blunt and rounded, margins usually entire, but some shallowly lobed. Acorns are annual, solitary or paired, mostly sessile or on stalks 3–8 mm long.

**Habitat:** Foothills of desert mountain ranges. 850–1,800 m (2,800–6,000 ft).

**Range:** Arizona (Cochise, Graham, Pima, and Santa Cruz counties), southern New Mexico, and western Texas; Mexico (Baja California Sur, Chihuahua, Coahuila, and Sonora).

**Seasonality:** Flowering April to May; acorns ripening in October.

**Status:** Native.

**Ecological significance:** This oak dominates Brown Canyon and is intermixed with *Q. emoryi* and, in lesser abundance, *Q. arizonica*, *Q. grisea*, and *Q. hypoleucoides*. Landrum (1994) records *Q. oblongifolia* from oak woodlands, as in Brown Canyon, and elsewhere extending into grasslands.

This tree is one of the species used by the Acorn Woodpecker. Scattered through the canyon are dead limbs with a polka-dot pattern of small holes that birds use to store acorns for later in the season. These birds are not as common in this canyon as some others, but are seen occasionally. Caterpillars of the Cecrops-Eyed Silk Moth (*Automeris cecrops*) eat the leaves.

**Human uses:** Akimel and Tohono O'odham are said to have considered the acorns of this oak a staple (Russell 1908, Castetter and Bell 1942). Because the trees are uncommon in the Akimel O'odham territory, these people traded for them with the Tohono O'odham (Castetter and Underhill 1935, Russell 1908). In the early 1900s, Hispanic and other people along the Mexican border harvested large quantities of acorns for food during the winter months. These acorns were kept free from insects by storing them in wood ashes (Wilbur-Cruce 1987).

According to most sources, the acorns of this species are like those of *Q. emoryi*, are not bitter, and may be eaten without preparation (e.g., Hodgson 2001). Those that I have tasted are certainly *not* like *Q. emoryi* and have an unpleasant bitter taste. Jesus Garcia, from Sonora, says they are called "*bellotas de cochi, porque solamente los cochis las comen por amargas*" (pig's acorns, because only pigs eat them they are so bitter) (personal communication, 2009). Felger (personal communication, 2008) has questioned that local people ever ate this species. Still, there are credible accounts of the Tohono O'odham having eaten the acorns (e.g., Rea 1997), although the Akimel O'odham did not share their southern neighbors' zeal for them.

There are two distinct "etymologies" of English "acorn," one supported by linguists and the other by "popular etymology" (Austin 2004, OED 2007). The popular version, dating from the 1400s, is that acorn came from the combination of Old English *ac* (oak) with *corn* (fruit), yielding "acorn." The now accepted version is that it came from Old English *æcern*, which is a cognate with Old Norse *akarn*, Danish *agern*, Norwegian *aakorn*, Dutch *aker*, Old High German *ackeran*, all meaning "oak or beech mast." Also related are Gothic *akran* (fruit), which is related to Gothic *akr-s*, Old Norse *akr*, and Old English *æcer* (field). The connection between *Quercus* and the Old English word is that "fields" retained oaks that provided acorns for humans and their animals.

That Germanic word is distinct from most of the other European names for the fruits, as are their words for the trees. Some have terms apparently related to the tree, as in Dutch *eikel*, German *Eichel*, and Norwegian *eikenøtt* (oak nut). Greeks say *balanos*. To Romans acorn was *glans* (Latin), and that remains in *ghiandi* (Italian). Spanish uses *bellota*, which is derived from Arabic, but they also use *glande*. The word "bellota" appeared in English in 1866 as "belloot" or "belote," in the *Treasury of Botany* that wrote, "The acorns of *Q. Ballota*, and of its variety *Q. Gramuntia*, are eaten . . . under the name of Belotes."

**Derivation of the name:** *Quercus* is the Latin name applied since the days of the Roman Empire. *Oblongifolia* refers to the leaf shape, which is more or less oblong.

**Miscellaneous:** *Quercus grisea* is called Gray or Arizona Oak in Arizona; it is known as *encino blanco* in Sonora. *Quercus grisea* has elliptic to ovate leaves, the veins not prominent on the lower surface, and it grows in dry habitats. *Quercus arizonica* often has larger, oblong to oblanceolate leaves, the veins more prominent below, and it grows in wetter habitats. Landrum (1994) put *Q. grisea* and *Q. arizonica* together; others disagree (e.g., Felger et al. 2001; Nixon and Muller 1997; Spellenberg, personal communication, Sept. 2008). The leaves of *Q. grisea* resemble those of *Q. oblongifolia*, but they are typically rough-pubescent and veins are slightly raised above. *Quercus arizonica* may also be rough-pubescent but the veins are slightly sunken above (Felger et al. 2001).

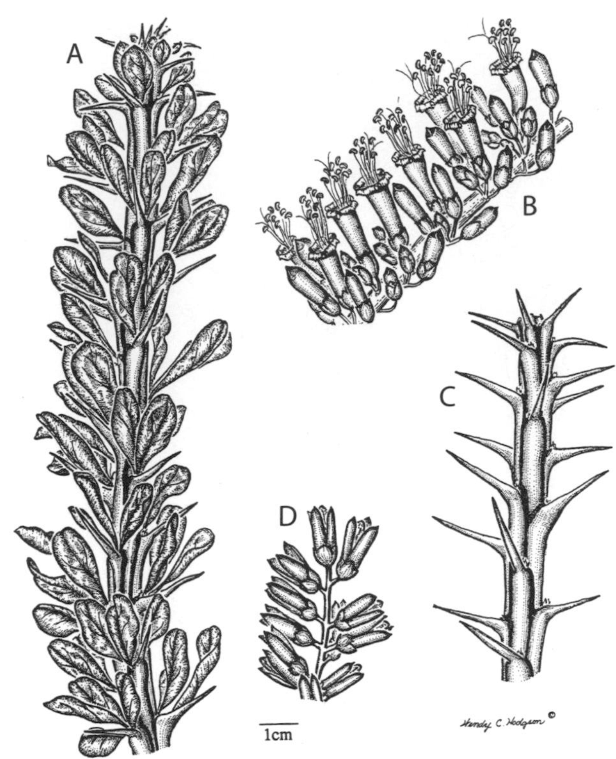


*Fouquieria splendens*. A. Stem with leaves of short shoots. B. Inflorescence. C. Stem in leafless condition. D. Capsules. Artist: W. C. Hodgson (courtesy Wendy Hodgson).

## *Fouquieria splendens* (Fouquieriaceae) Ocotillo, *ocotillo, melhog*

**Other names:** ENGLISH: Apache whipping stick, candlewood (Texas), candle bush, coach-whip (Arizona), Jacob's staff [wand], slimwood (Arizona), vine-cactus; SPANISH: *albarda* <*barda*> (pack-saddle, Sonora, Coahuila, Zacatecas), *barba* (beard, Coahuila), *chumari* (from Cahita *chunari*, Sonora), *cirio* (wax candle, Baja California, applied to another species), *colorín cimmarón* (wild red one, Mexico), *ocotillo* [*de corral*] ([corral] little torch, from Náhuatl *ocotl*, torch, New Mexico, Texas; Baja California, Chihuahua, Coahuila, Sonora, Zacatecas), *palo de Adán* (Adam's tree, Baja California, usually used for another species); ATHAPASCAN: *t'iis ts'ǫz* <*ges choze*> (*t'iis*, cottonwood leaf, *ts'ǫz*, ocotillo? Western Apache); HOKAN: *xomxéziz* <*xeshish*> (Seri), *xong* (Seri fide Martínez 1979, but not in Felger and Moser 1985); UTO-AZTECAN: *chimuchi chuwara* <*simuchi chuwara*> (*simuchi*, squirrel, *chuwara*, face, Tarahumara), *chumari* (Cahita), *cunuri* (Guarijío), *melhog* <*mïrok, mïro'k*> (Hiá Ceḍ O'odham and Tohono O'odham), *merihog* <*nuri'og*> (Onavas Pima; probably for *Fouquieria macdougalii*), *mureo* (Yaqui), *saar* (Mountain Pima), *tarákovara* (Northern Tepehuan), *utush* <*otosh*> (Cahuilla); YUMAN: *i'ikumadhí* (based on *atót*, spine, Maricopa), *í'i'qimie* <*igamye*> (Walapai), *ˀi:n$^{y}$áy* (Cocopa)

**Botanical description:** Shrub with several wand-like branches from a root crown, the branches erect- or sinuously-ascending, to 9 m tall, to 6 cm diameter, strongly grooved and ridged by decurrent spine bases, the bark grayish, the spines to 2 cm long. Leaves are sessile, obovate to spatulate, to 5 cm long, to 3 cm wide, entire, glabrous, early-deciduous. Inflorescences extend to 25 cm long. Sepals are almost orbicular, thin, 4–6 mm long. Corollas are scarlet, to 15 mm long. Capsules are ca. 15 mm long, 3-valved, persistent. Seeds have long whitish hairs.

**Habitat:** Desertscrub, hillsides, canyons. Sea level to 2,000 m (9,560 ft).

**Range:** Arizona (southern Apache, Coconino, northern Mohave, to Cochise, Pima, Santa Cruz, and Yuma counties), southeastern California, southern Nevada, New Mexico, and western Texas; Mexico (Baja California, Chihuahua, Sonora to Coahuila and Nuevo León, south to Durango, Hidalgo, San Luis Potosí, and Zacatecas). For details, see Henrickson (1972).

**Seasonality:** Flowering April to June.

**Status:** Native.

**Ecological significance:** This shrub is a dry-tropical species; it is most frequent on *bajadas*, but it grows elsewhere. On the slopes of Las Guijas Mountains and elsewhere, some stands have plants growing so close together that the stems are difficult to walk through. Typically, however, Ocotillos are more widely spaced. In southern Nevada, Ocotillo is probably near its climatic and ecologic limit (Turner et al. 1995).

Spring flowering lasts 50–60 days. On the slopes of the Sierrita Mountains in the Altar Valley, three species of hummingbirds visit flowers, and both Scott's (*Icterus parisorum*) and Hooded Orioles (*Icterus cucullatus*) drink nectar. Northern Cardinals (*Cardinalis cardinalis*), Pyrrholuxias (*Cardinalis sinuatus*), Lesser Goldfinches (*Carduelis psaltria*), and House Finches (*Carpodactus mexicanus*) also bite off flowers to get the nectar.

Plants are leafless for most of the year, but rains initiate growth within 24 hours, and fully mature leaves appear in five days (Turner et al. 1995). Five or six leafy periods can occur in one year, but two or three are more normal.

**Human uses:** Ocotillo has myriad uses. Tohono O'odham press the nectar from the flowers, let it harden, and chew it as a delicacy (Castetter and Underhill 1935). Cahuilla use the flowers to treat coughs, eat flowers and seed pods, and make a beverage of them (Bean and Saubel 1972, Standley 1920–1926).

The waxy coating on the stems makes them good tinder for starting campfires (Jaeger 1941). Tohono O'odham sometimes use the stems as the warp for weaving house frames (Castetter and Underhill 1935). At least the Seri and Northern Tepehuan also use the stems in houses (Pennington 1963, Felger and Moser 1985). Tohono O'odham employ the stems in making the framework of the structures used in the *wi:gida* <*viigida, vikita*> ("harvest" festival, based on *wi:gĭ*, down, originally from the Golden Eagle, Sheridan and Parezo 1996). Those structures are sometimes six feet square and represent the clouds and mountains (Castetter and Underhill 1935, Hayden 1987). Tarahumara formerly used *F. splendens* as a detergent (Pennington 1963). Northern Tepehuan use the stems to build fences, funnels for making soap, in constructing carrying baskets, and to make the hoop in the running game for women called *šibúbaruai* (circle of wood). Northern Tepehuan also use *Fouquieria* medicinally, as a poultice for aching gums, or eat the flowers to relieve throat problems (Pennington 1969). Pfefferkorn (1794–1795) gave details for treating wounds taught to him by a Sonoran Piman. Walapai too use the roots to reduce swelling (Watahomigie et al. 1982). Apache note that a tea of the roots will "clean you out top to bottom" (Gallagher 1976).

**Derivation of the name:** *Fouquieria* was named for physician Pierre Éloi Fouquier, 1776–1850, a professor of medicine in Paris. *Splendens* refers to striking or outstanding, probably alluding to the cluster of scarlet flowers.

## *Erodium cicutarium* (Geraniaceae) Filaree, *alfilerillo, hoho'ibaḍ*

**Other names:** ENGLISH: crane's bill, [red-stem] filaree (variation of *alfilaria*, from Spanish), pin-clover, red-stem, [red-stem] stork's bill [storkbill] (the name for *Erodium* dating to 1562 with Turner's *Herbal* and 1597 in John Gerarde's *Herbal*; akin to German *storchschnabel, schnabel*, beak, bill; *storkenebb*, stork's bill, Norwegian; Old Saxon *storkesnevel*); SPANISH: *aguaje del pastor* (shepherd's needle, Mexico), *agujitas* (little needle, Sonora, south), *alfilaria* [*alfilerillo, de pastor*] ([shepherd's] little needle; from Arabic *al-hillā*, the thorn, from *halla*, to pierce, California, New Mexico to Edo. México, Guerrero), *alfileres* [*alfileritos*] ([little] needles, Spain), *hierba de chuparrosa* [*yerba de chuparrosa*] (hummingbird

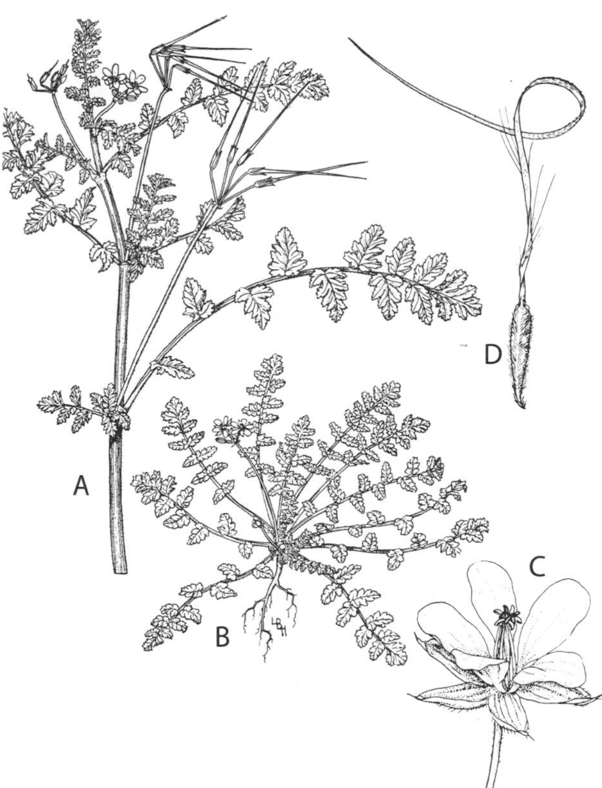


*Erodium cicutarium*. A. Branch with flowers and fruits. B. Basal rosette of young plant. C. Flower. D. Single fruit with corkscrew-like appendage. Artist: L. B. Hamilton (Parker 1958).

herb, Chihuahua), *herba de la coralina* (little pink herb, Mexico), *piene de bruja* (witch's comb, Edo. México), *pico de cigüeña* (crane's bill, Mexico), *tenedorcitos* (little forks, Spain); ATHAPASCAN: *chooyin 'azee'* <*čoyin ˀazeˀ*> (*chooyin*, menstruation, *'azee'*, medicine, Navajo), *dah yiitíhídą́ą́'* <*dahitíhídą́ˀ*> (hummingbird's food, Navajo), *tsís'ná dáá'* (bee food, Navajo), *dzílí bíláshgaan* <*tzílí pilackaan*> (chicken hawk [sic], its claw, Navajo); CHUMASH: *chikwi* (Barbareño Chumash), *kwł'łn* (Ventureño Chumash), *s'u'wlima'* (Ineseño Chumash); UTO-AZTECAN: *hawañ ta:tad* (*hawañ*, raven, *ta:tad*, feet, Tohono O'odham), *hoho'ibaḍ* (*ho'oipaḍ*, needle, based on *ho'i*, thorn, *bahidaj*, fruit, Akimel, Hiá Ceḍ, and Tohono O'odham), *ko:koḍ oipij* (*ko:koḍ* = ? *oipij*, awl, Tohono O'odham), *muutanavɨzivɨ* [*muutanamuzuvɨ*] (hummingbird beak, Kawaiisu), *pa'boiäts* (Ute), *pakhanat* (Cahuilla), *semuči* (Tarahumara), *yam'pagwanûp* (odor similar to yampah [*Perideridia gairdneri*], Shoshoni); YUMAN: *min*$^{a}$*m*$^{é}$*n'yá'* <*min'min'ya*> (Walapai)

**Botanical description:** Stems erect or finally prostrate to decumbent, 0.5–8 dm long or longer, rough-hairy and glandular. Leaves are 1–12 cm long, doubly pinnately dissected or parted, stipules lanceolate. Peduncles are 1–15 cm long. Inflorescences have pedicels 6–8 mm long. Sepals are 3–6 mm long, mucronate and bristle-tipped. Petals are pink or lilac, 5–7 mm long, spotted. Stylar beak 2–4 cm long, the carpel bodies 4–5 mm long, stiffly pilose.

**Habitat:** Grasslands, canyons, particularly in disturbed soils. Up to 2,130 m (7,000 ft).

**Range:** Arizona (throughout the state), California, New Mexico, Texas, east to Virginia, north to Quebec, Michigan, and Illinois; Mexico (Chihuahua, Sonora, and south to at least Edo. México).

**Seasonality:** Flowering February to July.

**Status:** Exotic; native of Europe.

**Ecological significance:** This garden weed from Spain was introduced to the New World in the sixteenth century with livestock (Dunmire and Tierney 1995, Dunmire 2004). Plants may have arrived as early as 1540 when Coronado drove his flocks up the Rio Grande to the Pecos pueblos. Even if *Erodium* came with Oñate almost 60 years later, it still arrived early. There was plenty of time for the species to be widely distributed after being brought, probably accidentally in saddle blankets. In the following years, plants thrived in the southwestern climates, which were similar to those of Spain, and the herbs soon colonized Arizona, New Mexico, and Texas, but did not stop there and continued spreading.

Dispersal and establishment is accomplished by the strip of the mericarp that breaks off with each seed. That tissue is sensitive to changes in moisture (hygroscopic) and curls and unfurls. This variation in shape literally screws the propagule into the ground. Sometimes these fruits are described as having "corkscrew" dispersal.

**Human uses:** This Old World species has long been used to treat throat and mouth problems (Martínez 1969). Whether or not the indigenous Americans learned uses from the Spanish or independently is uncertain.

Chumash, Jemez, Navajo, Ohlone, Tarahumara, and Zuni use the herbs as medicine (Bocek 1984, Pennington 1963, Mayes and Lacy 1989, Timbrook 1990, Dunmire and Tierney 1997, Moerman 1998, Wyman and Harris 1951). Chumash make a medicine for the blood. Tarahumara make an infusion of the entire plant to treat sore throat, cough, or stomachaches.

Leaves are crushed, dampened, and placed in the ear to relieve earaches. Navajo use *Erodium* on prayer sticks and in ceremonies.

Hopi, Isleta, Kawaiisu, Kumiai, Ohlone, and Navajo eat leaves, chew them as gum, or feed them to sheep (Bocek 1984, Moerman 1998, Vestal 1952, Zigmond 1981). Tarahumara cook the leaves for several hours and drain them before eating them as greens (Pennington 1963). Chumash were the only people found who eat the seeds (Timbrook 1990).

Tarahumara use Filaree to produce a yellow dye (Pennington 1963). Curiously, that dye comes from the pink flowers when boiled with alum.

**Derivation of the name:** *Erodium* is from Greek *erodios*, a heron, referring to the mericarps. These and *Geranium* were thought to resemble crane or heron heads with their long beaks. *Cicutarium* means that the leaves resemble *Cicuta*, the poison hemlock (Apiaceae).

**Miscellaneous:** A survey of plants for alkaloids found a weakly positive result for these herbs (Aplin and Cannon 1971).

## *Fendlera rupicola* (Hydrangeaceae) False Mock-Orange

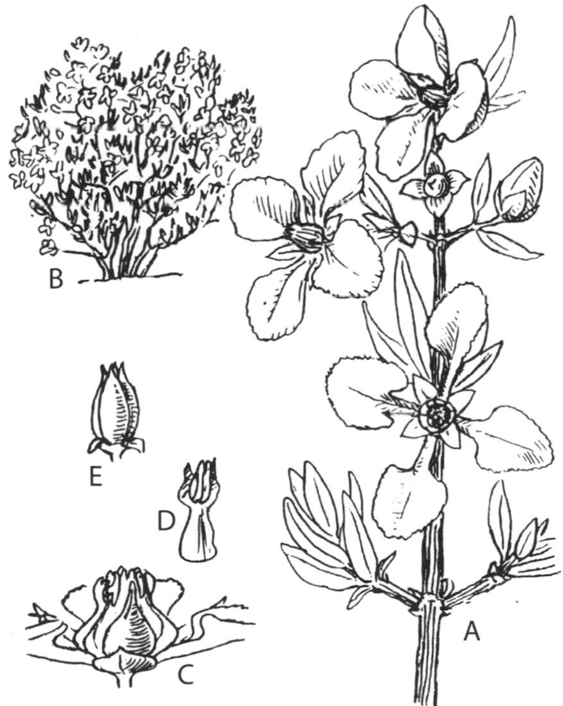


*Fendlera rupicola*. A. Flowering branch. B. Growth form. C. Flower, enlarged. D. Anthers. E. Fruit. Artist: R. D. Ivey (Ivey 2003).

**Other names:** ENGLISH: antelope brush (suggesting that Pronghorns browse the branches; since the habitat is wrong, probably a misnomer), [cliff] fendler-bush (a book name), false mock-orange (comparing the shrubs with the mock-orange, *Philadelphus*), fendlera (a book name), Fendler's buck-brush (a book name); ATHAPASCAN: *bį́į́hdą́ą́'* (*bį́į́h*, deer, *dą́ą́'*, food, Navajo), *tshińtł'zíh <ci 'iz>* (wood, hard, Navajo), *haashch'ééłti'í binát'oh <xašč'éłtiˀí binát'oh>* (Talking God's tobacco, Navajo)

**Botanical description:** Shrubs, much branched, 1–2 m tall (rarely taller), the bark of twigs longitudinally ridged and grooved, reddish to straw-colored, turning gray with age. Leaves opposite or appearing in bundles (fasciculate) and opposite, 9–30 mm long, 2–7 mm wide, lanceolate-linear, linear, elliptic, or sometimes ovate, entire, strongly rolled under (revolute), green and sparsely rough-hairy on both sides but becoming glabrous (glabrate), the midrib prominent, grooved above. Flowers are solitary or 2–3 together at the ends of short branches. Sepals are 3–5 mm long, to 8 mm in fruit, rough-hairy below, woolly above. Petals are white, 13–20 mm long, constricted to a narrow base, the blade to 11 mm wide. Capsules are 8–15 mm long.

**Habitat:** Along streams, on talus slopes, and on walls of canyons. 1,200–2,130 m (4,000–7,000 ft).

**Range:** Arizona (Apache to Hualapai, Mohave, south to Cochise, Pima, and Santa Cruz counties), southwestern Colorado, Nevada, New Mexico, Utah, and western Texas.

**Seasonality:** Flowering March to June (September).

**Status:** Native

**Ecological significance:** In April, the white flowers of these shrubs are favored by Two-Tailed Swallowtail Butterflies (*Papilio multicaudata*). A single stand of the trees below the arch may have three or more of these giant butterflies drinking nectar from the flowers at the same time.

This shrub is one of the four most important browse plants for the White-Tailed Deer (Anthony and Smith 1977). While *F. rupicola* is eaten by Mule Deer, it is not nearly as important in their diets. The shrubs are heavily browsed by deer, Bighorn Sheep, and goats (Kearney and Peebles 1951). When food is scarce, cattle browse on them.

These shrubs in the northern part of their range are found in Blackbrush (*Coleogyne*), mixed desert shrub, and pinyon-juniper communities (Welsh et al.

1987). In Utah, *Fendlera* grows from 1,370 to 1,710 m (4,500–5,610 ft). At Brown Canyon, the shrubs extend down into oak woodlands, at about 1,432 m (4,700 ft).

Plants in the Baboquivari Mountains represent a slight southwestern range extension for this species (Lamb 1975). Previously, the shrubs were known from the Coyote, Pajarito, Rincon, Santa Catalina, and Santa Rita Mountains in Pima and Santa Cruz counties (Toolin et al. 1979, Bowers and McLaughlin 1987). *Fendlera* has not been found in the Tucson Mountains (Rondeau et al. 1996) or in the Ironwood Forest National Monument (Dimmitt and Van Devender 2003).

**Human uses:** The Navajo consider *Fendlera* cathartic, and they use a medicine from it in their Male Shooting Way, Night Way (*Yeibichai*), Plume Way, and Wind Way ceremonies (Elmore 1944, Moerman 1998, Wyman and Harris 1951). Navajos drink an infusion of the bark as a remedy for "swallowed ants" (Elmore 1944). Navajo also boil *Fendlera* with juniper berries, pinyon buds, and cornmeal for the "mush-eating ceremonies," and they use the shrub to kill lice in the hair (Moerman 1998, Wyman and Harris 1951).

Anasazi in at least Colorado harvested the wood for fuel (Kohler and Matthews 1988). Havasupai and Navajo make the branches into arrow foreshafts, and Hopi use *F. rupicola* in religious ceremonies (Elmore 1944, Moerman 1998). The Home God rubs a smooth stick across a notched limb of *Fendlera* to make music during the Mountain Chant ceremony of the Navajo. Wood is also used to make weaving forks, planting sticks, and knitting needles (Elmore 1944). The wood is soft when green, it but becomes hard when seasoned.

**Derivation of the name:** *Fendlera* was created by George Engelmann and Asa Gray to commemorate German-born American explorer, traveler, and plant collector August (Augustus) Fendler, 1813–1883. *Rupicola* means rock-loving, in reference to the preferred habitat.

**Miscellaneous:** From the time the genus was described by Engelmann and Gray in the 1800s until well into the 1980s, *Fendlera* was assigned to the family Saxifragaceae. Molecular genetics studies in the early 1990s showed that the group was not monophyletic. Several genera, including Hydrangea (*Hydrangea*), Mock Orange (*Philadelphus*), and some others, were segregated into the family Hydrangeaceae. Now there are data which suggest that both *Fendlera* and *Jamesia* may represent a sister group to the remainder of the family (Judd et al. 2008). These two genera, in fact, may eventually be separated into a related family of their own, but they are now assigned to the Hydrangeaceae subfamily Jamesioideae.

## *Juglans major* (Juglandaceae)<br>Arizona Walnut, *nogal, u:pio*

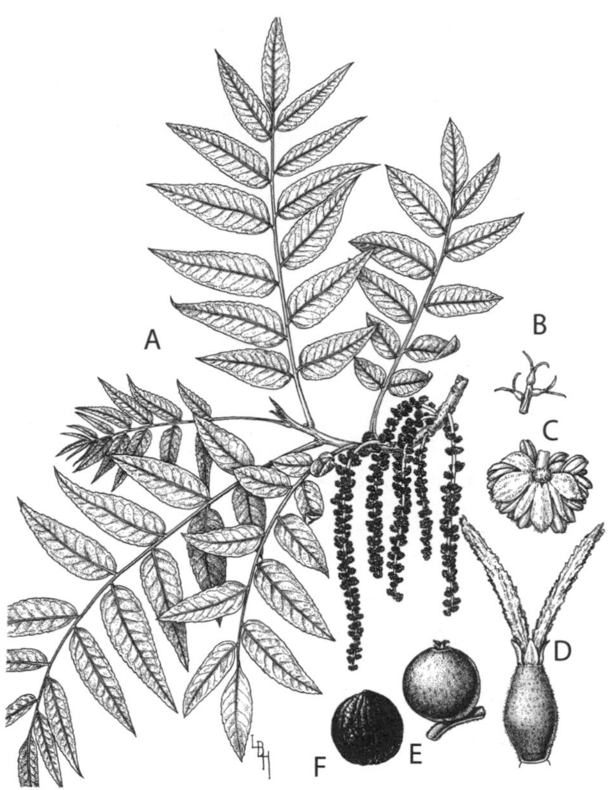


*Juglans major.* A. Male flowering branch. B. Female flowers, enlarged. C. Male flowers, enlarged. D. Female flower, enlarged. E. Fruit. F. Seed. Artist: L. B. Hamilton (Felger et al. 2001).

**Other names:** ENGLISH: [Arizona] walnut (by 1542, Leonard Fuchs said the walnut [*J. regia*] was cultivated throughout Germany, where it was called *Welchnusz*, meaning "Welch nut," and the Dutch said *nootboom*, nut tree; Italians were already saying *noce*, the French *noyer*, and the Spanish *nogal*, when *Juglans* finally reached England, where it first was called "Wealh nut" or "Gaul nut," and this term evolved into "walnut"); SPANISH: *nogal* (Chihuahua, Sonora; from Latin *nux*, through vulgar Latin *nucale*), *nogal silvestre* (wild walnut, Chihuahua, Sonora, Texas); ATHAPASCAN: *ch'iłdiiyé* [*ch'iłniiyé*] *<ch'il niyé>* (*ch'il*, plant/nuts, *niiyé*, you pound it, Western Apache), *hałsede <haltsede>* (that which is cracked, Chiricahua and Mescalero Apache), *ha'ałtsédii* <*xa'ałtsyétiih*> (that which is cracked, Navajo); UTO-AZTECAN: *ïpïvï* <*uupĭ*> (Onavas Pima; probably *epeve* or *upuvu̯*), *ïpïvï* <*ïpokai*> (Northern Tepehuan), *lačí* (Tarahumara), *mu̧ru̧kátu̧vu̧áci* (*tu̧vu̧áci*, nut, Ute), *noga'al u'sh* (from Spanish *nogal*, and O'odham *u:ṣ* or *u'us*, tree [plural], Mountain Pima), *súhūvi* (Comanche), *uup* [*uupio*] (Mountain Pima, Rea 1997), *uupai* (Northern Tepehuan), *uupio* (tree and nuts; also means skunk, Akimel O'odham), *u:pio* (Tohono O'odham); YUMAN: *kemtcūtek*[a] <*gamjudk*> (Walapai)

**Botanical description:** Trees to about 15 m tall, the bark deeply furrowed and ridged on older trees, branches forming a rounded crown. Leaves are pinnately compound, to 35 cm long, leaflets usually 9–13, rarely more, almost sessile, lanceolate, acuminate apically, more or less falcate, to 1 dm long and 35 mm wide below the middle, margins coarsely dentate-serrate, yellowish green, almost hairless with age. Male flowers are in drooping catkins. Fruits are spherical, 25–35 mm wide, the husk brown, thin, densely hairy. Nut has a thick, hard shell and a small edible kernel.

**Habitat:** Along streams and washes. 750–2,150 m (2,500–7,000 ft).

**Range:** Arizona (all counties except Apache), New Mexico, and Texas; Mexico (Chihuahua and Sonora south to Guerrero).

**Seasonality:** Flowering April to May.

**Status:** Native.

**Ecological significance:** These trees are more typical of the Ponderosa Pine zone but extend down into lower elevations along washes and canyons. The species is almost stopped by the Mogollon Rim in west-central Arizona, but it occurs locally in the Havasu Canyon in Coconino County. *Nogales* sometimes occur on drier sites than their usual associates of sycamores (*Platanus*), alders (*Alnus*), ash (*Fraxinus*), cottonwood (*Populus*), and willows (*Salix*) (Lamb 1975).

**Human uses:** Fruits are food to the Apache, Navajo, Northern Tepehuan, Tewa, Tohono O'odham, Walapai, and Yavapai (Castetter 1935, Castetter and Opler 1936, Elmore 1944, Gallagher 1976, Pennington 1969, Wilbur-Cruce 1987, Moerman 1998). *Juglans major* was the most important species among Mogollon Culture artifacts in Cordova Cave, New Mexico (Kaplan 1963). Leaves are made into a "beverage" in Chihuahua (Hocking 1997), but it must have been medicinal. The Apache derive fiber from the bark, and at least the Navajo, Northern Tepehuan, Tohono O'odham, and Walapai use the bark, leaves, twigs, and nuts as a dye source (Elmore 1944). The colors the dyes produced were different tones of brown or green (leaves).

Northern Tepehuan beat leaves and then put them into pools to catch fish (Pennington 1969). People in the eastern United States also use walnut bark to stupefy fish (Austin 2004).

The binational city of Nogales (Santa Cruz County, AZ, and Sonora, MX) has had that name since 1859, and it appeared on early maps of the area. This city was once called "*Los Dos Nogales*" because there were two trees, one on either side of the border (Granger 1960). The Nogales Wash "is now treeless, cement-lined, strewn with garbage, and subject to violent flash floods and criminal activities" (Felger et al. 2001).

There are two species in the eastern United States: *J. cinerea* (Butternut; New Brunswick to North Dakota, south to Georgia and Arkansas) and *J. nigra* (Black Walnut; northeastern states to Florida, west to the deciduous forests of Texas). In the West and Southwest, there is also *J. microcarpa* (River Walnut, Little Walnut; western Oklahoma, southern and western Texas, southeastern New Mexico, northern Mexico). Another species is endemic to California (*J. californica*). Nuts of all of these species are used as food, and all are used as medicine, for their wood, as dyes, and probably all are used to poison fish (Moerman 1998, Austin 2004).

**Derivation of the name:** *Juglans* was proposed by Linnaeus, who created the name from a contraction of *Jovis glans*, the nut of Jupiter—although *glans* has an alternative meaning, which he no doubt intended. Linnaeus delighted in words with dual meanings. *Major* means the largest, an odd name because it has the second-smallest fruits of the United States members of the genus. *Juglans microcarpa* fruits are slightly smaller.

**Miscellaneous:** *Juglans* contains 21 species, which grow in eastern Asia, the Mediterranean, North America, and the Andes of South America (Manning 1978).

## *Krameria erecta* (Krameriaceae) Range Ratany, *mezquitillo, edho*

**Other names:** ENGLISH: Pima [little-leaved, little-leaf, range] ratany (from *ratânia* or *ractania*, *rataña*, maybe from Quechua, meaning "ground-creeping";

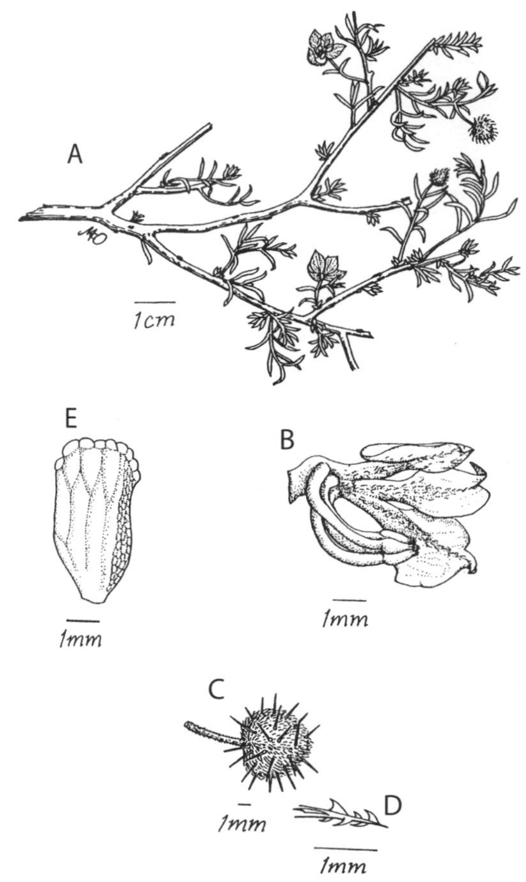


*Krameria erecta*. A. Flowering branch. B. Basally connate petaloid sepals with the four equal stamens arching upward from their point of insertion. C. Fruit. D. Fruit spine. E. Elaiphore. Artist: M. Ogorzaly (Simpson 1989).

in English by 1808), purple heather (superficially it resembles heather, Ericaceae); SPANISH: *mezquitillo* (little mesquite, comparing *Krameria* with *Prosopis*, Mexico), *zarsaparilla* (thorny vine, a name also given to *Rubus*, San Luis Potosí); HOKAN: *haxz iztim* (dog's hipbone, Seri); UTO-AZTECAN: *chacate* (Tohono O'odham; cognate with Náhuatl *chacatl*), *cósahui* (Yaqui, Hiá Ceḍ O'odham), *eḍho* (Hiá Ceḍ O'odham; used for *K. grayi*), *eḍho* <*'edho, e'eḍho*> (Tohono O'odham), *eeḍho* (Akimel O'odham), *'eḍho, he:ḍ* (Hiá Ceḍ O'odham), *kosawi* <*cosawi*> (Onavas Pima), *naka ˀbɔrïnanïmp* (Southern Paiute; for *K. grayi*), *tajimsi* (sun beard, Mayo), *tahué* <*tajué, tajuí*> (Guarijío), *tamichil* (Sonora), *wetahúpatci* (Tarahumara); YUMAN: *chacate* (Maricopa); (= *K. parvifolia* var. *imparata*)

**Botanical description:** Shrubs reaching 2–6 dm tall and about as wide, with numerous branches. Leaves are 3–15 mm long, 0.3–1 mm wide, linear to oblong, often spinulose-tipped, tomentose on both surfaces. Pedicels are glandular or not. Calices are 7–10 mm long, red-purple within. Upper petals are 2.5–2.8 mm long, retrorsely barbed along the rachis. Pods are almost globose, 5–9 mm in diameter, pilose-hirsute, spiny.

**Habitat:** Grasslands, ridges, canyons. Below 1,650 m (5,400 ft).

**Range:** Arizona (Coconino, Mohave, south to Cochise, Graham, Pima, and Santa Cruz, Yuma counties), California, Nevada, and Utah; Mexico (Baja California, Chihuahua, Coahuila, San Luis Potosí, Sonora, and south to Edo. México).

**Seasonality:** Flowering April to October.

**Status:** Native.

**Ecological significance:** There are two unusual traits about the family Krameriaceae, in addition to their zygomorphic flowers and spiny fruits (Austin 2004). Members of the genus are hemiparasitic, thus forming haustoria on the roots of a broad range of host plants. Flowers of *Krameria* are nectarless, and the rewards for visitors are fatty oils produced by modified external surfaces of the lower two petals (elaiphores). *Krameria* actually depends on oil-collecting bees for pollination, and only *Centris* (Anthophoridae) has been reported to visit their flowers (Simpson, personal communication, 2003). A given *Centris* species or even an individual, on the other hand, will visit other oil-producing plants, using the oils from the flowers for their larval food.

**Human uses:** According to Spanish explorer Hipólito Ruiz (1754–1815), he first found the South American *Krameria* plants being chewed by a Peruvian woman. She told him that *ratânia* was useful to whiten and strengthen the teeth (Simpson 1991). At that time, no other uses by indigenous people were known, but Ruiz began experimenting with *Krameria*. Later, he extolled *ratânia*'s virtues, and a variety of uses as an internal astringent were instigated in Peru and Spain. The medicine was introduced into England in 1806, and later into Germany in 1818.

Rea (1997) found *K. erecta* being made into astringent medicines and dyes by the Akimel and Tohono O'odham. Castetter and Underhill (1935) found *eḍho* in use by the Tohono O'odham to dye clothes a "terra cotta red." The Mayo also use the roots in astringent medicines and in dyeing (Yetman and Van Devender 2001). Guarijío boil *K. erecta* with *Eysenhardtia* (q.v.) to dye wool.

Guarijío use *tahué* for healing sores much as do their neighbors the Mayo (Yetman and Felger 2002). Seri have a name for *K. erecta*, but Felger and Moser (1985) record no use of this species. *Krameria erecta* is uncommon, while *K. grayi* is common in the Seri region, and this group does use that species. Tarahumara used an unidentified species as a toothache medicine, but the Northern Tepehuan did not (Pennington 1963, Pennington 1969). Onavas Pima brew the leaves into a hot tea (Pennington 1980). Tucson groceries and yerbarias still stock and sell *Krameria*, and many Hispanics use it (Kay 1996). In northeastern Mexico, Ratany is used to treat the blood, clean teeth, and heal infected gums (Ford 1975).

**Derivation of the name:** *Krameria* was named for the Austrian army physician, Johann Georg H. Kramer, d. 1742. *Erecta* indicates that the shrubs stand upright.

**Miscellaneous:** *Krameria* is a small genus, with 18 species found from the southern United States to northern Argentina and Chile (Simpson and Salywon 1999, Simpson et al. 2004). *Krameria* is the only genus in the Krameriaceae (Mabberley 1997).

## *Lamium amplexicaule* (Lamiaceae) Henbit, *lamio*

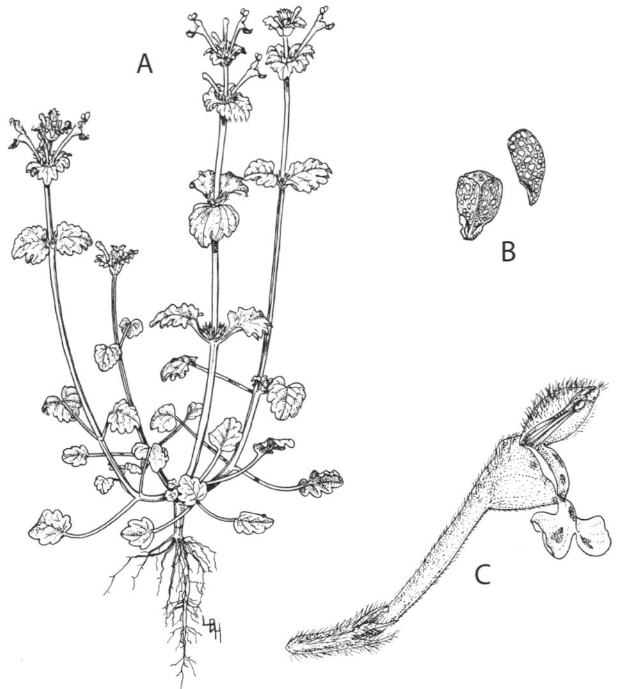


*Lamium amplexicaule*. A. Flowering plant. B. Nutlets. C. Side view of flower, enlarged. Artist: L. B. Hamilton (Parker 1958).

**Other names:** ENGLISH: archangel (the British name because it flowers about May 8, a feast day for the Archangel Michael), blind [dead] nettle ("nettle" from Old English *netele*, *netel*, *netle*, based on an Old Teutonic name for the genus *Urtica*, applied by ca. CE 705; it is "dead" or "blind" because it does not sting like *Urtica*), henbit (from Low German *Hoenderbeet*, or "hen's morsel"; in use by 1597 when John Gerarde called it "Great Henbit"); SPANISH: *lamio* (the Old World Spanish and Italian name derived from Latin), *ortiga muerta* (dead nettle, Mexico)

**Botanical description:** Annual or biennial herbs, sparsely pubescent, stems branched from the base and from leaf axils, ascending to decumbent, to 45 cm high. Leaves are broadly ovate to kidney-shaped or nearly orbicular, truncate to cordate basally, coarsely serrate, the basal leaves petioled. to 10 cm wide, the upper sessile or clasping and to 25 mm wide. Flowers are 6–10 in axillary or terminal clusters. Calices are pubescent, 4–6.5 mm long. Corollas are purple or reddish, 12–16 mm long, the tube slender. Seeds are smooth and shining, brown to olive-colored, pebbled or mottled with white, triangular, 2–2.4 mm long.

**Habitat:** Disturbed sites, washes, streams. About 30–2,740 m (100–9,000 ft).

**Range:** Arizona (Coconino and Pima counties), widespread in temperate United States; Mexico (Chiapas and Hidalgo); South America.

**Seasonality:** Flowering (February) March to May (November).

**Status:** Exotic. Introduced from Europe, probably as a medicine.

**Ecological significance:** Parker (1972) considered the herbs as "primarily a pest in lawns, especially new lawns; also found in gardens, flowerbeds, and plowed fields." The species can cause "staggers" in livestock.

In the canyons, *Lamium* is confined to streambeds and margins where there is enough water for the herbs to grow, flower, and fruit during the wet seasons. There is a colony below the rock dam crossed by the stream in Brown Canyon just below where the overflow pipe dumps water into a small pool. It has persisted through the wet seasons of 2001–2009.

*Lamium* produces non-opening (cleistogamous) flowers in the spring and fall, and may bloom at times when pollinators are not present. When insects are flying, the flowers are visited by bees, including the alien Honeybee (*Apis mellifera*).

In the Old World, *Lamium* is said to be dead, blind, deaf (Day-Nettle is a variant of *dea*, deaf), and dumb (Coffey 1993). All of these names are applied because Henbit does not have the stinging hairs of the real nettles (*Urtica*).

**Human uses:** The only uses found for *L. amplexicaule* were in Alabama as a stimulant, laxative, diaphoretic, and stomach tonic (Hocking 1997). No records of indigenous people using these plants have been found, and Moerman (1998) did not list the species. Records for *L. album* and *L. purpureum* are, however, markedly different.

These herbs, but particularly *L. album*, were used as a remedy in Roman times when they were mixed with axle grease, as mentioned by Pliny, CE 23–79 (Dobelis 1986). In early England, *Lamium* had a reputation as a cure for scrophula, a tuberculosis of the lymph nodes, known at the time as "King's Evil," because it was believed that it could be cured by the monarch's touch. Plants are rich in tannin and are considered by many as an astringent and as an anti-inflammatory dressing for cuts, wounds, and burns. A drink made from *Lamium* relieves diarrhea. The herb was eaten in the Old World as a vegetable or soup additive.

Purple Dead-Nettle (*L. purpureum*), sometimes hard to distinguish from *L. amplexicaule* without experience, is also native to Europe and the Mediterranean. That species is used as a diuretic, purgative, and styptic (Uphof 1968). Some think that this species is more effective in medicines than *L. album* (Hocking 1997).

**Derivation of the name:** *Lamium* is the classical Latin name of a nettle-like plant recorded by Pliny. The word is said to be based on *lamia*, a witch, or vampire, or a flatfish, in allusion to the throat or the resemblance of the flower and the fish. *Amplexicaulis* means stem-clasping, a reference to the leaves.

**Miscellaneous:** *Lamium* comprises about 30 species that are native to northern Africa and Eurasia.

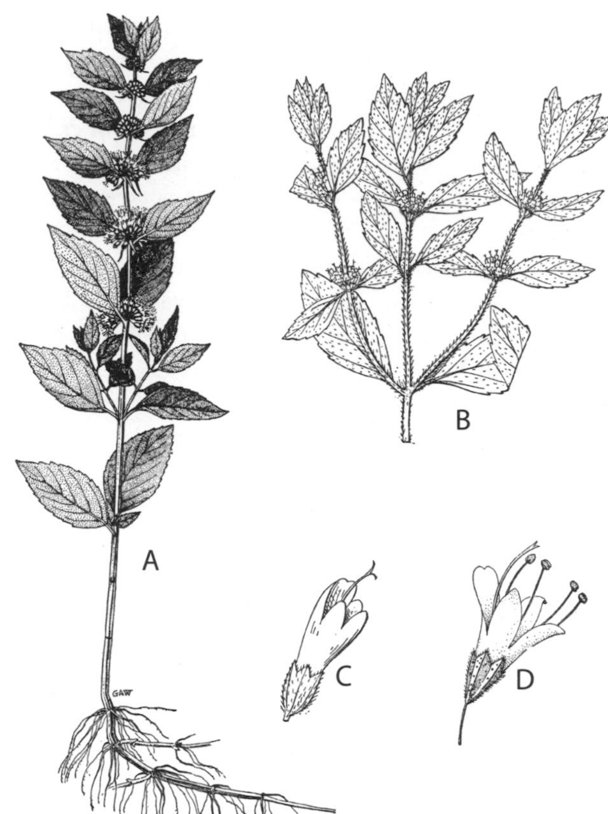


*Mentha arvensis*. A. Flowering plant showing roots. B. Enlarged branch of plant in flower. C. Flower in "female" stage with only stigma exposed. D. Flower in "male" stage with stamens. Artists: G. A. Walker, A, D (Dayton 1937); F. Emil, B, and A. Hollick, C (Britton and Brown 1896–1898).

## *Mentha arvensis* (Lamiaceae) Field Mint, *poléo*

**Other names:** ENGLISH: brook [American, field, Japanese, marsh, wild] mint (the ancient name, preceding Wm. Turner in 1548; from classical Latin *menta*, *mentha*, probably borrowed, like ancient Greek *minthe* and Hellenistic Greek *minthos*, from an unidentified source; cognate with Middle Dutch *minte*, Old Saxon *minta*, German *Minze*), corn-mint (akin to German *kornminze*, Dutch *corneminte*; from either Wm. Turner as a species of Calamint, *Calamintha acinos*, wild basil, or a book-name of *M. arvensis*), lamb's tongue, wild bergamont [bergamot] (from French *bergamote* for *Monarda*; originally from Turkish *bey-armuda*, prince's pear, for *Citrus bergamia*), water calamint (from *Calamintha*), wild pennyroyal (derived from Anglo-French *pulyole ryale*, and that from Latin *pulegiol*, little thyme, plus *real*, royal, sometimes rendered *ryal* or *rial*; originally applied to European *Mentha pulegium*), wild peppermint (comparing it to *M. piperita*); SPANISH: *hierbabuena* [*yerba buena*] (good herb, Old and New World Spanish), *menta* [*campestre*, *silvestre*] ([field, wild] mint, Mexico), *poléo* (Sonora and elsewhere in Mexico), *tabardillo* (an old Spanish name for several different plants, although originally applied to *Sanicula*, Apiaceae; now usually applied to *Calliandra* in Baja California and Sinaloa); ATHAPASCAN: *'azeeniłchin*

<*ˀazéˀ ńdoteži*> (knotted medicine, Navajo), *tséghą́ą́ʼ ʻadisxas nátʼoh* <*cékʼinˀałčízí nátʼoh*> (chuckwalla [lit. scrapes on rocks] tobacco, Navajo), *tłohntcine* (smelly plant, Chiricahua and Mescalero Apache), *txóltchiin* <*tolcin*> (water, smelly [fragrant plant that grows near water], Navajo); UTO-AZTECAN: *bawa-ne* (Tarahumara), *kwɨnānuva waíyanuva, tozitunaʼabe* (Northern Paiute), *pağóyˀna ˀnapu* (Ute), *paʼgwanûp* (*pa*, water, *gwanûp*, sweet odor, Shoshoni), *pakwana* (Shoshoni), *pawukɨtɨbɨ* [*powukɨtɨbɨ*] (Kawaiisu), *woléo* [*poléo*] (from Spanish, Guarijío)

**Botanical description:** Herbs from stolons, branching below or almost simple, to 8 dm tall, more or less retrorse-pubescent, at least on the angles. Leaves are oblong to ovate or lanceolate, rounded at the base into a distinct petiole, mostly serrate, minutely pubescent or short-villose, to 5 cm long, the upper smaller than the lower. Flowers are in axillary whorls. Calices are pubescent, 3 mm long, the triangular-subulate teeth about equaling the tube. Corollas are pink to violet or white, about twice as long as the calyx, mostly glabrous.

**Habitat:** Along streams. 1,500–2,740 m (5,000–9,000 ft).

**Range:** Arizona (Apache, Coconino, Navajo, Yavapai, south to Cochise, Greenlee, and Pima counties), New Mexico, and Texas; Mexico (Chihuahua and Sonora); widespread in North America and Old World.

**Seasonality:** Flowering June to October.

**Status:** Perhaps partly native and partly exotic; introduced by the Spanish in the 1500s, along with other medical herbs (Dunmire 2004). Tucker and Naczi (2007) argued that wild New World plants should be called *M. canadensis.*

**Ecological significance:** This mint is a circumboreal species that was first described from Europe (Tucker and Naczi 2007). Separation of native and introduced populations in the New World is not usually possible. Indeed, there seems to have been so much gene mixing that there is little more than an array of hybrids.

**Human uses:** Europeans were enamored of mints of various kinds well before the first known records during classical Greek and Roman times. Athenians perfumed their bodies with the fragrance and used the mint at feasts (Hart 1976). Greeks and Romans crowned themselves with mint foliage. Baucis and Philemon of Greek mythology scoured their festive board with mint before laying food on it for divine guests. Mint was used as payment to the Pharisees in biblical times. Pliny, CE 23–79, lauded mints, and their popularity continued through Elizabethan times, when more than 40 ailments were treated with mint extracts (Dobelis 1986). Numerous species were applied. Menthol is extracted from *M. arvensis* (Lawless 1995).

Indigenous Americans also use mints, and Hart (1976), Moerman (1998), Train et al. (1957), Wyman and Harris (1951), and Zigmond (1981) found records of *M. arvensis* being used as medicines by at least California tribes, Cherokee, Cheyenne, Flathead, Gros Ventre, Iroquois, Kawaiisu, Kutenai, Menomini, Navajo, Okanagan, Paiute, Salish, and Thompson. Tribes who use mints as food include Blackfoot, Cherokee, Cheyenne, Hopi, Kawaiisu, Lakota, Navajo, Okanagan, Paiute, Saanich, Sanpoil, Shuswap, and Thompson. Hart (1976) suggests that indigenous people long preferred *M. arvensis*, while those derived from the Old World favored *M. piperita* and *M. spicata.*

There are various other uses, including as perfumes, insecticides, and hair dressings. For example, the Paiute make a tea from the leaves (Kelly 1932). Gentry (1942, 1963) adds that the Mountain Guarijío made a beverage and medicine from the herbs. Mountain Pima make a tea to induce sleep (Pennington 1973). The Tarahumara eat the leaves as a *quelite*, and they make medicine for toothaches from the whole plant (Pennington 1969, Thord-Gray 1955). Navajo treat fevers and influenza with an infusion of *Mentha*, particularly when they think that those symptoms were caused by being struck by a whirlwind (Vestal 1952).

**Derivation of the name:** *Mentha* is a classical name, possibly from Minthe of Theophrastus (372–287 BCE), who was a nymph loved by Pluto. Legend says she was changed by Proserpine into the mint plant. *Arvensis* means "of the fields."

## *Mentha spicata* (Lamiaceae) Spear Mint, *yerba buena*

**Other names:** ENGLISH: bush mint (because the herbs grow into tall "bushes," Arizona), creek mint (Kentucky), Our Lady's Mint (England), spearmint (there are two versions of how the name arose, one from "spear" and the other from "spire"; "spear" is from Old English *spere*, cognate with West Frisian *spear*, Middle Dutch *spere*, *speer*, Dutch *speer*, Old

*Mentha spicata*. A. Flowering branch. B. Flower. C. Dissected flower. Artist: F. Emil, A; A. Hollick, B, C (Britton and Brown 1896–1898).

Saxon and Middle Low German, Old High German, and Middle High German *sper*, German *speer*, Old Norse plural *spjr*; Middle Swedish *spär*, and Danish *spær*; it is doubtful whether Latin *sparus*, hunting-spear, is related; perhaps the word was applied to these mints because of the leaf shape; in use by 1562 by William Turner for the common garden mint used in cookery; the other view is that the name came from "spire-mint," because of the flower clusters in "spires"); FRENCH: *baume* (balm, Quebec); SPANISH: *hierba [yerba] buena* (good herb, New Mexico, Mexico, Spain), *menta* (New Mexico, Mexico); UTO-AZTECAN: *pakwana* (Northern Paiute), *yerbagüen* (variant of *yerbawén*, loan from Spanish *hierba buena*, Mountain Pima); (= *Mentha viridis*)

**Botanical description:** Perennial herbs, from stolons, glabrous or sparingly pubescent at the nodes. Stems are erect, usually branched, to 1.2 m tall, often purplish. Leaves are sessile or short-petiolate, oblong-lanceolate to ovate-lanceolate, acute to acuminate apically, obtuse to rounded or almost cordate basally, unequally sharply serrate, the larger leaves 3–6 cm long. Flower whorls are in slender terminal leafless spikes, often 6–8 cm long in fruit. Bracts are subulate-lanceolate, equaling or surpassing the calyx, green, glabrous or ciliate. Calyx teeth are subulate, about equaling the tube, ciliate on the margins, the inflorescence otherwise glabrous. Corollas are pale lavender.

**Habitat:** Canyons, disturbed ground. 1,260–1,490 m (4,160–4,900 ft).

**Range:** Arizona (Cochise, Coconino, Gila, Graham, Maricopa, Mohave, Navajo, Pima, and Yavapai counties), extensively naturalized in North America.

**Seasonality:** Flowering July to September.

**Status:** Alien. Native to Europe.

**Ecological significance:** This mint was collected in Gila County in 1892, although it surely had been brought to Arizona earlier. We know that some "mint" was introduced by Padre Kino to the missions in Arizona between 1687 and 1706, but we cannot be sure of the species (Dunmire 2004). Given the Spanish taste for sheep and the ancient popularity of mints in Spain and other parts of southern Europe (Tardío et al. 2005), Spear Mint is probably the one introduced. Apparently, Parker (1972) did not consider Spear Mint a weed because she did not include the species.

Now Spear Mints are frequent on the floodplain near the canyon bottom above the junction with Jaguar Canyon and Arch Trail near 4,300 feet elevation. Winter cold kills back the parts above ground, but the herbs resprout in the spring to regrow into tall plants by the time of the monsoon rains. Soon after rains begin, the long clusters of blue-lavender flowers draw the attention of passing humans and a multitude of insect visitors.

**Human uses:** Spear Mint was mentioned by Dioscorides (fl. CE 40–80), Galen (ca. CE 129–ca. 199), and Pliny (CE 23–79). By 1542, the pharmaceutical name was *Mentha*. Because the Romans spread the mint through their empire (Rosengarten 1969), it began to be called *mente romaine* (Roman mint, 1550, French) and *menta romana* (Roman mint, 1551, Italian). Fuchs (1542) said that it was widely grown in European gardens. He listed it, along with four other kinds of mints, under "*De Hedyosmo*" (sweet-smelling, Greek).

In Elizabethan times, more than 40 ailments were treated with mints of various kinds. Surely that popularity led to settlers bringing Spear Mint to the New World. Peppermint (*M. piperita*) is now the best-known introduction, but Spear Mint was probably brought in earlier. We know for certain that Peppermint was in New England when John Josselyn wrote about it in the late 1660s (Coffey 1993).

Manasseh Cutler (1742–1823) thought that Spear Mint had a more agreeable flavor than the "Horse Mint" (*Mentha longifolia*) and that it was preferred during his time for culinary and medical purposes. This mint is still popular in "mint jelly" to accompany lamb, in *tzatziki* (eastern Europe), in *tabbouleh* (Middle East), in garnishing and flavoring, and for herbal teas and iced drinks (Bown 1995). In Mexico, *yerba buena* is used to cure *la cruda* or hangover (Small 1997).

**Derivation of the name:** *Mentha* is a classical name used by Theophrastus (372–287 BCE). He took the name from Minthe(s), daughter of Cocytus. She was a nymph who, legend says, was changed in a fit of jealousy by Proserpine [Proserpina], the suspicious wife of Pluto, into the mint plant. *Spicata* means that the flowers are in clusters without branches.

## *Stachys coccinea* (Lamiaceae) Scarlet Betony, *betónica*

*Stachys coccinea*. A. Flowering branches. B. Habit. Artist: B. Swarbrick (courtesy Bonnie Swarbrick).

**Other names:** ENGLISH: [scarlet, Texas] betony (from Latin *betonica*, which Pliny, CE 23–79, said was a Gaulish name from *vettonica*, derived from Vettones, people of Lusitania [Portugal], with cognates in Romance and Germanic languages), [scarlet] hedge nettle ("hedge nettle" is an allusion to where the herbs grew and the similarity of their leaves to nettle or *Urtica*; in English by 1678); SPANISH: *betónica* (cognates are Dutch and Norwegian *betoine*, French *bétoine* [in use by 1550], Italian *betonica* [in use by 1551], Spanish *betónica* [in use by 1557], German *Betonik* [in use by 1542]), *flor de chuparosa* (hummingbird flower, Mountain Pima; OTO-MANGUEAN: *mbarejnatr'eje* (Mazahua); UTO-AZTECAN: *bishish hióskem* <*bispshi hioshgama*> (maybe *bishish*, sneeze, *hióskem*, flower, Mountain Pima), *toi'yabagwanûp* [*pi'abagwanûp*] (Shoshoni; for *S. palustris*)

**Botanical description:** Perennial with assurgent to erect stems, to 1 m tall, pubescent with long, soft, spreading hairs below, less so above. Leaves are broadly to narrowly triangular-ovate to ovate-lanceolate, dentate-serrate, dark green and short-pubescent above, paler and densely pubescent below, acute apically, rounded to truncate or almost cordate basally, the lower leaves to 8 cm long, gradually smaller upward, passing into the sessile foliar bracts of the flower groups. Flower clusters interrupted spikes, whorls of flowers mostly 4. Calices are deeply and narrowly bell-shaped, with 5 triangular teeth about as long as the tube. Corollas are scarlet, 18–24 mm long, strongly 2-lipped, the upper lip kidney-shaped, the lower lip divided with the middle lobe ovate and the lateral lobes oblong-spatulate.

**Habitat:** Canyons, moist, rich soils of crevices, steep stony slopes. 450–2,440 m (1,500–8,000 ft).

**Range:** Arizona (southern Apache to Maricopa, south to Cochise, Pima, and Santa Cruz counties), New Mexico, and western Texas (Trans-Pecos region); Mexico (Baja California, Chihuahua, Sonora to Coahuila, and south to Oaxaca); Central America to Guatemala and Nicaragua.

**Seasonality:** Flowering March to October.

**Status:** Native.

**Ecological significance:** Hummingbirds love *Stachys*. Several species of these small nectar-feeding birds are resident in the canyons when Scarlet Betony is in flower. Particularly, the Broad-Billed Hummingbirds (*Cynanthus latirostris*) are frequent in the region, and these birds nest in trees near where these herbs grow. A short stop beside the flowering plants will reveal visits by these birds for nectar. Van Devender et al. (2005) reported 11 or 12 species of hummingbirds visiting the flowers in a three-day period in nearby Sonora.

Near the Baboquivari Mountains, *C. coccinea* is known from the Coyote, Pajarito, Quinlan, Rincon, Santa Catalina, San Luis, and Santa Rita Mountains. Dimmitt and Van Devender (2003) found it in the Ironwood National Monument, but it has not been found in the Tucson Mountains (Rondeau et al. 1996).

**Human uses:** No use found for Scarlet Betony, except as a cultivated plant.

Other species are used by people around the world. Hocking (1997) lists 17 species that are used. Moerman (1998) found in this region only *S. rothrockii*, which is used by the Navajo as a ceremonial medicine and chant lotion, and as a deodorant and cure for "deer infection" (Vestal 1952).

Not far away, in California, the Ohlone use *S. bullata* to treat sores, earaches, stomachaches, and as a gargle for sore throat (Bocek 1984). The Pomo heat the leaves to help cure boils (Moerman 1998). The Kawaiisu put a "cork" of leaves of *S. albens* in a basketry water bottle. These people say, "It gives a good taste to the water" (Zigmond 1981).

Pliny (CE 23–79) wrote of what we now call *Stachys officinalis*: "The Vettones in Spain discovered a mint called *vettonica* in Gaul, *serratula* in Italy, and *cestros* or *psychotrophon* by the Greeks, a plant more highly valued than any others" (Austin 2004). Along with that species, two others were famous among Europeans before they arrived in the New World—*S. palustris* and *S. sylvatica*. Two well-known English names of these were Betony and Woundwort, and all were renowned as treatments for sores and open wounds.

Most species have enlarged tubers that were used as food from Europe to Asia before the New World was discovered. John Lightfoot, in his *Flora Scotica* of 1777, wrote of *S. palustris* tubers that they were "sweet, and in times of necessity . . . eaten by men, either boiled, or dry'd, and made into bread." Fernald et al. (1958) had not tried that species, but speak highly of related *S. hyssopifolia* in New England. These authors found the nutty tubers "white and as good a nibble or salad as one could wish."

**Derivation of the name:** *Stachys* comes from Greek *stachyos*, a spike or ear of grain, thus referring to the flower cluster; akin to Latin *stachys*, the mint or horse-mint. *Coccinea*, red, refers to the flower color.

## ***Mentzelia albicaulis*** **(Loasaceae)** **White-Stem Blazing-Star,** ***pega-pega, ikus ho:ho'idam***

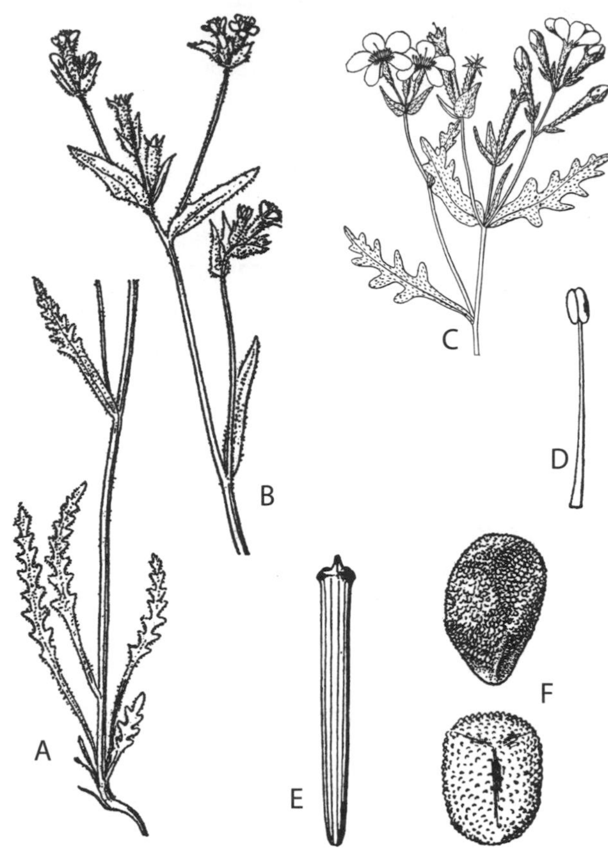


*Mentzelia albicaulis*. A. Basal part of plant. B. Flowering tip of plant. C. Flowering branch showing part of variation. D. Anther. E. Fruit. F. Seeds, in two views. Artists: J. R. Janish, A, B, F top (Abrams 1951); M. Timmerman, C; A. Hollick D–F lower (Britton and Brown 1896–1898).

**Other names:** ENGLISH: gravy plants (Great Basin), small-flowered [white-stem] blazing-star ("blazing-star" used for *Chamaelirium* by John Bartram in 1751, by 1789 for *Aletris*, and by 1837 for *Liatris*; later for other plants with star-shaped flowers), white-stem stick-weed ("stick-weed" alludes to the plant hairs); SPANISH: *pega pega* (stick-stick, a name also applied to *Boerhavia*, and used for several other unrelated species with sticky fruits, Mexico), *pegaropa* (clothes catcher, Mexico); ATHAPASCAN: *iiłtł'į́įh* <*ˀily.'ihi*> [*'iiłtł'įhįih*] ([plant whose leaves are] tenacious, Navajo); UTO-AZTECAN: *huwikaü* (*höwi*, mourning dove, *qa'ö*, corn, Hopi), *ku'hua* <*ku'hwa*> (Shoshoni), *kuha* (Panamint), *kuhu* <*kuhá*> (Northern

Paiute), *kuˀu* (Southern Paiute), *gu'ha* (Paiute), *iks s-hoohoidam* (*iks*, fabric, textile, *s-hoohoid*, it likes, *-dam*, attributive, or "when you get it and stick it on your clothes; it just sticks because they like it," Akimel O'odham), *ikus ho:ho'idam* (Tohono O'odham), *kul* <*ku-l*> (Tübatulabal), *kuˀuvɨ* (Kawaiisu; *mahavuˀuvɨ*, the name for *M. veachiana*), *sililitaqa* <*silílitaqa*> (*silí-lita*, crackle, *qa*, one, Hopi); YUMAN: *sele'* (probably grain(y) or sand(y), cf. Smith 1973, Walapai)

**Botanical description:** Annual herbs, forming a rosette, erect or spreading. Lower leaves are linear, usually deeply lobed; upper leaves are linear to ovate-lanceolate, mostly lobed, uppermost sometimes entire. Floral bracts are mostly linear-lanceolate, entire or angular. Flowers open in early morning. Calices have lobes 2–4 mm long. Petals are ovate, yellow, 3–5 mm long. Stamens are many, 3–5 mm long. Capsules are cylindrical, narrowed at base, erect or the first-formed sometimes recurved, 15–30 mm long, a "lid" falling when ripe. Seeds are 20–30 per capsule, irregularly angled, the surface with a checkered pattern (tesellate) and pointed papillae.

**Habitat:** Washes, canyons. 300–2,250 m (1,000–7,400 ft).

**Range:** Arizona (throughout state), California, New Mexico, Texas, north to Missouri, Nebraska, South Dakota, Wyoming, and Washington; British Columbia; Mexico (Baja California, Chihuahua, Coahuila, northern Sonora, and Tamaulipas).

**Seasonality:** Flowering February to August.

**Status:** Native.

**Ecological significance:** One story has it that the idea for "velcro" came from *Mentzelia*; however, the official Web site of that company says it was from *Xanthium* (q.v.). Either way, the "velcro" on the herbage and fruits serves to disperse Blazing-Star by sticking to passing animals, including humans. Once the fruits are moving, their "salt-shaker" design (topless, with numerous tiny seeds inside) makes for effective dispersal of the seeds.

**Human uses:** Indigenous people parched the seeds for meal (Kearney and Peebles 1951). There are records of Akimel O'odham, Cahuilla, Havasupai, Hohokam, Hopi, Kawaiisu, Klamath, Miwok, Montana tribes, Navajo, Paiute, Tübatulabal, Walapai, and Yavapai eating the seeds of *Mentzelia* (Bohrer et al. 1969, Castetter 1935, Dunmire and Tierney 1997, Elmore 1944, Fewkes 1896, Hodgson 2001, Hough 1898, Kelly 1932, Moerman 1998, Rea 1997, Smith 1973, Voegelin 1938, Watahomigie et al. 1982, Zigmond 1981).

Seeds recovered from a Hohokam site south of Tucson represent the earliest documented use in southern Arizona (Bohrer et al. 1969). About three pounds of seeds were also in a storage jar at Cordova Cave, New Mexico (Kaplan 1963). In the 1860s, the U.S. Cavalry reported hundreds of pounds of seeds traceable to this species stored in willow baskets among the Walapai (Hodgson 2001, Smith 1973). Clearly, *Mentzelia* was historically important food for indigenous peoples, although Rea (1997) almost missed finding out about the genus because cultural erosion was so advanced among the Akimel O'odham. Seeds of *M. albicaulis* should have been important food because they contain 6.4% fat, 3.9% protein, and 85.1% carbohydrates (Hodgson 2001).

At least the Shoshonean people of central Nevada burned off brush and sowed wild seeds of *Mentzelia* (Ebeling 1986). Cattle ranching stopped that practice and led to the cultural erosion of those people and others (Hodgson 2001).

Four tribes are known to have used *M. albicaulis* as medicine. Kawaiisu and Shoshoni treat burns with the seeds (Chamberlin 1911, Zigmond 1981). Hopi and Navajo use *M. albicaulis* to treat toothaches by crushing and soaking the plants or seeds and then applying them to the tooth (Moerman 1998, Vestal 1952). Navajo mix the leaves with other material to make a medicine to relieve snakebites (Vestal 1952). Apache use roots of *M. pumila* to relieve constipation (Reagan 1929).

**Derivation of the name:** *Mentzelia* is named for Christian Mentzel, 1622–1701, a German physician and philologist. *Albicaulis* means white-stemmed.

**Miscellaneous:** There are at least 100 species in the genus (Christy 1998). Two other species are also known from the Baboquivari Mountains. *Mentzelia isolata* is distinguished by having a broad placenta; it is almost always an erect herb, while *M. albicaulis* has a filiform placenta and the stems often spread to form a rosette. *Mentzelia montana* has bracts with whitish bases, mostly attached to the ovary, and straight fruits; *M. albicaulis* rarely has bracts with whitish bases, has them attached below the ovary, and has arched fruits.

## *Anoda cristata* (Malvaceae) Spurred Anoda, *pintapán*

**Other names:** ENGLISH: crested [spurred] anoda (Arizona, New Mexico); SPANISH: *aguatosa* (spiny, from Náhuatl *ahuatl*, spine, Oaxaca), *altea* (modifica-

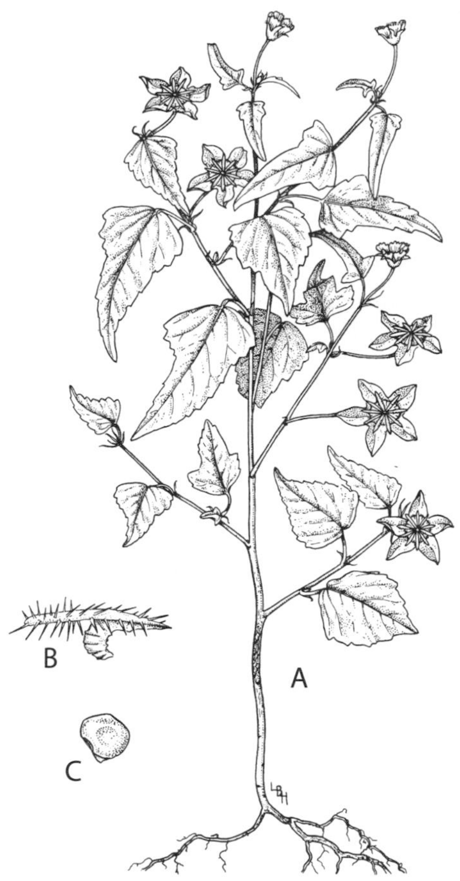


*Anoda cristata*. A. Habit of plant with flowers and fruits. B. Side view of carpel with spur on back. C. Seed. Artist: L. B. Hamilton (Parker 1958).

tion of Latin *althaea*, the hollyhock, Puebla), *amapola* [*amapolita*] [*del campo, morada*] ([little, wild, purple] poppy; *amapola* may be derived from Náhuatl *atl*, water, *mapaitl*, hand, but because it also occurs in European Spanish as *ababol*, Portuguese *papoila*, *papoula*, and French *pavot*, it probably came from Latin *papaver*, Chiapas, Veracruz, Distrito Federal, Edo. México, Jalisco, Puebla), *halache* <*halanche*> (maybe a variant of Mayan *balanche*, Puebla), *malva* [*chica, de Castilla*] ([little, Spanish] mallow, Aguascalientes, Guanajuato, Guerrero, Michoacán, Morelos, Jalisco, Sonora), *pintapán* (appears to be "paints the bread," but perhaps a loan word, Sonora), *quesitos* (little cheese, Hidalgo, Sonora), *tsayaltsay* <*tzalyaltzai*> (maybe from Náhuatl *tzalan*, between, *ializtli*, to leave, Yucatán), *violeta* [*del campo*] ([wild] violet, Edo. México, Veracruz to Oaxaca), *violeta silvestre* (wild violet, Sinaloa); MAYAN: *balanche* (*balan*, jaguar, *che'*, plant, Maya); OTO-MANGUEAN: *alachi* <*alache*> (Mixtec, Distrito Federal, to Guerrero, Puebla); TOTONACAN: *pax'tamac* (Totonac); UTO-AZTECAN: *rehué* (Tarahumara), *tlachpahuatla* (broom, based on the verb *chpa:na*, to sweep, Náhuatl, San Luis Potosí), *xihuitl* (herb, Náhuatl, Mexico), *shiipugi* (maybe from *sipuriki*, be piled or clustered, Mountain Pima), *tusi* (cross, cf. Névome *tusi viucama'*, cross-eyed, Mountain Pima)

**Botanical description:** Erect to decumbent herbs, the stems usually hispid. Leaves are ovate to hastate, dentate to almost entire, sparsely pubescent, often with purple blotch along midvein. Flowers are solitary in leaf axil on long stalk. Calices are 5–10 mm long. Petals are 8–16 mm long, lavender (rarely white). Fruits are disk-shaped, 8–11 mm in diameter; mericarps 10–19, with horizontal spines. Seeds are 3 mm long.

**Habitat:** Moist meadows, along streams. 1,100–1,800 m (3,000–5,900 ft).

**Range:** Arizona (southern Apache to Yavapai, south to Cochise, Pima, and Santa Cruz counties), California, east to Louisiana, North Carolina, Kentucky, and Missouri; Mexico (throughout); to South America.

**Seasonality:** (July) August through September (October).

**Status:** Native.

**Ecological significance:** Fryxell (1993) considered *A. cristata* a weedy species that grows from the United States south to Argentina, Bolivia, and Chile. Because the species has three ploidy levels, with diploid (2n = 30), tetraploid, and hexaploid, this diversity surely accounts for some of the known morphological variation.

*Anoda* comprises 23 species that grow in the southwestern United States, Mexico, and south. Arizona has six species in the genus, with *A. crenatiflora* being known from Santa Cruz County, *A. pentaschista* and *A. thurberi* from Cochise County, and *A. reticulata* from both Pima and Santa Cruz counties. All of these species are on the northern fringe of their ranges.

The other species in Brown Canyon is *Anoda abutiloides*. That species, with a narrower range (from Pima and Santa Cruz counties, AZ; Chihuahua, Jalisco, Sinaloa, Sonora, MX), grows in arid mountains and canyons, particularly in cracks in rocks, 1,100–1,500 m (3,500–5,000 ft). As the name indicates, it is more like an *Abutilon* with white pubescent leaves.

**Human uses:** Although *Anoda* is considered a medicine, the most frequent use, at least in central and southern Mexico, is as food. From at least the Tarahumara of Chihuahua to the Mixtecs of Guerrero and Oaxaca, young, tender leaves and buds are consumed as *quelites* (Linares and Aguirre 1992, Rendón et al. 2001). Casas et al. (1996) note that the Mixtec people of northeastern Guerrero recognize two variants, the *alache macho* (male anoda) and the *alache hembra* (female anoda). The "male" form has slender and pubescent leaves, high fiber content, and is not palatable. However, the "female" form has broader leaves without pubescence, low fiber, and is palatable. Seeds are actually spread in agricultural fields to increase abundance. Could it be that this is an incipient cultivar variation?

Leaves and flowers are emollient, are used against respiratory distress including whooping cough, and are taken to lower fever as far south as Oaxaca (Martínez 1969). Spurred Anodas are high in vitamin C, retinol, iron, and proteins, and are used as ornamentals, at least in Oaxaca (Angélica Bautista-Cruz and María R. Arnaud-Viñas, in litt. Nov. 2004).

**Derivation of the name:** *Anoda* is from a Sinhalese name for some species of *Abutilon* in Sri Lanka, as originally stated by Cavanilles, although alternative views have been given. Others have speculated that *Anoda* is from Greek *a*, without, and *odous*, *odontos*, a tooth, referring to the leaves, or that it came from Greek *a*, without, and *nodus*, a joint or node. *Cristata* means "crested." *Abutilon* has three seeds per mericarp, while *Anoda* has one seed per mericarp.

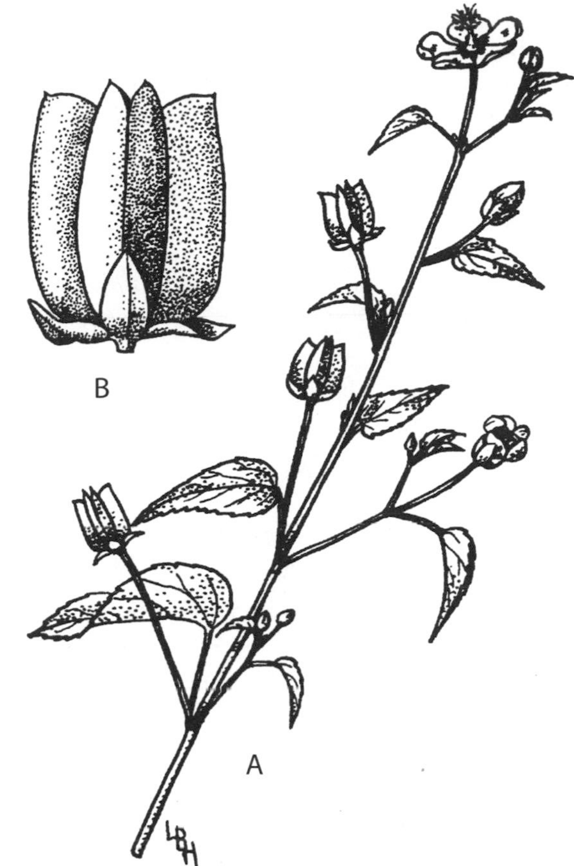

*Abutilon incanum*. A. Flowering and fruiting branch. B. Fruits, enlarged. Artist: L. B. Hamilton (Benson and Darrow 1945).

## *Abutilon incanum* (Malvaceae) Indian Mallow, *pelotazo*

**Other names:** ENGLISH: Indian mallow (from Latin *malva*); SPANISH: *escoba malva* (broom mallow, Sonora), *malva* (mallow, Sonora), *pelotazo* [*chico*] ([little] hairy one, Sinaloa), *tronadora* (based on *tronado*, poor or pauper, northern Mexico to Oaxaca); HOKAN: *caatc ipápl* (what grasshoppers are strung with, Seri), *hasla an ihoon* (ear is its place, Seri); UTO-AZTECAN: *jíchiquia to'ora cojuya* (ash broom, Mayo), *tosaporo* (Guarijío)

**Botanical description:** Subshrubs to ca. 1 m tall (rarely taller). Leaves are ovate, to 6 cm long, irregularly serrulate, densely tomentose. Flowers are solitary or in open panicles. Calices are 3–5 mm long. Corollas are yellow to pink with a dark red center, the petals reflexed, 4–6 mm long. Fruits are longer than the calyx, ca. 6 mm wide, tomentulose; mericarps 5, acute or apiculate at apex. Seeds are ca. 2 mm long, appearing glabrous but minutely puberulent.

**Habitat:** Well-drained slopes. 300–1,219 m (1,000–4,000 ft).

**Range:** Arizona (Cochise, Graham, Greenlee, La Paz, Maricopa, Mohave, Pima, Pinal, Yavapai, and Yuma counties) to western Texas; Mexico (Baja California, Nuevo León, San Luis Potosí, Sinaloa, and Sonora).

**Seasonality:** Flowering (April) August to September (October).

**Status:** Native.

**Ecological significance:** A plant of rocky slopes, gravelly plains, and along arroyos, the species is confined to warm, arid regions of North America. Distribution of the species reflects a need for warm-season rain and relatively mild winters (Turner et al. 1995). The southern limit is in Sinaloa and San Luis Potosí.

**Human uses:** Two unique names and uses are those among the Seri. *Hasla an ihoon* refers to putting small pieces of the stem through pierced ear holes to keep them from closing (Felger and Moser 1985). *Caatc ipápl* notes the seventeenth-century practice of stringing grasshoppers to bring them to camp for food. The species is considered a minor source of fiber in Sinaloa (Wiggins 1980).

The common recent use of most *Abutilon* is to make brooms. The Seri name *hant ipásaquim* (for *A. californicum*) means "broom." Mayo call the brooms *jíchiquia jérocha* (Yetman and Van Devender 2001). Guarijío still use *Abutilon* in medicines to help women who cannot conceive (Yetman and Felger 2002).

*Abutilon* is also notable for the mucilage content of many species. Several species are exploited, and the extract is used as a poultice for tumors (Austin 2004). The mucilage serves as an emollient, and many use a decoction of it to clean wounds. In the Bahamas and Yucatán, the heated leaves of *Abutilon permolle* are used to draw boils, and the decoction is used to bathe sores. Onavas Pima used *A. lignosum*, which they call *pintapán*, to relieve dysentery (Pennington 1980). Usage of mallows is a continuation of a medical application that goes back hundreds, perhaps thousands, of years in Europe and the Americas.

Using *Abutilon* is like the application of *Althea officinalis* (*Althea*, Greek, to heal; *officinalis*, of medicine), the mallow of Europe that also provided the sweet originally called Marsh-Mallow (Austin 2004). Emperor Charlemagne (742–814) thought so much of Marsh-Mallow that he ordered its cultivation. Doubtless, he had it used for both medicine and sweets. That original sweetmeat product was not much like the modern version—now the only plant extract that Marsh-Mallows contain is sucrose, and that sweetener does not come from mallows. Originally, the mucilage was extracted from Marsh-Mallow roots by boiling. Largely, it was used medicinally to soothe the skin and mucous membranes. That is essentially the use that *Abutilon* still has among some people.

**Derivation of the name:** British gardener Philip Miller created *Abutilon* by adapting the Arabic name, *aututilun*, used by physician Ibn Sina [Avicenna], who died in CE 1037. The name may be compounded from Arabic *abu*, father of, and Persian *tula* or *tulha*, mallow. *Incanum* means "white-pubescent," thus representing the thick covering of white trichomes on most plant parts.

**Miscellaneous:** There are six other *Abutilon* known from the Baboquivari canyons, including *A. abutiloides*, *A. malacum*, *A. mollicomum*, *A. reventum*, *A. sonorae* (Sonora Indian Mallow; *pintapán cimarrón*), and *A. thurberi*. Only three are known from Brown Canyon.

*Abutilon malacum* is similar to *A. incanum*, but the leaf tips distinguish the two. Leaves are blunt (obtuse) in *A. malacum* and pointed (acute-acuminate) in *A. incanum*. Stems and petioles are glabrous or puberulent in *A. reventum*, and the carpels are blunt (muticous). This morphology contrasts with that of *A. sonorae*, which has these organs sparsely to copiously hairy, with long, spreading or reflexed trichomes, and mucronate or cuspidate carpels.

## *Gossypium thurberi* (Malvaceae) Canyon Cotton, *algodoncillo, ban tokiga*

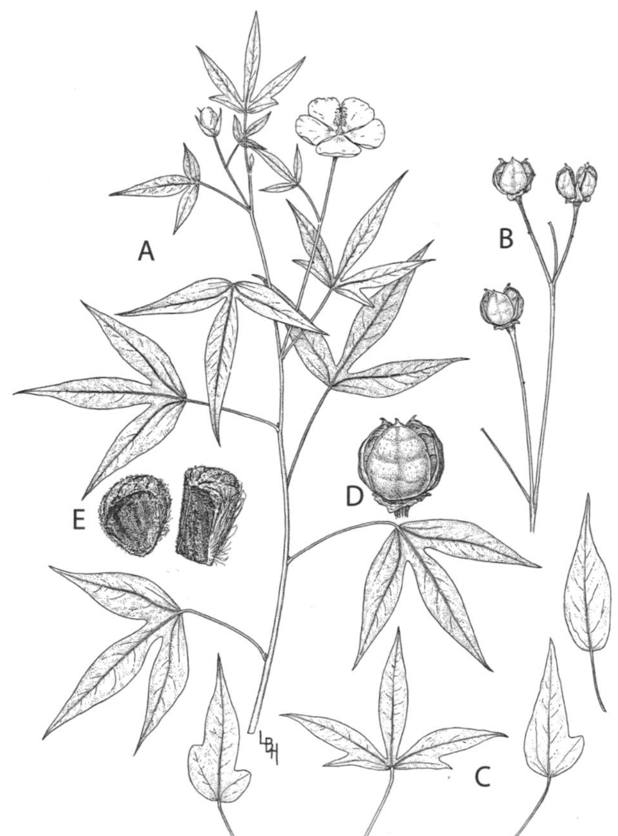


*Gossypium thurberi*. A. Flowering branch. B. Fruiting branch. C. Leaf variation. D. Fruit. E. Seeds, two views. Artist: L. B. Hamilton (from original artwork).

**Other names:** ENGLISH: canyon [desert, Thurber's] cotton ("cotton" in English and the Romance languages was derived from Arabic *al-qutn*, *qatn*, because these people brought the word and plant to Europe via Spain; cognates include Italian *cotone*, French *coton*, and Portuguese *cotão*; Spanish, misunderstanding the articles of Arabic, call it *algodón*, derived from *al-qutun* <*al-qoton*>, the cotton; English began using words *cotoun* and *coton* about CE 1300, and even in Gaelic it is *cotan*); SPANISH: *algodoncillo* [*del campo*, *del monte*] (little cotton [of the countryside, wild], Sonora); ATHAPASCAN: *ichoghąą* (Western Apache), *ndik'ą'* (Navajo); UTO-AZTECAN: *ban tokiga* (coyote's cotton, Tohono O'odham; *toki* <*to'ki*, *tokih*> is cultivated cotton), *tok* (Onavas Pima; for cultivated cottons), *toki* (Akimel O'odham; for cultivated cottons), *toˀsá* (Guarijío); YUMAN: *atcώ* <*xlitco*, *xotcώ*> (Maricopa), *hedjáwa* (Havasupai), *xsaw* [*xsa:w*] (Cocopa); LANGUAGE ISOLATE: *u'we* (down, Zuni for *G. hirsutum*)

**Botanical description:** Shrubs, erect to ascending, branched, 1–3.5 m tall, all parts gland-dotted. Leaves deeply 3–5 lobed, the lobes 1–3 cm wide, 3.5–10 cm long, lanceolate, entire, long-acuminate, glandular, glabrous petioles 2–8 cm long. Flowers are solitary or in corymbose cymes, on peduncles 1–3.5 cm long. Bracts are 3, ovate-lanceolate, entire or 3-toothed, 6–15 mm long. Calices are broadly cup-shaped, 4–5 mm high, truncate or 5-toothed. Petals are 2–3 cm long, white or cream with purple at base, turning purple with age, purple- or black-dotted. Capsules are broadly ovoid to almost globose, 1–2 cm long, apiculate, inner margins of sutures bearing tufts of long white hair. Seeds are 4–5 mm long, dark brown to almost black, top-shaped, finely pubescent with short hairs and a few long ones.

**Habitat:** Canyons, washes. 760–1,530 m (2,500–5,000 ft).

**Range:** Arizona (Cochise, Gila, Graham, Maricopa, Pima, Pinal, Santa Cruz, and Yavapai counties); Mexico (Sonora).

**Seasonality:** Flowering (July) August to October.

**Status:** Native.

**Ecological significance:** As the common name implies, these short shrubs are most common in canyons, particularly along streams. This shrub is a thornscrub species (Brown 1982).

Canyon Cotton is among the plants in the canyons that turn colors in the fall. When the first cool nights begin, the leaves turn from green to red. *Gossypium* is among the few plants that become red in the canyons, along with with *Rhus* (q.v.) and *Toxicodendron* (q.v.).

The structure of the flowers is so similar to *G. herbaceum* of the Caribbean and elsewhere (Austin 2004) that canyon cotton must also be pollinated by bees.

**Human uses:** According to Castetter and Bell (1942), Canyon Cottons were used as a source of fiber among the Tohono O'odham. Because the lint on the seed is insufficient for use, fiber must have been taken from the stems.

Indigenous people were growing cultivated cotton when Europeans first arrived in the Southwest. That species, *G. hirsutum*, had a long history of cultivation in the New World, and most tribes used it. Many indigenous people had used cottons for so long that they have simple terms (primary unanalyzable lexemes) for them (Austin 2004); others are probably compound. In the West Indies, the Arawak called it *ikálotopue*, and the Caribs say *maulu* <*mauru*> (Dominica, Surinam). The Carib word is considered a cognate with the South American Tupí name *amandiyu-b*. Mainland names include *dehti* (Otomí), *ma ˀhua* (Tarahumara), *odigé* (Hispaniola), *pishm* <*pishten*> (Mixe, Mexico; *pishten-kiup*, the plant; *pishten-puih*, the flower), *tüdy* (Otomí), and *upsana* (Cuna, Panama). However, a few languages have more interesting variants on their names. The Seri say *mooj*. The Maya say either *xchup* (stuff, as in pillows, Yucatán) or *taman* (Yucatán). After the Spanish arrived, the Maya began calling sheep and lambs *taman*, a name previously reserved for cotton. The Aztecs were more direct; they said *ichcatl* (cotton) or *ichcaxihuitl* (thread herb, Náhuatl).

**Derivation of the name:** *Gossypium* is from the Latin *gossypion*, used for the cotton plant by Pliny, CE 23–79. *Thurberi* commemorates George Thurber, 1821–1890, quartermaster of the Mexican Boundary Survey (1850–1853) and editor of the journal *American Agriculturist*.

**Miscellaneous:** There are 50 species in *Gossypium*, which are found in relatively arid tropical and subtropical parts of Africa, the Middle East, Australia, and North and South America (Fryxell 1993).

## ***Sida abutifolia*** **(Malvaceae)** **Prostrate Mallow, *malva***

**Other names:** ENGLISH: prostrate mallow ("mallow" is from Latin *malva*, a general term for members of the Malvaceae; akin to "maul" and "maw"), spreading fan-petals [sida] (Arizona, New Mexico);

*Sida abutifolia*. A. Flowering branch. B. Habit. Artists: unknown (Diggs et al. 1999), A; R. D. Ivey, B (Ivey 2003).

SPANISH: *hierba de la vieja* (old woman's herb, Durango), *malva* (Sonora), *yerba del buen día* (good day herb, San Luis Potosí, Nuevo León); MAYAN: *hauay-xiu* <*xauayxiu*> (*hauay*, expanded, *xiu*, herb, Maya); (= *S. procumbens*)

**Botanical description:** Perennial herbs from a woody rootstock. Stems extend to 50 cm long, stellate-pubescent. Leaf blades are linear to ovate or oblong, to 35 mm long, usually smaller, apically obtuse to acute, basally cuneate to somewhat cordate, crenate-dentate. Flowers are solitary in leaf axils. Sepals are ca. 5 mm long. Petals are much longer than sepals. Fruits have apiculate carpels, varying to having 2 prominent points.

**Habitat:** Grasslands, washes, canyons. 760–1,830 m (2,500–6,000 ft).

**Range:** Arizona (Cochise, Gila, Graham, Greenlee, Pima, Pinal, and Santa Cruz counties), New Mexico, Texas, and Florida Keys; Mexico (Chihuahua and Sonora); Caribbean; South America.

**Seasonality:** Flowering July to September.

**Status:** Native.

**Ecological significance:** This species is typically an herb hiding among grasses in the desert grasslands. Plants are inconspicuous until the flowers open, and then they are easily seen. Flowers do not open until midday or early afternoon, and then they close early; they are not open more than a few hours. Augochlorid bees are visitors. These herbs are scattered in open sites in both Brown and Jaguar Canyons, but they are nowhere as common as in the grasslands lower in the Altar Valley. *Sida abutifolia* is also documented in Baboquivari, Fresnal, and Moristo Canyons, and in the lowlands on both sides of the mountains (Appendix).

**Human uses:** *Sida abutifolia* is used in Nuevo León and San Luis Potosí to treat boils and kidney problems (Ford 1975).

Numerous other species have been used by a variety of peoples. Hocking (1997) lists 13 species for which he found recorded uses. Of those, four are known to have been used in Mexico. Others are used in the southeastern United States and the Caribbean.

Some, like *S. acuta*, are used to make brooms because of their slender, flexible stems (Austin 2004). Those same stems cause others, like *S. cordifolia*, to be used as a source of fibers (Hocking 1997). Many, however, are considered medicinal. Of those considered medicinal, only *S. rhombifolia* and *S. spinosa* occur in Arizona.

*Sida rhombifolia* is known as *axocatzin* (reference obscure, from Náhuatl *axoquetzin*, the young son of Tetzcuco, Arizona, Mexico), Indian Hemp, Indian Tea (Florida), and *malva de cochino* (pig mallow, Cuba). Branches are made into brooms, and leaves are smoked as a stimulant and adulterant for Marijuana; leaves contain ephedrine (Martínez 1969, Schultes and Hofmann 1979). Tarahumara use *S. rhombifolia* to wash their hair, saying it makes the hair grow (Pennington 1963). Similar to this species and often confused with it is *S. acuta*. That plant is called Wire-Weed (Florida), *malva de caballo* (horse mallow, Cuba), *chak' misbil* (red broom, Maya), and *chichipe* (something in the path that sticks, Maya).

The other Arizona species with records of uses, *S. spinosa* (Prickly Mallow, Arizona), is also called *t'iltha' thipon* (narrow *thipon*, Huastec, San Luis Potosí). While the Huastec have a name for *S. spinosa*, Alcorn (1984) was given no use. However, the common name compares *S. spinosa* with *thipon*

(*Malvastrum americanum*), and that is used to treat vomiting, swollen feet, and on external sores. Elsewhere in Mexico, *S. spinosa* is used to relieve dysentery (Hocking (1997). All of those uses are made of the species south of the border region.

Outside of the borderlands of Mexico and Arizona, species used include *S. ciliaris*, *S. pyramidata*, *S. triloba*, *S. cordifolia*, *S. glutinosa*, *S. ovalis*, *S. tomentosa*, and *S. urens* (Alcorn 1984, Hocking 1997).

**Derivation of the name:** *Sida* is the Greek name used by Theophrastus, 372–287 BCE, for some similar plant. *Abutifolia* means that it has leaves resembling those of *Abutilon*.

## *Sphaeralcea fendleri* (Malvaceae) Fendler's Globe-Mallow, *mal de ojo, haḍam tatk*

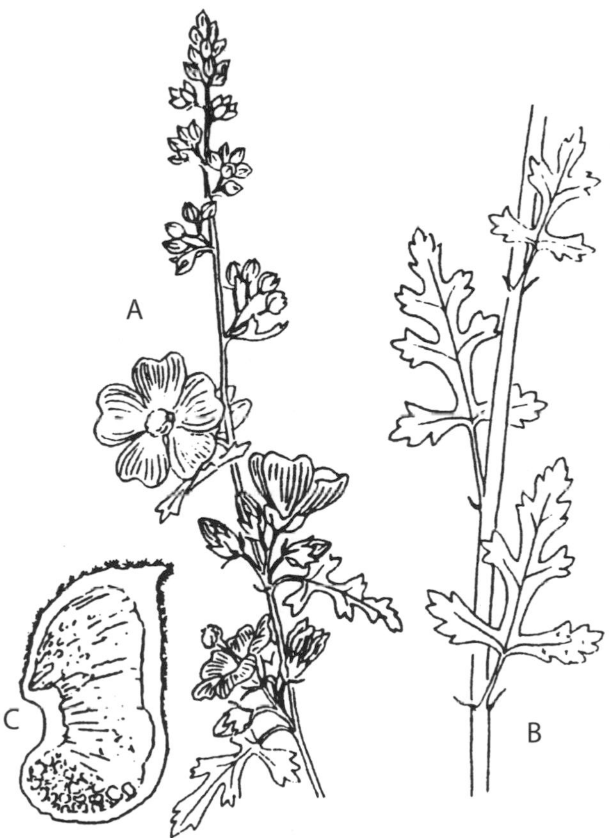


*Sphaeralcea fendleri*. A. Flowering branch. B. Segment of stem showing leaves and stipules. C. Fruits segment. Artist: R. D. Ivey (Ivey 2003).

**Other names:** ENGLISH: desert mallow ("mallow" is derived from the classical Latin *malva*, usually *Malva sylvestris*, and appeared in Old English before the 1300s), [Fendler's, scarlet, thicket] globe-mallow (the common name is a translation of the genus; which came first is problematical, Arizona, New Mexico), sore-eye poppies (Arizona); SPANISH: *mal de ojo* (evil eye, Arizona), *yerba de la negrita* (black woman's herb, New Mexico); ATHAPASCAN: *'azee'ntł'iní'* [*tsoh*] <*ˀazeˀ nƛ'íní*, *'azé' nini*, *'azee' ntł'inítsdh*> (*azee'*, medicine, *ntł'iní'*, gummy, *tsoh*, big [plant whose roots are gummy], Navajo), *azee' dijéé'* <*ˀazéˀ dit'iˀi*> (viscid medicine, Navajo), *izee dijééd* <*izee dit'iihee*> (*izee*, medicine, *dijééd*, viscid, Western Apache); KIOWA TANOAN: *'oḏa* (Tewa); UTO-AZTECAN: *haḍam tatk* (sticky roots, or *haḍam tatkam*, sticky roots it has, Akimel and Tohono O'odham), *heoko kuta* (*heo*, evening, *kuta*, stick, Yaqui), *koi'nakomp* (Shoshoni), *kopono* <*kapóna*> (Hopi); YUMAN: *jik buny* (Walapai)

**Botanical description:** Small shrubs or perennial herbs, often with several erect to ascending stems from a woody crown, green or white-pubescent, the stems to 14 dm tall. Leaves on slender petioles, the blades ovate-oblong to broadly ovate, cuneate basally, acute to obtuse and mucronulate apically, shallowly and deeply 3-lobed below the middle, lateral lobes narrow-triangular or rectangular, acute, the midlobe toothed or cleft, the margins crenate to crenate-dentate, 3–6 cm long. Inflorescences are a narrow many-flowered thyrse, the bractlets of the involucel thin, green. Calices are 4.5–6 mm long, with deltoid to ovate-lanceolate and acute to short-acuminate lobes. Petals are pinkish to lavender or mauve, 8–13 mm long. Fruits have 11–15 carpels, chartaceous walls, often connate when mature, 4–5 mm long.

**Habitat:** Along creeks. 900–2,440 m (3,000–8,000 ft).

**Range:** Arizona (Apache to Coconino, south to Cochise, Pima, and Santa Cruz counties), central Colorado, New Mexico, and Texas; Mexico (Chihuahua? and Sonora?).

**Seasonality:** Flowering August to October.

**Status:** Native. Our plants belong to var. *venusta*, restricted to southeastern Arizona and adjacent Mexico, exclusively a tetraploid with 2n = 20 chromosomes. The species also contains diploid (2n = 10) and triploid (2n = 15) races (La Duke 1986).

**Ecological significance:** This species is typical of Ponderosa Pine woodlands but it extends down from that habitat into the oak woodlands in the canyons. Although Fendler's Globe-Mallow grows down to

3,000 ft elsewhere in Arizona, it has not been found below about 4,700 ft in Brown Canyon.

This shrub is part of the diet of the Mule Deer (Krausman et al. 1997). The beetle *Agrilaxia flavimana* (Buprestidae) also feeds on Fendler's Globe-Mallows (MacRae and Nelson 2003).

The distribution of *S. fendleri* in Arizona follows the Mogollon Rim across the state from northwest to southeast, and it then follows the mountains down to the Mexican border. Although no specimens of var. *fendleri* have been located from Sonora, those from Sycamore Canyon in southern Santa Cruz County are so close to the border that it must be there. The Baboquivari Mountains are the western range limit (Felger et al. in prep.).

**Human uses:** *Sphaeralcea fendleri* is used by Navajo as medicine for external and mouth sores, and an infusion is drunk to relieve internal injury and hemorrhage (Moerman 1998, Vestal 1952, Wyman and Harris 1951). Curtin (1947), Hocking (1997), and Whiting (1939) record that *Sphaeralcea* is used to treat bruises, as an expectorant, a laxative, on tumors, and to promote hair growth. Apache boil the root to make a tea that halts diarrhea (Gallagher 1976). The remedy is sold for these purposes in Mexican-American stores (Kay 1996).

Jaeger (1941) notes that people in Baja California and elsewhere in the West consider Fendler's Globe-Mallows "*plantas muy malas*" because of the irritating hairs that sometimes cause sore eyes. Akimel O'odham and, presumably, Tohono O'odham are now mixed in their views of Globe-Mallows, although in 1908 they used *S. fendleri* to treat diarrhea or biliousness (Rea 1997). Now, some Akimel O'odham told Rea it would *cause* sore eyes, while others assured him it would *cure* those eyes. Others remember the older use and recommend it for relieving diarrhea—because it is *s-haḍam* (sticky).

Havasupai mix Fendler's Globe-Mallow juice with clay before molding it into ceramics (Moerman 1998). Seeds of *Sphaeralcea* occur in Hohokam sites with some regularity, and Gasser (1981) assumed that they were eaten. There may have been other uses.

**Derivation of the name:** *Sphaeralcea* comes from Greek *sphaera*, sphere or globe, from the spherical fruits, and the genus name *Alcea*, the Hollyhock, which it resembles. *Fendleri* commemorates German-born American August Fendler, 1813–1883, explorer and collector in North and South America; *venusta* means "charming" in Latin. *Sphaeralcea* is similar to other genera in the family, including *Sidalcea*. The two are distinguished by the reduced "second calyx" (involucel) in *Sphaeralcea* and lack of it in *Sidalcea* (Fryxell 1993).

## *Proboscidea parviflora* (Pedaliaceae) Annual Devil's Claw, *uña de gato, ban ihugga*

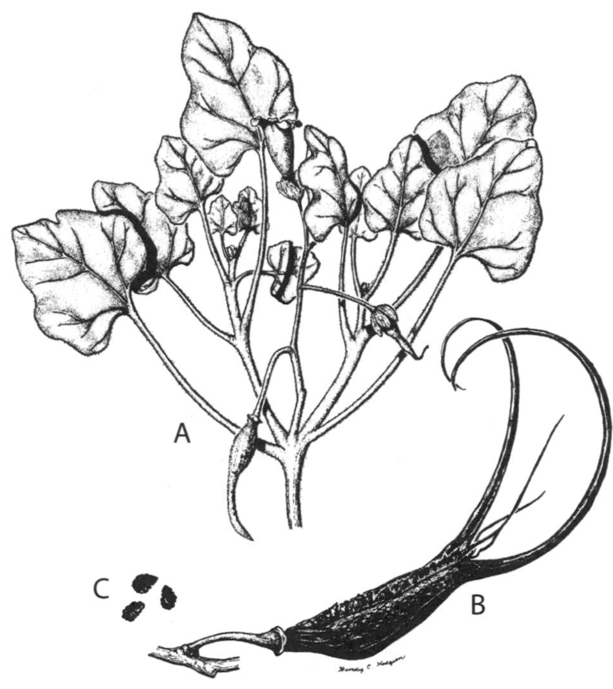


*Proboscidea parviflora.* A. Flowering and fruiting plant. B. Mature fruit. C. Seeds. Artist: W. C. Hodgson (courtesy Wendy Hodgson).

**Other names:** ENGLISH: double-claw (Arizona, New Mexico), devil's claw (a name applied to a European *Ranunculus* by 1878; later applied to other plants); SPANISH: *aguaro* (from Cahita *ahua*, horn, Chihuahua, Sonora), *catachio* (from Quechua *catachina*, to cover, in reference to the sticky pubescence, Guerrero to Oaxaca), *cuernitos* [*cuernatos*] (little horns, Sonora to central mesa of Mexico), *espuelito del diablo* (devil's little spur, Baja California), *garambullo* (a name also given to *Celtis*, *Cereus*, and *Pisonia*, Chihuahua), *gatito* (little cat, Sonora), *perritos* (little dogs, Sonora), *toro* (bull, Sonora), *torito*[*s*] (little bull, Sonora to central mesa of Mexico), *uña de gato* [*gatuño*] (cat's claw,

Sonora); ATHAPASCAN: *'akéshgaan* (claw, Navajo), *chogołshahé <chugoséhe, idághadé, itághadé>* (Western Apache), *daa'Yadebitabizaye* (devil's claw with small leaves, Chiricahua and Mescalero Apache); UTO-AZTECAN: *akawat* (Cahuilla), *ban ihugga <ban 'ihugga, ihu'k>* (coyote's devil's claw, Akimel O'odham and Tohono O'odham), *ban shu:shk* (coyote's sandals, Tohono O'odham), *čorí* [*čoríkari*] (Tarahumara), *'ihug* (devil's claw, Hiá Cеḍ O'odham), *<sahoobinump>* (Southern Paiute), *tamko'okochi* (Yaqui), *tankokochi <tancocochi>* (Guarijío? Gentry 1495 ARIZ), *tɨvoʾonɨbɨ* [*tɨvɨʾonɨbɨ*] (Kawaiisu), *tümuippüh* (Panamint), *tumo'ala <tumó'alá>* (*ala*, horn, Hopi); YUMAN: *gwóxtón* (Maricopa), *halák*[A] (Havasupai), *ʾi:cúc* (Cocopa), *mak ḍuny* (Walapai); (= *Martynia fragrans*)

**Botanical description:** Annual herbs, from a stout taproot, stems often hollow, branched and ascending or spreading, to 6 dm tall, and 2 m broad, sticky-pubescent. Leaves are cordate, entire to shallowly lobed or undulate, 15–25 cm long and almost as wide when mature, the petioles about as long as the blade. Racemes are 2–10 flowered. Calices are 9–12 mm long at anthesis, slightly accrescent and deciduous at maturity. Corollas are dirty yellow to pink or purplish, with a dark purple spot on the upper side, 24–35 mm long, the limb 15–35 mm broad. Fruits at maturity are strongly curved, 15–35 cm long, crested along the concave side at the broadened basal portion.

**Habitat:** Grasslands, desertscrub, along streams, washes. 300–1,500 m (1,000–5,000 ft).

**Range:** Arizona (Apache, Coconino, Gila, Graham, Greenlee, Maricopa, Mohave, to Cochise, Pima, Pinal, and Santa Cruz counties), southern California, New Mexico, Nevada, western Texas, and southern Utah; Mexico (Baja California, Chihuahua, Durango, Guerrero, Jalisco, Oaxaca, San Luis Potosí, Sinaloa, and Sonora).

**Seasonality:** Flowering (April) July to September (October).

**Status:** Native.

**Ecological significance:** This herb is the wild ancestor of the cultivated *ihug <'ihugga>* of the Tohono O'odham and Akimel O'odham. Plants are usually confined to wet sites, at least those that are wet during the summer rains. The herbs are also heliophilic and thrive only where there is abundant sunshine, languishing if they germinate under shaded places.

The long "claws" on the fruits are adept at spreading them. Anyone who has been walking carelessly through the region and had one of these fruits "grab" them by the ankle can attest to their efficiency. Probably, the fruits are also spread by Coyotes, deer, Javelina, and other animals.

**Human uses:** This wild *Proboscidea* probably gave rise to the domesticated form along the banks of the Gila River in Arizona (Rea 1997). The wild form, however, was already used as food and fiber by indigenous peoples. Seeds are eaten by at least the Akimel and Tohono O'odham, Apache, Havasupai, Hopi, Tarahumara, and Walapai, and probably also by the Cahuilla and Yaqui (Castetter and Opler 1936, Gallagher 1976, Hodgson 2001, Moerman 1998, Nabhan and Rea 1987, Pennington 1963, Whiting 1939). Tarahumara also boil the fresh leaves and add them to bean dishes (Pennington 1963). Oddly, the Northern Tepehuan claim no use of Devil's Claw (Pennington 1969).

At least the Akimel and Tohono O'odham, Kawaiisu, Panamints, and Western Apache make use of the fibers from the fruits to add the black color to their baskets, although basket makers prefer those from the cultivated form, *P. parviflora* ssp. *parviflora* var. *hohokamiana* (Bray 1998, Coville 1892, Hodgson 2001, Nabhan and Rea 1987, Rea 1997, Zigmond 1981). Onavas Pima (Sonora) use the seeds to polish fired *ollas* (pots) (Pennington 1980). Hopi believe that the long spines of the fruits draw lightning and rain; therefore, they never remove *Proboscidea* from cultivated fields (Whiting 1939).

**Derivation of the name:** *Proboscidea* is from Greek *proboskis*, snout, *-oidea*, resembling, thus alluding to the long curved beak of the fruit. *Parviflora* means small-flowered.

**Miscellaneous:** *Proboscidea althaeifolia* is the perennial species with yellow flowers. This perennial has been reported in Brown Canyon, but not verified. It has also been found in Baboquivari Canyon, in the Coyote and Sierrita Mountains, and 22 miles south of Robles Junction near Hwy. 286. There is a specialized bee (*Perdita hurdii*) that pollinates this species. *Proboscidea althaeifolia* is called *aguaro con camote* (devil's claw with a sweet potato, Sonora), *ban ihugga* (Akimel O'odham), *campanita* (little bell), *cuernito* (little horn, Sonora), *cuernos* [*espuela*] *del diablo* (devil's horns), Desert Unicorn Plant, *gato* (cat, Sonora), *tumo'ala* (Hopi), *uña de gato* (cat's claw, Sonora), *torito* (little bull, Sonora), and *uña del diablo* (devil's claw).

## *Morus microphylla* (Moraceae) Texas Mulberry, *mora, gohi*

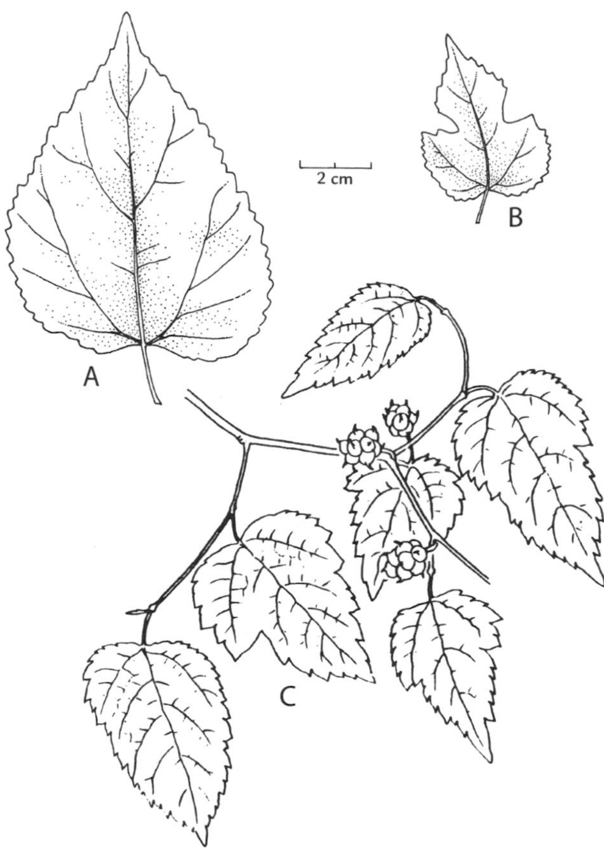


*Morus microphylla*. A. Unlobed leaf. B. Lobed leaf. C. Fruiting branch. Artist: M. B. Johnston, A–B (Felger et al. 2001); R. D. Ivey, C (Ivey 2003).

**Other names:** ENGLISH: little-leaf [Mexican, mountain, Texas] mulberry (in print in English by 1382; based on *morus* or *morum* in Latin, the fruits became *mora* in Italian, Spanish, and Portuguese, and *mûre* in French; *morus* plus berry was transcribed into *múlbere* in Old High German to become *Maulbeer* in modern German, *moerberie* in Dutch, and mulberry in English); SPANISH: *baya* (berry), *mora cimarróna* (wild mulberry, Arizona, Sonora), *moral* (mulberry grove, Chihuahua south), *salsa mora* (mulberry sauce, Sonora); ATHAPASCAN: *iłtį' tsį* <*its'in*> (*iłtį'*, gun, combined with *tsį* = bow, Western Apache), *tséłkani* (Western Apache), *tsełkane* (Chiricahua and Mescalero Apache); UTO-AZTECAN: *apurí* (Tarahumara), *gohi* (Hiá Ceḍ O'odham), *gohi* <*gōhi, gohih, goohi, gaw-hee, kohi, kóxi*> (var. *gohui*, Akimel O'odham and Tohono O'odham), *kohi* (Mountain Pima), *kóhi* <*kóji*> (Northern Tepehuan), *paboré* (Guarijío); YUMAN: *puima'á* (Havasupai), *puim'á* <*pu'i ma*> (Walapai)

**Botanical description:** Trees or shrubs to 7 m tall. Leaves with a petiole to 15 mm long, the blade ovate, 3–7 cm long, entire to 3–5 lobed, basally rounded to almost cordate, apically pointed, serrate or crenate-serrate, rough-hairy above, pubescent or glabrous below. Staminate catkins are 1–2 cm long. Fruits are ovoid, purple or blackish when ripe, 1–1.5 cm long.

**Habitat:** Washes, canyons. 1,100–1,500 m (3,500–5,000 ft).

**Range:** Arizona (Coconino, Greenlee, Yavapai, south to Cochise, Pima, and Santa Cruz counties), New Mexico, and western Texas; Mexico (Chihuahua, Durango, and Sonora).

**Seasonality:** Flowering March to April.

**Status:** Native.

**Ecological significance:** Mulberry is a temperate genus extending down toward the desert in canyons, where it is never abundant in the Baboquivari Mountains. The tree is mostly above 4,500 ft, but a few come down to 4,200–4,300 ft in the Baboquivari Mountains; it grows lower in other areas. Flowers are wind-pollinated, and the pollen is a major source of allergies in cities; planting *Morus* is no longer legal in Tucson. Fruits in *Morus* are termed "multiple," meaning that the single fruit is produced by numerous flowers. Fruits are avidly sought by birds and mammals. Martin et al. (1951) list about 30 species of birds and several mammals that feed on the fruits of *Morus*.

**Human uses:** Completely unlike the historical record of eating Mulberry fruits in the Southeast (Austin 2004), the record in the Southwest is sparse and fragmentary. Hodgson (2001) attributes that difference in part to the rarity of the trees, the low yield of fruits, and their popularity with birds and mammals. Still, the fruits are consumed by Apaches, Akimel and Tohono O'odham, Havasupai, Ópata, Tarahumara, Northern Tepehuan, Walapai, and Yavapai (Castetter 1935, Castetter and Opler 1936, Gallagher 1976, Pennington 1963, Pennington 1969, Moerman 1998, Hodgson 2001). At least the Tarahumara and Northern Tepehuan eat the young leaves cooked as *quelites*, but the Ópata apparently did not (Pennington 1963, Pennington 1969, Hodgson 2001). More recently, the Tarahumara use the sap for curdling milk to make cheese.

More important to most tribes than their use for edibility, the trees were valued for making bows. Akimel O'odham formerly made bows of Mulberry wood and were famous archers (Rea 1997). A dance-step, previously associated with rainmaking, is called *gohi-meli* (mulberry-run, also cripple dance). The wood was also prized for bows by at least the Apache, Tarahumara, Northern Tepehuan, Tohono O'odham, and Walapai (Castetter and Underhill 1935, Hodgson 2001, Moerman 1998, Pennington 1963, 1969). Western Apaches also prune trees in the fall to produce long, straight stems for making baskets (Gallagher 1976).

Like its eastern counterpart (Austin 2004), the western species is valued as a dye source. Yaqui use the dye, and early in the twentieth century, hundreds of carloads of wood were shipped from Mexico to the United States for dye making (Hodgson 2001).

**Derivation of the name:** *Morus* is from classical Latin *morum*, meaning "berry" or the Mulberry. *Microphylla* means having small leaves, although they are not particularly small compared to other species in the genus.

**Miscellaneous:** People have long thought the old nursery rhyme "Here we go round the Mulberry bush" referred to *Morus*. The rhyme always seemed odd to me because the Mulberry is a tree, not a bush. It turns out that the use of the word "Mulberry" in English is equivalent to Spanish where *mora* also means Blackberry (*Rubus*). Vickery (1995), who traced the English etymology, found that the "Mulberry bush" in the game was actually a *Rubus* and not a *Morus*; apparently, the mixing of their names goes back to Horace (65–8 BCE), Virgil (70–19 BCE), and Ovid (43 BCE–? CE 17). Indeed, a glossary published in the eleventh century by R. P. Wülcker recorded that *mora* was used as a name for all berries.

## *Allionia incarnata* (Nyctaginaceae) Wind-Mills, *hierba de la golpe*

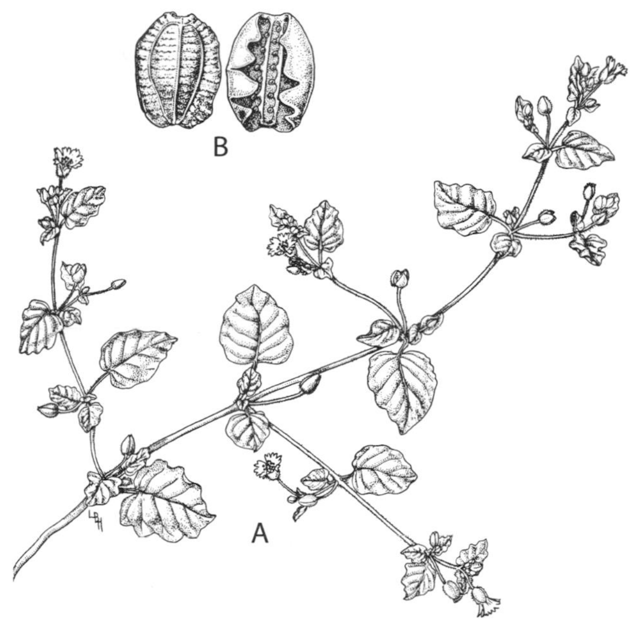


*Allionia incarnata*. A. Flowering plant. B. Enlargement of the two faces of the distinctive seeds. Artist: L. B. Hamilton (Parker 1958).

**Other names:** ENGLISH: pink three-flower (a reference to the flower cluster, Arizona), trailing four o'clock ("four o'clock" is a name given the related *Mirabilis jalapa* by Robert Browne in his *Flora of Jamaica* of 1756 noting the time when the flowers open), umbrella-wort ("wort" is an old English word meaning plant), [trailing] wind-mills (originally applied to wind-driven mills for grinding grain, pumping water, etc. in the Old World; in English by 1297; later applied to plants, Arizona, New Mexico); SPANISH: *guapile* (variant of *guapilla*, Sonora; also used for *Agave*, *Hechtia*), *hierba de la golpe* (bruise herb, Sonora), *hierba de la hormiga* [*mosca*] (ant [fly] herb, Durango, Nuevo León, Zacatecas), *Juan ematilli* (Onavas Pima); ATHAPASCAN: *tsét'ąą' ts'ósí* <*cedide.h c'o's*> (leaves like rock tea [*tsé*, rock, *'at'ąą'*, leaf], *ts'ósí*, slender, Navajo), *'ilt'ąą'* <*ˀilt'ąˀí*> (leaves like rock tea [lit. *'lį́*, valuable, *'at'ąą'*, leaf], Navajo); HOKAN: *hamíp cmaam* (female spiderling [*Boerhavia*], Seri); KIOWA TANOAN: *'okup'e* (*'oku*, turtle, *p'e*, plant, Tewa); UTO-AZTECAN: *totopwuvàapi* <*totópwuvápi*> (*totop*, fly, *wuvàapi*, whip, Hopi)

**Botanical description:** Perennial or winter annual herbs. Stems are prostrate, 1–8 dm long or more, radiating from the root crown. Leaf blades are 8–30(–55) mm wide, ovate to elliptic, entire or undulate to roundly lobed or toothed. Flowers on peduncles are 3–25 mm long, the perianth (corolla-like modified calyx) pink-purple to magenta, rarely white, 6–15 mm long. Fruits are 3–4 mm long, the inner side 3-veined, the margins usually toothed and incurved.

**Habitat:** Grasslands and mixed desert shrub communities. 800–1,830 m (2,624–6,000 ft).

**Range:** Arizona (throughout most of state, missing from some areas of southern Apache and Navajo counties), California, Colorado, Nevada, Texas, and Utah; south through Mexico; to South America.

**Seasonality:** Flowering April to October.

**Status:** Native.

**Ecological significance:** This species occurs in open areas, typically those places that have been disturbed in some way. Fruits of this genus, unlike some others in the area including *Boerhavia* (q.v.), do not have glandular plant-hairs along their margins. In *Allionia*, these glands are sunken within a cavity on the side of the fruits. In the other species, those sticky plant-hairs are efficient at attaching themselves to passing animals (including humans) that spread the propagules to new locations. It is uncertain what method is used to disperse *Allionia*.

This sun-loving plant (heliophile) thrives in the wet season, sometimes creating clumps over six feet (2 m) across. Bowers et al. (2004) found that *A. incarnata* appeared rarely or in small numbers. When the dry season begins, Wind-Mills quickly shrivel. The parts above ground disappear during the hottest seasons, and the entire plant "disappears." Instead of being dead, the herbs are simply dormant, living below ground until there is sufficient moisture to begin growing again. Leaves have specializations (Krantz anatomy) that are characteristic of C4 metabolism (Mulroy and Rundel 1977). That metabolism allows them to survive in hot climates.

Seeds are collected by four distinct species of rodents—Merriam's Kangaroo Rat (*Dipodomys merriami*), Arizona Pocket Mouse (*Perognathus amplus*), Bailey's Pocket Mouse (*P. baileyi*), and Rock Pocket Mouse (*P. intermedius*), but only *P. baileyi* uses the seeds "significantly" (Reichman 1975).

When the flower displays are at their peak, the showy pink corolla-like perianths are eye-catching. Each "flower" is actually a cluster of three, easily separable from the others and forming a wedge-shaped part of the group. These flowers open early in the morning and fade by late morning on hot days. Flowers develop into fruits that are unique in the Sonoran Desert (Felger 2000).

**Human uses:** The Seri consider *Allionia* "female" because it "spreads over the ground." *Allionia incarnata* is cooked in water, and the tea drunk to relieve diarrhea (Felger and Moser 1985). Hopi (Whiting 1939) use the same name for *Allionia* as for *Boerhavia*, q.v. Navajo treat swellings with an infusion of the roots (Vestal 1952).

**Derivation of the name:** *Allionia* was named for the Italian Carlo Ludovico Allioni, 1728–1804, a naturalist, physician, professor of botany, and correspondent of Linnaeus. *Incarnata* means "flesh-colored," thus alluding to the color of the corolla-like calices.

**Miscellaneous:** The genus *Allionia* contains two or three species, all of which occur in Arizona. Taxonomists generally agree that *A. choisyi* (Smooth Umbrella Wort) is a distinct species. However, *A. cristata* is controversial, with some putting it into synonymy.

Another member of the family in the canyons is *Commicarpus scandens* (*sonorita*, Spanish; *miona*, Mayo). *Commicarpus* is a larger, shrubby perennial that has smaller flowers (6–8 mm wide) than *Allionia*, with white to yellow-green petal-like bracts. Except for flower size, *Commicarpus* resembles *Boerhavia*. The two have been combined by some authors, but Spellenberg considers *Commicarpus* distinct (Spellenberg 2004).

## ***Boerhavia coccinea* (Nyctaginaceae) Red Spiderling, *jaunilipin***

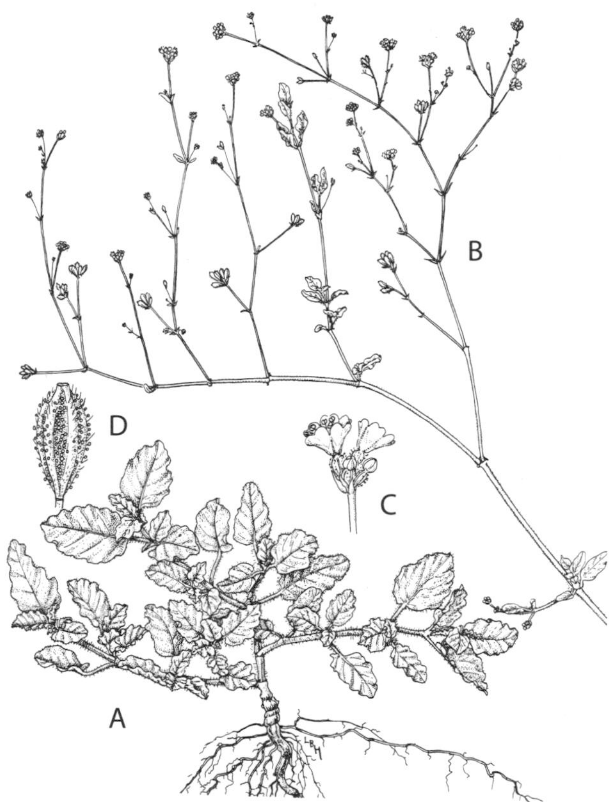


*Boerhavia coccinea*. A. Basal portion of plant. B. Flowering branch. C. Flower cluster. D. Gland-covered fruit. Artist: L. B. Hamilton (Parker 1958).

**Other names:** ENGLISH: Indian boerhavia (a book name), red [scarlet] spiderling ("spiderling" was not recorded in English until 1885 when it referred to a small spider; thus, its use in common names is much more recent and apparently an allusion to the similarity of the flowers to small arachnids; "spider" in English is much older, having appeared by 1340 in Old English as *spithre*, akin to Old English *spinnan*, to spin), wine-flower (a reference to flower color); SPANISH: *jaunilipin* (loan word in Spanish from Mayo? Sonora), *mochi(s)* (possibly from Guarijío *mochiná*, Sonora); ATHAPASCAN: *na'ashjé'ii dą́ą́'* <*naˀasjeˀi dáˀ*> (*na'ashjé'ii*, spider, *dą́ą́'*, food, Navajo); UTO-AZTECAN: *juana huipili* (Mayo, Sonora), *mochiná* (Guarijío), *totopwuvàapi* <*totópwuvápi*> (*totop*, fly, *wuvàapi*, whip, Hopi; for *B. erecta*); (= *B. diffusa* misapplied, *B. caribaea*)

**Botanical description:** Warm-weather annuals or perennials with thickened roots, sprawling, openly branched, often 1–2 m across. Stems are either hairy or glabrous. Flower clusters are diffuse, in umbrella-like clusters at ends of branchlets, much-branched, glabrous or glandular-hairy. Perianths are bright red-purple. Fruits are 2.8–3.5 mm long, sticky with exudate from short-stalked glandular hairs; round when immature, obovoid when mature, prominently ribbed, the tip rounded.

**Habitat:** Desertscrub, canyons, washes, disturbed places. Below 2,130 m (7,000 ft).

**Range:** Arizona (Mohave and Yavapai, south to Cochise, Pima, and Santa Cruz counties), southeastern California, New Mexico, Texas, and disjunct to Florida; Mexico (Baja California, Chihuahua, Coahuila, Nuevo León, Sonora, and Tamaulipas, south); Central America; South America; Bahama Islands, West Indies; and tropical Africa.

**Seasonality:** Flowering April to November.

**Status:** Native.

**Ecological significance:** *Boerhavia coccinea* are plants of disturbed sites, including along streams, in washes, roadsides, trails, and other places where something churns up the soils. Those same places are where passing animals also may drop the seeds of Red Spiderlings that cling to their feet or fur. Sticky seeds adhere to everything, and a person walking through a large patch when fruits are ripe will bring away dozens or even hundreds of adhesive propagules on clothing or skin.

Several species are food plants for the White-Lined Sphinx Moth (*Hyles lineata*). Although *B. coccinea* is not listed as one in most sources, the insects do use them. During the prime caterpillar season of August to November of 2002, both *Boerhavia* species in Brown Canyon were defoliated by those animals.

Bumblebees (*Bombus*) sip nectar from the flower clusters early in the morning. These bees are oddly mismatched in size for the tiny flowers.

**Human uses:** Guarijío, Mayo, and Seri take the roots, boil them, then drink the "tea" as a remedy for measles (Felger and Moser 1985, Yetman and Van Devender 2001, Yetman and Felger 2002). The Guarijío use *B. coccinea* or *B. erecta* interchangeably for this remedy and apply the same common names to both.

Seri call *B. erecta* the *hamíp caacöl* (large spiderling), based on their name *hamíp* for *B. coulteri*; they cook and eat the green herbage (Felger and Moser 1985). Mayo consider the herbage of *B. coccinea* and others to be good food for cows and goats (Yetman and Van Devender 2001).

Seri say that the edible caterpillar of the White-Lined Sphinx (*Hyles lineata*) is found more often on *B. erecta* than on the other species (Felger and Moser 1985). The Hiá Ceḍ O'odham use the name *makkumĭ ha-jeweḍ* (caterpillar their mother [lit. earth]) for *B. erecta*, *B. spicata*, and *B. wrightii*, but exclude *B. coccinea* (Felger 2000). Like their Hiá Ceḍ O'odham relatives, the Akimel O'odham call *B. erecta* and *B. intermedia* the *makkom jeej* or *makkum jeej*. Both names mean "mother of the caterpillar." O'odham did not eat the plant itself, but the moth larvae were formerly relished as food by both Akimel and Tohono and presumably Hiá Ceḍ O'odham (Rea 1997, Nabhan 2002). Rea (1997) was told that the species with "deep reddish or purple flowers" has no common names. That information conforms with that of Nabhan (1983b), Felger (2000), and Felger et al. (in prep.), who have not found names for *B. coccinea* among the Hiá Ceḍ or Tohono O'odham.

Northern Tepehuan make the leaves of *saranda*, an unidentified *Boerhavia*, into a tea to treat fevers (Pennington 1969). Hopi hang *B. erecta*, with its sticky leaves and stems, in the house as fly-paper (Hough 1898, Whiting 1939).

**Derivation of the name:** *Boerhavia* (also spelled *Boerhaavia*) commemorates the Dutch physician Hermann Boerhaave, 1668–1739. He was a professor of botany and medicine, and one of the most influential clinicians and teachers of the eighteenth century. Among his other accomplishments, he revived the Hippocratic method of bedside instruction. *Coccinea* refers to the red flowers.

**Miscellaneous:** Although several species grow in the mountains, *B. erecta* is the only other one noted in Brown Canyon.

## *Mirabilis longiflora* (Nyctaginaceae) Sweet Four-O-Clock, *maravilla, taṣ ma:had*

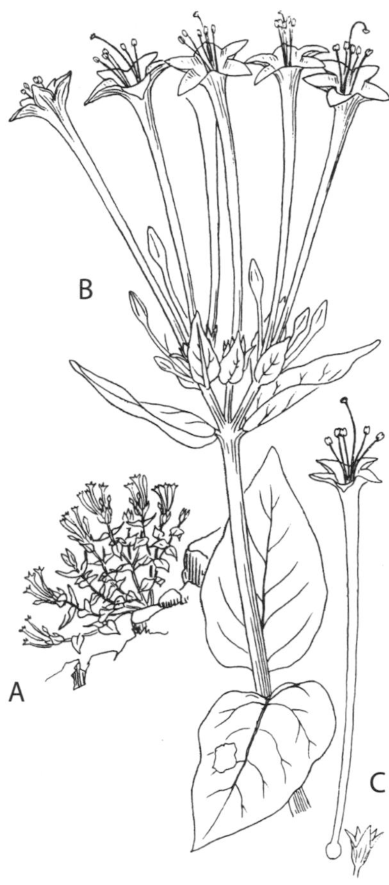


*Mirabilis longiflora*. A. Life-form and habitat. B. Flowering branch. C. Flower, with calyx separated. Artist: R. D. Ivey (Ivey 2003).

**Other names:** ENGLISH: [sweet] four-o-clock [o'clock] (named for the time at which the original species opened its flowers; in use by 1756); SPANISH: *maravilla* [*de jardin, del cerro*] ([garden, wild] marvel, wonderful, Mexico), *pebete* (little child, Oaxaca), *suspiro* (sigh, Jalisco); ATHAPASCAN: *tł'é'iigáhí* <*y'e-ˀgahi*> (white at night, Navajo); KIOWA TANOAN: *puhu* (Tewa); MAYAN: *xpak-u-pa* (sticks a little; from *pak*, to stick, *umpak*, a little, Maya); UTO-AZTECAN: *acxoyatic* <*acsoyate, atzayatl*> (Náhuatl, composed of *atl*, water, and *tzoucati*, hidden by clouds, applied to some plant with many long stems that are made into brooms; original identity uncertain, but applied also to *Abies* and *Ipomoea capillacea*), *maravii* (loan from Spanish, Mountain Pima), *piˀagabɨ* (*piˀaga*, an edible caterpillar of *Coloradia pandora*, the Pandora Moth [Saturniidae], Kawaiisu; for *M. laevis*), *taṣ ma:had* [*tash mahhad*] (*taṣ*, time, day, sun, god, *ma:had*, raised by hand, Tohono O'odham)

**Botanical description:** Perennial herbs, with erect stems 5–15 dm tall, branched, the branches erect or ascending, densely viscid-puberulent or short-villous. Leaves are opposite, with slender petioles less than 1 cm long on the lowest leaves, the upper sessile or nearly sessile, the blades cordate-ovate to narrowly deltoid-ovate or lanceolate-ovate, 6–11.5 cm long, 3–7 cm wide, basally cordate, apically acute to long-attenuate, pubescent like stems. Inflorescences of numerous dense axillary or terminal leafy clusters may be subtended by long linear bract-like leaves. Involucres on peduncles extend to 3 mm long, campanulate, 1–1.5 cm long, densely glandular with short hairs. Flowers are nocturnal, the perianth white or less often pink to magenta, 7–17 cm, long, densely viscid without, the slender tube ca. 2 mm in diameter, abruptly expanding into a limb 2–3 cm wide. Fruits are oblong to ellipsoid, 8 mm long, 5 mm wide, constricted at both ends, 5-angled, bumpy.

**Habitat:** Canyons, streamsides, washes. 760–2,130 m (2,500–7,000 ft).

**Range:** Arizona (Apache, Yavapai, south to Cochise, Pima, and Santa Cruz counties), New Mexico, and western Texas; Mexico (Chihuahua, Sonora? and south).

**Seasonality:** Flowering August to September.

**Status:** Native.

**Ecological significance:** *Mirabilis longiflora* has moth-flowers that open in the evening and close in the early morning. Their long, slender corolla tubes will permit access to the nectar only by an animal with a long proboscis like Sphinx Moths. Leaves are eaten during the monsoon by the larvae of the Two-Spotted Forester Moth (*Alypiodes geronimo*, Noctuidae, subfamily Agaristinae).

This *Mirabilis* usually grows along the sides of streams, where it has seasonally abundant moisture. Perhaps because plants receive more moisture, the herbs reappear at ca. 6,000 ft on Kitt Peak (*Ioligam Do'ag*, Jojoba [*Simmondsia*] mountain, Tohono O'odham) in the picnic grounds near the observatory. Sweet Four-O-Clock is often in clumps several feet across.

**Human uses:** Hocking (1997) says that the root is used as a drastic purgative in Mexico, but he gives no further details. That use would be in keeping with the most famous member of the genus, *Mirabilis jalapa*, which was long thought to be the true Jalap (*Ipomoea jalapa*). Roots of *M. jalapa* were substituted for Jalap, the preferred medicinal laxative of the 1800s (Lewis and Elvin-Lewis 2003). The Névome called some plant *vopoitudacarha* (*maravilla*), but its identity has been lost (Pennington 1980). Perhaps it was some *Mirabilis*.

Navajo chew several *Mirabilis* species to treat mouth sores and gum problems (Wyman and Harris 1941). Other uses are to treat rheumatism and arthritis, and to treat burns (called *kǫ̓ 'azee'* <*koˀ ˀaze.ˀ*>, fire medicine). Bohrer (1975b) reports roots of *M. multiflora* from an archaeological site in New Mexico and records multiple uses by several historic tribes.

**Derivation of the name:** *Mirabilis* is from Latin *mirabilus*, meaning wonderful. The flowers opening in the evening were remarkable to Linnaeus and others. One of the first species to reach Europe was *M. jalapa*. This species began to be called *maravilla* in Spanish, and by the time of John Gerarde in 1633, it was known as "beauty-of-the-night," "marvell of Peru," and "marvel of the World." Some of those names stuck because the flowers open in late afternoon and stay open all night. At the time, the Europeans were familiar with few such nocturnal blossoms. *Longiflora* means long-flowered because of the long-slender corollas.

**Miscellaneous:** Both *Mirabilis albida* (= *M. comata*, *Oxybaphus comatus*, *M. pumila*, *Oxybaphys pumilus*) and *M. linearis* (= *Oxybaphus linearis*) are frequent in Brown Canyon and adjacent Jaguar Canyon. *Mirabilis albida* is also recorded in Fresnal and Toro Canyons. These species are distinguished by the corolla being no more than 2.5 cm long and trumpet-shaped. According to Saxton and Saxton (1969), the Tohono O'odham called some species *taṣ ma:had*. Rea (1997) identified this as *Lupinus sparsiflorus*, but Shaul (2007) thought that it was both a *Mirabilis* and *Lupinus*.

*Mirabilis albida* has leaves with elongate petioles and blades that are cuneate, truncate, or cordate basally. By contrast, *M. linearis* has short or no petioles, and leaf blades that are acute to attenuate basally. This species has been found in Fresnal Canyon on the western slopes and in Toro Canyon on the eastern slopes. There is an old collection of *M. linearis* from an unidentified place in the Baboquivari Mountains, and it now grows in Brown Canyon.

## *Forestiera phillyreoides* (Oleaceae) Desert Olive, *palo blanco*

*Forestiera phillyreoides*. A. Fruiting branch. B. Habit. Artist: B. Swarbrick (courtesy Bonnie Swarbrick).

**Other names:** ENGLISH: desert olive (from Latin *oliva* and Greek *elaion*, through French and Anglo-Norman, appearing in English about CE 1100 for the Old World olive tree *Olea europea*; later extended to other members of the family), mountain privet ("privet" etymology unknown; in English by 1542), tangle-bush [brush] ("tangle" or encumber and hamper, in use by ca. 1340); SPANISH: *adelia* (an old generic name applied by T. S. Brandegee), *garapatillo* (little tick, San Luis Potosí), *palo blanco* (white wood), *palo de tecumblate* (see *Condalia* for etymology, Durango); ATHAPASCAN: *iYentłidzi* (hard seed, Chiricahua and Mescalero Apache); UTO-AZTECAN: *to(m)bovɨ* (Kawaiisu; for *F. neomexicana*); (= *F. shrevei*)

**Botanical description:** Shrubs, 2–3.5 m tall, branched. Leaves are oblong to oblong-oblanceolate or narrowly obovate, 1.5–2.5 cm long, 4–10 mm wide, rounded apically, with a small apical notch or slightly apiculate, broadly cuneate basally, the margins curled under (revolute). Inflorescences are of lateral racemes of fascicles. Scales of flower buds are 6–10 pairs, pale yellow, broadly ovate, about 1 mm long. Flowers are 2–6 in a bundle (fascicle). Corollas are absent. Stamens are usually 4, with purple anthers. Drupes are ellipsoid, 5–8 mm long, 2.5–3.5 mm wide, purple-black, usually asymmetrically curved. Seeds are striate.

**Habitat:** Canyons, rocky slopes. 600–1,380 m (2,000–4,500 ft).

**Range:** Arizona (Cochise, Pima, Santa Cruz, southern Maricopa, and Yuma counties); Mexico (Sonora).

**Seasonality:** Flowering December to March (August).

**Status:** Native.

**Ecological significance:** The species is on the northern limit of its range on the Gila Akimel O'odham Reservation in Maricopa County (Rea 1997).

Flowers are inconspicuous and seem to be built for wind pollination. However, the genus is reported as being visited by insects (Tomlinson 1980). These shrubs flower when almost no other plants have blossoms in the region, when rains and winter temperatures have been proper, in January. When in flower, the shrubs develop a yellowish tinge that contrasts with the grays and blacks of the other, typically leafless, shrubs. The purple anthers are obvious on the reduced male flowers.

Flowers are reduced to stamens, pistils, and scales. *Forestiera* is recorded as mostly having unisexual flowers, with only an occasional perfect blossom on individual plants (e.g., Correll and Johnston 1970, Felger et al. 2001). Plants examined in Brown Canyon from January to June of 2005 were covered with perfect flowers, and no unisexual blossoms were located among thousands. The anthers had shed pollen in January, and by Febuary the ovaries were beginning to swell. Fruits were mature by June. Plants in other localities have not yet been examined to determine if they are bisexual, similar to those in Brown Canyon. *Forestiera* missed the 2006 flowering and were just coming into flower in early August 2007. Zigmond (1981) noted that the Kawaiisu had no uses for the plants. However, he also commented that both bear and Coyotes ate the fruits.

**Human uses:** Although no records have been found of fruits of Desert Olive being used, the related genus *Olea* has a venerable history. The Olive is native from the Mediterranean to the Himalayas in Asia and southern Africa, and it was brought into domestication early in history. Indeed, the Olive, the Grape (*Vitis vinifera*), the Fig (*Ficus cairica*), and the Date (*Phoenix dactylifera*) comprise the oldest group of plants that founded horticulture in the Old World, dating from the Bronze Age at least 5,000 years ago (Zohary and Hopf 1993, Simpson and Conner-Ogorzaly 1995). The name given to the genus was derived from Latin and Greek and spawned similar names in most European languages. Examples include Spanish *olivo* (1147), Portuguese *oliva* (1200s), Italian *oliva* (1274), Middle Dutch *olve* (Dutch *olijf*), German *Olive*, and Old Swedish *oliva* (Swedish *oliv*). While fruits of the Olive produce an edible oil, none has been produced from other genera in the family.

The related *F. pubescens* is used to make digging sticks by the Hopi, and the fruits are reportedly eaten (Hocking 1997, Moerman 1998, Whiting 1939). Because fruits were also used in a medicine, the report of edibility is suspect. Leaves are an emetic to the Navajo; wood provides prayersticks, and fruits yield dye for Jemez and Navajo (Vestal 1952).

**Derivation of the name:** *Forestiera* was named by Jean Louis Marie Poiret to commemorate nineteenth-century French physician Charles Leforestier. *Phillyreoides* means that it resembles *Phillyrea*, another genus in the olive family that grows in southern Europe and Asia. Theophrastus (372–287 BCE) used the name *philyrea*, perhaps from Greek *philos* and *hiron*, for the "Mock Privet."

## *Fraxinus velutina* (Oleaceae) Velvet Ash, *fresno*, *bitoi*

**Other names:** ENGLISH: Arizona [desert, Toumey, velvet] ash (from Middle English *asshe*, Old English *æsc*, akin to Old Norse *askr*, in English by ca. 700); SPANISH: *botavaras* (to make sticks, Sonora), *fresno* [*terciopelo*] ([velvet]-ash, Arizona, New Mexico, Texas, Mexico); ATHAPASCAN: *dahba'* <*dabbaˀ*> (Navajo); UTO-AZTECAN: *awilibïl* (Tübatulabal; *F. oregona*), *bitoi* <*pitoi*> (Akimel, Hiá Ceḍ, and Tohono O'odham), *edɨvɨ* (*edɨ*, bow, Kawaiisu; for *F. latifolia*), *pávlas* (Luiseño), *piichai* (Mountain Pima), *pimaráakârâ* (Comanche), *pítai* <*petai*> (Northern Tepehuan), *pitai* <*potoi*> (Nevome), *uré* (Tarahumara); YUKI: *pök* (Yuki; for *F. latifolia*); YUMAN: *im'val* (Walapai), *mωRc* (Maricopa)

**Botanical description:** Small to medium-sized trees, mostly to 12 m tall, but ancient specimens may be taller, with spreading branches to form a rounded crown. Leaves are petioled, pinnately compound, 7.5–15 cm long, the leaflets 3–9, usually 5, short-petiolulate to almost sessile, varying greatly in shape, from elliptic to ovate, obtuse to long-pointed apically,

*Fraxinus velutina*. A. Fruiting branch. B. Male flower; note 2 stamens. C. Female flower. Artist: L. B. Hamilton (Benson and Darrow 1945).

25–75 mm long, almost entire, glabrous or densely short-pubescent. Flowers are small, yellow (staminate) or green (pistillate), appearing before leaves, many in clusters. Samaras are numerous, 2–3.5 cm long, wing oblong-obovate, 3–4 mm wide.

**Habitat:** Along streams and washes. 600–2,130 m (2,000–7,000 ft).

**Range:** Arizona (Apache to Coconino, south to Cochise, Pima, and Santa Cruz counties), southern California, southeastern Nevada, New Mexico, southwestern Utah, and Texas; Mexico (Baja California, Chihuahua? and Sonora).

**Seasonality:** Flowering March to May.

**Status:** Native.

**Ecological significance:** Where *Fraxinus* trees grow, there is assurance that water is near the surface all year, and may have been on the surface in the recent past. For example, along Arivaca Creek, there are enormous trees. That stream ran all year until about 1993, when urban use of water in the town of Arivaca grew so great that it ceased flowing.

**Human uses:** Although there is a pan-Piman folk genus for ash trees (Rea 1997), not many uses seem to have been recorded in Arizona. Surely, these groups used ash wood, as do others, for tools and firewood (Powell 1988). Moerman (1998) records only the Walapai, who use the wood to make bows, poles for gathering saguaro fruits and pine cones, and staffs. Guarijío make handles for tools from the wood (Yetman and Felger 2002). Tarahumara make violin bodies (Pennington 1963), while the Northern Tepehuan make the violin neck from it (Pennington 1969). Northern Tepehuan make arrows from the easily worked wood (Pennington 1969). Perhaps because the trees are not often as large as other species in the genus, less has been done with their wood.

Tarahumara and Northern Tepehuan seem to be unique in eating the leaves (Pennington 1963, Pennington 1969). Leaves are eaten as a condiment among the Tarahumara, or as a potherb by both groups after being repeatedly boiled.

*Fraxinus* is a more important tree in more northern cultures, where ashes are larger and more abundant. According to Norse legend, the world tree, an evergreen ash called *Yggdrasill*, overshadows the whole universe. *Yggdrasill*'s roots, trunk, and branches bind together Heaven, Earth, and the Netherworld. Rooted in the primordial abyss *Hel*, the subterranean source of matter, it bears three stems. The center stem runs up through *Midgard*, the earth, which it supports. It issues out of the mountain *Asgard*, where the gods assemble at the base of *Valhalla*. Heaven can be reached only by *Bifrost*, the bridge of the rainbow. The stem spreads its branches over the entire sky, the leaves are the clouds, and the fruits the stars. Four stags, *Dain*, *Dvali*, *Duneyr*, and *Durathor*, symbolizing the cardinal winds, live in these branches, feeding on the flower-buds and dripping dew from their antlers to earth. The second stem springs up in *Muspellsheim*, the warm South, where *Urth*, the Past; *Verdandi*, Present; and *Skuld*, the Future, dwell, and the gods sit in judgment. The third stem rises in *Nifleheim*, the cold North, where all knowledge of humans flows from the fountain of the Frost-Giant, *Mimir*, the personification of Wisdom. The tree is the Nordic Tree of Life, symbol of strength and vigor, because the first Norseman, *Ask*, sprang from the ash tree (Austin 2004).

**Derivation of the name:** *Fraxinus* is the classical Latin name for the ash tree and a spear or javelin made from it; akin to Akkadian *burasu*, Hebrew *beros*. *Velutina* means pubescent, in allusion to the plant-hairs on the leaf surface. At one time, there was a var. *toumeyi* recognized in honor of James William Toumey, 1865–1932, the first biology professor at the

University of Arizona, who began the herbarium in 1890, one year before the first class started. The variety is no longer recognized as distinct.

## *Oenothera primiveris* (Onagraceae) Yellow Desert-Primrose, *onagra, wipi si'idam*

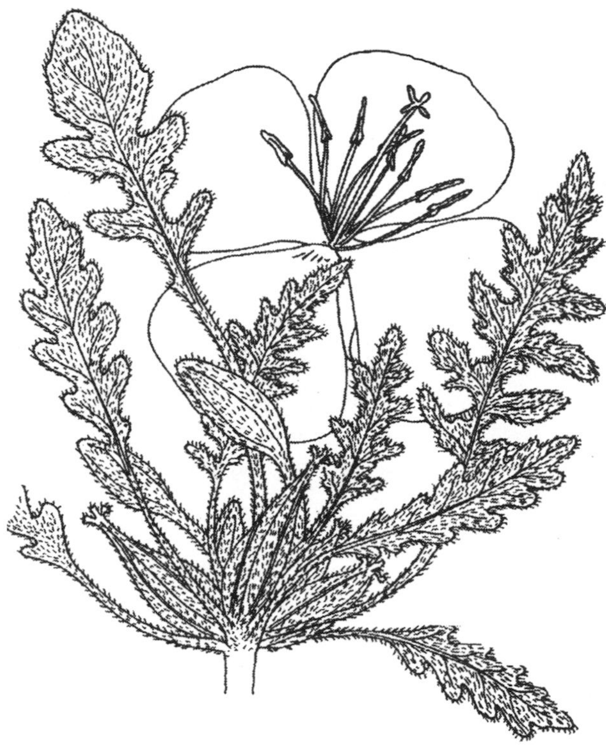

*Oenothera primiveris*. Flowering plant. Artist: unknown (Abrams 1951).

**Other names:** ENGLISH: [yellow] desert [evening] primrose ("primrose" from Latin *prima rosa*, the earliest rose, applied to *Primula* by 1400s; to indicate *Oenothera* since 1760s); SPANISH: *flor de San Juan* (St. John's flower, New Mexico, Mexico), *hierba de asno* (ass's herb, Spain), *onagra* (from Latin *onager*, Greek *onagros*, the wild ass, *Equinus onager*, Mexico; cognate with *enagra*, Italian, *onagre*, French); ATHAPASCAN: *tł'é'iigáhí* <*tł'éé'yigáahii, y'eˀígahi*> (one that becomes white at night, Navajo), *tł'é gogáhá* (*tł'é*, night, *gogáhá*, one that becomes white, Western Apache), *tłonaitsui* (plant with yellow flowers, Chiricahua and Mescalero Apache); UTO-AZTECAN: *ka'nagwana* (Shoshoni; for *O. caespitosa*), *polìisi* <*polí:si*> (*polìi*, butterfly, *si*, flower, Hopi; for white-flowered species), *vippi si'idam* (sucking at the breast, Akimel O'odham), *wipi si'idam* (breast sucker, Tohono O'odham); YUMAN: *su wá támp*$^{c}$ (Mohave or Chemehuevi, reported for *O. brevipes*)

**Botanical description:** Annuals, from a heavy taproot, caespitose, the herbage villous throughout or glabrate. Leaf blades are oblanceolate, 1–12 cm wide, 3–30 mm wide, rarely almost entire, usually deeply pinnatifid, lobes ovate to lanceolate, these again toothed or lobed, all lobes rounded to obtuse, the petioles one-half to as long as the blade. Hypanthium is 2–6 cm long. Sepals are linear-lanceolate, 12–28 mm long, 3–5 mm wide, usually reflexed in pairs at anthesis. Petals are yellow, aging orange-red, obovate, 2–3 cm long, notched apically (emarginate). Capsules are square in cross section, with a heavy median rib down each face, somewhat reticulate, gradually tapering to apex, 6–8 mm thick at base, 2–3.5 cm long. Seeds are in 2 rows in each locule, 2.5–3 mm long.

**Habitat:** Grasslands, desertscrub. Up to 1,370 m (4,500 ft).

**Range:** Arizona (Greenlee, Mohave, south to Cochise, Pima, Santa Cruz, and Yuma counties); Mexico (Baja California and Sonora).

**Seasonality:** Flowering (February) March to May.

**Status:** Native.

**Ecological significance:** The caespitose (growing in tufts) plants with yellow flowers and distinctive leaves pinnatifid into lanceolate or ovate lobes with rounded teeth or lobes are distinctive. Fruiting structures are also unique. Fruits remain as a rounded "hump" on the ground surface that is attached to the underground remains with a slender stalk. The surface is pitted with holes where the ovaries have dehisced. Fruits are more like some parasitic plant than remains of flowers. From fall through winter, these fruiting structures may be found scattered or even clumped in open sunny sites where Yellow Desert-Primrose flowered in the spring.

**Human uses:** Navajo use a medicine of *Oenothera primiveris* in ceremonies and for treating swelling (Vestal 1952). The Akimel O'odham name *vippi si'idam* is applied mostly to white-flowered species, but Rea (1997) records that the name also is given to *O. primiveris*. The flowers stick when applied to the breast; hence, the name. No one Rea talked with remembered any use.

Other species of *Oenothera* are widely used across North America. In the Southwest, those species are used by at least the Apache, Hopi, Jemez, Keres,

Kiowa, Navajo, Paiute, Pomo, and Zuni (Moerman 1998, Stevenson 1915, Vestal 1952, Whiting 1939). These people use Evening Primroses as food, medicine, dyes, and in ceremonies. Farther east, other tribes used other species (Austin 2004). As the derivation of the generic name shows, Europeans were using their species long before arriving in the New World.

**Derivation of the name:** *Oenothera* is said to be based on *oinos*, wine, and *thera*, catcher; a name used by Theophrastus, 372–287 BCE, for some Onagraceae. However, Pliny, CE 23–79, wrote, "*onothera, sive onear, hilaritatem afferens in vino*" (ass-catcher, or ass-hunter, a plant whose juice in wine causes happiness; or, as translated in 1601, "Oenothera, otherwise Onuris, an herb good also in wine to make the heart merry"). *Primiveris* means truly the first [in the spring], because of its early flowers.

**Miscellaneous:** A genus of 124 American species (Mabberley 1997). Although now known in the Americas as *hierba del asno* and *onagra*, those names were originally given to Old World natives. Several species are widely cultivated and naturalized in Europe. The genus is rich in gammalineolinic acid (GLA), which is important in human production of fatty acids and prostaglandins. This GLA was extracted from *O. biennis* and other species sold by a British company that earned £36 million in 1993 from the product (Mabberley 1997). Although the company claimed that the product was effective in treating premenstrual tension and other maladies, others noted that laboratory tests reveal that placebos are as good as this product. Perhaps that is why the company went out of business.

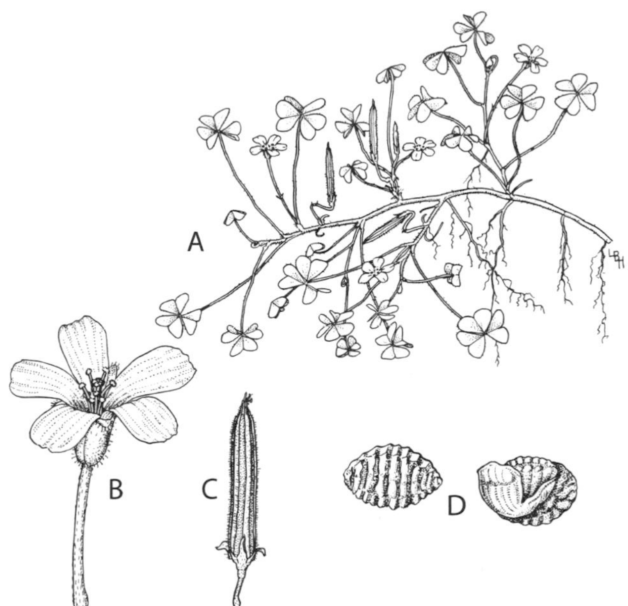


*Oxalis stricta*. A. Flowering and fruiting plant. B. Flower, enlarged. C. Fruit. D. Seeds, two views. Artist: L. B. Hamilton (Parker 1958).

## *Oxalis stricta* (Oxalidaceae) Yellow Wood Sorrel, *chanchaquilla*

**Other names:** ENGLISH: sour-grass, sheep-sorrel [sheep-showers] (sheep-sorrel in use by 1806; "sorrel" in English for *Rumex acetosa* by ca. CE 1400; William Turner in 1568 compared it to *Oxalis*; from the Old French *surele* in the CE 1100s, which is the diminutive of Germanic *sur*, sour; in modern German *sauer*); SPANISH: *agrito* (little sour one, San Luis Potosí), *chanchaquilla* (Arizona, sometimes used for *Cenchrus* and *Lantana*; *chancaco* is from Náhuatl *chancaca*, with an archaic meaning of "*mazapán de la tierra*," a confection of sugar, and a more recent application to herbs in the northern part of the country; the diminutive ending *-illa* is intended to separate *Oxalis* from the others), *hierba* [*zacate*] *alegre* (happy herb [grass], Mountain Pima), *limoncillo* (little lime, Chihuahua), *oreja de ratón* (mouse's ear, Mountain Pima), *socoyole* <*xocoyol, xocoyole, jocoyol*> (a term for sorrels, from Náhuatl *xoxocoyolin*, based on *xococ*, acid, sour, and *coyolli*, rattle, maybe comparing the fruit shape with a snake's rattle, New Mexico, Chihuahua south); ATHAPASCAN: *chąąt'inil* (Navajo), *itadnkodje* (sour plant, Chiricahua and Mescalero Apache); UTO-AZTECAN: *áli ïko* (little and sour, Northern Tepehuan), *čokóbari* (Tarahumara; for other species), *he'egtuli* <*hɨktɨlyi*> (*he'eg*, sour, *tuli*, fold/tuck [alluding to leaves], Mountain Pima)

**Botanical description:** Perennial herb without rhizomes, to ca. 25 cm tall, flowering the first year, the taproot slender or stout. Stems are erect to creeping and mat-forming. Leaves have 3 leaflets; these are glabrous to pilose above. Sepals are 3–7 mm long, without tubercules at tips. Petals are 5–12 mm long, yellow. Capsules are 8–25 mm long, glabrous to canescent.

**Habitat:** Along streams. 760–1,830 m (2,500–6,000 ft).

**Range:** Arizona (Cochise, Coconino, Graham, Maricopa, Pima, Santa Cruz, and Yavapai counties) and widespread in the United States; Mexico (Chihuahua and Sonora).

**Seasonality:** Flowering April to September.

**Status:** Native. The species complex, including *O. corniculata*, is spread around the world. It appears that there are native and exotic strains, and mixing of the two has contributed to the extreme variation within the yellow-flowered American plants.

**Ecological significance:** The long fruits are explosively dehiscent and expel seeds, sometimes for several meters from the parent plants. Because of this trait, the herbs quickly become spread through any new region where they occur. When the habitat is open, the seeds remaining in the seed bank take advantage and grow to add their own genetic mixture to the soil.

These small herbs are both weeds and part of the streamside ecosystem. In their native sites, *Oxalis* is part of the herb flora maintained by the unstable conditions of streams. Because *Oxalis* thrive on disturbance, moist microsites near people are prime places for them to grow.

**Human uses:** Northern Tepehuans cook *Oxalis* leaves as greens, and add them to cooked beans as a condiment (Pennington 1969). Mountain Pima eat the young stems (Laferrière et al. 1991). Tarahumara eat *O. decaphylla* and *O. divergens* in much the same way (Pennington 1963).

Other people who use sorrels, under the names *O. corniculata* and *O. stricta*, in ceremonies, medicine, and food include Cherokee, Iroquois, Kiowa, Menomini, Mesquakie, Mountain Pima, Omaha, Pawnee, Ponca, and Northern Tepehuan (Moerman 1998, Pennington 1969, 1973). The Northern Tepehuan boil *Oxalis* to make a beverage that is drunk to lower fever (Pennington 1969).

The "sour" taste of *Oxalis* and its relatives is attributable to the presence of oxalic acid (Hocking 1997). This compound is the most highly ionized organic acid known in nature, and it has been used to stop bleeding (as a hemostat).

The genus is probably the "shamrock" (from Old Irish *soamair*, diminutive *seamróg*, clover or little clover; Gaelic *seamrag*) of Ireland, although that name may have been applied to this and clovers (*Trifolium*). This *seamrag* was the lucky symbol of the sacred sun wheel to the Gaelic people (Dobelis 1986). Saint Patrick realized the importance of *Oxalis* to these people and used it to explain the Holy Trinity of Christianity, thus winning them to that religion and incorporating another pagan symbol into the fabric of Catholicism.

**Derivation of the name:** *Oxalis* is based on Greek *oxys*, sour. The taste has also given rise to the common name in several languages. *Stricta* means erect, upright.

**Miscellaneous:** *Oxalis corniculata* has been separated and lumped with *O. stricta* by a variety of authors. There is no current consensus on how species should be delimited, but there are technical differences. Unfortunately, each study of the group interprets those differences in distinct ways.

## *Argemone pleiacantha* (Papaveraceae) Bluestem Prickly Poppy, *chicalote, to:ta heosig*

*Argemone pleiacantha*. A. Growth form. B. Flowering and fruiting branch of similar *A. polyanthemos*. C. Fruits of *A. pleiacantha*. D. Spines on stem of *A. pleiacantha*. Artist: R. D. Ivey (Ivey 2003).

**Other names:** ENGLISH: cowboy's [fried] eggs (the white petals and yellow stamens resemble fried eggs, Arizona), [southwestern] prickly [thistle] poppy ("poppy" is from Middle English *popi*, Old English *popig, popaeg*, ultimately from Latin *papaver*, and perhaps from Sumerian *pa pa*, the noise made when chewing poppy seeds); SPANISH: *cardo* (Sonora; from Latin *carduus*; this was the name used by classical authors Virgil, 70–19 BCE, Pliny, CE 23–79, and by Linnaeus in 1753 for an array of bristly plants related to *Cirsium*; later extended to other spiny herbs), *chicalote* <*chilazotl, chichilotl, xicólotl*> (from Náhuatl *chicalotl*, spiny herb, Sonora); HOKAN: *xazácoz* (Seri); UTO-AZTECAN: *hipigdum* (Onavas Pima), *pa'ratĭtsĭnbogop*

(Shoshoni, questionably for *A. mexicana*), *so'lolopul* (Tübatulabal; for *A. polyanthemos*), *toi'yanabogop* (Shoshoni; for *A. mexicana*), *to:ta heosig* (*to:ta*, white, *heosig*, flower, Tohono O'odham), *tságida*[ʔa] (Northern Paiute; for *A. platyceras*)

**Botanical description:** Herbs, perennial, the stems to 10 dm tall, spiny, with yellow sap. Leaves are deeply lobed, to 20 cm long, basally clasping the stem, with abundant prickles on veins and margins. Flowers are solitary on the tips of branches. Corollas are white, to 12.7 cm wide, tissue-paper thin, with 4–6 wrinkled petals, numerous bright-orange stamens. Capsules are oblong, to 4 cm long, prickly. Seeds are numerous, small.

**Habitat:** Foothills and desert grasslands. 800–2,440 m (2,624–8,000 ft).

**Range:** Arizona (Cochise, Gila, Pima, and Santa Cruz counties) and New Mexico; Mexico (Chihuahua and Sonora).

**Seasonality:** Flowering (February) March to October.

**Status:** Native. Our plants are spp. *pleiacantha.*

**Ecological significance:** Although the white saucer-shaped flowers are not the "typical" moth-flower, they are visited by Sphinx Moths, particularly the Five-Lined Sphinx (*Hyles lineata*). Especially during the monsoon season when the *Boerhavia* is covered with the caterpillars of these moths, the adults are seen sipping nectar from the base of the petals in the early morning hours. Probably, moths also visit in the evenings as the flowers open then. As the day heats up, the flowers begin to wilt, and the moth visits terminate.

As grazing livetock will not eat these armed and poisonous plants, animals preferentially eat other species. Where there are large colonies of *Argemone* on rangeland, they are an indication of overgrazing.

In Brown and Jaguar Canyons, these plants are scattered as individuals. Although the canyons were grazed from the late 1870s or early 1880s until about 1997, the areas have not been degraded badly enough to cause this plant to dominate.

**Human uses:** The Seri make a tea from the leaves to relieve kidney pain, as a diuretic, and to clear the urine (Felger and Moser 1985). This herb is believed to "rest" the kidneys and relieve pain. The same tea was used by women after parturtion to "lose bad blood," and to help expel the placenta.

The most famous species in Mexico is *A. ocroleuca* (Martínez 1969). These herbs are called *chicalote*, *cardo*, or *hierba loca* (crazy herb, Mountain Pima) throughout the country. The Maya of Yucatán call that species *k'iix-k'anlol* (*k'iix*, spiny, *ka'an*, yellow, *lol*, flower), and the Náhuatl in Veracruz say *tlamexcaltzin* (from *tlalli*, earth, *mexkal*, agave liquor, *tsi:n*, diminutive). *Argemone ocroleuca* is called *tajíchuri* <*tachiná*> (Guarijío) or *táchino* <*táchiguo*> (Mayo) farther south in Sonora. The latex is used to clear spots from the cornea, and the flowers are considered narcotic. Some use the leaves to make a medicine to halt diarrhea. Guarijío and Mayo use the species to treat eye problems and as a laxative (Yetman and Felger 2002, Yetman and Van Devender 2001). Onavas Pima use *A. ocroleuca* and *A. mexicana* to make a purgative or to make one vomit after drinking too much *tesgüino* (home brewed liquor) (Pennington 1980). Mountain Pima treat wounds with *A. ocroleuca* (Pennington 1973).

The Spanish Dominican Fray Francisco Ximénez wrote in 1615, and later Francisco Hernández confirmed in 1651, that the Aztecs used *Argemone* as a medicine. Aztecs ground the seeds and put them in a drink that served as a laxative, the latex aided eye inflammations and cloudy corneas, and the flowers helped alleviate mange.

**Derivation of the name:** *Argemone* is from Greek *argemon*, or Latin *argema*, cataract of the eye, thus alluding to former use in treating that malady. *Pleiacantha* is from Greek *pleio*, more or full, *acantha*, spine or thorn, thereby alluding to the spiny nature of the herbs, although they are no more prickly than most other *Argemone*.

**Miscellaneous:** *Argemone* contains 31 species in North, Central, and South America, and in the West Indies (Ownbey et al. 1998). The genus is famous for its poisonous alkaloids (Nellis 1997).

## *Corydalis aurea* (Papaveraceae) Gold Smoke, *corídale*

**Other names:** ENGLISH: colic weed (unexplained; does it cure or cause colic?), golden corydalis (a book name), gold smoke (referring to the flower color and probably the odor of the herbage), scrambled-eggs (a reference to the yellow flowers, Arizona, New Mexico); SPANISH: *corídale* (from the Latin, Mexico; cognates occur in French and Italian), *fumaria* (from Latin *fumus*, smoke, Mexico); ATHAPASCAN: *bílátah łitso tsoh* <*bilátah łcoi coh*> (*bílátah*, flowers, *łitso*, yellow, *tsoh*, big, Navajo), *chooyin 'azee'* <*co'in ˀazéˀ*>

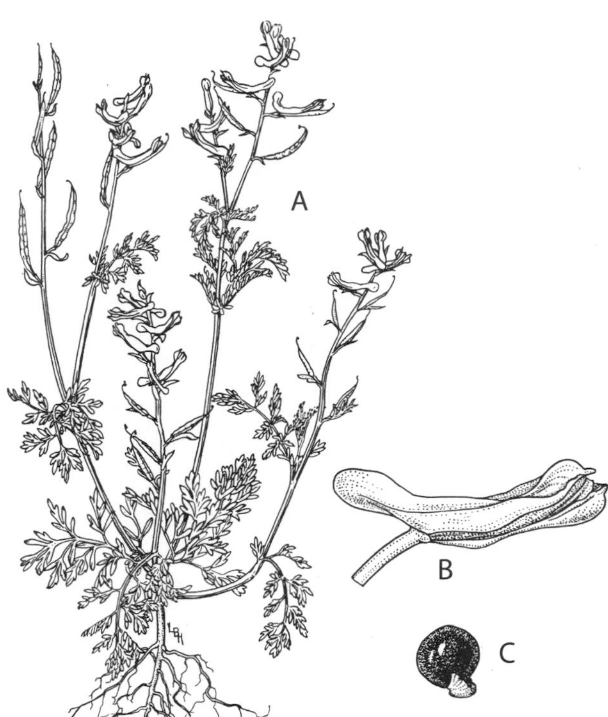


*Corydalis aurea*. A. Flowering plant. B. Flower with conspicuous spur at upper left. C. Seed. Artist: L. B. Hamilton (Parker 1958).

(menstruation medicine, Navajo), *gáagii binát'oh* <*gâgi binát'oh*> (raven's tobacco, Navajo), *hasbídídą́ą́'* <*hasbididá*ˀ> (*hasbídí*, mourning dove, *dą́ą́'*, food, Navajo), *nikookáá' łitso* <*naxoká*ˀ *łcoi*> (*nikookáá'*, earthly, *łitso*, yellow, Navajo), *tązhii halchiin ałts'íísígíí* <*tąžilčin* ˀ*ałc'ísí, tazhii yilchiin ałts'íísígíí*> (*ałts'íísígíí*, little, *tązhii*, turkey-like, *halchiin*, odor, Navajo; compare with *Delphinium*), *ts'yaa tł'ohdeeí* <*ciyahƛ' oh de*> (*ts'yaa*, under trees [contraction of *tsin 'ayaa*], *tł'ohdeeí*, grass whose seeds fall, Navajo; compare with *Chenopodium*)

**Botanical description:** Annual or biennial (or perennial), glaucous herbs, 6–40 cm long, from taproots. Stems are simple or with a few to several decumbent to ascending branches. Leaves are 1–4 times pinnately compound, the ultimate segments linear to oblong. Racemes are few- to many-flowered, the pedicels longer than the spur. Sepals are 1–3 mm long, yellowish to whitish. Corollas are yellow, 1–18 mm long, the spur 3–6 mm long. Capsules are 1.5–3 cm long, twisted (torulose), usually curved-ascending.

**Habitat:** Canyons, washes, grasslands. 450–2,895 m, elsewhere rarely to 3,355 m (1,500–9,500 ft).

**Range:** Arizona (throughout state except extreme western areas), California, north to Alaska, east to Quebec, and Pennsylvania; Mexico (Chihuahua and Sonora).

**Seasonality:** Flowering (February) March to May (June).

**Status:** Native.

**Ecological significance:** This northern-temperate-to-arctic species barely reaches into Mexico on the higher mountains, although it sometimes extends down to surprisingly low elevations. The genus, of perhaps 400 species, is best developed at higher latitudes, with only a few species being able to extend very far toward the tropics (Mabberley 1997). Europe has 11 species, and the Sino-Himalayas have ca. 280. Alaska supports 3 species, including *C. aurea*, and Arizona has this single species (Kearney and Peebles 1951, Hultén 1968). The genus is best developed in the Asian part of the Old World.

*Corydalis aurea* has not been studied, but some others in the genus are dispersed by ants. For example, European *C. cava* is spread by ants (Dunn 2005).

**Human uses:** Navajo use *Corydalis* to stop diarrhea, to help cure sores, including those associated with childbirth, and as a rheumatism remedy (Elmore 1944, Moerman 1998, Wyman and Harris 1951). An infusion is taken to relieve stomachaches, backaches, menstrual pains, injuries, and sore throats. A remedy of *Corydalis* is also put on domestic livestock to help cure snakebites (Mayes and Lacy 1989, Wyman and Harris 1951). Infusions are used to soak watermelon seeds to increase production, presumably because it serves as an insecticide (Vestal 1952). Farther east, the Ojibwa inhaled smoke from the burning plant to clear the head and revive a patient (Moerman 1998).

**Derivation of the name:** *Corydalis* is from Greek *korydallis* or *korydalos* (Aristotle), akin to *korys*, helmet, the name of the Crested Lark (*Galerida cristata*). Possibly, the name was given because of the resemblance of the spur to the hind claw of the bird. *Aurea* means golden, a reference to the yellow flowers.

**Miscellaneous:** For decades, *Corydalis* was segregated into the family Fumariaceae with the genus *Fumaria*. That second generic name is based on Latin *fumarium*, a smoke chamber for ripening wine and drying wood, from *fumus*, smoke, related to Greek *kapnos*, smoke. There, the views of why it is called that diverge. Fernald (1950) wrote: "presumably from the nitrous odor of the roots when first pulled from the ground."

Dobelis (1986) had another version. He wrote, "Like curls of smoke rising from the ground, fumitory's gray-green leaves have a ghostly appearance when seen from afar." In Shakespeare's time, 1564–1616, *Fumaria* was sold in apothecary shops as *fumus terrae* (earth smoke). Pliny, CE 23–79, considered the herb a wonderful medicine, and so did Dioscorides, fl. CE 40–80. Pliny wrote that it was called *kapnos* because "used as ointment for the eyes, it improves the vision, and, like smoke, produces tears." Dioscorides' and Pliny's version was probably closer to the original reason for naming, and more akin to the account written by Fernald (1950). In the Middle Ages, exorcists burned leaves because the smoke was thought to expel evil spirits (Coffey 1993). The *Grete Herball* of 1526 said the plant was "engendered of a coarse fumosity rising from the earth." That view is somewhat akin to the theory of spontaneous generation. Coffey (1993) gives yet other possible sources for the name, all more recent and less likely than the one given by Pliny.

Europeans knew *Fumaria* from Neolithic times, but *Corydalis* was not scientifically recognized as distinct until 1805, when DeCandolle described it in his third edition of the flora of France.

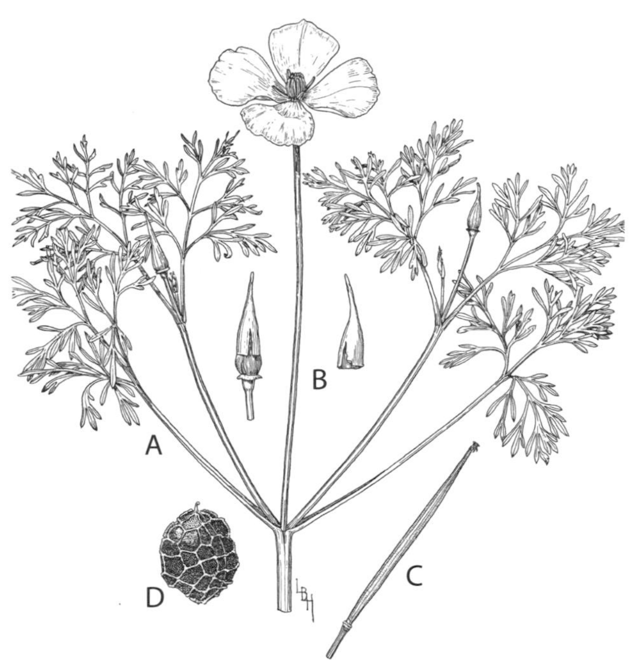


*Eschscholtzia californica.* A. Flowering plant. B. Splitting cap on and off bud. C. Fruit. D. Seed, enlarged. Artist: L. B. Hamilton (from original artwork).

## *Eschscholzia californica* (Papaveraceae) Mexican Poppy, *amapola del campo, ho:hĭ e'es*

**Other names:** ENGLISH: Mexican [gold-] poppy, California poppy ("poppy" entered English before the twelfth century, coming through Middle English *popi*, Old English *popaeg*, *popig*, modified from Latin *papaver*; the names allude to the similarity to the Old World *Papaver*); SPANISH: *amapola del campo* (wild poppy, Sonora); HOKAN: *hast ipénim* (splattered against [the] rock, Seri; for *E. parishii*); UTO-AZTECAN: *atóošanat* (Luiseño; *taróoshant* in Juaneño dialect), *hiyogʷivɨ* (Kawaiisu), *ho:hoi 'e'es* (mourning dove's plant, Hiá Ceḍ O'odham), *ho:hĭ e'es* <*ho:hī e'es, hahdkos*> (*ho:hĭ*, mourning dove, *e'es*, plants, Tohono O'odham), *hoohi e'es* (mourning dove's plant, Akimel O'odham, Arizona), *tesinat* (Cahuilla), *yogobul* (Tübatulabal); YUKI: *huicoñil* (Yuki)

**Botanical description:** Annual herbs, 1–3.5 dm tall. Leaves are basal and sometimes on the stem, erect or spreading, glabrous, sometimes glaucous, 3-times (ternately) dissected. Flowers have the bud erect, lanceolate-ovoid, acute to long-pointed, the outer rim (a funnel-shaped cup below the sepals) spreading to recurved, 0.5–2 mm wide, inner rim erect, hyaline, less than 1 mm tall. Sepals are glabrous, sometimes glaucous, 3–8 mm long. Petals are orange to yellow, sometimes with a darker basal spot, 1.5–7 cm long. Capsules are 3–11 cm long. Seeds are 1.5–1.8 mm wide, spherical to ellipsoid, net-ridged, brown to black.

**Habitat:** On mesas, bajadas, ridges, and dry washes, in desertscrub and grasslands. 300–1,400 m (1,000–4,500 ft).

**Range:** Arizona (all counties except Apache, Coconino, and Navajo), southeastern California, New Mexico, southern Nevada, Texas, and southwestern Utah; Mexico (northwestern Chihuahua and northern Sonora); subspecies *californica* extends from Baja California to Washington (Brasher and Clark in Ownbey et al. 1998, Felger 2000).

**Seasonality:** Flowering February to May.

**Status:** Native. Our plants are ssp. *mexicana*.

**Ecological significance:** The species is divided into two distinct ranges by the severe deserts of western Arizona and eastern California (Felger 2000). Apparently, the two were divided by the increasing aridity of the glacial times. Now ssp. *californica* has bifid cotyledons and is usually perennial, while ssp. *mexicana* has simple cotyledons and is annual. Subspecies *californica* is widely cultivated and used in reclamation seedlings throughout Arizona (Brasher and Clark in Ownbey et al. 1998).

Flowers are intensely sun-loving and do not open until the sun is fully up and has been on them for some time. During the day, if clouds intervene, the corollas again close.

The O'odham name for the herbs links it with the Mourning Dove (*Zanaida macroura*). Both the O'odham and English names for the dove link the birds with mourning. The O'odham apparently associate the bird's call with their own keening, repeating the kinship term of the deceased (Rea 1997). What Rea (1997) did not point out is that the name also serves to signify that the bird is one of the chief consumers of the seeds (Martin et al. 1951).

**Human uses:** Mexican Poppy was planted for use at Easter by the Akimel O'odham until "those educated men" (priests?) stopped it (Rea 1997). Rea was told that the species was part of Thin Leather's Creation Story, but he did not find it in the printed version. One of his consultants told him the tale: "It's a grandmother, the Hoohi; it's from the son's side, *Hoohi ka'akmaḍ*; and they were killed by the enemies—the grandsons. The messenger went and told Hoohi, and she said, 'Wait till my *e'es* [crops] are ripe. Then I'll cry.'. . . So, when the harvest is ready, then she start the song, and cry there, for her grandsons."

Plants are used as medicine by the Cahuilla, Chumash, Kawaiisu, Maidu, "Mendocino Indians," Ohlone, Pomo, and Yuki (Bean and Saubel 1972, Bocek 1984, Curtin 1957, Ebeling 1986, Garcia and Adams 2005, Moerman 1998, Timbrook 1990, Zigmond 1981). Ohlone and Chumash people also rub the flowers into the hair to kill lice.

Luiseño people chew the flowers with gum and cook the leaves as greens (Moerman 1998). The Mendocino tribes and Nisenan people also cook the leaves as greens (Ebeling 1986, Moerman 1998).

**Derivation of the name:** *Eschscholtzia* commemorates the Estonian physician Johann Friedrich Gustav von Eschscholtz, 1793–1831. He was also an explorer, naturalist, professor of medicine and anatomy, plant collector, and zoologist. Eschscholtz accompanied Captain Otto von Kotzebue (1787–1846) on his expeditions around the world on the *Rurik* in 1815–1818 and *Predpriatie* in 1823–1826. *Californica*, of California, means that the first specimens were collected in that state; *mexicana*, of Mexico, is the subspecies from Mexico and adjacent United States east of California.

## ***Passiflora arizonica*** **(Passifloraceae)** **Passion-Flower, *ojo de venado***

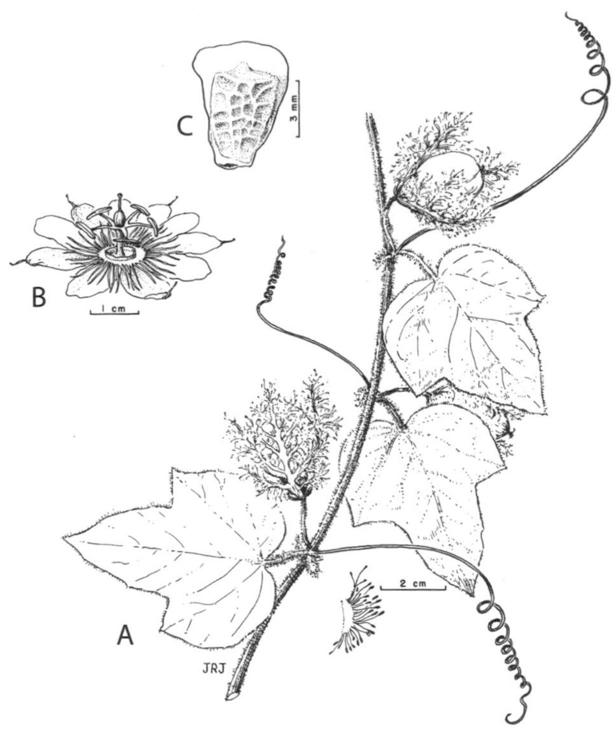


*Passiflora arizonica*. A. Flowering branch. B. Flower. C. Seed. Artist: J. R. Janish (Wiggins and Porter 1971).

**Other names:** ENGLISH: passion-flower ("passion-flower," referring to the passion of the crucifixion of Christ, see derivation of genus below); SPANISH: *amapola* (poppy, Mexico; see *Eschscholtzia* for etymology), *buli de venado* (deer's rattle, Sonora), *clavellín blanco* (white carnation, Sinaloa), *corona de cristo* (Christ's crown, Arizona, Texas), *jujito peludo* (little hairy passion-flower, Tabasco; *jujito* is based on Mayan *jujub*, the name for another *Passiflora*), *ojo de venado* (deer's eye, Sonora), *talayote* (from Náhuatl *tlal*, earth, *ayohtli*, squash, Sonora), *passionaria* (of the passion, New Mexico, Mexico); UTO-AZTECAN: *<boapkokama>* (Mountain Pima), *ma'aso alócosim* (deer's gourd, Mayo), *tonolochi* (*tonoro*, bright/shiny, *chi*, the one, Mountain Pima); (= *Passiflora foetida* var. *arizonica*)

**Botanical description:** Vines, perennial, the stems to 2–4 m, gray-villous throughout and viscous with glandular hairs. Leaves are palmately 3–5-lobed, to 5 cm long and wide, the lobes sinuate to again slightly lobed, denticulate. Bracts surrounding flowers are 2–3 cm long, 2-pinnatisect or 3-pinnatisect, the ultimate filiform segments gland-tipped, straight or nearly so. Flowers are 2–5 cm wide, white to purple. Sepals are ovate-oblong to ovate-lanceolate, awned dorsally near apex. Petals are oblong to oblong-lanceolate or oblong-spatulate, slightly shorter than sepals, white. Corona filaments are white to blue or purple, in several series, those of the outer 2 series filiform, about 1 cm long, the others capillary, 1–2 mm long. Fruits are 2–3.5 cm wide, green or greenish yellow. Seeds are ovate-cuneiform, about 5 mm wide, coarsely reticulate at center of each face.

**Habitat:** Canyons, bajadas, rocky slopes. 1,200–1,670 m (4,000–5,500 ft).

**Range:** Arizona (Pima and Santa Cruz counties); Mexico (Sonora).

**Seasonality:** Flowering August to September.

**Status:** Native. The Arizona plants are distinguished by having the basal lobes of the leaves again bilobed, and having the middle lobe narrowed toward the base. These traits, plus the nocturnal flowers, distinguish *P. arizonica* from the more tropical *P. foetida*, which they were thought to be until recently (Goldman 2003, MacDougal 2001).

**Ecological significance:** Southeastern Arizona is the northern limit of this species; it is a northwestern extension of the genus out of the tropics. This climber is known only in Pima County from the Baboquivari Mountains, Coyote Mountains, Las Guijas Mountains, Pozo Verde Mountains, and a few places in Santa Cruz County. The nocturnal flowers must be pollinated by moths.

This passion-flower often follows waterways, but it may grow well above the bottoms on the slopes of hills leading down into the valley. Considering the wet places where *P. foetida* grows in the southern and more tropical part of their range, it is difficult to believe that anyone ever thought the Arizona examples were that species.

Perhaps the bad odor of the leaves is an olfactory warning of the many poisonous chemicals they contain. Only caterpillars from butterflies in the Nymphalidae subfamily Heliconiineae are able to detoxify the *Passiflora* and eat them. This group of tropical animals is poorly represented in southeastern Arizona (Bailowitz and Brock 1991, Glassberg 2001). The Gulf Fritillary (*Agraulis vanillae*) may be frequent in the fall in the Baboquivari Mountains. *Passiflora* and *Agraulis* are known in the Santa Catalina and Galiuro Mountains, but are not recorded in the Peloncillo Mountains (Bailowitz and Brock 1991, SEINet 2009). The Variegated Fritillary (*Euptoieta cloudia*) is more common in the Baboquivari Mountains. Although it has been reported in the Baboquivari and Santa Catalina Mountains and eastward to the Chiricahua Mountains, I have not seen the Zebra Longwing (*Heliconius charitonius*).

**Human uses:** Fruits of related *Passiflora aridia* are eaten by Mayo (Yetman and Van Devender 2001), and farther south people eat fruits of *P. foetida*. Fruits of *P. arizonica* that I have tasted were terrible, although they might not have been completely ripe.

In Cuba, *P. foetida* is called *pasionaria hedonda* (stinking passion-flower). There and elsewhere in the Caribbean, *P. foetida* is medicinal, with uses as a sedative, to treat insomnia and convulsions in children, and for coughs, colds, and intestinal worms (Hocking 1997).

**Derivation of the name:** *Passiflora* is an adaptation of the Latin *flos passionis* (flower of the passion), comparing the flower parts with the Crucifixion of Jesus (3 stigmas = 3 nails, 5 anthers = 5 wounds, corona = crown of thorns). *Arizonica* means from Arizona.

**Miscellaneous:** *Passiflora mexicana* is the other species in Brown and Jaguar Canyons, and in other canyons in the vicinity (elevation 760–1,500 m [2,500–5,000 ft]). This plant is easily recognizable by its leaves, which are V-shaped, and by the odor of its diurnal flowers. While *P. mexicana* has pale lilac to lavender blossoms similar to those of *P. arizonica*, those of *P. mexicana* produce a foul odor resembling something that has been dead for a long time. This tropical species has an odd range in Arizona, extending north through Cochise, Graham, Pima, Pinal, and Santa Cruz counties. The oddity is that the climbers do not occur in eastern Cochise County, but are confined to the western part of that region, and extend down into Mexico through the western states only.

## *Rivina humilis* (Petiveriaceae)
## Rouge Plant, *coralillo cimarrón*

**Other names:** ENGLISH: pigeon berry (Arizona, Texas, Sonora), rouge-plant (from French *rouge*, based on Latin *rubeum*, *rubeus*, related to *ruber* and *rūfus*, Arizona; "rouge-berry" in Florida), wild ruby, wild tomato; SPANISH: *coral* (red one, Mexico), *coralillo cimarrón* (wild little red one, Sonora), *coralito* (little red one, Texas, Durango, Cuba), *hierba del zorrillo* (skunk herb; name also applied to *Petiveria alliacea*,

*Rivina humilis*. Flowering and fruiting branch. Artist: P. N. Honychurch (courtesy P. N. Honychurch-Billingham).

Mexico), *teihuist* (possibly from Guarijío *teywesí*, Sonora), *ucusuiro* (maybe from Guarijío *uruquiro*, Sonora), *achiotillo* (little achiote, *Bixa orellana*, Honduras); MAYAN: *akan t'ele'* (baby foot, Huastec, San Luis Potosí), *kuxubcan* <*coxubcan*, *cusucan*, *cuxuban*> (*ku'xub*, achiote [*Bixa orellana*], *kan*, snake, Maya, Belize), [*tsakam*] *taa' t'ele'* ([little] baby excrement, Huastec, San Luis Potosí); OTO-MANGUEAN: *skwam butz snya* (*skwam*, herb, Zapotec); UTO-AZTECAN: *bacot mútica* (snake's pillow, Mayo), *chilpantlaçolli* (*chilpan*, medicinal plant, *tlaçolli*, adultery, Náhuatl), *chilpanxuitl* <*chilpanxochitl*> (*chilpan*, wasp, *xiwitli*, herb, Náhuatl, San Luis Potosí), *tejocote* (Tarahumara; also used for *Crataegus* in Mexico; from Náhuatl *tetl*, something hard, *xocotl*, acid fruit), *teywesí* (Guarijío), *uruquiro* (Guarijío)

**Botanical description:** Herbs, with stems from perennial rootstocks, glabrate or rarely pubescent, erect or becoming vine-like to 15 dm tall. Leaves are bright green, ovate to ovate-lanceolate, acute to acuminate apically, broadly rounded to truncate basally, to 15 cm long, 9 cm wide. Flowers are in axillary or terminal panicles. Sepals are 4, white, greenish, or rose, 2–2.5 mm long. Berries are red or orange, 2–3.5 mm wide.

**Habitat:** Ravines, canyons, streamsides. 450–1,370 m (1,500–4,500 ft).

**Range:** Arizona (Greenlee to Maricopa, south to Cochise, Pima, and Santa Cruz counties) and Florida to Texas and Arkansas; Mexico (Baja California Sur, Chihuahua, Coahuila, Nuevo León, Sonora, Tamaulipas, and south); Caribbean; Central America; South America.

**Seasonality:** Flowering June to October (November).

**Status:** Native.

**Ecological significance:** These plants are tropical and reach their northern limit in the southern United States from Arizona to Florida. In Arizona, Rouge-Plants are confined to the margins of streams in canyons and washes, where it is seasonally wet. *Rivina humilis* is usually found in much wetter places in Florida and the Caribbean. In Arizona, the species extends west of the Baboquivari Mountains to the Ajo and Diablo Mountains.

The red fruits are visual cues to animals that a meal awaits, and many different avians eat the fruits with impunity. Among the birds feeding on the berries in Florida and Texas are Wild Turkeys (*Meleagris gallopavo*), Plain Chachalacas (*Ortalis vetula*), Mourning Doves (*Zanaida macroura*), White-Winged Doves (*Zanaida asiatica*), Northern Mockingbirds (*Mimus polyglottos*), Gray Catbirds (*Dumetella carolinensis*), and Northern Robins (*Turdus migratorius*) (Austin 2004). In the Baboquivari Mountains, some of those birds do not occur, but other fruit-eaters are present.

**Human uses:** People in western Chihuahua have eaten *Rivina humilis* fruits since at least the 1700s (Pennington 1963). Western Tarahumara collect the fruits in October for food (Pennington 1963). Tzotzil Maya eat the leaves (Breedlove and Laughlin 2000).

As is often the case, academics cannot agree on whether or not the fruits are edible for humans. Standley (1920–1926) considered Rouge Plant fruits edible. Nellis (1997) disagreed, saying that the berries were poisonous. The herbs are used to treat *susto* (fright) in Puebla and Veracruz, Mexico, and among the Huastecs of San Luis Potosí (Alcorn 1984, Vásquez-T. and Jácome-C. 1997).

Plants are touted as good fodder by some Mayo (Yetman and Van Devender 2001). Additionally, the fruits are used for food and dye in Sonora (Hodgson 2001), and for dye in Texas (Correll and Johnston 1970).

**Derivation of the name:** *Rivina* was named for Augustus Quirinus Rivinus [Bachman], 1652–1723, a German physician in Leipzig. *Humilis* means weak, thus alluding to the branches.

**Miscellaneous:** This genus is monotypic, with just this single species (Mabberley 1997).

Although historically placed in the Phytolaccaceae, Judd et al. (2008) and some others have put it in the Petiveriaceae with *Petiveria* and *Trichostigma*.

## *Pinus discolor* (Pinaceae)
## Border Pinyon, *piñon*, *huk*

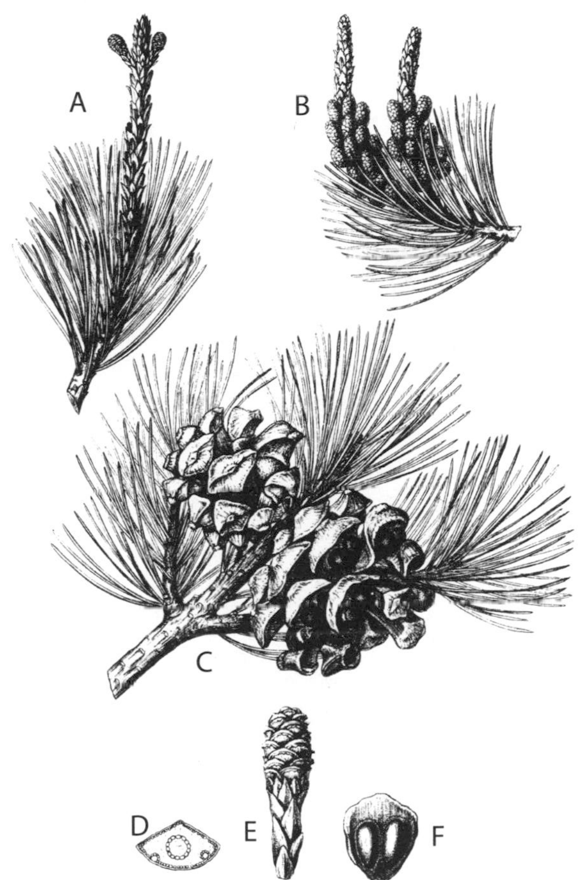


*Pinus discolor*. A. Female branch with young cones. B. Male branch with catkins. C. Adult cones. D. Cross section of leaf. E. Young female cone. F. Fertile scale. Artist: C. E. Faxon (Sargent 1890–1902).

**Other names:** ENGLISH: pine (from Latin *pinus*), pinyon (from Spanish); SPANISH: *pino* (Sonora; for pines), *piñon* (pinyon, Sonora), *piñonero* (pinyon tree, Sonora); ATHAPASCAN: *cha'oł* <*ca'ol, c'a o-l*> [*deetstsiin*] (Navajo; for pinyons), *īzēnchí* (Jicarilla Apache), *ništci* <*nictci*> (Chiricahua, Jicarilla, and Mescalero Apache), *obé*, *obé'tsin* <*obe'chin*> (*obé'*, pinyon nut, *tsįh*, tree, wood, Western Apache); CHUMASH: *tak* (Barbareño Chumash; for pines), *tomol* (Ineseño Chumash; for pines), *tsɨkɨnɨn* (Ventureño Chumash; for pines); HOKAN: *ṭlágū-m* (Washo); KIOWA TANOAN: *to* (Tewa); OTO-MANGUEAN: *xivatí* (Mazahua); UTO-AZTECAN: *huk* (Akimel and Tohono O'odham; for pines, Arizona), *huk* (singular), *hu'uk* <*hú-uk*> (plural) (Mountain Pima), *išíkuri* (used for *P. cembroides*, Tarahumara), *obi* (for pinyons, Paiute), *oco* (Tarahumara; for pines), *ocosaguat* (Ópata, Sonora; for pines), *ti'va*$^{g}$ (Southern Paiute), *tivat* <*téva-t, tewat*> (pinyon, Cahuilla), *tiva* (Ute, the seed of *P. monophylla*), *ti'bawara* (Shoshoni; *ti'ba*, the seed), *tïba-t* [*tɨba-t*] (Tübatulabal), *tɨvapɨ* (Kawaiisu), *tuba* <*tɨbá*> (Northern Paiute; for *P. monophylla*), *tūvāą́'*$^{a}$ [*tūvápi*] (Northern Paiute), *tuve'e* <*tuvé'e, tuvaü*> (from *tuva*, nut, Hopi; for *P. edulis* and *P. monophylla*), *wa'ápų* (Ute; see also *Juniperus*), *wahappin* (Panamint), *wohkó* (Guarijío), *woko* (Yaqui), *wónūp* (Mono); YUMAN: *a'ko* (Mohave, the seeds), *ˀi:x*$^{w}$*í:* <*ehwi*> (Cocopa; for *Pinus monophylla*; obtained by trade), *huwál* (Havasupai, Arizona; for pines), *hwío* (Yuma), *ī'kó'* (Walapai, seeds *ko'*; for *P. edulis*), *ixalúwi* (Maricopa, Arizona; for pines), *nnai* (Paipai; for pines); LANGUAGE ISOLATE: *he'sho tsi'tonné* (gum branch, Zuni, New Mexico)

**Botanical description:** Large shrubs or small trees, to 10 m tall, often branched near the base. Leaves are mostly 3, rarely 2 or 4, in a bundle, 2.5–5.2 cm long, stout, rigid, often curved upward, glaucous white on the inner face (bottom). Sheaths are curled into rosettes and eventually deciduous. Usually, plants are dioecious. Cones are small, 2–4 cm long, usually wider than long, light brown, deciduous soon after seeds ripen; scales relatively few, thick, the tips blunt, knobbed, and without spines, nearly sessile, the peduncles short. Seeds are obovoid 10–13 mm long, wingless.

**Habitat:** Slopes, ridges, canyons. About 1,370 m (4,500 ft) upward in Brown Canyon, elsewhere 1,525–2,490 m (5,000–8,170 ft).

**Range:** Arizona (Cochise, Pima, and Santa Cruz counties), southwestern New Mexico; Mexico (Chihuahua, Durango, and Sonora).

**Seasonality:** Pollen shed late May to June; cones with seeds falling mid-September to October.

**Status:** Native.

**Ecological significance:** This border-land pinyon is unique to the region. Although it has been confused with the more widespread Mexican Pinyon (*P. cembroides*), one of the major pinyon seed sources, this species is a unique kind that occurs from the Chiricahua Mountains to the Baboquivari Mountains. *Pinus cembroides* is distinguished partly by being entirely monoecious, having leaves in 2–3 per bundle, less distinctly 2-colored leaves with stomata on both surfaces, larger cones (to 5 cm), and generally larger trees, to 17 m tall (Felger et al. 2001, Malusa 1992). *Pinus discolor* has stomata on the inner surface, but not on the outer surface.

*Pinus discolor* is known from a few trees in the bottom of Brown Canyon, with at least four of them being reproductive females—the sexes are separate (dioecious). Some of the slopes of smaller side canyons, particularly those higher up, have numerous trees. The species is also found in Jaguar Canyon, Baboquivari Canyon, and Thomas Canyon. Border Pinyon also occurs in the Coyote and Quinlan Mountains, just to the north of the Baboquivari Mountains. Elsewhere in Pima County, *P. discolor* is recorded from the Santa Catalina, Santa Rita, Sierrita, and Rincon Mountains. Nearby in Santa Cruz County, the trees are known from the Pajarito and Patagonia Mountains. Farther east, *P. discolor* grows in the Chiricahua, Dragoon, Huachuca, Mule, Pelocillo, Swisshelm, Whetstone, and Winchester Mountains of Cochise County. Border Pinyon is also in a few mountains in Graham and Greenlee counties.

**Human uses:** Pines are arguably the most important kinds of North American trees. Wood from all species has found uses, varying in importance from one species to the other, for about every purpose for which wood is commonly employed (Standley 1920–1926). Wood is used, where available, for fuel. Bundles of pitch pine still are commonly found in Mexican markets, and various indigenous peoples have eaten the sapwood and inner bark of pines during famines (Austin 2004, Castetter and Opler 1936, Minnis 1992); others use the resin for waterproofing baskets and other wickerwork (Vestal 1952).

The Aztecs called pines *ocotl* (torch, Náhuatl) because of their use of the resinous stems in making torches (Siméon 1885). That word gave rise to *ocote* in Mexican Spanish. *Ocote* is sometimes substituted for *pino*, but *ocote* is often used for pine wood. The Aztec word for a pine tree is *oco-cuahuitli*, meaning "torch tree."

*Pinus discolor* is a pine nut producer, as is *P. cembroides*, and the source of commercial *piñones*, *P. edulis* (Felger et al. 2001). It would be surprising if all people living near Border Pinyon did not eat the seeds, as pine nuts were an essential food among the Great Basin peoples (Lanner 1981b), Apache (Castetter and Opler 1936, Gallagher 1976), and Navajo (Vestal 1952). Among other names, seeds are called pinenuts, *piñones* (Spanish), and *tü'ba* (Paiute). Along with other pines in Mexico, Border Pinyon is used to treat chest problems, fever, rheumatism, wounds, and sores (López-E. and Hinojosa-G. 1988).

**Derivation of the name:** *Pinus* is the classical Latin name of the pines. The Latin is a cognate with pine (English), *pin* (French), *pinho* (Portuguese), and *pino* (Italian, Spanish). The word is akin to Akkadian *pehum*, to caulk, Anglo-Saxon *pin*, *pinhnutu*, and Sanskrit *pitu-daruh*, a kind of pine. *Discolor* means two-colored, referring to the needles with green above and white below.

## *Maurandya antirrhiniflora* (Plantaginaceae) Blue Snapdragon Vine, *mipil*

*Maurandya antirrhiniflora*. A. Flowering branch. B. Lifeform. Artist: R. D. Ivey (Ivey 2003).

**Other names:** ENGLISH: chicka-biddy (a "pet" name for a favored young domestic chicken, allusion obscure), roving sailor (Arizona, New Mexico, Texas to Florida), [blue, little, violet, vine] snapdragon [vine] ("snapdragon" was in use by 1573, and defined by John Gerarde as "[t]he flowers [are] fashioned like a dragon's mouth; from whence the women haue taken the name Snapdragon"; in this sense, the word "snap" was more akin to Middle High German *snabel*, Middle Low German *snavel*, beak, bill); SPANISH: *mipil* (Hidalgo); ATHAPASCAN: *shį násdzid* <*si nal$_3$idi*> (*shį*, summer, *násdzid*, afraid of, Navajo), *tłonanesdidzi* (vine, Chiricahua and Mescalero Apache)

**Botanical description:** Perennial herbs, stems twining, much-branched, glabrous. Leaves have petioles 1–3 cm long, hastate-ovate, 15–25 mm long and wide, both main blade and lateral lobes acuminate, narrowly cordate at base. Flowers are on slender pedicels, 1–3 cm long. Sepals are 1–1.3 cm long, linear-lanceolate, entire. Corollas are 2–2.5 cm long, the tube pale and dull, the lobes violet to purple or whitish, with an upraised yellowish white dark-lined pubescent palate that partly closes the opening. Capsules are 5–6 mm long. Seeds are brown, about 1 mm long, irregularly corky-winged, the wings broken and some mere lines of tubercules.

**Habitat:** Stony slopes, canyons, washes. 450–1,830 m (1,500–6,000 ft).

**Range:** Arizona (Coconino and Mohave, south to Cochise, Pima, Santa Cruz, and Yuma counties), southern California, New Mexico, Texas, formerly in Utah, and Nevada; Mexico (Chihuahua, Coahuila, Sonora, to Tamaulipas, and south to Oaxaca); El Salvador and Honduras.

**Seasonality:** Flowering April to October.

**Status:** Native.

**Ecological significance:** These climbers almost always begin life in shaded spots, where there is more moisture than on exposed sites. *Maurandya* is often seen below shrubs, and it may climb up these supports to reach better light, although the climbers have been reported to form "curtains" in other parts of the region. Equally as often, *Maurandya* remains in the lower part of these shrubs

There must be a complicated pollination system for these vines, as the corollas, while comparatively small, are showy. The violet to purple contrasts markedly with the pale colors, and both stand out against the green of the foliage. With *Maurandya*, as with their namesake, *Antirrhinum*, there is a "crawl-space" in the "gullet flowers" that would force bees (presumably genera like *Bombus*) to contact stamens and stigmas on entering and leaving the flower. Elisens and Freeman (1988) show that the nectar composition is consistent with those kinds of long-tongued bees.

**Human uses:** None found for this species, except cultivation. This species appears to be a popular plant for gardens, particularly those with unusual flowers and those favoring twining plants (Brickell and Zuk 1997). Horticulturists and some botanists have decided that this species should be separated into the genus *Maurandella*, and that there is still a genus *Maurandya* with two other species. However, the latest study by Elisens (1985) puts the genus *Maurandella* in synonymy with *Maurandya*.

**Derivation of the name:** *Maurandya* is dedicated to physician Catalina Pancratia Maurandy, a botany teacher at Cartagena in the late 1800s, and the wife of A. J. Maurandy, director of the Cartagena Botanic Garden. *Antirrhinifolia* means that the vine has leaves like Old World *Antirrhinum*, the Snapdragon.

**Miscellaneous:** There are two species in this American genus, ranging through parts of western North America and South America. Oddly, professional botanists collecting plants often misidentify these vines as *Ipomoea* (Convolvulaceae). Although both are climbers, *Maurandya* has leaves opposite the bilobed flowers, as opposed to the alternate leaves and axillary radially symmetrical flowers of *Ipomoea*.

Some now put *Maurandya* in the Veronicaceae instead of the Scrophulariaceae (e.g., Wunderlin and Hansen 2003), although Judd et al. (2008) consider it part of the Plantaginaceae. Classification by Judd et al. (2008) dramatically alters the historical concept of the Plantaginaceae, which was considered to contain the genus *Plantago* and two closely allied genera, *Bougueria* (Andes) and *Littorella* (3 spp.—1 N. America, 1 S. America, 1 Europe and Azores). The old view was that the family was intermediate between the Scrophulariaceae and Rubiaceae, but it has long been considered ambiguous. The view of Judd et al. (2008) is that *Plantago* is highly reduced and adapted for wind pollination, while most of the remaining members of the family are animal-pollinated.

## *Plantago virginica* (Plantaginaceae) Pale-Seeded Plantain, *llantén, da:pk*

**Other names:** ENGLISH: Indian wheat (New Mexico, fide Hocking 1956), pale-seeded [Virginia] plantain (from Latin *plantagenim*, sole of foot, in reference to the large broad leaves; first used in English ca. CE 1255; cognates are *piantaggine*, Italian; *plantain*,

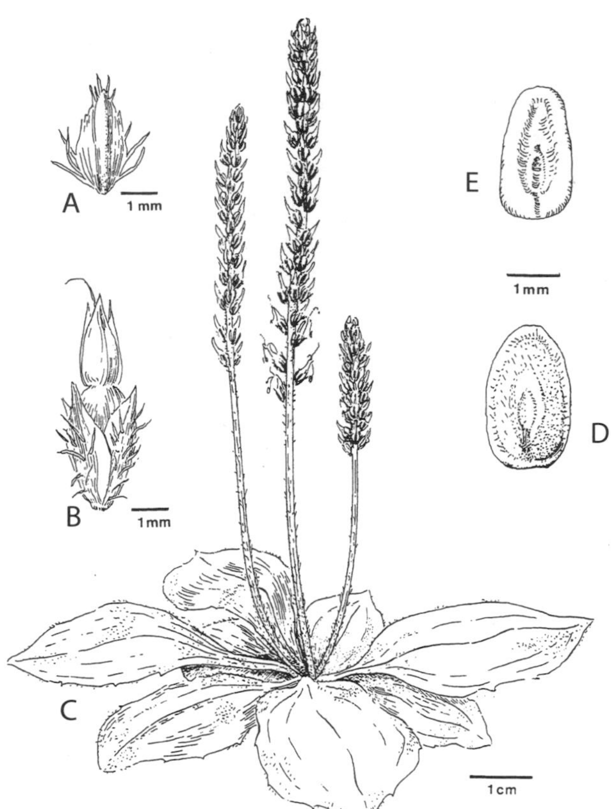


*Plantago virginica*. A. Bract. B. Flower. C. Habit. D. Seed of *P. rhodosperma*. E. Seed of *P. virginica*. Artist: M. Gregg (Huisinga and Ayers 1999, courtesy Tina Ayers).

French); SPANISH: *llantén* <*lantén, yanten*> (the broad one, from Latin *planus*; Mexicans sometimes shorten the "ll" to "l"), *pastora* [*pastorcita*] ([little] shepherd, Sonora); ATHAPASCAN: <*ˀalii: béyi.c'oi*> ("urine spurter," *łizh*, urine, and maybe *biyah chooʼį*, from *biyah*, supporting it, *chooʼį*, used or useful, Navajo), *bįįhjaaʼ* <*bíhi-ljáˀ*> (*bįįh*, deer, *jaaʼ*, ears, Navajo), *'ásaaʼ halts'aaʼ* <*c'aˀ xalc'aˀ*> (*'ásaaʼ*, bowl, *halts'aaʼ*, hollow [shaped seeds], Navajo); UTO-AZTECAN: *da:pk* (smooth/slippery, Tohono O'odham), *hahay'inga* <*hahaínga*> (*Hahay'i*, female kachina who represents ideals of womanhood, *nga*, medicine, Hopi), *mumsa* (Hiá CeḍO'odham; for seeds), *řorogoči* (Tarahumara), *šiñakali* (Northern Tepehuan), *toi'gûpagûnt* [*toígûpagûnt*] (*toi*, elevation [referring to the flower stalks], *gûp*, fruit, *a*, connecting vowel, *gûnt*, thorn, Shoshoni; for *P. major*); YUMAN: *ókʳīsa* (Havasupai, Arizona)

**Botanical description:** Annual herbs to 35.5 cm tall, although often shorter. Leaves petiolate, the blades lanceolate, 2.5–10 cm long, 0.8–2.7 cm wide, basally attenuate, apically acute, sparsely villose, distinctly 3-veined, sometimes with 2–4 widely spaced teeth. Peduncles are 1–20 cm long. Spikes are 1.5–17 cm long, sparsely to densely villous, the bracts subulate to narrowly or broadly triangular, 2–4.8 mm long, narrowly basally scarious-margined, ciliate. Flowers are more or less dioecious. Sepals are ovate, 2.5–3 mm long, broadly scarious-margined, acuminate apically, obtuse to acute basally, hirsute on midrib. Corollas have lobes erect, enclosing the capsule, lanceolate, 2–3 mm long; stamens 4. Capsules break in the middle. Seeds are 2, ellipsoid, 0.9–3 mm long, 0.8–1.6 mm wide, light brown, the inner surface deeply concave, the outer surface furrowed.

**Habitat:** Moist soil, canyons, along streams. 760–2,130 m (2,500–7,000 ft).

**Range:** Arizona (Cochise, Gila, Maricopa, Pima, and Santa Cruz counties), southern California, east to Florida, north to Connecticut, Michigan, and Missouri; Mexico (Chihuahua and Sonora).

**Seasonality:** Flowering March to September.

**Status:** Native.

**Ecological significance:** Plants disappear by the middle of November and reappear as seedlings by January or February. Pollen is spread by wind in all the species (Proctor et al. 1996). One British study found that *Plantago* pollen is the fifth most common in the air, behind grasses, ash, oaks, and elms. Flowers in plantains are partly unisexual and are gynodioecious (female flowers on one plant; hermaphroditic on another). *Plantago*, like a few other groups, is not fixed on one method, even if it is the most frequent. Insects do visit the flowers and pollinate them in some, perhaps all, species.

**Human uses:** Europeans had a long history of using *Plantago* when explorers arrived in the New World. To the Anglo-Saxons, *Plantago* was the "mother of herbs," and they had magical poems that spoke of them. One went, "Carts creaked over you, queens rode over you, brides bridled over you, bulls breathed over you; all these you withstand, so may you withstand poison and infection . . ." (Dobelis 1986). Gaelic-speakers called *Plantago* the *cuach Phàdruig* (*cuach*, drinking cup, *Phàdruig*, of Father Druid; for *P. major*) or the *slàn lus* (healthy herb) (Austin 2004). Scandanavian people said *kjempe* (giant, Norwegian) or *koempe* (warrior; because of a children's game; for genus, Danish). Other Germanic people said waybread [waybred] (from Old English *waybráde*, meaning "broad-leaved plant growing beside the way"), *vejbred* (waybread, Danish), *weegbree* (Dutch, from Middle Low German, *wegebrede*), *Wegebreite* (German, from Old High German *wegbrieta*), or *Wegerich* (of roads, German).

Partly because of their tannin content, the leaves help cure several problems. Leaves are applied to cuts, sores, burns, snake and insect bites, and inflammation (Dobelis 1986). A tea of the seeds is used to treat diarrhea, dysentery, and bleeding from mucous membranes. Psillium (seeds of *Plantago psyllium*) is still sold commercially to relieve constipation.

Acoma and Laguna eat young leaves (Castetter 1935). Tarahumara eat young leaves of several species raw, but older leaves are shredded and boiled before being consumed (Pennington 1963). Northern Tepehuan also eat the leaves of a *Plantago* as a *quelite* (Pennington 1969); they use two species medicinally, one to treat fever and the other for stomach cramps. Hohokam ate the seeds and, perhaps, also the leaves (Gasser 1981). Navajo eat the seeds of *P. patagonica* in mush (Wyman and Harris 1951).

**Derivation of the name:** *Plantago* is based on Latin *planta*, footprint or sole of the foot, plus *-ago*, from *agere*, to bear or resemble; first use by Pliny, CE 23–79. *Virginica* means of Virginia, where it was first collected.

**Miscellaneous:** *Plantago patagonica* is a narrow-leaved annual that appears with the spring wildflowers and is seasonally abundant. This species in Arizona is documented in all counties except La Paz and Yuma. The herb also grows from British Columbia to Saskatchewan south through Mexico to Argentina and Chile (Huisinga and Ayers 1999). It has been introduced into the southeastern United States.

## *Platanus wrightii* (Platanaceae) Arizona Sycamore, *aliso*

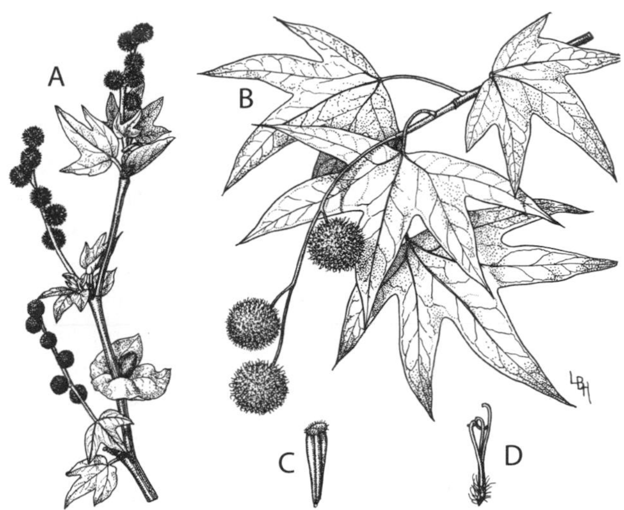


*Platanus wrightii.* A. Branch with young leaves and male flowers. B. Mature branch with female fruits. C. Single stamen. D. Pistillate flower. Artist: L. B. Hamilton (Benson and Darrow 1945).

**Other names:** ENGLISH: Arizona [white] sycamore ("sycamore" was originally applied to *Ficus sycomorus*, a fig in Egypt, Syria, and nearby countries; the word is a combination of Greek *sykos*, fig, and *moros*, the mulberry, and came into English ca. CE 1300; by ca. 1588, it was applied to European *Acer pseudoplatanus*, and by 1814 or perhaps before to American *Platanus*); SPANISH: *álamo* [*blanco*] ([white] *álamo* based on *ala*, wing, in both Spanish and Portuguese; from a resemblance of the fluttering leaves to wings on birds; the name is also used for *Populus*), *aliso* (Chihuahua, Sonora; originally Spanish for *Alnus*), *ciclamor* (Spanish for "sycamore"); ATHAPASCAN: *gashdla'é* <*gaastlae, k'ashdla'a*> (Western Apache); UTO-AZTECAN: *havatɨbɨa* (from *hava-*, shade? Kawaiisu), *ŕepogá* (Tarahumara), *sáoko* (Hopi), *ṣua'har* <*shua'jar*> (from **ṣuhar*, cognate with Northern Tepehuan *ṣohárat*, Mountain Pima), *sivíl* (Cupeño), *sivily* <*sivel, siví-l*> (Cahuilla), *sivé:-la* <*savé:-la, sevél*> (Luiseño), *ṣohárat* <*ṣojárat*> (Northern Tepehuan), *ušako* <*sako*> (Tarahumara); YUMAN: *ūhpúhl* (Kumiai)

**Botanical description:** Trees to 10–20 m, older individuals spreading, herbage generally woolly with branched (dendritic) hairs, becoming glabrous with age. Leaves are alternate, 24–38 cm long, 18–36 cm wide, simple, palmately lobed and veined, with 3–7 lobes, petioled. Inflorescences are axillary, unisexual, in dense, many-flowered, globose heads. Flowers have tufts of irritating, golden-brown hairs. Sepals and petals are mostly inconspicuous or reduced, soon deciduous. Male inflorescences are shorter than female inflorescences and soon falling. Fruiting heads are 2.5 cm wide, the individual fruits are achenes 6–7.5 mm long with persistent stigma bases, 4-angled, 1-seeded.

**Habitat:** Along streams in canyons and washes. 600–2,450 m (2,000–8,000 ft).

**Range:** Arizona (Cochise, Coconino, Gila, Greenlee, Maricopa, Mohave, Pima, Pinal, Santa Cruz, and Yavapai counties) and New Mexico; Mexico (Chihuahua, Sinaloa, and Sonora).

**Seasonality:** Flowering March to May.

**Status:** Native.

**Ecological significance:** These trees are confined to permanently moist sites, mostly as gallery forests along streams. Flowers are small and wind-pollinated; fruits are also wind-dispersed after the heads break apart. Surely, fruits are also water-dispersed.

Hollows in the branches and trunks are havens for many cavity-nesting birds, including Arizona Woodpeckers (*Picoides arizonae*), Gila (*Melanerpes uropygialis*), Lewis's (*Melanerpes lewis*), Western Screech Owls (*Megascops kinnecottii*), and Whiskered Screech Owls (*Megascops trichopsis*). Birds that build their nests in the branches include Cooper (*Accipiter cooperii*) and Gray Hawks (*Asturina nitida*), along with several hummingbirds.

The current prolonged drought is taking its toll. There are dead branches and large limbs on many trees, and some individuals are dying.

**Human uses:** Tarahumara and Northern Tepehuan crush the bark of Arizona Sycamore and decoct it into a drink (Pennington 1963, Pennington 1969). Thord-Gray (1955) said that only the inner bark is used. This beverage is considered a "general medicinal tea" among the Tarahumara, but simply a refreshing beverage among the Northern Tepehuan. The Tarahumara also remove the large knots on the lower trunks of trees and carve them into bowls. Probably, this is the species that Hopi called *sáoko* and use to dye leather (Whiting 1939). Western Apaches make a medicinal tea from the plant and use the bark for dyeing (Gallagher 1976).

No other records have been found of this species being used by other groups, except for being planted for shade. Trees are havens from the heat for hikers. There are dozens of places with "Sycamore" in their names in Arizona, including Sycamore Canyon, Sycamore Creek, Sycamore Mesa, Sycamore Pass, Sycamore Rim, and Sycamore Spring (Granger 1960). Of the Sycamore Canyon in Coconino County, Peattie (1953) wrote: "[I]t is certainly . . . romantic. . . . Indian caves are still found in it, and once, according to local legend, it was a hide-out for badmen and renegade Indians. Today, the great Sycamores throw their shadows on the canyon walls in peace."

**Derivation of the name:** *Platanus* was the original Latin; Greek *platanos*, broad, in reference to the leaves, a name for the Plane-Tree used by Pliny, CE 23–79; the Greek Theophrastus, 370–288 BCE, spelled it *platys*. *Wrightii* commemorates Charles Wright, 1811–1885, who is best known for his collections in Cuba and Texas (1837–1852).

**Miscellaneous:** There are two other native species, *P. occidentalis* and *P. racemosa*, in North America north of Mexico (Kartesz 1994). The eastern species, *P. occidentalis* (Sycamore, Plane-Tree, Buttonwood), ranges through the eastern deciduous forests and on to the Great Plains along waterways. That species probably was used by all people living within its range (Austin 2004). *Platanus racemosa* (California Sycamore) grows in California and down into Baja California in Mexico (Munz and Keck 1973, Wiggins 1980). There are perhaps eight species in the monotypic family (Mabberley 1997). Until recently, most have considered the relationships of the genus dubious. Now, molecular genetics suggests that *Platanus* belongs in the Proteales, along with Proteaceae and Nelumbonaceae (Judd et al. 2008).

## *Aristida ternipes* (Poaceae) Spider Grass, *zacate araña, waṣai*

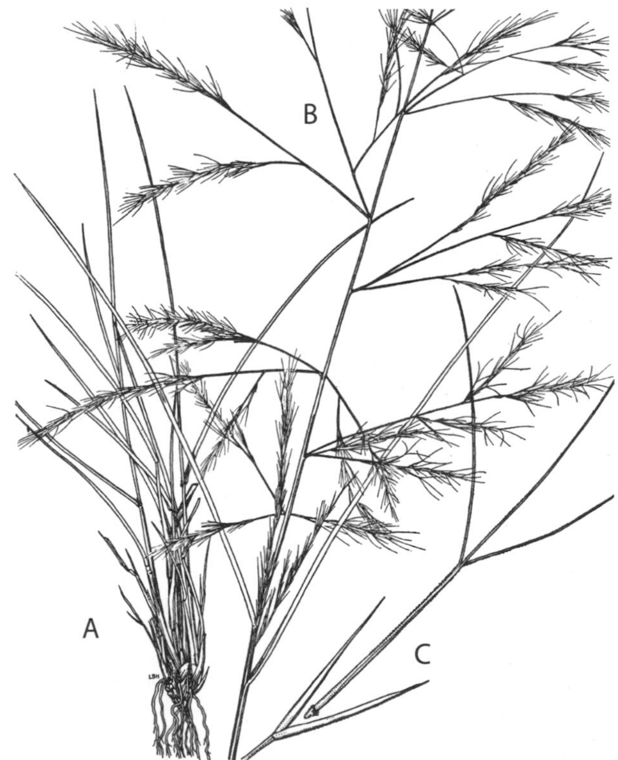

*Aristida ternipes*. A. Base of plant. B. Flowering tip of plant. C. Spikelet. Artist: L. B. Hamilton (Gould 1951).

**Other names:** ENGLISH: [poverty, six-weeks] three-awn (referring to the three needle-like outgrowths of the fruits; "awn" from Old Norse *ögn*, akin to Swedish *agn*, Danish *avn(e)*, Old High German *agana*, modern German *ahne*, Gothic *ahana*; the delicate spinous process, or "beard," that terminates the grain-sheath; in use by 1300); SPANISH: *otatillo* (from Náhuatl *otatl*, large grass, a name also given to some climbing grasses, genus *Lasiacis*, Mexico), *zacate* (from Náhuatl *zacatl*, grass), *tres barbas arqueado* (arched three barbs, Mexico), *zacate araña [de tres barbas]* ([three-awn] spider grass (applied to a *Panicum* by 1889; later extended to other grasses, Arizona, New Mexico, Sonora), *zacate barba* (barbed grass, Sonora); ATHAPASCAN: *tł'oh* (grass, Western Apache); MAYAN: *chak-suuk* <*tok-suuk*> (*chak*, red, *sak*, white, Maya); UTO-AZTECAN: *ba'aso* (Mayo), *guatoco* (Guarijío), *hahay'iqalmongwa* <*hahaí'iqálmongwa*> (*Hahay'i*, female kachina who represents ideals of womanhood, *qalmongwa*, bangs, because those of this kachina are red, Hopi), *o'gĭp [o'gwĭp, toi'yaogwĭp, yo'nĭp]* (Shoshoni; for *A. purpurea*), *wahái* (any grass, Northern Paiute), *waṣai* (any grass, Tohono O'odham); (= *Aristida hamulosa*)

**Botanical description:** Tufted perennials, the culms firm, wiry, 40–120 dm tall, often branched above the base. Leaf blades are firm, narrow, and involute on drying. Panicles are variable, 20–50 cm long, usually with a few stiffly spreading or drooping primary branches to 25 cm long, these mostly bare of spikelets on the lower one-third to one-half, branches rarely all short and slightly spreading. Glumes are about equal, the lemma tapering to a short, stout, scabrous, straight or slightly twisted awn column, 9–12 mm long; lateral awns absent or reduced to short stubs, rarely 2 mm or more long, the central awn 4–12 mm long.

**Habitat:** Desert grasslands, canyons, washes. To 1,800 m (5,905 ft).

**Range:** Arizona (Cochise, Coconino, Gila, Graham, Pima, Pinal, and Santa Cruz counties), southern California, New Mexico, and western Texas; Mexico (Chihuahua and Sonora); south to Honduras.

**Seasonality:** Flowering June to September.

**Status:** Native. Individuals with the single awns are var. *gentilis*.

**Ecological significance:** This grass has a triple method of dispersal. Early in the fall, when the awns are ripe, they stick to anything that passes. Humans and other animals are often adorned with the seeds and spread the propagules. People living near the grass find awns from the fruits stuck in socks, trousers, shirts, and even underwear. The needle-like barbs efficiently move the propagules.

Later in the season, the entire fruit-cluster breaks free from the stems and is rolled and blown across the grasslands and other habitats, where it drops the remaining fruit crop. Because autumn and early winter winds may be fierce and common, the second dispersal mechanism may be more effective than animal dispersal.

Seeds of *Aristida* are among the most common foods of Merriam's Kangaroo Rat (*Dipodomys merriami*). Hoffmeister (1986) recorded seeds of "annual three-awn," "perennial three-awn," and "annual grama" as the order of preference. No binomials were given.

**Human uses:** Mayo commonly add Spider Grass stems to adobe (Yetman and Van Devender 2001). These people also weave the stems into dense mats used in walls and roofs, even though it is a fire hazard with their cooking fires. A wall thus constructed will last for 15 years. Guarijío use it similarly (Yetman and Felger 2002). The Guarijío brew the grass into a tea that is drunk to cure spider (*turusi*) bites. Tarahumara use the grass to line the pit in which agave is being roasted (Pennington 1963). Tarahumara consider *A. ternipes* to be a good forage for livestock, and they search out meadows that contain it. Indeed, Johnson-G. and Carillo-M. (1977) say that the one-awned form is probably the best of the "*tres barbas*" for forage. The three-awned form is considered excellent in the thickets in pastures, but the *navajitas* (*Bouteloua*, q.v.) are better in open areas.

**Derivation of the name:** *Aristida* is from Latin *arista*, the awn, the beard of an ear of grain. *Ternipes* means in threes, alluding to the awns.

**Miscellaneous:** There are two named varieties of this species that were previously recognized as distinct species (*A. ternipes* vs. *A. hamulosa*). Variety *gentilis* is recognized by the solitary awns, while var. *ternipes* has the three awns characteristic of the genus. Typically, the grasses are stable in their awn size and number, but Felger (2000) has found the two variations within the same population, 3-awned during a wet year and 1-awned during a dry season. That change must be exceptional. For nine years, single-awned plants have remained stable near my home.

Other species of *Aristida* in the mountains are *A. adscensionis* (Six-Weeks Three-Awn, *zacate tres barbas*) and *A. laxa* (Beggar-Tick Grass, Three-Awn). Six-Weeks Three-Awn is an annual that has awns

1–2 cm long that are usually not widely divergent from the main axis. Beggar-Tick Three-Awn is a perennial; the lateral awns are reduced or absent with only one remaining, and they are rarely over 2 mm long.

### *Bothriochloa barbinodis* (Poaceae) Cane Beard-Grass, *zacate popotillo, waṣai*

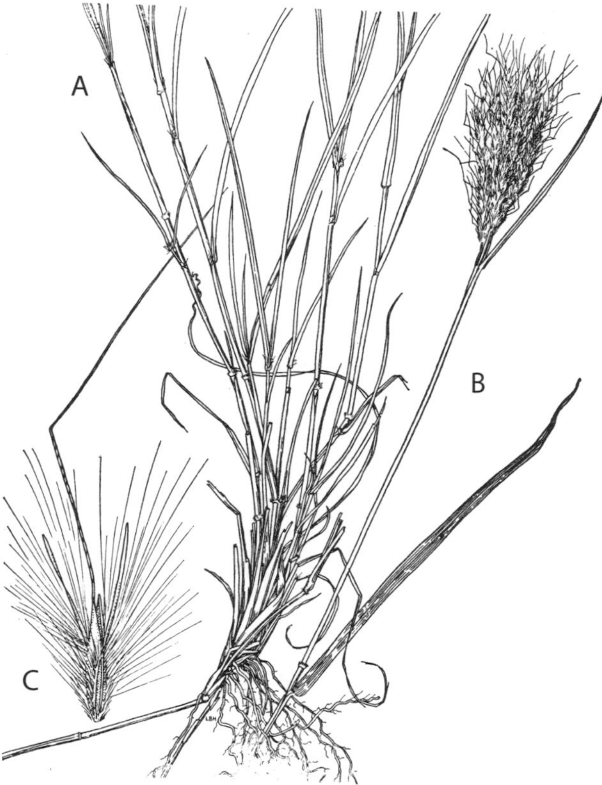


*Bothriochloa barbinodis*. A. Base of plant. B. Flowering tip of plant. C. Spikelet. Artist: L. B. Hamilton (Gould 1951).

**Other names:** ENGLISH: cane beard-grass (by 1841 "beard-grass" used for the European grass *Polypogon*, q.v.; later for other species), fuzzy-top, cane bluestem ("cane" is from Middle English *canne*, appearing in English by 1398; "bluestem" because of the bluish, glaucous leaf sheaths along the stems, in use by 1864 for *Andropogon gerardii*; subsequently, applied to *Bothriochloa* and *Schizachyrium*); SPANISH: *algodonoso* (cottony, Mexico; see *Gossypium* for etymology), *cola de coyote* (coyote's tail, Nuevo León), *popotillo* [*perforado, plateado*] ([perforated, folded] little broom, from Náhuatl *popotl*, broom, Sonora), *zacate popotillo* (little broom grass, Mexico); ATHAPASCAN: *tł'oh* (any grass, Western Apache); UTO-AZTECAN: *wahái̵* (any grass, Northern Paiute), *waṣai* (any grass, Tohono O'odham); (= *Andropogon barbinodis*)

**Botanical description:** Grasses with tufted stems, 6–13 dm tall, often branching below, the nodes glabrous to hispid. Leaf blades are commonly glaucous, almost glabrous, 3–6 mm wide. Panicles are long-exserted, silvery white, silky, dense, oblong, 7–15 cm long, the racemes 2–4 cm long, the rachis-joints and pedicels silky. Spikelets are 4 mm long, the awn bent like a knee (geniculate), twisted below, 1–1.5 cm long; pedicellate spikelets reduced.

**Habitat:** Desert grasslands, canyons, washes. 304–1,828 m (1,000–6,000 ft).

**Range:** Arizona (Navajo and Coconino south to Cochise, Graham, Greenlee, Pima, and Santa Cruz counties), California to Oklahoma, and Texas; Mexico (Chihuahua and Sonora).

**Seasonality:** Flowering July to October.

**Status:** Native.

**Ecological significance:** West of the Baboquivari Mountains, these grasses grow only near the Tule Tanks and Tinajas Altas tanks in Cabeza Prieta National Wildlife Refuge (Felger et al. in prep.). The grasses are more common near waterways in the Ajo Mountains. These authors further write, "This grass occurred in the Puerto Blanco Mountains about 5,200 years ago, and this or a similar grass grew in the Ajo Mountains about 8,000 to 9,500 years ago."

Cane Beard-Grass is abundant in both Brown and Jaguar Canyons. Elsewhere, it has been collected in Baboquivari and Thomas Canyons and surely grows throughout. The grass is abundant in most places in the Altar Valley from the Mexican border to at least the Sierrita Mountains (Austin 2003). It is frequent to uncommon in the Pajarito, Rincon, and Tucson Mountains (Toolin et al. 1979, Bowers and McLaughlin 1987, Rondeau et al. 1996).

There is growing evidence that this species declines rapidly with grazing. Bryant et al. (1990) found that the species indicated overgrazing in the Sonoran Desert. Felger et al. (in prep.) now finds remnant populations of these grasses in some of the dryer mountains of southwestern Arizona that have been protected from grazing. Betancourt et al. (1986) found this grass in Porcupine middens in the Sacramento Mountains of New Mexico dated at 1760 ± 100 years BP in the upper levels and 4540 ± 150 years BP in the lower levels.

**Human uses:** Cane Beard-Grass may be *toota muḍadkam* of the Akimel O'odham, but the identification is uncertain (Rea 1997). The people Rea talked with said this was "[l]ike a hygeria [a grain sorghum?] . . . likes moisture, lots of moisture." Rea speculates that the grass might have been *Bothriochloa barbinoides*, but he could not eliminate *Digitaria californica* or *Chloris crinata* from the possibilities. Only the name was remembered and no previous uses—except as an indicator of wet habitats and food for cattle.

Humphrey et al. (1952) consider this species a fair forage grass because the stems are coarse and the nutrients leach out after drying. However, plants are grazed readily by livestock during the summer growing season.

**Derivation of the name:** *Bothriochloa* comes from Greek *bothros*, a pit or hole, *bothrion*, a little hole, *chloa*, grass. When Kuntze named the genus, he wrote: "*Der Name 'Grubehengras' wegen der etwa 0.5 mm breiten und tiefen Grube auf der l. Spelze* (The name "pitted-grass" because of the nearly 0.5 mm wide and deep pit of the lower glume)." Not all species have the pit, but *barbinoides* does. *Barbinoides* means resembling a beard.

**Miscellaneous:** Species now placed in *Bothriochloa* were, for decades, uniformly included in the genus *Andropogon*. Otto Kuntze, a strict advocate of nomenclatural priority, created *Bothriochloa* in 1891—along with 29,999 other names. It was not until 1940 that the species was moved into that genus. Even then, neither Gould (1951) nor Kearney and Peebles (1951) called the grasses *Andropogon barbinoides*, and they did not even list the *Bothriochloa* binomial as a synonym. Later, Gould changed his mind, and Gould and Moran (1981) call them *Bothriochloa barbinoides*.

*Bothriochloa* is now considered by most to be distinct from *Andropogon*, and it contains 28 species that grow in the warm parts of the world (Zuloaga et al. 2003).

## *Bouteloua curtipendula* (Poaceae)
## Sideoats Grama, *navajita, dadpk waṣai*

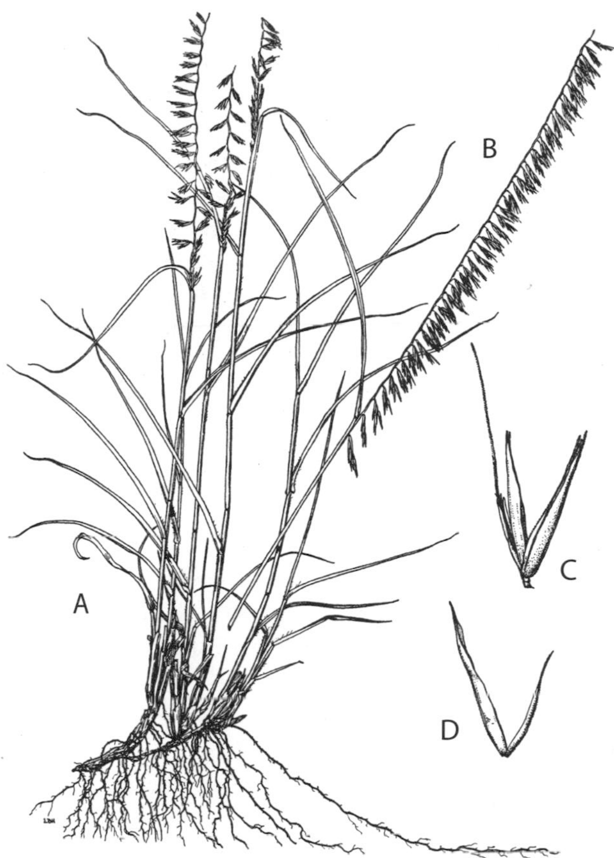


*Bouteloua curtipendula*. A. Flowering plant. B. Inflorescences, enlarged. C. Florets. D. Glumes. Artist: L. B. Hamilton (Gould 1951).

**Other names:** ENGLISH: Sideoats grama (sideoats, an oat-like grass with the flowers on one side of the stalk; "oat," cognate with West Frisian *oat*, Dutch *oot*, etc.; in use since Old English; "grama" from Spanish *grama*, grass; in English by 1844); SPANISH: [*pasto*] *banderilla* (little flag [grass], Chihuahua, Sonora), *gramilla* (little grass, Mexico), *navajita* [*banderilla*] (little knife [little flag], Baja California, Chihuahua, Sonora), *zacate de navaja* (knife grass, Sonora); ATHAPASCAN: *tłobindaiłkehntii* (grass with the seeds lying on top of one another, Chiricahua and Mescalero Apache), *tł'oh nástasí* (grass that bends back around, Navajo), *tł'oh* (any grass, Western Apache), *tł'oh łichíí'* <*y'oh lici*> (red grass, Navajo); HOKAN: *isnáap ic is* (whose fruit is on one side, also their name for *Lepidium*, Seri); KIOWA TANOAN: *ta tăn'iŋ* (*ta*, grass, *tăn*, seedy, Tewa), *tap'eñita* (*tap'eñi*, broom, *ta*, grass, Tewa); UTO-AZTECAN: *dadpk waṣai* <*da:pk washai, dadpk washai*> (smooth/slippery grass, Hiá Ceḍ and Tohono O'odham), *harushö* (*haru*, bean sprouts, *shö*, *Pleuraphis jamesii*, Hopi), *navaja sa'i* <*sha'i*> (*navaja*, knife in Spanish, *sa'i*, any grass, Mountain Pima), *owiv* (grass, Ute), *wahái* (any grass, Northern Paiute)

**Botanical description:** Tufted perennial, often with rhizomes by which the clumps expand. Culms are 3–11 dm long, 1–2 mm thick, erect, largely unbranched.

Ligules have a fringe or fringed scale 0.5–1 mm long. Leaf blades are 5–35 cm long, 2–6 mm wide, usually flat with the sides folded inward (involute). Spikes are 20–50, these 6–20 mm long (rarely longer), in 2 rows on one side of a panicle axis, 13–30 cm long, somewhat pendulous, at length each deciduous as a unit, the rachis prolonged beyond the most distal spikelet as a needle, sparsely appressed-strigose or glabrous. Spikelets are mostly 5–8 (as few as 3, as many as 13).

**Habitat:** Grasslands, canyons. Up to 2,133 m (7,000 ft).

**Range:** Arizona (throughout), Alabama, southern California, to Maine and Maryland, Montana, and Texas; Ontario; Mexico (Chihuahua and Sonora).

**Seasonality:** Flowering July to October.

**Status:** Native.

**Ecological significance:** Sideoats Grama was probably grazed by the Pronghorn (*Antilocapra americana*) that inhabited the Altar Valley. The Masked Bobwhite (*Colinus virginianus ridgewayi*) eats the seeds of the related *B. rothrockii* and probably those of *Bouteloua curtipendula*, too. Good stands of these grasses are indicators of healthy grasslands.

All *Bouteloua* species are important forage plants to livestock (Humphrey et al. 1952, Johnson-G. and Carillo-M. 1977) and constituted the favored species on the Buenos Aires land when it was a ranch. That grazing began in earnest in the 1850s when Pedro Aguirre, Jr., started a stagecoach and freight line between Tucson and the mining towns of Arivaca in Arizona, and Altar in Sonora, Mexico. He added a homestead in 1864 and named it *Buenos Ayres* ("good air"), for the winds there. Before cattle, the valley supported a large herd of Pronghorns and Masked Bobwhite Quail, but ranching and droughts eliminated both. The Buenos Aires National Wildlife Refuge was established to recover these endangered animals (USFWS 2003).

**Human uses:** Chiricahua and Mescalero Apache use the moist grass to lay on hot stones to prevent the loss of steam when cooking (Castetter and Opler 1936). Tewa bundle and dry the grass and make it into brooms to sweep floors, hearths, and metates, and also to brush the hair (Robbins et al. 1916). This species is probably the *dapk vashai* (naked or smooth/slippery grass) of the Akimel O'odham, but the identity is uncertain (Rea 1997).

The Spanish name *banderilla* is intriguing. Although literally the word means "little flag," it is rich with historical meaning dating back to pre-Columbian Spain. The flowering stalks of these grasses with the spikes hanging off one side reminded people of the "*banderilla*" of Spain. That instrument is the dart with a small flag that is used by the *picador* in bullfights. The *picador* enrages the bull with these *banderillas* by goading the animal and encouraging it to charge the *matador*. Indeed, this activity developed into an expression in Spanish used in other circumstances: *clavar a uno una banderilla*, to goad or taunt someone.

In a convergent analogy, a Kiowa warrior who had killed an enemy with a lance wore the grass. The flowering stem symbolized the feather-adorned spear he had used (Vestal and Schultes 1939).

**Derivation of the name:** *Bouteloua* is named for Spanish Claudio Boutelou, 1774–1842, a writer on floriculture and agriculture. *Curtipendula* means short-hanging, comparing the spikelets to other species in the genus.

**Miscellaneous:** The genus *Bouteloua* comprises 54 species that grow from Canada south to Argentina, although Mexico has more species than anywhere else (Peterson et al. 2001). Seven other species of *Bouteloua* have been found in the Baboquivari Mountains, and several of these extend up into the Brown Wash and lower Brown Canyon (Appendix).

Arguably, the most specialized of these grasses is *B. aristidoides*. This small annual grass is called Six-Weeks Grama (because of its short life-cycle after summer rains) and Needle Grama (because of the sharp awns that stick like needles) or *aceitilla* (little oily one). This grass is the favored food of the Rufous-Winged Sparrow (*Aimophila carpalis*) that just reaches the United States in southern Arizona.

## *Bromus anomalus* (Poaceae) Nodding Brome, *bromo, waṣai*

**Other names:** ENGLISH: brome (from the Latin *bromus* and Greek *bromos* used by Pliny, CE 23–79, for the genus; the word was also applied to oats, *Avena*), cheat (see below), chess (recorded for *Bromus* by 1736; referring to the parallel rows of grains in an ear of corn or grass; used in this sense by 1562 when William Turner wrote, "[Rice] hath comonly an Ear with ij chesses or orders of corn as barley hath"); SPANISH: *bromo dormilón* (nodding brome, Mexico), *bromo frondoso* (leafy brome, Mexico); ATHAPASCAN: *ye'i 'ac'osi ch'il* <*Yωóce čil*> (*ye'i*, deity, *'ac'osi*, impersonator, *ch'ił*, plant, Navajo), *hajíínáí ch'ił* <*xaǯínái čil*>

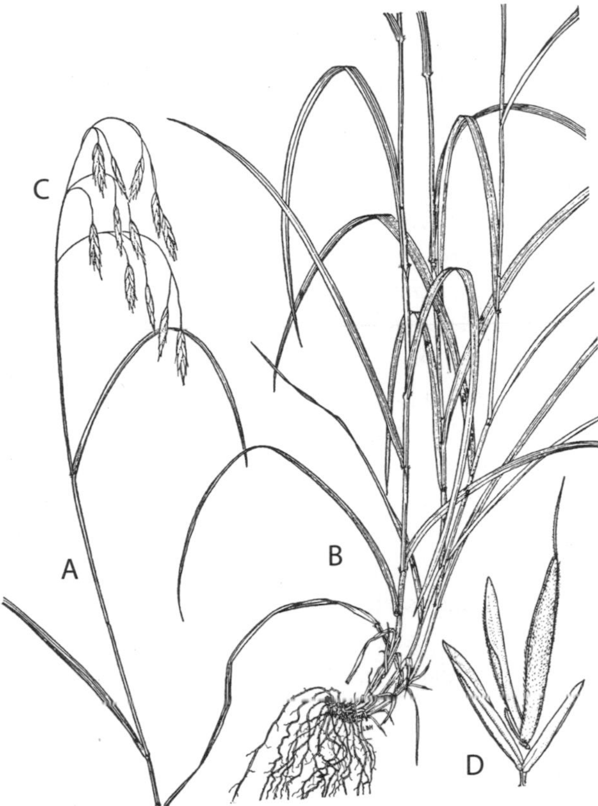


*Bromus anomalus*. A. Flowering tip. B. Base of plant. C. Floret. D. Glumes. Artist: L. B. Hamilton (Gould 1951).

(*hajíínáí*, emergence, *ch'ił*, plant, Navajo), *łį́į́tłts'oh* <*į́į́λokłicoh*> (*łį́į́'*, horse, *tłoh*, grass, *ts'oh*, big, Navajo); UTO-AZTECAN: *basi-awi* (Tarahumara; for *Bromus carinatus*), *owiv* (grass, Ute), *páshu* (any grass near water, Hopi), *wahái* (any grass, Northern Paiute), *waṣai* (any grass, Tohono O'odham)

**Botanical description:** Tufted perennials. Culms are 3–9 cm long, 1–2 mm thick, erect except shortly decumbent (somewhat rhizomatous) at the base, simple. Leaf sheaths are almost glabrous or shortly retrorse woolly-pubescent, the blades 2–5 mm wide, mostly flat. Panicles are nodding, 8–16 cm long, few-flowered. Spikelets are 5- to 9-flowered. First glume is 3-nerved, the second 5-nerved; lemmas with an awn 2–3 mm long, pubescent.

**Habitat:** Open woodlands. 609–3,048 m (2,000–10,000 ft)

**Range:** Arizona (Apache, Cochise, Coconino, Graham, Greenlee, Pima, Santa Cruz, and Yavapai counties), California, New Mexico, western Texas, and north to Idaho; Saskatchewan; Mexico (Chihuahua and Sonora south to Chiapas).

**Seasonality:** Flowering August to September.

**Status:** Native.

**Ecological significance:** Nodding Brome is a mountain plant in western Texas (Trans-Pecos) and in the Baboquivari Mountains. The species has not been found in southwestern Arizona west of the Baboquivari Mountains (Felger et al. in prep.). While the grass grows down to 2,000 feet within Arizona, it has not been found below 4,000 feet in the Baboquivari Mountains. Plants are seasonally frequent in both Brown and Jaguar Canyons. This *Bromus* is also recorded in Baboquivari Canyon.

This Mexican species barely reaches the United States between Texas and California; Wagnon (1952) thought that it was confined to Texas. Subsequently, it has been found farther west and north. There has been confusion of *B. anomalous* with other species, and both diploid and tetraploid races (2n = 14, 28) have been reported (Wagnon 1952, Reeder 1971).

**Human uses:** Nodding Brome is said to be the first plant the Navajo saw after the Emergence (Vestal 1952). The Navajo use Nodding Brome as horse food and in the Evil Way blackening (Vestal 1952).

The English common name applied to all members of *Bromus* is "cheat." That sounds like it was given because the grasses grew as weeds in fields and "cheated" the farmer by decreasing the crops through competition. That does not seem to be the case, although most English-speaking farmers will probably suggest that as a "folk etymology." The Oxford English Dictionary (OED 2007) provides this definition for cheat: "Wheaten bread of the second quality, made of flour more coarsely sifted than that used for MANCHET, the finest quality." That use of the term "cheat" was recorded in English literature by about 1450 and continued for wheat used to make loaves until at least 1861. Grains of several species of *Bromus* were used as human food in the Americas (Moerman 1998).

**Derivation of the name:** *Bromus* is the ancient Greek name for oats (*Avena sativa*), from *broma*, food. *Anomalus* means uneven or irregular, perhaps referring to the drooping branches of the flower cluster.

**Miscellaneous:** Brown Canyon has two other species of *Bromus* recorded, the native *B. carinatus* and the alien *B. rubens* (Foxtail Brome, Red Brome; *bromo rojo*). *Bromus anomalous* is distinguished by its glabrous or shortly pubescent stems and few-flowered inflorescences with drooping branches. *Bromus carinatus* is recognized by glabrous stems, open ascending to erect inflorescences, and the comparatively short awns (6–8 or rarely 11 mm). *Bromus rubens* is identified by its pubescent stems, erect, dense inflorescences, and the long awns (12–26 mm) on the florets.

A major problem caused by the Red Brome is that it forms dense colonies that promote fire in the desertscrub. Because that habitat is not adapted to burning, it may take centuries for it to recover (Esque and Schwalbe 2002). Red Brome is not yet as common in Brown Canyon as it is elsewhere in Arizona. There is a small colony near the entrance gate that was probably accidentally introduced as a weed on clothing or cars parked by hunters. Other colonies line the streambeds pretty much throughout in sunny places.

## *Cynodon dactylon* (Poaceae) Bermuda Grass, *zacate ingles, 'a'ai hihimdam waṣai*

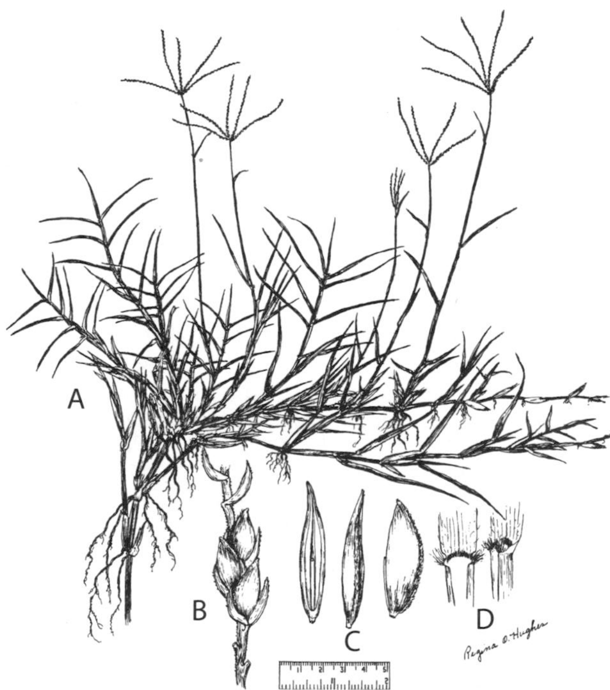


*Cynodon dactylon.* A. Habit of grass, minus the rhizomes. B. Spike. C. Florets. D. Ligule at base of leaf blade. Artist: R. O. Hughes (Reed 1971).

**Other names:** ENGLISH: Bermuda grass (in use by 1808, compared with "Cumberland grass"); SPANISH: *diente de perro* (dog's tooth), *gallito* (little rooster, Mexico), *grama* (grass, Spain), *pata de gallo* (rooster's foot, Sonora), *zacate bermuda* (bermuda grass, Sonora; from *zacatl*, grass, Náhuatl), *zacate conejo* (rabbit grass, Chihuahua), *zacata de lana* (wool grass, Mayo, Sonora), *zacate ingles* (English grass, Sonora); ATHAPASCAN: *tł'oh* (any grass, Western Apache); MAYAN: *kan-suuk* (*kan*, yellow, *sak*, white, Maya), *zarzuue* (Maya, Yucatán); UTO-AZTECAN: *'a'ai hihimdam vaṣai* (Hiá Ceḍ O'odham), *'a'ai himdam vashai* [*'a'ai hihimdam waṣai*] (grass that spreads in all directions, Akimel and Tohono O'odham), *acaxacahuitztli* <*acabacahuitztli*> (maybe from *acaxitl*, water deposit, *acaualliz*, dry weed, *itztetl*, obsidian, Náhuatl), *ki: weco vaṣai* (Hiá Ceḍ O'odham), *kii wecho vashai* [*ki: weco waṣai*] (grass around houses; used when first seen, Akimel and Tohono O'odham), *komal himdam* (spreads out flat grass, Akimel O'odham), *owiv* (grass, Ute), *váṣoi* <*vásoi*> (any grass, Northern Tepehuan), *wahái* (any grass, Northern Paiute); YUMAN: *xusí* (Cocopa)

**Botanical description:** Herbs, creeping with long stolons, scaly rhizomes, obvious internodes, often forming extensive mats. Leaves are mostly 2-ranked (distichous). Spikes are 4–7, digitate, slender, often 2–6 cm long, purplish to green. Spikelets are 1.7–2.5 mm long, awnless, numerous, crowded.

**Habitat:** Washes, *ciénegas*, disturbed sites. Up to ca. 1,830 m (6,000 ft).

**Range:** Arizona (throughout), southern California, east to Florida, and north to New Hampshire; Mexico (Chihuahua, Coahuila, Sinaloa, Sonora, Tamaulipas, and south); Central America; South America; Old World.

**Seasonality:** Flowering March to October.

**Status:** Exotic. This species was *anouphi* to the classical Egyptians (Manniche 1989) in its probable African home.

**Ecological significance:** This grass was not listed by Howard S. Gentry in 1942 in the Guarijío country of southern Sonora. Because the species was introduced from Africa, it probably had not yet arrived there. Since that time, it has garnered a number of names in Spanish and several indigenous languages. The grass remains a serious pest in agricultural and other areas.

Bermuda Grass spreads by seeds and rhizomes, and it is, therefore, almost impossible to eliminate. If the rhizomes are dug out completely, an almost hopeless task, seeds will sprout and re-establish the grasses. Several seed-eating birds consume the seeds, including sparrows and doves. Worse, for many people, the pollen is one of the most serious sources of hay fever in Arizona (Parker 1972).

According to Parker (1972), the grass cannot withstand freezing, shade, or frequent cultivation, but it can tolerate indefinite periods of drought. In spite of that statement, it persists in my yard, where it is frozen and dug out several times a year.

Farther west, in the Cabeza Prieta National Wildlife Refuge and the Organ Pipe Cactus National Monument, the grasses grow "near springs, alkaline seeps, tinajas, and other waterholes and canyon bottoms, where water may accumulate at least seasonally, in natural and disturbed habitats. Often at waterholes artificially 'enhanced' or 'improved' for wildlife" (Felger et al. in prep.). Bermuda Grass persists long after disturbances have been removed, as it does in patches in Brown Canyon. One of those spots is near the uppermost windmill; another is in oak woodlands higher up the canyon.

**Human uses:** Mayo used Bermuda Grass to relieve kidney pain (Yetman and Van Devender 2001). Maya use it to treat urinary problems and bad blood (Martínez 1969). The O'odham names reflect their recognition of this alien species as it appeared in their territory (Rea 1997).

Spencer (1957) maintained that dairymen of the South know that 10 or 12 acres of Bermuda Grass "are the equivalent of five or six times that acreage of Bluegrass [*Poa* spp.] or Red Top [*Agrostis gigantea*] or Timothy [*Phleum pratense*]." He recommended that farmers leave the grass in place and not buy hay for their animals. Unlikely as it seems, people in the province of Madrid, Spain, have long eaten the rhizomes raw as a snack (Tardío et al. 2005).

**Derivation of the name:** *Cynodon* is Greek for dog's tooth, the florets resembling them to Linnaeus. *Dactylon* means like a finger, referring to the flowering branches.

**Miscellaneous:** *Cynodon*, with its 17 species, is generally conceded to be native to Africa and other parts of the Old World (Mabberley 1997, Soreng et al. 2007). Several species in the African savanna have been there so long that plants have evolved in response to grazing by various species of antelope, particularly Gnu (*Connochaetes*).

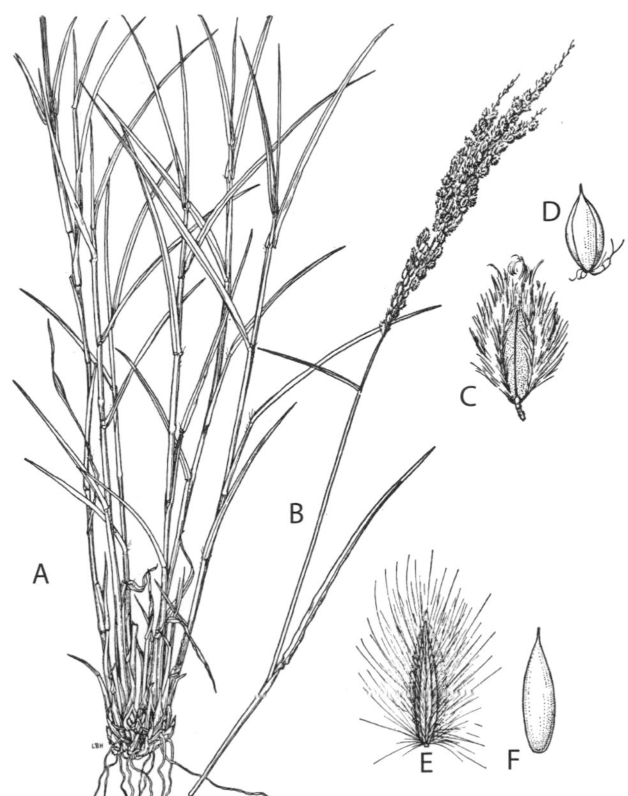


*Digitaria californica*. A. Base of plant. B. Flower cluster. C. Spikelet. D. Grain. *Digitaria insularis*. E. Spikelet. F. Grain. Artist: L. B. Hamilton (Gould 1951).

### *Digitaria californica* (Poaceae) Cotton-Top, *zacate punta blanca, waṣai*

**Other names:** ENGLISH: [Arizona, California] cotton-top (a reference to the white fruit cluster); SPANISH: *plumero blanco* (white feather duster), *zacate punta blanca* (white top grass, Chihuahua, Sonora); ATHAPASCAN: *tł'oh* (any grass, Western Apache); UTO-AZTECAN: *wahái* (any grass, Northern Paiute), *waṣai* (any grass, Tohono O'odham); (= *Trichachne californica*)

**Botanical description:** Perennials with hard, knotty bases. Stems are 50–90 cm tall, erect, with felt-like tricomes. Blades of largest leaves are 12–17 cm long, 4–6.5 mm wide. Panicles are 8–17 cm long, narrow, the branches alternate (racemose), appressed, densely flowered. One spikelet of the pair is long-pedicelled, the other shorter. Spikelets are 3–3.7 mm excluding the hairs. Lower glume is minute, membranous, glabrous, the upper glume slightly narrower and shorter than the sterile lemma. Upper glume and sterile lemma each have 3 major veins in addition to the marginal ones, with dense silky white or pale purplish marginal hairs overtopping the spikelet.

**Habitat:** Grasslands, rocky hills. 300–1,830 m (1,000–6,000 ft).

**Range:** Arizona (Coconino, Mohave, southern Navajo, south to Cochise, Pima, Santa Cruz, and

Yuma counties), southeastern Colorado, New Mexico, and Texas; Mexico (Baja California and Sonora to Puebla); South America; West Indies.

**Seasonality:** Flowering August to October.

**Status:** Native.

**Ecological significance:** This plant is the most widespread species in *Digitaria* section *Trichachne*, which comprises ca. 12 New World species. In the Altar Valley desert grasslands, this species is common. Plants thrive in any microsite that has enough moisture to support them, although they are typically scattered individual clump grasses and do not form dense colonies.

Felger et al. (in prep.) record *D. californica* in Alamo Canyon (Ajo Mountains) at 8,590 to 9,570 ybp, in Montezuma's Head dated 13,500 to 20,490 ybp, and in the Puerto Blanco Mountains at 7,560 to 7,970 ybp.

**Human uses:** Cotton-Top may be *toota mudhadkam* of the Akimel O'odham, but Rea (1997) was not able to find consultants who could identify the grass.

The grasses are considered "palatable forage" for livestock during a short period following spring or summer rains (Kearney and Peebles 1951).

The common name "witch-grass" for another *Digitaria* (see below) and for *Panicum hirticaule* in the lowlands is intriguing. "Witch" came into English from Old English *wicce*, which corresponds to *wicca*. By about 1000, the words meant a man who practiced witchcraft or magic; a magician, sorcerer, or wizard; in use about the same time for a female practitioner. Both *wicce* and *wicca* are related to *wiccian*, which corresponds to Middle and Low German *wikken*, *wicken*, of obscure origin. In the senses arising in Middle English and later, the words were probably shortened from *bewitch*. Either origin suggests magic.

The name "witch-grass" was recorded in 1840 when Jesse Buel wrote in the *Farmer's Companion*: "The quack, switch, or witch grass, a variety of the fiorin [a corruption of Irish *fiorthán*, an *Agrostis*], is highly nutritious, roots and all" [presumably for livestock]. Buel also wrote "switch-grass" for the first time, thereby comparing it to a riding whip or flexible branch used to beat something (or someone). "Quack" grass came earlier, being used by 1822 for a species of *Agropyron*. This word dates back to "quitch," which is from Old English *cwice* (akin to Middle Low German *kweke*; hence, German *quecke*, Dutch *kweek*). That word is thought to be related to *cwic* or "quick," in reference to its vitality. "Quick" dates to 1637 and refers to yet another genus of grasses. The impression is that all of these terms were originally for all grasses.

So, were any of these grasses used by witches? The question is unanswered by the etymology, but the answer is probably not. The various names are probably all corruptions and misunderstandings of "switch" because of the "beating" that the cultivated lands took from them. However, it is possible that the connotation came from flagellation that witches might have practiced.

**Derivation of the name:** *Digitaria* is from Latin *digitus*, a finger, referring to the finger-like branches of the spikes. *Californica* means from California, in this case meaning Baja California, as the species does not grow in the state of California.

**Miscellaneous:** Two other species are common, *D. cogata* (fall witch-grass) and *D. insularis* (cotton grass, *camalote*). Fall Witch-Grass is recognizable by its felty pubescence near the base of the stems. *Digitaria insularis* differs from *D. californica* by having tawny hairs on the spikelets rather than those that are white or purplish.

## *Elymus elymoides* (Poaceae) Squirrel-Tail, *zacate ladera, waṣai*

**Other names:** ENGLISH: [bottlebrush] squirrel-tail (originally applied to *Hordeum* by 1777); SPANISH: *triguillo desértico* (little desert wheat, Mexico), *zacate cebadilla* [*sevaidilla*] (little nourishing grass, Mexico), *zacate ladera* (slope grass, Sonora); ATHAPASCAN: *'zéé'iilwoii* <*ˀazeˀ i.l "oˀi*> (*'zéé'iilwoii*, one that runs into the, *'azéé'*, mouth, Navajo), *tł'oh* (any grass, Western Apache); UTO-AZTECAN: *$k^{w}$a-siyavɨ* (tail, Kwaiisu; for *E. multisetus*), *mono'pü* (Paiute), *o'ro* [*o'do, o'ro, o'rorop*], *ti'wabinĭp* (Shoshoni; for *E. canadensis*), *odorûmbĭv* (Ute), *pahankis* (Cahuilla), *pesru* <*pésru*> (*pe*, kangaroo rat, *sru*, tail, Hopi), *waiya* (Northern Paiute; for *E. condensatus*), *waṣai* (any grass, Tohono O'odham); (= *Sitanion hystrix*)

**Botanical description:** Culms are somewhat clumped, 15–50 cm tall. Leaves are glabrous or soft-pubescent, the blades 2–5 mm broad. Spikes are 5–10 cm long, bristly with long, spreading awns, densely flowered, often partially included in the upper sheaths. Rachises are fragmenting when dry, with internodes 4–8 mm long and spikelets regularly 2 at the nodes. Spikelets are mostly 2–4 flowered, disarticulating above the glumes and between the florets. Glumes are

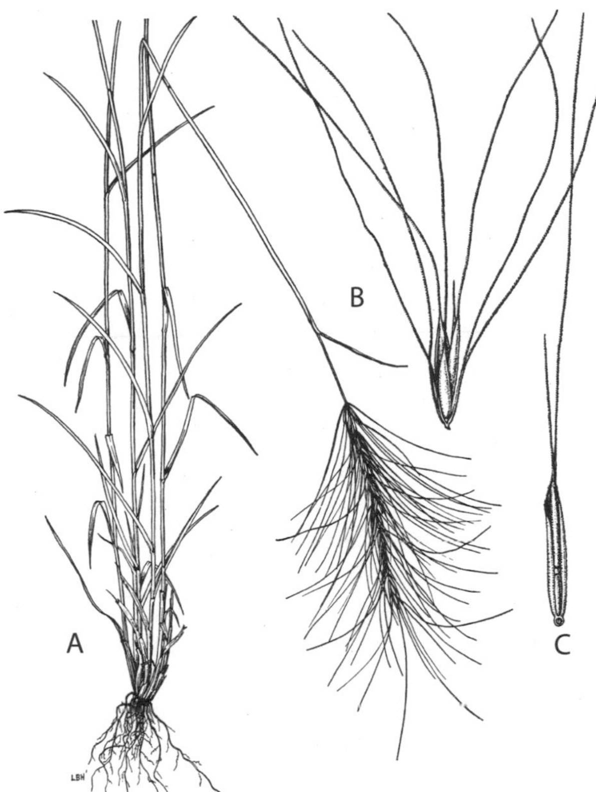


*Elymus elymoides*. A. Flowering plant. B. Spikelet. C. Floret. Artist: L. B. Hamilton (Gould 1951).

awn-like to the base, the awn stout, divergent, scabrous, 1.5–9 cm long.

**Habitat:** Rocky hillsides. 600–3,500 m (2,000–11,500 ft).

**Range:** Arizona (Apache to Mohave, south to Cochise, Pima, and Santa Cruz counties), California, New Mexico, Texas, east to Missouri, and north to South Dakota; British Columbia; Mexico (Chihuahua and Sonora).

**Seasonality:** Flowering March to September.

**Status:** Native.

**Ecological significance:** The bottle-brush-like fruiting clusters of Squirrel-Tail have a wonderful secondary (primary?) way of being dispersed. When the wind blows in open areas, fruit-heads roll across the ground like small tumbleweeds. Otherwise, the seeds are dispersed by adhering to passing animals, including humans. When the sharp awns penetrate the flesh, they can cause painful wounds that sometimes become inflamed in domestic stock and small mammals.

Squirrel-Tail was long considered to belong to the distinct genus *Sitanion*. That group was thought to be related to *Elymus* (wild-rye) and *Agropyron* (wheatgrass), and *Sitanion* was sometimes considered only a subdivision (section) of *Elymus*. Natural and artificial hybrids between those genera and *Hordeum* (barley) were studied by many (e.g., Stebbins et al. 1946, Dewey 1971) to understand their phylogeny and evolution (Wilson 1963). It turns out that part of the Squirrel-Tail genome came from *Hordeum* (Jones 1998, Jones and Larson 2005).

**Human uses:** Navajo use the young Squirrel-Tail plants as fodder for domestic animals (Moerman 1998). Seeds are eaten by the Acoma, Laguna, Shoshoni, and southern Paiute (Chamberlin 1911, Hodgson 2001, Kelly 1932). Without naming individual tribes, Ebeling (1986) lists Squirrel-Tail among the grasses that provide food for people in Arizona, New Mexico, southwestern Colorado, and the San Juan drainage area of Utah. This grass is a "cool-season" species that grows during the winter and spring rainy periods.

These grasses mature in early summer, when stored food supplies historically reached their lowest levels. Seeds harvested from four species of *Elymus*, in addition to those that come from *Agropyron repens*, *Agrostis* spp., *Festuca octoflora*, *Hordeum pusillum*, *Koeleria pyramidata*, *Oryzopsis hymenoides*, *Phalaris caroliniana*, *Poa* spp., and *Stipa speciosa*, provided the people in the Southwest with a nourishing food source (Bohrer 1975a). Moreover, when these grasses were at their peak of young growth, deer, Pronghorn, and Bison grazed them (Ebeling 1986). Those animals would have been easier to hunt when concentrated on these foods.

It is now known that *Hordeum pusillum* (Little Barley) and *Phalaris caroliniana* (May-Grass, *baabkam* <*papkam*> to Akimel O'odham) were domesticated 400–1450 years ago, the former by the Hohokam of Arizona (Austin 2004). *Hordeum* was in the Spiro site of Oklahoma, and it was cultivated in Illinois. Little Barley may be one of the dominant "seeds" (caryopses) in Late Woodland or Mississippian archaeological sites, and it dates to 2550–2770 ± 70 BP (600–800 BCE). More recently, Fritz (2000) has concluded it was domesticated by 2000 BP. May-Grass grains are dominant in Middle and Late Woodland archaeological sites (Austin 2004). The grains have been associated with people in Illinois, dating to 1700 ± 70 BP at one site and to 1400 ± 70 BP at another location. May-Grass was already an important cultivated food in Kentucky by 1000 BCE; Fritz (2000) also concluded that it was a domesticated crop by 3000 BP.

Both Little Barley and May-Grass were among several highly nutritious native foods (e.g., *Chenopodium*, *Helianthus*) in the Eastern Agricultural Complex that were abandoned when Maize became the primary American food. The generic name *Sitanion* (from Greek *sitos*, grain, *setaneios*, of this year or spring wheat; Latin *sitanius*, of this year, summer wheat), given by Rafinesque in 1819 to what we now call *Elymus elymoides*, also makes me wonder. Could these grasses also have been harvested in the eastern Great Plains when Rafinesque was there?

**Derivation of the name:** *Elymus* comes from Greek *elymos*, Millet (*Setaria italica*), probably based on *elyo*, to cover, as in fields. *Elymoides* means resembling Millet.

## *Eragrostis intermedia* (Poaceae) Plains Love-Grass, *zacate volador, waṣai*

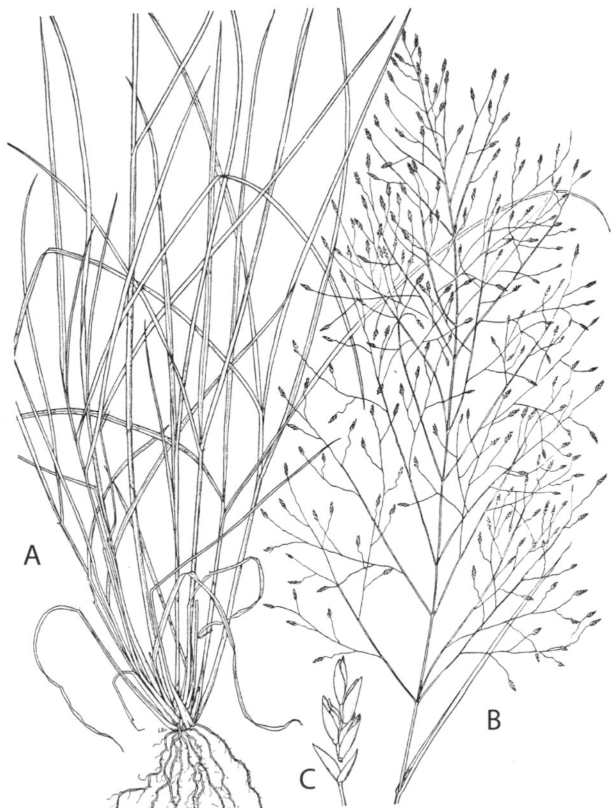


*Eragrostis intermedia*. A. Base of plant. B. Flowering apex of plant. C. Spikelet. Artist: L. B. Hamilton (Gould 1951).

**Other names:** ENGLISH: love-grass (a translation of *Eragrostis*; in use by 1702, when London apothecary James Petiver wrote: "What is peculiar in this Love-grass is its having just under each spike, its stalk clammy"); SPANISH: *zacate llanero* (prairie grass, Sonora), *zacate volador* (flying grass, Arizona, Sonora); ATHAPASCAN: *tł'oh* (any grass, Navajo), *tł'oh* (any grass, Western Apache); UTO-AZTECAN: *mṓnó* (Northern Paiute; for *E. secundiflora*), *waṣai* (any grass, Tohono O'odham); YUMAN: *kšam* <*košom, kwšam*> (Cocopa; reported as *E. mexicana*, but Felger 2000 thought it was *E. pectinacea*)

**Botanical description:** Tufted erect or ascending perennial, 25–90 cm tall. Leaf sheaths are glabrous except along the apex and the often pilose margins. Ligule is a dense row of white hairs ca. 0.4 mm long. Panicles are open, broadly pyramid-shaped, 20–40 cm long, 15–30 cm wide. Spikelets are 5–11 flowered, on peduncles longer than the spikelets. Caryopsis is oblong, striate, flattened, ca. 0.8 mm long.

**Habitat:** Sandy or rocky slopes and plains. 900–1,530 m (3,000–5,000 ft).

**Range:** Arizona (Coconino, Yavapai, south to Cochise, Gila, Maricopa, Pima, and Santa Cruz counties), east to Oklahoma, and Georgia; Mexico (Chihuahua and Sonora); to Central America.

**Seasonality:** Flowering June to September (October).

**Status:** Native.

**Ecological significance:** This species was historically the dominant native *Eragrostis* in the region. Plants are still widely scattered but fairly abundant in the canyons only where livestock is excluded. The alien African *E. lehmanniana* is abundant only in the lower parts of these canyons. That exotic grass was planted before 1989 on what is now the Buenos Aires National Wildlife Refuge when it was still a cattle ranch. In fact, *E. lehmanniana* and *E. curvula* (see below) were introduced into the region in the 1930s for "revegetation" on rangelands that were recovering from past episodes of widespread overgrazing by cattle. These introductions followed the philosophy that "if a green grass survives, it is useful" (Bock and Bock 2002). No thought was given to the impact of these alien species on indigenous organisms, and the Masked Bobwhite Quail (*Colinus virginianus ridgewayi*) and probably the Pronghorn (*Antilocapra americana*) in the region have suffered along with hundreds of kinds of plants.

Because Lehmann Love-Grass was "preadapted" to the climate of the region, which was similar to that of its native lands in southern Africa, because of its copious seed production, and because of its positive response to burning, it is widespread and dominant throughout the border region from the Sonoita Valley to the Altar Valley and beyond (Bock and Bock 2002). Following introduction, the African grass has almost excluded the native species in the Altar Valley. *Eragrostis lehmanniana* was growing in almost all of over 100 lowland quadrants (10 × 10 m) examined on the Buenos Aires National Wildlife Refuge in 2002. The native *E. intermedia* was growing in only a small percentage of those quadrants. In Brown and Jaguar Canyons, two other exotics, *E. cilianensis* and *E. curvula*, were present in all of the half-dozen plots examined, but *E. intermedia* remained more frequent than either of the alien species.

**Human uses:** Several species of Love-Grass are used as fodder. Plants on ranches near the Buenos Aires National Wildlife Refuge where *E. intermedia* persists are close-cropped, while *E. lehmanniana* is hardly touched by livestock. Johnson-G. and Carillo-M. (1977) said that, because *E. intermedia* is so palatable to livestock, it is one of the first grasses to be overgrazed.

**Derivation of the name:** *Eragrostis* is from *eros*, love, and *agrostis*, a field or country. *Intermedia* means intermediate, apparently between two other species.

*Eragrostis* was named in 1776 by Nathanael Matthaeus von Wolf (1724–1784). Charles L. Hitchcock (1902–1986) described *E. intermedia* in 1933, and the genus now contains 110 temperate and tropical plants (Peterson et al. 2001). Two of the most famous members of the genus are *E. tef* (*t'ef* or Teff), an important food plant in Ethiopia, and *E. pilosa*, also producing edible seeds, but best known as being a part of the bricks used to build Egyptian pyramids in the third millennium BCE (de Wet 1977).

**Miscellaneous:** There are three exotic *Eragrostis* species in the canyons, *E. cilianensis* (Stink-Grass, *zacate arestoso*), *E. curvula* (Boer's Love-Grass, Weeping Love-Grass), and *E. lehmanniana* (Lehmann Love-Grass). Lehmann Love-Grass is most likely to be confused with Plains Love-Grass, but it has peduncles shorter than its spikelets. The other two species have leaf sheaths densely villous on the back. Stink-Grass is most easily recognized by its sticky surface. Weeping Love-Grass has leaf blades 20–30 cm long, greater than those of the other three species.

## *Heteropogon contortus* (Poaceae) Tangle-Head, *zacate colorado*, *bihag waṣai*

*Heteropogon contortus*. A. Plant with immature florets. B. Branch with maturing florets. C. Spikelet. Artists: L. B. Hamilton, A, C (Gould 1951); E. M. Whitehorn, B (Hitchcock and Chase 1950).

**Other names:** ENGLISH: needle-grass (referring to the needle-like awns, New Mexico), tangle-head ("tangle" or encumber and hamper, in use by ca. 1340; the name alludes to the twisted awns of the fruits); SPANISH: *barba negra* (black beard, Mexico), *carrizo* (from Latin *carex*, a sedge, Sonora; also used for *Arundo* and *Phragmites*), *hierba negra de los prados* (black herb of the prairies, Mexico), *rabo de asno* (donkey's tail, Mexico), *retorcido moreno* (dark twisted, Mexico), *zacate colorado* (red grass, Arizona, Chihuahua, Sonora), *zacate retorcido* (twisted grass, Mexico), *zacate aceitillo* (oily grass, Chihuahua, Sonora); ATHAPASCAN: *tł'oh* (any grass, Navajo), *tł'oh* (any grass, Western Apache); UTO-AZTECAN:

*bihag waṣai* (wrap-around grass, Tohono O'odham), *biibhinol vashai* (wrap-around grass, Akimel O'odham, Arizona), *ujchú* (Guarijío), *wahái* (any grass, Northern Paiute)

**Botanical description:** Perennial, tufted grasses. Culms are 2–8 dm tall, branched above, the branches erect. Sheaths are smooth, compressed-keeled. Leaf blades are flat or folded, 3–7 mm wide. Racemes are 4–7 cm long, 1-sided. Sessile spikelets are ca. 7 mm long, slender, nearly hidden by the imbricate pedicellate spikelets, the awns 5–12 cm long, bent and flexuous, commonly tangled, at least at maturity. Pedicellate spikelets are ca. 1 cm long.

**Habitat:** Grasslands, canyons, slopes. 300–1,670 m (1,000–5,500 ft).

**Range:** Arizona (Mohave and Yavapai to Cochise, Pima, Pinal, Santa Cruz, and Yuma counties), New Mexico, Texas, and east to Florida; Mexico (Chihuahua and Sonora).

**Seasonality:** Flowering (May) August to October.

**Status:** Native.

**Ecological significance:** These grasses thrive on disturbance. In wild areas not disturbed by humans or livestock, or at least not recently, plants are confined to rockslides and waterway margins where the soils have been turned over. Around humans, the grasses are most common on trails, along roadsides, and in similar places.

**Human uses:** Guarijío in Sonora use Tangle-Head for roof construction when palm leaves are not available (Yetman and Felger 2002). Farther south, Mayan people similarly use the grass for thatch (Berlin et al. 1974, Lentz et al. 1996). Guarijío put a layer of *ujchú* in thick mats on the *latas* (small wooden pieces on the *vigas* or beams) and then cover it with dirt (Yetman and Van Devender 2001). Guarijío also stuff the grass into *angarillas* (woven rigid saddlebags) to transport it home for use in construction.

Watson and Dallwitz (1992) list *H. contortus* as both a significant weed species and an important native pasture plant. The utilization as food for livestock is confined to the period when Tangle-Head is young, tender, and without fruits (Johnson-G. and Carillo-M. 1977). Otherwise, the sharp fruiting parts cause damage to animals. Furthermore, Gould (1951) noted that the grasses have low palatability except during periods of vegetative growth, but that they are accepted by cattle when better forage is not available.

Tangle-Head is sometimes cultivated for ornament, although it is not listed in the *Sunset Western Garden Book* (Brenzel 1995). *Heteropogon* does provide a striking aisle of green texture during the rainy season, and one of browns contrasting with other tones during the fall and early winter. Those who grow Tangle-Head either do not have dogs, or have never spent hours extracting the awns and needle-sharp seeds from a dog's feet and fur. Indeed, unless caught early, these armed structures can cause nasty and painful sores in household pets and livestock. People in New Mexico told me that *Heteropogon* is called "needle grass," although none of the books I have seen use that name; instead, "needle grass" is confined to *Stipa*. Both have similarly awned seeds, and the use surely is more widespread than is evident from the literature.

**Derivation of the name:** *Heteropogon* is from Greek *hetero*, different, *pogon*, beard, because some spikelets have awns while others do not. *Contortus* alludes to the twisted awns that wrap around each other while still on the grasses. The English, Latin, and O'odham names all refer to the twisting (Rea 1997).

**Miscellaneous:** The other species in the canyons, although rare there, is *H. melanocarpus*, the Sweet Tangle-Head. That grass is distinguished by being an annual, and by having conspicuous glands on the inflorescences. Sweet Tangle-Head grows in Cochise, Pima, and Santa Cruz counties; east to Georgia, Alabama, and Florida; and south into the tropics. Apparently, the sweet odor contributes to this plant being unpalatable to cattle, and they will not eat it (Gould 1951).

## *Leptochloa dubia* (Poaceae) Green Sprangle-Top, *zacate gigante, waṣai*

**Other names:** ENGLISH: green sprangle-top ("sprangle" of obscure origin, to sprawl, in use by 1390 as "sprantle," by the early 1400s written "spranglynge" [sprangling]), Texas crow-foot (applied to buttercups [*Ranunculus*] with leaves resembling the feet of *Corvus* by ca. 1440; later extended to other plants); SPANISH: *desparramado dubia* (dubious sprangletop, Mexico; translation of scientific name), *hierba del hilo* (thread herb, Mexico), *pasto gigante* (giant grass, Mexico), *zacate gigante* (giant grass, Arizona, Chihuahua, Sonora); ATHAPASCAN: *tł'oh* (any grass, Navajo), *tł'oh* (any grass, Western Apache); UTO-AZTECAN: *wahái* (any grass, Northern Paiute), *waṣai* (any grass, Tohono O'odham); YUMAN: *kupo* (Mohave, reported as *L. viscida*); (= *Diplachne dubia*)

*Leptochloa dubia*. A. Base of plant. B. Apex of plant with flower cluster. C. Spikelet. Artist: L. B. Hamilton (Gould 1951).

**Botanical description:** Tufted perennials, culms 25–115 cm tall, erect, leafy, unbranched. Ligules are a dense row of eyelash-like hairs (cilia) 0.3–1.2 mm long. Blades are 5–30 cm long, 1–10 mm wide, flat or involute, the sheaths keeled and corners pilose. Panicle is 5–20 cm long, the racemes 3–13, widely spaced, ascending, 2–15 cm long, 3–6 mm thick. Spikelets are overlapping, 5–10 mm long, 3–9 flowered, the glumes 3–5 mm long, acute, the lemmas oblong, emarginate, sometimes mucronate, glabrous except the margins of the lower half.

**Habitat:** Ridges, slopes, grasslands to canyons. 760–1,800 m (2,500–6,000 ft).

**Range:** Arizona (Yavapai to Greenlee, south to Cochise, Maricopa, Pima, Pinal, and Santa Cruz counties), New Mexico, Oklahoma, Texas, and disjunct to Florida; Mexico (Baja California, Chihuahua, and Sonora, south); disjunct to South America.

**Seasonality:** Flowering July to October.

**Status:** Native.

**Ecological significance:** Green Sprangle-Top thrives on rocky, semiarid mountains. Plants are rarely in large colonies in the canyons, but are scattered through other native grassland species. Sprangle-Top is, like several others in the region, a tropical C4 grass that is on the northern limit of its range in the southern United States. Unlike many, this one extended up through Mexico into the border states, but not east into most of Florida. In Florida, the species is confined to the southern counties.

These grasses, like almost all of the family, are adapted for wind pollination. In spite of that, this species produces whole branches of flowers that never open (cleistogamous), instead remaining closed but still producing fruit (Hitchcock and Chase 1950).

**Human uses:** Gould (1951) considers Green Sprangle-Top a range grass "of considerable value" and notes that it was sometimes cut for hay in some regions. Leithead et al. (1971) add that it was readily grazed by all livestock, especially when green and succulent. Even during the dormant season, it provides good-quality forage. Leithead et al. (1971) consider its regions of importance to livestock to be central southern Florida and the central grasslands of Texas. Similarly, Johnson-G. and Carillo-M. (1977) compliment the grass in Sonora and say that it is a good pasture grass with high nutrition. Several of the species in *Leptochloa* are considered valuable fodder grasses.

**Derivation of the name:** *Leptochloa* is from Greek *lepto*, slender, *chloa*, grass, from the slender inflorescence branches. *Dubia* means doubtful. Apparently, Carl S. Kunth, who named the species in 1816, was uncertain what to do with it. As it was, Kunth put it in *Chloris*, the wrong genus. It was changed to *Leptochloa* by C. G. D. Nees von Esenbeck in 1824.

**Miscellaneous:** The genus *Leptochloa* comprises 17 species that are widely distributed through the Americas, and which also grow in Australia (Peterson et al. 2001).

The Red Sprangle-Top (*Leptochloa panicea* ssp. *brachiata*, formerly *L. mucronata* or *L. filiformis*), also known as *zacate salado* (salt grass) and *desparramo rojo* (red sprangle-top), may be locally more common. This species has shorter spikelets (2–3 mm long vs. 4–10 mm) and only 2–4 flowers per spikelet (vs. 4 to many flowers in *L. dubia*). Although common names are not always reliable, *L. panicea* has reddish flower and fruit clusters, while *L. dubia* is mostly green. Red Sprangle-Top, in its many forms, is spread across the southern states from California to Florida (Hickman 1993, Wunderlin and Hansen 2004). Similarly, the grasses extend into Mexico in Baja California,

Sonora, and south to Argentina and Peru (Wiggins 1980, Felger 2000, Peterson et al. 2001). Red Sprangle-Top is not considered good for forage; it is weedy (Whitson 1992).

### *Muhlenbergia dumosa* (Poaceae) Bamboo Muhly, *otatillo, waṣai*

*Muhlenbergia dumosa*. A. Flowering upper branches. B. Base of plant. C. Glumes. D. Floret. Artist: E. M. Whitehorn, A–B; A. Chase, C–D (Hitchcock and Chase 1950).

**Other names:** ENGLISH: bamboo muhly ("bamboo" perhaps from Malay, Sudanese, and Javanese *bambu*; maybe from Canarese *banbu* or *banwu*; appearing in English by 1586; "muhly" a book name based on the genus); SPANISH: *carricillo* (little sedge, Sonora), *otatillo* (little cane, from Náhuatl *otatli*, cane; a name also given to some climbing grasses, genus *Lasiacis*, Chihuahua, Sonora); ATHAPASCAN: *tło* (any grass, Chiricahua and Mescalero Apache), *tł'oh* (any grass, Navajo); UTO-AZTECAN: *áli tótoikami* (*áli*, little, *tótoikami*, reeds, Northern Tepehuan), *totchkam* <*totčkam*> (based on *totkam* [pl.], *tookam* [sing.], bundle of reeds, Mountain Pima, Sonora), *saawi* (any grass, Yaqui), *waṣai* (any grass, Tohono O'odham)

**Botanical description:** Perennial, with short, stout, creeping scaly rhizomes. Stems (culms) are robust, solid, thick, scaly at base, to 6 mm thick, the main stem erect or leaning, 1–3 m tall, the lower part with bladeless sheaths, branching at the middle and upper nodes, the branches numerous, bundled, spreading, twice-compound, the ultimate branchlets filiform. Leaf blades are flat or rolled inward (involute), smooth, those of the branches mostly less than 5 cm long and 1 mm wide. Flowering clusters (panicles) are numerous on the branches, often longer than the leaves, 1–3 cm long, narrow, somewhat zigzagged (flexuous). Spikelets, excluding the awn, are about 3 mm long, the glumes less than half as long, pale, with a green midnerve, usually minutely awn-tipped or the awn to 9 mm long.

**Habitat:** Rocky canyon slopes and valleys. ca. 1,200–1,490 m (ca. 4,000–4,900 ft). Down to 450 m (1,500 ft) in Sonora.

**Range:** Arizona (Gila, Maricopa, Pima, Pinal, Santa Cruz, and Yuma counties); Mexico (Baja California Sur, Chihuahua, Durango, Jalisco, Michoacán, Sinaloa, and Sonora); disjunct to Argentina.

**Seasonality:** Flowering March to May.

**Status:** Native.

**Ecological significance:** Bamboo Muhly was originally described from specimens collected by C. G. Pringle in 1884 in the Santa Catalina Mountains. West of the Baboquivari Mountains, these bamboo-like grasses are known in the Barry M. Goldwater Air Force Range and in Organ Pipe Cactus National Monument (Felger et al. in prep.). Although Kearney and Peebles (1951) seem to indicate that *Muhlenbergia dumosa* grows at low elevations, most of the records in Arizona are from mountain canyons.

In Brown Canyon, Bamboo Muhly is most obvious near the region of the arch. There, the grass spreads up the slope on the northern side to provide an eye-catching display. Unless one is familiar with the flora of the region, this species may be mistaken for "real" bamboo that has escaped and become naturalized in the site. Instead of being naturalized aliens, these are native North American grasses nearing the northern fringe of their range. There are several places in Jaguar Canyon where the grasses climb the slopes. The species also grows in several canyons on the western Baboquivari slopes and in the Coyote Mountains.

Fruits are sometimes infected with a smut fungus, *Ustilago muhlenbergiae*; Felger et al. (in prep.) reported this smut on *M. dumosa*. The only other smut reported for this grass is *U. sonoriana*, on plants from near Bavispe, Sonora. *Ustilago muhlenbergiae* is also known in southern Arizona from *M. pauciflora*, *M. porteri*, and *M. texana*.

**Human uses:** Northern Tepehuan brew Bamboo Muhly roots into a tea that is drunk to relieve stomach cramps (Pennington 1969). Mountain Pima make a tea to treat heart-related pains, premenstrual cramps, and pneumonia (Pennington 1973, Reina-G. 1993).

The only other use found for the species is as an ornamental. Gould (1951) records that Bamboo Muhly is being cultivated "to a limited extent" in Arizona and California. Brenzel (1995) adds that it is a "[s]plendid container plant," and she was of the opinion that it was an "odd but striking grass." People in the region of cultivation who may be inhibited from planting the overwhelming real bamboos might instead consider using this "well-behaved" native. In cultivation, plants are usually 1 m tall, but sometimes they reach 2 m

**Derivation of the name:** *Muhlenbergia* was named by J. C. D. von Schreber in honor of Gotthilf Heinrich [Henry] Muhlenberg, 1753–1815, a German-American Lutheran minister and amateur botanist who lived in Pennsylvania. *Dumosa* means bushy, shrubby.

**Miscellaneous:** *Muhlenbergia* is a fairly large and complex grass genus consisting of about 147 species (Peterson et al. 2001). The majority of species are in the tropical and warm parts of the Americas, especially North America, but some also grow in southern Asia. Charlotte O. Reeder (neé Goodding), a research associate at the University of Arizona, is the world authority on the genus. She is acknowledged in the introduction to Hitchcock and Chase (1950) as having helped with the genus.

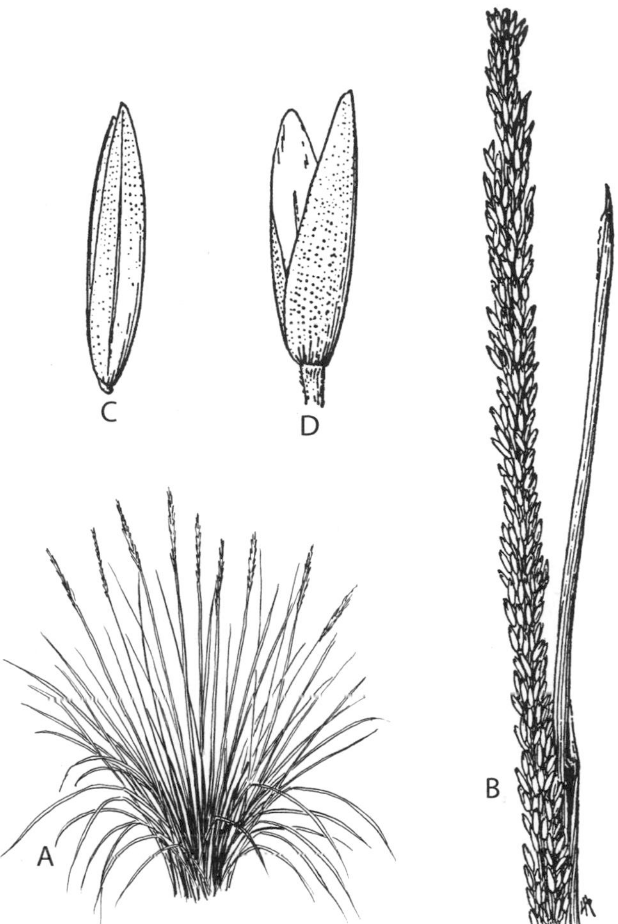


*Muhlenbergia rigens*. A. Habit. B. Flower cluster. C. Spikelet. D. Glumes. Artists: Bonnie Swarbrick, A; L. Hughey, B; A. Chase, C–D (Hitchcock and Chase 1950).

## *Muhlenbergia rigens* (Poaceae) Deer-Grass, *zacate venado, waṣai*

**Other names:** ENGLISH: deer-grass (a name applied to *Rhexia* by 1866; apparently later given to these grasses, Arizona); SPANISH: *escobón* (big brush, Sonora), *hierba del paisano* (country-man's herb, Sonora), *zacate venado* (deer grass, Sonora); ATHAPASCAN: *tło* (any grass, Chiricahua and Mescalero Apache), *tł'oh* (any grass, Navajo); UTO-AZTECAN: *mašil* (plant, Tübatulabal), *monopi* [*monope, mónop*] (Mono), *nor* <*norr*> <*nol*> (to turn [leaves], Mountain Pima), *sipu(m)bivɨ* [*šipu(m)bavɨ*] (Kawaiisu), *suul* (Cahuilla), *waṣai* (any grass, Tohono O'odham); LANGUAGE ISOLATE: *pi'shu li'awe* (*pi'shu*, come up quick, *li'awe*, tall, Zuni)

**Botanical description:** Tufted perennials, stems (culms) 10–15 dm long, 1.5–4 mm thick, erect, unbranched. Ligules are a firm blunt scale 1–2 mm long. Leaf blades are firm, 2–5 dm long, 3–4 mm wide, flat at the base, but mostly infolded (involute) above, nearly straight. Panicle is 2–6 dm tall, 4–13 mm wide, dense and spikelike. Glumes are 2–3 mm long, mostly mucronate or with a small awn.

**Habitat:** Washes, canyons near streams. 900–2,290 m (3,000–7,500 ft).

**Range:** Arizona (Apache, Coconino, south to Cochise, Pima, and Santa Cruz counties), California,

New Mexico, and Texas; Mexico (Baja California, Chihuahua, and Sonora).

**Seasonality:** Flowering September to October.

**Status:** Native.

**Ecological significance:** While these bunch grasses grow in California's valley grasslands, riparian areas, and meadows (Hickman 1993, Munz and Keck 1973), Deer-Grass is found only in canyons along streams in the Baboquivari Mountains. The species requires substantial amounts of water in comparison with some of the other grasses in the region, and it does not appear to be able to survive outside water courses. While the streams are only seasonal, that is sufficient to maintain the clumps. The grass is also sun-loving (heliophilic), and it presumably was not maintained by human-created fires in Arizona and elsewhere, as it was in California (Anderson 1996).

**Human uses:** Although at least 20 tribes in southern California use Deer-Grass in weaving baskets (Anderson 1997), there appear to be no records of it having been used that way in other states. This grass provides the most significant basketry material in California, and it is known as part of the flora of 15 of the southern California counties. By contrast, there are no records of the grass being used by other skilled and famous weaving tribes, such as the Akimel and Tohono O'odham, Guarajío, Mayo, or Seri (Felger and Moser 1985, Rea 1997, Yetman and Felger 2002, Yetman and Van Devender 2001). All use a coiling technique, and the reason for the difference is unclear (M. K. Anderson, personal communication, July 2007; J. Timbrook, personal communication, July 2008). Although the Tübatulabal told Voegelin (1938) that they had no distinct name for the grass, they did use Deer-Grass in making baskets.

Members of the Zuni Galaxy and *Shu'maakwe* fraternities (clowns or "delight makers") attach a piece of this grass to their prayer sticks (Stevenson 1915). Mountain Pima use the grass in teas to treat stomachaches and urinary problems (Reina-G. 1993). Apaches formerly ate the seeds (Reagan 1929).

**Derivation of the name:** *Muhlenbergia*, see *Muhlenbergia dumosa* for etymology. *Rigens* means stiff, referring to the compact, erect flower cluster.

**Miscellaneous:** At least six other *Muhlenbergia* species grow in the canyons. Most frequent is *M. emersleyi* (Bull-Grass; *zacate toro*, Sonora; *pičíraka*, Tarahumara, *cola de ratón*, Chihuahua), a clumpgrass of slopes, 0.5–1 m tall, which has characteristic nodding flower and fruit clusters. The Tarahumara cut the culms into about 10-inch segments to bundle into a brush. That brush was used to remove the corn meal from the *matáka* (*metate*) into the *pinole* container (Pennington 1963). Northern Tepehuan, on the other hand, made a broom of the culms to clean their houses (Pennington 1969).

*Muhlenbergia porteri* (Bush Muhly, Hoe Grass; *zacate aparejo*, Sonora) is locally frequent on a variety of sites, but most are well-drained ridges and some are fully exposed. Bush Muhly is a smaller spreading grass that grows to 50–100 cm tall. Tarahumara in Chihuahua called this species *bakú*, and they boiled and mashed the roots as a poultice to relieve backaches or rheumatic pains (Pennington 1963).

More difficult to distinguish are smaller *M. fragilis* (Fragile Muhly), *M. microsperma* (Little-Seed Muhly, *liendrilla chica*), and *M. monticola*. Both *M. fragilis* and *M. microsperma* are diminuitive annuals. The former is 10–30 cm tall and has lemmas essentially without awns, while *M. microsperma* is the same height but has awns 10–30 mm long. *Muhlenbergia monticola* is a perennial, 30–50 cm tall, and has densely tufted, slender culms.

## *Panicum bulbosum* (Poaceae) Bulb Panic-Grass, *panizo, waṣai*

**Other names:** ENGLISH: bulb panicum (a book name), panic grass (first written by John Gerarde in his *Herball* of 1597 as "Pannicke grass is garnished with chaffie and downie tufts"; "panic" from Latin *panicum*; "grass" from Old English *graes*, *gaers*, in use by about CE 725; akin to "green" and "grow"); SPANISH: *hierba de la flecha* (arrow herb, Mountain Pima), *panizo* (Mexico, Spain; akin to Italian *panico*, French *panic*; originally applied to *Panicum italicum* of Linnaeus, now *Setaria italica*, otherwise called Italian millet, and largely cultivated in Southern Europe; later extended to other species of the genus *Panicum*, many of which are grown in different parts of the world as cereal grains); ATHAPASCAN: *tł'oh* (any grass, Navajo), *tłołdei* (Chiricahua and Mescalero Apache); UTO-AZTECAN: *koomági yorádagi* (gray *yorádagi*, Northern Tepehuan), *nor ṣa'i* <sha'i> (*nor*, to turn [leaves], *ṣa'i*, grass, Mountain Pima, Sonora), *owiv* (grass, Ute), *waṣai* (any grass, Tohono O'odham)

**Botanical description:** Perennials from bulbous bases that are almost rhizomatous, the underground bases rounded, 1–2 cm thick. Culms are 50–120 cm tall, the base sometimes shortly decumbent but above

*Panicum bulbosum*. A. Base of plant. B. Spikelet. C. Grain. Artist: L. B. Hamilton (Gould 1951).

mostly erect, slightly compressed. Sheaths are glabrous to pilose near the summit. Ligule is membranous, ciliate. Leaf blades are 2–6 dm long, 3–17 mm wide, rough-hairy or glabrous above, glabrous below, flat. Panicle is 2–5 cm long, diffuse or open. Spikelets are 2.8–4.2 mm long, ellipsoid, faintly nerved, glabrous. First glume is about half the total length of the spikelet, usually blunt. Fertile lemma surface is wrinkled (rugose).

**Habitat:** Moist sites in canyons and open woodlands, often along banks. 1,370–2,440 m (4,500–8,000 ft).

**Range:** Arizona (Apache, Coconino, Navajo, south to Cochise, Pima, and Santa Cruz counties), New Mexico, and western Texas; Mexico (Chihuahua and Sonora south to Oaxaca).

**Seasonality:** Flowering July to October.

**Status:** Native.

**Ecological significance:** Bulb Panic-Grasses were in the Ajo Mountains in the ice age (14,500 to 32,000 ybp) (Felger et al. in prep.). Now, the species does not grow in that area or in any other part of arid southwestern Arizona west of the Baboquivari Mountains, and it has not been found in the Gran Desierto region of Sonora (Felger 2000).

*Panicum bulbosum* grows on the canyon slopes in rich detritus below trees, particularly in the oak woodland community. There, the tops are prone to leaning downhill, and the bulbous bases are often buried below several centimeters of water-retaining organic debris. There is at least one case on record of Montezuma Quail (*Cyrtonyx montezumae*) nesting below a clump of these grasses in the Huachuca Mountains (Cochise County, AZ) (Wallmo 1954). Because both the birds and grasses occur in Brown Canyon, they may use Bulb Panic-Grass there.

**Human uses:** Hopi grind Bulb Panic-Grass seeds with corn as food (Castetter 1935, Hough 1897, 1898). The other recorded use found for this grass is that the Northern Tepehuan seek out places where it grows because the roots are a favorite food for domestic pigs, *Sus* (Pennington 1969). These people themselves ate the roots of another kind of *yorádagi*, but the grass is not identified beyond its common name by Pennington (1969).

The genus *Panicum* has a rich history of different species being used by people, particularly the seeds providing food as flour or being ground for *atole*. The most famous member of the genus in the Southwest is *Panicum hirticaule* var. *millaceum* (incorrectly called *P. sonorum*, cf. Nabhan and De Wet 1984). This grass was cultivated by several indigenous groups including at least the Chemehuevi, Cocopa, Guarijío, and Yuma. The grass was called *ṣimca* *<shimcha>* (Tohono O'odham), *šmča:* *<heshmicha, šimča>* (Cocopa), and *sahuí* *<sauwi, sagui, sawi>* by the Guarijío (Felger 2000, Hodgson 2001, Nabhan 1985, Yetman and Felger 2002). This species is probably what the other Yuman speakers called *aksámta* (Mohave), *aksám* (Yuma), and *ikamacˀ* (Maricopa). Perhaps seeds of *P. bulbosum* were collected along with other foods by the prehistoric people who had villages near the mouth of Brown Canyon.

**Derivation of the name:** *Panicum* is from the classical Latin name for bread, *panis*, or Millet, *panus*; related to Akkadian *panu*, Italian, *pane*, bread. *Bulbosum* refers to the swollen, bulb-like underground base of the species.

**Miscellaneous:** Other species known in the mountains are *P. hirticaule* (not the cultivated variety), *P. hallii*, and *P. lepidulum*. *Panicum hirticaule* (Witch-Grass, see *Digitaria californica* for etymology) is annual, while the other three grasses are perennials.

Moreover, *P. hirticaule* has open, erect panicles with spikelets 2.7–4 mm long. Panicles are also open in *P. lepidulum* but the spikelets are 4–4.2 mm long. The culms of *P. lepidulum* are 50–80 cm tall, while they are 50–120 cm in *P. bulbosum*. In addition, *P. lepidulum* lacks the hardened corm-like base of *P. bulbosum*. Gould (1951) pointed out that *P. lepidulum* is similar to *P. hallei*. The curly basal leaves of *P. hallei* are usually characteristic enough to distinguish it, but when not, the glabrous culm internodes separate *P. hallei* from *P. lepidulum*, which is rough-hairy.

## ***Polypogon monspeliensis* (Poaceae)**
## **Rabbit-Foot Grass, *cola de zorro, waṣai***

*Polypogon monspeliensis*. A. Flowering plant. B. Spikelet. C. Spikelet of *D. interruptus* for comparison. Artist: L. B. Hamilton (Parker 1972).

**Other names:** ENGLISH: [rabbitfoot] beard-grass (first in the literature in *Withering's British Plants* of 1841, northeastern United States; the name is based on comparing the fruiting cluster with a "beard," from Old English *beard* [earlier **bard, *bærd*], cognate with Middle Dutch *baert*, Dutch *baard*, Old High German, modern German *Bart*, Old Norse **barr* retained only in compounds as *Langbarr* [but cognate with *bar*, "brim, edge, beak, prow," Old Teutonic **bardo-z*; and those words cognate with Old Slavic *barda*; kinship to Latin *barba* is, on phonetic grounds, doubtful]), [annual] rabbit('s)-foot grass [annual rabbitsfoot grass, rabbit'sfootgrass, rabbitfoot grass, rabbitfoot-grass] (apparently coined by A. S. Hitchcock in 1935 in the first edition of *Manual of the Grasses of the United States*), rabbitfoot polypogon (a book name); SPANISH: [*zacate*] *cola de zorro* (fox tail [grass], Sonora), *hierba de caso* (event herb, Sonora), *pata de conejo* (rabbit-foot grass, Sonora); ATHAPASCAN: *ch'ił ńdínísé* <*c'il dínesą́*> (*ch'ił*, plant, *ńdínísé*, growing prolongatively, Navajo), *dloziłgaii bitsee'* <*ƛozilgai biceˀ*> (*dloziłgaii*, white pine squirrel, *bi*, its, *'atsee'*, tail, Navajo), *'zéé'iilwoii* <*ˀazeˀ i.l "oˀi*> (runs into the mouth, Navajo); UTO-AZTECAN: *ban bahi* (*ban*, coyote, *bahi*, tail, Akimel O'odham), *pombikanan* (Tübatulabal), *ṣa'i* <*sa'e*> (also *vaṣa'i*, grass, Mountain Pima), *shelik bahi* <*sheshelik baabhai*, pl.> (*shelik*, round-tailed ground squirrel, *bahi*, tail, Akimel O'odham), *wahái̵* (any grass, Northern Paiute), *waṣai* (any grass, Tohono O'odham); YUMAN: *xṭpa nkʸšyułʸ* (*xṭpa*, white, *nkʸšyułʸ*, tail, Cocopa)

**Botanical description:** Annual with glabrous or scabrous foliage. Stems (culms) are (1–)2–6 dm tall, erect from a decumbent base, usually tufted. Leaf blades are linear, scabrous. Panicle is spike-like, dense, interrupted, 2–10(–15) cm long, pale and soft silky, bristly with yellow awns, often partly included in the uppermost sheaths. Spikelets are 2.5–3 mm long.

**Habitat:** Moist sites along streams, ditches, and seeps. Up to 2,440 m (8,000 ft).

**Range:** Arizona (Apache, Cochise, Coconino, Gila, Graham, Greenlee, La Paz, Maricopa, Mohave, Navajo, Pima, Pinal, Santa Cruz, Yavapai, and Yuma counties), New Mexico, Texas east to Georgia and Virginia, north to New Brunswick, and west to the Pacific states and Alaska; Mexico (Sonora). Also found in South America, Africa, Europe, and Asia.

**Seasonality:** Flowering April to October.

**Status:** Alien. Naturalized from the Old World.

**Ecological significance:** This Old World plant is crowding out native species that require wetlands. In the streambeds and moist parts of the canyons, this grass is sometimes the only species that may be found in June—except for sometimes being mixed with similarly exotic *Bromus rubens*. In the few places that there is permanent water, Rabbit-Foot Grass seems to be perennial. In those places, the grass remains green through the winter months. Perhaps the species is facultatively perennial and not obligately annual, as previously noted in the literature. However, in places that dry up completely, the beard-grass is annual.

Although Hitchcock did not apply a common name until the 1930s, records of the grass in the New World appeared much earlier. The first specimen found was of plants collected in 1805 in San Francisco (*H. Bolander* s.n. MO). Then, the grasses were discovered in Galveston, Texas, in 1843 (*F. Lindheimer* s.n. MO). By the 1890s, Rabbit-Foot Grass was known in at least Arizona, Colorado, Idaho, Iowa, Massachusetts, Oregon, Utah, Virginia, and Washington.

One early literature record of Rabbit-Foot Grass made in 1954 (C. E. Hubbard's book on *Grasses of the British Isles*) pointed out that the Old World "Annual Beard-grass" was the same plant as that "[k]nown in N. America as 'Rabbitfoot Grass.'"

**Human uses:** Presumably Rabbit-Foot Grass, *Elymus elymoides*, and *Setaria macrostachya* were all used by Navajo witches to kill a victim (Wyman and Harris 1941). All three grasses share the name *'zéé'iilwoii*. This death supposedly was achieved by dropping the fruit of the grass into the mouth of a sleeping person. Vestal (1952) simply said that *Elymus* and *Hordeum jubatum* (wild-barley) were feared because "it can kill a man if it gets into the mouth." Elmore (1944) said that the name referred to working its way down the throat. At the opposite extreme, the Tübatulabal burned off the awns and ate the seeds (Voegelin 1938).

Gould (1951) suggested that "[n]one of the four species introduced into the United States is of economic importance." If *Polypogon* was not important, the question arises as to how it arrived. Hitchcock and Chase (1950) suggested that the grasses arrived as contaminants in ballast from ships. Once in the Americas, these grasses, like many other Old World weeds in the New World, were probably spread as contaminants in food for domestic animals. Because the grasses grow with others that are preferred forage for livestock, it would be easy for them to become mixed with those plants.

**Derivation of the name:** *Polypogon* is Greek for "many beards," referring to the fruiting cluster. Because the genus compares the grasses to a beard, it has been "beard-grass" since 1798 when Desfontaines described it. However, Linnaeus, who described the species, put it in *Alopecurus* (Greek *alopex*, fox, *oura*, tail). *Monspeliensis* means that is was originally from Montpellier, France, whence Linnaeus and Desfontaines had specimens.

### *Setaria macrostachya* (Poaceae) Plains Bristle-Grass, *zacate tempranero, waṣai*

*Setaria macrostachya*. A. Base of plant. B. Flowering tip of plant, enlarged. C. Stalked spikelet with slender bristle at base. D. Fertile floret. E. Spikelet with fertile floret removed. Artist: L. B. Hamilton (Gould 1951).

**Other names:** ENGLISH: [plains, summer] bristle-grass ("bristle" from Middle English *brustel*, *brostle*; the Old Teutonic form of the root-syllable is **bors-*, pointing to Aryan **bhers-*, Sanskrit *bhṛshtí-s*, point, prong, edge; in English meaning one of the stiff hairs that grow on the back and sides of the hog and wild boar by ca. 1000; later applied to other materials; "bristle" was combined with "grass" to mean *Setaria* by 1841), foxtail [wild] millet ("fox-tail" applied to various grasses with brush-like spikes by 1552, originally the genus *Alopecurus*; "millet" from Latin *millium*, having a thousand grains); SPANISH: *zacate tempranero* [*temprano*] (early grass, Chihuahua, Sonora); ATHAPASCAN: *zéé'iilwoii* (one that goes into the throat, Navajo); HOKAN: *hasac* (Seri), *xica* *quiix* (globular things, Seri); MAYAN: *ne-kuuk-suuk* (*neh*, animal's tail, *k'uk'uk*, young plant part, *sak*, white, Maya); UTO-AZTECAN: *waṣai* (any grass, Tohono O'odham)

**Botanical description:** Tufted perennials, culms 2–10 dm tall, mostly simple or with a few branches on the lower part, ascending or erect. Leaf sheaths are mostly glabrous or the uppermost margin pilose, the blades 7–15 mm wide, glabrous or sparsely pilose near base. Panicles are 1–3 dm long, 1–2 cm thick, nearly cylindrical, with numerous sterile branchlets that appear as bristles subtending the spikelets, the spikelets sessile, 2–2.3 mm long, the second glume two-thirds to three-fourths as long as the fertile lemma whose tip is exposed, sterile palea as long as the convex fertile palea.

**Habitat:** Dry open ground, woods, hillsides. 600–2,130 m (2,000–7,000 ft).

**Range:** Arizona (all counties except Apache and Mohave), Colorado, New Mexico, and southeastern Texas; Mexico (Chihuahua, Sonora, east to Tamaulipas, south to Guanajuato, Queretaro, and Oaxaca).

**Seasonality:** Flowering May to October.

**Status:** Native.

**Ecological significance:** These clump-grasses grow from below the Harm House to above the Environmental Education Center in Brown Canyon, an elevation of 1,173–1,268 m (3,850–4,160 ft). Scattered plants are rare above that elevation. The species has also been found in Baboquivari and Fresnal Canyons, plus in several other localities, on the western slopes. Elsewhere in the region, *Setaria* grows on the slopes of Las Guijas Mountains east of the Altar Valley in the Buenos Aires National Wildlife Refuge. Fox-Tail Grasses are common in the Ajo Mountains and locally elsewhere in Organ Pipe Cactus National Monument, except in the more arid areas; also in canyons and on mountain slopes in the eastern part of Cabeza Prieta westward to the Cabeza Prieta Mountains, especially at higher elevations. The grass has been in the Ajo Mountains for at least 32,000 years (Felger et al. in prep.).

Unlike some of the other bristle-grasses in the area, this species is a strong heliophile that grows in open sunny spots along roads and trails. The seeds are eaten by a number of birds including doves, quails, sparrows, and many other songbirds (Martin et al. 1951).

**Human uses:** Oddly, few American species of *Setaria* have been recorded as being eaten. Those that have been listed are reported to produce meal resembling that from wheat (Ebeling 1986). Moerman (1998) does not list the genus, and Hocking (1997) mentions only the Old World species. In spite of that lack, American *Setaria* was eaten.

*Setaria macrostachya* was the primary starch source in the Tehuacán Valley for thousands of years, having been left in caves in both caches and in the refuse (Austin 2006a). The species was the only cereal present in levels radiocarbon dated at 5500 BCE, but it declined in importance after about 4500 BCE, following the rise of Maize (*Zea mays*) cultivation.

Maize did not become the main cereal grass in Mexico until about 4500 BCE. Before that, *Setaria* was the principal cereal over a large area for 1500 years (Callen 1967a, Prasada Rao et al. 1987). After Maize became dominant, Plains Bristle-Grass was eventually abandoned in most places. Consumption of *S. macrostachya* seeds by the Seri continues (Felger and Moser 1985). The Seri toast and grind the grain into flour, cook it with or without Green Turtle oil, and eat it as gruel; it was still a fairly important food in the 1980s.

Available data suggest that more people probably ate *Setaria* as a cereal before Maize became available. Since then, it appears that most of the knowledge of this old food has been lost (Austin 2006a).

**Derivation of the name:** *Setaria* comes from Latin *seta*, a bristle, referring to the bristles subtending the spikelets. *Macrostachya* means big spike or ear of grain.

## *Sporobolus airoides* (Poaceae) Sacaton, *zacatón, noḍ*

**Other names:** ENGLISH: dropseed (used in English for a *Muhlenbergia* by 1866 after Robert Brown had created *Sporobolus*, with the same meaning, in 1810), [big alkali] sacaton ("big grass," from Spanish); SPANISH: *zacatón* *<sacatón>* (augmentative of Spanish *zacate*, from Náhuatl *zacatl*, grass); ATHAPASCAN: *tłaltso* (big grass, Chiricahua and Mescalero Apache), *tł'oh ts'ósí* *<y'oh c'o's>* (slender grass, Navajo); UTO-AZTECAN: *noḍ* *<nawt, not>* (Akimel and Tohono O'odham), *nöönö* *<nɜ́:nɜ́>* (Hopi); (= *S. wrightii*, *S. airoides* var. *wrightii*)

**Botanical description:** Tufted perennial grasses, the culms numerous, 9–25 dm long, 2–9 mm thick, unbranched, erect, the ligule a fringe of eyelash-like hairs 2–10 mm long. Leaf blades are 2–7 dm long, 3–

*Sporobolus airoides* and *S. wrightii*. A. Base of plant. B. Flowering tip of plant. C. Tip of *S. airoides* for comparison. D. Glumes. E. Two views of florets. Artist: L. B. Hamilton (Gould 1951).

10 mm wide, flat or drying involute, summit of sheath glaucous. Panicle is 2–6 dm long, 12–26 cm wide, open, more or less pyramid-shaped when mature, the branches not whorled, spreading or slightly ascending, with short, densely flowered secondary branchlets. Spikelets are touching each other, the first glume 0.5–1 mm long, the second glume 0.8–1.8 mm long, the lemma 1.2–2.1 mm long.

**Habitat:** Washes, grasslands. 600–1,670 m (2,000–5,500 ft).

**Range:** Arizona (Cochise, Coconino, Navajo, Pima, and Santa Cruz counties), southern California, New Mexico, and western Texas; Mexico (Baja California, Chihuahua, Coahuila, Durango, Hidalgo, Nuevo León, Puebla, Sonora, and Zacatecas).

**Seasonality:** Flowering July to October.

**Status:** Native. Recent and most past agrostologists have recognized two species, *S. airoides* and *S. wrightii* (Soreng et al. 2007, TROPICOS 2008). The distinctions between those species are slight.

**Ecological significance:** Sacaton was formerly a dominant grass in the lowlands of the Altar Valley when the water table was higher. Now, healthy stands are confined to places where the water table is perched near the surface and remains comparatively moist through the year. Sites that are drying have plants that show stress, reduced fecundity, and wide individual spacing. Perhaps most remarkable is that the grasses persist at all. In the Tucson Valley, water tables are now more than 200 feet below the levels of just a century ago (Matlock and Davis 1972, Laney 1998).

Roseann Hanson (1997) recorded this grass in the lower Brown Canyon and wash. There are a few isolated plants remaining in the lower canyon between the Harm House and the Environmental Education Center.

**Human uses:** Akimel O'odham formerly made hairbrushes (called *gasvikuḍ*) from the grass before it disappeared following drainage of their region (Rea 1997). Tohono O'odham also used these brushes (Ortiz 1983), and they now use the word *gaswikuḍ* for combs. There was even a place on the Gila River reservation once called *Noḍ Ch-eḍ* ("place where sacaton grows"). Rea (1997) records that continuing drainage altered the locality that bore that name into an alkali salt flat. The site then became known as *Ongam Ch-eḍ* (salty place). More recently, it became an irrigated agricultural field where Alfalfa and Cotton were grown, and then a housing development.

North of the Coyote Mountains, there is a village called *Nawt Vaya* on road maps, and Granger (1960) lists it as *Not Vaya*. Both spellings are the same words, meaning "sacaton well." *Nawt*, *Not*, and *Noḍ* are transcription variants of the grass name; *Vaya* is a corruption of *Wahia* or *Vahia*, well. Lumholtz (1912) found a village with that name in the northern part of the Comobabi Mountains northwest of the Coyotes.

Navajo have ground the seeds and used them alone or mixed with Maize to make flour for bread or mush (Vestal 1952). Historically, the Tohono O'odham also used the seeds for food (Castetter and Underhill 1935). Apache, like the Akimel O'odham, made brushes of the roots, but used them to clean the spines off cacti before eating them. Western Apaches used *Sporobolus* for thatching (Reagan 1929).

Today, the only use of Sacaton seems to be as forage and hay for livestock (Hocking 1997). Apparently, wherever the grass remains is considered good winter range.

**Derivation of the name:** *Sporobolus* was named from Greek *spora*, seed, *bolos*, casting. The scientific and common name "dropseed" both refer to the seeds being dispersed, at least in part, by simply falling off the fruiting stalk; also, perhaps, to free seeds being forcibly ejected when the mucilaginous pericarp dries. *Airoides* presumably comes from resemblance to *Aira*, Hair Grass. *Wrightii* commemorates Charles Wright, 1811–1885, a collector in the American Southwest.

**Miscellaneous:** *Sporobolus* is a genus of 160 species that grow in the Americas (77 species), Asia, and Africa. There is a single species in Europe (Mabberley 1997, Peterson et al. 2001). At least Apache, Hopi, Keres, Kiowa, and Navajo have eaten the seeds of *S. cryptandrus* (Castetter and Opler 1936, Elmore 1944, Moerman 1998, Vestal 1952), a species that grows in riparian areas on the lower Buenos Aires National Wildlife Refuge.

*Sporobolus contractus* (Spike Dropseed) grows in the Altar Valley and west of the Baboquivari Mountains. When the rains have been good, this perennial clump-grass is frequently in flower in the valley. Apache, Hopi, and Navajo, and probably other peoples, ate seeds of this species (Whiting 1939, Moerman 1998, Steggerda 1941).

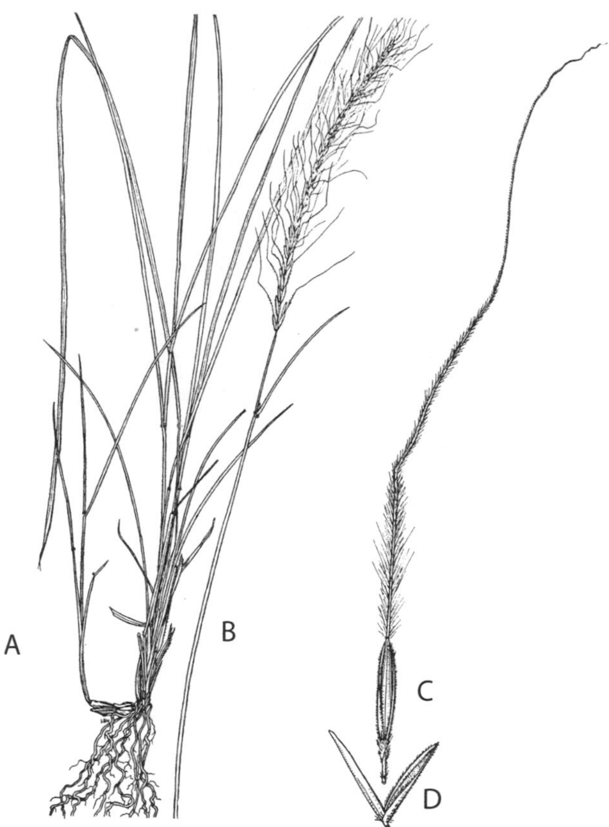

*Trachypogon spicatus*. A. Base of plant. B. Flowering tip. C. Spikelet. D. Glumes. Artist: L. B. Hamilton (Gould 1951).

## *Trachypogon spicatus* (Poaceae) Crinkle-Awn, *zacate barba larga, waṣai*

**Other names:** ENGLISH: crinkle-awn (Arizona; "crinkle" probably from crinkle as a verb, but the noun may be the earlier; akin to Dutch and Low German *krinkel*, curve, flexure, crookedness, curvature, diminutive of *kring, krink*, circle; in use by 1596; "awn," see *Aristida* for etymology), grey tussock grass ("tussock" was in use by 1550 and originally referred to a tuft of hair; later applied to bunch-grasses; by 1842 the term "tussock-grass" had appeared); SPANISH: [*zacate*] *barba larga* (big beard [grass], Mexico); MAYAN: *alcapajac* (Chontal, Oaxaca); UTO-AZTECAN: *waṣai* (any grass, Tohono O'odham); LANGUAGE FAMILY UNKNOWN: *polmuc* (Mexico), *quimec* (Mexico); (= *Trachypogon montufari*; = *Trachypogon secundus*)

**Botanical description:** Tufted perennials 6–12 dm tall, the nodes appressed-hirsute, the sheaths with an erect auricle 2–5 mm long. Leaf blades are flat to almost involute, 3–8 mm broad. Racemes are spike-like, solitary, terminal, dense, 10–18 cm long, the rachis glaucous, remaining intact. Spikelets are in pairs, 6–8 mm long, one spikelet of each pair nearly sessile, staminate, awnless, the other pedicellate, perfect, the pedicel disarticulating near its base and constituting a sharp-barbed callus below the detached spikelet, the awn stout, bent and flexuous, 4–6 cm long, short-plumose below and nearly glabrous toward the tip.

**Habitat:** Rocky hillsides. 300–1,830 m (1,000–6,000 ft).

**Range:** Arizona (Cochise, Gila, Graham, Mohave, Pima, Pinal, and Santa Cruz counties), southwestern New Mexico, and southern Texas; Mexico (Chihuahua and Sonora); south to Panama; and Africa.

**Seasonality:** Flowering September to October.

**Status:** Native.

**Ecological significance:** These grasses have not been found west of the Baboquivari Mountains in southwestern Arizona or in the Gran Desierto Region of Sonora (Felger 2000, Felger et al. in prep.). Within the canyons, Crinkle-Awn is typically found on rocky hillsides, and it occurs in both Brown Canyon and Jaguar Canyon. The species is, for example, on the slope of the upper trail not far above the Environmental Education Center. The grasses are also known in Thomas Canyon, on the western side of the Baboquivari Mountains, and in the Coyote Mountains to the northeast.

This widespread species reaches its northern limit in southeastern Arizona. West of the Baboquivari Mountains, the climate is too arid, and much of the elevation is too low for the species to occur (Felger et al. in prep.). East of those mountains, the species goes north to, but not upon, the Mogollon Rim, probably because of climate limitations.

Proctor et al. (1996) indicate that the grass family requires cross-pollination because of self-incompatibility. In the Poaceae there is a special kind of incompatibility where the stigma "recognizes" the genome of the pollen tube growing from the grains on the stigmatic surface. Following recognition the pollen tube is blocked as it trys to penetrate the stigmatic surface. There seem to be two genes (loci) that interact to achieve this termination.

**Human uses:** Crinkle-Awn plants were used prehistorically as thatch in El Salvador (Lentz et al. 1996). The species is also widely used for thatch in southern Mexico and Central America (Berlin et al. 1974, Williams 1981, Lentz et al. 1996). In addition, the grasses are used to make brushes for whitewashing walls there. Hocking (1997) reports that the root was considered antibiotic, and the leaves were used as a tea substitute. Hocking does not say, but these uses were probably in Africa (Stiven 1952).

**Derivation of the name:** *Trachypogon* is from Greek *trachys*, rough, and *pogon*, a beard, the spikelet having a plumose awn. *Spicatus* is from Latin, and means that the flowers are in a usually unbranched, elongated, simple cluster, with flowers having no stalks.

**Miscellaneous:** This genus comprises 3–13 species (Mabberley 1997, Watson and Dallwitz 1992). Zuloaga et al. (2003) have further altered the number of species to 5; these grow in tropical America and in Africa.

*Trachypogon* is susceptible to infection by rusts (*Puccinia*) in some areas; smuts (*Sphacelotheca, Ustilago*) also grow on them (Watson and Dallwitz 1992). The genus has been placed in the subfamily Panicideae, tribe Andropogonodae, and subtribe Andropogoneae by Zuloaga et al. (2003). Thus, Alexander Humboldt, Aimé Bonpland, and Carl Ludwig von Willdenow placed the species near some of its relatives when these three called it *Andropogon* in 1806.

## *Gilia mexicana* (Polemoniaceae) Gilia, *alhelilla*

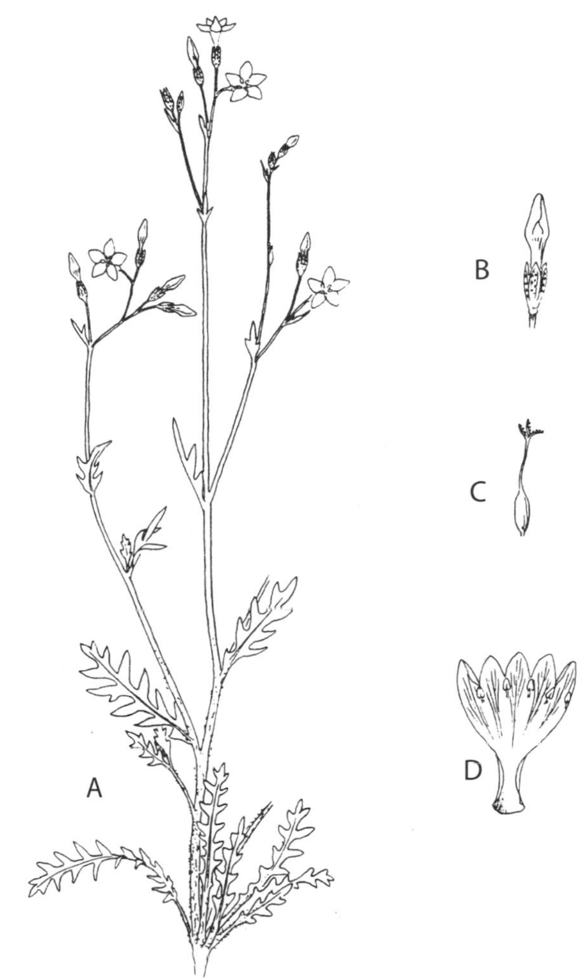


*Gilia mexicana*. A. Plant in flower. B. Unopened bud. C. Pistil showing 3-lobed style. D. Dissected corolla. Artist: R. D. Ivey (Ivey 2003).

**Other names:** ENGLISH: [El Paso] gilia (a book name); SPANISH: *alhelilla* (resembling *alhelí*, a name for the genus based on *alhelí*, the crucifer *Matthiola*, Mexico); ATHAPASCAN: *dlǫ́ii 'azee'* <*ƛǫˀiˀazeeˀ*> (weasel medicine, Navajo; for *G. gunnisoni*); UTO-AZTECAN: <*'a'nu'ngahu*> (Hopi; for *Ipomopsis multiflora*), *iambĭp* (*i'a*, wound, *ûm*, possessive, *bĭp*, plant, Shoshoni), *yogûmsĭtagwĭv* (*yo'guvĭtc*, coyote, *ŭm*, possessive, *sĭtagwĭv*, medicine, Ute; for *G. gracilis*)

**Botanical description:** Herbs, annuals, erect, with 1 to several leafy stems 10–33 cm tall from a basal rosette, the stems slender, gray-pubescent below the middle, with some minute glandular hairs on the inflorescence. Basal leaves are pinnately or bipinnately lobed, rachis and lobes 0.8–1.3 mm wide, the primary lobes 4.5–6.5 mm long, lower cauline leaves similar to the basal ones, progressively smaller upward, the uppermost bract-like, linear. Calices are 2.5–4 mm long, glabrous or sparsely glandular or cobwebby pubescent, the lobes acute. Corollas are 4.5–8 mm long, the tube included within calyx, or throat excluded, or twice as long as calyx, tube and throat white with yellow spots, lobes white to pale blue with violet dots along midveins. Capsules are narrowly ovoid, 3.5–5 mm long.

**Habitat:** Rocky ridges, hillsides, rocky ledges. 900–1,220 m (ca. 3,000–4,000 ft).

**Range:** Arizona (Cochise, Maricopa, Pima, Pinal, Santa Cruz, and Yavapai counties) and New Mexico; Mexico (Chihuahua to Edo. México).

**Seasonality:** Flowering in (February) March to April.

**Status:** Native.

**Ecological significance:** This small herb is a Chihuahuan Desert species that barely enters the Sonoran Desert (Shreve and Wiggins 1964). The species is rare where it just reaches the western tip of Texas (Correll and Johnston 1970).

These wildflowers are abundant in the cool seasons. Especially when there has been sufficient winter rain, the minute herbs may blanket the open spaces on the ground. Flowers are probably pollinated by long-tongued bees and butterflies. However, the corolla shape does not exclude pollination by bee-flies, as occurs in related *G. splendens* (Proctor et al. 1996). *Ipomopsis multiflora* and *I. thurberi* are both known from the western slopes of the mountains. *Ipomopsis*, too, is notable for having multiple pollination types within its species. Some are bird-pollinated (*I. aggregata*), while others are adapted for pollination by night-flying Sphinx Moths, and others by bee-flies.

**Human uses:** No uses were found for this species, but other species are used. Navajo use *G. inconspicua* to treat fevers (Vestal 1952), put *G. leptomeria* on scorpion stings and other bites, employ *G. subnuda* to aid childbirth (Wyman and Harris 1951), and use *G. gunnisoni* to treat sores and the blood (Wyman and Harris 1951). Keres use *G. rigidula* on muscle cramps, and Havasupai parch the seed of *G. sinuata* for food (Moerman 1998).

Pennington (1958) found that some species were used to poison fish, particularly *G. macombii* (*maté-šuwa*, Tarahumara). That name is also used for *Ipomopsis thurberi*, another fish poison. He found that several species of *Gilia* contain saponins that would be effective at poisoning fish. Chamberlin (1911) found *G. gracilis* being put on wounds by Shoshoni. Train et al. (1957) note several other species being used to clean wounds, to treat venereal diseases, and for other problems among the Paiute, Shoshoni, and Washo of Nevada.

**Derivation of the name:** *Gilia* honors Italian astronomer Filippo Luigi Gilii [Gilij], 1756–1821, who wrote on Roman agriculture and the delineation of natural genera (*Delineazione dei generi naturali*). *Mexicana* refers to Mexico, where the species was first collected.

**Miscellaneous:** There is one other species recorded in the Baboquivari Mountains, *Gilia flavocincta*. Those plants are like *G. mexicana*, but they have sepals that are purple-striped. Also, *G. flavocincta* has a more open basal rosette, a simpler stem, and longer corollas (8–15 mm). *Gilia flavocincta*, too, flowers in the spring and may be intermixed with *G. mexicana*. This species is either more common or more often collected than *G. mexicana* in the state; it occurs in Apache, Cochise, Gila, Graham, Maricopa, Mohave, Pima, Pinal, Santa Cruz, Yavapai, and Yuma counties.

*Allophyllum gilioides* (Straggling Gilia) is probably the most abundant member of the family in the region. The herb grows on rocky ridges and flowers from March to April. This herb has a tight cluster of flowers at the apex that is surrounded by bracts and sepals. The diminutive annual is known in Coconino, Gila, Graham, Maricopa, Pima, Pinal, and Yavapai counties. This species occurs mostly below the Mogollon Rim and extends into southeastern Arizona along the mountain chains. Straggling Gilia seems to be disjunct between the Baboquivari and Sierrita Mountain region to the Santa Catalina and Rincon Mountains.

## *Eriogonum abertianum* (Polygonaceae) Wild Buckwheat

**Other names:** ENGLISH: [Abert's] buckwheat (alluding to relationship and similarity with the true "buckwheat," *Fagopyrum*, in the same family; the name is related to Dutch *boekweit* [*bockweydt*] or German *Buchweize*, "beech-wheat," from the shape of the three-angled seeds); ATHAPASCAN: *łe'azee'* (*łeezh*, earth, *'azee'*, medicine, Navajo); UTO-AZTECAN: *hulaqal* (Cahuilla), *powáwi* (Hopi), *pu'iwanûp* (Shoshoni, for *E. brevicaule*; several other species with

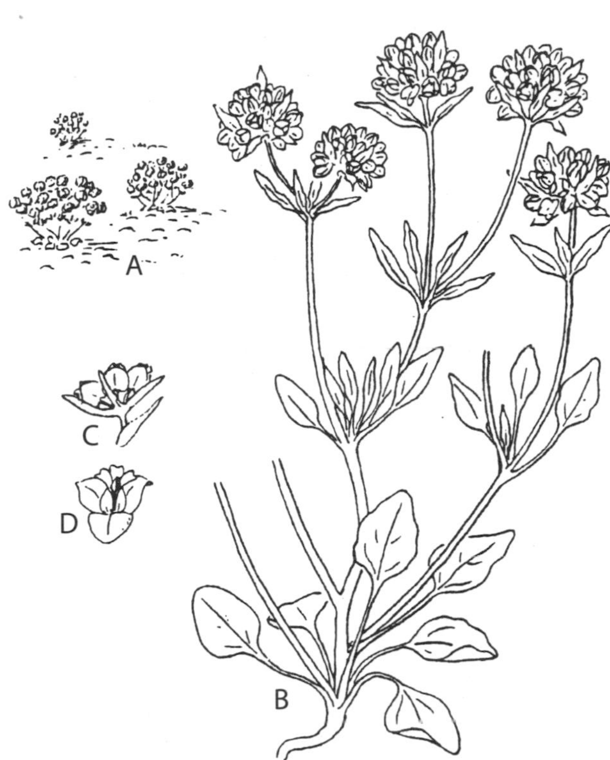


*Eriogonum abertianum*. A. Life-form of plant. B. Flowering branch. C. Flower cluster. D. Flower detail. Artist: R. D. Ivey (Ivey 2003).

distinct names), *tunabol* (Tübatulabal, several species recognized)

**Botanical description:** Annual herb, spreading or erect, branched, to 5 dm tall. Leaves are basal and cauline, all loosely villose or white pubescent, the basal ovate to oblong, 1–4 cm long, 1–3 cm wide, the petioles to 6 cm long, the cauline sessile or nearly so, obovate-lanceolate to linear. Peduncles when present extend to 6 cm long, hirsute. Involucres are broadly campanulate, the tubes 2–3 mm long, reflexed at maturity, villous-canescent. Perianths are white to pale yellow, often tinged with rose or reddish, 3–4.5 mm long. Achenes are dark brown, 0.6–1 mm long.

**Habitat:** Canyon slopes and ridges. 460–2,130 m (1,500–7,000 ft).

**Range:** Arizona (Coconino and Yavapai south to Cochise, Graham, Greenlee, Pima, and Santa Cruz counties), New Mexico, and western Texas; Mexico (Chihuahua and Sonora to central Mexico).

**Seasonality:** Flowering March to September.

**Status:** Native.

**Ecological significance:** These plants grow in rocky areas, often being found among the boulders and stones beside the creek beds within canyons and washes. Wild buckwheats occur in both grassland and oak woodland communities, and they are often found associated with Mesquite (*Prosopis*).

The small white-to-pinkish flowers are visited by insects, especially bees and flies. Probably, these are the pollinators. The butterfly genus *Euphilotes* (called "Blues") is restricted to *Eriogonum*, with larvae of *Euphilotes rita* in the Baboquivari Mountains feeding exclusively on *Eriogonum wrightii*. Not only do the larvae feed exclusively on single *Eriogonum* species, but also the adults drink nectar only from its flowers, and they often remain faithful to individual populations of one species. This fidelity limits gene exchange and results in an array of variation (Glassberg 2001).

**Human uses:** Navajo use *Eriogonum abertianum* in a medicine to treat skin cuts (Vestal 1952). There are another 36 species listed by Moerman (1998) as used in North America, excluding Mexico. Others that grow in the Baboquivaris are *E. polycladon*, *E. thurberi*, and *E. wrightii*. Of these, only *E. wrightii* is listed as being used. This species, too, is used by the Navajo, but as an emetic (Wyman and Harris 1951). Hopi use several species for women's problems (Whiting 1939).

The seeds of many species have been pounded into a meal, eaten dry or mixed with water, and drunk as *atole*. Dunmire and Tierney (1995) found that the Anasazi ate the seeds of wild buckwheats, and carbonized seeds and flower parts have been found in eastern Arizona, southern Colorado, and in dried human feces at Chaco Canyon and in the Four Corners area.

Some species used by other tribes in this vicinity are *E. alatum* (medicine; roots and seeds for food, Navajo), *E. cernuum* (medicine; seeds for food, Navajo), *E. corymbosum* (medicine, Havasupai; food, Hopi), *E. divaricatum* (medicine, Navajo), *E. hookeri* (food condiment, Hopi), *E. inflatum* (medicine, Navajo; food, stems as tobacco pipe stems, Havasupai, Kawaiisu, Yavapai), *E. jamesii* (medicine, Apache, Navajo, Zuni; mythological, Zuni), *E. leptophyllum* (medicine, Navajo), *E. racemosum* (medicine; food, Navajo), *E. rotundifolium* (medicine, Keres, Navajo; food, Navajo), *E. tenellum* (medicine, Keres), and *E. umbellatum* (medicine, Kawaiisu, Navajo).

**Derivation of the name:** *Eriogonum* comes from *erion*, wool, and *gonum*, joint or knee, referring to the downy nodes of the stems. *Abertianum* is probably named for James William Abert, 1820–1897, son of John James Abert, 1788–1863. The species was named by John Torrey in 1848, when John James Abert was chief of the Topographical Bureau of the War Department of the United States. However, the son was attached in the summer of 1845 to the third expedition of John Charles Frémont, whose assignment was "to make reconnaissance southward and eastward along the Canadian River through the country of Kiowa and Comanche." Frémont took the main party to California, and he gave command of the Canadian River mission to J. W. Abert, with an assistant, Lt. William G. Peck. In his report, Abert described the geology, flora, and fauna of the Canadian River valley. Later, Abert and Peck accompanied Gen. Stephen W. Kearny's Army of the West to New Mexico. The two lieutenants conducted a survey of New Mexico as far south as Socorro. The surveyors visited each of the Rio Grande pueblos and took note of the geology and wildlife of the new American territory, as well as of the habits and customs of their native residents.

## *Polygonum persicaria* (Polygonaceae) Smartweed, *hierba pejiguera*

**Other names:** ENGLISH: gander-grass (eaten by male geese? South), heart's ease (akin to the name for *Viola*, in reference to treating heart problems with them, Texas to NE United States), [spotted] lady's thumb (a usage in the United States, postdating 1888; said to have been applied after "Our Lady pulled up a plant, leaving her thumb print on the leaf"; another version is that it was in Gaelic the *lus chann ceusaidh*, the herb of the Crucifixion tree, and that it grew under the cross and was spotted with Christ's blood), peachwort (derived from *persicaria*, peach-like), red-shanks [red-leg] (an allusion to the red stems), smartweed (applied to *P. hydropiper* by 1787; that same year, W. H. Marshall, writing on the rural economy of Norfolk, England, said: "*Smartweed*, biting and pale-flowered persicarias; arsmart"; indeed, the OED lists "arse-smart" as the common name), willow-weed (leaves somewhat resemble those of willow, *Salix*); SPANISH: *chillero* (little chile, *Capsicum*), *duraznillo* (little peach, a comparison of its leaves and those of the peach tree, applied to numerous unrelated plants in Mexico; originally used in medieval times for knotweed [*Polygonum*]), [*hierba*] *pejiguera* (peach-like [herb], Mexico), *moco de guajolote* [*quajolote*] (turkey snot,

*Polygonum persicaria*. A. Plant. B. Flowering spike. C. Achenes, two views. D. Nodal stipules (ocrea). Artist: R. O. Hughes (Reed 1971).

Texas, Mexico), *pimpenilla* (from *pimpernilla*; "pimpernel" came from Medieval Latin *pipinella*, which perhaps came from *bipinella*, a diminutive of *bipennis*, two-winged, apparently in reference to the umbel, *Pimpinella*, Apiaceae); UTO-AZTECAN: *tamandiŋ tibohišŋ* (*tamandiŋ*, tooth, *tibohišŋ*, medicine, Tübatulabal), *tiɨkarāŋɨva* (Northern Paiute)

**Botanical description:** Annual herbs, stems branched, ascending, glabrous or nearly so, green or sometimes marked with red, 1–10 dm tall. Leaf blades are lanceolate, nearly glabrous, 3–15 cm long, 5–18 mm wide, the ochrea brittle, 2–4 times as high as wide, ciliate, the bristles mostly less than 3 mm long. Peduncles are usually glabrous, racemes numerous, mostly less than 3 cm long, some to 5 cm, dense. Calices are without glands, white, pink, or purple, 2.2–3.2 mm long. Achenes are dark brown or black, lustrous, lens-shaped or 3-angled, 2–2.7 mm long.

**Habitat:** Along streams, in ponds. 900–2,130 m (3,000–7,000 ft).

**Range:** Arizona (Coconino, Navajo, Yavapai, south to Cochise, and Pima counties), widespread in the United States; Mexico (Chihuahua and Sonora).

**Seasonality:** Flowering July to October.

**Status:** Exotic; introduced from Europe.

**Ecological significance:** These herbs were introduced from the Old World and quickly spread into the wetter areas of the southwestern United States. Birds doubtless helped in their spread by eating the seeds and planting them when the avians stopped to drink at pools, streams, and other wet spots. Now, these alien herbs may be the only plant growing in temporary ponds and pools at lower elevations on the Buenos Aires National Wildlife Refuge (Austin 2003).

**Human uses:** People in the Old World knew *Polygonum* when they arrived in the Americas. Among their names for the genus are the following: Arssmerte (for *P. hydropiper*, by Wm. Turner in 1548), *corriola bastarda* (false morning glory, Portuguese), *glùineach* (having large knees, Gaelic), *Knöterich* (knotted, German), Knotgrass (from the knotted or jointed stem of *Polygonum aviculare*, in English by 1500), *lus an fhògair* (banishment herb, Gaelic for *P. hydropiper*), and *renouée* (knotted, French). Several of their species were used for food and medicine (Dobelis 1986). Coffey (1993) notes that *P. persicaria* "will dye woollen cloth yellow," after having been dipped in alum.

No records have been found of *P. persicaria* being used in southern Arizona or New Mexico, but farther east, it is used as medicine by Algonquians, Cherokee, Iroquois, and Ojibwa (Merrill and Feest 1975, Moerman 1998). Algonquians call this species *wisakon*, and they use it to treat poisoned and green (i.e., infected, putrid) wounds (Merrill and Feest 1975). In Mexico, the Guarijío boil the leaves and eat them as greens (Yetman and Felger 2002). Hocking (1997) wrote that *P. persicaria* is used in home remedies similarly to *P. hydropiperiodes*, that is, to treat kidney stones, gout, rheum (watery matter secreted by the mucous membranes), hemorrhoids, icterus (jaundice), and as a vulnerary on wounds and eczema.

Tarahumara use *P. punctatum*, known as *korísowa*, to catch fish. Bundles of *korísowa* were put in baskets, crushed, and dipped in pools; fish would then rise almost immediately (Pennington 1958, 1963). Tarahumara use *P. pensylvanicum* (*watonáka*, *yerba del pescado*) the same way.

**Derivation of the name:** *Polygonum* is from Greek *poly*, many, *gonu*, knees or nodes, referring to the swollen joint of the stems. *Persicaria* is an old generic name, said to be a comparison of the leaves with those of the peach (*Prunus persica*) or *persica* in Latin.

### *Rumex hymenosepalus* (Polygonaceae) Wild Rhubarb, *cañaigre*, *siwidculis*

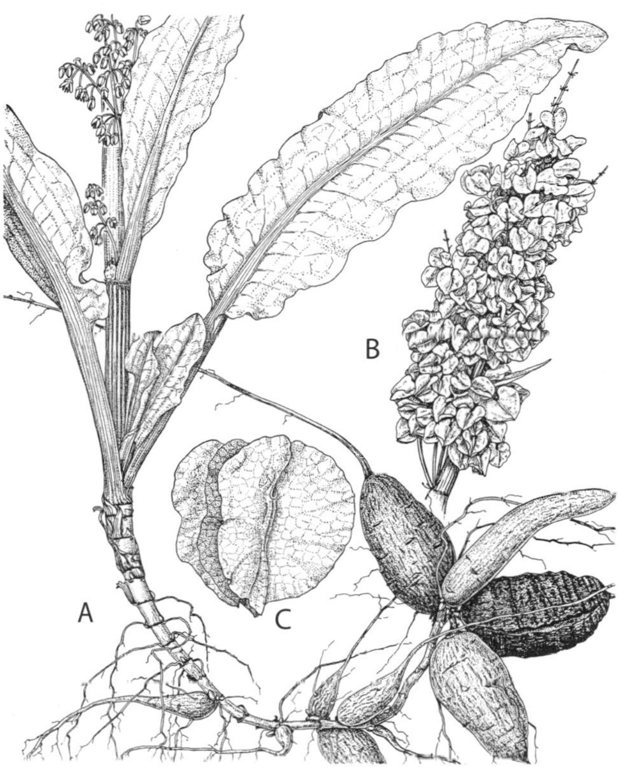


*Rumex hymenosepalus*. A. Basal part of plant showing tubers. B. Fruiting branch. C. Fruiting calyx. Artist: L. B. Hamilton (Parker 1958).

**Other names:** ENGLISH: desert [wild] rhubarb (a comparison with *Rheum*), [red, sand] tanner's dock (docken, singular, from Old English *docce*, in use by the CE 1100s; related to Middle Dutch *docke*, Old Danish *adokke*, Old French *éadocce*; Gaelic *dogha*, cognate with English "dock," but originally designating "burdock," *Arctium lappa*), wild pie plant (a comparison with *Rheum*; the petioles of both are used to make pies); SPANISH: *cañaigre* (Spanish from *caña agria*, sour cane), *hierba colorada* (red herb, Baja California, Sonora), *hierba de la mula* (mule herb, Coahuila), *raíz colorada* (red root, Sonora), *raíz del indio* (Indian root, Chihuahua, Coahuila); ATHAPASCAN: *'asdzą́ą́ nádleehébishéé'* <*ˀasʒą́náƛehébižéˀ*> (*'asdzą́ą́nádleehé*, Changing Woman, *bi*, her, *shéé'*, saliva, Navajo), *ch'ilt'ozhé* <*jił dozhe*> (Western Apache), *chaad'iniih* <*chaat'inii, čaˀt'íní, tchąat'iniih*> (hidden one [the roots], Navajo), *tjiłt'oo'íh* <*jilt'o'í, jil't'o'í*> (which is sucked [the stem or petioles?], Navajo); CHUMASH: *alaqpɨi* (Ventureño Chumash), *sha'w* (Barbareño Chumash, Ineseño Chumash); UTO-AZTECAN: *abanal* (Tübatulabal), *aingappawaia* (Shoshoni; for *R. crispus*), *äñ'kapadjarûmp* [*äñ'kapaidjarûmp, äñ'kapatsarûmp*] (Shoshoni; for *R. salicifolius*), *ɨtsākān*$^{va}$ (Northern Paiute; for *R. crispus*), *avaanarɨbɨ* (Kawaiisu), *maalval* (Cahuilla), *pawai* (Northern Paiute), *siwidculis* <*s-hiwiculs, siwidculs*> (probably meaning "it is bitter," Tohono O'odham), *sivijuls* (either from *siv*, bitter, or *hiv*, rubbing against, as in a scraping stick of musical instruments, Akimel O'odham), *sayávi* (Hopi), *vakas nener* <*vakashiniñir*> (*vakas*, cow, *nener*, tongue, Mountain Pima; for *R. obtusifolius*), *wakondam* (washer, launderer, Tohono O'odham); YUMAN: *akyésa* (Mohave), *akyés* (Maricopa, Yuma), *ki:š* <*kíš*> (Cocopa), *thi'hach* (Walapai)

**Botanical description:** Perennial herbs from deeply seated tuberous roots. Stems are 2–10 cm tall, lower leaves long-petiolate, the blades 8–25 cm long, 2–12 cm wide, elliptic to lanceolate or oblanceolate, cuneate basally, acute to acuminate apically, more or less fleshy, cauline leaves reduced, stipular sheaths (ochreas) 1–4 cm long. Panicles are compact, 10–40 cm long, usually pinkish. Perianth is 2–4 mm long at anthesis, the valves 8–18 mm long in fruit, cordate-ovate to almost orbicular, reticulate, rounded apically. Fruits are 3-angled, 4–7 mm long.

**Habitat:** Grasslands, desertscrub, ridges, canyons, stream beds. Up to 1,830 m (6,000 ft).

**Range:** Arizona (Coconino, Mohave, Navajo, Yavapai, south to Graham, Pima, Pinal, and Santa Cruz counties), California, Colorado, New Mexico, Nevada, Texas, Utah, and Wyoming; Mexico (Baja California, Coahuila, Chihuahua, and northern Sonora).

**Seasonality:** Flowering (February) March to April.

**Status:** Native.

**Ecological significance:** *Rumex hymenosepalus* was described by John Torrey in his *Report on the United States and Mexican Boundary* in 1859. The species is more common near lower grasslands and desertscrub washes, but it remains in the lower parts of the canyons. Plants are abundant in the area between the Harm House (3,855 ft) and the Environmental Education Center (4,160 ft), but become rare and then disappear above that.

As the soil dries up, Wild Rhubarb dies back to the perennial roots after flowering in the summer. Leaves come back with winter rains and the cooler weather, and *R. hymenosepalus* blossoms with other spring wildflowers. Leaves appear as early as December in years with good rains, but by February plants are prominent in the understory.

**Human uses:** Wiggins (1980) reports that in Baja California Wild Rhubarb roots, containing "over 30% tannin," are dug for that compound. Roots are soaked in water and the liquid used to tan skins. Others using *R. hymenosepalus* for tanning include the Akimel O'odham, Cahuilla, Navajo, and Walapai (Bean and Saubel 1972, Curtin 1949, Moerman 1998, Vestal 1952). This technique was learned from Europeans (Dunmire and Tierney 1995).

At least the Akimel O'odham, Arapaho, Cahuilla, Hopi, Kickapoo, Navajo, Paiute, Pawnee, Tohono O'odham, and Western Apache use Wild Rhubarb medicinally (Bray 1998, Castetter and Underhill 1935, Curtin 1949, Gallagher 1976, Moerman 1998, Romero 1954, Train et al. 1957, Vestal 1952, Whiting 1939). Tohono O'odham powder the root, put it on external sores, and eat it to treat a sore throat (Castetter and Underhill 1935). Kickapoo cut the roots into pieces, put it in mescal, steep it for several days, and apply it to rheumatic joints (Latorre and Latorre 1977). The *Rumex* is not native where the Kickapoo live but is cultivated in their gardens.

Brown, green, gold, orange, and red dyes are obtained from the roots by Akimel O'odham, Chumash, Hopi, Navajo, and Walapai (Curtin 1949, Deschinny 1984, Timbrook 1990, Rea 1997, Moerman 1998, Whiting 1939).

Although most people did not eat the roots, they were consumed by some. The most commonly eaten parts are the flower stalks, leaves, and seeds. These parts are food to Akimel O'odham, Cahuilla, Chumash, Cocopa, Kawaiisu, Maricopa, Navajo, Tohono O'odham, and Walapai (Curtin 1949, Castetter and Bell 1951, Hodgson 2001, Rea 1997, Moerman 1998, Timbrook 1990). Castetter and Bell (1951) reported *R. crispus*, but Felger et al. (in prep.) say it was *R. hymenosepalus*.

**Derivation of the name:** *Rumex* is the classical Latin name for Sorrel, *R. acetosella*; akin to Akkadian *ramaku*, to pour, Hebrew *romak*, spear or javelin.

*Hymenosepalus*, meaning having membranous sepals, refers to the wings on the fruit.

### ***Claytonia perfoliata*** **(Portulacaceae)** **Miner's Lettuce, *verdolaga de invierno***

*Claytonia perfoliata*. A. Flowering plant. B. Flowering branch tip. C. Seed. Artist: J. R. Janish (Wiggins 1980).

**Other names:** ENGLISH: Indian lettuce (for etymology, see *Lactuca*), miner's lettuce (Coffey 1993 said that the name was given because of its use during the gold rush of 1849 in California), winter purslane (for etymology, see *Portulaca*); SPANISH: *petota* (California; a loan word in Spanish? it looks suspiciously like Caribbean *batata*, the name for *Ipomoea batatas* and the word that gave rise to potato; in fact, there is a *Solanum* section *Petota* based on that word), *verdolaga de invierno* (winter purslane, Mexico); CHUMASH: *shilik'* (Barbareño Chumash), *shilik* (Ineseño Chumash); UTO-AZTECAN: *pa'gwodzûp* (Shoshoni), *palsingat* (Cahuilla), *uutukʷa'aribɨ* (Kawaiisu); (= *Montia perfoliata*)

**Botanical description:** Annual herbs, the flowering stems few to several from the root crown, are 5–20 cm tall. Basal leaves are 2–15 cm long, somewhat sheathing basally, the blades linear to spatulate or broader. Cauline leaves are 2, opposite, usually connate-perfoliate and forming a disk-like structure 0.8–3 cm wide. Inflorescences are racemose, terminal, and sometimes also axillary. Flowers are mostly 3–8, on pedicels 2–10 mm long or longer, nodding, only the lower subtended by a bract. Sepals are 2, rounded, 1.5–3 mm long. Petals are 5, white or pinkish, 3–5 mm long. Capsules are 2–4 mm long. Seeds are usually 3, black, lustrous.

**Habitat:** Among rocks, near seeps and streams. 760–2,290 m (2,500–7,500 ft), sometimes higher (to 3,300 m) outside Arizona.

**Range:** Arizona (Coconino to Mohave south to Pima and Santa Cruz counties), California, southwestern Colorado, Nevada, Utah, and east to South Dakota; north to British Columbia; Mexico (Baja California, Sonora? Chiapas, Jalisco, Michoacán, and Edo. México); El Salvador, Guatemala; and Argentina.

**Seasonality:** Flowering February to May.

**Status:** Native.

**Ecological significance:** Through most of its range Miner's Lettuce is a spring wildflower and herb that grows in pinyon-juniper, riparian, and spruce-fir communities. Where there is locally abundant water, it may extend down into Creosote-Bush, Blackbrush, and Sagebrush communities (Welsh et al. 1987). This herb is a temperate and boreal species that reaches south across the Mexican border along the mountain chains.

**Human uses:** Miner's Lettuce is used as a salad plant and potherb by Anglos, Hispanics, and indigenous peoples. Coffey (1993) provides a famous old-time recipe for miner's lettuce: peeled *tuna* (*Opuntia* fruits) nestled among *petota* leaves, served with a dressing of olive oil, salt, pepper, and vinegar. The herbs are used as food by at least the Cahuilla, Ohlone, Kumiai, Kawaiisu, Luiseño, "Mendocino people," Miwok, "Montana people," Nisenan, and Paiute (Bean and Saubel 1972, Bocek 1984, Moerman 1998, Zigmond 1981). Chumash ate not only the leaves as greens but also the seeds (Timbrook 1990). Powers (1877) reports that the Miwok (or Washo?) people living in the Sierra Nevada Mountains in Amador County, California, put the fresh plants on ant nests (probably *Formica rubra* group fide Ebeling 1986), and ate them after the insects had crawled on them. The formic acid sprayed by the agitated ants gave Miner's Lettuce a sour taste similar to that of vinegar.

Because *Claytonia* are confined to higher regions in the mountains, perhaps the O'odham did not use them. However, it is also likely that O'odham collected and ate the herbs seasonally, but abandoned that use because of the introduction of more easily available Old World greens. Dunmire (2004) has elegantly summarized the changes in their culture with the advent of Old World winter crops. Although Old World greens are used, native species are still harvested by the O'odham (cf. Rea 1997).

Shoshoni put mashed plants on rheumatically painful places (Train et al. 1957). Cahuilla consider the juice an appetite restorer, and Thompson make an eye medicine of them (Moerman 1998, Romero 1954).

**Derivation of the name:** *Claytonia* commemorates John Clayton, 1686 (or 1694)–1773, physician, plant collector, and immigrant to Virginia from England. He was the clerk of Gloucester County, VA, and he collected many of the specimens that formed the basis of Jan Gronovius's *Flora Virginica* of 1739–1743. *Perfoliata* means that the stems appear to grow through the leaves. In reality, the opposite leaves are sessile and partly fused to form what appears to be a single leaf around the node.

**Miscellaneous:** For many years, the species was placed in the genus *Montia* (named for Giuseppe Monti, 1682–1760, professor and director of the Bologna Botanical Garden from 1722 to 1760). *Montia* and *Claytonia* were separated because of life-form (annual vs. perennials with corms or tuberous roots), number of cauline leaves (several pairs or if one pair, then connate-perfoliate vs. one pair that is not perfoliate), and number of ovules (3 vs. 6). Most students of the group now agree that the annual plants are simply derived from the perennial.

## *Portulaca umbraticola* (Portulacaceae) Purslane, *verdolaga, ku'ukpalk*

**Other names:** ENGLISH: Chinese hat (Texas), [wing-pod, wingpod] purslane (from Latin, *porcil(l)ca*, used by Pliny, CE 23–79, as an alternative spelling for *portulaca*; in English by ca. 1387 as "purcelan," derived from Old French *porcelaine*; akin to Italian *porcellana*, Arizona, New Mexico, Texas), pusley (a corruption of "purslane" appearing by 1861); SPANISH: *verdolaga* [*de la sierra*] ([wild] purslane, Sonora; derived from Arabic *burd(u)lagá*); ATHAPASCAN: *ch'í'atsii' ch'ił* <*ci-aji c'il, ci-yajilc', tséghánłch'i*> (*chił*, plant, *ch'í*, its *'atsii'*, hair, Navajo), *tsi'Yalcide* (red hair, Chiricahua and Mescalero Apache); UTO-AZTECAN:

*Portulaca umbraticola*. A. Flowering and fruiting plant. B. Inset of petal. Artist: R. D. Ivey (Ivey 2003).

*guaro* (Mayo), *ku'ukpaḍ* (Akimel O'odham; derived from *kupal*, upside-down or face down, describing the growth form; for *P. oleracea* and *P. retusa*), *ku'ukpalk* <*ki'ukpalk*> (Hiá Ceḍ O'odham; used for *P. oleracea*), *ku'ukpalk* <*ku'ukpàlk*> (Tohono O'odham, Arizona), *ku'ukpalk* <*kumpuri*> (Onavas Pima), *pihala* <*pihála*> (Hopi), *saˀrúci* (Guarijío), *verdolaaga* (loan from Spanish, Mountain Pima).

**Botanical description:** Herbs, annual, fleshy, glabrous, the stems prostrate to erect or ascending. Leaves are few, the blade flat, sessile, the lower spatulate or obovate to obtuse and rounded, the upper oblanceolate to oblong and often acute, 1–3 cm long, 1–11 mm wide. Flowers are in clusters at the ends of branches. Sepals are ovate, obscurely keeled. Corollas are yellow or orange to partly red, the petals spatulate or obovate, acute to cuspidate. Stamens are 7–27. Capsules are circumscissile at the middle or above, the rim crowned by a narrow wing, the lid flattish. Seeds are gray, tuberculate.

**Habitat:** Desertscrub, rocky ridges. 760–1,830 m (2,500–6,000 ft).

**Range:** Arizona (Cochise, Gila, Graham, Greenlee, Pima, Pinal, and Santa Cruz counties), New Mexico, and western Texas to Georgia; Mexico (Baja California, Chihuahua, and Sonora); Cuba and Jamaica.

**Seasonality:** Flowering July to September.

**Status:** Native.

**Ecological significance:** These herbs are typical of dry sites on ridges where *Portulaca* appears after the summer rains begin. There, *verdolaga* flourishs until the moisture fails, but by then it has flowered and fruited and left thousands of seeds behind to germinate with the next season's rains.

Fruits have numerous seeds, presumably about the same as *P. oleracea*. In that species, 52,300 seeds have been counted on a single plant (Martin et al. 1951). Common Ground Doves (*Columbina passerina*), several sparrows, other songbirds, and a number of mammals, including rabbits and kangaroo rats, are recorded as eating *Portulaca* seeds.

Flowers do not open early in the morning. Indeed, it is often 10 a.m. or later before they open.

**Human uses:** By far the most commonly used species of *Portulaca* in the New World is *P. oleracea* (Hocking 1997, Moerman 1998). That species was found in the Salts Cave in Kentucky and radiocarbon dated at 3000 BCE (Chapman et al. 1974). Similarly, the species has been dated at 2000 BCE in Asia (Dunmire 2004). *Portulaca* was a popular source of greens eaten in ancient Egypt and classical Greece and Rome (Davidson 1999). Some think *P. oleracea* arrived late in Europe, perhaps about 1582 (Hedrick 1919). This garden plant from Spain probably was introduced to the New World between 1540 and 1600 (Dunmire 2004). Subsequently, the Old World and New World genomes mixed.

*Portulaca oleracea* is uncommon in the canyons, but two other species, *P. suffrutescens* and *P. umbraticola*, occur more frequently. Most similar to *P. oleracea* is *P. umbraticola*, and people do use it. Mayo gather and eat it as a source of greens when the succulent leaves appear in the summer (Yetman and Van Devender 2001). Onavas Pima cook leaves as greens (Pennington 1980). Akimel O'odham mostly use *P. oleracea*, and Rea (1997) made no mention of *P. umbraticola*. Probably its growth form is different enough that these Pimans would not include it in *ku'ukpaḍ*. Moreover, *P. umbraticola* does not seem to grow along the Gila River (SEINet 2009).

**Derivation of the name:** *Portulaca* is perhaps a diminutive of the Latin *porta*, a gate or door, from the lid of the fruit; others suggest that it derives from *portu-laca*, "milk carrier," or even from *porca*, defined as "*Porcelle*, as *Porche*, the fine Cockle or Muscle shels which Painters put their colors in." The name was used by Pliny (CE 23–79) for *P. oleracea*. An unexpected interpretation of *porcella*, a sow, is that it is a "vulgar term for the pudendum" (Coffey 1993). Moreover, "purslane," derived from it, is said to mean "herb of the womb." *Umbraticola* means growing in shady places, a misnomer for Sonoran area plants.

## *Androsace occidentalis* (Primulaceae) Western Rock-Jasmine

*Androsace occidentalis*. A. Life-form. B. Fruit, enlarged. Artist: F. Emil (Britton and Brown 1896–1898).

**Other names:** ENGLISH: western fairy candelabra (from Latin *candelabrum*, in classical Greek and Roman times, a candle-stick; entering English by 1815; by 1834, the word was applied to plants that were branched like a candelabrum), [western] rockjasamine (jasmine or jessamine, is derived from Persian *yasmin*, *yasman*; plants of the "real" jasmine are *Jasminum* in the Oleaceae; typically the name is applied to flowers with a strong evening scent); ATHAPASCAN: *'azee' dilkǫǫh* <*ˀazeˀ dilkǫhí*> (*'azee'*, medicine, *dilkǫǫh*, smooth, Navajo; for *A. septentrionalis*); UTO-AZTECAN: *ka'na* (Shoshoni; questionably *Androsace*, used for *Lewisia*, cf. Chamberlin 1911)

**Botanical description:** Annual herbs. Leaves in a basal rosette, 6–30 mm long, lanceolate to oblanceolate or spatulate, not differentiated into a blade and petiole, entire or toothed, puberulent with simple hairs. Stalks (scapes) are 1 to many, 2–10 cm tall, puberulent with forked hairs. Bracts are lanceolate to elliptic, pubescent with forked hairs. Flower clusters are umbels, 2–10-flowered, the pedicels slender, 3–30 mm long. Calices are top-shaped (turbinate) to bell-shaped (campanulate), 3.8–5 mm long, puberulent, keeled below each lobe, the lobes about equal to the tube. Corollas are white, included in the calyx. Capsules are globose, about equal to the calyx in length, opaque. Seeds are brown, ca. 1 mm long.

**Habitat:** Along streams, canyons, rocky north-facing slopes. 300–1,500 m (1,000–5,000 ft); 1,280–2,135 m in Utah (Welsh et al. 1987).

**Range:** Arizona (all counties except Navajo), California, New Mexico, Texas, north to Indiana, Michigan, and Ohio; British Columbia to Manitoba; and Mexico (Sonora).

**Seasonality:** February to April (May).

**Status:** Native.

**Ecological significance:** In Arizona, these herbs grow from the Sonoran Desert to open pine forests (Cholewa 1992). Western Rock-Jasmines are typically not abundant below about 4,300 feet in the Baboquivari Mountains but reappear at the Arivaca Ciénega farther east at 3,626 feet. The species is also known from the nearby Coyote and Quinlan Mountains. Only the similar *A. septentrionalis* is recorded from west of the Baboquivari Mountains (Felger and Broyles 2007).

In the canyons, these herbs become apparent in late January or early February, and their basal rosettes of leaves are often tinged reddish. The flowers are tiny and so inconspicuous that the leaves are usually noticed first. Because the area is often dry, Western Rock-Jasmine rarely reaches its full height potential; the plants are often only 2–4 cm tall. Moreover, *Androsace* tends to be interspersed with other spring plants and may not be easily seen. However, examination of north-facing slopes, particularly where there is water seeping, will usually reveal these small plants. Once they are found, it may be surprising how many of them are tucked among other species in an area that at first seemed to lack them.

On the eastern part of its range, Western Rock-Jasmine is known from a single site in Ohio where it is considered endangered, and it is similarly endangered in Indiana and Michigan (Cusick 1989, Anonymous 1999, Michigan Botanical Club 2001). That view of one group of scientists contrasts with another who consider it a "weed" to be controlled in agricultural fields (Blackshaw 2003). Because the seedlings may grow in densities of over 200 per meter (Karl et al. 1999), the "weed" concept may be understandable. Both views emphasize variable frequencies in different parts of the range of *A. occidentalis*.

Although this species is listed on the national registry of wetland plants, few in Arizona would consider the places *Androsace* grows as "wetlands." However, Western Rock-Jasmine plants in Brown Canyon do typically grow in streambeds and in "seeps" on north-facing slopes where groundwater comes to the surface. Based on its overall distribution, Stevens (1920) considered this plant a "Sonoran life zone" species that extended north and east into North Dakota.

**Human uses:** Navajo make a compound decoction of *A. occidentalis* to stop postpartum hemorrhage, to prevent birth injury, and to treat venereal disease and pain (Vestal 1952, Moerman 1998). A decoction is also drunk before the sweat bath (Vestal 1952). In addition, Navajo use *A. septentrionalis* as protection from bewitchment and pain from witches' arrows (Wyman and Harris 1951).

**Derivation of the name:** *Androsace* is derived from Greek *andros*, a man, male, and *sakos*, a shield; akin to Latin *androsaces*. The name originally was applied by Dioscorides (fl. CE 40–80) to some unknown plant. *Occidentalis* means "western," in comparison with another species, which is *septentrionalis* or "northern."

**Miscellaneous:** *Androsace*, with 100–150 species, is a member of the largely temperate and boreal family Primulaceae (Robbins 1944, Cholewa 1992, Mabberley 1997). This species was the first recorded for North America and was described in 1814 by Thomas Nuttall. Indeed, Nuttall collected the specimen in 1811 near a "Maha" [Omaha] village on the Missouri River in what is now the state of Nebraska (Robbins 1944).

## *Astrolepis sinuata* (Pteridaceae) Wavy Cloak-Fern, *helecho*

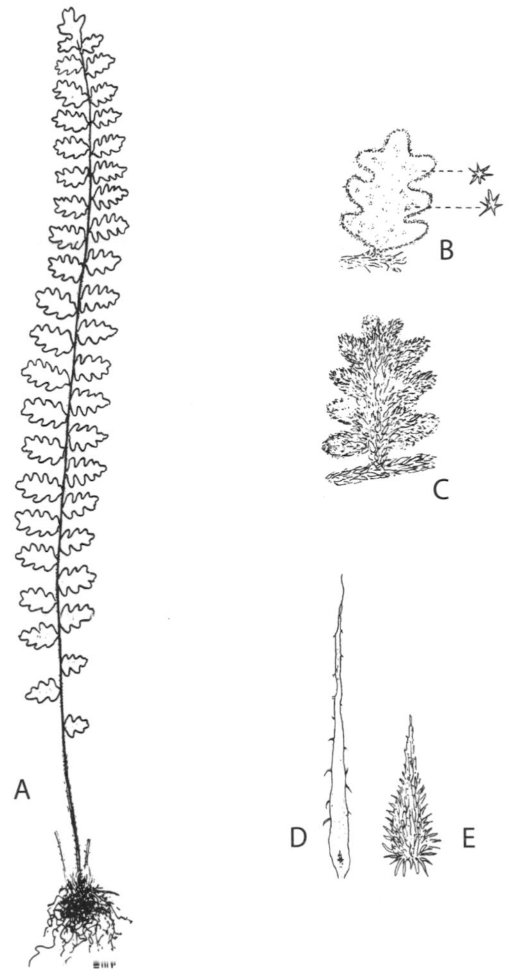


*Astrolepis sinuata*. A. Frond with base. B. Upper surface of pinna with scales, enlarged. C. Lower surface of pinna. D. Rhizome scale. E. Pinna rachis scale. Artist: E. M. Paulton (Mickel 1979).

**Other names:** ENGLISH: star-scaled [wavyleaf, wavy, scaly] cloak-fern ("star-scale" because of the unique star-like pubescence; "cloak fern" is a translation of a former generic name, *Notholaena*, from Greek where *chalena* means cloak; "cloak" is from Old French *cloke*, in turn from medieval Latin *cloca*, *clocca*, a cape worn by horsemen and travellers, in use by ca. 1275, and figuratively as something that covers or conceals by 1526); SPANISH: *calaguala* (Chihuahua; from Quechua *kalla-hualla*, a widespread name for ferns adapted into Mexico for an array of medical species), *cañahuala* (variant of *calaguala*, Chihuahua), *candelilla* (little candle, Coahuila), *doradillo* (little golden one, Coahuila), *helecho* (fern, Mexico; a general fern name), *nacahuela* (maybe from Náhuatl *nacaztli*, ear, Edo. México); UTO-AZTECAN: *kalawala* (Tarahumara; adapted from Spanish *calaguala*), *mási-l*$^{y}$ (fern, Cahuilla), *máṣ-la* (fern, Luiseño); (= *Notholaena sinuata*, *Cheilanthes sinuata*)

**Botanical description:** Rhizomes with linear scales, these of uniform color (concolorous) or weakly of two distinct colors (bicolorous), pale reddish brown, sometimes with a darker central stripe, many short marginal teeth. Leaf stalks (stipes) are 1.5–14 cm long, round or slightly flattened, reddish brown or brown, with lanceolate, toothed, appressed scales. Blades are linear, 10–45 cm long, 2–7 cm wide, pinnate, tapering gradually at the base and apex, the ultimate segments oblong, ovate, or nearly lanceolate, with 3–8 pairs of narrow lobes extending 1/3–1/2 way to the costa (midrib), curled upward to reveal the lower (abaxial) surface when dry, the segment margins underolled.

**Habitat:** Dry, rocky slopes and crevices, often on limestone. 304–2,133 m (1,000–7,000 ft).

**Range:** Arizona (Cochise, Coconino, Graham, Pima, Santa Cruz, and Yuma counties), and California to western Oklahoma and Texas; Mexico to Chile; Jamaica and Hispaniola.

**Seasonality:** Reproductive during wet seasons.

**Status:** Native.

**Ecological significance:** Usually on limestone rocks or among them in crevices in other places. No limestone is known in Brown Canyon. The species is on the northwestern part of its range near the Baboquivari Mountains, although it goes much farther north in New Mexico.

This species is perhaps the most common fern in the canyons. Particularly during rainy winter months when the majority of plants are leafless, these green ferns show up against the gray background. When winter rains have been generous, the ferns are fully uncurled and photosynthetic.

When the dry season approaches, the ferns become less conspicuous. The pinnae curl in on themselves and eventually become whitish from exposure of the scaly lower surface.

Unlike *Notholaena standleyi* (q.v.), *A. sinuata* is not often hidden below overhanging rocks.

**Human uses:** People in Chihuahua make an infusion of the entire Wavy Cloak-Fern plant to treat stomach problems (Ford 1975).

In Chihuahua, ferns called *Notholaena tomentosa* (= *Cheilanthes tomentosa*) are called *doradilla* (little golden one) and are used as a diuretic and depurative (Hocking 1997). This species may be the *doradilla* that the Névome called *babahuhogisa*, but there is insufficient evidence to be certain (Pennington 1979).

**Derivation of the name:** *Astrolepis* combines Greek *astron*, a star, and *lepis*, a scale, to refer to the scales on the frond surface. *Sinuata* alludes to the wavy frond margins.

**Miscellaneous:** The genus *Astrolepis* contains eight species. The combination of a chromosome base number of x = 29, pinnate leaves, two vascular bundles in the petioles, unique stellate or coarsely ciliate scales on the lower blade surface, and other traits indicate that *Astrolepis* is a monophyletic group (Bentham and Windham 1993).

The similar *A. cochisensis* (*helechillo*, little fern) has been found only a single time in Brown Canyon. It also grows in the Coyote Mountains and in nearby mountains, west into the Pinacates, east into the Trans-Pecos part of Texas, and south into Mexico. This fern extends in Texas northeast into the Plains Country (Motley County) and east to the Edwards Plateau (Sterling and Edwards counties). Although *A. cochisensis* has not been found in nearby Thomas Canyon, perhaps the species occurs elsewhere in the mountain range. *Astrolepis cochisensis* has leaf pinnae typically 4–7 mm long, entire or asymmetrically lobed, with nearly circular to elliptical scales. The largest pinnae on *A. sinuata* are usually 7–35 mm long, symmetrically lobed, with elongate scales. (According to the OED the word "elongate" means "[l]engthened, prolonged, extended; esp. in *Bot.* and *Zool.* that is long in proportion to its breadth; that has a lengthened, slender, or tapering form." "Elongated" has a slightly different meaning.)

## *Cheilanthes fendleri* (Pteridaceae) Golden Lip-Fern, *helecho*

**Other names:** ENGLISH: [Fendler's] lip-fern (any fern of the genus *Cheilanthes*; an allusion to the lip-like covering of the reproductive structures [indusium]; in use by 1890); SPANISH: *helecho* (fern); ATHAPASCAN: *łogaxe* (coffee plant, Chiricahua and Mescalero Apache); KIOWA TANOAN: *'ok'up'e'ñaebi* (*p'e'ñaebi*, plant, Santa Clara Tewa) *papi'e* (San Ildefonso Tewa); UTO-AZTECAN: *ti(m)binawivɨ* (rock apron, Kawaiisu)

**Botanical description:** Fronds scattered on a slender creeping rhizome, to 3 dm tall, but usually

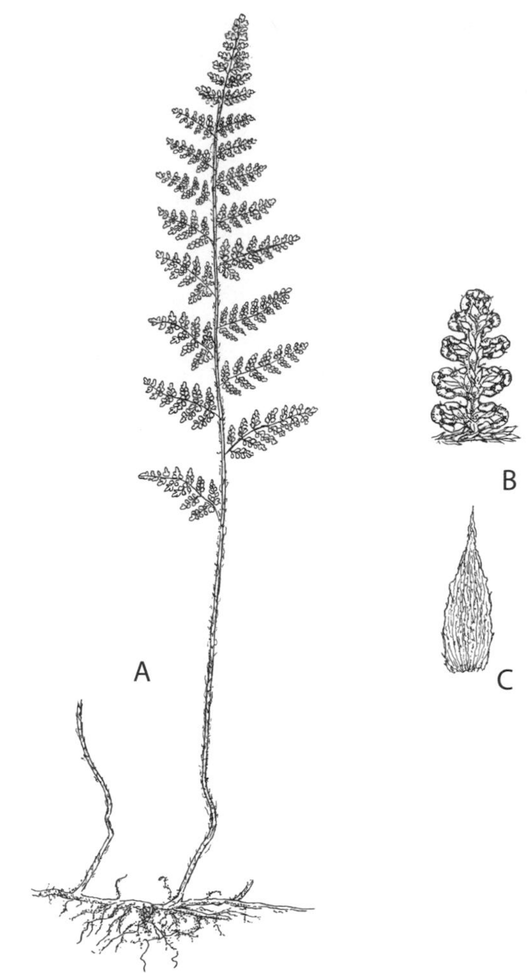


*Cheilanthes fendleri*. A. Frond with fern base. B. Lower surface of pinnule. C. Blade scale. Artist: E. M. Paulton (Mickel 1979).

much smaller. Scales are concolorous to weakly 2-colored, pale brown, and sometimes with a dark brown central stripe, entire. Leaves are with stalks (stipes) 3–17 cm long, brown to purplish black, with narrow scales having long teeth near their base plus linear, hair-like scales. Blades are narrowly ovate-lanceolate to oblong-lanceolate, 4–15 cm long and 1–4 cm wide, 3-pinnate, basally obtuse, apically acuminate, rigidly herbaceous, glabrous above (adaxially) and below but with scales and a few hairs below (abaxially), with hair-like scales on the axes, the segment margins strongly folded under.

**Habitat:** Growing among rocks on slopes and cliffs. 1,219–2,895 m (4,000–9,500 ft).

**Range:** Arizona (Coconino, Mohave south to Cochise, Greenlee, Pima, and Santa Cruz counties), California, Colorado, New Mexico, and western Texas

(Hudspeth and Jeff Davis counties and disjunct to Crosby County); Mexico (Baja California, Chihuahua, and Sonora).

**Seasonality:** Reproducing during the rainy seasons.

**Status:** Native.

**Ecological significance:** This fern is another that grows in rock crevices, at the base of boulders, and along ledges, particularly in shaded canyons. The species was named by J. D. Hooker in 1852 based on specimens collected in New Mexico by A. Fendler, and it was though to be endemic to the southwestern United States until after the middle 1980s. Although C. G. Pringle collected the species in Chihuahua in 1885, the specimen was either misidentified or unknown to fern specialists until after 1985. This fern is thought to hybridize with *C. wootonii* (Windham and Rabe 1993).

**Human uses:** Chiricahua and Mescalero Apaches boil the leaves and young stems to make a beverage (Castetter and Opler 1936). Keres use an infusion of *C. fendleri* as a douche after childbirth (Moerman 1998). Tewa grind Golden Lip-Fern and put it on the lips to treat cold sores (Robbins et al. 1916). Navajo also use the related *C. wootonii* as a medicine to treat gunshot wounds, and as a Life Medicine (Vestal 1952).

**Derivation of the name:** *Cheilanthes* is from Greek *cheilos*, lip or margin, and *anthos*, a flower, in allusion to the marginal indusia or membrane covering the sporangia. *Fendleri* commemorates German-born American August Fendler, 1813–1883, explorer and collector in North and South America.

**Miscellaneous:** *Cheilanthes wootonii* also occurs in Brown Canyon. This fern is similar to *C. fendleri*, but has stalk scales with eyelash-like structures along the margins (ciliate). This species is able to reproduce without sexual gametes (apogamous), and it is a triploid whose ancestors are unknown.

Also found in the Baboquivari Mountains are *Cheilanthes eatoni* (commemorates E. A. Eaton, 1865–1908, a student of ferns, who was a lecturer and writer in New York), *C. lindheimeri* (commemorates Ferdinand Jakob Lindheimer, 1801–1879, German-born collector of Texas plants, 1836–1879, and resident of New Braunfels, Texas), *C. pringlei* (commemorates C. G. Pringle, 1838–1911, plant collector in the Pacific states and Mexico; his diary was published in 1918 as *The Record of a Quaker Conscience*), *C. tomentosa* (pubescent, but most are, so this one is not unique), and *C. wrightii* (commemorates Charles Wright, 1811–1885, a collector in the American Southwest). All of these except *C. tomentosa* have been identified in Brown Canyon. These species are difficult to distinguish, the differences being largely based on microscopic traits. *Hierba de la peña* (sorrow herb) is a name applied to *C. lindheimeri* in San Luis Potosí.

*Cheilanthes* comprises perhaps 150 species that grow mostly in the Western Hemisphere, but a few are known in Europe, Asia, Africa, the Pacific Islands, and Australia (Windham and Rabe 1993). This genus is the largest and most diverse of the xeric-adapted ferns. *Cheilanthes* is considered related to both *Notholaena* and *Pellaea*, and it has at one time or another been considered part of one or both of those genera.

## ***Notholaena standleyi*** **(Pteridaceae)** **Star Cloak-Fern, *helecho***

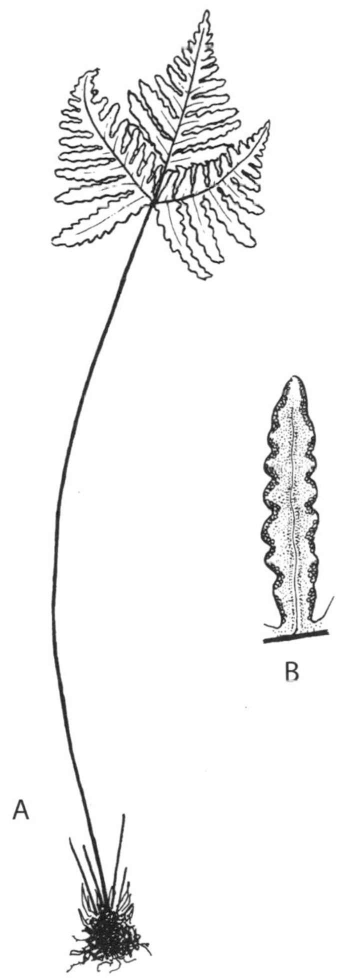


*Notholaena standleyi*. A. Leaf with part of basal rhizome. B. Fertile leaf segment (pinnule). Artist: E. M. Paulton (Mickel 1979).

**Other names:** ENGLISH: [star] cloak fern ("star" is unexplained; perhaps it is from the more or less star shape of the scales in some species, or perhaps from the star-shaped leaves; "cloak" from the generic name where *chalena* means cloak), rock fern; SPANISH: *helecho* (fern); SERI: *hehe quina* (hairy plant, Seri); (= *Cheilanthes standleyi*)

**Botanical description:** Rhizomatous ferns, scales with a dark red-brown center, the margins membranous to orange-brown, entire to slightly ragged, the center turning black with age. Petioles are 3–13 cm, long, dark brown, glabrous or with felt-like hairs when young. Leaf blades are 5-angled in outline, about as wide as long, 3–7 cm wide, divided into 5 major, deeply cleft pinnate segments (pinnae), the upper surface green, lower surface somewhat obscured by golden yellow glands.

**Habitat:** Often among broken rocks on slopes, in canyons, washes. 300–1,980 m (1,000–6,500 ft).

**Range:** Arizona (Cochise, Gila, Graham, Greenlee, Maricopa, Pima, Pinal, and Santa Cruz counties), New Mexico, Texas to western Oklahoma, and southeastern Colorado; Mexico (Baja California, Chihuahua, and Sonora to Tamaulipas to Durango and Puebla).

**Seasonality:** Reproductive after rains.

**Status:** Native.

**Ecological significance:** The Sonoran Desert plants are smaller than those in mesic regions and are probably sexual diploids (Felger 2000). This species is perhaps the second most common fern in the canyons, with *Astrolepis sinuata* (q.v.) being the most frequent. The 5-angled leaves of these small ferns are often in patches among rock clusters and at the bases of boulders where the ferns take advantage of moisture retained by the stones after it has been lost in more open sites.

When the 1.9-inch rain of mid-March temporarily broke the winter drought of 2006, these ferns were among the first to awaken. Earlier, their leaves had been folded and curled so that fronds were comparatively inconspicuous. After that rain, fronds unfolded and became obvious below their protecting boulders and logs.

**Human uses:** The Seri make a tea from the fronds to help women conceive (Felger and Moser 1985). More important to them perhaps is the fern's value with regard to the supernatural. Various preparations of leaves bring "luck" to the person carrying them. All of the uses involve putting pieces of the fern in a small cloth bag, and these help in bringing luck in gambling, causing rain, and protecting the wearer from floods.

**Derivation of the name:** *Notholaena* is from Greek *nothos*, false, and *chlaena*, cloak, thereby referring to the leaf margins and the incomplete indusium (covering of the spore-producing structures). *Standleyi* is dedicated to Paul Carpenter Standley, 1884–1963, a student of the flora of Central America and the southwestern United States, particularly New Mexico (e.g., Wooton, E. O., and P. C. Standley. 1915. Flora of New Mexico. *Contributions from the U.S. National Herbarium* 19: 514–519).

**Miscellaneous:** The other two common species in the area are *N. aschenborniana* (= *Cheilanthes aschenborniana*) and *N. grayi*. Leaves on these species are not 5-angled as in *N. standleyi*, but longer than wide and feather-shaped. These two species are distinguishable by examining the margins of the scales on the leaf stalk. If the scales are dissected or have long eyelash-like margins (ciliate) and the undersides of the fronds are obscured by scales, the individuals are *N. aschenborniana*. Plants that have smooth-margined scales and undersides of the fronds evident without having to remove the scales to see the leaf surface are *N. grayi*.

*Notholaena aschenborniana* grows from southern Arizona, across the tip of southwestern New Mexico, through southwestern Texas, and south into Mexico (Chihuahua, San Luis Potosí, Tamaulipas, Sonora, and south to Distrito Federal). These ferns are found among rocks and cliffs from 4,500 to 9,000 feet in elevation, particularly on limestone substrates.

*Notholaena grayi* is known in almost the same region as *N. aschenborniana*, southern Arizona, across the tip of southwestern New Mexico, through southwestern Texas, and south into Mexico (Chihuahua, Coahuila, and Sonora). *Notholaena grayi* grows at somewhat lower elevations, from 3,000 to 5,000 feet. Both species are recorded by Mickel (1979) as "frequent."

## ***Pellaea truncata*** **(Pteridaceae)**
## **Spiny Cliff Brake, *calaguala***

**Other names:** ENGLISH: spiny cliff brake ("spiny" refers to sharp points on the leaflets; "brake" is perhaps from Scandinavian, akin to Old Swedish *braekne*, fern); SPANISH: *calaguala* (Mexico, from Quechua *kallahualla*, a widespread name for ferns adapted into Mex-

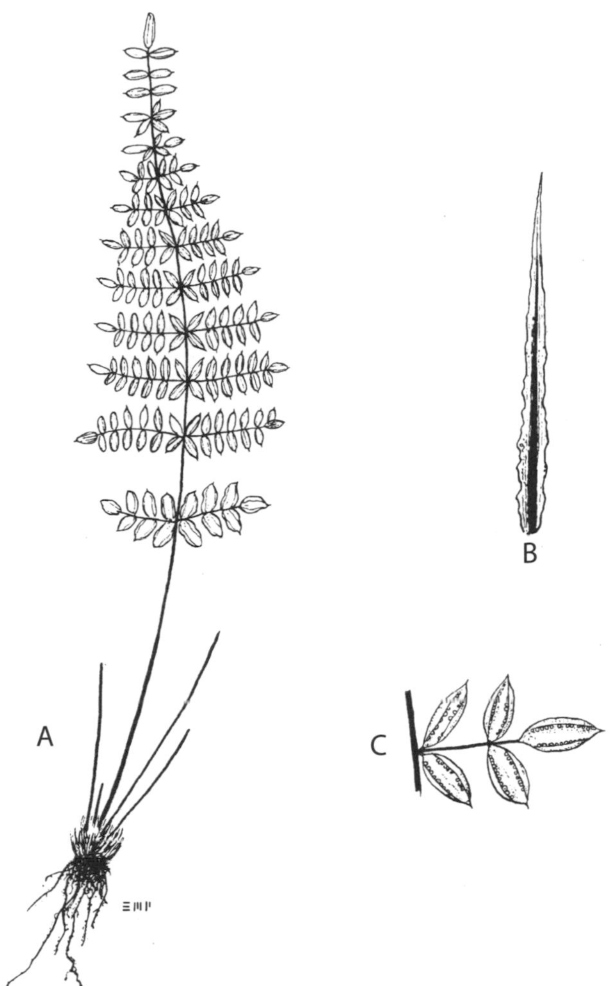


*Pellaea truncata*. A. Frond with basal segment with roots. B. Scale. C. Fertile pinnule. Artist: E. M. Paulton (Mickel 1979).

ico for an array of medical species); UTO-AZTECAN: *tča* (Tübatulabal; for *P. mucronata*), *ti(m)binawivɨ* (rock apron, Kawaiisu; for *P. mucronata*)

**Botanical description:** Rhizomes with a bundle of erect branches, or short-creeping, 2–6 mm in diameter, usually much-branched, scales 2-colored, brown with a black central stripe, sometimes somewhat contorted, distally toothed. Fronds are crowded, with stipes 3–20 cm long, adaxially (ventrally) flattened or shallowly grooved, reddish brown to purplish black, glabrous. Leaf blades are mostly lanceolate, 8–22 cm long, 2.5–10 cm wide, 2-pinnate, the pinnules elliptic, 5–12 mm long, sessile, not articulate, strong rolled under, obtuse to truncate basally, mucronate apically.

**Habitat:** Among rocks, cliffs, canyons. 600–1,830 m (2,000–6,000 ft).

**Range:** Arizona (Coconino, Mohave, south to Cochise, Pima, Santa Cruz, and Yuma counties), southern California, eastern Colorado, southern Nevada, New Mexico, western Texas, and southwestern Utah; Mexico (Chihuahua and Sonora).

**Seasonality:** Reproductive after rains, particularly in the late spring and fall.

**Status:** Native.

**Ecological significance:** Spiny Cliff Brake is a mountain species. Although the fern has been reported as low as 2,000 feet within the state, it has not been recorded that low in the Baboquivari Mountains. There, the plants are nestled among rocks beginning about 4,000 feet elevation and continuing upward. Spiny Cliff Brake has a curious distribution; it is found in the Sierra San Pedro Mártir of Baja California, in the Ajo Mountains of Organ Pipe Cactus National Monument in southwestern Arizona, and then reappearing in the Baboquivari Mountains, to extend from there to the east and north. The entire genus is adapted for living in drought areas (Windham and Rabe 1993).

All of the *Pellaea* species in North America outside Mexico are an apparently monophyletic alliance usually called section *Pellaea* (Windham and Rabe 1993). That group of species may actually be distinct from the ferns in Asia that have been put in the genus. The closest relatives of *Pellaea* in the United States seem to be the *Cheilanthes alabamensis* complex and the distinct genus *Argyrochosma*.

**Human uses:** None found for this species. However, *P. atropurpurea* is used by the Cahuilla of southwestern California in a medicine to thin the blood and to prevent sunstroke (Romero 1954). *Pellaea truncata* and *P. atropurpurea* are considered closely related (Felger 2000). Tübatulabal in California eat the leaves and stalks of *P. compacta* (Voegelin 1938).

**Derivation of the name:** *Pellaea* is from Greek *pellos*, dark, possibly referring to the bluish-gray leaves or perhaps the stalks (rachises). *Truncata* means cut off.

**Miscellaneous:** *Pellaea* is a genus of about 40 species that are widely distributed in the tropics and warm temperate regions (Windham and Rabe 1993). There is a single species in Europe (Mabberley 1997). Most of the species grow in the southwestern United States and adjacent Mexico, with 15 known from the United States and Canada. All of the species in the border region of Arizona are known in the Baboquivari Mountains.

*Pellaea atropurpurea* (Purple Cliff Brake) is frequent. This species is distinguished by the stipes and rachises without grooves on the top and the purple color. By contrast, *P. truncata* has grooves.

Also present are *P. intermedia* and *P. wrightiana*. *Pellaea intermedia* is distinctive because of its petioles and rachises, which are straw-colored, tan or gray, and rarely lustrous. All the others have petioles and rachises that are dark brown to black and usually lustrous. Scales must be examined to separate the other species. Scales are uniformly reddish brown or tan in *P. atropurpurea*. At least some stem scales are 2-colored, with a dark central region and lighter, brown margins in both *P. truncata* and *P. wrightiana*. The pinna costa is much longer than the ultimate segments, the largest pinnae are divided into 11 or more segments, and the blades are ovate-deltate to lanceolate and usually more than 4.5 cm wide in *P. truncata*. By contrast, the pinna costa is usually shorter than or equal to the ultimate segments, the largest pinnae are divided into 3–11 segments, and the blades are linear-oblong, and usually less than 4.5 cm wide in *P. wrightiana*.

## *Anemone tuberosa* (Ranunculaceae) Desert Wind-Flower

**Other names:** ENGLISH: desert anemone [windflower] ("anemone" appeared in English with William Turner's *Herbal* of 1551; he wrote, "Anemone hath the name because the floure neuer openeth it selfe, but when the wynde bloweth"), tuber anemone (a book name, New Mexico); UTO-AZTECAN: *toi'yamohagûp* [*toi'yamogup*] (Shoshone; for *A. multifida*)

**Botanical description:** Perennial herbs, the stems 8–50 cm tall, sparingly silky or glabrous. Basal leaves 6–16 cm long, 3-foliolate, the leaflets deeply cleft or parted, the ultimate segments obovate to cuneate. Peduncles are with 1–5 flowers, pilose to glabrous. Flowers are 18–36 mm wide, with 9–18 sepals, white to rose-pink. Achenes are 3.5–4.5 mm long, woolly-pubescent.

**Habitat:** Among rocks on mesas and foothills. 762–1,524 m (2,500–5,000 ft).

**Range:** Arizona (Coconino and Mohave counties south to Cochise, Greenlee, Pima, and Santa Cruz counties), California, New Mexico, extreme western Texas, and Utah. Found mostly below the Mogollon Rim.

**Seasonality:** Flowering February to April.

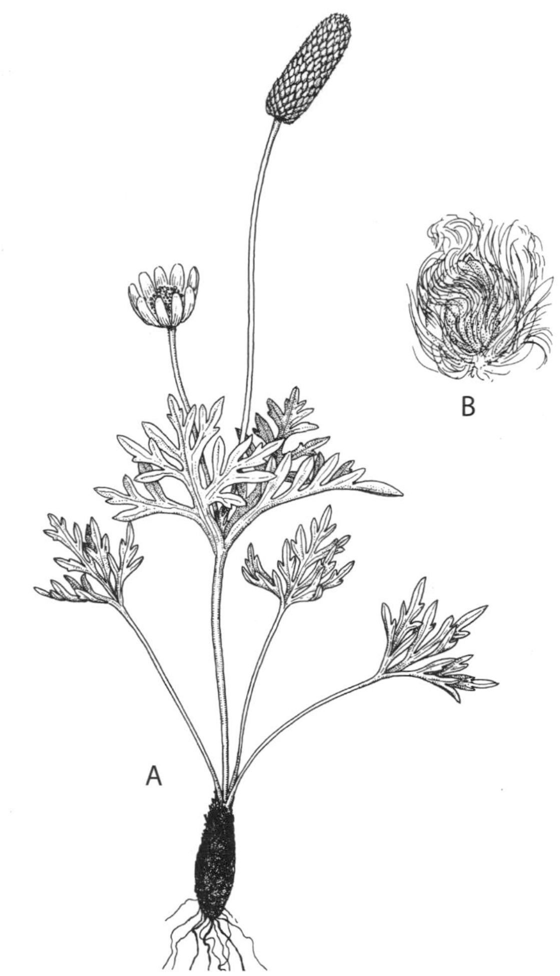


*Anemone tuberosa*. A. Flowering and fruiting plant. B. Detail of seed. Artist: L. B. Hamilton (from original artwork).

**Status:** Native.

**Ecological significance:** These plants are among several kinds of perennial herbs with above-ground parts that disappear during the dry season. In the spring, particularly when there has been sufficient rain, the above-ground parts appear and produce patches of showy flowers. This ephemeral existence mimics that of the annual species, but it has been achieved through a different life-strategy.

*Anemone tuberosa* is one of the first native herbs to flower in the canyons in the spring. Buds are obvious and prevalent by the last two or so weeks in January. In the first week or two of February, the flowers open. Flowers are first white, and they become tinged with pink or rose with age. Neither leaves nor flowers appeared after the almost rainless winter of 2006.

West of the Baboquivari Mountains, this species has been found in the Barry M. Goldwater Air Force Range, Organ Pipe Cactus National Monument, and the Sonoran Desert National Monument (Felger et al. in prep.).

**Human uses:** No records were found of people using *Anemone tuberosa*.

Moerman (1998) lists 9 of the 25 species recognized in the *Flora of North America* (including *Pulsatilla*) that have been used to treat a variety of ailments. One must wonder, in the absence of uses among most indigenous American peoples, if some natives learned of these plants from Old World settlers. Among the Europeans, *Anemone hepatica*, also called *Hepatica*, was known as the *erba trinita* (Italian), the herb trinity, and *trinitaire* (of the Trinity, France). That name came from the classical pharmaceutical name *Herba Trinitalis.* Sir Thomas Browne wrote in 1646 that *Hepatica* "obtaineth that name onely from the figure of its leaves, and is one kind of liverworte or Hepatica."

**Derivation of the name:** *Anemone* is from Greek *anemone.* There are two views of its meaning: (1) "daughter of the wind" (from *anemos*, wind, and the patronymic suffix *-one*) or (2) "red" (from *amona*, the root word being *mon*, "to be red"). The latter view would be akin to Sanskrit *hema* (red), Akkadian *sama*, *samtu*, *samat* (red), *damu* (blood), and Greek *haima*, *haimatos* (blood). Linguist Henry Bradley (1845–1923), an editor of the Oxford English Dictionary, objected to the first derivation because the suffix was not "exclusively patronymic." Coffey (1993) also supported the latter view by saying that the word *Anemone* is derived from *Namaan*, the Semitic name for Adonis, from whose blood the flower supposedly arose, or that it grew from Aphrodite's tears as she wept for the slain Adonis. *Tuberosa* refers to the tuber-like caudex (underground stem) that allows the species to survive drought periods.

The Crown Anemone (*Anemone coronaria*) is the "Lily-of-the-Field" of the Bible. The reference is in Matthew (6:28–29), as: "Consider the lilies of the field, how they grow; they toil not, neither do they spin: And yet I say unto you, That even Solomon in all his glory was not arrayed like one of these."

**Miscellaneous:** *Anemone* comprises 144–150 species (Mabberley 1997, Dutton et al. 1997). There are 17 species in Europe, and Kartesz (1994) admitted 27 for all of North America (including two he put in *Pulsatilla*). Dutton et al. (1997) recognize only 25 for the *Flora of North America North of Mexico.* Those species put in *Pulsatilla* have historically been separated because of their long plumose achene beaks. Other morphological and phylogenetic analyses suggest that *Pulsatilla* should be included within *Anemone*, as it is not a natural phylogenetic grouping.

## *Clematis drummondii* (Ranunculaceae) Virgin's Bower, *barba de chivato, keli ciñwo*

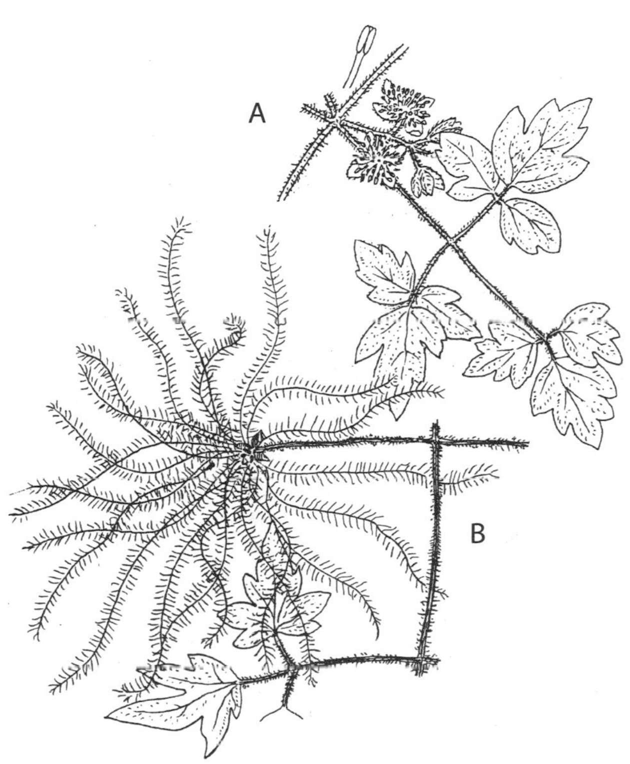


*Clematis drummondii.* A. Branch of male plant. B. Branch of female plant. Artist: R. D. Ivey (Ivey 2003).

**Other names:** ENGLISH: Drummond's clematis (a book name, New Mexico), old-man's-beard (referring to the white-plumed fruits and their resemblance to a white beard), [Texas] virgin's bower ("virgin's bower" was applied to the European climber *Clematis vitalba* by John Gerarde in 1597; later extended to American species); SPANISH: *barba de chivato [chivo]* (goat's beard, Chihuahua, Coahuila, San Luis Potosí, Sonora, Tamaulipas, Zacatecas), *chilillo* (branch used as a whip, probably originally from Mayan), *hierba de los averos* (herb of the disgraceful ones, San Luis Potosí), *redadura de nopal* (wraps around cactus, Mountain Pima); ATHAPASCAN:

*ch'ił na'atł'o'ii* [*ts'oh, ts'ósí*] <*č'il na'aƛ'o'i 'ałc'ósí, c'il naʾar-ʾóʾi* [*coh, c'o's*]> (*ch'ił*, plant, *na'atł'o'ii*, weaving, *ts'oh*, big, *ts'ósí*, slender, Navajo), *nanisdiz* (Western Apache); UTO-AZTECAN: *kava vopar* <*kaava boporo*> (*kava*, horses', *vopar*, hairs, Mountain Pima), *keli ciñwo* (*keli*, old man, *ciñwo*, whiskers, Tohono O'odham), *keri tenvo* <*kuri tunvo*> (old man's whiskers, Onavas Pima, Sonora), *o'bĭndamanûmp* (Shoshoni; for *C. douglasii* and *C. ligustricifolia*), *pogʷitina hɨapiina* (Grizzly Bear's trap) (Kawaiisu)

**Botanical description:** Shrubs, the stems scrambling to climbing with tendril-like petioles and leaf-rachises. Leaf blades are odd-pinnate, usually 5-foliolate, the leaflets deltoid to ovate, strongly 3-parted to 3-cleft, 1.5–5.5 cm long, membranous to leathery, margins dentate. Inflorescences are usually axillary, 3–12-flowered simple cymes or compound. Flowers are unisexual. Sepals are wide-spreading, white to cream, oblong to obovate or oblanceolate, 9–13 mm long, and pubesent. Achenes are elliptic to ovate, with a plume-like beak 4–9 cm long.

**Habitat:** Canyons, washes. Up to 2,200 m (7,200 ft).

**Range:** Arizona (Cochise, Gila, Greenlee, Maricopa, Pima, Pinal, and southern Yavapai counties), southeastern California, Colorado, New Mexico, Oklahoma, and Texas; Mexico (Sonora to Tamaulipas and south).

**Seasonality:** Flowering (March) June to September (October).

**Status:** Native.

**Ecological significance:** Virgin's Bower grows only in places where there is some extra moisture, particularly in canyons and along protected washes. This species is a northern relative of *C. dioica* from farther south in Mexico and extending to Argentina (Moreno and Essig 1997).

The stigmas are effective in dispersing the fruits. As the fruits mature, the stigmas elongate to become plumose. When fully mature, the parachute-like stigmas with attached ripened ovary (achene) are dispersed by the wind. It takes only a slight puff to lift and carry these propagules considerable distances. That particular mechanism is effective, because the winds may blow fiercely, sometimes for days. An indication of that effectiveness is that the plumose stigmas have evolved in two distinct families, Ranunculaceae and Rosaceae (see also *Cercocarpus*).

**Human uses:** Poultices of Virgin's Bower leaves are used to treat irritations of the skin in humans and other animals (Moreno and Essig 1997), as they have been for *C. dioica* and other members of the subgenus. Martínez (1969) was particularly complimentary to the related *C. dioica*, which has more common names. Perhaps the name *cabeza de viejo* (old man's head) could be applied to *C. drummondii*. All of the names allude to the notable white elongated stigmas of the fruits.

*Clematis* contains acrid, toxic compounds that are highly irritating to skin and mucous membranes. These toxins include anemonin, a dilactone derived from protoanemonin (Austin 2004). Rafinesque noted in 1830, "Bark and blossoms acrid, raising blisters on the skin; a corrosive poison internally, loses the virulence by cooking." The French (*herbe aux gueux*) and Spanish (*hierba de los pordioseros*), names referring to beggars, are references to their past use to irritate the skin to simulate sores and induce sympathy from potential donors. Ingestion of any part may cause bloody vomiting, severe diarrhea, and convulsions (Foster and Duke 1990). In spite of those drastic symptoms, the Kickapoo use the leaves to treat toothaches, and they add *C. drummondii* to another plant to stop excessive bleeding at childbirth (Latorre and Latorre 1977). The Western Apache used *C. drummondii* roots in making *tułbáí* (corn liquor) (Bray 1998).

**Derivation of the name:** *Clematis* comes from Greek *klematis*, a shoot or vine-branch or twig. The genus was created by Linnaeus, but originally used by Dioscorides, fl. CE 40–80, for a climbing plant with long and lithe branches. *Drummondii* signifies that the discoverer of the species was Thomas Drummond, 1780–1835, a Scottish botanist and collector in America.

**Miscellaneous:** *Clematis* is a genus of ca. 300 species, with 32 in North America north of Mexico (Moreno and Essig 1997). The genus is found worldwide, mostly in temperate zones, although a few are subarctic, subalpine, and tropical. Of the five species in Arizona, *C. drummondii* belongs with *C. ligusticifolia* in subgenus *Clematis*.

## *Delphinium scaposum* (Ranunculaceae) Bare-Stem Larkspur, *espulita cimarrona, kukṣo wu:plim*

**Other names:** ENGLISH: [tall mountain, bare-stem] larkspur ("larkspur" in use by 1578 when Henry Lyte translated Rembert Dodoens's *Cruydeboek* to compare the floral spur with the spur of a

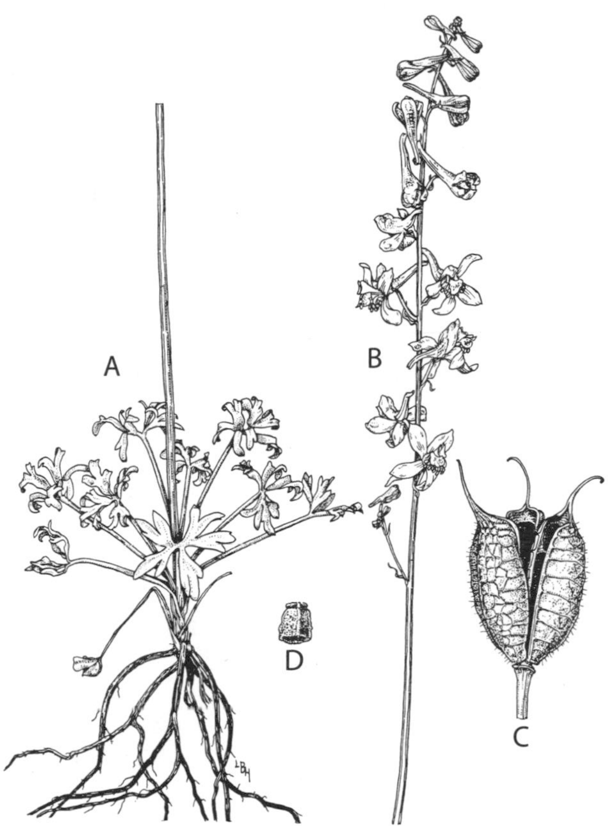


*Delphinium scaposum*. A. Basal portion of plant. B. Flowering stalk. C. Fruit. D. Seed. Artist: L. B. Hamilton (Parker 1958).

lark); SPANISH: *espuelita cimarrona* (wild little spurs, Arizona, Sonora); ATHAPASCAN: *'akee' 'ąą <k'ey ahi'>* (*'akee'*, foot, *'ąą*, spreading [under], Navajo), *bik'íhoochįįh nteel <k'ixwootxyeelih>* (*bik'íhoochįįh*, covers it [covers the ground], *nteel*, widely, Navajo), [*biką'*] *tádidíín dootł'izh <tádídín doƛ 'iš, tádidíín dootł'izhii,* [*biką'í*] *tididi'n do'y'is, txatitįiootl'ij>* (*biką'*, male, *tádidíín*, pollen, *dootł'izh*, blue, Navajo), *tł'ízí 'azee' <ƛ'ízí ˀaze·ˀ>* (*tł'ízí*, goat, *'azee'*, medicine, Navajo); UTO-AZTECAN: *chiinō hiitpa* (*chiinō*, Chinese, *hiitpa*, queue or braid, Akimel O'odham; for *D. parishii*), *cucul i'ispul <cu:cul, chuchul-i'spul, cuculi 'i'ispul, kuksho-wuuplim>* (*cucul*, rooster, *i'ispul*, spurs, from Spanish *espuela*, Tohono O'odham), *kukṣo wu:plim <kukṣo wu:pulim>* (*kukṣo*, neck, *wu:plim*, alike here and there, Tohono O'odham), *motoobɨ* (Kawaiisu; for *D. parryi*), *sawarĭnt* (Ute; for *D. menziesii*), *teoro'si <tcorosi>* (bluebird flower, from *tecoro*, bluebird, *sihu*, flower, Hopi), *tu'kubagûmp* [*pa'gasauwinoûp*] (*tu'kûm*, the sky or blue, Shoshoni), *tukymsi <tukyámsi>* (*tukya*, prairie-dog, *si*, flower, Hopi)

**Botanical description:** Perennials from thickened, fibrous-fleshy roots, with 1 to several erect stems 36–85 cm tall, herbage glabrous or soft pubescent. Leaves are basal, somewhat fleshy, the petioles 2–13.5 cm long, the blades 2.5–6 cm wide and about as long, with deeply cleft primary divisions, the lobes broad, each with a nipple-like tooth. Racemes are 5–12-flowered. Flowers are showy, 2.5–3 cm wide. Sepals are deep azure blue, upper pair of petals white with blue markings, lower petals dark blue like the sepals.

**Habitat:** Rocky slopes. 600–1,500 m, rarely 2,440 m (2,000–5,000 ft, rarely 8,000 ft).

**Range:** Arizona (Apache, Coconino, Gila, Graham, Maricopa, Mohave, Navajo, Pima, Pinal, and Yavapai counties), southwestern Colorado, New Mexico, southern Nevada, and Utah; Mexico (northern Sonora).

**Seasonality:** Flowering March to April (May, June).

**Status:** Native.

**Ecological significance:** Although these herbs will grow down in the desert, larkspur rarely thrives there. *Delphinium* is largely a temperate circumboreal group, with 25 species in Europe and 61 in North America (Mabberley 1997). Warnock (1997) reported *D. scaposum* as endemic, but this is a near-endemic; it also occurs in Sonora, Mexico (Felger 2000). Flowers are pollinated by long-tongued bees, particularly bumblebees (*Bombus*, cf. Macior 1975).

In the western United States, *Delphinium* is second to locoweeds (*Astragalus*) in causing the death of livestock that accidentally eat it while grazing (Parker 1972, Dobelis 1986). This poisoning typically happens only in areas where the herbs are abundant, as smaller quantities only sicken and do not kill. Plants contain delphiline, delphinine, and similar alkaloids with a curare-like action (Mabberley 1997).

**Human uses:** Hopi grind Bare-Stem Larkspur flowers with corn to make "blue pollen" (called *cakwatalasi*) for the flute altar (Fewkes 1896). Navajo also use the petals to make a blue dye (Elmore 1944). Both tribes use the plant as an emetic in rituals, and petals and pollen are used by Navajo medicine men to make *tádidíín dootł'izh* (blue pollen) (Elmore 1944, Kearney and Peebles 1951, Whiting 1939, Wyman and Harris 1941). Navajo use the *Delphinium* to aid new mothers, and it is eaten by women to increase fertility (Wyman and Harris 1951); they also give it to female goats for the same purpose, hence, the name *tł'ízí 'azee'*. These practices are dangerous because the plants are poisonous.

Larkspurs have long been used in the Old World to kill parasites that live on humans, such as lice or itch mites. Larkspur was considered the choice plant for delousing people in medieval times (Dobelis 1986). The British issued *Delphinium* to Wellington's troops at the Battle of Waterloo in 1815. Similarly, during the American Civil War, Union troops are said to have used larkspur to kill body lice.

**Derivation of the name:** *Delphinium* is from Greek *delphinion*, the name for the larkspur or dolphin-flower, from *delphis*, a dolphin, referring to the form of the flowers, which are sometimes like the classical figures of dolphins. *Scaposum* means having a leafless stem, an uncommon trait in the genus.

**Miscellaneous:** The other species of *Delphinium* that has been reported in the Baboquivari Mountains is *D. parishii* (Parish's Larkspur, Apache Larkspur, Clary's Larkspur, or Mohave Larkspur). Warnock (1997) considered all *Delphinium* to the west of the Baboquivari Mountains to be *D. parishii* spp. *parishii*, but Felger et al. (in prep.) do not consider the plants distinct from *D. scaposum*. The Baboquivari Mountains are the region where Warnock (1997) shows the division of these two species. Warnock considers those with sepals that are light blue to white and fruits 3 or fewer times longer than wide to be *D. parishii* and those with fruits usually more than 3 times longer than wide and sepals blue to dark blue to be *D. scaposum*. There reportedly are also differences in the cells of the seed coats.

*Delphinium* is a genus of ca. 300 species (Mabberley 1997, Warnock 1997).

## *Thalictrum fendleri* (Ranunculaceae) Meadow Rue, *ruda cimarron*

**Other names:** ENGLISH: Fendler's meadow rue ("meadow rue" in use by 1653 when Culpeper gave the modifiers "meadow or wild" to distinguish between wild plants and the cultivated rue, *Ruta graveolens*); SPANISH: *paloma consulta* (consulted dove, Chihuahua), *ruda cimarron* [*de la sierra*] (wild [mountain] rue, New Mexico), *tanito* ("little Tano," maybe a reference to the Tano or Jemez people of New Mexico; or could it be from *tăŋ*, the Tewa word for seed? Chihuahua); ATHAPASCAN: *tazi'lci'n* <*tazi.lci.n*> (turkey odor, Navajo), *diłhił bitsighaá* <*nahił bichuwa*> (*diłhił*, black, *bitsighaá*, hair, Western Apache); KIOWA TANOAN: *tăŋ su'iŋ* (*tăŋ*, seed, *su*, to smell, Tewa); YUKI: *kinkuhach* (Yuki)

*Thalictrum fendleri*. A. Base of plant. B. Male flowering branch. C. Male flower. D. Female flower. E. Achene. Artist: Leta Hughey (Dayton 1937).

**Botanical description:** Herbs, perennial, dioecious, the stems mostly 2–10 cm tall, often purplish, often branching above. Leaves are mostly along stems, but larger toward the base, usually in two groups of threes (biternate), the ultimate segments 5–20 mm long, ovate to obovate or orbicular, thin, dark green above, pale and often glandular below. Panicles are small to large, leafy bracted, the pedicels straight. Sepals are white or greenish, 2–4 mm long. Stamens are prominent, more obvious than sepals, 18–25. Achenes are 5–11 mm long, sessile or nearly so, spreading and often appearing as a spheroidal cluster, the body 4–7 mm long, strongly flattened, glabrous or glandular, obliquely ellipsoid-obovate in outline.

**Habitat:** Canyons, streamsides, washes; oak woodlands. 1,200–2,900 m (4,000–9,500 ft).

**Range:** Arizona (Apache to Mohave, south to Cochise, Pima, and Santa Cruz counties), California,

New Mexico, north to Colorado, Wyoming, Oregon, and South Dakota; Mexico (Chihuahua).

**Seasonality:** Flowering April to August.

**Status:** Native.

**Ecological significance:** *Thalictrum* is a temperate genus that barely reaches south into this region. The species is more typical of pine forests than the oak woodlands where it grows in the Baboquivari Mountains. Farther north, it is associated with willow and birch forests, aspen, and even spruce-fir communities (Park and Festerling 1997).

Flowers of this species are small, not showy, unisexual (dioecious), and are surely wind-pollinated (Proctor et al. 1996). Indeed, this is one of the few members in a largely animal-pollinated family that has developed wind pollination. Not all members of *Thalictrum* are wind-pollinated, as a few, such as cultivated Eurasian *T. flavum* and *T. aquilegifolium*, have conspicuous feathery flower clusters with cream- or lilac-colored stamens.

**Human uses:** Meadow Rue was a widely used plant in the Old World, where it was known as *rù beag* (*rù*, rue, *beag*, little, Gaelic), *Wiesenraute* (meadow rue, German), Meadow-rhubarb, *rubarbio de los pobres* (poor people's rhubarb, Spanish) and *ruibarba dos pobres* (poor people's rhubarb, Portuguese), *pigamo* (from Greek *peganon*, the name of *Ruta*, Italian), and *pigamon* (Quebec, France). These names compare Meadow Rues with *Ruta* (rue) and *Rheum* (rhubarb). Both *Rheum* and *Thalictrum* are purgative, and *Ruta* and *Thalictrum* have been used as spices (Austin 2004).

Navajo and Shoshoni use the root infusion to treat venereal diseases (Train et al. 1957, Vestal 1952), while Keres and Washo treat colds and sores with it (Hocking 1997, Moerman 1998, Train et al. 1957). The medicine has an emetic, digitalis-like activity. Navajo also use the ash for a black dye (chant lotion, *béé'di*, <*keX'o*>) during the Enemy Way ceremony, and they take it as a tea and bath on the fifth night after the blackening ceremomy of the War Dance (Elmore 1944, Wyman and Harris 1941). Tarahumara use a decoction of *T. fendleri* to relieve heat prostration (Pennington 1963). Northern Tepehuan make a medicine of it to relieve stomach cramps and to reduce fever (Pennington 1969); Western Apache make a tea of the bark (Gallagher 1976).

**Derivation of the name:** *Thalictrum* is from Greek *thaliktron*, a plant with coriander-like leaves; used by Dioscorides, fl. CE 40–80. Pliny, CE 23–79, first applied the Latin *thalictrum* to Meadow Rue. The name *thaliktron* may be based on *Thalia*, "the blooming one," the Greek muse of pastoral poetry and comedy. *Fendleri* commemorates German-born American August Fendler, 1813–1883, explorer and collector in North and South America.

**Miscellaneous:** There are 120–330 northern temperate, New Guinea, tropical American, and African species in the genus, with 15 in Europe (Mabberley 1997, Park and Festerling 1997). *Flora of North America* recognized 22 species.

## ***Condalia warnockii*** **(Rhamnaceae)** **Mexican Buck-Thorn, *bindó, u:spaḍ***

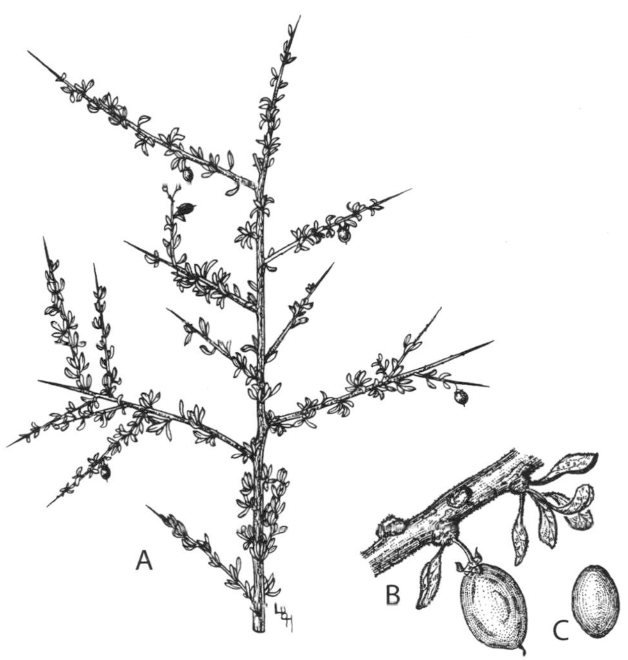


*Condalia warnockii*. A. Fruiting branch. B. Fruiting segment of branch. C. Seed. Artist: L. B. Hamilton, A (Benson and Darrow 1945); W. Hodgson, B–C (courtesy Wendy Hodgson).

**Other names:** ENGLISH: [Mexican] buck-thorn (now applied to several genera, but first applied to *Rhamnus cathartica* by Dodoens in 1554; the Latin *cervi spina* was applied to *Rhamnus* by Valerius Cordus, 1514?–1544), lote-bush (based on *lotos*, the Greek name used for a plant with sweet berries which was "the food of the Lotophagi, which Heroditus, Dioscorides, and Theophrastus describe as sweet, pleasant and wholesome, and which Homer says was so delicious as to make those who ate it forget their native country"), Mexican crucillo, squaw-bush (Arizona, New Mexico), [Warnock's] snakewood (a plant use to treat snakebite, used for *Strychnos* by 1598, for *Colubrina* by 1832; later applied to *Condalia*, New Mexico); SPANISH: *bindó* (from Zapotec *bindó*, the name of another *Condalia*, San Luis Potosí), *guichutilla* (Sonora, possibly from Cahita *güichura*, the name of a milkweed, but any relationship is obscure), *teconblate* [*tecomblate*] (probably from Náhuatl *teçonpatli*, *teçontli*, stone, *patli*, medicine, New Mexico); UTO-AZTECAN: <*balchata*> (Onavas Pima), *ho'i* (sticker, thorn, Mountain Pima; for *C. brandegeei* fide Pennington 1973), *kavk kuavuli* (*kavk*, dry, hard, *kuavulĭ*, *Lycium*, Hiá Ceḍ O'odham; for *C. globosa*), *u:sbaḍ* <*'u:padh*, *u'usbaḍ*, *u:spa't*> (a contraction of *u:s jeweḍbaḍ*, *u:s*, bush/sticks, *jeweḍbaḍ*, abandoned land, Tohono O'odham; compare with *Ziziphus*); (= *C. spathulata* of authors, not A. Gray)

**Botanical description:** Shrubs grow to 2 m tall. Leaves are in fascicles of 2–5 at the short shoots, spatulate, 3–7(–10) mm long, 1–3(–4.5) mm wide, apically acute, basally acuminate, margins entire, hispidulous, 2–4 pairs of lateral veins, these obvious. Flowers are solitary or fascicled. Petals are absent. Drupes are nearly globose, red to reddish black, 4–6 mm long.

**Habitat:** Canyons, especially bajadas. 500–1,700 m (1,640–5,600 ft)

**Range:** Arizona (Cochise, Graham, Greenlee, Maricopa, Pima, and Pinal counties), New Mexico, and western Texas; Mexico (Chihuahua, Coahuila, Hidalgo, Sonora, and Zacatecas).

**Seasonality:** Flowering July to September.

**Status:** Native. Baboquivari plants are var. *kearneyana*. This variety is endemic to the Sonoran Desert (Johnston 1962). Our variety grows only in southern Arizona and northern Sonora; variety *warnockii* is confined to New Mexico, Texas, and northern Mexico. The range of variety *kearneyana* does not overlap with that of var. *warnockii*.

**Ecological significance:** Flowers appear in response to summer rains and are scented and attract insects (Turner et al. 1995). Leaves are usually evergreen, but may be shed during unusual droughts. The species seems to be restricted to climates with mild winters and biseasonal rainfall dominated by summer storms. This shrub is a dry-tropical species.

**Human uses:** Bohrer (1991) found seeds of *Condalia warnockii* in a Hohokam archaeological site called La Ciudad (current Phoenix). These seeds, dated from CE 800 to 1100, presumably had been selected by people so that they were larger than those of wild plants. Those she found were 4–7 mm long; modern seeds are 3–6 mm. *Condalia warnockii* is not now known in the Salt River Valley, and its seeds have not been recovered from within other Hohokam sites within its current range. What these pre-European people were using the fruits for is unknown, but food or medicine seems likely.

Kearney and Peebles (1951) simply say that the stones of the fruits "have a soapy taste and are only slightly bitter." The name *teconblate* implies that the Aztecs used *Condalia* for medicine, but no other sources seem to have recorded that. Perhaps the long persistence of the Náhuatl name in Mexico records that history, as does the Mountain Pima use of *C. brandegeei* to treat exhaustion (Pennington 1973).

Hocking (1997) records that *C. hookeri* of Texas and Mexico (to Nuevo León) has edible fruits that have been made into jelly. Given the comment by Kearney and Peebles, perhaps *C. warnockii* has been similarly used. The Texas species is a source of blue dye, and that makes me wonder if several species may not have had that use.

**Derivation of the name:** *Condalia* was named for the Spanish physician and explorer of South America, Antonio Condal, 1745–1804, a native of Barcelona. *Warnockii* is dedicated to Barton H. Warnock, 1911–1998, a botanist who worked at Sul Ross State University in Alpine, TX. *Kearneyana* commemorates Thomas H. Kearney, one of the authors of the *Arizona Flora* (Kearney and Peebles 1951).

**Miscellaneous:** *Condalia* is a genus of 18 species that grows in warm parts of the Americas (Christie et al. 2006). This genus is related to *Rhamnus* (q.v.) and *Ziziphus* (q.v.). Indeed, what is now *Ziziphus canescens* was still being called *Condalia lycioides* when Kearney and Peebles (1951) wrote the *Arizona Flora*. Part of that confusion arose because it was thought at the time that *Ziziphus* was a genus of the Old World.

A second species in Brown Canyon is *C. correllii* (Mexican Buck-Thorn, *bindó*, *kawk kuavulĭ; kawk*, dry, *kuavulĭ*, includes both *Celtis* and *Lycium*, q.v.). This species is distinguished from *C. warnockii* by its wider (mostly 4–10 mm vs. 1.5–3 mm), obovate-to-elliptic leaves, with inconspicuous veins on the lower leaf surfaces (Christie et al. 2006). Additionally, the black fruits are on pedicels 1–2.5(–8) mm long in *C. warnockii*, while pedicels are up to 0.5 mm long in *C. correllii*, and the red to black fruit is nearly sessile. *Condalia correllii* is known in Arizona (Cochise, Pima, and Santa Cruz counties) and Mexico (Chihuahua, Coahuila, and Sonora).

Also in the canyons is *Sageretia wrightii*. This sprawling, straggling, vine-like shrub is similar to *Condalia* in many ways, but the life-form is distinctive. Moreover, it has fruits with 2–4 distinct stones, while *Condalia* has a single stone. These shrubs grow between about 4,500 and 5,000 feet in the canyons and may be in flower from March to September.

### *Rhamnus californica* (Rhamnaceae) California Buck-Thorn, *cáscara sagrada*

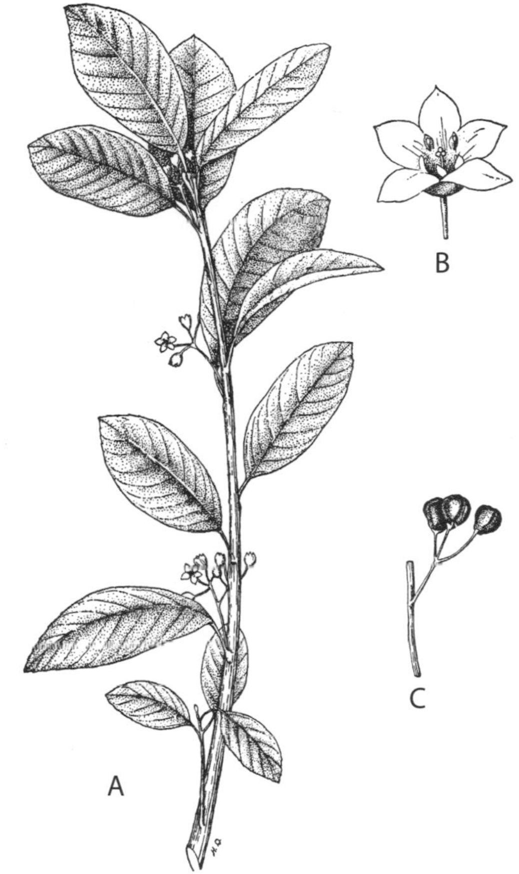


*Rhamnus californica*. A. Flowering branch. B. Flower. C. Fruiting cluster. Artist: H. Dreja (Dayton 1937).

**Other names:** ENGLISH: bitter-bark (California), California [southern] buck-thorn (from Latin *cervi spina*; applied to *Rhamnus cathartica* by German pharmacists and one of the fathers of pharmacognostics, Valerius Cordus, 1514?–1544), chittemwood (based on Hebrew *shittim* or *shittah*, originally a name for *Acacia*, California), coffee-berry (California), Indian cherry, pigeon-berry (fruits eaten by birds, California), wild coffee (resemblance of seeds to coffee beans, California), yellow wood; SPANISH: *cacachilla* (based on slang *caca*, feces, northern Mexico), *capulincillo* (little cherry; based on Náhuatl *capuli* or *capolin*, northern Mexico, also applied to *Karwinskia*, another member of Rhamnaceae), *cáscara sagrada* (sacred bark, comparing *R. californica* with better-known medicinal *R. purshiana*, New Mexico), *espino sagrado* (holy spine, Mexico), *hierba del oso* (bear berry, used for several genera in Mexico); ATHAPASCAN: *hidáń shash <bida są>* (*hidáń*, food, *shash*, bear, Western Apache), *shash dą́ą́' <šaš dą́ˀ>* (*shash*, bear, *dą́ą́'*, food, Navajo, for *R. betulifolia*); UTO-AZTECAN: *hunwet qwa'i'va'a <hoon-wet-que-wa>* (Cahuilla), *kapí* (Tarahumara; for an unidentified species), *opobol* [*opobul*] (Tübatulabal), *sinaˀahyaavibɨ* [*sinaˀayhaavibɨ*] (*sinaˀa*, coyote, Kawaiisu); YUMAN: *i'kai* (Paipai); (= *Frangula californica*)

**Botanical description:** Shrubs, 1–4 m tall, upright or spreading, bark of young twigs usually reddish. Leaves are persistent, oblong to elliptic, flat or somewhat curled under (revolute), entire to serrate, acute to obtuse, 3–8 cm long, usually shiny and glabrous above, glabrous or somewhat pubescent below. Inflorescences are umbellate, on peduncles 4–18 mm long, 6–50 flowered. Flowers are perfect, 5-parted, rarely 4-parted, 2–3 mm long. Berries are black or red when ripe, almost globose or somewhat elongate, 10–12 mm long.

**Habitat:** Canyons. 1,100–1,980 m (3,500–6,500 ft).

**Range:** Arizona (southern Coconino, Mohave to Cochise, Graham, Pima, and Santa Cruz counties), California, and New Mexico to southern Oregon; Mexico (Baja California and Sonora).

**Seasonality:** Flowering May to July.

**Status:** Native.

**Ecological significance:** These shrubs typically appear in the understory of the oak woodlands and are almost always in deep shade. Fruits are sought out by birds, packrats (*Neotoma* ssp.), and by Coyotes (*Canis latrans*); elsewhere by bears (Ebeling 1986, Quinn and Keeley 2006). There are no longer bears in the Baboquivari Mountains (Hoffmeister 1986). La Osa ranch was supposedly named in the 1800s when a cowboy roped and killed a Silvertip Grizzly (*Ursus arctos horribilis*) (Granger 1960). Black Bears (*Ursus americanus*) formerly ranged west into these mountains, and the last known Grizzly Bears in Arizona were killed in the Mogollon area in the 1920s and 1930s.

**Human uses:** Bark from California Buck-Thorn has been used as a laxative, as the names *cacachilla* and *cáscara sagrada* suggest. Fruits historically were eaten by Kawaiisu and Paiute (Ebeling 1986). These people ate fruits sparingly because they are strong, bitter, and laxative. Given the comparison with *Karwinskia* in the common names, *R. californica* probably produced some of the same problems along with both uses. Fruits of *Karwinskia* are considered edible and sweet, but if eaten in excess, they cause paralysis and trembling in the legs (Hodgson 2001). The toxic chemicals are the anthracenones, which cause "progressive bilateral ascending paralytic neuropathy (demyelination)" that may "terminate in respiratory paralysis. Weakness occurs only after a latent period of several weeks, and paralysis may progress for a month or more" (Lampe and McCann 1985). The anthracenones are derived from anthrones (10H-anthracen-9-ones), as are anthraquinones (anthracen-9,10-diones). All of these chemicals are common in the family, but particularly in *Rhamnus.*

Hodgson (2001) found records that Cochimís did not know that the poisonous compounds were concentrated in the seed, but another tribe did. Cochimís tried to keep their children from eating the poisonous fruits, but the Pericúe, or perhaps the Guaicura, simply removed the stones before eating the fruits and experienced no problems.

*Rhamnus* is a well-known laxative because it contains different anthracene derivatives (Eckart Eich, personal communication, Dec. 2004). It was after learning to speak Spanish that I connected "casscáre-ah" of my parents with *cáscara* (bark) and *cáscara sagrada* (holy bark). My parents' generation was dosed with that plant weekly as a general "preventative" for everything. Literature at the time calls *cáscara* a "tonic." However, those who actually were obliged to take it have other, more derogatory terms for *cáscara*—and its stimulatory impact on the alimentary canal.

**Derivation of the name:** *Rhamnus* was applied to Mediterranean plants by Theophrastus, 372–287 BCE, and Pliny, CE 23–79. The ancient name is said to be based on Greek *rhamnos.* See *R. ilicifolia* for other views. *Californica,* of California, means that it was first collected in what is now that state.

**Miscellaneous:** Some put *Rhamnus californica* in the genus *Frangula* (e.g., Kartesz 1994, Christie et al. 2006). By this definition, plants with 4-parted imperfect flowers are *Rhamnus,* while those with 5-parted and perfect flowers are *Frangula.* Judd et al. (2008) notably recognize only *Rhamnus,* not even listing *Frangula* as a synonym.

## *Rhamnus ilicifolia* (Rhamnaceae) Holly-Leaf Buck-Thorn

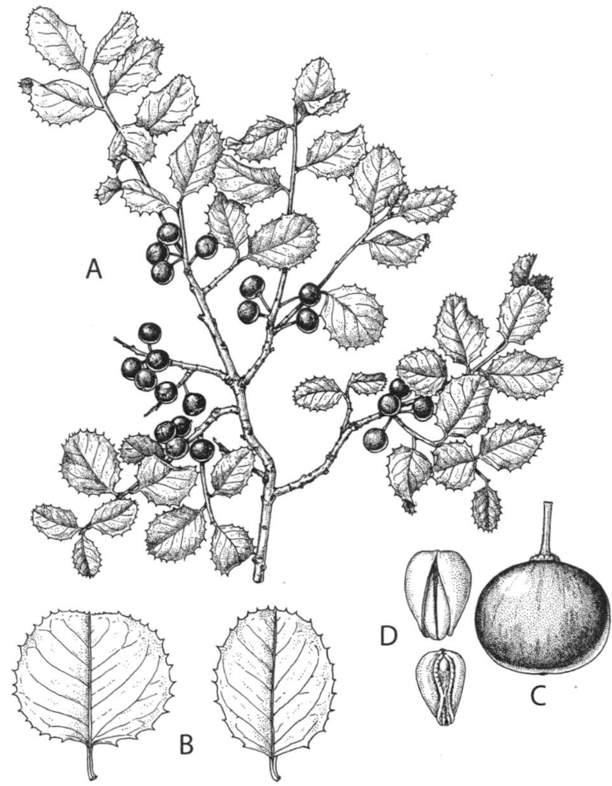


*Rhamnus ilicifolia.* A. Fruiting branch. B. Leaves, showing some variation. C. Berry. D. Seeds. Artist: L. B. Hamilton (from original artwork).

**Other names:** ENGLISH: [holly-leaf, hollyleaf] buck-thorn (from Latin *cervi spina;* applied to *Rhamnus cathartica* by German pharmacists and one of the fathers of pharmacognostics, Valerius Cordus, 1514?–1544), mountain holly-tree ("holly" from Old English *holgen,* a comparison with the genus *Ilex,* in use by 1150; related *holen* or *hollin* used by 725), [holly-leaf] red-berry (California); ATHAPASCAN: *bį it'ą* (*bi,* his, a circumlocution for "bear"? *it'ą',* spinach, Western Apache); CHUMASH: *suqup'i* (Barbareño Chumash); UTO-AZTECAN: *čual* (Tübatulabal, for *R. crocea*), *huˀupiyagadɨbɨ* (having berries like *Lycium andersonii* or *huˀupi,* Kawaiisu)

**Botanical description:** Shrubs, branched, spreading, gray-green, 1–2 m tall, with rigid often spinescent branchlets, evergreen. Leaves are often fascicled, glabrous or slightly puberulent, shining, rigidly leathery, oval to almost orbicular, and truncate to notched (emarginate) at the apex, 10–25 mm long, spiny-dentate, on petioles 1–4 mm long. Flowers are unisexual, 4- or less often 5-parted, on pedicels 1–4 mm long. Petals are absent. Berries are red, 5–6 mm long, obovoid, 2-seeded. Seeds are 4 mm long, finely reticulate, rounded at apex, the outer side deeply grooved.

**Habitat:** Canyons, mountain slopes. 900–2,130 m (3,000–7,000 ft).

**Range:** Arizona (Coconino, Gila, Graham, Maricopa, Mohave, Pima, Pinal, and Yavapai counties), California, Nevada, and Oregon; Mexico (Baja California and Sonora).

**Seasonality:** Flowering March to May.

**Status:** Native. Arizona plants are morphologically and taxonomically distinct from those in California and Baja California (Hickman 1993, Wiggins 1980). The population in south-central Arizona has been called *R. crocea*, but it is now assigned to *R. ilicifolia*, which differs by having leaves that are spinose along the margins, oval to almost orbicular, and truncate to notched (emarginate) at the apex (Felger et al. in prep.). *Rhamnus crocea*, which grows west of the Baboquivari Mountains, has leaves that are smaller, narrower, and merely serrulate to almost entire (Hickman 1993, Christie et al. 2006).

**Ecological significance:** This species is typically associated with chaparral and open, coniferous forest (Kearney and Peebles 1951). In the Baboquivari Mountains, the shrub occurs in canyons within oak woodlands. The species is in the mountains of southeastern Arizona, and it follows the highlands up the Mogollon Rim into northwestern Arizona.

Holly-Leaf Buck-Thorns are browsed by deer and Bighorn Sheep (Kearney and Peebles 1951). Berries are also eaten by doves and other birds (Ebeling 1986).

At first glance, this shrub may be confused with *Quercus turbinella*. Fruits and flowers are totally distinct in these two, being insect flowers and berries in *Rhamnus* and wind-pollinated catkins and acorns in *Quercus*. Sterile plants are more difficult to identify, but the two genera are separable by a number of traits. Bark is light gray, rough, and fissured in *Q. turbinella*; it is gray-green and smooth in *R. ilicifolia*. Leaves are densely to sparsely glandular and stellate above and below, although glabrate in *Q. turbinella*; they are glabrous to slightly puberulent in *R. ilicifolia*. Leaf blades are blue-green in the oak and green in *R. ilicifolia*.

**Human uses:** Apache eat the fruits of *R. ilicifolia* with meat (Kearney and Peebles 1951, Ebeling 1986). Cahuilla use the berries of Holly-Leaf Buck-Thorn for food (Bean and Saubel 1972). Hocking (1997) says that indigenous people added the fruits to pemmican.

The bark extract has a red color and pleasant odor and is used like *cáscara sagrada* (*R. purshiana*) as a laxative, but it is said to be milder (Hocking 1997). Pioneers made a sour summer beverage from the berries, and they added it to gin drinks. Chumash use the roots to make a yellow dye (Timbrook 1990). California tribes also use their *cáscara sagrada* (*R. purshiana*) as a dye (Moerman 1998). For *Rhamnus*, Cannon and Cannon (2003) note the presence of flavonols, including kaempferol, rhamnetin, rhamnocetrin, and quercetin, along with the anthraquinones, alaterine, frangularoside, and rhamnicoside, often with a dominance of emodin.

**Derivation of the name:** *Rhamnus* was applied to Mediterranean plants by Theophrastus, 372–287 BCE, and Pliny, CE 23–79. The ancient name is said to be based on Greek *rhamnos*. According to Millspaugh (1892), the name is from Gaelic *ràmh*, branching. Quattrocchi (1999) thought the name was from Greek *rhabdos*, a stick; akin to Akkadian *rapasu*, to beat, Sumerian *rab*, stick, branch. Jaeger (1941) thought that Theophrastus used *rhamnus* for some crocus, perhaps because of the branched stigmas. *Ilicifolia* means having leaves similar to Holly (*Ilex*) or to Holm Oak (*Quercus ilex*).

## *Ziziphus obtusifolia* (Rhamnaceae) Gray-Thorn, *abrojo, u:spaḍ*

**Other names:** ENGLISH: gumdrop tree (Texas), lotebush (see *Condalia* for etymology), thorn (Arizona), white crucillo; SPANISH: *abrojo* (bur, Mexico), *bachata* (from Eudeve *va:tsa'a* <*vātzá*> or Ópata *batzat*, Sonora), *barabachatas* (dearest bearded one; from *barba*, beard, *chata*, a term of endearment, Sonora), *chaparro prieto* (black thicket, Tamaulipas), *ciruela de monte* (wild cherry, Sonora), *crucillo blanco* (little white cross, Sonora), *garambullo* (spiny plant, Mayo, Sonora), *garrapata* (tick, Mexico), *palo blanco* (white tree, Mexico); ATHAPASCAN: *ch'il ńłdzig* <*chi gatoiłjit*> (*ch'il*, plant, *ńłdzig*, rotten, Western

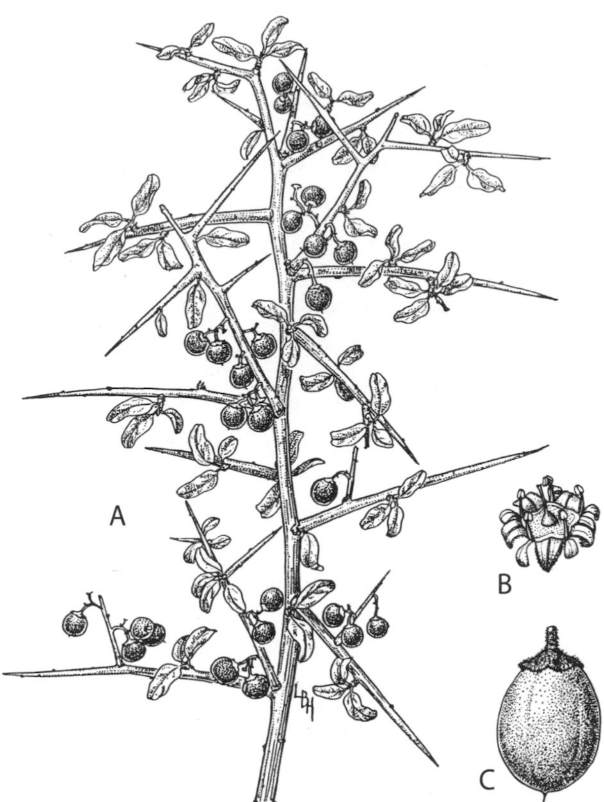


*Ziziphus obtusifolia*. A. Fruiting branch. B. Flower, enlarged. C. Fruit, enlarged. Artist: L. B. Hamilton (Humphrey 1960, Benson and Darrow 1945).

Apache); UTO-AZTECAN: *huichilame* (Mayo), *jó'otoro* (Mayo), *hutki* <*jutuqui*> (Mayo), *jeweḍbaḍu:s* <*duwastbaḍ uus*> (tall, dead-looking bush, Onavas Pima), *'u:spaḍ* <*u:supaḍ*> (Hiá Ceḍ O'odham), *u:spaḍ* <*'uspaḍ*> (contraction of *u:s jeweḍbaḍ*, Tohono O'odham), *'us jeveḍpaḍ* (Hiá Ceḍ O'odham), *u:s jeweḍbaḍ* <*'us jewedhpadh, u:s tcui'tpa't*> (*u:s*, sticks, *jeweḍbaḍ*, abandoned land, Tohono O'odham), *u'us chevaḍbaḍ* <*ositc u'wutpat, u-us dji-wuht-paht*> (*u'us*, bush, *cheveḍ*, to become tall, *paḍ*, dead, Akimel O'odham fide Rea 1997); YUMAN: *uwé* (Maricopa); (= *Condalia lycioides* var. *canescens*)

**Botanical description:** Shrubs 1–4 m tall, with straight, stout, thorn-tipped branches, the twigs and thorns grayish or whitish, with a wax-like covering (bloom). Leaves are grayish green, 8–19 mm long, 5–9 mm wide, variable from deltoid to ovate or oblong or nearly linear, deltoid or ovate serrate in the upper half. Inflorescences of axillary umbels are present. Flowers are inconspicuous, with 5 sepals and petals. Drupe is dark blue or black, ellipsoid, 6–8 mm long, with a single woody 2-celled stone.

**Habitat:** Canyons, washes. 300–1,500 m (1,000–5,000 ft).

**Range:** Arizona (Cochise, Coconino, Gila, Greenlee, Maricopa, Mohave, Pima, Pinal, Santa Cruz, Yavapai, and Yuma counties), southeastern California, southern Nevada, New Mexico, and western Texas; Mexico (Baja California and Sonora).

**Seasonality:** Flowering (May) July to September, but intermittently throughout the year in response to rains.

**Status:** Native. Our plants are var. *canescens*. Variety *obtusifolia* has been found in Cochise County, but it is a Chihuahuan Desert taxon (Christie et al. 2006).

**Ecological significance:** Birds eat the insipid, blue fruits, and they take refuge in the thorny branches. Although Gray-Thorns are found on open flats, slopes, mesas, and other habitats, they are most common in bottomlands along streams (Turner et al. 1995). This abundance perhaps reflects the importance of those sites to the animals that disperse the seeds. Leaves are drought and winter deciduous.

**Human uses:** Akimel O'odham, Maricopa, Mayo, Mohave, Seri, Tohono O'odham, Yavapai, and probably other tribes eat fresh Gray-Thorn fruits (Castetter and Underhill 1935, Castetter and Bell 1951, Felger and Moser 1985, Moerman 1998, Rea 1997, Russell 1908, Yetman and Van Devender 2001). Formerly, the Tohono O'odham fermented the fruits for a beverage (Castetter and Underhill 1935). Russell (1908) recorded that the Akimel O'odham beat the berries down with a stick and gathered them in baskets. The pulp was separated from the seeds and eaten; seeds were discarded. Onavas Pima also eat the fruits (Rea 1997). Akimel O'odham children were told that, if they ate the seeds, they would "grow up to be bald—like those seeds" (Mary Narcia Juan in Rea 1997).

Roots are used in Tamaulipas and in New Mexico (by Apache) for soap, and a decoction is put on sores (Standley 1920–1926). Akimel O'odham treat sore eyes with a decoction of the roots, and they also use it as soap (Curtin 1949, Kearney and Peebles 1951, Rea 1997). Seri use the powdered roots on skin and scalp sores (Felger and Moser 1985). Yetman and Van Devender (2001) record that their Mayo consultants held Gray-Thorns in high regard for healing several maladies, and that the "plants seem to exude an aura of healing."

The Havasupai use the branches to make planting sticks (Moerman 1998). Formerly, the Akimel O'odham did the same (Rea 1997). The Akimel

O'odham once also made a musical instrument of them, a notched stick (*hivkuḍ*) that was played over a basket.

**Derivation of the name:** *Ziziphus* was taken from the Persian *zizfum* or *zizafun*, or the Arabic *zizouf*, the name for the *Ziziphus lotus*, the Jujube-Tree of Pliny and Columella. *Obtusifolia* means that the leaves have rounded apices; *canescens* indicates that the leaves are gray-pubescent.

## *Cercocarpus montanus* (Rosaceae) Alder-Leaf Mountain-Mahogany, *palo duro*

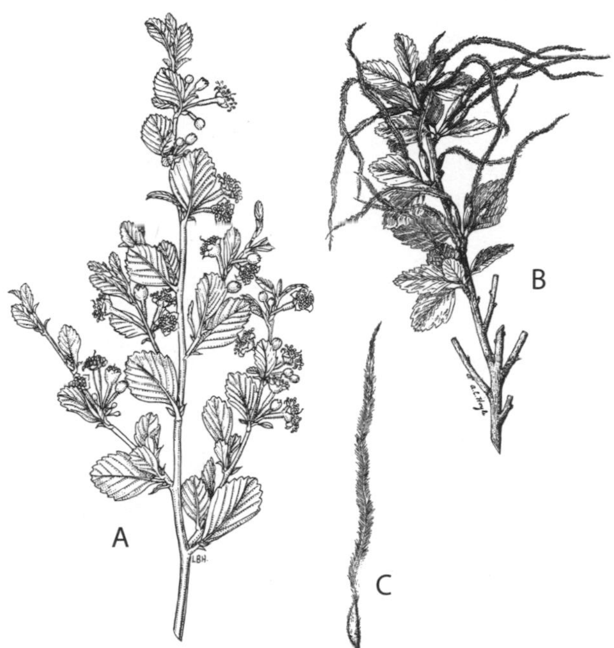


*Cercocarpus montanus*. A. Flowering branch. B. Fruiting branch. C. Achene with feathery "tail" (style). Artists: A. C. Hoyle, B, C (Dayton 1937); L. B. Hamilton, A (from original artwork).

**Other names:** ENGLISH: feather-brush (in English by 1856 in another context: "He was dusting . . . with a feather-brush"; later used as a plant name), [alder-leaf, hardtack, true] mountain mahogany (Arizona, New Mexico; "mahogany" perhaps from Taino *maga* or *magua*, meaning fertile lowland or plains; in Spanish by 1528, English by 1660; later used for other plants with similar wood; see below for more details); SPANISH: *palo duro* (hard wood, New Mexico, Chihuahua); ATHAPASCAN: *tsé'ésdaazii* <*ce'esda-zi'*, *tshé'estaazih*> (*tsé*, stone, *'ánísdáás*, heavy [comparative], Navajo), *tsį ntł'izí* <*ges ndazhe*> (*tsį*, stone, *ntł'izí*, tough, Western Apache); KIOWA TANOAN: *p'eke'in* (*p'e*, wood, *ke*, hard, Tewa), *qwae* (Tewa); UTO-AZTECAN: *čikáka* (Tarahumara), *piáitcampi* (Southern Paiute probably belongs here), *putsivi* <*putcívi*> (Hopi), *sinaˀaruubɨ* (possibly *sɨnaˀa*, coyote, *tuu(m)bɨ*, *Ceanothus cuneatus*, Kawaiisu), *tubi* (Northern Paiute, for *C. ledifolius*), *tunampi* [*tu'nûmp*, *tu'hinûp*] (Shoshoni, for *C. ledifolius* and *C. parvifolius*); YUKI: *umse* (Yuki); (= *C. betuloides*, *C. breviflorus*)

**Botanical description:** Shrubs, less commonly small trees, 1.2–4 m tall. Leaves are short-petiolate. Blades are obovate, oblanceolate, or orbicular, 6–44 mm long, 5–23 mm wide, crenate-serrate, glabrous above, pubescent below but sometimes glabrous, deciduous. Flowers are 9.5–17.5 mm long, inconspicuous. Sepals are 0.9–1.7 mm long. Achenes have tails 3–10 cm long.

**Habitat:** Mountain slopes, canyons. 1,280–1,438 m (4,200–8,000 ft).

**Range:** Arizona (southern Apache, Greenlee, Coconino, and Yavapai to Cochise, Pima, and Santa Cruz counties), New Mexico, and western Texas; Mexico (Chihuahua, Coahuila, Sonora, south to San Luís Potosí, and Hidalgo).

**Seasonality:** Flowering March to November.

**Status:** Native. Our plants are var. *paucidentatus*.

**Ecological significance:** Fruits of *Cercocarpus* have styles that develop into plume-like structures that are effective in dispersing the propagules. In this regard, Mountain Mahogany is similar to Apache Plume (*Fallugia paradoxia*) of the same family and also to *Clematis* (virgin's bower, q.v.) of the Ranunculaceae. This convergence of unrelated families to a similar dispersal method is evidence of both the effectiveness of the system and the strong selective pressures imposed by the wind.

Although the species usually comes down to only 4,500 feet within its range, it is below about 4,200 feet in Brown Canyon. In that canyon, the plants are largest just at the Arch at about 4,900 feet; they become younger and smaller below that point.

This species and others are valuable browse plants for wildlife. *Cercocarpus* is a component of wild seed mixtures in reclamation attempts (Welsh et al. 1987). Deer like the plant so well that Navajo hunters sometimes chew the leaves from shrubs browsed by them to increase their hunting luck (Mayes and Lacy 1989).

Mountain Mahogany has the actinomycete fungus *Frankia* living as a symbiont on its roots. That fungus fixes nitrogen that Mountain Mahogany is able to utilize (Mabberley 1997).

**Human uses:** Mormon settlers in Utah began using the name "Mountain Mahogany" sometime between the late 1840s and 1860s. These people built the Mormon Tabernacle in Salt Lake City in the 1860s, bringing logs by ox team across 300 miles of desert from Pine Valley Mountain in Washington County. Among the logs workers carried were those called "Mountain Mahogany" that they used to build the organ in the Tabernacle (Benson and Darrow 1981). John C. Frémont and John Torrey used the Mormon name in the botanical literature, and it has since become the most widely used name of the genus.

Apaches, Keres, Navajo, and Tewa make medicine from *C. montanus* (Elmore 1944, Galligher 1976, Moerman 1998, Reagan 1929, Robbins et al. 1916). This species is one of the Navajo Life Medicines. Tarahumara make a tea from the leaves to treat colds, constipation, and dysentery (Pennington 1963). Roots are ground and used to help sores heal. The species figures prominently in Navajo ceremonies, including the Mountain Chant (in which its wood is used to build the sweathouse), Plume Way, and Chiricahua Wind Way (Elmore 1944, Mayes and Lacy 1989, Vestal 1952).

Apaches make bows of *Cercocarpus* (Reagan 1929). Keres and Navajo make brooms and tools of the branches and wood (Elmore 1944, Moerman 1998). Tarahumara make arrows from the branches (Pennington 1963). Hopi, Isleta, Keres, and Navajo make dye from the bark (Hocking 1956, Moerman 1998, Vestal 1952, Whiting 1939). The Paiute use another species to make digging sticks (Kelly 1932), calling the implement *tupi* <*tu·pi*>.

**Derivation of the name:** *Cercocarpus* is from Greek *kerkos*, a tail, and *karpos*, fruit, referring to the tail-like plume of the fruits. *Montanus* notes the habit in the mountains. *Paucidentatus* means that the leaves have few teeth.

**Miscellaneous:** *Cercocarpus* has eight species in western and southwestern North America (Mabberley 1997). Lehr (1978) and Kartesz (1994) recognized four species in Arizona. Most of these kinds are well north of the Mexican border region and the Baboquivari Mountains. The genus is not recorded for the Pajarito Mountains (Santa Cruz County), the Rincon Mountains, or the Tucson Mountains (Toolin et al. 1979, Bowers and McLaughlin 1987, Rondeau et al. 1996). Similarly, the genus is not present in the area west of the Baboquivari Mountains (Felger et al. 2001, Felger et al. in prep.). Within Pima County, *C. montanus* is also present in the Empire, Quinlan, Santa Rita, and Santa Catalina Mountains.

## *Prunus serotina* (Rosaceae) Choke Cherry, *capulín*

*Prunus serotina*. A. Leafy branch. B. Individual flower. C. Flower cluster. D. Fruits cluster. E. Fruit, enlarged. Artist: L. B. Hamilton (from original artwork).

**Other names:** ENGLISH: choke cherry (refers to the astringent fruits; dating from about 1796 in the northeastern United States; "cherry" derived in the 1300s from Old English *ceris*, in turn, from Latin *cerasus*), wild black cherry; SPANISH: *capulín* [*grande, pequeño*] ([big, small] cherry, Chihuahua; taken from Náhuatl *capulí*), *cerezo* [*americano, criollo, negro americano, negro serótino, negro silvestre, negro, silvestre*] ([American, creole, American black, black serotinous, wild black, black, wild] cherry; from Latin *cerasus*); ATHAPASCAN: *tł'oh tłíishi* (*tł'oh*, plant, *tłíishi*, snake, Western Apache), *tłoh dotł'izhé* (*tłoh*, herb, *dotł'izhé*, green, Western Apache), *didzé* <*dzidzé, titzéh,* $di_3e$>

(berry, Navajo), *didzé dík'ǫ́zhí* <*k'įicjínih*> (*didzé*, cherry, *dík'ǫ́zhí*, sour, Navajo), *dze*, *dzedeyui* (*dze*, chokecherry, *dzedeyui*, sour berry, Chiricahua and Mescalero Apache), *mą'iidą́ą́'* (*mą'ii*, coyote, *dą́ą́'*, food, Navajo); HOKAN: *tsámṭu* (Washo); KERES: *apu* (Acoma, Cochiti, Laguna, San Felipe); KIOWA TANOAN: *abè* (Tewa); MAYAN: *tup* (Quiché, Guatemala); OTO-MANGUEAN: *detze* (Otomí), *taunday* (Zapotec); TARASCAN: *xengua* (Purépecha); UTO-AZTECAN: *aatuuvɨ* (Kawaiisu; for *P. virginiana*), *ˀátu-t* <*a-tut*, *atut*> (Cahuilla, Luiseño), *aguasiqui* <*aguasique*, *ahuasiqui*> (Mountain and Onavas Pima, Sonora; loan from Tarahumara fide Thord-Gray 1955), *bachí* (Tarahumara), *čámi-š* (Cahuilla), *chaamch* (Luiseño, Juaneño dialect; for *P. ilicifolia*), *cháamich* (Luiseño), *do'·-icabui* (Paiute), *eˀkó* (Guarijío), *heco* <*jeco*> (Guarijío), *humi* <*jumpil*, *húmpely*> (cognate with Hopi *sïpi*, berry; proto-Uto-Aztecan **sepi*, Mountain Pima), *mo'oshgama* (Mountain Pima), *táishavui* (Western Paiute), *tóǫ̂ˀnapi* (Ute), *toonampü* (Panamint), *to'onûmp* [*toñ'gicip*] (Shoshoni), *toshebuı* [*toishapui*] <*tó'isabui*> (Northern Paiute), *tûghávuhya* (Mono), *túkuši* (Northern Tepehuan), *usábi* <*usábiki*, *cusabi*> (Tarahumara), *úvagi* (Tarahumara), *wasiki* (Mountain Pima; loan from Tarahumara or Guarijío fide Laferrière 1991); YUKI: *chulmäm* (Yuki, for *P. virginiana*); YUMAN: *ahkaí* (Kumiai)

**Botanical description:** Trees, to 10 m, young leaves pubescent, otherwise glabrate. Leaves are mostly 3.5–8.7 cm long, blade elliptic to sometimes ovate or obovate, the apex acute to obtuse, shiny green above, paler below, the margins finely serrated with forward-pointing and mostly gland-tipped teeth, petioles 5–15 mm long. Racemes are at ends of leafy branchlets, the branchlet and its raceme 3–12 cm long. Calices are persistent in fruit. Corollas are white. Fruits are drupes, 8–12 mm diameter, black. Seeds are 6.5–7.8 mm long.

**Habitat:** Usually along stream banks, canyon banks. 1,100–1,830 m (3,500–6,000 ft).

**Range:** Arizona (Apache to Mohave, south to Cochise, Pima, and Santa Cruz counties), New Mexico, and Texas; Mexico (Baja California, Sonora east to Tamaulipas, and south to Oaxaca and Chiapas); south through Central America; and South America.

**Seasonality:** Flowering (March) April to May.

**Status:** Native.

**Ecological significance:** These temperate trees follow the mountains from North to South America. The small white flowers are pollinated by bees and probably flies. Although within Arizona the species extends down to 3,500 feet, Choke-Cherries have not been found below about 4,500 feet in Brown Canyon.

**Human uses:** The Mexican and tropical American plants are *Prunus serotina* ssp. *capuli* (Felger et al. 2001). Fruits have long been eaten fresh and preserved through the Americas, and they are made into alcoholic and nonalcoholic drinks (Felger et al. 2001). Hernán Cortés's men first noticed this cherry in Mexico in 1519. Its fruits were an important food for Spanish soldiers during the siege of Mexico City (Standley and Steyermark 1946). Later, Francisco Ximénez said in 1615 that the fruits "*no son nada inferiores a nuestras cerezas*" (are not inferior to our cherries). Fernald et al. (1958) thought that indigenous people first pounded the dried fruits (including pits) and then leached out the poisonous chemicals before using them in foods. An alternative view is that, once pounded, the poisons volatilize to render them harmless (Dunmire and Tierney 1997).

Guarijío still use the trees, particularly as fuel (Yetman and Van Devender 2001). Others use the wood for cabinets, furniture, handles, interior trim, panels, patterns, printer's blocks, scientific instruments, toys, veneer, and woodenware (Austin 2004).

Tarahumara and Northern Tepehuan use the leaves and bark to catch fish, and the Tarahumara use both in a tea to treat whooping cough (Pennington 1958, 1963, 1969). Tarahumara also include young leaves in their corn dish *esquiate* (similar to cornmeal mush) to add spice, and they use the branches like *Polygonum* (q.v.) to catch fish. Paiute make a tea from the leaves (Kelly 1932) and treat several maladies with it (Train et al. 1957). Kickapoo allow only members of the Buffalo clan to use the *Prunus* as a medicine (Latorre and Latorre 1977). Those individuals make a decoction of the bark, which is drunk to stop excessive menstruation. Mexicans make a decoction of the bark that is used to treat a number of maladies (Martínez 1969).

This tree is sacred to the Navajo, serving as an emetic, *ˀiiłkóóh*, and being used in the *Hóchxǫ́'íji* (Evil Way) and Mountain Top Way (Wyman and Harris 1941, Elmore 1944). Navajo use the wood for dance implements, for prayersticks to the north, and represent it in Mountain Chant sand painting, among other applications.

**Derivation of the name:** *Prunus* is the classical Latin name of the Plum, *P.* × *domestica*; Greek *proumne. Serotina* means late-ripening, a reference to the fruits.

**Miscellaneous:** Wilted foliage contains cyanogenic or prussic acid (cyanide or HCN) that is poisonous to livestock and humans, although fresh herbage is eaten by animals with impunity.

## *Vauquelinia californica* (Rosaceae) Arizona Rose-Wood, *árbol prieto*

*Vauquelinia californica*. A–B. subsp. *pauciflora* A. Branch. B. Leaf shape and margin. C–D. subsp. *sonorensis* C. Branch. D. Leaf shape and margin. Artist: N. Bartels (Hess and Henrickson 1987, courtesy James Henrickson).

**Other names:** ENGLISH: Arizona rose-wood ("rose-wood" has been applied since about 1660 to legumes with valuable, close-grained wood and the trees yielding it, originally legumes such as *Dalbergia*; the name was given because of the similarity of the wood's fragrance to that of roses, *Rosa*), few-flower vauquelinia (a book name), Torrey vauquelinia (a book name); SPANISH: *árbol* [*palo*] *prieto* (black tree, Durango), *guayul* [*guayule*] (from Nahuatl *cuahitl*, tree, *uli*, rubber, a name applied to *Parthenium*, and *Vauquelinia*; Standley 1920–1926 said it should be *Vauquelinia*, but that is in Coahuila; in Durango it is exclusively for *Parthenium*, cf. Santamaría 1959), *palo verde* (green tree, Durango); LANGUAGE FAMILY UNKNOWN: *ucas* (Mexico)

**Botanical description:** Large shrubs or small trees, evergreen, with a hard wood, multiple-stemmed, the herbage with short crinkled to matted hairs, sometimes glabrate with age. Leaves are firm (sclerophyllous), short petiolate, the blades narrow, lanceolate to narrowly elliptic to linear, leathery, margins finely wavy (crenulate) with small, gland-tipped teeth and glands often between the teeth. Inflorescences are of terminal, compound corymbs. Flowers are about 1 cm wide, the petals white. Capsules are 4.5–6.5 mm long, firm and somewhat woody at maturity, with 5 carpels. Seeds have a terminal wing.

**Habitat:** Rocky slopes, extending down into canyons. 760–1,500 m (2,500–5,000 ft).

**Range:** Arizona (Cochise, Gila, Maricopa, Pima, and Pinal counties) and southwestern New Mexico; Mexico (Baja California, Chihuahua, Coahuila, Durango, and Sonora).

**Seasonality:** Flowering May to August.

**Status:** Native.

**Ecological significance:** This mountain tree grows in the Ajo, Baboquivari, Sierra Estrella, Superstition, Whetstone, Guadalupe, and several other lower mountain ranges. The distribution of the species suggests that it is a relict of a more widespread range from during the Pleistocene and Holocene epochs (Turner et al. 1995, Rea 1997). Rondeau et al. (1996) considered the species a relict chaparral plant, along with *Quercus turbinella* and *Rhus aromatica*. The *Rhus* appears throughout both Brown and Jaguar Canyons, but the oak has been found only on the western slopes of the Baboquivari Mountains. The *Quercus* also occurs in Las Guijas Mountains across the Altar Valley.

There are now four subspecies recognized within the species, three of them occurring in Arizona (Hess and Henrickson 1987). Only ssp. *californica* has been reported in the Baboquivari Mountains (Felger et al. 2001), but there are plants in Brown Canyon that better fit ssp. *sonorensis*. Plants that grow in shaded sites even key to ssp. *pauciflora*, but that is surely because the lower leaf surface lacks the white ("shade-leaves"). Subspecies *pauciflora* has leaves green or yellow on both surfaces and grows in extreme southeastern Arizona and nearby New Mexico. Subspecies *californica* has broader (10–20 mm) leaves that are lanceolate to oblong-lanceolate or narrowly elliptic. Subspecies *sonorensis* has narrow leaves (7–10 mm wide) that are linear to linear-lanceolate. This subspecies is supposed to grow from the Ajo Mountains westward, being disjunct into Baja California.

Leaves on the trees in Brown Canyon are less than 10 mm wide, linear to linear-lanceolate, and white below. Felger (personal communication, 9 Feb. 2005), who examined Arizona Rose-Woods with me, says leaves are "not white enough," while he acknowledges that the leaves are too narrow to be ssp. *californica*.

**Human uses:** Wood and bark are used for dyeing goatskins yellow (Standley 1920–1926, Santamaría 1959). Plants are cultivated as ornamentals in Arizona and recommended by Brenzel (1995).

**Derivation of the name:** *Vauquelinia* commemorates the French pharmacist Louis Nicolas (Nicolas Louis) Vauquelin, 1763–1829, professor of chemistry at the Collège de France and at the Muséum d'Histoire Naturelle, Paris. *Californica* refers to Baja California where it was first collected; the species does not grow in the state of California. The original specimen was named by John Torrey in Emory's *Notes of a Military Reconnoissance* published in 1848. The editor of the book, Major William Hemsley Emory (1811–1887), was an army officer and director of the Mexican Boundary Survey.

**Miscellaneous:** Although the relationships of this genus have been debated for decades, new research suggests that it is actually part of the lineage including apples and pears (Morgan et al. 1994, Campbell et al. 1995, Evans and Campbell 2002, Verbylaitė et al. 2006).

## *Galium aparine* (Rubiaceae) Bed-Straw, *cuajaleche*

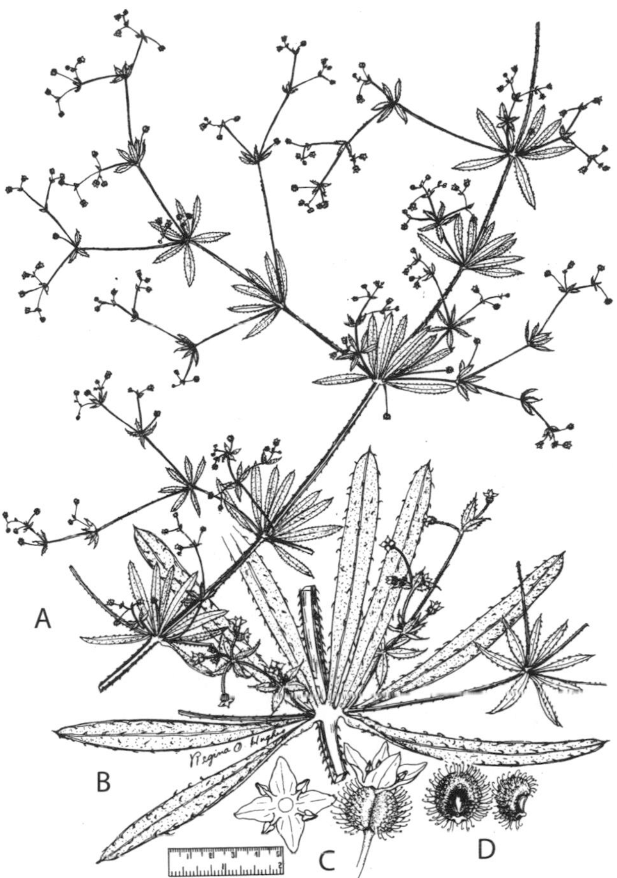


*Galium aparine*. A. Branch of a flowering plant. B. Enlarged leaf whorl. C. Flowers. D. Fruits. Artist: R. O. Hughes (Reed 1971).

**Other names:** ENGLISH: beggar's lice, catchweed, [common] bedstraw ("bedstraw," by ca. 1527 applied to *Galium* because they resembled straw that was formerly used to stuff mattresses; originally applied to one species, *G. verum*, Our Lady's Bedstraw, and later as a name for all species in the genus; early Christian tradition had it that *Galium* filled the manger in Bethlehem), cleavers (from Old English *clife*, Old High German *chliba*, from root of *clifian*, to stick, adhere), cleaver-wort ("wort" is an old English word meaning plant), goose-grass (since at least 1538 applied to this species and others because branches were used to feed geese), sticky-willy (Arizona, New Mexico); SPANISH: *amor de hortelano* (love of gardens; recorded as *amor del ortelano* in 1542), *aspérgula* (scratchy one; a pharmaceutical name from 1542), *azotalengua* (whip tongue), *cuajaleche* (curdles milk), *lapa* (clinging person/thing), *lárgalo* (makes it bigger), *presera* (guard of the dam or irrigation canals), *yerba del coyote* (coyote's herb, Northern Tepehuan and Mountain Pima); ATHAPASCAN: *'atą̨ą' ts'ósí <ˀat'wˀc'o-s>* (slender leaves, Navajo); KIOWA TANOAN: *'awi* (Tewa); UTO-AZTECAN: *pûñ'gonatsu* (horse medicine, Shoshoni)

**Botanical description:** Herbs with weak or reclining stems, annual from a slender taproot, retrorsely hispid on the angles, hairy above the joints, to 1 m long. Leaves are mostly 6–8 in a whorl, linear-oblanceolate, tapering to the base, mostly 2–7 cm long, bristle-tipped, the margins and lower midrib retrorsely hispid. Peduncles are 1–3 flowered. Corollas white. Fruits are bristly, 2–4 mm wide.

**Habitat:** Along streams and in canyons. 600–2,440 m (2,000–8,000 ft).

**Range:** Arizona (all counties except La Paz and Yuma), from Newfoundland to Alaska, south to Florida, Texas, and west to California; Old World.

**Seasonality:** Flowering March to May.

**Status:** Reportedly native and exotic.

**Ecological significance:** In the springtime, these sprawling herbs blanket the slopes above the streambed with a green coat. As the season progresses, their small flowers appear to punctuate the covering with white. Then, as the weather warms up and the moisture dries, *Galium* withers and disappears.

**Human uses:** Europeans knew *Galium aparine* when they arrived in the New World since the species is native to all parts of Europe. John Gerarde wrote in 1597 that "Women do usually make pottage of Cleavers with a little mutton and Otemeal, to cause lankness, and keep them from fatnesse." John Evelyn wrote in 1699 that "Clavers, *Aparine*; the tender winders with young Nettle-Tops, are us'd in Lenten pottages." Another British name for the herbs is Clabber-Grass. Jonathan Swift wrote in 1730 that the Irish country people "live with comfort on potatoes and bonny clabber." The word "clabber" is from Irish *claba*, thick, in reference to curdling milk, as also noted in the Spanish name *cuajeleche*.

Lady's Bed-Straw contains coumarin and has been used to treat an array of maladies, including bladder and kidney inflammations, dropsy, kidney stones, and fever (Austin 2004). The juice contains citric acid, which may have antitumor activity. Some think that *G. verum* lowers blood pressure, and it is known to contain asperuloside, an anti-inflammatory compound. Juice is used to prevent scurvy, but it may cause contact dermatitis. Mountain Pima use *G. aparine* to treat sprains, scrapes, and inflammation (Pennington 1973). Northern Tepehuan make a lotion of *Galium* to treat aching limbs (Pennington 1969). Northern Tepehuan also use *G. triflorum* (*yerba del coyote de la barranca*) to treat aching limbs and colic.

Moreover, Lady's Bed-Straw is the source of important dyes. Among other uses, it was one of those dyes used to color Scottish tartans, and in several countries, to color cheeses. Roots contain several anthraquinones, including alizarin, purpuroxanthin, rubiadin, purpurin, and lucidin (Cannon and Cannon 2003). Most species of *Galium* contain these pigments.

**Derivation of the name:** *Galium* is from Greek *gala*, milk, alluding to the European *G. verum*, whose flowers were used to curdle milk for cheese-making or to color cheese. *Aparine* is an old generic name, meaning to catch, cling, or scratch.

**Miscellaneous:** Other species in the canyons are *G. microphyllum* (Small-Leaf Bed-Straw), *G. proliferum*, *G. stellatum*, and *G. wrightii* (Cleavers). *Galium microphyllum* is unique in having solitary sessile or nearly sessile flowers surrounded by a cluster of leaf-like bracts and fleshy mature fruits. Only *G. proliferum* is annual; the others are perennial and either shrubby or at least woody at the base. *Galium proliferum* is rough-hairy, has no more than 4 leaves per whorl, and almost sessile solitary (or in 2s) flowers subtended by two reduced leaves. *Galium stellatum* has white flowers and rough-pubescent herbage; *G. wrightii* has purple or brown flowers and soft pubescence.

## ***Bouvardia ternifolia* (Rubiaceae)**
## **Trumpet Flower, *trompetilla, wipismal je:j***

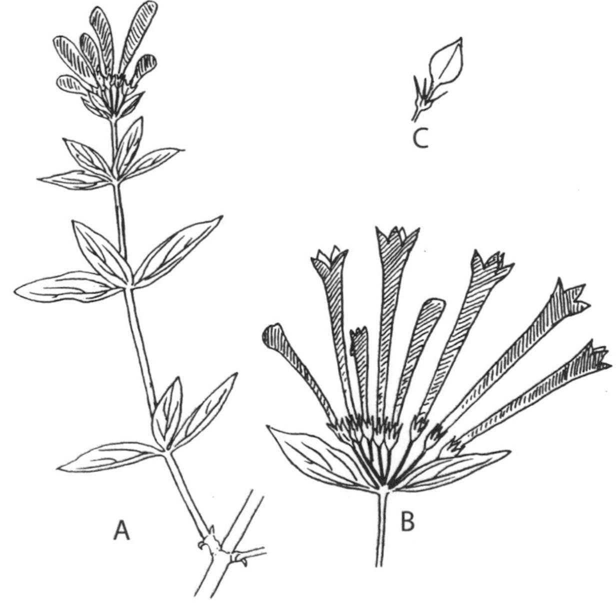


*Bouvardia ternifolia*. A. Flowering branch of shrub. B. Detail of flower cluster. C. Fruit. Artist: R. D. Ivey (Ivey 2003).

**Other names:** ENGLISH: firecracker bush (New Mexico), scarlet bouvardia (a book name); SPANISH: *chuparrosa* (rose-sucker, Sonora), *clavillo* (little carnation, Arizona; typically applied to the first root that emerges from the seed of corn or beans in Mexico), *contrahierba* [*colorado*] ([red] counter-poison, antidote, Edo. México), *doncellita* (little lady, Oaxaca), *hierba del indio* (Indian's herb, Sinaloa, Sonora), *hierba del pasmo* (herb for *pasmo*, from Latin *spasmus*, Sinaloa), *indita* (little Indian girl, Chihuahua), *mirto* [*del campo*] ([wild] myrtle, Coahuila, Durango), *tabaquillo* (little tobacco, Michoacán), *trompetilla* [*rosa*] (little [red] trumpet, Hidalgo, Edo. México, Oaxaca, Sonora, Veracruz), *yerba de zorrillo* (little skunk's herb, Mountain Pima); OTO-MANGUEAN: *rnanta* (Mazahua); UTO-AZTECAN: *expatli* (*ex* [*e:š*], three, *patli*, medicine, Náhuatl), *rurikuči* (Tarahumara), *tlacoxiuitl* (*tlacotl*, slave or servant, *xiuitl*, herb, Náhuatl), *tlacoxóchitl* *<tlacosúchil>* (*tlacotl*, slave or servant, *xóchitl*, flower, Náhuatl), *wipismal je:j* *<wipismal jehj>* (hummingbird's mother, Tohono O'odham); LANGUAGE FAMILY UNKNOWN: *tonati-sochit* (Hidalgo); (= *Bouvardia glaberrima*)

**Botanical description:** Shrubs or herbaceous above and woody below, to 1 m tall, the herbage glabrous or minutely rough-hairy, the bark pale when young but turning brown with age. Leaves are lanceolate to ovate-lanceolate, typically whorled in 3s around the stems, although sometimes in 2s or 4s, to 8 cm long, basally cuneate, apically acute, and pubescent. Inflorescences of cymose clusters are on the ends of branches. Calices are of 4 lobes, these subulate. Corollas are narrowly tubular, red to reddish orange, rarely pink or white, 2–3 cm long, to 7 mm wide at the slightly flaring apex. Capsules are subglobose, paired (didymous). Seeds are flat, peltate, winged.

**Habitat:** Canyons and slopes. 750–2,743 m (2,500–9,000 ft).

**Range:** Arizona (southern Apache, Cochise, Graham, Greenlee, Pima, Pinal, and Santa Cruz counties) and New Mexico; Mexico (Chihuahua and Sonora to Durango).

**Seasonality:** Flowering May to October.

**Status:** Native.

**Ecological significance:** The tubular red to orangish flowers on these shrubs are visted by hummingbirds. In Brown Canyon, the Broad-Billed Hummingbird (*Cynanthus latirostris*) is among the common species that drink nectar from Trumpet Flower where individuals grow on slopes leading down toward the streambeds.

Flowers are of two kinds, varying in the length of the stamens and styles (distylous). That arrangement is a specialization for outcrossing.

Within Arizona, these shrubs are known in pine-oak, juniper-oak, and oak chapparal communities (Dempster and Terrell 1995). In the Baboquivari Mountains, Trumpet Flower grows in oak woodlands with scattered pines and junipers.

**Human uses:** Mexicans use the Trumpet Flower root as a cardiac stimulant and to treat dysentery, hydrophobia, and heat exhaustion (Standley 1920–1926, Hocking 1997). As examples, the Mountain Pima and Tarahumara decoct the stems and leaves to make a medicine for heart trouble (Pennington 1963, 1973).

The root is considered a styptic, and it was used by Mexicans at the time of contact to stop bleeding that had been initiated as part of a medical cure. *Bouvardia ternifolia* was discussed by Fray Bernardino de Sahagún in the 1500s under the name *tlacoxochitl*. He wrote, "The flavor of the root is both bitter and sweetish. They are good for heat and exhaustion of the heart. Ground with about 15 grains of maize and as much cacao, and mixed with water, they should be taken several times on an empty stomach or after meals." Later, Trumpet Flower was described and illustrated by Francisco Hernández in 1651.

Northern Tepehuan gather the flowers and attach them to the *olla* (jar) containing fermenting *tesgüino* (home-brewed liquor) (Pennington 1969). This decoration has an undisclosed meaning to them.

**Derivation of the name:** *Bouvardia* commemorates Charles Bouvard, 1572–1658, superintendent of the Jardin du Roi, Paris. *Ternifolia* refers to the three leaves in whorls around the stems.

**Miscellaneous:** *Bouvardia* is a genus of 20–50 species that grows from Arizona, New Mexico, and Texas south through Mexico to Central and South America (Dempster and Terrell 1995, Mabberley 1997).

## *Ptelea trifoliata* (Rutaceae) Hop-Tree, *cola de zorrillo*

**Other names:** ENGLISH: hop-tree (from ca. 1877, with the substitution of the fruits of *Ptelea* for those of hops, *Humulus lupulus*, for making malt liquor, and as a tonic and soporific), skunk-bush (both leaves

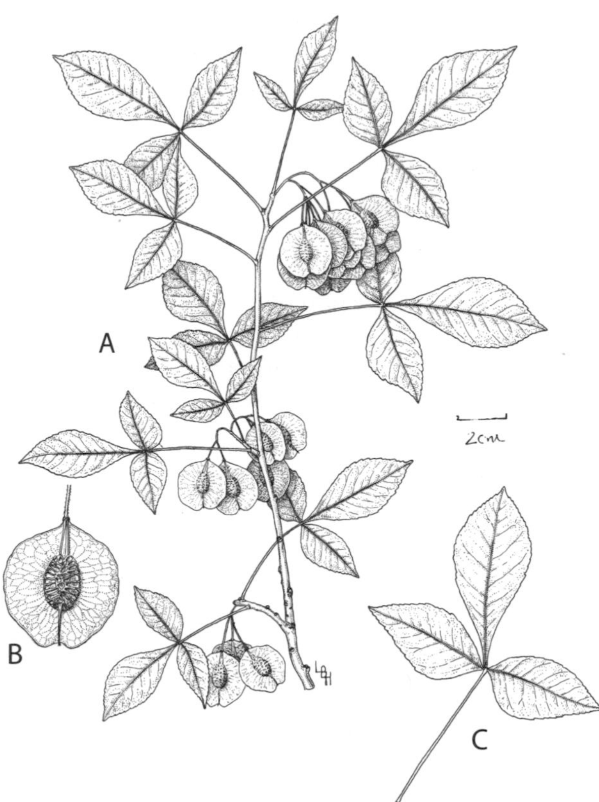


*Ptelea trifoliata*. A. Fruiting branch. B. Fruit, enlarged. C. Leaf, enlarged. Artist: L. B. Hamilton (from original artwork).

and mammal have a similar odor), wafer-ash; SPANISH: *cola de zorrillo* (skunk tail, Arizona, Texas, Chihuahua, Sonora, Tamaulipas, Veracruz), *fresno* (ash, northern Mexico), *pinacatillo* (comparing the smell with the pinacate beetle, Coahuila; from Náhuatl *pinacatl*, the Aztec name for the beetle *Elodes* spp., Tenebrionidae, a black insect that sprays an irritating, stinking chemical for defense), *telea* (variant of Greek *ptelea*), *vara de zorro* (fox bush, northern Mexico to Veracruz), *zorillo* (skunk, Hidalgo); KERES: *chibanini* (San Felipe); UTO-AZTECAN: *ápago <apaga, upaga>* (Tarahumara); (= *P. angustifolia*)

**Botanical description:** Shrubs or, less often, small trees; herbage aromatic. Leaves are winter deciduous, alternate, glabrous or at least glabrous with age, the leaflets 3, these 3.5–10 cm long, 1.5–5 cm wide, lanceolate to obovate, thin, lateral leaflets unequal, the margins minutely toothed to entire. Panicles are terminal on branches. Flowers are greenish white, unisexual (female) and bisexual, 11–15 mm wide, 4–5 merous. Sepals are 1–2 mm long. Petals are 4–5 mm long. Fruits are flattened, papery, disk-shaped samaras, 1–2.3 cm wide and about as long, with a conspicuously veined, broad wing encircling the fruit.

**Habitat:** Canyons, washes. 1,100–2,590 m (3,500–8,500 ft).

**Range:** Arizona (all counties but Yuma), New Mexico, and Texas; Mexico (Chihuahua, Sonora to Tamaulipas, and south to Oaxaca).

**Seasonality:** Flowering May to June (July).

**Status:** Native. Baboquivai plants are subspecies *angustifolia*; ssp. *trifoliata* grows in much of the eastern United States.

**Ecological significance:** These temperate trees belong to the Rocky Mountain (Petran) Province and reach down toward the desert and grasslands in the canyon bottoms. Because the shrubs need abundant water, they are never far from the moisture remaining in the bottoms of seasonal streams.

As the common names indicate, the foliage smells like a skunk. This odor is released when branches are only bruised or even lightly brushed in passing. Crushed leaves have a different, citrus-like fragrance. Plants contain an array of alkaloids, including dictamnine, pteleine, ptelecultinium, pteleatinium chloride, and several others. In addition to the alkaloids, Hop-Trees produce coumarins (Austin 2004).

**Human uses:** According to Diggs et al. (1999), *P. trifoliata* contain a poisonous saponin that may cause photodermatitis. Leaves may cause dermatitis in some people, but they have been used as a substitute for Hops (*Humulus*) in brewing beer. Fruits have been substituted for Hops in making beer since at least 1858, when the U.S. Department of Agriculture examined possible commercial use (Bailey 1960).

The wood is carved into beads and crosses by the Tarahumara (Pennington 1963).

Constantine F. Rafinesque considered *Ptelea* vulnerary and vermifuge in his *Medical Botany* of 1830. Menominis hold the root bark to be a sacred medicine and panacea to season and render other medicines potent (Austin 2004). Mesquakies use it similarly, especially to treat lung problems, often in tea with other barks. Havasupai use Hop-Trees for medicine, but they also made things of the wood, and San Felipe Pueblo people eat the seeds (Castetter 1935, Moerman 1998).

In Mexico, *Ptelea* is still used in remedies for dyspepsia, as a mild tonic, in a bath of an alcoholic infusion of leaves against chills, and against rheumatism (Martínez 1969). Tarahumara make a decoction of leaves and roots to treat rheumatic pain (Pennington 1963).

**Derivation of the name:** *Ptelea* is from *ptao*, to fly, the Greek name for the elms (*Ulmus*), reapplied by Linnaeus to this genus with a similar fruit; akin to Akkadian *petelu*, to wind, entwine. *Trifoliata* means that it has three leaflets per leaf.

**Miscellaneous:** There are from 3 to 11 species in *Ptelea*, all endemic to North America (Mabberley 1997, Felger et al. 2001). Some of them contain alkaloids that have been shown to be antifungal.

## *Nolina microcarpa* (Ruscaceae) Bear-Grass, *sacahuista, moho*

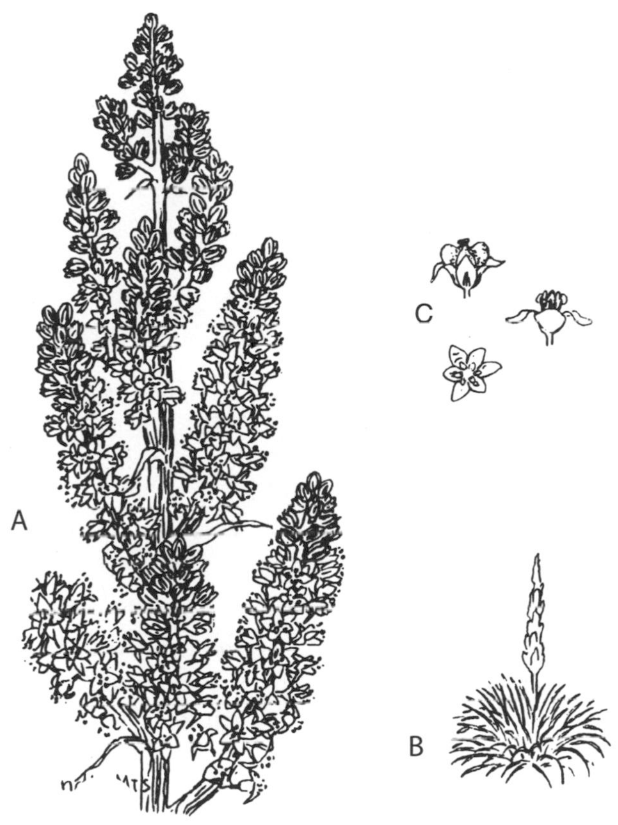


*Nolina microcarpa*. A. Flower cluster. B. Aspect of plant. C. Detail of flowers. Artist: R. D. Ivey (Ivey 2003).

**Other names:** ENGLISH: bear-grass (in use by 1750, presumably because bears use them; originally applied in Kentucky, surely to some real grass, Poaceae; later extended to other plants), saw-grass (a comment on the cutting leaf margins; in use by 1822 for *Cladium* and later extended to other plants), squaw-flowers; SPANISH: *palma* (palm, Chihuahua), *palmilla* (little palm, Sonora), *sacahuista* (Arizona, Sonora, probably from Náhuatl *zacatl*, grass, *cuahuitl*, tree), *sotol chiquito* (little sotol, because *Dasylirion* is *sotol*, Sonora), *zacate* [*cortador*, *de aparajo*, *de armazón*] ([cutting, packsaddle, framework] grass, from Náhuatl *zacatl*), *soyate* (from Náhuatl *zoyatl*, palm); ATHAPASCAN: *bį', gogisa* (*bį'*, the name for *Nolina*, *gogisa*, sharp, Western Apache), *etłodeitsa* (yucca, fringed, Chiricahua and Mescalero Apache), *hogéesh* <*ho.gisi'*> (cutting leaves, Navajo); UTO-AZTECAN: *duya* <*ruya, guru*> (Tarahumara), *kukt* (Tübatulabal; for *N. parryi*), *kurú* [*wirúku*] (Guarijío), *moho* (Hiá Ceḍ O'odham), *moho* <*mōhō*> (Tohono O'odham, probably from *mo'o*, hair, head; apparently a pan-Tepiman name), *moho* <*moh, moj*> (Mountain and Onavas Pima), *silíomóhu* (*móhu*, hair, Hopi); YUMAN: *ŏk$^r$ĭnyúda* (Havasupai), *qanyud* (Walapai)

**Botanical description:** Perennial, with a large woody, mostly underground caudex, the flowering stem stout, to 18 dm long. Leaves are concavo-convex, thick and keeled, about 1 m long, 6–12 mm wide, raggedly denticulate to strongly serrulate, fibrous and torn (lacerate) apically. Panicle is narrow, branched at base, primary floral branches ca. 3 dm long, with ascending branchlets 5–7.5 cm long. Bracts are small. Fruits are thin, inflated, notched on both ends, ca. 6 mm wide. Seeds are 2–2.5 mm in diameter.

**Habitat:** Rocky slopes, grasslands. 900–1,980 m (3,000–6,500 ft).

**Range:** Arizona (Apache, Cochise, southwestern Coconino, Gila, Graham, Greenlee, Maricopa, Pima, Pinal, and Santa Cruz counties), New Mexico, Nevada, western Texas, and southwestern Utah; Mexico (Chihuahua, Durango, and Sonora).

**Seasonality:** Flowering May to June.

**Status:** Native.

**Ecological significance:** This species barely extends west of the Baboquivari Mountains (Felger et al. in prep.). There, the species is disjunct from farther east into the higher elevations of the Ajo Mountains in western Pima County. Thus, this shrub is more a Chihuahuan Desert species than a Sonoran Desert plant. Although Kearney and Peebles (1951) recorded the species as "almost throughout the state," it is most frequent in southeastern Arizona, and it extends northwest along the Mogollon Rim.

The extended drought has been taking its toll on *Nolina*. After the almost rainless winter was finally broken in March of 2006, scattered plants of *Nolina* and *Dasylirion* were dying throughout the lower parts of Brown Canyon. Although these and other plants are adapted to desert conditions, there is a limit to how much time any organism can live without water.

**Human uses:** *Sotol* (from Náhuatl *tzotolli*) is both the name of the plant and the potent alcoholic beverage made from the "heart" (apical meristematic region). All of the Arizona species are reportedly used similarly in making this brew (Hocking 1997, Moerman 1998).

Apaches of several bands eat the roasted young flower stalks (Castetter and Opler 1936, Hodgson 2001). Bean and Saubel (1972) also found that the Cahuilla bake the stalks of *N. bigelovii* in rock-lined pits. Similarly, the Tübatulabal and Walapai eat the stalks of *N. parryi* (Moerman 1998, Voegelin 1938). Several tribes, including Apache and Isleta, eat the fruits, either fresh or cooked (Castetter and Opler 1936, Moerman 1998, Hodgson 2001). Fresh fruits have an awful astringent taste; I consider them inedible.

The Tohono O'odham use the leaves of Bear-Grass to weave baskets, while Akimel O'odham use the flower stalks of *uḍvak* <*oodvak, otoxak*> (*Typha domingensis*) (Rea 1997). Similarly, *Nolina* fibers and leaves are used in weaving and building materials by Apache, Havasupai, Hopi, Isleta, Jemez, Keres, and Yavapai (Castetter and Opler 1936, Gallagher 1976, Moerman 1998, Whiting 1939). Tohono O'odham and Walapai also use the leaves of *N. bigelovii* and *N. parryi* in weaving.

Navajo make dye from *Nolina* (Moerman 1998). Apache use the roots for soap (Moerman 1998); they contain the sapogenin yuccagenin (Hocking 1997).

**Derivation of the name:** *Nolina* commemorates P. C. Nolin, an eighteenth-century French agriculturist. *Microcarpa* means having small fruits.

**Miscellaneous:** *Nolina* consists of about 20 species, being best developed in Mexico but with species extending into the southern United States, from California to Florida (Gentry 1972). There are four species in Sonora and four in Arizona; however, they are not all the same. Sonora has *N. bigelovii*, *N. matapensis*, *N. microcarpa*, and *N. texana*. Arizona lacks *N. matapensis* but has *N. parryi*. Texas has *N. arenicola*, *N. erumpens*, *N. lindheimeriana*, and *N. micrantha* (Correll and Johnston 1970). The genus was formerly put in the Nolinaceae.

## *Dodonaea viscosa* (Sapindaceae) Hop-Bush, *jarilla*

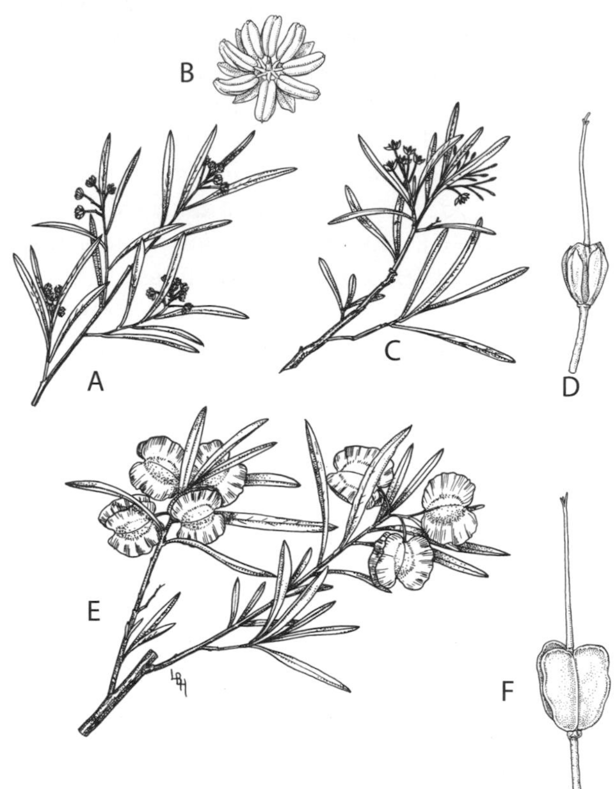


*Dodonaea viscosa*. A. Male flowering branch. B. Male flower, enlarged. C. Female flowering branch. D. Female flower, enlarged. E. Fruiting branch. F. Developing fruit. Artist: L. B. Hamilton (Benson and Darrow 1945).

**Other names:** ENGLISH: hop-bush (Arizona to Florida; used in place of hops, *Humulus lupulus*, in making beer), hopseed bush, switch sorrel (Arizona); SPANISH: *aría* <*airía*> (error for *jarilla*? Mexico), *alamillo* (little winged one, Chihuahua, Sonora), *chupulitztle* <*chapuliz*> (Spanish from Náhuatl *chapulichtli*, grasshopper fiber, as these insects eat the leaves), [*hierba de la*] *cucaracha* (cockroach [herb], Durango), *cuerno de cabra* (goat's horn, Oaxaca), *gitarán* (probably based on Spanish *gitano*, gypsy, clever, Mexico, Puerto Rico), *granadina* (maybe leaves resemble *Psidium*, the *granada*, Baja California), *guachomó* (from Maya *huach*, person born in Yucatán, Mexico), *guayabillo* (little guava [*Psidium*], Baja California), *huesito* (little bone, Chiapas), *jarilla* [*de loma*] ([wild] little arrow, Chihuahua, Sonora to Oaxaca), *jirimú* (corruption of Purépecha *pirimu*? Michoacán), *mundito*(little world, perhaps a reference to the fruits, Hidalgo), *ocotillo* (little torch, from Náhuatl *ocotl*, Guanajuato, Hidalgo), *palomilto* (*palo*, tree, *milto*, wild, from Náhuatl, San Luis Potosí), *tapachile* (maybe from Náhuatl *tapachtli*, sea shell or fish scale, in reference to the fruits), *tarachique* [*tarachico*] (from Ópata, Sonora; *Tarachi* is the Yaqui region of the state), *varal* (branch thicket, Arizona, Hidalgo, Sonora); TARASCAN: *pirimu* (Purépecha); UTO-AZTECAN: *maˀcikári* (Guarijío), *taratsike* (*tara*, foot, *tsi*, possessive/object, *ke*, little, Ópata, Sonora), *tonalcotl-xihuitl* [*toñalokotl*] (*toñal*, sun, *okotl*, pine pitch, *xihutil*, herb, Náhuatl); (= *D. angustifolia*)

**Botanical description:** Shrubs 1.5–2 m tall. Branches are mostly erect-ascending. Herbage is densely resinous-glutinous. Leaves are simple, 3–9 cm long, 3–9 mm wide, linear to linear-oblanceolate, narrowed at base and essentially sessile, with margins entire. Flowers are mostly dioecious but some bisexual, pedicellate, in small clusters, yellow-green. Sepals are 2 mm long. Capsules are 1–2 cm wide, papery, 3-winged, often pinkish red, drying straw-colored. Seeds are 3–3.5 mm across, black, lens-shaped.

**Habitat:** Rocky slopes, canyons, often on limestone. 600–1,500 m (2,000–5,000 ft).

**Range:** Arizona (southern Yavapai to Cochise, Pima, and Santa Cruz counties) and New Mexico; Mexico (Baja California and Sonora to Tamaulipas south); Central America; South America, West Indies, and worldwide.

**Seasonality:** Flowering February to October.

**Status:** Native. Baboquivari plants belong to var. *angustifolia*.

**Ecological significance:** These bushes are tropical thornscrub species (Brown 1982) that barely reach into this region. Where Hop-Bush is found, it is an indication of a warmer microclimate during the winter. Plants in the open are killed by cold weather on the western slopes of the Sierrita Mountains, while plants are barely touched in Brown Canyon where it is over 1,000 feet higher in elevation.

Flowers are probably wind-pollinated (Turner et al. 1995). Although Turner et al. (1995) suggest that fruits are wind-dispersed, they usually fall apart before dropping. I suspect that seeds are dispersed by birds.

**Human uses:** Given the many uses in Hawaii and Asia (Hocking 1997, Moerman 1998), and the numerous names applied in the Sonoran region, it is surprising that no uses have been found in this area. Farther south in Mexico, the species is used as a medicine (Martínez 1969).

The leaves and their resinous exudate are aromatic, bitter, astringent, sudorific, purgative, and febrifuge (Austin 2004). Both leaves and/or wood decoctions are used against fevers (Mexico, Panama, and Brazil). The leaf decoctions are also used against rheumatism and venereal disease (Mexico and Brazil). Bark extracts are used in astringent baths and fomentations (Mexico and Brazil) to provide relief from several problems, including rheumatism. Although people in some areas (Mexico) eat the seeds, they and the entire plant contain enough saponin that people also use them as *barbasco* to poison fish (Panama and Colombia). Leaves are chewed as a stimulant—maybe that is why Colombians confuse it with coca—or to treat toothaches (Panama). The sap is applied to clear tumors (Brazil), and sap with other plant parts is used in cataplasms in the treatment of flatulence (Brazil), gout, and venereal diseases (Mexico and Brazil). Leaf infusions are hemostatic for people and animals (Colombia and Panama).

**Derivation of the name:** *Dodonaea* was created by Philip Miller in 1754 to honor Dutchman Rembert Dodoens (1517?–1585), one of the foremost physicians and herbalists of his day. *Viscosa* means sticky, an allusion to the waxy surfaces; *angustifolia* indicates that this variant has narrow leaves.

## *Sapindus drummondii* (Sapindaceae) Soap-Berry, *amolillo*

**Other names:** ENGLISH: soap-berry (in use by 1693; "soap" was in English by ca. CE 1000, from Latin *sapo* by Pliny, CE 23–79; cognates are present in most European languages); SPANISH: *amole* <*yamole, yamolli*> (soap, from Náhuatl), *amole de bolita* (soap balls, Mexico), *amolillo* (little soapy one, Sonora), *arbolio* (little tree, Sonora), *cirioni* <*cherioni*> (maybe from Latin *cereus*, for wax, Arizona), *jaboncillo* (little soap, Nuevo León, San Luis Potosí, Sonora, Tamaulipas and south), *matamuchacho* (boy killer, Sonora), *palo blanco* (white tree, Chihuahua); OTO-MANGUEAN: *bibí* <*pipe, pipal*> (fruit, Zapotec); UTO-AZTECAN: *jutuhui* (Guarijío), *tehistle* <*tehoitzli, tehuixtle, tehuitle*> (sharp rock, Náhuatl; allusion abstruse), *tubchi* <*tupchi*> (Mayo, Sonora); LANGUAGE FAMILY UNKNOWN: *boliche* (also the name of a ball game, Sinaloa); (= *Sapindus saponaria* var. *drummondii*)

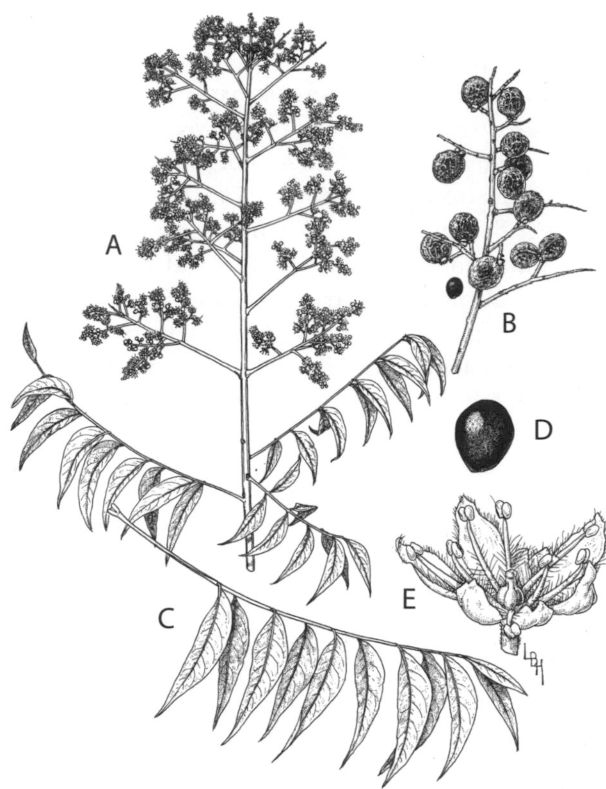


*Sapindus drummondii*. A. Flowering branch. B. Fruiting branch. C. Leaf, enlarged. D. Seed, enlarged. E. Flower, enlarged. Artist: L. B. Hamilton (from original artwork).

**Botanical description:** Trees or small shrubs 5–15 m tall. Leaves are drought- and cold-deciduous, 12–45 cm long, the leaflets 5–10, usually 6–8 pairs, these offset and alternate on the rachis, elliptic-lanceolate, larger leaflets 5.5–20 cm long, sometimes longer, 1–6 cm wide, leafstalk sometimes narrowly winged. Flowers are 4–5 mm wide, with long, white hairs, especially near the base. Sepals are 1.5–1.8 mm long. Petals are white, 2.2–2.5 mm long, cupped, obovate to orbicular. Fruits (or each lobe) are 12.5–18.5 mm in diameter, each lobe 1-seeded.

**Habitat:** Canyons, streamsides, washes. 760–1,500 m (2,500–5,000 ft).

**Range:** Arizona (Coconino, Mohave, south to Cochise, Pima, and Santa Cruz counties), New Mexico, Texas, east to Kansas, Missouri, and Louisiana; Mexico (Chihuahua, Sinaloa, Sonora to Tamaulipas, and south to Oaxaca).

**Seasonality:** Flowering May to August.

**Status:** Native.

**Ecological significance:** Flowers attract numerous flies and medium-sized wasps that are presumably the pollinators. Plants always grow near water, but they are typically enough upslope that Soap-Berry roots are rarely flooded. Trees are usually on inclines leading down to streams, ponds, and similarly wet sites. Tree location suggests that the seeds might be dispersed by water during floods. However, it is known that Raccoons eat the fruits, at least in small quantities (Tyler et al. 2000). That consumption is consistent with the seeds germinating better after acid treatment (Vora 1989).

Birds also eat fruits, particularly during fall migration. As with other species that produce during this season, these "berries" are relatively large, and the pericarps are rich in lipids. Stiles (1980) classifies these fruits as "low-quality" because they remain on the trees, and low rates of seed dispersal continue into the winter.

**Human uses:** Gonzalo Fernández de Oviedo y Valdés called the Mexican trees *cuentas del xabón* (bead soap) in 1535. That name records two of the uses that Americans have for the seeds, as beads in necklaces and rosaries and as soap. The first idea persists today in *palo de cuentas* (bead tree, Oaxaca). People throughout the range of these trees use the soapy traits for washing (Austin 2004). The fruits contain 30% saponins, thereby making them useful for cleaning. However, the chemicals cause dermatitis in some individuals (Hocking 1997). Fruits formerly were used in adding a head to beer, and the fatty oils of seeds have been used in manufacturing soap.

Kiowa use the sap in a poultice to treat wounds (Vestal and Schultes 1939). At least the Comanche and Tohono O'odham use the wood to make arrows (Castetter and Underhill 1935, Moerman 1998). The Tohono O'odham use the wood only to make the foreshaft of the arrow where the head was attached (Castetter and Underhill 1935).

**Derivation of the name:** *Sapindus* comes from Latin *sapo*, soap, and *indicus*, of the Indians. The name *drummondii* commemorates Thomas Drummond, 1780–1835, a Scottish botanist and collector in North America. The other specific name, *saponaria*, is from Latin *sapo*, soap. Thus, both the genus and the name of the second species emphasized previous use as soap.

**Miscellaneous:** There has been confusion and disagreement about species limits in this genus since *Sapindus* was named by Hooker and Arnott in 1839 (*The Botany of Captain Beechey's Voyage*). Salywon (1999) thought that *S. drummondii* was the same as

S. *saponaria*. There are certainly two kinds of plants that grow sympatrically throughout much of the Americas. *Sapindus drummondii* typically lacks the wings on the compound leaf axis. By contrast, the eastern species, *S. saponaria*, is characterized by those very wings. Fruits also tend to be smaller in *S. drummondii* than in *S. saponaria*.

## *Sideroxylon lanuginosum* (Sapotaceae) Gum Belly, *tempixtle*

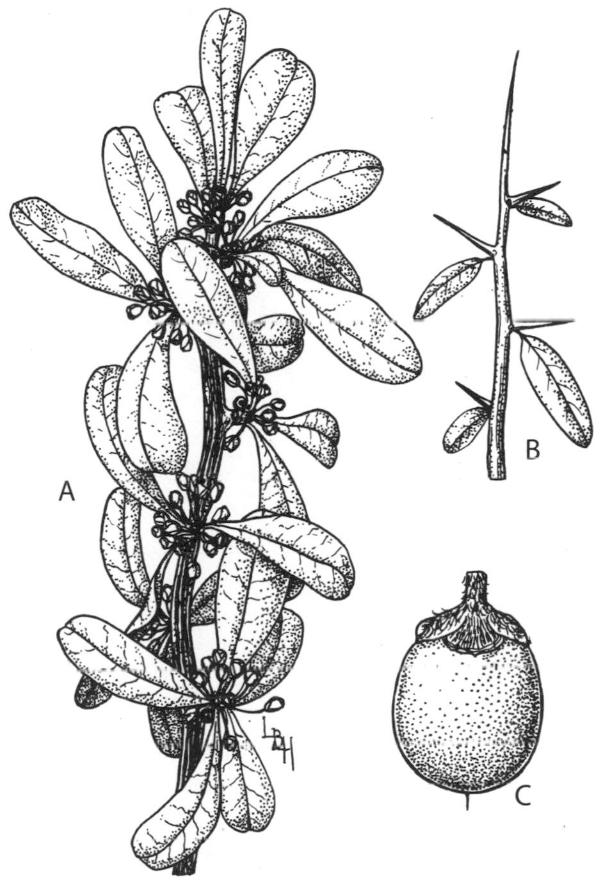


*Sideroxylon lanuginosum*. A. Flowering branch. B. Young twig with spines. C. Fruit, enlarged. Artist: L. B. Hamilton (Benson and Darrow 1945).

**Other names:** ENGLISH: chittam-wood <shittam-wood> (based on Hebrew *shittim* or *shittah*, originally the name of *Acacia*, Arizona, Sonora), black haw (from Old English *haga*, hedge; akin to Dutch *haag*, German *hage*, Danish *have*, garden), gum belly [billy] (Arizona, New Mexico, Florida), gum bumelia (Arizona), gum elastic (Texas), ironwood (name given, often locally, to various trees with extremely hard wood; dating from at least the 1650s), [false] woolly buckthorn (from Latin *cervi spina*; applied by Valerius Cordus, 1514?–1544, to *Rhamnus cathartica*; later used for other genera, Texas), woolly bucket-wood; SPANISH: *coma* (from Náhuatl *comal*, alluding to the shape of the fruit and the cooking utensil, Texas, Nuevo León), *tempixtle* <*temisque, tempistle*> (from Náhuatl *tentli*, lip, *pixtle*, seed of the sapote [*Pouteria sapota*], Mexico); MUSKOGEAN: *afoló intató (afoló*, screech owl, *in*, its, *tató*, honey-locust, Koasati), *coyyí notá okwołí* (*coyyí*, pine, *notá*, underside, *okwołí*, dewberry, Koasati, Texas); OTO-MANGUEAN: *mulché* (Zapotec, Mexico); UTO-AZTECAN: *bebelama* (juice, Ópata, Mexico), *chu-ga-ka* (Tarahumara), *hake* (Guarijío); (= *Bumelia lanuginosa*)

**Botanical description:** Shrubs or small trees 2–15 m tall, sometimes taller, the trunk usually 1–2 dm in diameter, the bark gray, and twigs brown and spiny. Leaves are oblanceolate to obovate or elliptic, broadly rounded to acute at the apex, 2–10 cm long, 1–3.5 cm wide, leathery, loosely woolly-villous when young, soon glabrate above, persistently hairy to sometimes glabrate below, reticulate-veined on both surfaces, sometimes in groups (fascicled). Flower clusters have 10–15 blossoms, the hairy or almost glabrous pedicels to 15 mm long. Sepals are strongly hairy or almost glabrous, 1.5–3.2 mm long. Corollas are 3–4.7 mm long, the tube 1.3–2 mm long. Sterile stamens (staminodes) are deltoid-ovate, nearly equaling the corolla lobes. Fruits are obovoid to broadly ellipsoid to almost globose, purplish black when ripe, 7–12 mm long.

**Habitat:** Canyons, washes, particularly along seasonal or permanent streams. 1,100–1,830 m (3,500–6,000 ft).

**Range:** Arizona (Cochise, Pima, and Santa Cruz counties), New Mexico, Texas, to Kansas, Oklahoma, Missouri, Illinois, Georgia, and Florida; Mexico (Chihuahua and Sonora to Tamaulipas).

**Seasonality:** Flowering June and July.

**Status:** Native.

**Ecological significance:** In Arizona, this shrub is considered a thornscrub species (Brown 1982); however, farther east in the southern United States, it is not. The species has been divided into more or less geographically distinct varieties. The western variety that grows from the Trans-Pecos region of Texas to Arizona and Sonora is var. *rigidum*. This form has more densely pubescent leaves than the others. Birds eagerly eat the fruits, and probably also mammals eat them (Vines 1977).

**Human uses:** Gum Belly gum is chewed by children in Texas and in northern Mexico along the Rio Grande (Bourke 1895, Kearney and Peebles 1951). Fruits are eaten by indigenous people throughout the southern states (Austin 2004). This plant is probably the one that Thord-Gray (1955) said was eaten by the Tarahumara. His description of the sticky black fruits puckering "the mouth in an unpleasant manner" is consistent with several members of the family. The fruit is edible "but not tasty and causes stomach upset and dizziness if eaten in quantity" (Vines 1977).

Bark is pounded by the Kiowa and used as chewing gum (Vestal and Schultes 1939). The tree was the source of an orange dye in the 1800s (Hocking 1997). Wood is used for tools and cabinets.

**Derivation of the name:** *Sideroxylon* is from Greek *sidero*, iron, and *xylon*, wood, a reference to the hard wood. *Lanuginosum* means woolly-pubescent.

**Miscellaneous:** As the common name implies, these shrubs were at one time put in a distinct genus, *Bumelia*. That name, the classical Greek name of an ash (*Fraxinus*), was given by Theophrastus, 372–287 BCE, as *boumelios*, and became *bumelia* in Latin. Olaf Peter Swartz (1760–1818) reapplied the name to American plants in 1788, and the genus was considered separate from others in the family until the group was studied for the *Flora Neotropica* (Pennington 1990).

As now understood, *Sideroxylon* comprises 75 species, most of which grow in the tropics of the world (Mabberley 1997). There are more species in the southeastern United States than in the Southwest.

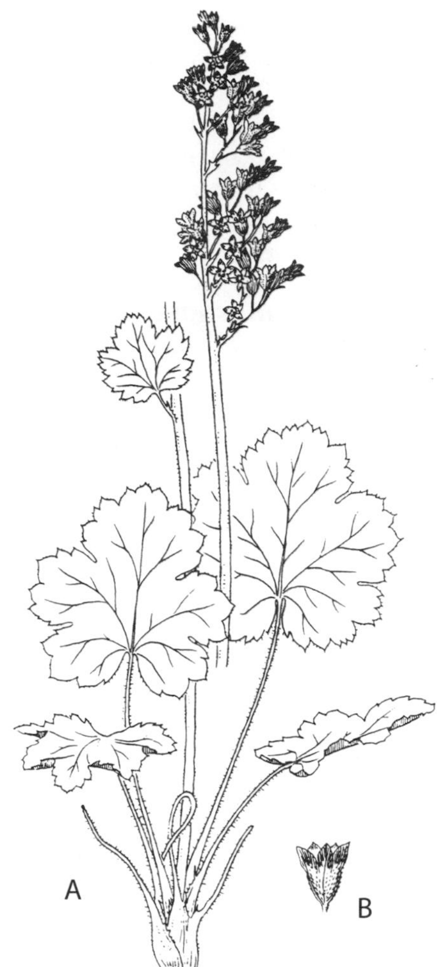


*Heuchera sanguinea*. A. Flowering plant. B. Detail of calyx. Artist: R. D. Ivey (Ivey 2003).

## *Heuchera sanguinea* (Saxifragaceae) Coral Bells, *campanilla de coral*

**Other names:** ENGLISH: alum root (name given to the astringent roots of various plants; applied to *Heuchera* by Benjamin Barton in 1798), coral bells (from ca. 1305 applied in English to red corals; "coral" from Old French *coral*, *coural*, later *corail*, cognate with Portuguese *coralh*, Spanish *coral*, Italian *corallo*, all ultimately derived from Latin *corallum*, *coralium*, and Greek *korallion*, red coral); SPANISH: *campanilla de coral* (little coral bells, Mexico), *flor de piedra* (rock flower, Mountain Pima); ATHAPASCAN: *tséyi' łé'étsoh binát'oh* <*céyiˀ leˀécoh binát'oh*> (*tséyi'*, canyon, *łé'étsoh*, rat, *binát'oh*, its tobacco, Navajo; for *H. parviflora*); UTO-AZTECAN: *sáaperek hióskem weg* <*hoda hioshgara, jod yorsh'cum wig*> (*sáaperek*, rock, *hiosgam*, flower, *vegi*, red, Mountain Pima), *wi'gûndaza* [*pa'sawi'gûnza*] (Shoshoni; for *H. rubescens*)

**Botanical description:** Perennial herbs without above-ground stems. Leaves are basal, the blade ovate, 1–7 cm long, the base cordate, the margins shallowly 7–10 lobed, the lobes dentate to aristate, on pubescent petioles 1.5–20 cm long. Flower clusters are open, cylindrical to pyramidal panicles with 0–2 leaf-like bracts below, 20–60 cm tall. Flowers are bright pink to deep red, glandular-hairy. Sepals are equal, ovate-oblong, erect, 1–4 mm long. Petals are shorter than sepals, 0.5–3.5 mm long, oblanceolate, white to pink. Fruits are capsular, 2-valved. Seeds are many, small.

**Habitat:** Moist, shaded cracks in rocks, particularly where there are seeps of water. 1,050–2,900 m (3,500–9,500 ft).

**Range:** Arizona (southern Apache to Cochise, Pima, and Santa Cruz counties) and southwestern

New Mexico; Mexico (Chihuahua and Sonora to Coahuila,).

**Seasonality:** Flowering (February) March to May (October).

**Status:** Native.

**Ecological significance:** Wherever these herbs occur is an indication that there is permanent water, at least just below the surface. Plants are restricted to a few seeps in Brown Canyon and some of its side canyons. The only other collection seen of this plant from the Baboquivari Mountains was made in Fresnal Canyon on the Tohono O'odham side in the 1970s.

This herb is a Mexican species that barely reaches the United States in Arizona and New Mexico; it occurs in no other states. The species was probably first collected in Chihuahua by German-born Friedrich (Frederick) Adolph(us) Wislizenus, 1810–1889, as indicated by his specimen collected in 1846 (MO).

The red flowers are not typical of those adapted morphologically for bird pollination. However, the red of the flower-clusters make *H. sanguinea* stand out when in blossom, and hummingbirds are notoriously opportunistic. These small nectar-feeders will feed on them.

**Human uses:** Moore (1979, 2003) says that all *Heuchera* species may be used interchangeably as medicines and that *H. sanguinea* is one such example. No other source has been found that includes this species as a medicine.

In 1798, physician Benjamin Barton wrote of *H. americana*, "This [plant] is sometimes called American Sanicle. It is more commonly called Alum-Root. The root is a very intense astringent. It is the basis of a powder which has lately acquired some reputation in the cure of cancer." Constantine Samuel Rafinesque wrote in 1828 that "it was used by the Indians, and is still used in Kentucky and the Alleghany Mountains, in powder, as an external remedy for sores, wounds, ulcers, and even cancers." That last use did not persist, but the astringent nature has.

Hart (1976) and Moerman (1998) list 12 other *Heuchera* that are used. Of those listed, most were used outside the Southwest and Mexico region. However, *H. bracteata* is used by Navajos to relieve indigestion and toothaches (Elmore 1944). *Heuchera bracteata* yields a pink tan dye (Elmore 1944). Similarly, *H. novomexicana* is used by the Navajo to treat internal pain, external sores and swelling, and in their Life Medicine (Vestal 1952). Navajo also use *H. parvifolia* to treat bites and stomachaches, to ease delivery during childbirth, as a Life Medicine, to treat venereal disease, and for rat bites and lightning infection (Moerman 1998, Vestal 1952, Wyman and Harris 1951). Other western tribes that use the genus include Arapaho, Blackfoot, Cheyenne, Shoshoni, Kutenai, Miwok, Okanagan, Paiute, Salish, Shuswap, Skagit, Thompson, and Tlingit (Moerman 1998, Train et al. 1957). Several eastern tribes use other species.

A common name for eastern *H. americana* is "alum root," and that and many other species still are used as an astringent and styptic (Hocking 1997, Moerman 1998). At least some species contain as much as 20% tannin (Moore 1979).

**Derivation of the name:** *Heuchera* commemorates Johann Heinrich Heucher, 1677–1747, a German physician and professor of medicine. *Sanguinea* means red-colored, in reference to the flowers.

**Miscellaneous:** *Heuchera* comprises 55 species that are endemic to North America. The United States and Canada have 35 species and 2 hybrids (Kartesz 1994). Although Kearney and Peebles (1951) and Lehr (1978) recognized 9 species in Arizona, the latest study accepted only 6 (Elvander 1992). By comparison, there are 8 species in all of the northeastern United States (Gleason and Cronquist 1963).

## *Castilleja austromontana* (Scrophulariaceae) Indian Paint-Brush, *cu:wĭ taḍpo*

**Other names:** ENGLISH: [Rincon Mountain] Indian paint-brush (recorded by 1892 for *Castilleja*), painted-cup (used by 1776 for European *Bartsia*; later for *Castilleja*); SPANISH: *flor de la piedra* (rock flower, Mexico for *C.* cf. *tenuifolia* fide Bye 1986). ATHAPASCAN: *dah yiitíhídą́ą́' <dahi-tihidáˀ>* (*dahyiitíhí*, hummingbird [lit. it hangs suspended], *dą́ą́'*, food, Navajo), *nát'oh nił chiin <nát'oh nił chiiní>* (*nát'oh*, tobacco, *nił chiin*, smelly, Navajo), *na'ashjé'ii dą́ą́' <naˀašʒᵛeˀi dą́ˀ>* (*na'ashjé'ii*, spider, *dą́ą́'*, food, Navajo; for *C. applegatei* [as *C. chromosa*]); UTO-AZTECAN: *agakidɨbɨ* (*agakidɨ*, red, *bɨ*, plant, Kawaiisu), *cu:wĭ taḍpo* (*cu:vĭ*, rabbit, *taḍpo*, eyelash, Hiá Ceḍ and Tohono O'odham; for *C. exserta*), *koi'digĭp* (Shoshoni; for *C. miniata*), *mo'tendĭt* (Ute; for *C. miniata*), *mansi*, *palámansi* (*pala*, red, *mansi*, flower, Hopi; for *C. lineariifolia*), *pią́sǫ́ǫˀmiˀnapų* *to'goûngona* (*to'goa*, snake, rattlesnake, *ûn*, gun, *gu'na*, fire, or snake fire, Shoshoni; for *C. parviflora*, also *C. miniata*), *pĭ'tcĭnūpūva* (Northern Paiute), *rosábochi* (Tarahumara for *C.* cf. *tenuifolia* fide Bye 1986)

*Castilleja austromontana*. Flowering branch of herb. Artist: R. D. Ivey (Ivey 2003).

**Botanical description:** Herbs, perennial. Stems are 25–35 cm tall or larger, usually several, slender, simple or branched, pubescent. Leaves are linear to linear-lanceolate, the uppermost with 3 obvious nerves, the lower with a single nerve, green, acute, narrowed to partly clasping basally, glabrous above, rough-hairy below. Flowers are numerous, in a terminal spike to about 10 cm long, on pedicels 1–2 mm long. Bracts are entire or with 2 broadly linear lateral teeth, or sometimes with a few more teeth, lanceolate to broadly oblong, obtuse to acute, 15–30 mm long, the upper scarlet for most of the length, light green at base, sparsely pubescent along margins, with yellow-green veins. Calices are 10 mm long, finely glandular-pubescent, green but tinged with yellow and tipped with red. Corollas are 20–25 mm long, yellow-green. Capsules are lanceolate, acuminate, glabrous, 10–12 mm long. Seeds are gray-brown, triangular, small.

**Habitat:** Mountainous oak woodlands, pinelands, Sycamore galleries. 1,371–2,895 m (4,500–9,500 ft).

**Range:** Arizona (in White Mountains of Apache and Greenlee counties, and in Cochise and Pima counties) and New Mexico (Catron); Mexico (Sonora).

**Seasonality:** Flowering April to September.

**Status:** Native.

**Ecological significance:** All species are obligate hemiparasites, and they attach themselves to their host's roots (Crosswhite 1983). As hemiparasites, they live on the sugars and other nutrients of their hosts, but they still retain chlorophyll. No one knows what they are using the chlorophyll to produce if not sugars for food. Many concentrate selenium when growing in soils containing that element, and they have levels that cause poisoning in livestock that graze on them.

Although *Castilleja* appears to have red flowers, corollas are usually some shade of green or yellowish. It is actually the enlarged bracts surrounding the flowers that are red, and other colors range from yellow to purple in a few species. These colors attract hummingbirds that pollinate the flowers, while the herbs provide the birds with needed food (nectar).

*Castilleja austromontana* was named by Paul Standley and Jacob C. Blumer in 1911, based on specimens collected by Blumer in the Rincon Mountains in 1909. The original plants were growing at 7,900 feet (2,420 m) in the pine forests near a mining camp. Bowers and McLaughlin (1987) recorded the species from pine forests and mixed coniferous forests in that range.

**Human uses:** No use was found for *Castilleja austromontana*, but Moerman (1998) lists 17 species used in the United States. At least 13 of those are used by people in the Southwest.

Kearney and Peebles (1951) record that the genus is used medicinally and ceremonially by the Hopi. To that information, Moerman (1998) and Whiting (1939) add that Hopi use *C. affinis* and *C. linariifolia*. Hopi maidens use *C. affinis* to adorn their hair on holidays (Fewkes 1896). Additionally, Hopi eat the flowers, use them ceremonially to symbolize the southeastern direction, and they mix them with paint. Tewa also use the flowers to symbolize the southeastern direction, and they, too, mix them with paint. The flower decorates pottery, is painted and carved on wood, and is included in colored yarn weaving among the Tewa (Robbins et al. 1916). Hispanics call this Indian PaintBrush the *flora de Santa Rita* (New Mexico). Hopi, Navajo, Shoshoni, and Tewa make *C. linariifolia* into medicine to prevent conception, and to treat menstruation problems, stomach disorders, and several other maladies (Robbins et al. 1916, Train et al. 1957, Vestal 1952, Whiting 1939). Other Navajos use *C. applegatei* to treat spider bites (Wyman and Harris 1951). Farther west, the Ohlone treat sores with a powder of the entire plant (Bocek 1984).

The other two species in Brown Canyon are used. *Castilleja lineata* is used by the Navajo to relieve stomach problems, and they suck the sweet nectar from the flowers (Elmore 1944). Apache mix *Castilleja minor* with added materials to color deer and other skins (Reagan 1929). Deschinny (1984) records the Navajo name *dah yiitíhídą́ą́'* <*dahitxįhidaą'*> for *C. integra* and said it is used to make brown dye. Northern Tepehuan and Hispanic people in Chihuahua use *C. indivisa* as a diuretic and purgative (Pennington 1969). This herb is known as *chupón* (big sucker).

**Derivation of the name:** *Castilleja* was named for the eighteenth-century Spanish botanist Domingo Castillejo of Cádiz. *Austromontana* means from southern mountains.

**Miscellaneous:** The two other species in the canyons are *C. lineata* and *C. minor.* Of the three species, only *C. minor* is an annual; it also has sticky herbage. Bracts and flowers on *C. lineata* are not highly colored, and they are never bright red. Instead, corollas are greenish or yellowish and almost the same length as the calyx.

## *Mimulus guttatus* (Phrymaceae) Monkey-Flower, *lama*

**Other names:** ENGLISH: common [round-leaf, seep, spring, spotted, yellow] monkey-flower (the name "monkey-flower" was in use by 1786 and refers to the colored markings on the yellow flower that resemble the face of that mammal); SPANISH: *almizcle amarillo* (yellow *almizcle* or bezoar, Mexico; see below), *berro* (water cress), *lama* (mud; surely a comment on the wet places where the herbs grow, although the word is used in Spanish much as "moss" is applied in English to anything green growing in water, Chihuahua, Sonora), *llantén* <*lantén*> *cimmarón* (wild *Plantago*, Chihuahua), *mímulo* (translation of the Latin genus, Mexico); UTO-AZTECAN: *antapittsehkwana* (Shoshoni), *baseró* (Tarahumara, Chihuahua), *paakoribi* (*paa*, water, Kawaiisu), *suudági mamaradï* (*suudági*, water, *mamaradï*, children, Northern Tepehuan, Chihuahua), *tocasoiahui* (Guarijío), *tokasoiawi* (Mayo)

*Mimulus guttatus*. A. Flowering branch tip. B. Segment of stems showing leaf variations. C. Gynoecium. D. Seed. Artists: F. Emil, A–B; A. Hollick C–D (Britton and Brown 1896–1898).

**Botanical description:** Annuals or perennials from stolons, 0.5–9 dm tall (when well watered), glabrous or pubescent, the stems stout, erect or decumbent, simple or branched, often succulent. Leaves are irregularly dentate, broadly ovate to obovate or kidney-shaped, palmately 5–9 veined, petiolate below and sessile above. Inflorescences are of terminal long-pedicellate racemes or solitary. Calices are accrescent, campanulate, 6–16 mm long in flower, longer and inflated in fruit. Corollas are yellow, 9–30 mm long, falling soon after opening, 2-lipped, the throat flaring, closed by the palate, spotted. Capsules are cylindrical, many-seeded.

**Habitat:** Streams, seeps, springs. 152–2,900 m (500–9,500 ft).

**Range:** Arizona (Apache to Mohave, south to Cochise, Pima, Santa Cruz, and Yuma counties), California, north to Montana and Alaska; Mexico (Baja California, Chihuahua and Sonora).

**Seasonality:** Flowering March to September, infrequently October, November, December.

**Status:** Native.

**Ecological significance:** These herbs always grow where there is a water supply. Monkey-Flowers are most common in permanent streams or pools, seeps, and springs. Although this species ranges well into the boreal zones, Monkey-Flowers are sensitive to both frost and drought and often disappear during the colder or dryer seasons.

*Mimulus guttatus* was first recorded in Britain in 1830 (Clapham et al. 1987). It is now widely naturalized there and elsewhere in Europe and the eastern United States. Plants are still widely cultivated for ornament.

**Human uses:** Kearney and Peebles (1951) record that *Mimulus* is used for salads or greens but do not mention the people involved. Moerman (1998) names the people who lived in Mendocino County, California, and the Miwok (Sierra Nevada, California), as some who used the herbs this way. The Tarahumara cook the leaves with beans (Pennington 1963), and Northern Tepehuan cook the leaves and eat them as a potherb (Pennington 1969).

The herbs are used as medicine by Kawaiisu, Shoshoni, and Yavapai (Moerman 1998, Train et al. 1957). Onavas Pima and Northern Tepehuan boil Monkey-Flower to make a febrifuge (Pennington 1969, 1979). Plants also are used to treat a sore chest or back, wounds or rope burns, and gastrointestinal upsets. Hocking (1997) says that *M. guttatus* is used in homeopathic medicine in Great Britain.

The name *almizcle amarillo* is particularly intriguing. In Spanish, the word *almizcle* refers to what is known in English as "bezoar" (from Arabic *bzahr* or *bdizahr* and Persian *pd-zahr*, counterpoison, antidote). These stones form in the stomachs of ruminants; they were held to be cures for an array of maladies from at least the 1400s through the 1700s. *Almizcle amarillo* suggests some relationship between the herbs and the antidote idea.

**Derivation of the name:** *Mimulus* is from Latin *mimus*, mime, buffoon, or comic, from the face-like corolla of some species. *Guttatus* is Latin, meaning spotted or covered with small drop-like glands.

**Miscellaneous:** *Mimulus cardinalis* (crimson monkey-flower) is rare in the upper reaches of Jaguar Canyon and elsewhere in the Baboquivari Mountains. This species has larger red flowers (3–5.5 cm long). Crimson Monkey-Flower, too, was eaten by the Kawaiisu in California, and it was used medicinally by the Karok (Moerman 1998).

## *Nuttallanthus texanus* (Scrophulariaceae) Toad-Flax, *linaria*

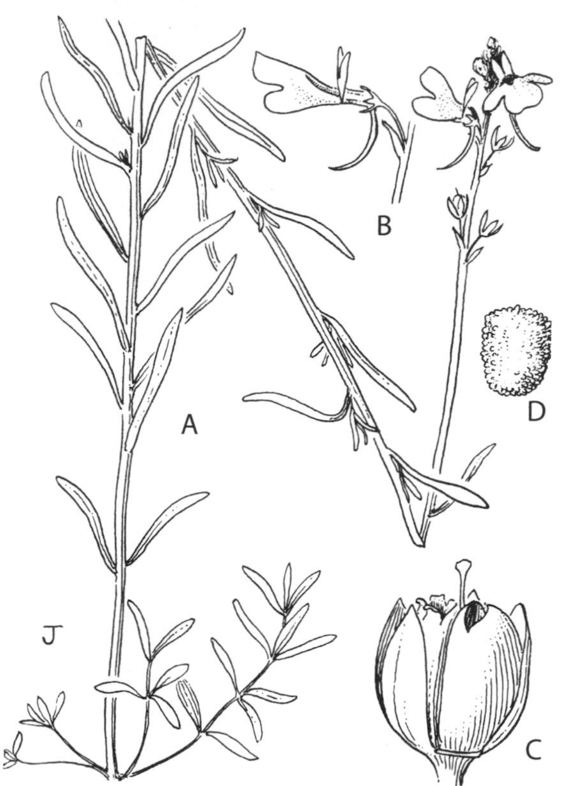


*Nuttallanthus texanus*. A. Flowering plant. B. Flower, enlarged. C. Fruit. D. Seed. Artist: J. R. Janish (Wiggins 1980).

**Other names:** ENGLISH: [Texas] toad-flax ("toad" in this context means "wild"; the other context is as an amphibian [*Bufo*]; "flax," probably from Greek *plek-ein* or *flah-*, akin to Old Aryan **plak-* as in flay, the etymological notion being connected with the process of "stripping," by which the fiber is prepared; the combination used in English by 1578 when Henry Lyte translated Rembert Dodoen's Dutch herbal of 1554); SPANISH: *linaria* (from Latin, with cognates in French and Italian, Mexico); (= *L. canadensis* var. *texana*, *Linaria texana*)

**Botanical description:** Annual herbs, the flowering stems erect, 1–8 dm tall, with short spreading prostrate branches, often at the base. Leaves are narrowly linear, sparse near the apex. Corollas are violet or bluish, 1–1.4 cm long, not including the 5–9-mm-long spur. Capsules are 2.5–3.5 mm long, about as wide, dehiscing 1/3–1/2 its depth. Seeds are cylindrical, with rounded densely tuberculate angles.

**Habitat:** Grasslands, canyons, washes. 450–1,500 m (1,500–5,000 ft).

**Range:** Arizona (Cochise, Gila, Graham, Greenlee, Maricopa, Pima, and Pinal counties), east to South Carolina, and north to British Columbia; Mexico (Baja California and Sonora, south); Central America; and South America.

**Seasonality:** Flowering February to May.

**Status:** Native.

**Ecological significance:** Although *N. texanus* grows in numerous habitats, it almost always lives in sandy soils. *Nuttallanthus* appears most often after something has disturbed the soil and scratched or notched the seed coat so that it is easier for the embryo to absorb water and germinate.

Flowers are pollinated by long-tongued bees and butterflies (Hill 1909). Hill also found flies (Syrphidae) visiting and commented that flower visitors of any kind were uncommon. Although cleistogamous flowers apparently have not been found in *N. texanus*, they are in *N. canadensis* (Hill 1909). Indeed, the earliest record of *N. canadensis* having self-fertilizing flowers seems to have been published by P. A. Rydberg in his *Flora of the Black Hills* in 1896. This account was followed soon after by Brandegee (1900) and Webster (1900) and then by Hill (1909). Recent study by Crawford and Elisens (2006) confirms that this species is mostly self-pollinating and genetically isolated from the others.

**Human uses:** No uses were found for *N. texanus.* Hocking (1997) reports that similar *N. canadensis* is used as a diuretic and laxative, and that the crushed plant has been applied to relieve problems with hemorrhoids. That species grows from Canada through much of the eastern United States to Texas, and it has been recorded in Mexico.

Farther east, two large ethnic groups had several uses for the European-introduced species *Linaria vulgaris*, called Butter and Eggs (Moerman 1998; see below for relationships between *Linaria* and *Nuttallanthus*). The Iroquois use Butter and Eggs to stop diarrhea, and in a remedy to remove a spell involving a "love medicine." Ojibwa use *L. vulgaris* as an inhalant in the sweat lodge. Because these herbs contain linarin, pectolinarin (a flavonoid glycoside), and aureusin (a glycoside), the uses are doubtless effective in some cases (Hocking 1997). Homeopathic physicians have used the species to promote purging. In Europe, this species is called "Cancerwort," although it is unclear if that name originally was applied to *Nuttallanthus* or *Veronica* (Austin 2004).

**Derivation of the name:** *Nuttallanthus* is a patronym based on Thomas Nuttall (1786–1859), plus *anthus*, flower. *Texanus* means from Texas, where it was first collected. The alternative generic name *Linaria* compares these wild herbs with *Linum*, the Flax, some of them having similar leaves.

**Miscellaneous:** Depending on how the genus is defined, *Linaria* comprises 150 species, mostly Eurasian. There are 77 species in Europe. The New World species are now considered part of the endemic American *Nuttallanthus.* As segregated, there are three species of *Nuttallanthus* in North America (*N. canadensis*, *N. floridanus*, *N. texana*), one in South America (*N. subandinus*), and none native of *Linaria*, although there are introduced Old World species.

The technical name of this herb remains problematical, as authorities cannot agree how many species there are. For example, Kartesz (1994) and Diggs et al. (1999) called them *Nuttallanthus texanus*, following in the species definitions of Correll and Johnston (1970), who used *Linaria texana.* Felger et al. (in prep.) use the name *Linaria canadensis* var. *texana*, preferring to recognize the similarity with the widespread species. Differences are admittedly minor, including whether or not the corolla is over or under 1 cm long, and seed traits (cf. Crawford and Elisens 2006, Fassett 1942).

## *Simmondsia chinensis* (Simmondsiaceae) Jojoba, *jojoba, ho:howai*

**Other names:** ENGLISH: coffee bush (see uses), deer [goat, pig, sheep]-nut (these animals apparently eat the fruits), jojoba (English, from Spanish, ultimately from indigenous *ho:howai*), quinine plant, wild hazel; SPANISH: *jojoba* (from indigenous *ho:howai*); HOKAN: *pnaacöl* (Seri); UTO-AZTECAN: *hohoova* (Yaqui), *ho:hovai* (Hiá Ceḍ O'odham), *hohowai* [*ho:howai*, pl.; *hohwi*, sing.] (Tohono O'odham), *qawnaxal* <*kowanukal*> (Cahuilla)

**Botanical description:** Shrubs 1–2 m tall, with stiff branches, pubescent on younger growth. Leaves are oblong-ovate, almost sessile, 2–4 cm long, dull green, more or less canescent-pubescent. Peduncles are 3–10 mm long. Sepals of male flowers are greenish, 3–4 mm long, those of the female flowers becoming 10–20 mm long. Capsules are leathery, nut-like, ovoid,

*Simmondsia chinensis*. A. Male branch. B. Individual male flower. C. Female flower. D. Fruiting branch. Artist: L. B. Hamilton (Benson and Darrow 1945).

obtusely 3-angled, 1.5–2 cm long, filled by the large oily puberulent seeds.

**Habitat:** Canyons, slopes, washes. 300–1,500 m (1,000–5,000 ft).

**Range:** Arizona (Greenlee to southern Yavapai, south to Cochise, Pima, and Yuma counties) and southern California; Mexico (Baja California and Sonora).

**Seasonality:** Flowering December to July.

**Status:** Native.

**Ecological significance:** Jojoba is a dioecious shrub, with male and female plants being sometimes widely separated. This adaptation is associated with obligate outcrossing.

In a paper with a spectacular series of color photographs, Buchmann (1987) provided detailed information on the wind pollination of Jojoba. *Simmondsia* sheds 523 grams of pollen per plant, with the peak in the afternoon between 1:00 p.m. and 4:00 p.m. During that time period, the stigmas captured 60–63 grains per cubic meter. Buchmann further found that the leaves are arranged around the female flowers to maximize the movement of pollen toward the stigmas. Removal of leaves decreased the efficiency of the flowers, dropping the capture as much as 50%. Although bees, including the introduced Honeybee (*Apis melifera*), visit flowers, they collect pollen from only males and never visit female blossoms.

The region of the Baboquivari Mountains is typically considered too cold for Jojoba, but the shrubs are tucked into protected sites. Mostly, individuals grow on the south-facing slopes, where they continue getting winter sun. *Simmondsia* are also often protected either by other plants or by rocky outcrops within valleys and canyons.

Antelope Squirrels (*Ammospermophilus harrisii*) store fresh seeds and aid dissemination (Jaeger 1941).

**Human uses:** Jojoba seeds are eaten raw or parched, but contain too much tannin for most tastes when raw. The seeds are eaten by at least the Cahuilla, Cochimí, Cocopa, Hiá Ceḍ and Tohono O'odham, Kiliwa, Seri, and Yavapai (Bean and Saubel 1972, Castetter and Bell 1951, Castetter and Underhill 1935, Felger and Broyles 2007, Felger and Moser 1985, Hodgson 2001, Moerman 1998). Cahuilla make the seeds into a coffee-like beverage but the use must have been much more widespread (Bean and Saubel 1972). Hispanics made the beverage along Arivaca Creek in the early 1900s (Wilbur-Cruce 1987). Kiliwas eat the seeds in *pinole*, but the Cochimí did not use them until after contact with Jesuits (Hodgson 2001). Seri consider the seeds survival food only, and said they sometimes cause diarrhea (Felger and Moser 1985).

Seeds are used by the Seri in preparations to treat sores, eye problems, sore throat, colds, and at childbirth (Felger and Moser 1985). Hiá Ceḍ O'odham, Tohono O'odham, and Yavapai use them similarly (Castetter and Underhill 1935, Felger and Broyles 2007, Moerman 1998). Fr. Martín del Barco, born in 1535, a priest in Baja California, wrote that the Jojoba was "extraordinarily effective" against a number of maladies (Hodgson 2001).

Seri also string the seeds for necklaces, and because of the ca. 50% liquid wax, use them to shampoo the hair (Felger and Moser 1985). These Mexican people make from the wood skewers for cooking meat.

Jojoba wax has been substituted for beeswax, and it was considered a substitute for olive oil by Spanish missionaries (Hocking 1997, Hodgson 2001). More recently, it has been touted as a replacement for whale oil, and hundreds of thousands of acres have been planted within and outside its native range.

**Derivation of the name:** *Simmondsia* is dedicated to physician Thomas William Simmonds, who died in Trinidad in 1804. He was described as the "ardent Botanist and Naturalist who accompanied Lord Seaforth to Barbadoes about 1804, and died soon after while engaging in exploring the island of Trinidad" (Jaeger 1941). *Chinensis* means from China, under the mistaken idea that it was from that part of the world.

## *Datura wrightii* (Solanaceae) Thorn Apple, *toloache, kotaḍopi*

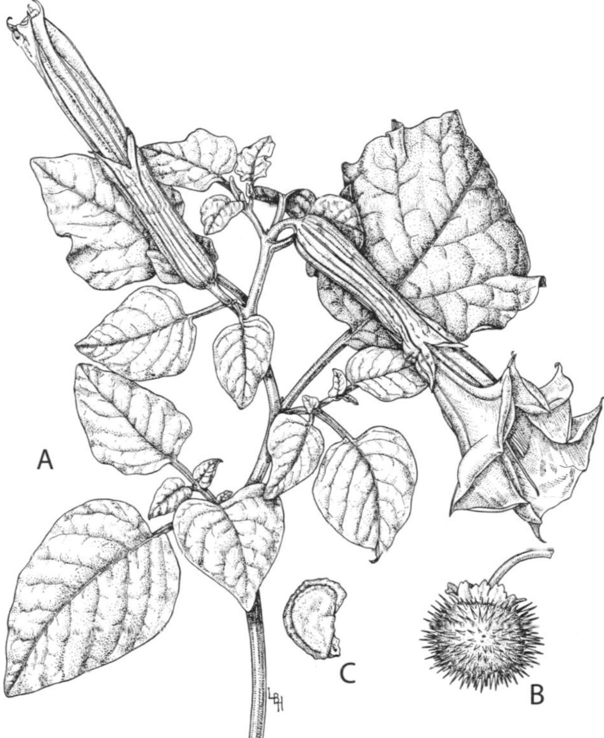


*Datura wrightii*. A. Leafy branch with flowers. B. Spiny capsule. C. Seed. Artist: L. B. Hamilton (Parker 1958).

**Other names:** ENGLISH: angel's-trumpet, Indian apple, jimson weed (from Jamestown, noting a poisoning incident in 1676), sacred datura, [sacred] thorn apple (allusion to spiny fruits); SPANISH: *chamico* (from Mayan *sham*, to retard, *mi*, negative, *ik*, respiration, Yucatán), *estramonio* (from Latin *struma* or *strama*, swelling, from medical use of *D. stramonium*), *tlapa* (from Náhuatl name *tlapatl*), *toloache* <*toluache, tolguacha*> (from Náhuatl, *tolohua*, to incline or bow the head); ATHAPASCAN: *ch'óhojilyééh* <*č'óxoʒ*$^{v}$*ilyêi, c'oxojiléi*> (madness producing, Navajo), *itanasbase* (round leaf, Chiricahua and Mescalero Apache), *ndíyíliitsoh* <*ntíGíliitshoh*> (*ndíyílii*, sunflower, *tsoh*, large, Navajo), <*jaa ilgodó*> (forget yourself, from *yaa yisdag*, he forgot it? Western Apache); CHUMASH: *mo'moy* (Barbareño Chumash), *momoy* (Ineseño and Ventureño Chumash); KIOWA TANOAN: *saemp'e* (*saen*, porcupine, *p'e*, plant, Tewa); UTO-AZTECAN: *dekúba* <*deku-ba, reku-ba*> (Tarahumara), *hakatdam* <*hakandam*> (Onavas Pima), *gegeda a'gama* <*gugudua'gcama, gugurha agama*> (the one with big horns or big horned one, Nevome, Sonora), *kiksawva'al* <*kikisowil*> (Cahuilla), *kookivuri* <*kokovuri*> (*koko*, snakes, *uri*, tied [tied with snakes], Mountain Pima fide Pennington 1973), *kotadopi* (Hiá Ceḍ O'odham; for *D. discolor*), *kotaḍopi* <*kotata'p*> (Tohono O'odham), *kotḍopi* <*kotodopi, kodop, kododophi, kotobi, kotdobi*> (Akimel O'odham), *máanet* (Luiseño, Juaneño dialect), *main-oph-weep* (Paiute), *mïmïp* [*manopweep, manop*$^{h}$*weep*] (Southern Paiute), *moopɨ* (Kawaiisu), *momoht* (Tübatulabal), *muipə* <*muipe*> (Northern Paiute), *muippüh* (Panamint), *navamutuda* <*nabamutuda*> (from *atosigar*, to poison or intoxicate, Nevome, Sonora), *táguaro* (an Ópata dance and the name of *Datura*, Sonora), *taŋaniva* (Northern Paiute), *tebwi* (Yaqui), *tecuyani* (a mythological animal that eats people, Náhuatl; from *te*, some, *cuani*, the one who eats), *tecuyaui* (cognate with Náhuatl? Guarijío), *tikúwari* (cognate with Náhuatl? Tarahumara), *tokorakai* (Northern Tepehuan), *tokorep* <*tókorew*> (Mountain Pima), *tókocovi* <*tokorhobi*> (related to *tonoro*, to shine, and *tóko*, spider, Nevome, Sonora), *tolohua-xíhuitl* <*tologuaxíhuitl*> (*tolohua*, to incline, *xíhuitl*, herb, Náhuatl), *tsimona* <*tcimóna*> (Hopi), *ʼųnų́pųvų* (Ute); YUMAN: *cmalgapít* (ear-deaf, Maricopa), *smalgatúʼ* (ear-something inside, based on *smalga*, ear, Havasupai), *ṣmal ka:pí:ṭ* (*ṣmal*, plant, *ka:pí:ṭ*, crazy, Cocopa), *shmalk tuch* (Paipai), *smalk*$^{a}$*tûʼ* (Walapai), *malyakatu'* (Mohave); LANGUAGE ISOLATE: *a'neglakya* (named for a legendary boy, Zuni); (= *D. inoxia*, *D. meteloides* of authors)

**Botanical description:** Annual or perennial herbs. Stems 3–10 dm tall, erect or sprawling, white-puberulent with dense, fine, gray hairs. Leaves are 5–25 cm long, the blades ovate, repand-dentate to entire, often uneven at base, white-pubescent like stems. Calices are 7–13 cm long, the lobes lanceolate, ca. 2 cm long. Corollas are 15–23 cm long, white to violet. Fruits are a nodding capsule, subglobose, opening irregularly when mature.

**Habitat:** Canyons, washes. 300–1,980 m (1,000–6,500 ft).

**Range:** Arizona (all counties, although sparse in northeastern and southwestern parts of state), southern California, Colorado, New Mexico, Texas, and Utah; Mexico (Baja California, Chihuahua, Sonora, and south to Edo. México).

**Seasonality:** Flowering (April) May to October.

**Status:** Native.

**Ecological significance:** These herbs thrive along washes, where there is extra moisture, and they bloom there when plants higher up the slope are barely in leaf. This species formerly was in the Organ Pipe Cactus National Monument, but it has not recently been found there (Felger et al. in prep.). Seeds found in Alamo Canyon have been dated at 9,570 ybp, while those in the Puerto Blanco Mountains were 9,070 ybp. Aridity in southwestern Arizona and cold in the northeastern parts of the state seem to limit the distribution.

The noctural flowers are visited by native Sphinx Moths during the night, and alien Honeybees visit during the early morning before corollas close. Flowers are adapted for moth pollination; bees are probably just robbers.

**Human uses:** Dunmire and Tierney (1997) found records of Thorn Apple seeds and pollen throughout the prehistoric Greater Pueblo area, and the western pueblos are known to have used it within historic times as an analgesic. Historic and modern records show that, in addition to its uses by Anglos and Hispanics, *toloache* is used by Akimel O'odham, Apache, Aztecs, Cahuilla, Chumash, Gabrielino, Guarijío, Havasupai, Hopi, Kawaiisu, Keres, Kumiai, Luiseño, Mayo, Mountain Pima, Navajo, Névome, Ohlone, Ópata, Paipai, Paiute, Seri, Shoshoni, Tarahumara, Northern Tepehuan, Tohono O'odham, Tübatulabal, Ute, Yaqui, Yavapai, Yoruk, Yuma, and Zuni (Bean and Saubel 1972, Bocek 1984, Castetter and Opler 1936, Castetter and Underhill 1935, Elmore 1944, Gallagher 1976, Havard 1896, Kay 1996, Moerman 1998, Palmer 1878, Pennington 1963, 1973, Rea 1997, Reagan 1929, Romero 1954, Train et al. 1957, Vestal 1952, Voegelin 1938, Whiting 1939, Wyman and Harris 1941, 1951, Yetman and Felger 2002, Yetman and Van Devender 2001). Roots are sometimes added by the Western Apache to *tułbáí* [*tulipi, tulpi, tulapai, tulipai*] (*tú*, water, *łibaa*, gray) or among the Chiricahua Apaches as *tiswin* [*teswin, tesvino, tizwin, tesgüino*] (from Náhuatl *teyhuinti*, intoxicating, Hrdlička 1908, La Barre 1938) to make it more potent (Havard 1896, Hrdlička 1904, Reagan 1929).

All plant parts contain dangerous belladonna alkaloids, and even touching them may be dangerous by way of moving chemicals that can dilate pupils or interfere with mucus membranes. Northern Tepehuan rub the crushed seeds and leaves onto a fishline. Once fish have gathered around the line and hook, the fisherman snags an animal (Pennington 1969).

**Derivation of the name:** *Datura* may be from Arabic *Tatorah* or Hindustani *Dhatura*, although Linnaeus said he took it from Latin *dare*, to give, because it was "given to those whose sexual powers are weak or enfeebled." *Wrightii* commemorates Charles Wright, 1811–1885, a collector in the American Southwest.

## ***Lycium exsertum* (Solanaceae)**
## **Wolf-Berry, *tomatillo, kuawul***

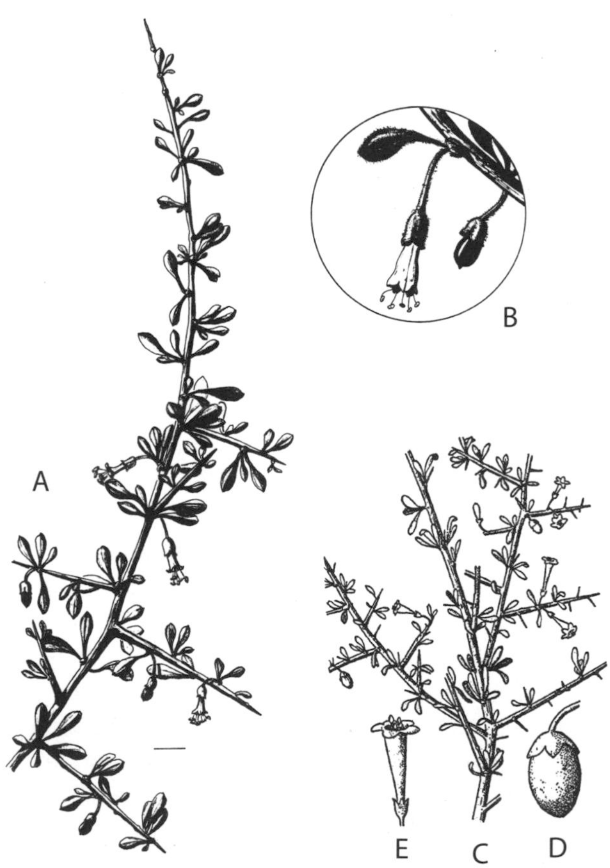


*Lycium exsertum*. A. Flowering branch. B. Inset of flower. *Lycium andersonii*. C. Flowering branch. D. Fruit. E. Flower. Artists: W. C. Hodgson, A–B (courtesy Wendy Hodgson); J. R. Janish, C–E (Abrams 1951).

**Other names:** ENGLISH: box [desert] thorn (Arizona), narrow-leaf thorn-bush, squaw-berry (Arizona), tomato berry, wolf-berry (applied by 1834 to *Symphoricarpus*, an allusion comparing that plant to the fierceness or rapacity of the beast; later extended to *Lycium*); SPANISH: *manzanita* (little apple, usually used for *Arctostaphylos*, Mexico), *tomatillo* (little tomato, Arizona, Sonora); ATHAPASCAN: *haashch'éédą́ą́'* <*hašč'éˀdą́', hásčédáˀ*> (*haashch'ééh*, deity, *dą́ą́'*, food, Navajo; for *L. pallidum*); KERES: *dyakuna* (Acoma; for *L. pallidum*); UTO-AZTECAN: *at wusha'i* (Onavas Pima), *hupui* (Northern Paiute; for *L. andersonii*), *huˀupivɨ* (Kawaiisu; for *L. andersonii*), *ˀici-s* (Luiseño), *kuávul* <*kuáwul*> (Akimel O'odham; see also *Celtis*), *kuavulĭ* <*kuavuli*> (Hiá Ceḍ O'odham; for other species), *kuawul* <*kuawur*> (Tohono O'odham), *kyeeva* <*ké:ve*>, *kyeptsoki* (*kyeve*, *Lycium*, *tsoki*, upright [plant], Hopi), *pi'ict* <*piˀis-t*> (Tübatulabal; for *L. torreyi*); YUMAN: *axtó* (Maricopa, Yuma), *axto't* <*a tu't*> (Mohave), *xcuc* (Cocopa)

**Botanical description:** Shrubs, much-branched and often rounded, the stems commonly 2.5 m tall, glandular-pubescent. Thorns are slender, needle-like. Leaves are spatulate to obovate, thickened, 1.5–4 cm long. Flowers are 1 per leaf axil, on pedicels 4–12 mm long. Calices are tubes 3.5–4 mm long, the lobes 0.5–2 mm long, unequal, tubular. Corollas are white, tinged with lavender, tubular-funnelform, the tube 7.5–8.5 mm long or more, hairy inside, narrowed below, moderately flared above, the lobes 1.5 mm long, spreading, with stamens and style exserted on staminate plants. Fruits are ellipsoid to ovoid, orange, fleshy, ca. 1 cm long.

**Habitat:** Slopes above washes. Up to ca. 1,310 m (4,300 ft).

**Range:** Arizona (Coconino, Gila, Graham, La Paz, Maricopa, Pinal, Yavapai, south to Cochise, Pima, Santa Cruz, and Yuma counties); Mexico (Baja California, Sinaloa? and Sonora).

**Seasonality:** Flowering February to April.

**Status:** Native. The type specimen was collected in Sonora by C. G. Pringle in 1884.

**Ecological significance:** These shrubs are most common along the margins of washes, where there is more moisture. Birds relish the fruits and often strip the bushes of the berries as soon as they ripen. Some plants have fruits remaining long after the season is over; individuals vary in taste, with some being "good" and others not. Plants are functionally dioecious (Felger 2000).

**Human uses:** The red or orange fruited species of *Lycium* play important roles in the lives of southwestern people (Hodgson 2001). People eat the fully ripe fruits raw, sun-dried, or cooked in water to make soups, stews, syrups, sauces, or beverages, even though eating them temporarily blackens the teeth (Rea 1997). Fruits are low in phosphorus, calcium, magnesium, and sodium.

Palmer (1878) wrote of *L. andersonii* that "[t]he berries . . . are eaten by Indians of Arizona and California; in fact, Whites relish them also. They are quite agreeable to the palate, being of a sweet, mucilaginous substance . . . berry . . . and when dried, resembles in taste dried currants." There are many more names for this species, including *hahöj-enej* (empty *Lycium*, Seri), *hahöj ináil coopol* (black-barked *Lycium*, Seri), *salicieso* [*saliciesa*] (Sonora), *sigreropo* (Mayo), *s-toa kuavulĭ* <*kuavuli*> (white wolfberry, Hiá Ceḍ O'odham), *s-toha koawul* <*kuawur*> (white wolfberry, Tohono O'odham). Fruits of *L. exsertum* are variable, but not that good.

Seri eat the fruits of *L. andersonii*, but prefer other species in their area (Felger and Moser 1985). Others that eat *Lycium* fruits are Acoma, indigenous Baja California people, Cahuilla, Cocopa, Hiá Ceḍ O'odham, Hopi, Jemez, Kawaiisu, Laguna, Tohono O'odham, Mohave, Paiute, Quechan, and Zuni (Bean and Saubel 1972, Castetter 1935, Castetter and Bell 1951, Castetter and Underhill 1935, Felger and Broyles 2007, Hodgson 2001, Moerman 1998, Vestal 1952, Whiting 1939). The Tohono O'odham were selling 10-pound bags of the fruits in the 1980s (Hodgson 2001).

Many other species of *Lycium* have edible fruits, and even those that are not good are sometimes eaten. Dunmire and Tierney (1997), Vestal (1952), and Whiting (1939) record that Hopi and Navajo sometimes mixed marginally edible fruits such as *Lycium* with white clay to remove the bitterness and fill an empty stomach. *Lycium pallidum* was a famine food during the 1863 drought when Hopi had to leave their homes temporarily and live among the Rio Grande Puebloans. Hopi still use that species in ceremonies, and they do not forget the debt owed the ancient food plants (Whiting 1939).

**Derivation of the name:** *Lycium* from Greek *lykion*, a name used by Dioscorides, fl. CE 40–80, and Pliny, CE 23–79, for the thorny shrub called Dyer's Buck-Thorn, perhaps a species of *Rhamnus*. The name is said to have been used because the shrub was growing in *Lycia*, a southwestern region of Asia Minor. *Exsertum*, protruding, means that the stamens and style and stigma extend outside the corolla mouth.

**Miscellaneous:** There are three species of *Lycium* in the canyons. *Lycium andersonii* and *L. berlandieri* grow in the lower part of the canyons and wash region. *Lycium berlandieri* is called *bachata* (see *Ziziphus*) and *salicieso* in Arizona and Sonora. While *L. exsertum* has glandular-pubescent herbage and flowers, the other two species are glabrous. Flowers are needed to distinguish the other two. Corolla tubes are cylindrical, or nearly so, and the lobes are lavender in *L. andersonii*, while tubes are campanulate and conspicuously expanded above and white in *L. berlandieri*.

## *Nicotiana obtusifolia* (Solanaceae) Desert Tobacco, *tabaquillo, ban wiwga*

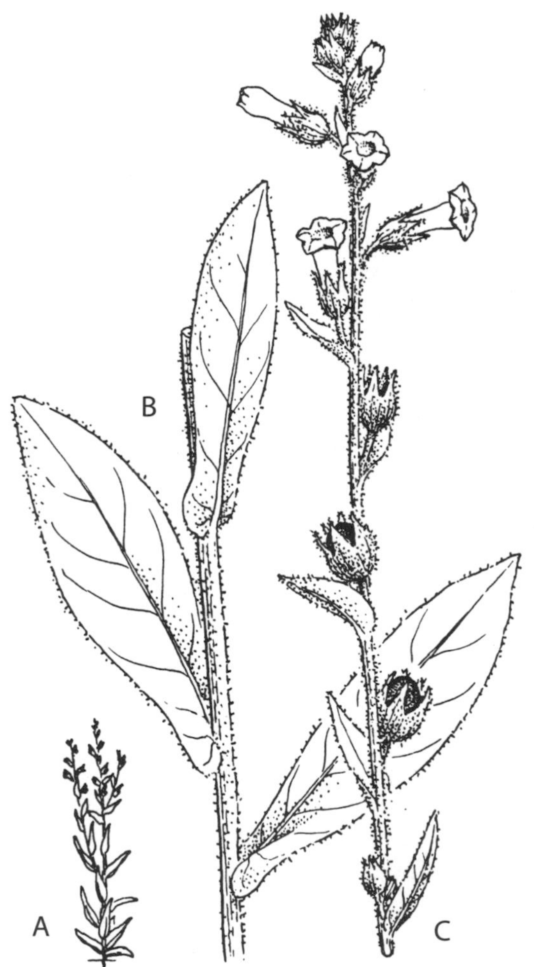


*Nicotiana obtusifolia*. A. Habit. B. Stem leaves. C. Flower and fruit cluster. Artist: unknown (courtesy James Henrickson).

**Other names:** ENGLISH: coyote [desert] tobacco (to the Taino of the Caribbean, *tabaco* meant what we today call a cigar; "tobacco" was applied to the leaves prepared for smoking or chewing after *Nicotiana* was introduced to Europe by the 1500s); SPANISH: *tobaco cimarrón* (wild tobacco, Sonora), *tobaco* [*de*] *coyote* [*loco*] (coyote [crazy] tobacco, Chihuahua, San Luis Potosí, Sonora), *tabaquillo* [*de coyote*] (little [coyote] tobacco, Texas to Arizona, Sonora); ATHAPASCAN: *nát'oh* (for all tobaccos, Navajo), *dziłnát'oh* <*$_{3}$il mit'oh*> (mountain tobacco, Navajo; for *N. attenuata*), *naayízí binát'oh* <*nayáši binát'oh*> (translated as Mountain sheep tobacco by Wyman and Harris 1951; error for *naayízí*, squash, *binát'oh*, its tobacco, Navajo; for *N. attenuata*), *nátotē* (Jicarilla Apache); HOKAN: *hapis casa* (putrid tobacco, Seri); UTO-AZTECAN: *ban vivga* <*ban vi:v*> (coyote tobacco, Akimel O'odham), *ban wiwga* (Tohono O'odham), *biy, biba-ta* (Ópata), *goy biba* (Mayo), *soˀo(n)dɨ* <*soódá*> (Kawaiisu), *o'odham ha vivga* (people's tobacco, Hiá Ceḍ O'odham, Arizona, Sonora), *ha wiwga* (their tobacco, Tohono O'odham), *isily piv'a* <*pivat-isil*> (coyote's tobacco, Cahuilla), *pamu* (Mono), *pā́müpi* [*pā́mūbī, pá$^h$mü, pá$^h$mū*] (Northern Paiute; for *N. attenuata*), *pahompin* <*pāmüpi*> (Panamint), *pahmóbi* (Mono), *pahmú* (Western Paiute), *piiva* <*piva, pi'va, pí:wa*> (Hopi), *pí:va-t* [*píivat*] (Luiseño), *pu'ibau* (Shoshoni; for *N. attenuata*), *qoˀápų* (Ute), *qɔ'ápI* (Southern Paiute), *šoˀogonht* <*šoogonhi*> (Tübatulabal; for *N. attenuata, N. bigelovii*), *tsawawap* (Southern Paiute), *vivai* (Northern Tepehuan), *vivam* (Yaqui), *vivá-t* (Eudeve), *wiopuli* <*wiopulĭ, wiupuri, víopoli*> (Tohono O'odham), *wipá* (Guarijío), *wipáka* <*auraka, bawa-ra-ka, huipá, pawa-ra-ka*> (*pewa*, to smoke, Tarahumara), *viv* (Onavas Pima), *wiw* <*viva*> (Mountain Pima); YUMAN: *kaθódn$_y$iúva* (Havasupai), *mela' ū'v* (coyote tobacco, Yuma), *ˀu:p* <*op*> (Cocopa), *hatalewah ū'v* <*a'uv, aúva*> (coyote tobacco, Mohave), *'úva* <*u:v*> (Walapai), *uvaanál$_y$a* (Maricopa; based on *u'úvac*); YUKI: *intelwayok* (old time tobacco, Yuki); (= *Nicotiana trigonophylla*)

**Botanical description:** Biennial or perennial herbs, sticky-pubescent, to 9 dm tall, with a simple or wand-like branched stem. Leaves are all sessile or the lower tapering to a winged petiole and obovate-oblong, the upper elliptic-lanceolate to oblong-elliptic with a broader cordate half-clasping base, the largest 22 cm long, 6 cm wide, apically rounded to acuminate. Inflorescences are loosely paniculate-racemose. Flowers are open throughout day, persisting into night.

Calices have lobes shorter than or equaling the tube. Corollas are tubular-campanulate, greenish white or yellowish, 12–23 mm long, the limb to 8 mm wide. Capsules are ovoid, 8–11 mm long. Seeds are brown, shining, pitted-reticulate.

**Habitat:** Grasslands, sandy washes. Up to 1,830 m (6,000 ft).

**Range:** Arizona (almost throughout state), southern California, Nevada, New Mexico, western Texas, and Utah; Mexico (Baja California, Sonora to Tamaulipas, and south to Oaxaca).

**Seasonality:** Flowering through year, mostly during rainy seasons.

**Status:** Native.

**Ecological significance:** These herbs have a patchy distribution within the Altar Valley and the Baboquivari Mountains. Within grasslands, this tobacco is usually confined to areas with a little more moisture, such as depressions and swales. Otherwise, *N. obtusifolia* usually grows along washes and streams. Flowers stay open throughout the day, but they are fragrant only at night when Sphinx Moths like *Hyles lineata* visit.

**Human uses:** As with many of the native tobaccos, *N. obtusifolia* was smoked by indigenous people, including at least the Akimel and Tohono O'odham, Apache, Cahuilla, Havasupai, Hopi, Mohave, Navajo, Tarahumara, Northern Tepehuan, Walapai, and Yuma (Bean and Saubel 1972, Castetter and Bell 1951, Castetter and Underhill 1935, Moerman 1998, Pennington 1963, 1969, Rea 1997, Reagan 1929, Russell 1908, Thord-Gray, 1955, Vestal 1952, Whiting 1939). Tohono O'odham tried to obtain the herb for smoking from "Elder Brother's Mountain" (Baboquivari; "Elder Brother" is *I'itoi*) (Castetter and Underhill 1935). These people considered *Nicotiana* a "purifying agent." The pan-Piman word for tobacco is *viv* <*wiw*>; Onavas Pima now apply this name to *N. tabacum*, but not to *N. obtusifolia* (Pennington 1980). Havasupai and Yuma cut mesquite trees, burn them on unbroken soil, and scatter seeds in the cold ashes (Jaeger 1941).

Although some historical literature says that tobacco was simply "used for smoking," the potent chemicals in this species were usually saved for ceremonial contexts. Hocking (1997) does not even mention the species, and Turner and Szczawinski (1991) say Desert Tobacco is poisonous. Of course, all tobaccos are poisonous because of nicotine and other alkaloid content. At the Hopi Niman Kachina ceremony, the kachinas or spirits are blessed by priests who blow smoke on them (Engard 1989). Indigenous peoples uniformly revered and respected tobacco and its power, while the Anglos and Hispanics rarely understood (Felger and Moser 1985, Winter 2001). Similarly respectful, the Navajo put powdered leaves in ceremonial prayer sticks (Elmore 1944).

Cahuilla use the leaves as a medicine (Bean and Saubel 1972), as essentially all species were used (Winter 2001, Austin 2004). The Tohono O'odham now run a store called *Wiwpul Du'ag* (wild tobacco mountain) just north of what Anglos call the Coyote Mountains. The village called San Pedro on maps of the reservation is called *Wiwpul* (wild tobacco village) by the Tohono O'odham.

**Derivation of the name:** *Nicotiana* was named by Linnaeus after the French diplomat Jean Nicot, 1530–1600, ambassador to Portugal in 1560, who promoted *N. tabacum* to the Portuguese and French courts. Nicot considered tobacco of great medicinal value. *Obtusifolia* means having leaves with rounded apices.

## ***Physalis pubescens*** **(Solanaceae)**
## **Hairy Ground-Cherry, *tomatillo***

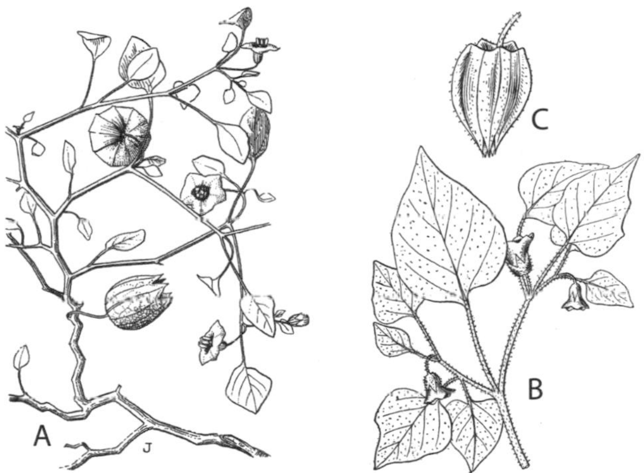


*Physalis*. A. *P. crassifolia*. B. *P. pubescens* plant. C. Fruit, enlarged. Artists: J. R. Janish, A (Abrams 1951); M. Timmerman, B; A. Hollick, C (Britton and Brown 1896–1898).

**Other names:** ENGLISH: [downy, low hairy] hairy ground cherry (the first English comparisons of *Physalis* with the cherry was "winter chirrir" by 1568, and "red winter cherries" by 1597; Germans were saying *Judenkirsen* or "Jew's cherry" by 1542; other names in the United States and Europe are "cherry tomato," "wild cherry," and "winter cherry," applied because the enlarged calyx turns red on the European *P. alkekengi*), husk-tomato (because of the enlarged calyx), pop vine, strawberry tomato; SPANISH: *tomate fresadilla* (little strawberry tomato, Sonora), *tomatillo* (little tomato, Sonora south); ATHAPASCAN: *bį̨į̨h yiljaa'í <pį̨i yil tee'íh, bį̨į̨h yildee'í>* (*bį̨į̨h*, deer, *yiljaa'í*, tongue, Navajo; for *P. longifolia*, and for *Frasera*), *gáán bidáá* [*bidáhá, bináhá*] (*gáán*, crown dancer, *bidáá*, his eyes, Western Apache), *nestani Ygati* (fruit that you can see through, Chiricahua and Mescalero Apache), *<tadílcosí>* (pops on forehead, an allusion to children popping the fruits, Navajo); HOKAN: *insáacaj* (Seri); UTO-AZTECAN: *cocostomatl* (yellow tomato, Náhuatl), *lali tomaata* (little tomato, Mountain Pima), *rurusí* (Tarahumara), *tulumisi* (Guarijío), *woghoína* (Mono), *xaltotompil* (tomato between sour and spicy, Náhuatl); YUMAN: *ḍamets* (Walapai), *kúm chulk* (Mohave)

**Botanical description:** Annual herb, the stems 8–90 cm tall, villous and sometimes sticky-puberulent (viscid), to more or less glabrous. Leaf blades are ovate, acuminate, the margins usually irregularly toothed, sometimes entire, 4–9 cm long, 2–4 cm wide, on petioles 2–7 cm long. Flowers are solitary, on slender pedicels, 1–2 cm long. Calices are campanulate, 4–10 mm long, 3–12 mm wide at anthesis, in fruit 18–30 mm long, the lobes ovate-deltoid to lanceolate. Corollas are yellowish, dark-spotted, 7–10(–12) mm long, 10–15 mm wide. Anthers are green, blue to violet. Berry is 10–18 mm in diameter, greenish, globose.

**Habitat:** Canyons, rocky slopes, washes. 914–1,830 m (3,000–6,000 ft).

**Range:** Arizona (Apache, Coconino, Graham, Greenlee, Yavapai, south to Cochise, Pima, and Santa Cruz counties), California, north to Colorado, and east to Pennsylvania and Florida; Mexico (Baja California, Chihuahua, Sonora, Sinaloa, east to Tamaulipas, and south to Yucatán); south to Panama; South America; and Caribbean.

**Seasonality:** Flowering August to September.

**Status:** Native to the eastern United States and Latin America. Introduced into western North America.

**Ecological significance:** Although these herbs may grow in open sites on ridges, Hairy Ground-Cherries are most frequent in the canyons below the canopy of the oak woodlands and along the gallery woodlands of the streams. There, their flowers are inconspicuous below their own herbage and other plants.

**Human uses:** Aztecs called *Physalis* fruits *tomatl.* The cultivated plant we now call "tomato" (*Solanum* [*Lycopersicon*] *esculentum*), the Aztecs called *xitomatl* *<jitomatl>* (red tomato, Náhuatl). Spanish-speakers in the New World generally call wild *Physalis* either *tomatillo* (little tomato) or *tomate del campo* (wild tomato).

In 1615, Francisco Ximénez published his book *Naturaleza y virtudes de las plantas* (Nature and Virtues of Plants) that included another kind called *coztomatl* (yellow tomato, Náhuatl). Martínez (1969) was still recommending that species (*P. coztomatl* in his book) as a medicine. Both authors consider the leaves diuretic, and the roots carminative and antidiarrheal. Unripe fruits and young leaves of all species contain solanine, and they are poisonous (Turner and Szczawinski 1991). This toxicity disappears as the *tomatillos* mature and the fruit becomes ripe. Today we know and eat fruits of the Mexican species (*P. philadelphica* or *P. ixocarpa*) as *tomatillo* in *salsas verdes.* Uncooked fruits are used to make the *salsas*, while they may also be cooked in a variety of dishes.

Other North Americans who use different species of *Physalis* include the Akimel O'odham, Apache, Cherokee, Dakota, Ho-Chunk, Hohokam, Iroquois, Isleta, Keres, Kumiai, Mesquakie, Mohave, Navajo, Omaha-Ponca, Pawnee, Seri, Tarahumara, Tohono O'odham, Walapai, Yuma, and Zuni (Castetter and Bell 1951, Castetter and Opler 1936, Felger and Moser 1985, Gasser 1981, Hodgson 2001, Rea 1997, Reagan 1929, Moerman 1998, Stevenson 1915, Vestal 1952). All of these people eat the fruits and use the herbs as medicines (Austin 2004, Curtin 1947, Felger and Moser 1985, Moerman 1998, Rea 1997, Yetman and Van Devender 2001). Onavas Pima use the seeds of some species inside rattles (Pennington 1980).

**Derivation of the name:** *Physalis* is from the Greek word for bladder, meaning the inflated calyx that grows around the berries. *Pubescens* means hairy.

**Miscellaneous:** There are several *Physalis* species in the canyons. *Physalis acutifolia* (= *P. wrightii*) is an annual, usually taller than wide, with glabrous or sparsely pubescent leaves. *Physalis pubescens* is also an annual but has pubescent leaves. Both have bluish anthers. *Physalis hederifolia* (*P. hederaefolia* in Kearney and Peebles 1951) is a perennial like *P. crassifolia*. *Physalis hederifolia* is distinguished from *P. crassifolia* by being inconspicuously pubescent or glabrate and not being sticky with glands.

*Margaranthus solanaceus* may be confused with *Physalis*, and some people put them together. Historically, this small herb has been separated from *Phy-*

*salis* by its urn-shaped (urceolate) corolla. The small flowers may be pollinated by bees, but the corollas are so small and obscured among the other plants that there is surely some interesting interaction veiled from normal observation.

### *Solanum elaeagnifolium* (Solanaceae) Silver Horse-Nettle, *mala mujer, wako hahaisa*

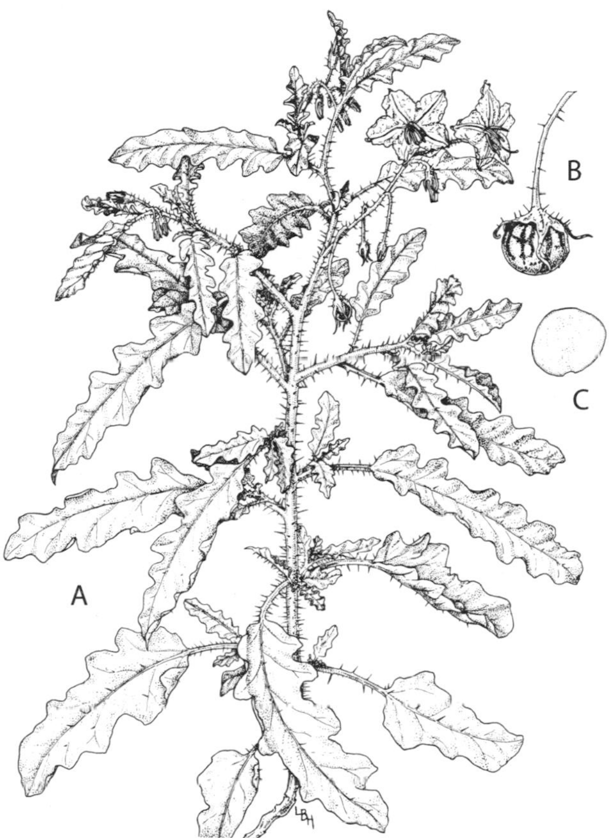


*Solanum elaeagnifolium*. A. Plant with flowers and fruits. B. Berry. C. Seed. Artist: L. B. Hamilton (Parker 1958).

**Other names:** ENGLISH: bull-nettle (bull = big, Germanic from ca. 1200; "nettle," Germanic, dating to the 400s; it alludes to stinging or prickly plants), iron-weed (Texas), silver [-leaf] nightshade ("nightshade," alluding to the narcotic properties of some European species; in English before CE 1000, of Germanic origin), white horse-nettle (horse = large, New Mexico, Texas; "horse" is a Teutonic word dating to ca. 825 in Old English), white-weed (Texas); SPANISH: *buena* [*mala*] *mujer* (good [bad] woman, Sonora), *pera* (pear, Coahuila), *trompillo* (little top, diminutive of *trompo*, New Mexico, Texas, Chihuahua, San Luis Potosí, Sonora), *rosillo* (dialectic for grayish, Sonora), *saca manteca* (butter puller, Arizona, Sonora), *tomatillo de buena mujer* (good woman's little tomato, Sonora); ATHAPASCAN: *gáán bidáá* [*bináá*] (see *Physalis*, Western Apache), *nááłtsoí* <*náałshoih*, *ʾanatco-i*> (*náá*, eyes [yellow seed (*bináá'*) by Elmore, in error], *łitso*, yellow, Navajo); KERES: *ashika* (Cochiti); UTO-AZTECAN: *vakoa hai* (broken gourd, Akimel O'odham), *vakoa hahaisig* (gourd broken into pieces, Akimel O'odham), *vakoa hahaiñig* (cracked gourd, Akimel O'odham), *vi'ul* (Hiá Ceḍ O'odham; fruits), *wako hahaisa* (*wako*, gourd, *hahaisa*, broken or chips of squash, Tohono O'odham); LANGUAGE ISOLATE: *ha'watapa* (*ha'li*, leaf, *wa'tapa*, prickly, Zuni)

**Botanical description:** Herbs, perennial, to ca. 1 m tall, with deep-running rootstock, silvery canescent all over, spines small and needle-shaped (acicular), frequent to sparse. Leaves have a petiole to 5 cm long, blades oblong to linear or oblong-lanceolate, to 15 cm long, usually tapered at base, mostly obtuse, entire to sinuate-wavy (repand). Cymes are terminal, short-peduncled, few-flowered, the pedicels long, recurved or reflexed in fruit. Calices are 5 angled, with slender lobes as long as tube. Corollas are violet (rarely white), shallowly 5-lobed, 2–2.5 cm wide, the lobes triangular-ovate. Berry is globose, to ca. 15 mm wide, yellow, eventually black.

**Habitat:** Roadsides. 300–1,670 m (1,000–5,500 ft).

**Range:** Arizona (throughout state), California, New Mexico, east to Colorado, Kansas, Missouri, and Texas; Mexico (Chihuahua, Coahuila, Nuevo León, northern Sonora, Tamaulipas, and south to Oaxaca); and Central and South America.

**Seasonality:** Flowering May to October, occasionally into November.

**Status:** Correll and Johnston (1970) say that the species is native from Missouri and Kansas south to Louisiana, Texas, Arizona, and adjacent Mexico; exotic elsewhere in the United States.

**Ecological significance:** These herbs thrive only in dry but disturbed soils, where Silver Horse-Nettles are pioneers. As the soils are colonized by other species and shaded, *S. elaeagnifolium* begins to die out. Once the site is completely shaded, there are rarely any of this species remaining.

Felger et al. (in prep.) found that Silver Horse-Nettles formerly were in Organ Pipe Cactus National Monument 9,570–20,490 years ago. The *Solanum* was in the Tinajas Altas 10,950 ybp. *Solanum elaeagnifolium* is no longer in those areas.

**Human uses:** Navajo use *S. elaeagnifolium* as a remedy for sore eyes, and for nose and throat trouble (Elmore 1944, Mayes and Lacy 1989). Similarly, other tribes that use Silver Horse-Nettles for a variety of medicines include the Akimel O'odham, Apache, Keres, and Zuni (Curtin 1949, Moerman 1998, Rea 1997, Reagan 1929, Stevenson 1915).

As with several *Solanum*, the fruits are used to curdle milk by Akimel O'odham, Cochití, Navajo, Hispanics, and Zuni (Castetter 1935, Curtin 1949, Ford 1975, Hocking 1997, Stevenson 1915). This use was reported by Curtin (1949), but the technique must have been learned from Hispanics because cows were unknown before being introduced by Spaniards. Rea (1997) reports that *vakoa hai* contains a protein-digesting enzyme resembling papain. Regarding similar uses, *S. erianthum* is called *hoja de manteca* (butter leaf, Oaxaca) and *saca manteca* (butter puller, Sinaloa). Knowledge of this use of *Solanum* fruits is widespread in the Americas.

Hopi use the fruits in necklaces worn by clowns (Whiting 1939). Keres make them into necklaces for women (Moerman 1998).

**Derivation of the name:** *Solanum* is from Latin *solamen*, a comfort, or *sol*, *solis*, the sun. *Elaeagnifolium* means having leaves like *Elaeagnus*, the genus of Oleaster and Russian Olive.

**Miscellaneous:** *Solanum douglasii* is found in Brown Canyon, and it flowers from July to September. This plant has white flowers instead of purple, and it lacks spines.

## *Celtis pallida* (Cannabaceae) Desert Hackberry, *granejo, ko:m*

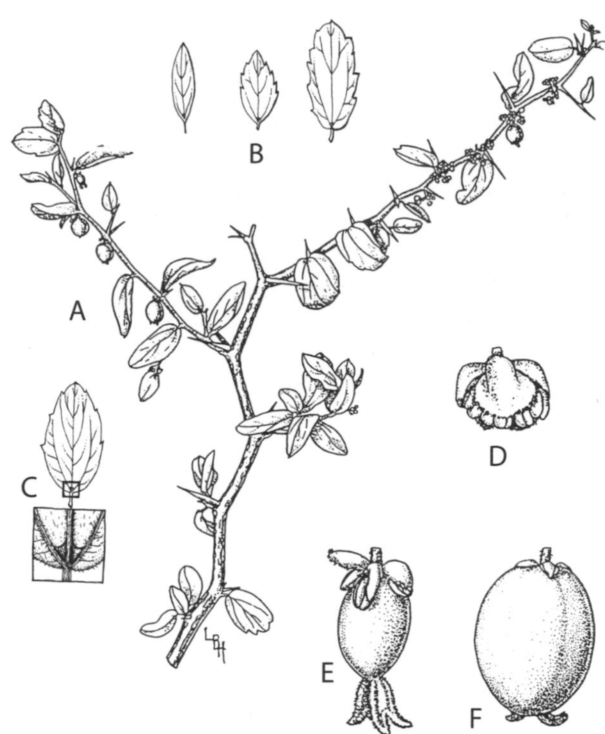


*Celtis pallida*. A. Flowering and fruiting branch. B. Leaf variations. C. Detail of leaf underside at base. D. Male flower. E. Perfect flower with stamens and pistil. F. Fruit. Artists: L. B. Hamilton, A, D–F (Benson and Darrow 1945); Felicia Bond, B, C (courtesy James Henrickson).

**Other names:** ENGLISH: [desert, spiny] hackberry [hagberry, hegberry] (appearing in English with John Gerarde's herbal in 1597, but based on Old Norse, *heggr*; the names are still given to *Prunus padus* and *P. avium*, as Danish *haegge-baer*, Norwegian *heggebär*, and Swiss *hägg-bär*; "hedge" was the original reference with *hæg*, derived from *haek*; still, an alternate meaning, at least originally, was to the female personification of evil we call a "hag"), spiny [shiny] hackberry; SPANISH: *acebuche* (comparing it with *Forestiera* or wild olive, Coahuila, Sonora), *capul* <*capui*> (cherry or *capulí*, Sonora, Durango, Texas), *garabato* (iron hook, Sinaloa), *garambullo* (spiny plant, Mayo, Sonora), *granejo* [*amarillo*] ([yellow] little seed, Chihuahua, Durango, Nuevo León, Sonora, Tamaulipas, Texas), [*granejo*] *huasteco* (Huastec [seeds], Tamaulipas), *palo de águila* (eagle's tree, Sonora), *rompecapa* (cape tearer, Sonora, Oaxaca); ATHAPASCAN: *jiłhazhí* <*jiłhazhi*> (*ch'il*, plant, *hazhí*, biting, Navajo; see also *Celtis reticulata*); HOKAN: *ptaacal* (Seri); OTO-MANGUEAN: *gec cehd* (*gec*, thorny, Zapotec), *guichi-bezia* (Zapotec, maybe akin to Náhuatl *huitztli*, spine); UTO-AZTECAN: *bainora* <*vainora*> (Cahita, probably because of similarity with *baiguo*, *Pisonia capitata*, Sonora), *kunwo* (Yaqui), *gumbro* <*cumbro*, *cúmero*> (Cahita, Mayo, Onavas Pima; Sobarzo says it is from Ópata), *ko:m* <*kohm*> (Akimel O'odham, Tohono O'odham), *kuavulĭ* <*kókauli*> (Akimel O'odham, Hiá Cẹd O'odham), *kuwavul* <*ku'avor*> (Tohono O'odham, Onavas Pima; Rea 1997:161 was given *k:om*, *kuwavul*, and *wusha'i* by Onavas Pima; see also *Celtis reticulata*), *suhtú* (Guarijío)

**Botanical description:** Shrubs 2–2.5 m tall, the young twigs, thorns, and inflorescences with stiff, appressed white hairs, becoming glabrous with age, twigs often zigzag, with single or paired thorns, these 1–3 per node, axils bearing flowers or a small leaf. Leaves are drought deciduous, 2.2–7 cm long, the blades ovate to broadly elliptic, dotted with hard, minute glands often with a stiff forward-pointing hair (scabrous), rough like sandpaper, margins with forward-pointing teeth or entire, petiole short. Flowers are green, inconspicuous, in small axillary clusters shorter than leaves. Fruits grow to 8 mm broad, orange, the pericarp thin, fleshy. Seeds are hard.

**Habitat:** Drainages, canyons. 457–1,828 m (1,500–6,000 ft).

**Range:** Arizona (Cochise, Gila, Greenlee, Pima, and Pinal counties), New Mexico, Nevada, western Texas, and Utah; Mexico (Baja California, Sonora east to Tamaulipas, and south to Oaxaca).

**Seasonality:** Flowering July to September.

**Status:** Native.

**Ecological significance:** The species is common along washes and/or rocky and gravelly slopes. This shrub is one of the few species in the genus adapted to xeric habitats. In the southeastern United States, all others are mesic site plants, except for *C. iguanaea* (Austin 2004). Like *C. pallida*, *C. iguanaea* is a tropical species adapted to dry places that is on the northern fringe of its range in Florida and in Sonora (Felger et al. 2001).

Fruits ripen during the summer and fall and are consumed by a variety of birds, Coatimundis (*Nasua narica*), Coyotes (*Canis latrans*), foxes, and Javelinas (*Pecari tajacu*) (Turner et al. 1995). The branches are browsed by deer, and they serve as nesting sites for White-Winged Doves (*Zanaida asiatica*) and cover for quail and mammals. Many birds, including Northern Mockingbirds (*Mimus polyglottos*), thrashers, and others, eat and distribute the fruits. Often young plants appear where these birds have a "favorite" perch in another species.

**Human uses:** Fruits on *C. pallida* are edible and reasonably sweet and tart, but a little waxy. Mayo, Onavas Pima, Seri, and probably others eat the raw fruits (Felger and Moser 1985, Pennington 1980, Yetman and Van Devender 2001).

Seri use the wood to make bows because of its flexibility (Felger 1977). Surely, other people living near Desert Hackberry similarly use the wood. Onavas Pima drag heavy logs of the trees to break up clods in plowed fields, and they weave saplings into wire fences to strengthen them (Pennington 1980).

**Derivation of the name:** *Celtis* is the Greek name used by Pliny, CE 23–79, for a lotus with sweet berries; also described by Herodotus, fl. 484 BCE, Dioscorides, fl. CE 40–80, Theophrastus, 372–287 BCE, and Homer, ninth–eighth? century BCE. *Pallida* means pale.

**Miscellaneous:** *Celtis* is comprised of 60–100 species (Mabberley 1997, Felger et al. 2001). There are three species in Sonora (Felger et al. 2001). Although it was formerly included within the Ulmaceae, recent studies have established that *Celtis* is not closely related to that group. Their shared traits appear to be resultant from convergence (Judd et al. 2008).

## *Celtis reticulata* (Cannabaceae) Net-Leaf Hackberry, *palo blanco, ko:m*

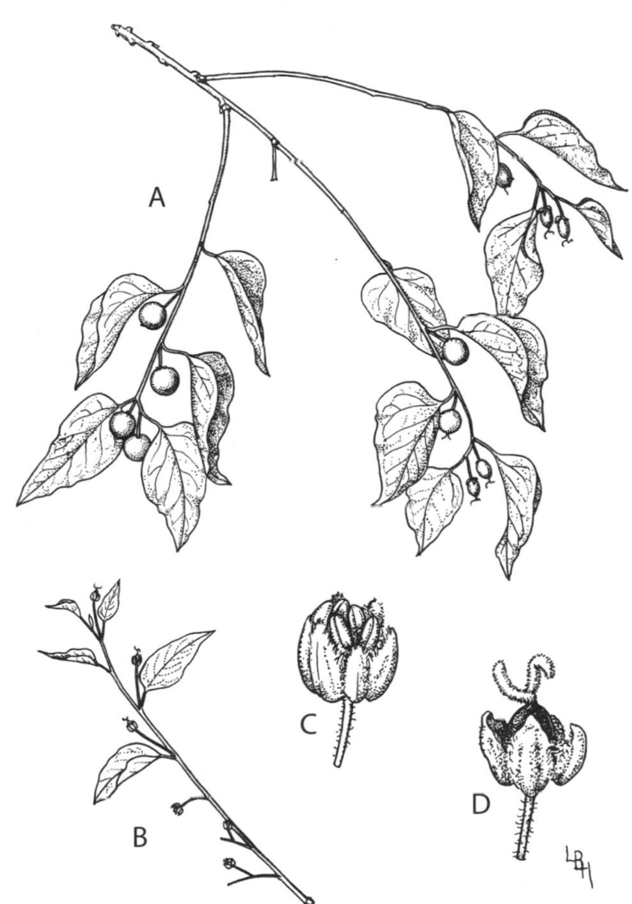


*Celtis reticulata*. A. Fruiting branch. B. Young branch with immature leaves; female flowers above, male below. C. Male flower. D. Female flower. Artist: L. B. Hamilton (Benson and Darrow 1945).

**Other names:** ENGLISH: [net-leaf] hackberry ("net-leaf" is a translation of the specific name *reti-*

*culata*; see *C. pallida* for derivation of "hackberry"), sugar-berry, western hackberry; SPANISH: *acibuche* <*acebuche*> (comparing it with *Forestiera* or wild olive, Chihuahua), *aceituna* (olive), *bainoro* <*vainora*> (probably because of similarity with *baiguo*, *Pisonia capitata*, Sonora), *garabato blanco* (white iron hook, Baja California), *membrillo* (comparing it with quince, *Cydonia oblonga*, Rosaceae, San Luis Potosí), *palo blanco* (white tree, Coahuila, Durango, Tamaulipas, Arizona, Texas), [*palo*] *cumbro* (Sinaloa), *palo duro* (hard tree, New Mexico), *palo mulato* (mulato tree, Durango), *vaior* (maybe akin to *bainora*, Mexico); ATHAPASCAN: *didzé bik'ǫǫdlizí* <*diʒé bekǫ'ƛizí*> ("hard seed," *didzé*, berry, *bi*, its, *'at'ǫǫ*, seed, *dlizí*, hard? Navajo), *iYntłidz* (hard seed, Chiricahua and Mescalero Apache), *jiłhazhí* <*jilxazi*, *tjiłxájih*> (chewing plant, Navajo; see also *Sambucus*), *jiłhááze* (cognate with Navajo *jiłhazhí*, Western Apache); KERES: *shikai-shikai-ka* (Acoma, Laguna); UTO-AZTECAN: *cúmero* <*combro*, *cumaro*, *cumbro*> (Cahita, Mayo, Sonora, Sinaloa), *gumbro* (Onavas Pima), *keˀmoci* (Guarijío), *ko:m* <*kom*> (Tohono O'odham), *kumar* (Onavas Pima; akin to *ku'avor* for *C. pallida*), *machaquí* <*uchieá*> (from *machacar*, to pound, mash, Guarijío, Sonora); YUMAN: *ᵃ'qwá'* <*aqwa'*> (Walapai)

**Botanical description:** Small trees or sometimes shrubby, to ca. 5 m tall, with a spreading canopy almost as broad, the mature bark reticulate and with corky ridges, branchlets usually hairy. Leaf blades are 2–8 cm long, 1.5–4 cm wide, obliquely ovate to lanceolate, obtuse to acute, rounded to cordate basally, entire or sparingly serrate, coriaceous, scabrous, reticulately veined, usually pale below. Fruits are spherical, reddish to orange or dark red, the flesh thin and sweet, 8–9 mm wide, on pedicels 6–12 mm long.

**Habitat:** Watercourses. 762–1,828 m (2,500–6,000 ft).

**Range:** Arizona (almost throughout state; not in high-elevation NE or arid SW), California to Oklahoma and Texas, and north to Colorado; Mexico (Baja California, Chihuahua, and Sonora to Tamaulipas).

**Seasonality:** Flowering March to April.

**Status:** Native.

**Ecological significance:** While *C. reticulata* may grow as shrubs on higher sites, these trees reach their best development along washes and canyons that are at least seasonally wet, with consistently high water tables. Brown (1982) considered this a thornscrub species, although farther east it grows in wetter habitats. This species suffered considerably in the drought of the early 2000s. After the long rainless period during the winter of 2005–2006, several of the larger trees were dead and winds toppled them.

Fruits of this species are eaten and dispersed by the same animals that spread *C. pallida*. Trees are an especially good food source for fruit-eating birds. Martin et al. (1951) recorded 31 species of birds and 14 mammals that consume the fruits of hackberry species throughout the range of the genus in the United States. Martin et al. further comment that the greatest importance of the trees to wildlife is in the West.

Snout Butterflies (*Libytheana carinenta*), Emperors (*Asterocampa*), and other genera require some type of hackberry for their larvae (Bailowitz and Brock 1991, Glassberg 2001). During some, but not all, wet seasons, Snout Butterflies may be the most common lepidopteran in the region. During the fall of 2001, this species was abundant in the Altar Valley, and flocks of hundreds "puddled" on wet sand after rains to garner minerals.

**Human uses:** At least the Acoma, Apache, Laguna, Mayo, Navajo, Tohono O'odham, Tewa, and Walapai eat berries of Net-Leaf Hackberry (Castetter 1935, Castetter and Opler 1936, Castetter and Underhill 1935, Elmore 1944, Moerman 1998, Robbins et al. 1916, Yetman and Van Devender 2001). Wood is burned and used to make tools by at least the Havasupai, Navajo, and Tewa, but surely by all tribes in its range. Tohono O'odham formerly used the bark to make sandals (Castetter and Underhill 1935). Mayo make a medicine for tonsilitis from the bark (Yetman and Van Devender 2001).

Several *Celtis* species may be used to make a tan dye. Navajos use it to make red-brown dye (Elmore 1944). At least the nonthorny kinds may be used in making baskets (Tull 1999), and the slender, flexible stems were favorites among some weavers.

**Derivation of the name:** *Celtis*, see *C. pallida*. *Reticulata* means with net-veining obvious on the leaves.

**Miscellaneous:** *Celtis* is comprised of 60–100 species (Mabberley 1997, Felger et al. 2001). There are three species in Sonora—*C. iguanaea*, *C. pallida*, and *C. reticulata* (Felger et al. 2001).

## *Hybanthus attenuatus* (Violaceae) Green Violet

**Other names:** ENGLISH: green violet (this name is applied to all members of the genus in the United States; sometimes *H. verticillatus* is called "nodding

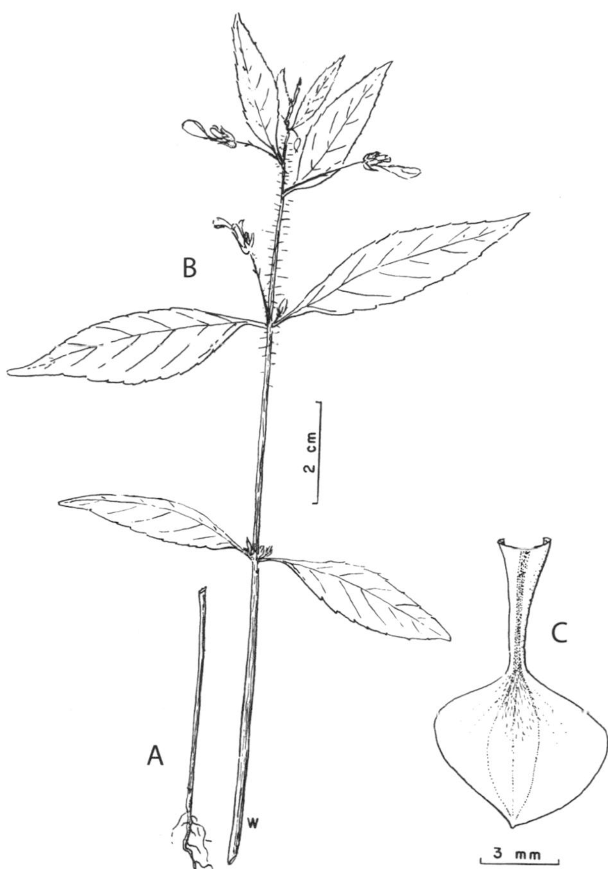


*Hybanthus attenuatus*. A. Base of plant. B. Flowering plant tip. C. Lower petal. Artist. I. L. Wiggins (Wiggins 1980).

green violet"); SPANISH: *hierba del rosario* (rosary herb, El Salvador, Honduras), *ipecacuanilla* (little ipecac [emetic], El Salvador, Honduras), *malva amarilla* (yellow mallow, a misnomer, Oaxaca)

**Botanical description:** Annual herbs 11–46 cm tall, the stems erect, with few branches, glabrous or pilose, often purple-mottled. Leaves are opposite below, alternate above, with petioles 4–7 mm long, the blades 1.5–10.5 cm long, elliptic to lanceolate, pubescent to glabrous, apex and base attenuate, the margins crenate to serrate, ciliate. Flower clusters are solitary or more often 2–4-flowered, in axillary, erect to horizontal clusters. Flowers have narrow sepals, these acuminate, entire, ciliate, 2–5 mm long, 0.5 mm wide at the base. Corollas are purple to blue, with lower petals 7–12 mm long, spatulate, the tip 4–7 mm wide above a median constriction, glabrous to pubescent on upper surface, lateral petals narrower, white to purple; cleistogamous flowers on short pedicels. Fruits are capsular, ovoid, 3–5.5 mm long, glabrous. Seeds are often 6 per capsule, globose to somewhat flattened with angular edges, dark brown to black with white to gray patches.

**Habitat:** Canyons, washes. 914–1,524 m (3,000–5,000 ft).

**Range:** Arizona (Cochise, Pima, and Santa Cruz counties); Mexico (Chihuahua, Sonora south through Guerrero, Guanajuato, Jalisco, Michoacán, and Veracruz); Belize, Costa Rica, El Salvador, Honduras, Nicaragua, and Panama; and Colombia, Ecuador, and Peru.

**Seasonality:** Flowering August to September.

**Status:** Native.

**Ecological significance:** This species is at the northern limit of its range in Arizona. Within the state, the herb is known in the Mule and Huachuca Mountains in Cochise County, in the Pajarito Mountains of Santa Cruz County, and in the Baboquivari Mountains in Pima County (Kearney and Peebles 1951, Toolin et al. 1979). At the onset of the summer rains, flowering begins. Several *Hybanthus* have self-fertilizing flowers that never open; however, those of *H. attenuatus* that open are probably pollinated by insects. Based on the flower structure, the blossoms must be pollinated by long-tongued bees.

Another species, *H. verticillatus*, grows in the Rincon and Tucson Mountains in Pima County (Bowers and McLaughlin 1987, Rondeau et al. 1996). This second species has a more northern distribution from Arizona and Mexico (Chihuahua to Coahuila, south to Puebla, San Luis Potosí, Veracruz, Zacatecas, and Oaxaca) to southeastern Colorado, Kansas, northeastern New Mexico, western Oklahoma, and Texas. Little (2001) illustrated the separate ranges of the two *Hybanthus* in Arizona.

*Hybanthus* and *Viola* are oddities in the family. The family is largely tropical, and these northern variations have biased temperate biologists into thinking that their traits are the "norm" for the family. Instead, the family is comprised largely of trees, shrubs (some twining), and lianas with actinomorphic flowers (Smith et al. 2004). Those plants form a stark contrast to the herbs *Hybanthus* and *Viola* with zygomorphic flowers.

**Human uses:** No uses were found for *Hybanthus attenuatus*, although other similar species have uses. Navajos say that *H. verticillatus* is the *'áłts'óózi <ˀa-lc' o-sigi>* (slender [medicine]), and they use it medicinally. Wyman and Harris (1941) do not say how the plant is used. The name may suggest a use for losing weight or it might simply be descriptive.

The Mayo use *H. mexicanus*, called *jépala* (Mayo) and *jarial* (Spanish), for fodder (Yetman and Van Devender 2001). *Hybanthus yucatenensis* is used in Mexico and Guatemala to treat snakebites (Hocking 1997). Other species used medicinally include *H. calceolaris* (*poaya branca, purga do campo*, white ipecacuanha, wild purge, Brazil) and *H. parviflorus* (*cuichunchulli*; *maytecillo* <*maintecillo*>, Argentina), both used as emetics, and *H. lanatus* (*pacacomha*, Brazil), used to treat cough and flu (Hocking 1997, Mabberley 1997).

**Derivation of the name:** *Hybanthus* is from Greek *hybos*, hump-backed, and *anthos*, a flower, referring to the anterior spurred petal. *Attenuatus* means narrowed, with reference to the leaves.

**Miscellaneous:** There are four species of *Hybanthus* in the United States, but only three of them are native (Kartesz 1994). The single eastern species is *H. concolor*, and the most widespread is *H. verticillatus*. *Hybanthus attenuatus* has the most restricted range of the native species. The genus consists of 150 species, found mostly in tropical and warm regions of the world (Mabberley 1997).

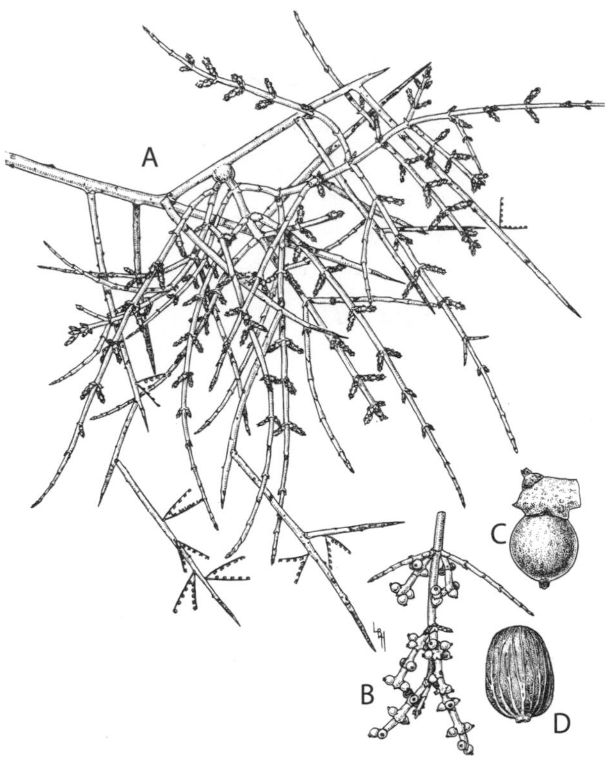


*Phoradendron californicum*. A. Branch of female flowers. B. Branch with fruits. C. Fruit, enlarged. D. Seed, enlarged. Artist: L. B. Hamilton (Benson and Darrow 1981).

## ***Phoradendron californicum*** **(Santalaceae) Desert Mistletoe, *toji, ha:kwaḍ***

**Other names:** ENGLISH: desert [mesquite] mistletoe ("mistletoe" is from *mistle*, dung, and *tan*, twig, a word from Old English in use by CE 1100; akin to Old High German *mistil*, dung, and *zein*, twig; the name came into existence because people believed that the Mistle Thrush, *Turdus viscivorous*, excreted the seeds on limbs; it actually scrapes them off its bill); SPANISH: *chile de espino* (spiny chile, Sonora), *toji* (from O'odham *toa*, oak, or *to:kĭ*, mistletoe); UTO-AZTECAN: *chayal* (Cahuilla), *haakvoḍ* (Akimel O'odham; probably based on the kinship term *hakit*, younger uncle, as an expression of the host-parasite relationship), *ha:kvaḍ* (Hiá Ceḍ O'odham), *ha:kwaḍ* <*hakowa't*> (Tohono O'odham), *haramkulyi* (*haram*, sticky, *kulyi*, old man, Mountain Pima), *o'ka* (Shoshoni; for *P. juniperinum*), *pohótela* (Phainopepla, Mayo, so called because that bird is the major disperser of the seeds), *to:kĭ* (Hiá Ceḍ O'odham, Arizona), *to(a)ker* <*toc'guer*> (on the oak, Mountain Pima); YUMAN: *kamúc* (Maricopa), *sxacál* [*sxyacál*] (Cocopa)

**Botanical description:** Parasites on tree branches, dioecious or sometimes monoecious. Flowers are fragrant. Calices are thick, fleshy, yellow-green. Anthers are short and yellow. Fruits are globose, 4.5–5.5 mm in diameter when fresh, the fresh pulp viscid and translucent white, salmon-colored on exposed surfaces and whitish to yellow-white where not exposed to sunlight. Seeds are 4 mm long, oblong, the surface white, the seedling root that attaches to the host stem (radicle haustorium) 0.5 mm long, green, and starting to protrude as the fruit ripens.

**Habitat:** Epiphytic on a variety of legumes.

**Range:** Arizona (Cochise, Coconino, Gila, Graham, Greenlee, La Paz, Maricopa, Mohave, Pima, Pinal, Santa Cruz, Yavapai, and Yuma counties), southern California, southern Nevada, and southern Utah; Mexico (Baja California's Cape Region, Chihuahua, Sinaloa, and Sonora).

**Seasonality:** Flowering January to November, largely during the fall monsoon season. However, plants may flower and fruit with little regard to season, and the fruits ripen in late winter and early spring.

**Status:** Native.

**Ecological significance:** *Phoradendron californicum* grows on legumes like *Acacia*, *Parkinsonia*, and *Prosopis*. Sometimes it grows on Rhamnaceae and even on *Larrea* (Kearney and Peebles 1951). Mistletoes put specialized roots (haustoria) into the vascular systems of their host plants and rob them of liquid and the nutrients dissolved in it. The epiphytes retain chlorophyll and use the nutrients and liquids to make their own food through photosynthesis.

Fruits of this epiphyte are extremely important to several birds during the winter months when few other plant foods are available (Jaeger 1941). Most important among the avians is the Phainopepla (*Phainopepla nitens*). These crested birds move seasonally from the lowlands in the summer to the uplands in the winter, where there is an abundance of mistletoe berries. In the uplands, the birds guard areas with mistletoe and consume the fruits with apparent delight (Alcock 1990). Because birds wipe the excess fruits and seeds from their bills, new *Phoradendron* are "planted" on other places on the same trees or on new individuals. While guarding trees with mistletoe that the birds use as their major food, Phainopeplas collect insects to support their brood of two nestlings. Each pair defends about an acre during nesting, as that provides enough food to feed the adults and young.

Northern Mockingbirds (*Mimus polyglottos*) also focus on these mistletoe fruits during the winter. These mimics (family Mimidae) fly to places with water and leave the semidigested fruits in their feces along the edges. Like the Phainopepla, these birds also wipe the sticky fruits and seeds from their bills at waterholes and on branches, where many sprout and grow to create new food sources for them.

The large old growth of these parasitic plants on legumes provides another resource for birds—nesting sites. Among the birds that take advantage of the protection for their nests are the Verdins (*Auriparus flaviceps*) and the White-Winged Dove (*Zanaida asiatica*). In addition, Gambel's Quail (*Callipepla gambelii*), Mourning Doves (*Zanaida macroura*), and others roost in the comparative security of mistletoe clusters.

**Human uses:** The Tohono O'odham boil Desert Mistletoe "leaves" and drink the liquid to treat colds (Castetter and Underhill 1935). Castetter (1935), Curtin (1949), and Jaeger (1941) found the Akimel O'odham eating the berries. However, the people with whom Rea (1997) talked were "afraid to eat" the berries, even though they had a good taste; at least fruits from plants growing on Mesquite did. Mayo make a tea of Mistletoes to treat blood pressure problems (Yetman and Van Devender 2001).

**Derivation of the name:** *Phoradendron* was named by Thomas Nuttall with the Greek *phora*, a thief, and *dendron*, tree, because of the parasitic habit. *Californicum* means from California, where the species was first discovered by the scientific community.

## *Phoradendron serotinum* (Santalaceae) Mistletoe, *injerto, ha:kwaḍ*

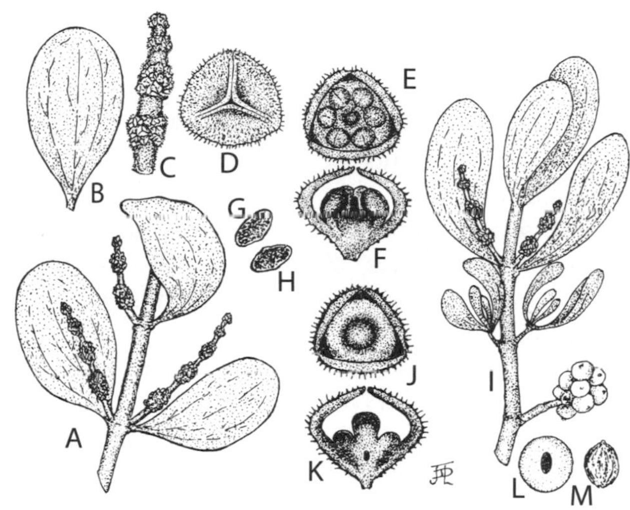


*Phoradendron serotinum*. A. Male plant. B. Leaf, enlarged. C. Part of a male flower cluster. D. Male flower from above. E. Cross section of male flower. F. Longitudinal section of male flower. G–H. Pollen grains. I. Female plant. J. Female flower in cross section. K. Longitudinal section of female flower. L. Fruit. M. Seed. Artist: J. D. Laudermilk (Benson 1957).

**Other names:** ENGLISH: big-leaf [Christmas, hairy] mistletoe (see *P. californicum* for etymology); SPANISH: *cuashilaca* (Hidalgo), *injerto* (grafted, Arizona to Texas, Durango), *liga* (tied, Mexico), *muérdago* (bird feces, California Spanish), *seca palo* (tree killer, Mexico), *silmo* (Sinaloa), *tohi* <*toje, toji*> (see *P. californicum* for origin, Sonora); ATHAPASCAN: *dahts'aa'* (Navajo), *t'iis bidahs'ą'* [*gad bidahs'ą'*] (*gad*, juniper, *t'iis*, cottonwood leaf, *bidahs'ą'*, its mistletoe, Western Apache); CHUMASH: *shlamulasha'w* (Barbareño Chumash), *stumuku'n* (Ineseño Chumash); OTO-MANGUEAN: *búngu* (Mazahua; for *P. velutinum*);

UTO-AZTECAN: *chichiham* (Yaqui), *haakvoḍ* (see *P. californicum* for etymology, Akimel O'odham), *ha:kwaḍ* (Tohono O'odham), *haramkulyi* (Mountain Pima), *kuchó'oko* (Tarahumara fide Bye 1986), *lomápi* (Hopi), *o'ka* (Shoshoni; for *P. juniperinum*), [*havatɨbɨa, maˀahnidɨbɨa*] *sanapiceeka* (mistletoe [growing on *Platanus racemosa, Quercus douglasii*], Kawaiisu), *to(a)ker* <*toc'guer*> (on the oak, Mountain Pima), *to:kĭ* (Hiá Ceḍ O'odham; for *P. californicum*); YUMAN: *kamúc* (Maricopa), *sxacál* [*sxyacál*] (Cocopa); (= *P. flavens, P. flavescens, P. tomentosum* var. *macrophyllum*)

**Botanical description:** Parasitic plants, stout, dioecious, 3–6 dm tall, the internodes 3–5 cm long, minutely white-pubescent to white-tomentose, becoming glabrous with age. Leaves are thick, elliptic-obovate to oblong-oblanceolate, obtuse, 1–5 cm long, glabrous with age, cuneately petioled. Spikes are 1.5–5 cm long. Male flower clusters are 20–30 flowered, the female 6–12. Fruits are white or tinged pink, globose, 4–5 mm in diameter.

**Habitat:** Parasitic on broadleaf trees, particularly cottonwoods (*Populus*), but also on ash (*Fraxinus*), hackberry (*Celtis*), oaks (*Quercus*), walnuts (*Juglans*), and others.

**Range:** Arizona (Yavapai to Graham, south to Cochise, Pima, Santa Cruz, and Yuma counties), California, and New Mexico; Mexico (Baja California, Chihuahua, Sonora to Coahuila, Durango, and south-central Mexico).

**Seasonality:** Flowering March, sometimes again August to October.

**Status:** Native. Our plants are ssp. *tomentosum*.

**Ecological significance:** These are hemiparasites, retaining their chlorophyll and ability to make food, but stealing their liquid nutrients from their hosts. Fruits from mistletoes everywhere are important bird foods. In Brown Canyon, Western Bluebirds (*Sialia mexicana*) have been seen feeding on the fruits in April. Phainopeplas (*Phainopepla nitens*), Northern Mockingbirds (*Mimus polyglottos*), and probably other kinds of birds also eat them.

**Human uses:** Probably *P. serotinum* is the species recorded as being used as food, a dye source, medicine, and as ceremonial plants among at least the Cahuilla, Kumiai, Hopi, Akimel and Tohono O'odham, and Navajo (Bean and Saubel 1972, Castetter and Underhill 1935, Moerman 1998, Russell 1908, Whiting 1939). Various medicines were made to treat several maladies (e.g., Hough 1897). Cahuilla use the leaves to make a tea, and the berries are boiled and eaten (Bean and Saubel 1972). Chumash consider a decoction contraceptive (Timbrook 1990).

Cahuilla also use the leaves to dye baskets black (Bean and Saubel 1972). Navajos use Mistletoe in the War Dance liniment, and they hang it over hogan doorways as protection from lightning (Elmore 1944).

*Phoradendron serotinum* resembles the Old World Mistletoe (*Viscum album*) that played such an important role in the history of several countries. Their species was known as *druidh lus* (*druidh*, Druid, *lus*, herb, Gaelic), *gui* (French), *guis* (sticky, Irish), *muérdango* (bird feces, Spanish), *uil'-ioc* (all-heal, Gaelic), *vischio* (from Latin *viscus*, sticky, Italian), *visco* (Portuguese), and *Vogel-liem* <*Vogellym*> (birdlime, German) among its other names (Austin 2004). Beliefs in the mystical powers of Mistletoes were widespread, but those among the Gaelic people are the best known. Less well known is that Mistletoes were important medicines (Austin 1998b).

All Mistletoes with wide leaves like these have figured prominently in people's beliefs. These plants are now used particularly during the Christmas season, when the branches are hung prominently to invite a kiss. This custom actually originally came into practice from the ancient Scandinavians, although Celts, Greeks, and Romans had similar views. Scandinavians hang branches of Mistletoe in their houses to ensure good luck. The Scandinavians believe that good luck will ensue because the Mistletoe, associated with Fragga, the goddess of love, was powerful and lucky. Thus, a house bearing Mistletoe was assured good luck and freedom from evil spirits. The Church adopted *Phoradendron* because it helped the priests enlist pagans from among all these cultures to Christianity by using a plant already sacred to them.

**Derivation of the name:** *Phoradendron*, see *P. californicum* for etymology. *Serotinum* refers to late flowering or fruiting. *Tomentosum* means densely woolly, matted with hairs.

**Miscellaneous:** The genus comprises 190 American species, and it is most diverse in the tropics (Mabberley 1997). Dried material is sold as *flores de palo* (tree flowers) in Central America.

## *Vitis arizonica* (Vitaceae) Canyon Grape, *parra del monte, u:ḍwis*

**Other names:** ENGLISH: canyon [Arizona, gulch, wild] grape ("canyon" came from Spanish *cañon*, tube, pipe, conduit, barrel, cannon [augmentative of *caña*,

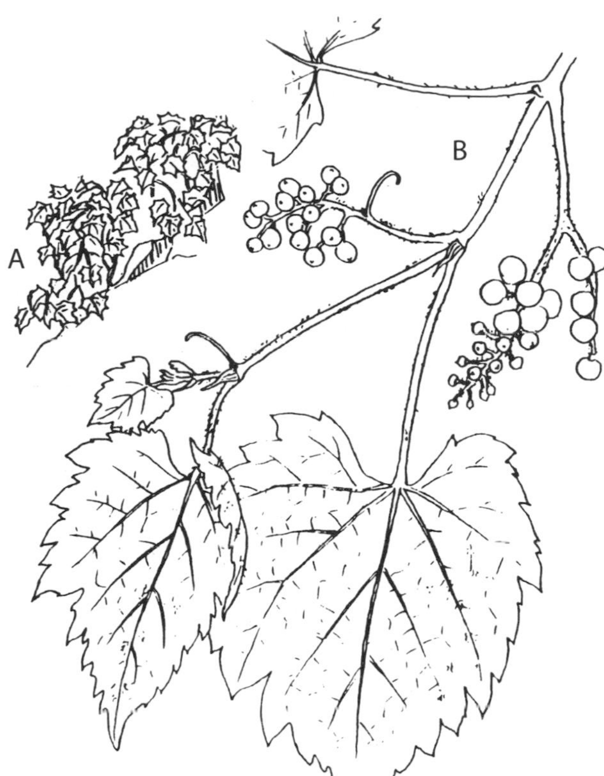


*Vitis arizonica*. A. Life-form. B. Fruiting branch. Artist: R. D. Ivey (Ivey 2003).

from Latin *canna,* reed, pipe, quill, cane, thus the same word as Italian *cannone*, Portuguese *canhão*, Provençal and French *canon*, English *cannon*], but specifically applied by the Spaniards of New Mexico to gulch, and into English by 1834; a "gulch" is a narrow and deep ravine, with steep sides, marking the course of a torrent; used in the United States by 1832; our English word "grape" came into the language before CE 1290 with "A luytel foul . . . brochte a gret bouch Fol of grapus swyte rede" [a little fowl . . . brought a great bunch full of sweet red grapes]; prior to CE 1290, the word was *krappen* in Teutonic, *krappa* in Old High German, and *grap, grappe* in Old French, all meaning a hook; presumably, the hook was the one used to harvest fruit from the vine); SPANISH: *parra* (vine, Tamaulipas), *parra del monte* [*silvestre*] (wild grape, Arizona, Texas, Chihuahua), *uva* [*cimarrón*] (wild grape, Chihuahua, Sonora; from Latin *uva*), *vid* (vine); ATHAPASCAN: *ch'il na'atł'o'ii* (weaving plant, Navajo), *dahts'aa'* <*dasts'aa, dahts'aa' benanisdizí, tach'aa*> (*dasts'aa'*, grape, Western Apache), *dastsa* <*dastasa*> (Chiricahua and Mescalero Apache), *tutzé* (Jicari-lla Apache); UTO-AZTECAN: *bakámai bišáparagai* (barn enveloper [sic, fide Pennington 1969], Northern Tepehuan), *isampu* (Panamint), *jeyulí* (Guarijío), *ó:va* (from Spanish *uva*, Hopi), *mákwit* (Luiseño), *mischiñ uuḍvis* <*mischiñ huuḍvis*> (*mischiñ*, something that is wild but should be tame, *uuḍvis*, grape, Akimel O'odham), *u:dvis* (Hiá Ceḍ O'odham), *shohar u'ushi* (*shohar*, crying, *u'ushi*, bushes, Mountain Pima), *sonótova* (Mono), *sųų'rǫ'oˀnapų* (Ute), *u:ḍwis* (Tohono O'odham), *uirí* (Guarijío), *urí* <*uli*> (Tarahumara), *uuva* (Yaqui), *u:va* <*uuwa*> (from Spanish *uvas*, Onavas Pima), *wolont* [*wolon-t*] (Tübatulabal; for *V. californica*); YUKI: *mö'mäm* <*mú'mám*> (Yuki; for *V. californica*); YUMAN: *idjérk*$^{r}$*a* (Havasupai), *i'icamác* (Maricopa), *itcêq*$^{a}$ <*i'je:qa*> (Walapai), *ˀu:vs* (Cocopa, from Spanish *uvas*)

**Botanical description:** Shrubs, stems climbing, branched, the tendrils falling early if not attached to a support. Leaves are broadly cordate-ovate, to almost kidney-shaped, apically triangular, 5–12 cm long, usually slightly broader than long, the margins with large teeth, both sides cottony when young, becoming glabrous or almost glabrous with age except along veins, the petioles often pink-tinged, pubescent or glabrous. Inflorescences are compound (thyrsiform), 5–10 cm long, slender-stalked. Berries arc 6–10 mm thick, black or sometimes with a thin bloom (glaucous), the skin thick, the pulp juicy and sweet or tart. Seeds are 4–6 mm long, 3–4 mm wide, short-beaked.

**Habitat:** Canyons, streamsides, washes. 600–2,130 m (2,000–7,000 ft).

**Range:** Arizona (Coconino, Mohave, Navajo, south to Cochise, Greenlee, Pima, and Santa Cruz counties), New Mexico, Nevada, western Texas, and southern Utah; Mexico (Baja California, Chihuahua, Coahuila, Durango, Guerrero, Sonora, and Tamaulipas).

**Seasonality:** Flowering (March) April to July.

**Status:** Native. Our plants are var. *arizonica*.

**Ecological significance:** Plants often twine over trees, shrubs, and boulders in mountain canyons; hence, the common name. Roots and stems are efficient at stopping erosion along waterways (Kearney and Peebles 1951).

Animals of various kinds, including birds and mammals, are fond of the fruit. Indeed, animals like the fruits so much that humans often have a hard time finding grapes on the plants.

**Human uses:** Canyon Grapes are eaten by at least Acoma, Apache, Guarijío, Havasupai, Isleta, Jemez, Laguna, Navajo, Paiute, Tarahumara, Akimel O'odham, Tohono O'odham, and Yavapai (Castetter

1935, Castetter and Opler 1936, Gallagher 1976, Hodgson 2001, Moerman 1998, Palmer 1878, Pennington 1963, Rea 1997, Yetman and Felger 2002). Palmer (1871) reports that grapes were formerly cultivated by Pueblo people. Fruits have been dried to eat like raisins, eaten fresh, or juiced to ferment into wine. Opinion varies about how good these grapes are, but many people were like the modern Akimel O'odham, who consider the fruits almost too sour to eat (Rea 1997). Perhaps opinion changed with time and availability of other foods, as Palmer (1878) records indigenous people "consuming them in large quantities." The Tarahumara also were eating the fruits in the late nineteenth century (Pennington 1963). The Northern Tepehuan were still eating them when visited by Pennington (1969).

Rea (1997) thought that *uuḍvis* (*u:ḍwis* in Tohono O'odham) was part of the postcontact combination in the Akimel O'odham name and that it came from Spanish *uvas*. Shaul (personal communication, 2007) agrees that the Onavas Pima *u:wa* is a loan word from Spanish. Shaul does not think that the Northern Tepehuan, Guarijío *uirí*, or Tarahumara *urí* are from the same base; he thinks they are indigenous names.

Kearney and Peebles (1951) suggest that the leaves allay thirst when chewed. Guarijío plant the seeds near their houses to provide shade (Yetman and Felger 2002). Maybe that is the reference in the Northern Tepehuan name.

Guarijío rub the leaves on Poison Ivy (q.v.) rash as a treatment (Yetman and Felger 2002). Northern Tepehuan make a remedy for scratches from the leaves (Pennington 1969). Navajos use the stems as part of an offering during courtship (Moerman 1998). Havasupai use the stems to make the hoop in the hoop and pole game. Jemez mix the juice with white clay as body paint for dancers.

**Derivation of the name:** *Vitis* is Latin for bend, plat, or weave. *Arizonica*, from Arizona, refers to the place where the species was first collected.

# Appendix: Plants of the Baboquivari Mountain Chain

This inventory is a preliminary list because there are added species being found regularly within the mountain ranges. The localities cited are based on specimens from mountain ranges to the north and south of the Baboquivari Mountains that are rarely distinguished. Although the floras of these four ranges (Coyote, Quinlan, Baboquivari, and Pozo Verde Mountains) overlap greatly, none are thoroughly collected and studied. There are probably species in the list that occur in some but not other ranges of these mountains. Species that are known in the valleys that extend upward into the lower reaches of canyons are also included. Collections are from the Baboquivari Mountains unless the other range is named. Specimens are deposited at ARIZ, ASU, and BANWR (Buenos Aires National Wildlife Refuge) unless otherwise noted (acronyms follow the system developed for *Index Herbariorum* and now available online at http://sciweb.nybg.org/science2/IndexHerbariorum.asp). Names are current as of July 23, 2008.

## ACANTHACEAE

***Anisacanthus thurberi*** (Torrey) A. Gray
Baboquivari Canyon, Brown Canyon, Thomas Canyon; Coyote Mountains
***Carlowrightia arizonica*** A. Gray
Quinlan Mountains
***Dicliptera resupinata*** (Vahl) Jussieu [*Dicliptera pseudoverticillaris* A. Gray]
Baboquivari Canyon, Brown Canyon, Fresnal Canyon, Toro Canyon
***Elytraria imbricata*** (Vahl) Persoon
Brown Canyon
***Justicia californica*** (Bentham) D. Gibson [*Beloperone californica* Bentham]
Fresnal Canyon, South Canyon; Coyote Mountains
***Justicia longii*** Hilsenbeck [*Siphonoglossa longiflora* (Torrey) A. Gray]
Baboquivari Canyon, Canyon Wash; Coyote Mountains
***Ruellia nudiflora*** (Engelmann & A. Gray) Urban
Baboquivari Canyon, Toro Canyon
***Tetramerium nervosum*** Nees [*T. hispidum* Nees]
Baboquivari Canyon, Brown Canyon, Fresnal Canyon, South Canyon, Thomas Canyon, Toro Canyon

## AGAVACEAE

***Agave palmeri*** Engelmann
Brown Canyon, Thomas Canyon
***Agave parviflora*** Torrey subsp. ***parviflora***
Brown Canyon
***Agave schottii*** Engelmann
Baboquivari Canyon, Brown Canyon
***Dasylirion wheeleri*** S. Watson
Baboquivari Canyon, Brown Canyon
***Nolina microcarpa*** S. Watson
Brown Canyon, Thomas Canyon
***Yucca baccata*** Torrey var. ***brevifolia*** (Schott ex Torrey) L. D. Benson & R. A. Darrow [*Yucca arizonica* McKelvey; *Yucca ×schottii* Engelmann (pro parte)]
Baboquivari Canyon; Pozo Verde Mountains; Quinlan Mountains

## ALLIACEAE

***Allium plummerae*** S. Watson
Baboquivari Canyon
***Nothoscordum texanum*** M. E. Jones
Brown Canyon, Fresnal Canyon

## AMARANTHACEAE

***Alternanthera pungens*** Kunth [*Alternanthera repens* (Linnaeus) Link non J. F. Gmelin]
Brown Canyon
***Amaranthus fimbriatus*** (Torrey) Bentham
Brown Canyon; Coyote Mountains
***Amaranthus palmeri*** S. Watson
Brown Canyon; Quinlan Mountains
***Amaranthus powellii*** S. Watson
Coyote Mountains, Quinlan Pass
***Amaranthus torreyi*** (A. Gray) Bentham ex S. Watson
Bear Canyon
***Froelichia arizonica*** Thornber
Thomas Canyon
***Gomphrena caespitosa*** Torrey
Brown Canyon, Thomas Canyon
***Gomphrena sonorae*** Torrey
Baboquivari Canyon, Bear Canyon, Brown Canyon, Thomas Canyon, Toro Canyon; Coyote Mountains

***Guilleminea densa*** (Willdenow) Moquin [*Brayulinea densa* (Willdenow) Small]
Baboquivari Canyon, Brown Canyon, Thomas Canyon
***Iresine heterophylla*** Standley
Baboquivari Canyon, Brown Canyon; Coyote Mountains
***Tidestromia lanuginosa*** (Nuttall) Standley subsp. ***eliassonii*** Sánchez-del Rio & Flores Olivera
Brown Canyon; Coyote Mountains; Quinlan Mountains

## ANACARDIACEAE

***Rhus aromatica*** Aiton var. ***trilobata*** (Nuttall ex Torrey & A. Gray) A. Gray [*Rhus trilobata* Nuttall ex Torrey & A. Gray, *Rhus trilobata* Nuttall ex Torrey & A. Gray var. *pilosissima* Engler, *Rhus emoryi* (Greene) Wooton]
Baboquivari Canyon, Brown Canyon, Thomas Canyon; Quinlan Mountains
***Rhus virens*** Lindheimer ex A. Gray [*Rhus virens* var. *choriophylla* (Wooton & Standley) L. D. Benson, *Rhus choriophylla* Wooton & Standley]
Jaguar Canyon
***Toxicodendron radicans*** (Linnaeus) O. Kuntze [*Rhus radicans* Linnaeus]
Brown Canyon

## APIACEAE

***Bowlesia incana*** Ruiz & Pavón
Brown Canyon, Fresnal Canyon
***Daucus pusillus*** Michaux
Brown Canyon, Fresnal Canyon, Toro Canyon; Coyote Mountains; Pozo Verde Mountains
***Lomatium nevadense*** (S. Watson) Coulter & Rose var. ***parishii*** (Coulter & Rose) Jepson
Brown Canyon
***Spermolepis echinata*** (Nuttall ex De Candolle) Heller
Baboquivari Canyon, Brown Canyon, Thomas Canyon; Coyote Mountains; Quinlan Mountains
***Yabea microcarpa*** (Hooker & Arnott) Koso-Polák [*Caucalis microcarpa* Hooker & Arnott]
Baboquivari Canyon, Brown Canyon, Fresnal Canyon, Toro Canyon

## APOCYNACEAE

***Amsonia kearneyana*** Woodson
Brown Canyon, Baboquivari Peak, South Canyon
***Amsonia palmeri*** A. Gray
Sycamore Canyon
***Haplophyton crooksii*** (L. D. Benson) L. D. Benson) [*H. cimicidum* of authors not A. De Candolle, *H. cimicidum* A. De Candolle var. *crooksii* L. D. Benson]
Baboquivari Canyon, Baboquivari Peak, Brown Canyon, Sabino Canyon; Coyote Mountains

## ARALIACEAE

***Aralia humilis*** Cavanilles
Baboquivari Peak, Brown Canyon, Forest Cabin Canyon, South Canyon, Sycamore Canyon, Thomas Canyon, Toro Canyon

## ARISTOLOCHIACEAE

***Aristolochia watsonii*** Wooton & Standley
Baboquivari Canyon, Brown Canyon; Coyote Mountains; Quinlan Mountains

## ASCLEPIADACEAE

***Asclepias asperula*** (Decaisne) Woodson
Brown Canyon, Thomas Canyon
***Asclepias brachystephana*** Engelmann ex Torrey
Baboquivari Mountains
***Asclepias glaucescens*** Kunth [*Asclepias elata* Bentham]
Brown Canyon
***Asclepias involucrata*** Engelmann ex Torrey
Baboquivari Peak, Brown Canyon
***Asclepias lemmoni*** A. Gray
Brown Canyon
***Asclepias linaria*** Cavanilles
Baboquivari Canyon, Brown Canyon, Forest Cabin Canyon, Fresnal Canyon, Spring Canyon; Coyote Mountains; Pozo Verde Mountains; Quinlan Mountains
***Asclepias quinquedentata*** A. Gray
Quinlan Mountains
***Cynanchum arizonicum*** (A. Gray) Shinners [*Metastelma arizonicum* A. Gray]
Baboquivari Canyon, Brown Canyon, Fresnal Canyon, Ventana Peak, E of San Miguel, Virtud Tank area; Coyote Mountains; Pozo Verde Mountains
***Funastrum crispum*** (Bentham) Schlechter [*Sarcostemma crispum* Bentham]
Brown Canyon

***Funastrum cynanchoides*** (Decaisne) Schlechter [*Sarcostemma cynanchoides* Decaisne]
Brown Canyon Wash
***Gonolobus arizonicus*** (A. Gray) Woodson [*Matelea arizonica* (A. Gray) Shinners]
Baboquivari Canyon, Brown Canyon, Forest Cabin Canyon, South Canyon; Coyote Mountains

## ASPLENIACEAE

***Asplenium dalhousiae*** Hooker
Baboquivari Canyon, Moristo Canyon
***Asplenium palmeri*** Maxon
Baboquivari Canyon, Brown Canyon, north of Moristo Canyon
***Asplenium resiliens*** Kunze
Baboquivari Peak

## ASTERACEAE

***Acourtia wrightii*** (A. Gray) Reveal & King [*Perezia wrightii* A. Gray]
Brown Canyon, Forest Cabin Canyon; Coyote Mountains
***Adenophyllum porophylloides*** (A. Gray) Strother [*Dyssodia porophylloides* A. Gray]
Coyote Mountains
***Ageratina herbacea*** (A. Gray) King & H. E. Robinson [*Eupatorium herbaceum* (A. Gray) Greene]
Baboquivari Peak, Brown Canyon, Fresnal Canyon
***Ageratina paupercula*** (A. Gray) King & H. E. Robinson [*Eupatorium pauperculum* A. Gray]
Baboquivari Canyon, Moristo Canyon, Spring Canyon S of Fresnal Canyon; Coyote Mountains
***Ambrosia ambrosioides*** (Cavanilles) Payne
Coyote Mountains
***Ambrosia confertiflora*** De Candolle
Brown Canyon; Quinlan Mountains
***Ambrosia cordifolia*** (A. Gray) Payne
Pozo Verde Mountains
***Ambrosia monogyra*** (Torrey & A. Gray) Strother & B. G. Baldwin [*Hymenaclea monogyra* Torrey & A. Gray]
Brown Canyon, Fresnal Canyon
***Artemesia dracunculoides*** Pursh
Thomas Canyon; Quinlan Mountains
***Artemisia ludoviciana*** Nuttall
Baboquivari Canyon, Brown Canyon; Coyote Mountains; Quinlan Mountains
***Baccharis brachyphylla*** A. Gray
Baboquivari Canyon, Brown Canyon, Sabino Canyon
***Baccharis emoryi*** A. Gray
Jaguar Canyon
***Baccharis pteronioides*** De Candolle
Pozo Verde Mountains
***Baccharis salicifolia*** (Ruiz & Pavón) Persoon [*B. glutinosa* Persoon]
Baboquivari Canyon, Brown Canyon
***Baccharis sarothroides*** A. Gray
Brown Canyon
***Baccharis thesioides*** Kunth
Baboquivari Peak, Brown Canyon, Fresnal Canyon, Jaguar Canyon, Virtud Tank area; Pozo Verde Mountains
***Bahia absinthifolia*** var. ***dealbata*** (A. Gray) A. Gray
Brown Canyon
***Bidens aurea*** (Aiton) Sherff
Brown Canyon, Moristo Canyon; Quinlan Mountains
***Bidens bigelovii*** A. Gray
Brown Canyon, South Canyon
***Bidens leptocephala*** Sherff
Brown Canyon; Coyote Mountains
***Brickellia amplexicaulis*** B. L. Robinson
Brown Canyon, Sycamore Canyon
***Brickellia baccharidea*** A. Gray
Baboquivari Mountains, without locality
***Brickellia betonicaefolia*** A. Gray
Baboquivari Canyon, Brown Canyon
***Brickellia californica*** (Torrey & A. Gray) A. Gray
Baboquivari Canyon, Brown Canyon, Thomas Canyon, Virtud Tank area
***Brickellia coulteri*** A. Gray var. ***coulteri***
Brown Canyon, Toro Canyon, Ventana Peak E of San Miguel; Pozo Verde Mountains
***Brickellia eupatorioides*** (Linnaeus) Shinners var. ***chlorolepis*** (Wooton & Standley) B. L. Turner [*B. chlorolepis* (Wooton & Standley) Shinners, *Kuhnia rosmarinifolia* Ventenat]
Baboquivari Canyon, Brown Canyon, Jaguar Canyon
***Brickellia grandiflora*** (Hooker) Nuttall
Brown Canyon, Thomas Canyon
***Brickellia parvula*** A. Gray
Baboquivari Canyon, Fresnal Canyon
***Brickellia venosa*** (Wooton & Standley) B. L. Robinson
Baboquivari Canyon, Brown Canyon, Fresnal Canyon, Sycamore Canyon, Toro Canyon

***Carminatia tenuiflora*** De Candolle
Baboquivari Canyon, Brown Canyon, South Canyon
***Carphochaete bigelovii*** A. Gray
Baboquivari Canyon, Thomas Canyon; Quinlan Mountains
***Chaenactis stevioides*** Hooker & Arnott
Brown Canyon
***Chrysopsis fulcrata*** Greene
South Canyon
***Chrysopsis villosa*** (Pursh) Nuttall
Coyote Mountains
***Cirsium arizonicum*** (A. Gray) Petrak
Brown Canyon, Thomas Canyon
***Cirsium neomexicanum*** A. Gray
Baboquivari Canyon, Brown Canyon; Coyote Mountains
***Conyza canadensis*** (Linnaeus) Cronquist [*Erigeron canadensis*]
Brown Canyon; Quinlan Mountains
***Coreocarpus arizonicus*** (A. Gray) Blake
Baboquivari Canyon, Brown Canyon, Thomas Canyon; Coyote Mountains; Quinlan Mountains
***Diaperia verna*** (Rafinesque) Moerfield [*Evax verna* Rafinesque]
Pozo Verde Mountains
***Encelia farinosa*** Torrey
Brown Canyon
***Ericameria cuneata*** (A. Gray) McClatchie [*Applopappus cuneatus* var. *spathulatus* (A. Gray) Blake]
Baboquivari Canyon, Brown Canyon, Moristo Canyon, Toro Canyon; Coyote Mountains; Quinlan Mountains
***Ericameria laricifolia*** (A. Gray) Shinners [*Haplopappus laricifolius* A. Gray]
Baboquivari Canyon, Brown Canyon, Fresnal Canyon, Thomas Canyon, Toro Canyon; Coyote Mountains; Quinlan Mountains
***Erigeron arisolius*** Nesom
Baboquivari Canyon
***Erigeron colomexicanus*** A. Nelson
Brown Canyon, Toro Canyon; Quinlan Mountains
***Erigeron divergens*** Torrey & A. Gray
Brown Canyon, Thomas Canyon; Quinlan Mountains
***Erigeron lobatus*** A. Nelson
Quinlan Mountains
***Erigeron oreophilus*** Greenman
Baboquivari Canyon, Brown Canyon, Thomas Canyon, Toro Canyon, Virtud Tank area; Coyote Mountains
***Eriophyllum lanosum*** A. Gray
Brown Canyon; Coyote Mountains; Quinlan Mountains
***Fleischmannia pycnocephala*** (Lessing) King & H. E. Robinson [*Eupatorium pycnocephalum* Lessing]
Baboquivari Canyon, Brown Canyon, Canyon north of Moristo Canyon, South Canyon, Sycamore Canyon, Thomas Canyon, Toro Canyon; Coyote Mountains
***Fleischmannia sonorae*** (A. Gray) King & H. E. Robinson [*Eupatorium sonorae* A. Gray]
Canyon north of Moristo Canyon; Coyote Mountains
***Galinsoga parviflora*** Cavanilles
Baboquivari Peak, Toro Canyon
***Gamochaeta stagnalis*** (I. M. Johnston) A. Anderberg [*Gnaphalium purpureum* Linnaeus, *Gamochaeta purpurea* (Linnaeus) Cabrera misapplied, cf. Nesom 2004]
Baboquivari Canyon, Brown Canyon; Coyote Mountains
***Geraea canescens*** Torrey & A. Gray
Quinlan Mountains
***Gnaphalium wrightii*** A. Gray
Baboquivari Canyon, Brown Canyon, Fresnal Canyon, Thomas Canyon, Ventana Peak E of San Miguel
***Guardiola platyphylla*** A. Gray
Baboquivari Canyon, Brown Canyon, Fresnal Canyon; Quinlan Mountains
***Gutierrezia arizonica*** (A. Gray) Lane
W of Baboquivari Peak; Quinlan Mountains
***Gutierrezia microcephala*** (De Candolle) A. Gray
Baboquivari Canyon, Brown Canyon; Quinlan Mountains
***Gutierrezia sarothrae*** (Pursh) Britton & Rusby
Baboquivari Canyon, Bear Canyon, Brown Canyon, Fresnal Canyon, Jupiter Canyon, Moristo Canyon, South Canyon, Toro Canyon; Quinlan Mountains
***Heliomeris longifolia*** (B. L. Robinson) Cockerell var. ***annua*** (M. E. Jones) Yates [*H. annua* (M. E. Jones) Cockerell, *Viguiera annua* (M. E. Jones) Blake]
Baboquivari Canyon, Brown Canyon, between Fresnal Canyon and Toro Canyon; Pozo Verde Mountains

***Heliomeris multiflora*** Nuttall [*Viguiera multiflora* (Nuttall) Blake]
Baboquivari Peak, Brown Canyon, Cave Canyon, Thomas Canyon; Coyote Mountains
***Heliopsis parvifolia*** A. Gray
Baboquivari Canyon, Toro Canyon
***Heterosperma pinnatum*** Cavanilles
Brown Canyon
***Heterotheca fulcrata*** (Greene) Shinners
South Canyon; Coyote Mountains
***Heterotheca grandiflora*** Nuttall
Toro Canyon; Coyote Mountains; Quinlan Mountains
***Heterotheca subaxillaris*** (Lamarck) Britton & Rusby [*H. psammophila* B. Wagenknecht]
Brown Canyon
***Heterotheca villosa*** (Pursh) Shinners [*Heterotheca viscida* (A. Gray) Harms]
Coyote Mountains; Quinlan Mountains
***Hymenothrix wislizenii*** A. Gray
Brown Canyon, Ventana Peak, E of San Miguel
***Hymenothrix wrightii*** A. Gray
Baboquivari Canyon, Fresnal Canyon; Quinlan Mountains
***Isocoma coronopifolia*** (A. Gray) E. Greene
Fresnal Canyon
***Isocoma tenuisecta*** Greene [*Haplopappus tenuisectus* (Greene) S. F. Blake ex L. D. Benson]
Brown Canyon; Quinlan Mountains
***Koanophyllon solidaginifolium*** (A. Gray) King & H. E. Robinson [*Eupatorium solidaginifolium* A. Gray]
Baboquivari Canyon, Brown Canyon, Dumosa Canyon, Fresnal Canyon, South Canyon, Toro Canyon
***Lactuca graminifolia*** Michaux
Brown Canyon
***Lactuca serriola*** Linnaeus
Brown Canyon
***Laënnecia sophiifolia*** (Kunth) G. L. Nesom [*Conyza sophiifolia* Kunth (as *sophiaefolia*)]
Brown Canyon
***Lagascea decipiens*** Hemsley
Baboquivari Canyon, Brown Canyon, South Canyon, south end of Baboquivari Mountains; Pozo Verde Mountains; Quinlan Mountains
***Lasianthaea podocephala*** (A. Gray) K. Becker
Baboquivari Canyon, Thomas Canyon
***Lasthenia californica*** DeCandolle ex Lindley [*L. chrysostoma* (Fischer & C. A. Meyer) Greene]
Pozo Verde Mountains; Quinlan Mountains
***Logfia californica*** (Nuttall) Holob [*Filago californica* Nuttall]
Baboquivari Mountains
***Machaeranthera asteroides*** (Torrey) Greene
Below Moristo Canyon
***Machaeranthera gracilis*** (Nuttall) Shinners
Brown Canyon, South Canyon
***Machaeranthera tagetina*** Greene
Brown Canyon, Fresnal Canyon, South Canyon, Ventana Peak; Coyote Mountains
***Machaeranthera tanacetifolia*** (Kunth) Nees
Sabino Canyon
***Malacothrix clevelandii*** A. Gray
Brown Canyon; Coyote Mountains
***Malacothrix glabrata*** A. Gray
Ronstadt Ranch; Quinlan Mountains
***Malacothrix stebbinsii*** W. S. Davis & P. H. Raven
Baboquivari Peak, Brown Canyon, Fresnal Canyon, Toro Canyon; Coyote Mountains
***Melampodium longicorne*** A. Gray
Brown Canyon
***Packera neomexicana*** (A. Gray) W. A. Weber and A. Löve [*Senecio neomexicanus* A. Gray]
Brown Canyon
***Parthenice mollis*** A. Gray
Baboquivari Canyon, Brown Canyon, Fresnal Canyon, Thomas Canyon, Toro Canyon, Virtud Tank area; Coyote Mountains
***Pectis rusbyi*** Greene ex A. Gray
Brown Canyon
***Pericome caudata*** A. Gray
Baboquivari Mountains, without locality
***Perityle coronopifolia*** A. Gray
Baboquivari Canyon, Fresnal Canyon, Jaguar Canyon, Jupiter Canyon; Quinlan Mountains
***Porophyllum gracile*** Bentham
Coyote Mountains; Pozo Verde Mountains
***Porophyllum ruderale*** (Jacquin) Cassini subsp. ***macrocephalum*** (De Candolle) R. R. Johnson
Baboquivari Canyon, Brown Canyon, South Canyon, Toro Canyon
***Pseudognaphalium canescens*** (De Candolle) W. A. Weber
Fresnal Canyon, Thomas Canyon
***Pseudognaphalium leucocephalum*** (A. Gray) A. A. Anderberg [*Gnaphalium leucocephalum* A. Gray]
Brown Canyon Wash, Thomas Canyon
***Pseudognaphalium pringlei*** (A. Gray) A. Anderberg [*Gnaphalium pringlei* A. Gray]
Baboquivari Canyon, South Canyon

***Pseudognaphalium stramineum*** (Kunth) Anderberg [*Gnaphalium chilense* Sprengel]
Brown Canyon
***Psilostrophe cooperi*** (A. Gray) Greene
Coyote Mountains
***Rafinesquia californica*** Nuttall
Baboquivari Canyon, Brown Canyon; Coyote Mountains
***Roldana hartwegii*** (Bentham) H. E. Robinson & Brettell [*Senecio carlomasonii* B. L. Turner & T. M. Barkley]
Baboquivari Canyon, Brown Canyon, Moristo Canyon
***Sanvitalia abertii*** A. Gray
Baboquivari Canyon, Brown Canyon, Fresnal Canyon, Jaguar Canyon
***Senecio flaccidus*** Lessing var. ***monoensis*** (Greene) B. L. Turner & T. M. Barkley [*Senecio monoensis* Greene]
Thomas Canyon; Coyote Mountains; Quinlan Mountains
***Senecio lemmoni*** A. Gray
Brown Canyon, Toro Canyon; Coyote Mountains; Pozo Verde Mountains
***Solidago altissima*** Linnaeus
Thomas Canyon
***Solidago canadensis*** Linnaeus
Baboquivari Mountains, without locality
***Solidago velutina*** De Candolle subsp. ***sparsiflora*** (A. Gray) Semple [*Solidago sparsiflora* A. Gray]
Baboquivari Canyon, Brown Canyon, Thomas Canyon; Pozo Verde Mountains; Quinlan Mountains
***Sonchus asper*** (Linnaeus) Hill
Baboquivari Canyon, Brown Canyon; Coyote Mountains
***Sonchus oleraceus*** Linnaeus
Coyote Mountains
***Stephanomeria pauciflora*** (Torrey) A. Nelson in J. M. Coulter and A. Nelson
Brown Canyon
***Stevia lemmoni*** A. Gray
Quinlan Mountains
***Stevia micrantha*** Lagasca
Baboquivari Peak, Toro Canyon
***Stevia serrata*** Cavanilles
Brown Canyon; Quinlan Mountains, Kitt Peak
***Tagetes lemmonii*** A. Gray
Baboquivari Canyon, Brown Canyon, Thomas Canyon; Quinlan Mountains
***Tetradymia canescens*** De Candolle
Baboquivari Canyon
***Thymophylla pentachaeta*** (De Candolle) Small var. ***pentachaeta*** [*Dyssodia pentachaeta* (De Candolle) B. L. Robinson subsp. *pentachaeta*]
Quinlan Mountains
***Tithonia thurberi*** A. Gray
Baboquivari Canyon, Brown Canyon, Toro Canyon
***Trixis californica*** Kellogg
Baboquivari Canyon, Brown Canyon, Fresnal Canyon, Virtud Tank area; Coyote Mountains; Pozo Verde Mountains
***Uropappus lindleyi*** (De Candolle) Nuttall [*Microseris linearifolia* (De Candolle) Schultz-Bipontinus]
Brown Canyon, Fresnal Canyon
***Verbesina rothrockii*** B. L. Robinson and Greenman
Brown Canyon
***Viguiera dentata*** (Cavanilles) Sprengel var. ***lancifolia*** Blake
Baboquivari Canyon, Brown Canyon, South Canyon, Thomas Canyon; Coyote Mountains; Pozo Verde Mountains
***Xanthisma spinulosum*** (Pursh) D. R. Morgan & R. L. Hartman var. ***gooddingii*** (A. Nelson) D. R. Morgan & R. L. Hartman [*Machaeranthera pinnatifida* (Hooker) Shinners]
Brown Canyon, Sabino Canyon
***Xanthium strumarium*** Linnaeus
Brown Canyon; Coyote Mountains
***Zinnia acerosa*** (De Candolle) A. Gray [*Zinnia pumila* A. Gray]
Brown Canyon; Quinlan Mountains

## BERBERIDACEAE

***Berberis haematocarpa*** Wooton
Brown Canyon, south end of Baboquivari Mountains
***Berberis repens*** Lindley
Near base of Baboquivari Peak
***Berberis wilcoxii*** Kearney
Baboquivari Canyon

## BIGNONIACEAE

***Tecoma stans*** (Linnaeus) A. Jusssieu ex Kunth
Baboquivari Canyon, Brown Canyon, Sycamore Canyon; Quinlan Mountains

## BORAGINACEAE

***Amsinckia menziesii*** (Lehmann) A. Nelson & J. F. Macbride var. ***intermedia*** (Fischer & C. A. Meyer) Ganders [*Amsinckia intermedia* C. F. Fischer & C. A. Meyer]
Brown Canyon; Quinlan Mountains
***Cryptantha angustifolia*** (Torrey) Greene
Baboquivari Mountains, without locality; Coyote Mountains
***Cryptantha barbigera*** (A. Gray) Greene
Baboquivari Canyon, Brown Canyon, Fresnal Canyon; Quinlan Mountains
***Cryptantha micrantha*** (Torrey) I. M. Johnston
Coyote Mountains; Pozo Verde Mountains
***Cryptantha pterocarya*** (Torrey) Greene
Toro Canyon; Coyote Mountains; Pozo Verde Mountains
***Heliotropium fruticosum*** Linnaeus
Brown Canyon
***Pectocarya heterocarpa*** (I. M. Johnston) I. M. Johnston
Baboquivari Peak
***Pectocarya recurvata*** I. M. Johnston
Brown Canyon, Thomas Canyon
***Plagiobothrys arizonicus*** (A. Gray) Greene ex A. Gray
Brown Canyon, Fresnal Canyon, Thomas Canyon; Coyote Mountains
***Plagiobothrys pringlei*** Greene
Baboquivari Peak
***Tiquilia canescens*** (De Candolle) A. T. Richardson
Baboquivari Mountains, without locality

## BRASSICACEAE

***Boechera perennans*** (S. Watson) W. A. Weber [*Arabis perennans* S. Watson]
Brown Canyon, Thomas Canyon
***Caulanthus lasiophyllus*** (Hooker & Arnott) Payson [*Thelypodium lasiophyllum* (Hooker & Arnott) Greene]
Brown Canyon; Coyote Mountains
***Descurainia pinnata*** (Walter) Britton
Brown Canyon, Thomas Canyon; Coyote Mountains
***Draba cuneifolia*** Nuttall ex Torrey & A. Gray [*Draba cuneifolia* Nuttall ex Torrey & A. Gray var. *integrifolia* S. Watson, *Draba cuneifolia* Nuttall ex Torrey & A. Gray var. *platycarpa* (Torrey & A. Gray) S. Watson, *Draba platycarpa* Torrey & A. Gray]
Brown Canyon, Thomas Canyon, Toro Canyon
***Draba helleriana*** Greene
Baboquivari Canyon
***Draba petrophila*** Greene
Baboquivari Peak
***Dryopetalon runcinatum*** A. Gray
Baboquivari Peak, Brown Canyon, South Canyon, Thomas Canyon, Toro Canyon; Coyote Mountains; Quinlan Mountains
***Erysimum asperum*** (Nuttall) De Candolle
Brown Canyon; Quinlan Mountains
***Erysimum capitatum*** (Douglas ex Hooker) Greene
Fresnal Canyon, South Canyon; Quinlan Mountains
***Halimolobos diffusa*** (A. Gray) O. E. Schulz
Baboquivari Canyon, Brown Canyon, South Canyon
***Lepidium lasiocarpum*** Nuttall
Coyote Mountains
***Lepidium oblongum*** Small
Thomas Canyon; Coyote Mountains
***Lepidium thurberi*** Wooton
Coyote Mountains; Quinlan Mountains
***Lepidium virginicum*** Linnaeus [*Lepidium medium* Greene, *Lepidium virginicum* var. *medium* (Greene) C. L. Hitchcock]
Brown Canyon, Thomas Canyon, Toro Canyon; Coyote Mountains
***Physaria gordonii*** (A. Gray) O'Kane & Al-Shehbaz [*Lesquerella gordonii* (A. Gray) S. Watson var. *gordonii, Lesquerella tenella* A. Nelson]
Baboquivari Peak
***Schoencrambe linearifolia*** (A. Gray) Rollins [*Thelypodium integrifolium* (Nuttall) Endlicher, *Hesperidanthus linearifolius* (A. Gray) Rydberg]
Baboquivari Peak, Brown Canyon, South Canyon, Thomas Canyon; Quinlan Mountains
***Sisymbrium irio*** Linnaeus
Brown Canyon
***Thlaspi montanum*** Linnaeus [*Thlaspi montanum* Linnaeus var. *fendleri* (A. Gray) P. Holmgren]
Baboquivari Canyon, Brown Canyon
***Thysanocarpus curvipes*** Hooker
Baboquivari Peak, Brown Canyon, Thomas Canyon, Toro Canyon; Coyote Mountains; Quinlan Mountains

## BURSERACEAE

***Bursera fagaroides*** (Kunth) Engelmann var. ***elongata*** McVaugh & Rzedowski
Fresnal Canyon

## CACTACEAE

***Carnegiea gigantea*** (Engelmann) Britton and Rose
Brown Canyon
***Coryphantha vivipara*** (Nuttall) Britton & Rose
Baboquivari Canyon
***Cylindropuntia arbuscula*** (Engelmann) F. M. Knuth [*Opuntia arbuscula* Engelmann]
Quinlan Mountains
***Cylindropuntia fulgida*** (Engelmann) F. M. Knuth var. ***fulgida*** [*Opuntia fulgida* Engelmann var. *fulgida*]
Near Sasabe
***Cylindropuntia fulgida*** (Engelmann) F. M. Knuth var. ***mamillata*** (A. Schott ex Engelmann) Backeberg [*Opuntia fulgida* Engelmann, *Opuntia fulgida* Engelmann var. *mamillata* (A. Schott ex Engelmann) Coulter]
Brown Canyon
***Cylindropuntia leptocaulis*** (De Candolle) F. M. Knuth [*Opuntia leptocaulis* De Candolle]
Brown Canyon
***Cylindropuntia spinosior*** (Engelmann) F. M. Knuth [*Opuntia spinosior* (Engelmann) Toumey]
Brown Canyon; Quinlan Mountains
***Cylindropuntia versicolor*** (Engelmann) F. M. Knuth [*Opuntia versicolor* Engelmann]
Brown Canyon
***Echinocereus coccineus*** Engelmann sensu lato [*Echinocereus polyacanthus* Engelmann, *Echinocereus coccineus* Engelmann, *Echinocereus triglochidiatus* var. *polyacanthus* (Engelmann) L. D. Benson]
Baboquivari Canyon, Brown Canyon; Quinlan Mountains
***Echinocereus fendleri*** (Engelmann) F. Seitz var. ***fasciculatus*** (Engelmann ex B. D. Jackson) N. P. Taylor
Fresnal Canyon; Quinlan Mountains
***Echinocereus fendleri*** (Engelmann) F. Seitz var. ***rectispinus*** (Peebles) L. D. Benson
Fresnal Canyon
***Echinocereus rigidissimus*** (Engelmann) Haage f.
Brown Canyon, Fresnal Canyon
***Ferocactus cylindraceus*** (Engelmann) Orcutt [*Ferocactus acanthodes* sensu Britton & Rose non Lemaire]
Coyote Mountains
***Ferocactus emoryi*** (Engelmann) Orcutt [*Ferocactus covillei* Britton & Rose]
Brown Canyon, Fresnal Canyon, Thomas Canyon
***Ferocactus wislizeni*** (Engelmann) Britton & Rose
Brown Canyon, Fresnal Canyon
***Mammillaria grahamii*** Engelmann var. ***grahamii*** [*Mammillaria microcarpa* Engelmann]
Brown Canyon
***Mammillaria macdougalii*** Rose [*M. heyderi* Muehlenpfordt var. *macdougalii* (Rose) L. D. Benson]
Brown Canyon, Fresnal Canyon; Pozo Verde Mountains
***Opuntia chlorotica*** Engelmann & Bigelow
Brown Canyon
***Opuntia engelmannii*** Salm-Dyck [*O. phaeacantha* var. *discata* (Griffiths) L. D. Benson & Walkington]
Brown Canyon, Fresnal Canyon, South Canyon, Toro Canyon
***Opuntia macrocentra*** Engelmann [*O. violacea* Engelmann ex B. D. Jackson var. *macrocentra* (Engelmann) L. D. Benson]
Coyote Mountains
***Opuntia phaecantha*** Engelmann var. ***major*** Engelmann [*Opuntia gilvescens* Griffiths]
Brown Canyon, Fresnal Canyon, South Canyon, Toro Canyon
***Opuntia santa-rita*** (Griffiths & Hare) Rose
Brown Canyon, Baboquivari Canyon
***Peniocereus striatus*** (Brandegee) Buxbaum
Near San Miguel

## CAMPANULACEAE

***Lobelia cardinalis*** Linnaeus
Brown Canyon, Thomas Canyon
***Triodanis biflora*** (Ruiz & Pavón) Greene
Baboquivari Canyon, Brown Canyon; Coyote Mountains
***Triodanis holzingeri*** McVaugh
South Canyon
***Triodanis perfoliata*** (Linnaeus) Nieuwland
Baboquivari Canyon

## CANNABACEAE

***Celtis pallida*** Torrey
Baboquivari Foothills, Brown Canyon, Fresnal Canyon
***Celtis reticulata*** Torrey
Baboquivari Canyon, Brown Canyon, Fresnal Canyon, Toro Canyon

## CAPRIFOLIACEAE

***Sambucus nigra*** Linnaeus subsp. ***cerulea*** (Rafinesque) R. Bolli [*Sambucus mexicana* C. Presl ex De Candolle]
Brown Canyon
***Symphoricarpos longiflorus*** A. Gray
Baboquivari Peak
***Symphoricarpos rotundifolius*** A. Gray
Quinlan Mountains

## CARYOPHYLLACEAE

***Arenaria lanuginosa*** (Michaux) Rohrbach subsp. ***saxosa*** (A. Gray) Maguire
Baboquivari Peak
***Cerastium brachypodum*** (Engelmann) Robinson [*C. nutans* Rafinesque]
Brown Canyon
***Silene antirrhina*** Linnaeus
Baboquivari Canyon, Brown Canyon, Toro Canyon; Coyote Mountains

## CHENOPODIACEAE

***Atriplex canescens*** (Pursh) Nuttall
Brown Canyon; Coyote Mountains
***Chenopodium berlandieri*** Moquin-Tandon [*C. murale* of literature]
Brown Canyon, Thomas Canyon
***Chenopodium fremontii*** S. Watson
Brown Canyon; Coyote Mountains
***Chenopodium inamoenum*** Standley
Baboquivari Canyon, South Canyon
***Chenopodium neomexicanum*** Standley
Brown Canyon, Sabino Canyon, Toro Canyon
***Dysphania graveolens*** (Willdenow) Mosyakin & Clemants [*Chenopodium graveolens* Willdenow]
Baboquivari Peak, Fresnal Canyon
***Salsola tragus*** Linnaeus
Baboquivari Canyon

## COCHLOSPERMACEAE

***Amoreuxia gonzalezii*** Sprague & Riley
Thomas Canyon
***Amoreuxia palmatifida*** Mociño & Sessé ex De Candolle
Brown Canyon, Fresnal Canyon

## COMMELINACEAE

***Commelina erecta*** Linnaeus
Baboquivari Canyon, Brown Canyon, Thomas Canyon
***Tradescantia occidentalis*** (Britton) Smyth [*Tradescantia occidentalis* var. *scopulorum* (Rose) Anderson & Woodson]
Baboquivari Canyon, Brown Canyon, Fresnal Canyon, South Canyon, Thomas Canyon
***Tradescantia pinetorum*** Greene
Between Baboquivari Canyon and the canyon to the south

## CONVOLVULACEAE

***Cuscuta erosa*** Yuncker
Baboquivari Canyon, Sabino Canyon, Toro Canyon
***Evolvulus alsinoides*** (Linnaeus) Linnaeus
Baboquivari Canyon, Brown Canyon, Thomas Canyon; Coyote Mountains
***Evolvulus arizonicus*** A. Gray
Baboquivari Peak, Brown Canyon, South Canyon
***Evolvulus nuttallianus*** J. A. Schultes [*E. pilosus* Nuttall]
Brown Canyon; Pozo Verde Mountains
***Evolvulus sericeus*** Swartz
Brown Canyon
***Ipomoea barbatisepala*** A. Gray
Brown Canyon, South Canyon, Thomas Canyon
***Ipomoea costellata*** Torrey
Baboquivari Canyon, Brown Canyon, South Canyon, Thomas Canyon, Toro Canyon
***Ipomoea cristulata*** Hallier f. [*Ipomoea coccinea* misapplied]
Baboquivari Canyon, Brown Canyon, Fresnal Canyon, Thomas Canyon, Toro Canyon; Coyote Mountains
***Ipomoea hederacea*** Jacquin
Baboquivari Canyon, Brown Canyon, Thomas Canyon, Toro Canyon; Coyote Mountains
***Ipomoea pubescens*** Lamarck
Brown Canyon, Fresnal Canyon, South Canyon, Toro Canyon
***Ipomoea ternifolia*** Cavanilles var. ***leptotoma*** (Torrey) J. A. McDonald
Baboquivari Camp, Toro Canyon; Coyote Mountains
***Jacquemontia agrestis*** (Martius ex Choisy) Meisner
Baboquivari Canyon, Sabino Canyon

***Jacquemontia pringlei*** A. Gray
Pozo Verde Mountains

## CRASSULACEAE

***Crassula connata*** (Ruiz & Pavón) Berger [*Tillaea erecta* Hooker & Arnott]
Brown Canyon, Fresnal Canyon; Coyote Mountains
***Graptopetalum rusbyi*** (Greene) Rose [*Cotyledon rusbyi* Greene, *Dudleya rusbyi* (Greene) Britton & Rose, *Echeveria rusbyi* (Greene) A. Nelson & J. F. Macbride, *Graptopetalum orpetii* E. Walther]
Brown Canyon
***Sedum cockerellii*** Britton
Baboquivari Canyon, South Canyon #2

## CROSSOSOMATACEAE

***Crossosoma bigelovii*** S. Watson
Baboquivari Peak; Coyote Mountains; Quinlan Mountains

## CUCURBITACEAE

***Apodanthera undulata*** A. Gray
Brown Canyon, Jaguar Canyon
***Cucurbita digitata*** A. Gray
Baboquivari Canyon, Brown Canyon; Quinlan Mountains
***Cyclanthera dissecta*** (Torrey & Gray) Arnott
Moristo Canyon, South Canyon, Sycamore Canyon, Toro Canyon
***Echinopepon wrightii*** (A. Gray) S. Watson
Baboquivari Camp, Brown Canyon, Fresnal Canyon, Moristo Canyon, Toro Canyon; Coyote Mountains
***Marah gilensis*** Greene
Brown Canyon, South Canyon; Coyote Mountains; Quinlan Mountains
***Sicyos ampelophyllus*** Wooton & Standley
Baboquivari Canyon
***Sicyosperma gracile*** A. Gray
Baboquivari Camp, Brown Canyon, South Canyon, Sycamore Canyon, Thomas Canyon, Toro Canyon

## CUPRESSACEAE

***Juniperus deppeana*** Steudel
Brown Canyon

## CYPERACEAE

***Bulbostylis capillaris*** (Linnaeus) Kunth ex C. B. Clarke
Baboquivari Canyon, between Fresnal and Toro Canyons
***Bulbostylis juncoides*** (Vahl) Kükenthal
Between Fresnal and Toro Canyons
***Carex agrostoides*** Mackenzie
Moristo Canyon
***Carex petasata*** Dewey
Baboquivari Peak
***Carex spissa*** Kük
Brown Canyon, Jaguar Canyon
***Cyperus aggregatus*** (Willdenow) Endlicher [*C. huamarensis* (Kunth) M. C. Johnston, *Cyperus flavus* (Vahl) Nees]
Brown Canyon, Thomas Canyon; Quinlan Mountains
***Cyperus aristatus*** Rottboell var. ***inflexus*** Rottboell
Coyote Mountains
***Cyperus esculentus*** Linnaeus [*Cyperus esculentus* Linnaeus var. *leptostachyus* Boeckeler]
Fresnal Canyon; Coyote Mountains
***Cyperus flavicomus*** Michaux [*Cyperus albomarginatus* (Martius & Schrader ex Nees) Steudel]
Brown Canyon; Coyote Mountains
***Cyperus mutisii*** (Kunth) Grisebach [*Cyperus pringlei* Britton]
Baboquivari Peak, Brown Canyon, Moristo Canyon, South Canyon, Thomas Canyon, Toro Canyon; Coyote Mountains
***Cyperus odoratus*** Linnaeus
Sabino Canyon; Coyote Mountains
***Cyperus squarrosus*** Linnaeus [*C. aristatus* Rottboell]
Baboquivari Canyon, Brown Canyon, Sabino Canyon; Coyote Mountains
***Lipocarpha micrantha*** (Vahl) G. Tucker [*Hemicarpha micrantha* (Vahl) Pax]
Brown Canyon

## DRYOPTERIDACEAE

***Dryopteris filix-mas*** (Linnaeus) Schott
Thomas Canyon
***Phanerophlebia auriculata*** Underwood
Baboquivari Canyon, Brown Canyon

## EPHEDRACEAE

***Ephedra trifurca*** Torrey ex S. Watson
Brown Canyon

## EQUISETACEAE

***Equisetum hiemale*** Linnaeus
Thomas Canyon

## EUPHORBIACEAE

***Acalypha californica*** Bentham
South end of Baboquivaris, west slope of Baboquivaris
***Acalypha neomexicana*** Mueller-Argoviensis
Baboquivari Canyon, Brown Canyon, Sabino Canyon, Toro Canyon
***Chamaesyce albomarginata*** (Torrey & A. Gray) Small [*Euphorbia albomarginata* Torrey & A. Gray]
Brown Canyon
***Chamaesyce arizonica*** (Engelmann) Arthur [*Euphorbia arizonica* Engelmann]
Baboquivari Peak, South Canyon, Toro Canyon
***Chamaesyce capitellata*** (Engelmann) Millspaugh [*Euphorbia capitellata* Engelmann]
Baboquivari Canyon, between Fresnal and Toro Canyons; Coyote Mountains
***Chamaesyce florida*** (Engelmann) Millspaugh [*Euphorbia florida* Engelmann]
Baboquivari Canyon; Coyote Mountains; Quinlan Mountains
***Chamaesyce gracillima*** (S. Watson) Millspaugh [*Euphorbia gracillima* S. Watson]
Between Fresnal and Toro Canyons
***Chamaesyce hirta*** (Linnaeus) Millspaugh [*Euphorbia hirta* Linnaeus]
Brown Canyon, Fresnal Canyon, Thomas Canyon
***Chamaesyce hyssopifolia*** (Linnaeus) Small [*Euphorbia hyssopifolia* Linnaeus]
Baboquivari Camp, Brown Canyon, Santa Margarita Ranch
***Chamaesyce melanadenia*** (Torrey) Millspaugh [*Euphorbia melanadenia* Torrey]
Baboquivari Canyon, Brown Canyon, Sabino Canyon, Thomas Canyon; Quinlan Mountains
***Chamaesyce polycarpa*** (Bentham) Millspaugh ex Parish [*Euphorbia polycarpa* Bentham]
Coyote Mountains
***Euphorbia exstipulata*** Engelmann
Baboquivari Peak, Fresnal Canyon
***Jatropha cardiophylla*** (Torrey) Mueller-Argoviensis
Baboquivari Canyon, Brown Canyon; Quinlan Mountains
***Jatropha macrorhiza*** Bentham
Brown Canyon, Thomas Canyon
***Manihot angustiloba*** (Torrey) Mueller-Argoviensis
Baboquivari Canyon, Brown Canyon, Fresnal Canyon, Thomas Canyon; Coyote Mountains; Quinlan Mountains
***Manihot davisiae*** Croizat
Baboquivari Canyon, Brown Canyon
***Poinsettia cuphosperma*** (Engelmann) Small [*Euphorbia cuphosperma* (Engelmann) Boissier]
Brown Canyon, Toro Canyon
***Poinsettia heterophylla*** (Linnaeus) Klotzsch & Garcke [*Euphorbia heterophylla* Linnaeus]
Baboquivari Canyon, Brown Canyon, Thomas Canyon, Toro Canyon; Coyote Mountains
***Tragia laciniata*** (Torrey) Mueller-Argoviensis
Brown Canyon
***Tragia nepetaefolia*** Cavanilles
Baboquivari Canyon, Brown Canyon, Thomas Canyon, Toro Canyon; Quinlan Mountains

## FABACEAE

***Acacia angustissima*** (Miller) Kuntze var. ***suffrutescens*** (Rose) Isely
Baboquivari Canyon, Brown Canyon, Fresnal Canyon, Moristo Canyon, South Canyon, Thomas Canyon, Toro Canyon; Coyote Mountains; Quinlan Mountains
***Acacia farnesiana*** (Linnaeus) Willdenow
Baboquivari Canyon, Brown Canyon, Thomas Canyon
***Acacia greggii*** A. Gray
Brown Canyon; Quinlan Mountains
***Amorpha fruticosa*** Linnaeus
Brown Canyon
***Astragalus allochrous*** A. Gray var. ***playanus*** Isely [*Astragalus wootonii* Sheldon]
Coyote Mountains
***Astragalus arizonicus*** A. Gray
Brown Canyon, Toro Canyon; Coyote Mountains; Quinlan Mountains
***Astragalus lentiginosus*** Douglas ex Hooker
Coyote Mountains; Quinlan Mountains
***Astragalus nothoxys*** A. Gray
Brown Canyon, Thomas Canyon, Toro Canyon

***Astragalus nuttallianus*** De Candolle
Thomas Canyon, Toro Canyon
***Calliandra eriophylla*** Bentham
Baboquivari Canyon, Brown Canyon, Fresnal Canyon; Quinlan Mountains
***Chamaecrista absus*** (Linnaeus) H. S. Irwin & Barneby [*Cassia absus* Linnaeus]
Cave Canyon, Baboquivari Canyon, Montezuma Bluff
***Chamaecrista nicitans*** (Linnaeus) Moench [*Cassia leptadenia* Greenman]
Fresnal Canyon, Thomas Canyon; Coyote Mountains
***Clitoria mariana*** Linnaeus
Baboquivari Canyon, Montezuma Bluff
***Cologania angustifolia*** Kunth
Brown Canyon
***Coursetia caribaea*** (Jacquin) Lavin var. ***sericea*** (A. Gray) Lavin
Brown Canyon, Jaguar Canyon
***Coursetia glandulosa*** A. Gray [*C. microphylla* A. Gray]
Baboquivari Peak, Brown Canyon, Fresnal Canyon; Coyote Mountains; Pozo Verde Mountains
***Dalea albiflora*** A. Gray
Bear Canyon, Fresnal Canyon, Thomas Canyon; Quinlan Mountains
***Dalea grayi*** (Vail) L. O. Williams
Fresnal Canyon
***Dalea lumholtzii*** B. L. Robinson & Fernald
Baboquivari Canyon, Toro Canyon
***Dalea pogonathera*** A. Gray
Brown Canyon, Forest Cabin Canyon, Fresnal Canyon, Sabino Canyon
***Dalea pringlei*** A. Gray
Bear Canyon, Brown Canyon, Fresnal Canyon
***Dalea pulchra*** Gentry
Brown Canyon, Fresnal Canyon, south end of Baboquivaris; Coyote Mountains; Pozo Verde Mountains; Quinlan Mountains
***Dalea tentaculoides*** Gentry
Baboquivari Canyon, Toro Canyon; Coyote Mountains, W Mendoza Canyon; Quinlan Mountains
***Dalea versicolor*** Zuccarini
Thomas Canyon; Coyote Mountains; Quinlan Mountains
***Desmanthus covillei*** (Britton & Rose) Wiggins ex B. L. Turner var. ***arizonicus*** B. L. Turner
Fresnal Canyon
***Desmodium angustifolium*** (Kunth) De Candolle
Baboquivari Canyon, Fresnal Canyon
***Desmodium batocaulon*** A. Gray
Brown Canyon, South Canyon; Quinlan Mountains
***Desmodium grahami*** A. Gray
Brown Canyon
***Desmodium procumbens*** (Miller) Hitchcock
Brown Canyon
***Desmodium psilocarpum*** A. Gray
Baboquivari Canyon, Brown Canyon, Fresnal Canyon, Sabino Canyon, Toro Canyon
***Desmodium psilophyllum*** Schlect
Baboquivari Canyon, Brown Canyon
***Desmodium rosei*** Schubert
Brown Canyon, Toro Canyon
***Erythrina flabelliformis*** Kearney
Baboquivari Canyon, Brown Canyon; Coyote Mountains; Quinlan Mountains
***Eysenhardtia orthocarpa*** (A. Gray) S. Watson
Baboquivari Canyon, Brown Canyon, Stone House Canyon, Thomas Canyon; Coyote Mountains
***Galactia wrightii*** A. Gray
Baboquivari Canyon, Brown Canyon, South Canyon, Thomas Canyon; Coyote Mountains
***Indigofera sphaerocarpa*** A. Gray
Baboquivari Peak, Thomas Canyon; Quinlan Mountains
***Lotus humistratus*** Greene [*Hosackia rosea* Eastwood]
Baboquivari Canyon, Fresnal Canyon, Thomas Canyon
***Lotus plebeius*** (Brand) Barneby [*Lotus oroboides* (Kunth) Ottley, *Hosackia puberula* Bentham var. *nana* A. Gray]
Baboquivari Peak, Brown Canyon, Fresnal Canyon, South Canyon, Toro Canyon; Quinlan Mountains
***Lotus rigidus*** (Bentham) Greene [*Hosackia rigida* Bentham]
Thomas Canyon
***Lotus wrightii*** (A. Gray) Greene [*Hosackia wrightii* A. Gray]
Thomas Canyon
***Lupinus concinnus*** Agardh
Brown Canyon, Thomas Canyon; Coyote Mountains; Quinlan Mountains

***Lupinus sparsiflorus*** Bentham
Baboquivari Canyon, Brown Canyon, Fresnal Canyon, Toro Canyon; Pozo Verde Mountains; Quinlan Mountains
***Macroptilium atropurpureum*** (Mociño & Sessé ex De Candolle) Urban
Fresnal Canyon; Coyote Mountains
***Macroptilium gibbosifolium*** (Ortega) A. Delgado
Brown Canyon
***Mimosa aculeaticarpa*** Ortega var. ***biuncifera*** (Bentham) Barneby [*M. biuncifera* Bentham]
Brown Canyon
***Mimosa dysocarpa*** Bentham var. ***wrightii*** (A. Gray) Kearney & Peebles
Brown Canyon, Fresnal Canyon, Sycamore Canyon, Toro Canyon
***Mimosa grahamii*** A. Gray var. ***lemmonii*** (A. Gray) Kearney & Peebles
Baboquivari Canyon, Brown Canyon
***Nissolia schottii*** (Torrey) A. Gray
Baboquivari Canyon, Brown Canyon; Coyote Mountains
***Parkinsonia aculeata*** Linnaeus
Brown Canyon, Fresnal Canyon; Coyote Mountains; Quinlan Mountains
***Parkinsonia florida*** (A. Gray ex A. Gray) S. Watson [*Cercidium floridum* A. Gray ex A. Gray]
West side on Baboquivari Road
***Parkinsonia microphylla*** Torrey [*Cercidium microphyllum* (Torrey) Rose & I. M. Johnston]
West side on Baboquivari Road; Quinlan Mountains
***Phaseolus acutifolius*** A. Gray
Baboquivari Canyon, Brown Canyon, Fresnal Canyon, Thomas Canyon, Toro Canyon; Coyote Mountains; Quinlan Mountains
***Phaseolus maculatus*** Scheele
Baboquivari Canyon; Quinlan Mountains
***Prosopis velutina*** Wooton
Brown Canyon
***Psoralidium tenuiflorum*** (Pursh) Rydberg [*Psoralea tenuiflora* Pursh]
Cave Canyon
***Robinia neomexicana*** A. Gray
Quinlan Mountains
***Senna lindheimeriana*** (Scheele ex Schlectendal) H. S. Irwin & Barneby [*Cassia leptocarpa* Bentham, *Senna hirsuta* (Linnaeus) H. S. Irwin & Barneby]
Brown Canyon, Thomas Canyon, Toro Canyon
***Sphinctospermum constrictum*** (S. Watson) Rose
Fresnal Canyon; Coyote Mountains
***Tephrosia leiocarpa*** A. Gray
Baboquivari trail, Brown Canyon, Virtud tank area
***Tephrosia tenella*** A. Gray
Baboquivari Peak, Fresnal Canyon; Coyote Mountains
***Vicia exigua*** Nuttall
Baboquivari Peak, Fresnal Canyon, Thomas Canyon; Coyote Mountains
***Vicia ludoviciana*** Nuttall
Baboquivari Peak, Brown Canyon
***Zapoteca formosa*** (Kunth) H. M. Hernández ssp. ***schottii*** (Torrey ex S. Watson) H. M. Hernández [*Calliandra schottii* Torrey]
Baboquivari Canyon, Fresnal Canyon, Moristo Canyon, South Canyon; Coyote Mountains
***Zornia diphylla*** (Linnaeus) Persoon
Baboquivari Canyon, Brown Canyon; Quinlan Mountains

## FAGACEAE

***Quercus arizonica*** Sargent
Brown Canyon; Quinlan Mountains
***Quercus chrysolepis*** Liebmann
Quinlan Mountains
***Quercus emoryi*** Torrey
Baboquivari Canyon, Brown Canyon, Spring Canyon, Thomas Canyon
***Quercus grisea*** Liebmann
Baboquivari Canyon, Brown Canyon, Thomas Canyon; Quinlan Mountains
***Quercus hypoleucoides*** A. Camus
Baboquivari Canyon, Brown Canyon; Quinlan Mountains
***Quercus oblongifolia*** Torrey
Baboquivari Canyon, Brown Canyon, Moristo Canyon, Spring Canyon, Thomas Canyon; Coyote Mountains; Quinlan Mountains
***Quercus turbinella*** Greene
Spring Canyon

## FOUQUIERIACEAE

***Fouquieria splendens*** Engelmann subsp. ***splendens***
Brown Canyon

## FUMARIACEAE

***Corydalis aurea*** Willdenow subsp. ***occidentalis*** (Engelmann ex A. Gray) G. B. Ownbey
Baboquivari Canyon, Brown Canyon, Moristo Canyon, Thomas Canyon, Toro Canyon

## GARRYACEAE

***Garrya wrightii*** Torrey
Baboquivari Canyon, Brown Canyon; Coyote Mountains; Quinlan Mountains

## GENTIANACEAE

***Centaurium calycosum*** (Buckley) Fernald
Brown Canyon; Coyote Mountains
***Centaurium nudicaule*** (Engelmann) B. L. Robinson
Baboquivari Mountains, without locality

## GERANIACEAE

***Erodium cicutarium*** (Linnaeus) L'Héritier ex Aiton
Brown Canyon
***Erodium texanum*** A. Gray
Brown Canyon

## HYDRANGEACEAE

***Fendlera rupicola*** A. Gray
Baboquivari Canyon, Brown Canyon, South Canyon, Spring Canyon, Thomas Canyon; Coyote Mountains; Quinlan Mountains

## HYDROPHYLLACEAE

***Eucrypta micrantha*** (Torrey) Heller
Brown Canyon; Quinlan Mountains
***Nama hispidum*** A. Gray
Coyote Mountains
***Phacelia affinis*** A. Gray
Brown Canyon, Fresnal Canyon, Toro Canyon
***Phacelia arizonica*** A. Gray
Coyote Mountains; Quinlan Mountains
***Phacelia coerulea*** Greene
Brown Canyon, Toro Canyon
***Phacelia congesta*** Hooker
Quinlan Mountains
***Phacelia distans*** Bentham
Baboquivari Canyon, Brown Canyon, Fresnal Canyon, Toro Canyon; Coyote Mountains; Quinlan Mountains
***Phacelia ramosissima*** Douglas ex Lehmann
Brown Canyon, Fresnal Canyon; Coyote Mountains; Quinlan Mountains

## JUGLANDACEAE

***Juglans major*** (Torrey) Heller
Baboquivari Canyon, Brown Canyon

## JUNCACEAE

***Juncus bufonius*** Linnaeus
Coyote Mountains; Pozo Verde Mountains
***Juncus marginatus*** Rostkovius
Coyote Mountains
***Juncus saximontanus*** A. Nelson [*J. ensifolius* Wikström var. *brunnescens* (Rydberg) Cronquist]
Coyote Mountains
***Juncus tenuis*** Willdenow
Coyote Mountains; Quinlan Mountains

## KRAMERIACEAE

***Krameria erecta*** Willdenow ex J. A. Schultes [*K. parvifolia* Bentham var. *imparata* J. F. Macbride]
Baboquivari Canyon, Brown Canyon, Fresnal Canyon; Quinlan Mountains

## LAMIACEAE

***Agastache breviflora*** (A. Gray) Epling
Brown Canyon
***Agastache rupestris*** (Greene) Standley
Baboquivari Peak; Quinlan Mountains
***Agastache wrightii*** (Greenman) Wooton & Standley
Baboquivari Canyon, Brown Canyon, Sycamore Canyon; Coyote Mountains
***Hedeoma dentatum*** Torrey
Baboquivari Peak, Brown Canyon, Moristo Canyon, South Canyon, Thomas Canyon, Toro Canyon
***Hedeoma oblongifolium*** (A. Gray) Heller
Fresnal Canyon, Jupiter Canyon, Thomas Canyon
***Hyptis emoryi*** Torrey
Baboquivari Canyon; Coyote Mountains; Quinlan Mountains

***Lamium amplexicaule*** Linnaeus
Brown Canyon
***Mentha arvensis*** Linnaeus
Brown Canyon
***Mentha spicata*** Linnaeus [*M. viridis* (Linnaeus) Linnaeus]
Brown Canyon
***Monarda citriodora*** Cervantes ex Lagasca subsp. ***austromontana*** (Epling) Scora [*Monarda austromontana* Epling]
Baboquivari Canyon, Brown Canyon
***Monarda pectinata*** Nuttall
Baboquivari Canyon
***Salvia columbariae*** Bentham
Coyote Mountains
***Salvia subincisa*** Bentham
Fresnal Canyon
***Scutellaria potosina*** T. S. Brandegee [*Scutellaria potosina* var. *tessellata* (Epling) B. L. Turner]
Baboquivari Canyon, Brown Canyon, Toro Canyon
***Stachys coccinea*** Jacquin
Baboquivari Canyon, Brown Canyon, Fresnal Canyon, Thomas Canyon; Coyote Mountains; Quinlan Mountains

## LEMNACEAE

***Lemna aequinoctialis*** Philippi
Coyote Mountains
***Lemna gibba*** Linnaeus
Coyote Mountains
***Lemna minor*** Linnaeus
Coyote Mountains
***Lemna valdiviana*** Philippi
Coyote Mountains

## LILIACEAE

***Calochortus ambiguus*** (M. E. Jones) Owenby
Pozo Verde Mountains
***Calochortus kennedyi*** Porter var. ***munzii*** Jepson
Brown Canyon

## LOASACEAE

***Mentzelia albicaulis*** (Douglas ex Hooker) Douglas ex Torrey & A. Gray
Brown Canyon, Thomas Canyon; Coyote Mountains; Quinlan Mountains
***Mentzelia asperula*** Wooton & Standley
Brown Canyon
***Mentzelia isolata*** H. S. Gentry
Baboquivari Canyon, Fresnal Canyon, South Canyon, Toro Canyon
***Mentzelia montana*** (A. Davidson) A. Davidson
Baboquivari Canyon, Fresnal Canyon, Toro Canyon
***Mentzelia pumila*** Nuttall ex Torrey & A. Gray
Brown Canyon

## LOGANIACEAE

***Buddleja sessiliflora*** Kunth
Coyote Mountains; Quinlan Mountains

## LYTHRACEAE

***Cuphea wrightii*** A. Gray
Brown Canyon
***Lythrum californicum*** Torrey & A. Gray
Thomas Canyon; Coyote Mountains

## MALPIGHIACEAE

***Cottsia gracilis*** (A. Gray) W. R. Anderson & C. Davis [*Janusia gracilis* A. Gray]
Brown Canyon, Moristo Canyon

## MALVACEAE

***Abutilon abutiloides*** A. Gray
Baboquivari Canyon, Bear Canyon, Fresnal Canyon; Coyote Mountains; Quinlan Mountains
***Abutilon incanum*** (Link) Sweet
Baboquivari Canyon, Brown Canyon, Fresnal Canyon, south end of Baboquivaris, Virtud Tank area; Coyote Mountains; Quinlan Mountains
***Abutilon malacum*** S. Watson
Brown Canyon
***Abutilon mollicomum*** (Willdenow) Sweet
Baboquivari Mountains, without locality; Coyote Mountains
***Abutilon parishii*** S. Watson
Coyote Mountains
***Abutilon reventum*** S. Watson
Brown Canyon, South Canyon; Coyote Mountains
***Abutilon sonorae*** A. Gray
Baboquivari Canyon, Brown Canyon, South Canyon, Toro Canyon; Coyote Mountains

***Abutilon thurberi*** A. Gray
Baboquivari Canyon, Fresnal Canyon
***Anoda abutiloides*** A. Gray
Baboquivari Canyon, Brown Canyon, Fresnal Canyon, South Canyon #2, South Canyon, Sycamore Canyon, Thomas Canyon, Toro Canyon
***Anoda cristata*** (Linnaeus) Schlechtendal
Baboquivari Canyon, Brown Canyon
***Ayenia filiformis*** S. Watson [*A. compacta* Linnaeus, *A. pusilla* Linnaeus]
Brown Canyon, South Canyon, Thomas Canyon; Quinlan Mountains
***Gossypium thurberi*** Todaro
Brown Canyon, Fresnal Canyon; Quinlan Mountains
***Herissantia crispa*** (Linnaeus) Brizicky
Baboquivari Canyon, Brown Canyon; Coyote Mountains
***Hibiscus biseptus*** S. Watson
Baboquivari Canyon, Brown Canyon, Fresnal Canyon, Sabino Canyon, South Canyon, Thomas Canyon, Toro Canyon; Coyote Mountains
***Hibiscus coulteri*** Harvey ex A. Gray
Baboquivari Canyon, Brown Canyon, Fresnal and Toro Canyons; Coyote Mountains; Pozo Verde Mountains
***Hibiscus denudatus*** Bentham var. ***denudatus***
Coyote Mountains
***Malva parviflora*** Linnaeus
Thomas Canyon
***Rhynchosida physocalyx*** (A. Gray) Fryxell
Baboquivari Canyon, Fresnal Canyon, Sycamore Canyon, Toro Canyon
***Sida abutifolia*** P. Miller [*S. procumbens* Swartz]
Baboquivari Canyon, Brown Canyon, Fresnal Canyon, Moristo Canyon, Rondstadt Ranch
***Sphaeralcea fendleri*** A. Gray var. ***venusta*** Kearney
Brown Canyon; Quinlan Mountains
***Waltheria detonsa*** A. Gray
Baboquivari Canyon above Forest Cabin Canyon, Fresnal Canyon
***Waltheria indica*** Linnaeus [*W. americana* Linnaeus]
Baboquivari Canyon, Fresnal Canyon, Ventana Peak, E of San Miguel; Coyote Mountains; Quinlan Mountains

## MARSILEACAEAE

***Marsilea vestita*** Hooker & Greville
Presumido Pass

## MARTYNIACEAE

***Proboscidea althaeifolia*** (Bentham) Decaisne
Baboquivari Canyon; Coyote Mountains
***Proboscidea parviflora*** (Wooton) Wooton & Standley [*Martynia fragrans* Lindley]
Brown Canyon, Fresnal Canyon

## MENISPERMACEAE

***Cocculus diversifolius*** De Candolle
Baboquivari Canyon, Brown Canyon, Fresnal Canyon, South Canyon, Thomas Canyon

## MOLLUGINACEAE

***Mollugo cerviana*** (Linnaeus) Seringe
Coyote Mountains
***Mollugo verticellata*** Linnaeus
Baboquivari Canyon, Brown Canyon, Sabino Canyon, Sycamore Canyon; Coyote Mountains

## MORACEAE

***Morus microphylla*** Buckley
Baboquivari Canyon, Brown Canyon, Fresnal Canyon, Moristo Canyon; Coyote Mountains

## NYCTAGINACEAE

***Allionia incarnata*** Linnaeus [*A. incarnata* Linnaeus var. *nudata* (Standley) Munz, *A. incarnata* Linnaeus var. *villosa* (Standley) B. L. Turner]
Brown Canyon; Coyote Mountains; Quinlan Mountains
***Boerhavia coccinea*** Miller [*B. caribaea* Jacquin, *B. diffusa* misapplied]
Baboquivari Canyon, Brown Canyon; Coyote Mountains; Quinlan Mountains
***Boerhavia coulteri*** (Hooker) S. Watson var. ***palmeri*** (S. Watson) Spellenberg [*B. watsonii* Standley]
Coyote Mountains
***Boerhavia erecta*** Linnaeus
Baboquivari Canyon, Brown Canyon, Sabino Canyon
***Boerhavia gracillima*** Heimerl
Baboquivari Canyon; Coyote Mountains
***Boerhavia megaptera*** Standley
Baboquivari Canyon

***Boerhavia spicata*** Choisy
Baboquivari Canyon
***Commicarpus scandens*** (Linnaeus) Standley [*Boerhavia scandens* Linnaeus]
Baboquivari Canyon, Brown Canyon, Fresnal Canyon, South Canyon; Coyote Mountains
***Mirabilis albida*** (Walter) Heimerl, sensu lato
Brown Canyon, Fresnal Canyon, Toro Canyon; Coyote Mountains; Quinlan Mountains
***Mirabilis laevis*** (Bentham) Curran var. ***villosa*** (Kellogg) Spellenberg
Coyote Mountains
***Mirabilis linearis*** (Pursh) Heimerl [*Oxybaphus linearis* (Pursh) B. L. Robinson]
Baboquivari Mountains, without locality, Brown Canyon
***Mirabilis longiflora*** Linnaeus
Baboquivari Canyon, Brown Canyon, Fresnal Canyon, Thomas Canyon, Toro Canyon; Quinlan Mountains

## OLEACEAE

***Forestiera phillyreoides*** (Bentham) Torrey [*F. shrevei* Standley]
Baboquivari Canyon, Brown Canyon, Sabino Canyon, Toro Canyon
***Fraxinus velutina*** Torrey [*F. pennsylvanica* Marsh var. *velutina* (Torrey) G. N. Miller]
Baboquivari Canyon, Brown Canyon, Fresnal Canyon

## ONAGRACEAE

***Camissonia californica*** (Nuttall ex Torrey & A. Gray) P. H. Raven [*Oenothera leptocarpa* Greene]
Brown Canyon; Coyote Mountains; Pozo Verde Mountains; Quinlan Mountains
***Camissonia chamaenerioides*** (A. Gray) P. H. Raven
Pozo Verde Mountains; Quinlan Mountains
***Camissonia claviformis*** (Torrey & Frémont) P. H. Raven subsp. ***peeblesii*** (Munz) P. H. Raven [*Oenothera claviformis* subsp. *peeblesii* (Munz) P. H. Raven, *Oenothera claviformis* var. peeblesii Munz]
Coyote Mountains; Quinlan Mountains
***Epilobium canum*** (Greene) P. H. Raven subsp. ***latifolium*** (Hooker) P. H. Raven [*Zauschneria californica* J. Presl subsp. *latifolium* (Hooker) D. D. Keck]
Baboquivari Canyon, Brown Canyon, Sycamore Canyon
***Oenothera elata*** Kunth subsp. ***hirsutissima*** (A. Gray ex S. Watson) W. Dietrich [*O. hookeri* Torrey & A. Gray]
Baboquivari Canyon, South Canyon, Thomas Canyon
***Oenothera hexandra*** (Ortega) W. L. Wagner & Hock subsp. ***gracilis*** (Wooton & Standley) W. L. Wagner & Hock [*Gaura hexandra* Ortega subsp. *gracilis* Wooton & Standley]
South Canyon
***Oenothera primiveris*** A. Gray
Brown Canyon, Fresnal Canyon, Thomas Canyon

## OROBANCHACEAE

***Orobanche cooperi*** (A. Gray) Heller
Baboquivari Canyon

## OXALIDACEAE

***Oxalis albicans*** Kunth [*O. corniculata* sensu authors]
Baboquivari Canyon, Brown Canyon; Coyote Mountains
***Oxalis alpina*** (Rose) Knuth
Brown Canyon
***Oxalis stricta*** Linnaeus [*O. corniculata* sensu authors]
Brown Canyon, Thomas Canyon

## PAPAVERACEAE

***Argemone gracilenta*** Greene
Coyote Mountains
***Argemone pleiacantha*** Greene subsp. ***pleiacantha***
Baboquivari Canyon, Brown Canyon, Thomas Canyon
***Eschscholzia californica*** Chamisso subsp. ***mexicana*** (Greene) C. Clark [*Eschscholzia mexicana* Greene]
Brown Canyon; Quinlan Mountains
***Platystemon californicus*** Bentham
Brown Canyon, Sabino Canyon; Quinlan Mountains

## PASSIFLORACEAE

***Passiflora arizonica*** (Killip) D. H. Goldman [*P. foetida* Linnaeus var. *arizonica* Killip]
Baboquivari Canyon, Brown Canyon, Forest Canyon, Fresnal Canyon, Thomas Canyon, Virtud Tank area; Coyote Mountains; Pozo Verde Mountains
***Passiflora mexicana*** Jussieu
Brown Canyon; Coyote Mountains; Quinlan Mountains

## PHRYMACEAE [sensu Scrophulariaceae of authors]

***Mimulus cardinalis*** Douglas
Baboquivari Canyon, Jaguar Canyon, Thomas Canyon, Moristo Canyon, Toro Canyon
***Mimulus guttatus*** De Candolle
Brown Canyon, Moristo Canyon, Spring Canyon south of Fresnal Canyon, Thomas Canyon; Quinlan Mountains
***Mimulus rubellus*** A. Gray
Baboquivari Canyon, Brown Canyon, Thomas Canyon, Toro Canyon; Coyote Mountains; Pozo Verde Mountains; Quinlan Mountains

## PHYTOLACCACEAE

***Rivina humilis*** Linnaeus
Baboquivari Canyon, Brown Canyon, Fresnal Canyon, Jupiter Canyon, Thomas Canyon, Toro Canyon

## PINACEAE

***Pinus discolor*** D. K. Bailey & Hawksworth [*P. cembroides* of authors not Zuccarini]
Baboquivari Canyon, Brown Canyon, Thomas Canyon; Coyote Mountains; Quinlan Mountains

## PLANTAGINACEAE

***Mecardonia procumbens*** (P. Miller) Small [*Mecardonia vandellioides* (Kunth) Pennell]
Brown Canyon, Fresnal Canyon, Toro Canyon
***Penstemon barbatus*** (Cavanilles) Roth
Baboquivari Canyon, Brown Canyon, Thomas Canyon
***Penstemon parryi*** A. Gray
Brown Canyon, Forest Cabin Canyon, Fresnal Canyon, Moristo Canyon, Thomas Canyon
***Plantago patagonica*** Jacquin [*P. purshii* Roemer & Schultes]
Brown Canyon, Thomas Canyon; Coyote Mountains; Pozo Verde Mountains; Quinlan Mountains
***Plantago virginica*** Linnaeus
Brown Canyon; Coyote Mountains
***Schistophragma intermedia*** (A. Gray) Pennell [*Leucospora intermedia* (A. Gray) Keil]
Brown Canyon, Baboquivari Canyon, Fresnal Canyon, Toro Canyon
***Stemodia durantifolia*** (Linnaeus) Swartz
Brown Canyon; Mendoza Canyon, Coyote Mountains

## PLATANACEAE

***Platanus wrightii*** S. Watson
Bear Canyon, Brown Canyon, Toro Canyon

## PLUMBAGINACEAE

***Plumbago zeylanica*** Linnaeus [*P. scandens* Linnaeus]
Baboquivari Canyon, Brown Canyon, Fresnal Canyon, Mendoza Canyon; Coyote Mountains; Quinlan Mountains

## POACEAE

***Agrostis exarata*** Trinius
Forest Cabin Canyon
***Agrostis scabra*** Willdenow
Forest Cabin Canyon
***Aristida adscensionis*** Linnaeus
M. S. Ranch (north end of Baboquivari Mountains), Sabino Canyon
***Aristida californica*** Thurber ex S. Watson var. ***glabrata*** Vasey
Quinlan Mountains
***Aristida laxa*** Cavanilles
Thomas Canyon; Quinlan Mountains
***Aristida purpurea*** Nuttall var. ***parishii*** (A. S. Hitchcock) Allred
Coyote Mountains
***Aristida ternipes*** Cavanilles var. ***ternipes***
Baboquivari Peak, Brown Canyon, South Canyon; Coyote Mountains; Quinlan Mountains

***Blepharoneuron tricholepis*** (Torrey) Nash
Toro Canyon
***Bothriochloa barbinoides*** (Lagasca) Herter [*Andropogon barbinodis* Lagasca]
Baboquivari Canyon, Brown Canyon, Sabino Canyon, Thomas Canyon; Coyote Mountains
***Bouteloua aristidoides*** (Kunth) Grisebach
Brown Canyon, between Fresnal and Toro Canyons
***Bouteloua barbata*** Lagasca var. ***barbata***
Baboquivari Mountains, without locality
***Bouteloua barbata*** Lagasca var. ***rothrockii*** (Vasey) Gould
Baboquivari Canyon, Brown Canyon, M. S. Ranch (north end of the Baboquivari Mountains), S of Ash [Fresnal] Canyon, Ventana Peak E of San Miguel; Coyote Mountains
***Bouteloua chondrosioides*** (Kunth) Bentham
Brown Canyon, Hwy 286–Arivaca Rd. Jct., Virtud Tank area; Quinlan Mountains
***Bouteloua curtipendula*** (Michaux) Torrey
Baboquivari Canyon, Brown Canyon, Thomas Canyon
***Bouteloua eriopoda*** (Torrey) Torrey
Ventana Peak, E of San Miguel
***Bouteloua hirsuta*** Lagasca
Brown Canyon, Thomas Canyon; Coyote Mountains; Quinlan Mountains
***Bouteloua radicosa*** (Fournet) Griffiths
Coyote Mountains
***Bouteloua repens*** (Kunth) Scribner & Merrill
Baboquivari Canyon, Thomas Canyon; Coyote Mountains; Quinlan Mountains
***Brachiaria arizonica*** (Scribner & Merrill) S. T. Blake [*Panicum arizonicum* Scribner & Merrill, *Urochloa arizonica* (Scribner & Merrill) Morrone & Zuloaga]
South Canyon, Toro Canyon
***Bromus anomalus*** Ruprecht ex Fournet [*B. porteri* (Coulter) Nash]
Baboquivari Canyon, Brown Canyon
***Bromus arizonicus*** (Shear) Stebbins
Coyote Mountains; Quinlan Mountains
***Bromus carinatus*** Hooker & Arnott
Brown Canyon
***Bromus catharticus*** Vahl
Brown Canyon
***Bromus ciliatus*** Linnaeus
Without locality (*L. N. Goodding s.n.*, 27-Dec-36); Quinlan Mountains
***Bromus frondosus*** (Shear) Wooton & Standley
Baboquivari Peak; Quinlan Mountains
***Bromus mucroglumis*** Wagnon
Baboquivari Canyon, Toro Canyon
***Bromus porteri*** (Coulter) Nash var. ***lanatipes*** Shear [*B. lanatipes* (Shear) Rydberg]
Baboquivari Canyon
***Bromus rubens*** Linnaeus [*Bromus madritensis* subsp. *rubens* (Linnaeus) Husnot]
Brown Canyon
***Cottea pappophoroides*** Kunth
Baboquivari Canyon, Bear Canyon, Brown Canyon, Window Butte
***Cynodon dactylon*** (Linnaeus) Persoon
Brown Canyon; Coyote Mountains; Quinlan Mountains
***Digitaria californica*** (Bentham) Henrard [*Trichachne californica* (Bentham) Chase]
Brown Canyon, Fresnal Canyon, Moristo Canyon
***Digitaria ciliaris*** (Retzius) Koeler
Coyote Mountains
***Digitaria cognata*** (Schultes) Pilger [*Leptotoma cognatum* (Schultes) Chase]
Baboquivari Peak, Brown Canyon; Pozo Verde Mountains
***Digitaria cognata*** var. ***pubiflora*** (Vasey ex L. H. Dewey) Wipff
Baboquivari Peak
***Digitaria insularis*** (Linnaeus) Fedde [*Trichachne insularis* (Linnaeus) Nees]
Baboquivari Canyon, Brown Canyon, Fresnal Canyon, Moristo Canyon, Toro Canyon
***Echinochloa colona*** (Linnaeus) Link
Fresnal Canyon, Virtud Tank area
***Elionurus barbiculmis*** Hackel ex Scribner
Baboquivari Canyon, Thomas Canyon
***Elymus arizonicus*** (Scribner & J. G. Smith) Gould
Without locality (*H. S. Haskell, F. W. Gould & R. A. Darrow 2767*, 7-Oct-44)
***Elymus elymoides*** (Rafinesque) Swezely [*Sitanion hystrix* J. G. Smith]
Baboquivari Peak, Brown Canyon, Fresnal Canyon, Sycamore Canyon, Thomas Canyon; Quinlan Mountains
***Enneapogon desvauxii*** Desvaux ex Beauvois
Between Fresnal and Toro Canyons, Canyon north of Moristo Canyon, Sabino Canyon, Rondstadt Ranch; Quinlan Mountains

***Eragrostis cilianensis*** (Allioni) Link
Baboquivari Canyon, Brown Canyon, Sycamore Canyon, Thomas Canyon
***Eragrostis curvula*** (Schrader) Nees var. ***conferta*** Nees
Brown Canyon, Sabino Canyon; Coyote Mountains; Quinlan Mountains
***Eragrostis intermedia*** A. S. Hichcock
Baboquivari Canyon, Brown Canyon, Forest Cabin Canyon, Thomas Canyon; Coyote Mountains; Quinlan Mountains
***Eragrostis lehmanniana*** Nees
Baboquivari Canyon, Brown Canyon; Coyote Mountains; Quinlan Mountains
***Eragrostis mexicana*** (Horneman) Link
Baboquivari Peak
***Eragrostis pectinacea*** (Michaux) Nees [*E. pectinacea* (Michaux) Nees var. *miserrima* (E. Fournier) J. Reeder]
Baboquivari Canyon, Toro Canyon
***Eragrostis superba*** Peyritsch
Sabino Canyon; Pozo Verde Mountains; Quinlan Mountains
***Eriochloa acuminata*** (J. Presl) Kunth [*E. gracilis* (Fournet) A. S. Hitchcock; *E. lemmoni* Vasey & Scribner var. *gracilis* (Fournet) Gould]
Brown Canyon; Coyote Mountains
***Eriochloa acuminata*** (J. Presl) Kunth var. ***minor*** (Vasey) R. B. Shaw
Brown Canyon; Coyote Mountains
***Eriochloa aristata*** Vasey
Without locality (*L. N. Goodding s.n.*, 14-Oct-36); Coyote Mountains
***Heteropogon contortus*** (Linnaeus) Beauvois
Brown Canyon, Thomas Canyon; Coyote Mountains
***Heteropogon melanocarpus*** (Elliott) Bentham
Brown Canyon, Cave Canyon
***Hordeum murinum*** Linnaeus subsp. ***glaucum*** (Steudel) Tzvelev
Brown Canyon
***Hordeum pusillum*** Nuttall
Fresnal Canyon
***Koeleria macrantha*** (Ledebour) J. A. Schultes
Baboquivari Canyon; Quinlan Mountains
***Leptochloa dubia*** (Kunth) Nees [*Diplachne dubia* Scribner]
Baboquivari Canyon, Brown Canyon, Thomas Canyon, Virtud Tank area; Coyote Mountains
***Leptochloa fusca*** (Linnaeus) Kunth subsp. ***fascicularis*** (Lamarck) N. Snow
Coyote Mountains
***Leptochloa panicea*** (Retzius) Ohwi subsp. ***brachiata*** (Steudel) N. Snow [*L. filiformis* (Persoon) P. Beauvois; previously also known as *L. mucronata* (Michaux) Kunth or *L. panicea* subsp. *mucronata* (Michaux) Nowack (in part); subsp. *mucronata* now restricted to SE and central U.S.]
Baboquivari Canyon, Brown Canyon, Cave Canyon; Coyote Mountains
***Leptochloa uninervia*** (J. Presl) A. S. Hitchcock & Chase
Coyote Mountains
***Lycurus setosus*** (Nuttall) C. G. Reeder
Baboquivari Canyon, Brown Canyon, Thomas Canyon; Coyote Mountains; Quinlan Mountains
***Melica porteri*** Scribner
Baboquivari Peak
***Melinis repens*** (Willdenow) Zizka subsp. ***repens*** [*Rhynchelytrum repens* (Willdenow) C. E. Hubbard]
Brown Canyon; Coyote Mountains
***Microchloa kunthii*** Desvaux [This rare species has only been collected 5 times in AZ; in Cochise, Pima, and Santa Cruz counties]
Aros Ranch; Las Guijas Mountains
***Muhlenbergia arizonica*** Scribner
Baboquivari Canyon, Canyon north of Moristo Canyon, Dumosa Canyon, Fresnal-Toro Canyons; Coyote Mountains
***Muhlenbergia dubioides*** C. O. Goodding
Moristo Canyon
***Muhlenbergia dumosa*** Scribner
Baboquivari Peak, Brown Canyon, between Fresnal and Toro Canyons, Virtud Tank area; Coyote Mountains
***Muhlenbergia elongata*** Scribner ex Beal [*M. xerophila* C. O. Goodding]
Forest Cabin Canyon, Moristo Canyon
***Muhlenbergia emersleyi*** Vasey [*M. gooddingii* Soderstrom]
Baboquivari Canyon, Brown Canyon, Canyon north of Moristo Canyon, Fresnal Canyon, Moristo Canyon, Thomas Canyon; Coyote Mountains; Quinlan Mountains
***Muhlenbergia fragilis*** Swallen
Baboquivari Canyon, Brown Canyon, Fresnal Canyon, Thomas Canyon
***Muhlenbergia longiligula*** A. S. Hitchcock
Moristo Canyon

***Muhlenbergia microsperma*** (De Candolle) Kunth
Baboquivari Canyon, Brown Canyon, Sabino Canyon, Window Butte; Coyote Mountains; Quinlan Mountains
***Muhlenbergia minutissima*** (Steudel) Swallen
Baboquivari Peak
***Muhlenbergia pauciflora*** Buckley
Baboquivari Peak, Sycamore Canyon, Thomas Canyon
***Muhlenbergia polycaulis*** Scribner
Baboquivari Peak, Moristo Canyon, Sycamore Canyon; Coyote Mountains
***Muhlenbergia porteri*** Scribner
Baboquivari Mountain Road, Brown Canyon; Coyote Mountains
***Muhlenbergia rigens*** (Bentham) A. S. Hitchcock
Baboquivari Canyon, Brown Canyon, Thomas Canyon; Coyote Mountains; Pozo Verde Mountains; Quinlan Mountains
***Muhlenbergia sinuosa*** Swallen
Baboquivari Canyon; Quinlan Mountains
***Muhlenbergia tenuifolia*** (Kunth) Kunth [*M. monticola* Buckley]
Baboquivari Canyon, Brown Canyon; Coyote Mountains; Quinlan Mountains
***Panicum bulbosum*** Kunth
Baboquivari Canyon, Brown Canyon, Jupiter Canyon, Toro Canyon
***Panicum hallei*** Vasey
Brown Canyon; Quinlan Mountains
***Panicum hirticaule*** J. Presl var. ***hirticaule*** [*P. capillare* Linnaeus var. *hirticaule* (J. Presl) Gould]
Baboquivari Canyon, M. S. Ranch, northern end of Baboquivaris, Thomas Canyon; Quinlan Mountains
***Panicum hirticaule*** J. Presl var. ***verrucosum*** Zuloaga & O. Morrone
South Canyon
***Panicum lepidulum*** A. S. Hitchcock & Chase
Quinlan Mountains
***Paspalum distichum*** Linnaeus
Fresnal Canyon
***Phalaris caroliniana*** Walter
Chango [Chivo] Pond
***Piptochaetium fimbriatum*** (Kunth) A. S. Hitchcock
Baboquivari Peak, Canyon north of Moristo Canyon
***Piptochaetium pringlei*** (Beal) Parodi
Quinlan Mountains
***Poa bigelovii*** Vasey & Scribner
Brown Canyon; Coyote Mountians
***Poa compressa*** Linnaeus
Brown Canyon
***Poa fendleriana*** (Steudel) Vasey var. ***fendleriana***
Without locality (*L. N. Goodding s.n.*, 14-Apr-36)
***Polypogon monspeliensis*** (Linnaeus) Desfontaines
Brown Canyon
***Polypogon viridis*** (Gouan) Breistroffer [*Agrostis semiverticillata* (Forsskal) C. Christensen; *Polypogon semiverticillata* (Forsskal) Hylander]
Jaguar Canyon, Thomas Canyon; Coyote Mountains
***Schizachyrium cirratum*** (Hackel) Wooten & Standley [*Andropogon cirratus* Hackel]
Baboquivari Canyon, Brown Canyon; Coyote Mountains
***Schizachyrium hirtiflorum*** Nees [*Andropogon hirtiflorus* (Nees) Kunth]
Baboquivari Canyon; Coyote Mountains
***Setaria arizonica*** Roeminger
Brown Canyon, west slope of Baboquivaris
***Setaria grisebachii*** E. Fournier
Virtud Tank area; Quinlan Mountains
***Setaria leucopila*** (Scribner & Merrill) Schumann
Dumosa Canyon; Coyote Mountains
***Setaria macrostachya*** Kunth
Baboquivari Canyon, Brown Canyon, Fresnal Canyon, Ventana Peak east of San Miguel; Coyote Mountains
***Sorghum halepense*** (Linnaeus) Persoon
Brown Canyon; Coyote Mountains
***Sphenopholis obtusata*** (Michaux) Scribner
Forest Cabin Canyon
***Sporobolus airoides*** Torrey [*S. wrightii* Munro ex Scribner, *S. airoides* Torrey var. *wrightii* (Munro ex Scribner) Gould]
Brown Canyon; Coyote Mountains
***Sporobolus contractus*** A. S. Hitchcock
Mesas W of Baboquivari
***Sporobolus cryptandrus*** (Torrey) A. Gray
Coyote Mountains
***Trachypogon spicatus*** (Linnaeus f.) Kuntze [*T. montufarii* (Kunth) Nees; *T. secundus* (J. Presl) Scribner]
Brown Canyon, Jaguar Canyon, Thomas Canyon, Virtud Tank area; Coyote Mountains
***Tridens eragrostoides*** (Vasey & Scribner) Nash [*Triodia eragrostoides* Vasey & Scribner]
Coyote Mountains
***Trisetum interruptum*** Buckley
Aguirre Lake

## POLEMONIACEAE

***Allophyllum gilioides*** (Bentham) A. & V. Grant
Brown Canyon, Fresnal Canyon, Sycamore Canyon, Toro Canyon
***Eriastrum diffusum*** (A. Gray) H. Mason
Bajillo de Las Chivas; Coyote Mountains; Quinlan Mountains
***Gilia flavocincta*** A. Nelson
Brown Canyon, Thomas Canyon; Coyote Mountains
***Gilia mexicana*** A. & V. Grant
Brown Canyon
***Ipomopsis macombii*** (Torrey ex A. Gray) V. Grant
Quinlan Mountains
***Ipomopsis multiflora*** (Nuttall) V. Grant
Baboquivari Canyon
***Ipomopsis thurberi*** (Torrey) V. Grant
Fresnal Canyon, Sycamore Canyon, Toro Canyon
***Linanthus aureus*** (Nuttall) Greene
Thomas Canyon
***Linanthus bigelovii*** (A. Gray) Greene
Fresnal Canyon, Toro Canyon
***Loeselia*** aff. ***ciliata*** Linnaeus
Brown Canyon

## POLYGALACEAE

***Monnina wrightii*** A. Gray
Toro Canyon
***Polygala barbeyana*** Chodat [*Polygala reducta* Blake]
Pozo Verde Mountains
***Polygala macradenia*** A. Gray
Brown Canyon
***Polygala obscura*** Bentham
Baboquivari Canyon
***Polygala orthotricha*** Blake
Brown Canyon, Thomas Canyon

## POLYGONACEAE

***Eriogonum abertianum*** Torrey
Baboquivari Canyon, Bear Canyon, Brown Canyon, Fresnal, Thomas Canyon; Coyote Mountains; Quinlan Mountains
***Eriogonum deflexum*** Torrey
Brown Canyon
***Eriogonum polycladon*** Bentham
Baboquivari Canyon, Brown Canyon, Toro Canyon
***Eriogonum thurberi*** Torrey
Baboquivari Canyon; Coyote Mountains
***Eriogonum wrightii*** Torrey var. ***wrightii***
Baboquivari Canyon, Jaguar Canyon, Thomas Canyon, Ventana Peak east of San Miguel
***Polygonum pensylvanicum*** Linnaeus
Aguirre Lake
***Polygonum persicaria*** Linnaeus
Brown Canyon
***Rumex hymenosepalus*** Torrey
Brown Canyon, Fresnal Canyon; Coyote Mountains

## POLYPODIACEAE

***Polypodium hesperium*** Maxon
Baboquivari Peak
***Polypodium thyssanolepis*** Klotzsch
Sycamore Canyon

## PORTULACACEAE

***Calandrinia ciliata*** (Ruiz & Pavon) De Candolle
Baboquivari Canyon, Brown Canyon
***Cistanthe monandra*** (Nuttall) Hershkovitz [*Calyptridium monandrum* Nuttall]
Brown Canyon; Quinlan Mountains
***Claytonia perfoliata*** Donn [*Montia perfoliata* (Donn) Howell]
Brown Canyon, Thomas Canyon
***Phemeranthus aurantiacus*** (Engelmann) Kiger [*Talinum aurantiacum* Engelmann]
Brown Canyon
***Portulaca oleracea*** Linnaeus
South Canyon
***Portulaca suffrutescens*** Engelmann
Baboquivari Canyon, Brown Canyon, Fresnal Canyon, Thomas Canyon; Coyote Mountains; Quinlan Mountains
***Portulaca umbraticola*** Kunth
Brown Canyon, Fresnal Canyon; Coyote Mountains
***Talinum paniculatum*** (Jacquin) Gaertner
Baboquivari Canyon, Brown Canyon; Coyote Mountains

## PRIMULACEAE

***Androsace occidentalis*** Pursh
Brown Canyon, Thomas Canyon; Coyote Mountains; Quinlan Mountains

## PTERIDACEAE

***Astrolepis cochisensis*** (Goodding) D. M. Benham & Windham subsp. ***cochisensis*** [*Notholaena cochisensis* Goodding]
Brown Canyon; Coyote Mountains
***Astrolepis integerrima*** (Hooker) D. M. Benham & Windham [*Notholaena sinuata* var. *integerrima* Hooker, *Notholaena integerrima* (Hooker) Hevly, *Cheilanthes integerrima* (Hooker) Mickel]
Coyote Mountains
***Astrolepis sinuata*** (Lagasca ex Swartz) D. M. Bentham & Windham subsp. ***sinuata*** [*Notholaena sinuata* (Lagasca ex Swartz) Kaulfuss, *Cheilanthes sinuata* Swartz]
Baboquivari Canyon, Brown Canyon, Thomas Canyon; Coyote Mountains
***Bommeria hispida*** (Mettenius) Underwood
Baboquivari Canyon, Brown Canyon, Thomas Canyon; Coyote Mountains; Quinlan Mountains
***Cheilanthes bonariensis*** (Willdenow) Proctor [*Notholaena aurea* (Poiret) Desvaux]
Baboquivari Canyon, Mendoza Canyon, Moristo Canyon, Thomas Canyon
***Cheilanthes eatoni*** Baker
Brown Canyon, Moristo Canyon, South Canyon, Thomas Canyon
***Cheilanthes fendleri*** Hooker
Baboquivari Peak, Brown Canyon, Moristo Canyon, Thomas Canyon
***Cheilanthes lindheimeri*** Hooker
Baboquivari Canyon, Brown Canyon, Fresnal Canyon, Thomas Canyon, Toro Canyon; Coyote Mountains; Quinlan Mountains
***Cheilanthes pringlei*** Davenport
Fresnal Canyon
***Cheilanthes tomentosa*** Link
Baboquivari Canyon, Moristo Canyon, Thomas Canyon
***Cheilanthes wootoni*** Maxon
Brown Canyon, Moristo Canyon, Thomas Canyon; Coyote Mountains
***Cheilanthes wrightii*** Hooker
Brown Canyon, Canyon north of Moristo Canyon, Fresnal Canyon; Coyote Mountains; Pozo Verde Mountains
***Notholaena aschenborniana*** Klotzsch [*Cheilanthes aschenborniana* (Klotzsch) Mettenius]
Brown Canyon
***Notholaena grayi*** Davenport
Baboquivari Canyon, Brown Canyon, Moristo Canyon, Thomas Canyon; Coyote Mountains
***Notholaena lemmoni*** D. C. Eaton
Coyote Mountains
***Notholaena limitanea*** Maxon
Coyote Mountains
***Notholaena standleyi*** Maxon [*Cheilanthes standleyi* (Maxon) Mickel]
Baboquivari Peak, Brown Canyon; Coyote Mountains
***Pellaea atropurpurea*** (Linnaeus) Link
Baboquivari Canyon, Brown Canyon, Moristo Canyon, Sabino Canyon, Thomas Canyon
***Pellaea intermedia*** Mettenius ex Kuhn
Baboquivari Canyon, Brown Canyon, South Canyon, Thomas Canyon; Coyote Mountains
***Pellaea truncata*** Goodding
Baboquivari Canyon, Brown Canyon, Sabino Canyon, South Canyon, Ventana Peak E of San Miguel; Coyote Mountains; Quinlan Mountains
***Pellaea wrightiana*** Hooker
Brown Canyon, Thomas Canyon
***Pityrogramma triangularis*** (Kaulfuss) Maxon
Baboquivari Canyon, Cave Canyon, Mendoza Canyon, South Canyon; Coyote Mountains

## RANUNCULACEAE

***Anemone tuberosa*** Rydberg
Brown Canyon, Thomas Canyon; Coyote Mountains
***Aquilegia chrysantha*** A. Gray
Bear Canyon, Moristo Canyon
***Aquilegia longissima*** A. Gray ex S. Watson
Baboquivari Peak, Brown Canyon, Thomas Canyon
***Clematis drummondii*** Torrey & A. Gray
Brown Canyon
***Clematis ligusticifolia*** Nuttall
Baboquivari Canyon, Sycamore Canyon, Thomas Canyon; Quinlan Mountains
***Delphinium parishii*** A. Gray subsp. ***parishii***
Forestry Cabin Canyon; Coyote Mountains
***Delphinium scaposum*** Greene
Baboquivari Canyon, Brown Canyon; Coyote Mountains
***Myosurus cupulatus*** Watson
Baboquivari Canyon, Thomas Canyon, Toro Canyon; Coyote Mountains
***Myosurus minimus*** Linnaeus
Brown Canyon

***Thalictrum fendleri*** Engelmann
Brown Canyon, Sycamore Canyon

## RHAMNACEAE

***Condalia correllii*** M. C. Johnston
Brown Canyon, Toro Canyon; Pozo Verde Mountains
***Condalia warnockii*** M. C. Johnston [*C. spathulata* of authors, not A. Gray]
Baboquivari Canyon, Brown Canyon, Fresnal Wash
***Rhamnus californica*** Eschscholtz [*Frangula californica* A. Gray]
Brown Canyon, Moristo Canyon
***Rhamnus ilicifolia*** Kellogg [*R. crocea* Nuttall subsp. *ilicifolia* (Kellogg) C. B. Wolf]
Brown Canyon; Coyote Mountains; Quinlan Mountains
***Sageretia wrightii*** S. Watson
Brown Canyon, without locality (*R. H. Peebles, G. J. Harrison & T. H. Kearney 3836*, 29-Mar-27)
***Zizyphus obtusifolia*** (Hooker ex Torrey & A. Gray) A. Gray var. ***canescens*** (A. Gray) M. C. Johnston [*Condalia lycioides* (A. Gray) Weberbauer var. *canescens* (A. Gray) Trelease]
Baboquivari Canyon, Brown Canyon, Forest Cabin Canyon, Thomas Canyon

## ROSACEAE

***Cercocarpus montanus*** Rafinesque var. ***paucidentatus*** (S. Watson) F. L. Martin [*C. breviflorus* A. Gray, *C. betuloides* Nuttall, *C. breviflorus* A. Gray]
Baboquivari Canyon, Brown Canyon, Sycamore Canyon, Thomas Canyon; Quinlan Mountains
***Holodiscus dumosus*** (Nuttall ex Hooker) Heller
Quinlan Mountains
***Prunus serotina*** Ehrhart
Baboquivari Canyon, Brown Canyon, Thomas Canyon
***Rubus arizonensis*** Focke
Brown Canyon, Moristo Canyon, Thomas Canyon
***Rubus neomexicanus*** A. Gray
Toro Canyon
***Vauquelinia californica*** (Torrey) Sargent subsp. ***sonorensis*** W. J. Hess and Henrickson
Baboquivari Canyon, Brown Canyon, south end Baboquivari Mountains; Coyote Mountains

## RUBIACEAE

***Bouvardia ternifolia*** (Cavanilles) Schlechtendal [*B. glaberrima* Engelmann]
Baboquivari Canyon, Brown Canyon, Fresnal Canyon, Sycamore Canyon, Thomas Canyon; Coyote Mountains; Quinlan Mountains
***Diodia teres*** Walter
Aros Ranch road, Toro Canyon
***Galium aparine*** Linnaeus
Baboquivari Canyon, Brown Canyon, Thomas Canyon
***Galium mexicanum*** Kunth
Baboquivari Canyon, South Canyon, Thomas Canyon
***Galium microphyllum*** A. Gray
Brown Canyon; Coyote Mountains; Quinlan Mountains
***Galium proliferum*** A. Gray
Brown Canyon, Fresnal Canyon, Toro Canyon
***Galium stellatum*** Kellogg
Jaguar Canyon
***Galium wrightii*** A. Gray
Baboquivari Canyon, Brown Canyon, South Canyon; Quinlan Mountains

## RUTACEAE

***Ptelea trifoliata*** Linnaeus subsp. ***angustifolia*** (Bentham) V. Bailey [*P. angustifolia* Bentham]
Baboquivari Canyon, Brown Canyon, South Canyon, Thomas Canyon, Toro Canyon

## SALICACEAE

***Populus fremontii*** S. Watson subsp. ***fremontii***
South end of Baboquivari Mountains
***Salix exigua*** Nuttall
Moristo Canyon, South Canyon, Toro Canyon; Coyote Mountains; Quinlan Mountains
***Salix gooddingii*** Ball
Brown Canyon
***Salix lasiolepis*** Bentham
Sycamore Canyon, Thomas Canyon
***Salix taxifolia*** Kunth
Thomas Canyon

## SANTALACEAE

***Comandra umbellata*** (Linnaeus) Nuttall
Coyote Mountains; Quinlan Mountains
***Phoradendron californicum*** Nuttall
Brown Canyon; Coyote Mountains
***Phoradendron serotinum*** (Rafinesque) M. C. Johnston subsp. ***tomentosum*** (De Candolle) Kuijt [*P. coryae* Trelease, *P. villosum* var. *coryae* (Trelease) Wiens, *P. flavens* [as "*flavescens*"] (Pursh) Nuttall, in part), *P. tomentosum* (de Candolle) Engelmann ex A. Gray var. *macrophyllum* (Engelmann ex Rothrock) L. D. Benson]
Brown Canyon, Thomas Canyon; Coyote Mountains; Quinlan Mountains

## SAPINDACEAE

***Cardiospermum corindum*** Linnaeus
Coyote Mountains
***Dodonaea viscosa*** Jacquin var. ***angustifolia*** (Linnaeus f.) Bentham [*D. angustifolia* Linnaeus f.]
Baboquivari Canyon, Brown Canyon; Coyote Mountains; Quinlan Mountains
***Sapindus drummondii*** Hooker & Arnott [*S. saponaria* Linnaeus var. *drummondii* Hooker & Arnott]
Brown Canyon; Coyote Mountains; Quinlan Mountains

## SAPOTACEAE

***Sideroxylon lanuginosum*** Michaux [*Bumelia lanuginosa* Persoon]
Baboquivari Canyon, Brown Canyon

## SAXIFRAGACEAE

***Heuchera sanguinea*** Engelmann
Brown Canyon, Spring Canyon, S of Fresnal Canyon; Coyote Mountains; Quinlan Mountains

## SCROPHULARIACEAE

***Castilleja austromontana*** Standley & Blumer
Brown Canyon
***Castilleja exserta*** (Heller) Chuang & Heckard [*Orthocarpus purpurascens* Bentham]
Baboquivari Canyon, Brown Canyon; Coyote Mountains; Quinlan Mountains
***Castilleja lanata*** A. Gray
Brown Canyon; Quinlan Mountains
***Castilleja minor*** (A. Gray) A. Gray
Brown Canyon; Coyote Mountains
***Castilleja tenuiflora*** Bentham
Baboquivari Canyon, Jupiter Canyon, South Canyon, Thomas Canyon
***Maurandya antirrhiniflora*** Humboldt & Bonpland [*Maurandella antirrhiniflora* (Humboldt & Bonpland ex Willdenow) Rothmaler]
Baboquivari Canyon, Brown Canyon
***Nuttallanthus texanus*** (Scheele) D. A. Sutton [*Linaria texana* Scheele, *L. canadensis* (Linnaeus) Dumortier var. *texana* (Scheele) Pennell]
Brown Canyon, Fresnal Canyon, Thomas Canyon; Quinlan Mountains
***Sairocarpus nuttallianus*** (Bentham ex A. De-Candolle) D. A. Sutton [*Antirrhinum nuttallianum* Bentham]
Brown Canyon, South Canyon, Toro Canyon; Pozo Verde Mountains
***Scrophularia parviflora*** Wooton & Standley
Toro Canyon

## SELAGINELLACEAE

***Selaginella arizonica*** Maxon
Coyote Mountains; Quinlan Mountains
***Selaginella rupincola*** Underwood
Coyote Mountains
***Selaginella underwoodii*** Hieronymus
Thomas Canyon

## SIMMONDSIACEAE

***Simmondsia chinensis*** (Link) C. K. Schneider
Baboquivari Canyon, Brown Canyon, Forest Cabin Canyon, Fresnal Canyon; Quinlan Mountains

## SOLANACEAE

***Capsicum annuum*** Linnaeus var. ***glabriusculum*** (Dunal) Heiser & Pickersgill
Baboquivari Canyon, Moristo Canyon, Sycamore Canyon

***Datura discolor*** Bernhardi
Jaguar Canyon
***Datura wrightii*** Regel [*D. inoxia* Miller, *D. meteloides* of authors, not Dunal]
Brown Canyon; Quinlan Mountains
***Jaltomata procumbens*** (Cavanilles) J. L. Gentry
Baboquivari Canyon
***Lycium andersonii*** A. Gray var. ***andersonii***
Brown Canyon
***Lycium berlandieri*** Dunal
Brown Canyon, Fresnal Canyon
***Lycium exsertum*** A. Gray
Brown Canyon, Fresnal Canyon; Pozo Verde Mountains
***Margaranthus solanaceus*** Schlechtendal
Brown Canyon
***Nicotiana obtusifolia*** Martens & Galleotti [*N. trigonophylla* Dunal]
Brown Canyon, Sabino Canyon, Thomas Canyon, Ventana Peak, E of San Miguel
***Physalis acutifolia*** (Miers) Sandwith [*P. wrightii* A. Gray]
Brown Canyon
***Physalis crassifolia*** Bentham
Brown Canyon; Coyote Mountains
***Physalis hederifolia*** A. Gray var. ***fendleri*** (A. Gray) Waterfall
Brown Canyon, Baboquivari Canyon, Toro Canyon, Thomas Canyon
***Physalis latiphysa*** Waterfall
Brown Canyon, Ronstadt Ranch
***Physalis pubescens*** Linnaeus
Brown Canyon, Fresnal Canyon, South Canyon, Toro Canyon; Coyote Mountains
***Solanum adscendens*** Sendtner
Brown Canyon, Fresnal Canyon, Toro Canyon
***Solanum deflexum*** Greenman
Brown Canyon
***Solanum douglasii*** Dunal
Baboquivari Canyon, Brown Canyon, Moristo Canyon, Thomas Canyon; Coyote Mountains
***Solanum elaeagnifolium*** Cavanilles
Brown Canyon, Sabino Canyon

## TAMARICACEAE

***Tamarix ramosissima*** Ledebour
Brown Canyon (single tree removed just below Environmental Education Center in 2002; another in side canyon in 2006)

## THEMIDACEAE

***Dichelostemma pulchellum*** (Salisbury) Heller
Brown Canyon

## TYPHACEAE

***Typha domingensis*** Persoon
Brown Canyon, Jaguar Canyon

## URTICACEAE

***Parietaria floridana*** Nuttall
Brown Canyon
***Urtica gracilenta*** Greene
Baboquivari Canyon, Brown Canyon

## VERBENACEAE

***Aloysia gratissima*** (Gill & Hooker) Troncoso
Pozo Verde Mountains
***Aloysia wrightii*** (A. Gray) Heller
Brown Canyon, Fresnal Canyon
***Glandularia bipinnatifida*** (Nuttall) Nuttall [*Verbena bipinnatifida* Nuttall]
Baboquivari Canyon, Brown Canyon; Pozo Verde Mountains
***Glandularia gooddingii*** (Briquet) Solbrig [*Verbena gooddingii* Briquet]
Baboquivari Canyon, Brown Canyon, Fresnal Canyon, Thomas Canyon; Coyote Mountains; Quinlan Mountains
***Glandularia wrightii*** (A. Gray) Umber [*Verbena wrightii* A. Gray]
Thomas Canyon
***Verbena ambrosifolia*** Rydberg
Without locality (*L. N. Goodding s.n.*, 17-Apr-35)
***Verbena neomexicana*** (A. Gray) Small
Fresnal Canyon, Toro Canyon

## VIOLACEAE

***Hybanthus attenuatus*** (Humboldt & Bonpland) C. K. Schulze
Brown Canyon, Toro Canyon

## VITACEAE

***Cissus trifoliata*** Linnaeus
Coyote Mountains

***Vitis arizonica*** Engelmann
Brown Canyon; Coyote Mountains; Quinlan Mountains

## WOODSIACEAE

***Cystopteris fragilis*** (Linnaeus) Bentham
Baboquivari Peak
***Woodsia mexicana*** Fée
Brown Canyon
***Woodsia phillipsii*** Windham [*W. mexicana* of authors, not Fée]
Brown Canyon
***Woodsia plummerae*** Lemmon
Baboquivari Peak, Toro Canyon
***Woodsia*** aff. ***scopulina*** D. C. Eaton
Brown Canyon

## ZYGOPHYLLACEAE

***Kallstroemia grandiflora*** Torrey ex A. Gray
Brown Canyon; Coyote Mountains; Pozo Verde Mountains; Quinlan Mountains
***Tribulus terrestris*** Linnaeus
Coyote Mountains

# REFERENCES CITED


Abrams, Le Roy. 1951. *Illustrated Flora of the Pacific States*. Vol. 3. Stanford University Press, Stanford, CA.

Abrams, Le Roy, and Roxana S. Ferris. 1960. *Illustrated Flora of the Pacific States*. Vol. 4. Stanford University Press, Stanford, CA.

Ahmed, W., Z. Ahmed, and A. Malik. 1992. Stigmasteryl galactoside from *Rhynchosia minima*. *Phytochemistry* 31: 4038–4039.

Ajilvsgi, Geyata. 1984. *Wildflowers of Texas*. Shearer Publishers, Bryan, TX.

Alcock, John. 1990. *Sonoran Desert Summer*. University of Arizona Press, Tucson.

Alcorn, Janis B. 1984. *Huastec Mayan Ethnobotany*. University of Texas Press, Austin.

Al-Shehbaz, Ihsan. 2003. Transfer of most North American species of *Arabis* to *Boechera* (Brassicaceae). *Novon* 13: 381–391.

Anderson, Edward F., José Cauich Canul, Arora Dzib, Salvador Flores Guido, Gerald Islebe, Felix Medina Tzuc, Odilón Sánchez Sánchez, and Pastor Valdez Chale. 2003. *Those Who Bring the Flowers: Maya Ethnobotany in Quintana Roo, Mexico*. El Colegio de la Frontera Sur, Chiapas, Mexico.

Anderson, M. Kat. 1996. The ethnobotany of Deergrass, *Muhlenbergia rigens* (Poaceae): Its uses and fire management by California Indian tribes. *Economic Botany* 50: 409–422.

Anderson, R. Scott, and Thomas R. Van Devender. 1991. Comparison of pollen and macrofossils in packrat (*Neotoma*) middens: A chronological sequence from the Waterman Mountains of southern Arizona, U.S.A. *Review of Paleobotany and Palynology* 68: 1–28.

Anonymous. 1999. *Endangered, threatened, and rare species documented from Lake County, Indiana*. http://www.in.gov/dnr/naturepr/species/lake.pdf.

Anthony, Robert G., and Norman S. Smith. 1977. Ecological relationships between mule deer and white-tailed deer in southeastern Arizona. *Ecological Monographs* 47 (3): 255–277.

Aplin, T. E. H., and R. Cannon. 1971. Distribution of alkaloids in some western Australian plants. *Economic Plants* 25: 366–380.

Arber, Agnes R. 1986. *Herbals, Their Origin and Evolution: A Chapter in the History of Botany 1470–1670*. Third Edition, with an introduction and annotations by William T. Stearn. Cambridge University Press, Cambridge.

Arizona Rare Plant Committee. 2001. *Arizona Rare Plant Field Guide*. Arizona Game and Fish Commission, Tucson.

Austin, Daniel F. 1979. Plants without beginning or end. *Bulletin of the Fairchild Tropical Garden* 34 (3): 17–19.

Austin, Daniel F. 1986. Nomenclature of the *Ipomoea nil* complex (Convolvulaceae). *Taxon* 35 (2): 355–358.

Austin, Daniel F. 1990. Comments on southwestern United States *Evolvulus* and *Ipomoea* (Convolvulaceae). *Madroño* 37 (2): 124–132.

Austin, Daniel F. 1998a. Convolvulaceae Morning Glory Family. In: A New Flora of Arizona. *Journal of the Arizona-Nevada Academy of Sciences* 30 (2): 61–83.

Austin, Daniel F. 1998b. Christmas botany: or how reindeer learned to fly. *The Palmetto* 17 (3): 12–14, 23.

Austin, Daniel F. 2002. Roots as a Source of Food. Pp. 1025–1043. In: Waisel, Yoav, Amram Eshel, and Uzi Kafkaki, eds. *Plant Roots. The Hidden Half*, Third Edition, Revised and Expanded. Marcel Dekker, Inc., New York.

Austin, Daniel F. 2003. Vascular Plants of the Buenos Aires National Wildlife Refuge. Unpublished report to BANWR.

Austin, Daniel F. 2004. *Florida Ethnobotany*. CRC Press, Boca Raton, FL.

Austin, Daniel F. 2006a. Fox-tail millets (*Setaria*: Poaceae)—abandoned food in two hemispheres. *Economic Botany* 60 (2): 143–158.

Austin, Daniel F. 2006b. Noteworthy distributions and additions in southwestern Convolvulaceae. *Canotia* 2 (3): 79–106. http://lifesciences.asu.edu/herbarium/canotia.html.

Austin, Daniel F., Kaoro Kitajima, Yoshiaki Yoneda, and Lianfen Qian. 2001. A putative tropical American plant, *Ipomoea nil* (Convolvulaceae), in Pre-Columbian Japanese art. *Economic Botany* 55 (4): 515–527.

Ayensu, Edward S. 1975. *Underexploited Tropical Plants with Promising Economic Value*. National Academy of Sciences, Washington, DC.

Bahre, Conrad J. 1984. Effects of historic fuelwood cutting on the semidesert woodlands of the Arizona-Sonora borderlands. Pp. 101–110. In: *History of Sustained-Yield Forestry: A Symposium*, Harold K. Steen, ed. Forest History Society of America, Portland, OR.

Bahre, Conrad J. 1991. *A Legacy of Change. Historic Human Impact on Vegetation of Arizona Borderlands*. University of Arizona Press, Tucson.

Bahre, Conrad J., and Charles F. Hutchinson. 1985. The impact of historic fuelwood cutting on the semidesert woodlands of southeastern Arizona. *Journal of Forest History* 29: 175–186.

Bailey, Flora L. 1940. Navajo food and cooking methods. *American Anthropologist* 42 (2): 270–290.

Bailey, Virginia L. 1960. Historical review of *Ptelea trifoliata* in botanical and medical literature. *Economic Botany* 14: 180–188.

Bailowitz, Richard A., and James P. Brock. 1991. *Butterflies of southeastern Arizona*. Sonoran Arthropod Studies, Inc., Tucson, AZ.

Baldwin, Bruce G., Bridget L. Wessa, and José L. Panero. 2002. Nuclear rDNA evidence for major lineages of Helenioid Heliantheae (Compositae). *Systematic Botany* 27 (1): 161–198.

Balick, Michael J., Michael H. Nee, and Daniel E. Atha. 2000. *Checklist of the Vascular Plants of Belize. With Common Names and Uses*. New York Botanical Garden Press, Bronx.

Barkley, Theodore M., ed. 1977. *Atlas of the Flora of the Great Plains*. Iowa State University Press, Ames.

Barkley, Theodore M., ed. 1986. *Flora of the Great Plains*. University Press of Kansas, Lawrence.

Barrows, David P. 1900. *The Ethnobotany of the Coahuilla Indians of Southern California*. Including a Cahuilla bibliography and introductory essays by Harry W. Lawton, Lowell John Bean, and William Bright. Reprinted 1967 by Malki Museum Press, Banning, CA.

Basehart, Harry W. 1974. *Mescalero Apache Subsistence Patterns and Sociopolitical Organization*. Garland Publications Inc., New York.

Bean, Lowell J., and Katherine S. Saubel. 1972. *Temalpakh (from the Earth). Cahuilla Indian Knowledge and Usage of Plants*. Malki Museum Press, Morongo Indian Reservation, Banning, CA.

Bean, Travis, Martin Karpiscak, and Raymond Turner. 2004. Desert plant community development following retirement from agriculture. Oral Session 94: Disturbance Ecology VI: Anthropogenic Disturbance. 2004 Annual Meeting of the ESA, Portland, Oregon. http://abstracts.co.allenpress.com/pweb/esa2004/document/36423.

Bennett, Bradley S. 2007. Doctrine of Signatures: An explanation of medicinal plant discovery or dissemination of knowledge? *Economic Botany* 61 (3): 246–255.

Benson, Lyman. 1957. *Plant Classification*. D. C. Heath & Co., Boston.

Benson, Lyman. 1969. *The Cacti of Arizona*. University of Arizona Press, Tucson.

Benson, Lyman, and Robert A. Darrow. 1945. A Manual of Southwestern Desert Trees and Shrubs. *University of Arizona Bulletin* 15, no. 2. Tucson.

Benson, Lyman, and Robert A. Darrow. 1981. *Trees and Shrubs of the Southwestern Deserts*. University of Arizona Press, Tucson.

Bentham, Dale M., and Michael D. Windham. 1993. *Astrolepis*. Pp. 140–143. In: Editorial Committee. *Flora of North America North of Mexico*. Vol. 2. Oxford University Press, New York.

Berlin, Brent, Dennis E. Breedlove, and Peter H. Raven. 1974. *Principles of Tzeltal Plant Classification: An Introduction to the Botanical Ethography of a Mayan-Speaking People of Highland Chiapas*. Academic Press, New York.

Betancourt, Julio L., Thomas R. Van Devender, and Martin Rose. 1986. Comparison of plant macrofossils in Woodrat (*Neotoma* sp.) and Porcupine (*Erethizon dorsatum*) middens. *Journal of Mammalogy* 67 (2): 266–273.

Blackshaw, Robert E. 2003. Soil temperature and soil water effects on pygmyflower (*Androsace septentrionalis*) emergence. *Weed Science* 51 (4): 592–595.

BLM. 2005. *Baboquivari Wilderness*. http://www.az.blm.gov/rec/baboquiv.htm.

Bocek, Barbara R. 1984. Ethnobotany of the Costanoan Indians, California, based on collections by John P. Harrington. *Economic Botany* 38 (2): 240–255.

Bock, Jane H., and Carl E. Bock. 2002. Exotic species in grasslands. Pp. 147–164. In: Tellman, Barbara, ed. *Invasive Exotic Species in the Sonoran Region*. University of Arizona Press, Tucson.

Bohlman, Ferdinand, and Christa Zdero. 1979. Naturlich vorkommende Terpen-Derivate. 156. Uber eine neue Gruppe von Sesquiterpenlactonen aus der Gattung *Trixis* [Naturally occurring terpene derivatives. 156. On a new group of sesquiterpene lactones from the genus *Trixis*]. *Gesellschaft Deutscher Chemiker* 112 (2): 435–444.

Bohrer, Vorsila L. 1975a. The prehistoric and historic role of the cool-season grasses in the Southwest. *Economic Botany* 29: 199–207.

Bohrer, Vorsila L. 1975b. Recognition and interpretation of prehistoric remains of *Mirabilis multiflora* (Nyctaginaceae) in the Sacramento Mountains of New Mexico. *Bulletin of the Torrey Botanical Club* 102 (1): 21–25.

Bohrer, Vorsila L. 1991. Recently recognized cultivated and encouraged plants among the Hohokam. *The Kiva* 56: 227–235.

Bohrer, Vorsila L., Hugh C. Cutler, and Jonathan D. Sauer. 1969. Carbonized plant remains from two Hohokam sites. Arizona BB:13:41, 13:50. *The Kiva* 35 (1): 1–10.

Bolton, Herbert E. 1936. *Rim of Christendom: A Biography of Eusebio Francisco Kino, Pacific Coast Pioneer*. Macmillan Company, New York. Reissued, copyright 1984, University of Arizona Press, Tucson.

Bourke, John G. 1890. Notes upon the gentile organization of the Apaches of Arizona. *The Journal of American Folklore* 3 (9): 111–126.

Bourke, John G. 1895. The folk-foods of the Rio Grande Valley and of northern Mexico. *The Journal of American Folklore* 8 (28): 41–71.

Bowers, Janice E. 1980. Flora of Organ Pipe Cactus National Monument. *Journal of the Arizona-Nevada Academy of Sciences* 15: 1–11, 33–47.

Bowers, Janice E. 1981. Local floras in the Southwest, 1920–1980: An annotated bibliography. *Great Basin Naturalist* 42: 105–112.

Bowers, Janice E., and Steven P. McLaughlin. 1982. Plant species diversity in Arizona. *Madroño* 29: 227–233.

Bowers, Janice E., and Steven P. McLaughlin. 1987. Flora and vegetation of the Rincon Mountains, Pima County, Arizona. *Desert Plants* 8 (2): 51–94.

Bowers, Janice E., and Steven P. McLaughlin. 1996. Flora of the Huachuca Mountains, a botanically rich and historically significant sky island in Cochise County, Arizona. *Journal of the Arizona-Nevada Academy of Sciences* 29 (2): 68–107.

Bowers, Janice E., Raymond M. Turner, and Tony L. Burgess. 2004. Temporal and spatial patterns in emergence and early survival of perennial plants in the Sonoran Desert. *Plant Ecology* 172 (1): 107–119.

Bown, Deni. 1995. *The Herb Society of America Encyclopedia of Herbs and Their Uses*. Dorling Kindersley Publishing Inc., New York.

Bradley, G., R. Glock, T. H. Noon, and C. Reggiardo. 1998. Equine. *Arizona Veterinary Diagnostic Laboratory Newsletter* 3 (4): 2.

Brandegee, Townshend S. 1900. Cleistogamous flowers in Scrophulariaceae. *Zoe* 5: 13.

Bray, Deni, ed. 1998. *Western Apache-English Dictionary: A Community-Generated Bilingual Dictionary*. Bilingual Press/Editorial Bilingüe, Tempe, AZ.

Breedlove, Dennis E., and Robert M. Laughlin. 2000. *The Flowering of Man: A Tzotzil Botany of Zinacantan*. Smithsonian Institution Press, Washington, DC.

Brenzel, Kathleen N., ed. 1995. *Sunset Western Garden Book*. Sunset Publishing Corporation, Menlo Park, CA.

Bretting, Peter. 1986. Folk names and uses for Martyniaceous plants. *Economic Botany* 38: 452–463.

Brickell, Christopher, and Judith D. Zuk, eds. 1997. *The American Horticultural Society A–Z Encyclopedia of Garden Plants*. DK Publishing Inc., New York.

Britton, Nathaniel L., and Addison Brown. 1896–1898. *An Illustrated Flora of the Northern United States, Canada, and the British Possessions*. Charles Scribner's Sons, New York.

Britton, Nathaniel L., and Joseph N. Rose. 1937. *The Cactaceae: Descriptions and Illustrations of Plants of the Cactus Family*. Reprinted 1963 by Dover Publ. Inc., New York.

Brown, David E., ed. 1982. Biotic communities of the American Southwest—United States and Mexico. *Desert Plants* 4 (1–4): 1–342.

Bryant, Nevin A., Lee F. Johnson, Anthony J. Brazel, Robert C. Balling, Charles F. Hutchinson, and Louisa R. Beck. 1990. Measuring the effect of overgrazing in the Sonoran Desert. *Climatic Change* 17 (2–3): 243–264.

Buchmann, Stephen L. 1987. Floral biology of jojoba (*Simmondsia chinensis*), an anemophilous plant. *Desert Plants* 8 (3): 111–124.

Burkill, Isaac H. 1966. *A Dictionary of the Economic Products of the Malay Peninsula*. Ministry of Agriculture and Cooperatives, Kuala Lumpur.

Bye, Robert A., Jr. 1972. Ethnobotany of the Southern Paiute Indians in the 1870s: With a note on the early ethnobotanical contributions of Dr. Edward Palmer. Pp. 87–104. In: Fowler, D. Don, ed. *Great Basin Cultural Ecology: A Symposium*. Desert Research Institute Publications in the Social Sciences No. 8. University of Nevada, Reno.

Bye, Robert A., Jr. 1979a. Hallucinogenic plants of the Tarahumara. *Journal of Ethnopharmacology* 1: 23–48.

Bye, Robert A., Jr. 1979b. An 1878 ethnobotanical collection from San Luis Potosí: Dr. Edward Palmer's first major Mexican collection. *Economic Botany* 33: 135–162.

Bye, Robert A., Jr. 1986. Comparative study of Tarahumara and Mexican market plants. *Economic Botany* 40: 103–124.

Callen, Eric O. 1963. Diet as revealed by coprolites. Pp. 186–194. In: Brothwell, Don R., and Eric Higgins, eds. *Science and Archaeology*. Basic Books Inc., New York.

Callen, Eric O. 1967a. The first New World cereal. *American Antiquity* 32: 535–538.

Callen, Eric O. 1967b. Food habits of some pre-Columbian Mexican Indians. *Economic Botany* 19 (4): 335–343.

Campa, Arthur L. 1970. *Hispanic Culture in the Southwest*. University of Oklahoma Press, Norman.

Campbell, Christopher S., Michael J. Donoghue, Bruce G. Baldwin, and Martin F. Wojciechowski. 1995. Phylogenetic relationships of Maloideae (Rosaceae): Evidence from sequences of internal transcribed spacers of nuclear ribosomal DNA and its congruence with morphology. *American Journal of Botany* 82 (7): 903–918.

Cannon, John, and Margaret Cannon. 2003. *Dye Plants and Dyeing*, Revised Edition. Timber Press, Inc., Portland, Oregon, in association with Royal Botanic Gardens, Kew, England.

Carter, Jack L. 1988. *Trees and Shrubs of Colorado*. Johnson Books, Boulder.

Casas, Alejandro, María del Carmen Vázquez, Juan L. Viveros, and Javier Caballero. 1996. Plant management among the Nahua and the Mixtec in the Balsas River Basin, Mexico: An ethnobotanical approach to the study of plant domestication. *Human Ecology* 24 (4): 455–478.

Castetter, Edward F. 1935. Ethnobiological studies in the American Southwest I. Uncultivated native plants used as sources of food. *University of New Mexico Bulletin* 4 (1): 1–44.

Castetter, Edward F., and Willis H. Bell. 1942. *Pima and Papago Indian Agriculture*. University of New Mexico Press, Albuquerque.

Castetter, Edward F., and Willis H. Bell. 1951. *Yuman Indian Agriculture: Primitive Subsistence on the Lower Colorado and Gila Rivers*. University of New Mexico Press, Albuquerque.

Castetter, Edward F., and Morris E. Opler. 1936. Ethnobiological studies in the American Southwest III. The ethnobiology of the Chiricahua and Mescalero Apache. *University of New Mexico Bulletin* 4 (5): 1–63.

Castetter, Edward F., and Ruth M. Underhill. 1935. The ethnobiology of the Papago Indians. Ethnobiological Studies in the American Southwest II. *University of New Mexico Bulletin, Biological Series* 4 (3): 1–84.

Center for Desert Archaeology. 2005. Native American Lifeways (11,000 B.C. to Present). http://www.cdarc.org/pages/what/current/SCNHA/ch04_e.pdf.

Chamberlin, Ralph V. 1909. Some plant names of the Ute Indians. *American Anthropologist* 11 (1): 27–40.

Chamberlin, Ralph V. 1911. The ethnobotany of the Gosiute Indians of Utah. *Memoirs of the American Anthropological Association* 2 (5): 331–405.

Chapman, Jefferson, Robert B. Stewart, and Richard A. Yarnell. 1974. Archaeological evidence for pre-Columbian introduction of *Portulaca oleracea* and *Mollugo verticillata* into eastern North America. *Economic Botany* 28: 411–412.

Cholewa, Anita F. 1992. Primulaceae. Primrose Family. *Journal of the Arizona-Nevada Academy of Science* 26 (1): 17–21.

Christie, Kyle, Michael Currie, Laura Smith Davis, Mar-Elise Hill, Suzanne Neal, and Tina Ayers. 2006. Vascular plants of Arizona: Rhamnaceae. *Canotia* 2 (1): 23–46. http://collections.asu.edu/herbarium/canotia.html.

Christy, Charlotte M. 1998. Vascular plants of Arizona: Loasaceae. In: A New Flora of Arizona. *Journal of the Arizona-Nevada Academy of Sciences* 30 (2): 96–111.

Clapham, A. R., T. G. Tutin, and D. M. Moore. 1987. *Flora of the British Isles*. Cambridge University Press, Cambridge.

Coffey, Timothy. 1993. *The History and Folklore of North American Wildflowers*. Houghton Mifflin Co., Boston.

Cohan, Dan. 2002. Map of land ownership adjacent to the Buenos Aires National Wildlife Refuge, Arizona. Unpublished.

Corn, Joseph L., and Robert J. Warren. 1985. Seasonal food habits of the collared peccary in south Texas. *Journal of Mammalogy* 66 (1): 155–159.

Correll, Donovan S., and Helen B. Correll. 1972. *Aquatic and Wetland Plants of Southwestern United States*. Environmental Protection Agency, Water Pollution Control Research Series 16030 DNL 01/72, Washington, DC.

Correll, Donovan S., and Helen B. Correll. 1982. *Flora of the Bahama Archipelago*. J. Cramer, Vaduz.

Correll, Donovan S., and Marshall C. Johnston. 1970. *Manual of the Vascular Plants of Texas*. Texas Research Foundation, Renner.

Coville, Frederick V. 1892. The Panamint Indians of California. *American Anthropologist* 5 (4): 351–362.

Crawford, James M. 1989. *Cocopa Dictionary*. University of California Publications in Linguistics Volume 114. University of California Press, Berkeley.

Crawford, Phillip T., and Wayne J. Elisens. 2006. Genetic variation and reproductive system among North American species of *Nuttallanthus* (Plantaginaceae). *American Journal of Botany* 93: 582–591.

Crispens, Charles G., Jr., Irven O. Buss, and Charles F. Yocom. 1960. Food habits of the California Quail in eastern Washington. *The Condor* 62 (6): 473–477.

Crosswhite, Frank S. 1980. The annual saguaro harvest and crop cycle of the Papagos, with special reference to ecology and symbolism. *Desert Plants* 2: 2–61.

Crosswhite, Frank S. 1981. Desert plants, habitat, and agriculture in relation to the major pattern of cultural differentiation in the O'odham people of the Sonoran desert. *Desert Plants* 3 (2): 47–76.

Crosswhite, Frank S. 1983. Selenium and *Castilleja*. *Desert Plants* 5: 96.

Crosswhite, Frank S., and Carol D. Crosswhite. 1984. The southwestern pipevine (*Aristolochia watsonii*) in relation to snakeroot oil, swallowtail butterflies, and ceratopogonid flies. *Desert Plants* 6 (4): 203–207.

Culpeper, Nicholas. 1653. *Complete Herbal: Consisting of a comprehensive description of nearly all herbs with their Medicinal Properties and directions for compounding the medicines extracted from them.* W. Foulsham & Co., Ltd., London.

Curtin, Leonora S. M. 1947. *Healing Herbs of the Upper Río Grande*. Laboratory of Anthropology, Santa Fe.

Curtin, Leonora S. M. 1949. *By the Prophet of the Earth: Ethnobotany of the Pima*. Reprinted 1984 by University of Arizona Press, Tucson.

Curtin, Leonora S. M. 1957. Some plants used by the Yuki Indians of Round Valley, Northern California. *The Masterkey* 31 (1): 85–94.

Curtis, Edward S. 1907–1930. *The North American Indian: Being a Series of Volumes Picturing and Describing the Indians of the United States and Alaska*. Published by the author; edited by Frederick Webb Hodge.

Cusick, Allison W. 1989. *Androsace occidentalis*. http://www.dnr.state.oh.us/Portals/3/Abstracts/Abstract_pdf/A/ANDROSACE_OCCIDENTALIS.pdf.

Daniel, Thomas F. 1984. The Acanthaceae of the southwestern United States. *Desert Plants* 5 (4): 162–179.

Daniel, Thomas F. 2004. Acanthaceae of Sonora: Taxonomy and phytogeography. *Proceedings of the California Academy of Sciences* 55 (35): 690–805.

Daniel, Thomas F., Lucinda A. McDade, Mariette Manktelow, and Carrie A. Kiel. 2008. The "*Tetramerium* Lineage" (Acanthaceae: Acanthoideae: Justicieae): Delimitation and intra-lineage relationships based on cp and nrITS sequence data. *Systematic Botany* 33 (2): 416–436.

Davidson, Alan. 1999. *The Oxford Companion to Food*. Oxford University Press, Oxford.

Dayley, Jon P. 1989. *Tümpisa (Panamint) Shoshone Dictionary*. University of California Publications in Linguistics Volume 116. University of California Press, Berkeley.

Dayton, William A. 1937. *Range Plant Handbook*. United States Department of Agriculture, Forest Service, Washington, DC.

Dempster, Lauramay, and Edward T. Terrell. 1995. Rubiaceae. In: A New Flora of Arizona. *Journal of the Arizona-Nevada Academy of Sciences* 29 (1): 29–38.

Deschinny, Isabell. 1984. *Native Plant Dyes. Series 1: Introduction*. Published by author, Window Rock, AZ.

Desowitz, Robert S. 1987. *New Guinea Tapeworms and Jewish Grandmothers*. W. W. Norton and Co., New York.

DeVlaming, Victor, and Vernon W. Proctor. 1968. Dispersal of aquatic organisms: Viability of seeds recovered from the droppings of captive killdeer and mallard ducks. *American Journal of Botany* 55 (1): 20–26.

De Wet, Jan M. J. 1977. Domestication of African cereals. *African Economic History* 3: 15–32.

Dewey, Douglas R. 1971. Synthetic hybrids of *Hordeum bogdanii* with *Elymus canadensis* and *Sitanion hystrix*. *American Journal of Botany* 58 (19): 902–908.

Diggs, George M. J., Barney L. Lipscomb, and Robert J. O'Kennon. 1999. *Shinner's & Mahler's Illustrated Flora of North Central Texas*. Center for Environmental Studies and Department of Biology, Austin College, Sherman, Texas, and Botanical Research Institute of Texas (BRIT), Fort Worth.

Dimmitt, Mark A. 2000. *Aristolochia*, *Ferocactus*, and *Echinocactus*. Pp. 168–169; 202–207. In: Phillips, S. J., and P. W. Comus, eds. *A Natural History of the Sonoran Desert*. University of California Press, Berkeley, and Arizona-Sonora Desert Museum Press, Tucson.

Dimmitt, Mark A., and Thomas R. Van Devender. 2003. Biological Survey of the Ironwood Forest National Monument. Final Report to the Bureau of Land Management, Tucson Field Office.

Di Peso, Charles C. 1953. The Sobaipuri Indians of the Upper San Pedro River Valley, Southeastern Arizona. Ph.D. Dissertation, University of Arizona, Tucson.

Dixon, D. M. 1969. A note on cereals in ancient Egypt. Pp. 131–142. In: Ucko, Peter J., and G. W. Dimbleby, eds. *The Domestication and Exploitation of Plants and Animals*. Aldine, Chicago.

Dobelis, Inge N., ed. 1986. *Magic and Medicine of Plants*. Reader's Digest Association, Pleasantville, NY.

Dodge, Natt N. 1976. *Flowers of the Southwest Deserts*. Southwest Parks and Monuments Association, Globe.

Donovan, A. J., and R. Topinka. 2004. The pollination ecology of Kearney's bluestar (*Amsonia kearneyana*: Apocynaceae). Poster at Fourth Southwestern Rare and Endangered Species Conference. http://nmrareplants.unm.edu/conference/abstracts.htm.

Doolittle, William W. 2000. *Cultivated Landscapes of Native North America*. Oxford University Press, New York.

Douglas, Marjory S. 1947. *The Everglades: River of Grass*. Hurricane House Publishing, Inc., Coconut Grove, FL.

Downie, S. R., G. M. Plunkett, M. F. Watson, K. Spalik, D. S. Katz-Downie, C. M. Valiejo-Roman, E. I. Terentieva, A. V. Troitsky, B.-Y. Lee, J. Lahham, and A. El-Oqlah. 2001. Tribes and clades within Apiaceae subfamily Apioideae: The contribution of molecular data. *Edinburgh Journal of Botany* 58: 301–330.

Duke, James A., with Mary J. Bogenschutz-Godwin, Judi duCellier, and Peggy-Ann Duke. 2002. *Handbook of Medicinal Herbs*. Second Edition. CRC Press LLC, Boca Raton, FL.

Dunmire, William W. 2004. *Gardens of New Spain. How Mediterranean Plants and Foods Changed America*. University of Texas Press, Austin.

Dunmire, William W., and Gail D. Tierney. 1995. *Wild Plants of the Pueblo Province: Exploring Ancient and Enduring Uses*. Museum of New Mexico Press, Santa Fe.

Dunmire, William W., and Gail D. Tierney. 1997. *Wild Plants and Native Peoples of the Four Corners*. Museum of New Mexico Press, Santa Fe.

Dunn, Rob R. 2005. Jaw of life. *Natural History* 114 (7): 30–35.

Dutton, Bryan E., Carl S. Keener, and Bruce A. Ford. 1997. *Anemone*. Pp. 139–158. In: Editorial Committee. *Flora of North America North of Mexico*. Vol. 3. Oxford University Press, New York.

Ebeling, Walter. 1986. *Handbook of Indian Foods and Fibers of Arid America*. University of California Press, Berkeley.

Editorial Committee. 1993. *Flora of North America North of Mexico*. Vol. 2. Oxford University Press, New York.

Editorial Committee. 2006. *Flora of North America North of Mexico*. Vol. 19. Oxford University Press, New York.

Ehrenfeld, Joan. 1976. Reproductive biology of three species of *Euphorbia* subgenus *Chamaesyce* (Euphorbiaceae). *American Journal of Botany* 63 (4): 406–413.

Elisens, Wayne J. 1985. Monograph of the Maurandyinae (Scrophulariaceae-Antirrhineae). *Systematic Botany Monographs*. Vol. 5. Ann Arbor, MI.

Elisens, Wayne J., and C. Edward Freeman. 1988. Floral nectar sugar composition and pollinator type among New World genera in tribe Antirrhineae (Scrophulariaceae). *American Journal of Botany* 75 (7): 971–978.

Elmore, Francis H. 1944. Ethnobotany of the Navajo. *University of New Mexico Bulletin* No. 392 1 (2): 1–136.

Elvander, Patrick. 1992. Saxifragaceae. In: A New Flora of Arizona. *Journal of the Arizona-Nevada Academy of Sciences* 26 (1): 36–41.

Engard, Rodney G. 1989. *The Flowering Southwest: Great Impressions*, Tucson, AZ.

Escalante-H., Roberto, and Zarina Estrada-F. 1993. *Textos y gramática del pima bajo*. Universidad de Sonora, Hermosillo, Sonora, Mexico.

Esque, Todd C., and Cecil R. Schwalbe. 2002. Alien annual grasses and their relationships to fire and biotic changes in Sonoran desertscrub. Pp. 165–194. In: Tellman, Barbara, ed. *Invasive Exotic Species in the Sonoran Region*. University of Arizona Press, Tucson.

Evans, Rodger C., and Christopher S. Campbell. 2002. The origin of the apple family (Maloideae; Rosaceae) is clarified by DNA sequence data from duplicated GBSSI genes. *American Journal of Botany* 89 (9): 1478–1484.

Everett, J. R., and C. L. Gonzalez. 1981. Seasonal nutrient content in food plants of white-tailed deer on the South Texas Plains. *Journal of Range Management* 34 (6): 506–510.

Ezell, Paul H. 1983. History of the Pima. Pp. 149–160. In: Ortiz, Alfonso, ed. 1983. *Southwest. Handbook of North American Indians*. Vol. 10. Smithsonian Institution, Washington, DC.

Faden, Robert B. 1992. Floral attraction and floral hairs in the Commelinaceae. *Annals of the Missouri Botanical Garden* 79 (1): 46–52.

Farish, Thomas E. 1915–1918. *History of Arizona*. Phoenix, AZ. Online version, http://www.library.arizona.edu/swetc/projects.html.

Fassett, Norman C. 1942. Populations of *Linaria* on the Gulf Coast. *American Journal of Botany* 29 (5): 351–352.

Felger, Richard S. 1977. Mesquite in Indian cultures of southwestern North America. Pp. 150–176. In: Simpson, Beryl B., ed. *Mesquite: Its Biology in Two Desert Scrub Ecosystems*. Dowden, Hutchinson & Ross, Stroudsburg, PA.

Felger, Richard S. 1999. The flora of Cañón del Nacapule: A desert-bounded tropical canyon near Guaymas, Sonora, Mexico. *Proceedings of the San Diego Society of Natural History* 35: 1–42.

Felger, Richard S. 2000. *Flora of the Gran Desierto and Río Colorado Delta*. University of Arizona Press, Tucson.

Felger, Richard S. in preparation. *Grasses of the Sonoran Desert*.

Felger, Richard S., and Bill Broyles, eds. 2007. *Dry Borders: Great Natural History Reserves of the Sonoran Desert*. University of Utah Press, Salt Lake City.

Felger, Richard S., Matthew S. Johnson, and Michael F. Wilson. 2001. *The Trees of Sonora, Mexico*. Oxford University Press, New York.

Felger, Richard S., and Mary B. Moser. 1974. Columnar cacti in Seri Indian culture. *The Kiva* 39 (3–4): 257–275.

Felger, Richard S., and Mary B. Moser. 1976. Seri Indian food plants: Desert subsistence without agriculture. *Ecology of Food and Nutrition* 5: 13–27.

Felger, Richard S., and Mary B. Moser. 1985. *People of the Desert and Sea: Ethnobotany of the Seri Indians*. University of Arizona Press, Tucson.

Felger, Richard S., and Gary P. Nabhan. 1976. Agroecosystem diversity: A model from the Sonoran Desert. Pp. 129–149. In: *Social and Technological Management in Dry Lands: Past and Present, Indigenous and Imposed*. Gonzalez, G., ed. AAAS Selected Symposium no. 10. Westview Press, Boulder, CO.

Felger, Richard S., Susan Rutman, Thomas R. Van Devender, and Michael F. Wilson. in preparation. *Ajo Peak to Tinajas Altas: Flora of Southwestern Arizona*.

Felger, Richard S., Susan Rutman, Michael F. Wilson, and Kathryn Mauz. 2007. Botanical diversity of southwestern Arizona and northwestern Sonora. Pp. 202–271. In: Felger, Richard S., and Bill Broyles, eds. *Dry Borders: Great Natural Reserves of the Sonoran Desert*. University of Utah Press, Salt Lake City.

Felger, Richard S., Peter L. Warren, Susan A. Anderson, and Gary P. Nabhan. 1992. Vascular plants of a desert oasis: Flora and ethnobotany of Quitobaquito, Organ Pipe Cactus National Monument, Arizona. *Proceedings of the San Diego Society of Natural History* 8: 1–39.

Fernald, Merritt L. 1950. *Gray's Manual of Botany*. American Book Co., New York.

Fernald, Merritt L., Alfred C. Kinsey, and Reed C. Rollins. 1958. *Edible Wild Plants of Eastern North America*. Harper and Row Publishers, New York.

Fewkes, J. Walter. 1896. A contribution to ethnobotany. *American Anthropologist* 9 (1): 14–21.

Fish, Suzanne K., and Paul R. Fish. 1992. Prehistoric landscapes of the Sonoran Desert Hohokam. *Population and Environment* 13 (4): 269–283.

Fisk, Erma J. 1983. *The Peacocks of Baboquivari*. W. W. Norton & Co., New York.

Fontana, Bernard L. 1980. Ethnobotany of the saguaro, an annotated bibliography. *Desert Plants* 2: 62–78.

Fontana, Bernard L. 1983a. Pima and Papago: Introduction. Pp. 125–136. In: Ortiz, Alfonso, ed. 1983. *Southwest. Handbook of North American Indians*. Vol. 10. Smithsonian Institution, Washington, DC.

Fontana, Bernard L. 1983b. History of the Papago. Pp. 137–148. In: Ortiz, Alfonso, ed. 1983. *Southwest. Handbook of North American Indians*. Vol. 10. Smithsonian Institution, Washington, DC.

Ford, Karen C. 1975. *Las Yerbas de la Gente: A Study of Hispano-American Medicinal Plants*. Anthropological Papers, Museum of Anthropology, University of Michigan, No. 60, Ann Arbor.

Foster, Steven, and James A. Duke. 1990. *A Field Guide to Medicinal Plants*. Houghton Mifflin Co., Boston.

Fowler, Catherine S. 1972. Some ecological clues to Proto-Numic homelands. Pp. 105–122. In: Fowler, Don D., ed. *Great Basin Cultural Ecology: A Symposium*. Desert Research Institute Publications in the Social Sciences No. 8. University of Nevada, Reno.

Fowler, Catherine S., ed. 1989. Willard Z. Park's ethnographic notes on the Northern Paiute of Western Nevada, 1933–1940. *University of Utah, Anthropological Papers*, 114.

Fowler, Catherine S., and Joy H. Leland. 1967. Some Northern Paiute native categories. *Ethnology* 6 (4): 381–404.

Franciscan Fathers. 1910. *An Ethnologic Dictionary of the Navaho Language*. Navajo Indian Mission, Saint Michaels, AZ.

Freeman, Donald C., E. Durant McArthur, and Kimball T. Harper. 1984. The adaptive significance of sexual lability in plants using *Atriplex canescens* as a principal example. *Annals of the Missouri Botanical Garden* 71: 265–277.

Fritz, Gayle J. 2000. Levels of native biodiversity in eastern North America. Pp. 223–247. In: Minnis, Paul E., and Wayne J. Elisens, eds. 2000. *Biodiversity and Native America*. University of Oklahoma Press, Norman.

Fryxell, Paul A. 1993. Malvaceae Mallow Family Part One: All genera except *Sphaeralcea* St. Hil. *Journal of the Arizona-Nevada Academy of Science* 27 (2): 222–236.

Gallagher, Marsha V. 1976. Contemporary ethnobotany among the Apache of the Clarkdale, Arizona. Coconino and Prescott National Forests. Report No. 14. U.S. Department of Agriculture, Forest Service, Southwestern Region.

Garcia, Cecilia, and James D. Adams. 2005. *Healing with Medicinal Plants of the West*. Abedus Press, La Crescentia, CA.

Gasser, Robert E. 1981. Hohokam use of desert plant foods. *Desert Plants* 3 (4): 216–234.

Gentry, Howard S. 1942. *Rio Mayo Plants*. Carnegie Institution, Washington, DC.

Gentry, Howard S. 1963. The Warihio Indians of Sonora-Chihuahua: An ethnographic survey. Smithsonian Institution Bureau of American Ethnological Bulletin No. 186. Anthropology Papers No. 65. U.S. Government Printing Office, Washington, DC.

Gentry, Howard S. 1972. The Agave Family in Sonora. Agricultural Research Service, USDA, Agric. Handbook No. 399. Washington, DC.

Gentry, Howard S. 1982. *Agaves of Continental North America*. University of Arizona Press, Tucson.

Getachew, G., H. P. S. Makkar, and K. Becker. 2000. Tannins in tropical browses: Effects on *in vitro* microbial fermentation and microbial protein synthesis in media containing different amounts of nitrogen. *Journal of Agricultural and Food Chemistry* 48 (8): 3581–3588.

Glassberg, Jeffrey. 2001. *Butterflies through Binoculars. The West*. Oxford University Press, New York.

Gleason, Henry A., and Arthur Cronquist. 1963. *Manual of Vascular Plants of Northeastern United States and Adjacent Canada*. D. Van Nostrand Company, Inc., Princeton, NJ.

Goldberg, Deborah E., and Raymond M. Turner. 1986. Vegetation changes and plant demography in permanent plots of the Sonoran Desert. *Ecology* 67 (3): 695–712.

Goldman, Douglas H. 2003. Two species of *Passiflora* (Passifloraceae) in the Sonoran Desert and vicinity: A new taxonomic combination and an introduced species in Arizona. *Madroño* 50: 243–264.

Gordon, Raymond G., Jr., ed. 2005. *Ethnologue: Languages of the World*, Fifteenth Edition. SIL International, Dallas, TX. http://www.ethnologue.com.

Gottlieb, L. D. 2006. *Rafinesquia* Pp. 348–349; *Stephanomeria* Pp. 350–359. In: Editorial Committee, eds. *Flora of North America North of Mexico*. Vol. 19. Oxford University Press, New York.

Gould, Frank W. 1951. Grasses of southwestern United States. *University of Arizona Bulletin* 22 (1): 1–352.

Gould, Frank W., and Reid Moran. 1981. The Grasses of Baja California, Mexico. *Memoirs of the San Diego Society of Natural History*. Vol. 12. San Diego, CA.

Granger, Byrd H. 1960. *Will C. Barnes' Arizona Place Names*. University of Arizona Press, Tucson.

Hamilton, Patrick. 1881. *Resources of Arizona: Its Mineral, Farming, and Grazing Lands, Towns, and Mining Camps; Its Rivers, Mountains, Plains, and Mesas; With a Brief Summary of Its Indian Tribes, Early History, Ancient Ruins, Climate, etc. etc. A Manual of Reliable Information Concerning the Territory. Under Authority of the Legislature*. Prescott, Arizona. This electronic copy was produced in February 2002, http://southwest.library.arizona.edu/reaz/body.1_div.9.html#page41.

Hanson, Roseann B. 1997. Checklist of Brown Canyon. Unpublished manuscript at Buenos Aires National Wildlife Refuge, Sasabe, AZ.

Hardin, James W., and Jay M. Arena. 1974. *Human Poisoning from Native and Cultivated Plants*. Duke University Press, Durham, NC.

Harrington, John P. 1933. Preliminary Juaneño vocabulary. Notes at http://www.ncidc.org/bright/Juaneno.doc.

Hart, Jeff. 1976. *Montana: Native Plants and Early Peoples*. Montana Historical Society, Helena.

Haufler, Christopher H., and Douglas E. Soltis. 1986. Genetic evidence suggests that homosporous ferns with high chromosome numbers are diploid. *Proceedings of the National Academy of Sciences, United States* 83: 4389–4393.

Havard, Valéry. 1896. Drink plants of the North American Indians. *Bulletin of the Torrey Botanical Club* 23 (2): 33–46.

Hayden, Julian D. 1987. The Vikita ceremony of the Papago. *Journal of the Southwest* 29: 273–324, plates i–xi.

Hedrick, Ulysses P., ed. 1919. *Sturtevant's Notes on Edible Plants*. J. B. Lyon Co., Albany, NY.

Heinrich, Michael. 2002. Ethnobotany, phytochemstry, and biological/pharmacological activities of *Artemisia ludoviciana* ssp. *mexicana* (estafiate). Pp. 107–117. In: Wright, Colin W. Artemisia. *The Genus* Artemisia: *Medicinal and Aromatic Plants—Industrial Profiles*. Vol. 18. Taylor & Francis, London.

Heiser, Charles B., Jr. 1979. *The Gourd Book*. University of Oklahoma Press, Norman.

Heiser, Charles B. 2008. The sunflower (*Helianthus annuus*) in Mexico: Further evidence for a North American domestication. *Genetic Resources and Crop Evolution* 55: 9–13.

Henrickson, James. 1972. A taxonomic revision of the Fouquieriaceae. *Aliso* 7: 439–537.

Henrickson, James, and Marshall C. Johnston. n.d. A Flora of the Chihuahuan Desert Region. Unpublished manuscript.

Hess, William J., and James Henrickson. 1987. A taxonomic revision of *Vauquelinia* (Rosaceae). *Sida* 12: 101–163.

Hickman, James C., ed. 1993. *The Jepson Manual. Higher Plants of California*. University of California Press, Berkeley.

Hicks, Sam. 1966. *Desert Plants and People*. The Nalor Company, San Antonio, TX.

Hill, Ellsworth J. 1909. Pollination in *Linaria* with special reference to cleistogamy. *Botanical Gazette* 47 (6): 454–466.

Hill, Jane H. 2001. Proto-Uto-Aztecan: A community of cultivators in Central Mexico? *American Anthropologist* 103 (4): 913–934.

Hill, Jane H. 2002. Toward a linguistic prehistory of the Southwest: "Azteco-Tanoan" and the arrival of maize cultivation. *Journal of Anthropological Research* 58 (4): 457–475.

Hill, Kenneth C., Emory Sekaquaptewa, Mary E. Black, Ekkehart Malotki, and Michael Lomatuway'ma. 1998. *Hopi Dictionary/Hopìikwa lavàytutuveni: A Hopi-English Dictionary of the Third Mesa Dialect with an English-Hopi Finder List and a Sketch of Hopi Grammar.* University of Arizona Press, Tucson.

Hilton, K. Simón. 1993. *Diccionario Tarahumara de Samanchique, Chihuahua, México.* Instituto Linguístico de Verano. Tucson, AZ.

Hitchcock, Albert S., and Agnes Chase. 1950. *Manual of the Grasses of the United States.* U.S. Department of Agriculture, Miscellaneous Publication No. 200. Reprinted 1971 by Dover Publications, Inc., New York.

Hocking, George M. 1956. Some plant materials used medicinally and otherwise by the Navaho Indians in the Chaco Canyon, New Mexico. *Palacio* 63: 146–165.

Hocking, George M. 1997. *A Dictionary of Natural Products: Terms in the Field of Pharmacognosy Relating to Natural Medicinal and Pharmaceutical Materials and the Plants, Animals, and Minerals from Which They Are Derived.* Plexus Publishing, Inc., Medford, NJ.

Hodgson, Wendy C. 1999. Agavaceae Agave Family Part One. *Agave* L. Century Plant, Maguey. *Journal of the Arizona-Nevada Academy of Science* 32 (1): 1–21.

Hodgson, Wendy C. 2001. *Food Plants of the Sonoran Desert.* University of Arizona Press, Tucson.

Hoffmeister, Donald F. 1986. *Mammals of Arizona.* University of Arizona Press and Arizona Game and Fish Department, Tucson.

Holm, Richard W. 1950. The American species of *Sarcostemma* R. Br. (Asclepiadaceae). *Annals of the Missouri Botanical Garden* 37 (4): 470–560.

Holmgren, Noel H., Patricia K. Holmgren, and Arthur Cronquist. 2005. *Intermountain Flora: Vascular Plants of the Intermountain West, U.S.A.* Volume Two, Part B. *Subclass Dilleniidae.* The New York Botanical Garden Press, Bronx.

Hough, Walter. 1897. The Hopi in relation to their plant environment. *American Anthropologist* 10 (2): 33–44.

Hough, Walter. 1898. Environmental interrelations in Arizona. *American Anthropologist* 11 (5): 133–155.

Hrdlička, Ales. 1904. Method of preparing tesvino among the White River Apaches. *American Anthropologist* 6: 190–191.

Hrdlička, Ales. 1908. Physiological and medical observations among the Indians of southwestern United States and northern Mexico. Bureau of American Ethnology Bulletin 34. Smithsonian Institution, Washington, DC.

Hsu, Hong-Yen. 1986. *Oriental Materia Medica.* Keats Publishing Inc., New Canaan, CT.

Huisinga, Kristin D., and Tina J. Ayers. 1999. Plantaginaceae. *Journal of the Arizona-Nevada Academy of Sciences* 32 (1): 62–76.

Hultén, Eric. 1968. *Flora of Alaska and Neighboring Territories.* Stanford University Press, Stanford, CA.

Humphrey, Robert R. 1960. Forest production on Arizona ranges: Pima, Pinal, and Santa Cruz counties. University of Arizona, *Agricultural Experiment Station Bulletin* 302: 1–138.

Humphrey, Robert R., Albert L. Brown, and A. C. Everson. 1952. Common Arizona Range Grasses. *Agricultural Experiment Station Bulletin* 243.

Hurd, Paul, Jr., and Earle G. Linsley. 1971. Squash and gourd bees (*Peponapis, Xenoglossa*) and the origin of the cultivated *Cucurbita. Evolution* 25 (1): 218–234.

Ivey, Robert D. 2003. *Flowering Plants of New Mexico,* Fourth Edition. RD & V Ivey Publishers, Albuquerque, NM.

Jackson, Benjamin D. 1928. *A Glossary of Botanic Terms.* Gerald Duckworth & Co., Ltd., London; Reprinted 1965 by Hafner Publishing Co. Inc., New York.

Jaeger, Edmund C. 1941. *Desert Wild Flowers.* Stanford University Press, Stanford, CA.

Jeffrey, Charles. 2001. *Cucurbitaceae.* Pp. 688–717. In: Stevens, W. Douglas, Carmen Ulloa Ulloa, Amy Pool, and Olga M. Montriel, eds. 2001. *Flora de Nicaragua. Introducción, Gimnospermas y Angiospermas.* Tomo I. Missouri Botanical Garden Press, St. Louis.

Jöel, J. 1976. Some Paipai accounts of food gathering. *Journal of California Anthropology* 3 (1): 59–71.

Johnson-G., D., and L. Carillo-M. 1977. Algunos zacates de Sonora. Boletin de Comite de Fomentos Ganadero, Sonora. 58 pp.

Johnston, Marshall C. 1962. Revision of *Condalia* including *Microrhamnus* (Rhamnaceae). *Brittonia* 14: 332–368.

Jones, J. R. 1975. Official Pima County Map. County of Pima, Arizona.

Jones, Thomas A. 1998. Viewpoint: The present status and future prospects of squirreltail research. *Journal of Range Management* 51 (3): 326–331.

Jones, Thomas A., and Stephen R. Larson. 2005. Status and use of important native grasses adapted to sagebrush communities. USDA Forest Service Proceedings RMRS-P 38: 49–55.

Judd, Walter S., Christopher S. Campbell, Elizabeth A. Kellogg, Peter F. Stevens, and Michael J. Donoghue. 2008. *Plant Systematics: A Phylogenetic Approach.* Third Edition, Sinauer Associates, Sunderland, MA.

Kaplan, Lawrence. 1963. Archaeoethnobotany of Cordova Cave, New Mexico. *Economic Botany* 17 (4): 350–359.

Kaplan, Lawrence, and Lucille N. Kaplan. 1992. Beans of the Americas. Pp. 61–79. In: Foster, Nelson, and Linda S. Cordell, eds. *Chilies to Chocolate: Food the Americas Gave the World.* University of Arizona Press, Tucson.

Karis, Per O., Mari Kallersjo, and Kate Bremer. 1992. Phylogenetic analysis of the Cichorioideae (Asteraceae), with emphasis on the Mutisieae. *Annals of the Missouri Botanical Garden* 79 (2): 416–427.

Karl, Michael G., R. K. Heitschmidt, and Marshall R. Haferkamp. 1999. Vegetation biomass dynamics and patterns of sexual reproduction in a northern mixed-grass prairie. *American Midland Naturalist* 141 (2): 227–237.

Kartesz, John T. 1994. *A Synonymized Checklist of the Vascular Flora of the United States, Canada, and Greenland.* Vol. 1. Checklist. Timber Press, Portland, OR.

Kay, Margarita A. 1996. *Healing with Plants in the American and Mexican West.* University of Arizona Press, Tucson.

Kearney, Thomas H., and Robert H. Peebles. 1951. *Arizona Flora.* University of California Press, Berkeley.

Keck, David D. 1946. A revision of the *Artemisia vulgaris* complex in North America. *Proceedings of the California Academy of Sciences* (Series 4) 25: 421–468.

Keil, David J., Melissa A. Luckow, and Donald J. Pinkava. 1988. Chromosome studies in Asteraceae from the United States, Mexico, the West Indies, and South America. *American Journal of Botany* 75 (5): 652–668.

Kelly, Isabel T. 1932. Ethnography of the Surprise Valley Paiute. *University of California Publications in Archaeology and Ethnology* 31 (3): 67–210.

Kelly, Isabel T. 1939. Southern Paiute Shamanism. *Anthropological Records* 2 (4): 151–167.

Kelly, William H. 1977. Cocopa Ethnography. *Anthropological Papers of the University of Arizona No.* 29. University of Arizona Press, Tucson.

Keys, Roy N., Stephen L. Buchmann, and Steve E. Smith. 1995. Pollination effectiveness and pollination efficiency of insects foraging *Prosopis velutina* in south-eastern Arizona. *Journal of Applied Ecology* 32 (3): 519–527.

Kimball, Rebecca T., and Daniel J. Crawford. 2004. Phylogeny of Coreopsideae (Asteraceae) using ITS sequences suggests lability in reproductive characters. *Molecular Phylogenetics and Evolution* 33 (1): 127–139.

Kimball, Rebecca T., Daniel J. Crawford, and Edwin B. Smith. 2003. Evolutionary processes in the genus *Coreocarpus*: Insights from molecular phylogenetics. *Evolution* 57 (1): 52–61.

Kissell, Mary L. 1916. Basketry of the Pima-Papago. *American Museum of Natural History* 17 (4).

Kitt, Mrs. G. F. 1926–1929. Reminiscences of R. C. Brown. Notes in the Arizona Historical Society, Tucson.

Kohler, Timothy A., and Meredith H. Matthews. 1988. Long-term Anasazi land use and forest reduction: A case study from southwest Colorado. *American Antiquity* 53 (3): 537–564.

Kotowicz, Claudia, Luis R. Hernández, Carlos M. Cerda-García-R., Margarit B. Villecco, César A. N. Catalan, and Pedro Joseph-N. 2001. Absolute configuration of trixanolides from *Trixis pallida*. *Journal of Natural Products* 64 (10): 1326–1331.

Krausman, Paul R., Amy J. Kuenzi, Richard C. Etchberger, Kurt T. Rautenstrauch, Leonard L. Ordway, and John J. Hervert. 1997. Diets of desert mule deer. *Journal of Range Management* 50: 513–522.

Kuijt, Job. 2003. Monograph of *Phoradendron* (Viscaceae). *Systematic Botany Monographs* Vol. 66, 643 pp.

La Barre, Weston. 1938. Native American beers. *American Anthropologist* 40 (2): 224–234.

La Duke, John C. 1986. Chromosome numbers in *Sphaeralcea* section *Fendlerianae*. *American Journal of Botany* 73 (10): 1400–1404.

Ladyman, Juanita A. R. 2005. *Isocoma tenuisecta* agricultural circular. http://www.fs.fed.us/global/iitf/pdf/shrubs/Isocoma tenuisecta.pdf.

Laferrière, Joseph E. 1991. Optimal Use of Ethnobotanical Resources by the Mountain Pima of Chihuahua, Mexico. Dissertation, University of Arizona, Tucson.

Laferrière, Joseph E., Charles W. Weber, and Edwin A. Kohlhepp. 1991. Use and nutritional composition of some traditional Mountain Pima plant foods. *Journal of Ethnobiology* 11 (1): 93–114.

Lamb, Samuel H. 1975. *Woody Plants of the Southwest*. The Sunstone Press, Santa Fe, NM.

Lampe, Kenneth F., and Mary A. McCann. 1985. *AMA Handbook of Poisonous and Injurious Plants*. American Medical Association, Chicago.

Landrum, Leslie R. 1994. Fagaceae. In: A New Flora of Arizona. *Journal of the Arizona-Nevada Academy of Sciences* 27 (2): 220–214.

Laney, Nancy R. 1998. *Desert Waters: From Ancient Aquifer to Modern Demands*. Arizona-Sonora Desert Museum Press, Tucson.

Lankford, George E., ed. 1987. *Native American Legends: Southeastern Legends*. August House, Little Rock, AR.

Lanner, Ronald M. 1981a. Fuel for a Silver Empire. Pp. 117–130. In: Lanner, Ronald M. *The Piñon Pine*. University of Nevada Press, Reno.

Lanner, Ronald M. 1981b. *The Piñon Pine*. University of Nevada Press, Reno.

Latorre, Dolores L., and Felipe A. Latorre. 1977. Plants used by the Mexican Kickapoo Indians. *Economic Botany* 31: 340–357.

Lawless, Julia. 1995. *The Illustrated Encyclopedia of Essential Oils: The Complete Guide to the Use of Oils in Aromatherapy and Herbalism*. Element Books, Inc., Rockport, MA.

Lawrence, George H. M. 1951. *Taxonomy of Vascular Plants*. Macmillan Publ. Co., New York.

Lehr, J. Harry. 1978. *A Catalogue of the Flora of Arizona*. Desert Botanical Garden, Phoenix, AZ.

Leithead, Horace L., Lewis L. Yarlett, and Thomas N. Shiflet. 1971. *100 Native Forage Grasses in 11 Southern States*. USDA, Soil Conservation Service, Agriculture Handbook 389, Washington, DC.

Lentz, David L., Mary E. D. Pohl, Kevin O. Pope, and Andrew R. Wyatt. 2001. Prehistoric sunflower (*Helianthus annuus* L.) domestication in Mexico. *Economic Botany* 55: 370–376.

Lentz, David L., Maria L. Reyna de Aguilar, Raul Villacorta, and Helen Marini. 1996. *Trachypogon plumosus* (Poaceae, Andropogoneae): Ancient thatch and more from the Ceren site, El Salvador. *Economic Botany* 50 (1): 108–114.

Lewis, Walter H., and Memory P. F. Elvin-Lewis. 2003. *Medical Botany: Plants Affecting Human Health*, Second Edition. John Wiley & Sons, Inc., Hoboken, NJ.

Linares, Edelmira, and Judith Aguirre, eds. 1992. *Los Quelites, un Tesoro Culinario*. Universidad Nacional Autónoma de México, D. F., México.

Linnaeus, Carolus. 1749. *Flora Oeconomica or Household Uses of Wild Plants in Sweden*. Reprinted 1979 by Rediviva Publishing House, Stockholm.

Lionnet, André. 1985. Relaciones internas de la rama sonorense. *Amerindia* 10: 25–58. http://celia.cnrs.fr.

Little, R. John. 2001. Violaceae. In: A New Flora of Arizona. *Journal of the Arizona-Nevada Academy of Sciences* 33 (1): 74–82.

Loayza, Ingrid, David Abujder, Rosemary Aranda, Jasmin Jakupovic, Guy Collin, Hélène Deslauriers, and France-Ida Jean. 1995. Essential oils of *Baccharis salicifolia*, *B. latifolia*, and *B. dracunculifolia*. *Phytochemistry* 38 (2): 381–389.

López-E., Rigoberto, and Alicia Hinojosa-G. 1988. *Católogo de Plantas Medicinales Sonorenses*. Universidad de Sonora, Hermosillo.

Lumholtz, Carl. 1912. *New Trails in Mexico*. Charles Scribner's Sons, New York. Reprinted 1990 by University of Arizona Press, Tucson.

Mabberley, David J. 1997. *The Plant-Book*. Cambridge University Press, Cambridge.

MacDougal, John M. 2001. Passifloraceae. *Journal of the Arizona-Nevada Academy of Science* 33 (1): 41–45.

Macior, Lazarus W. 1975. The pollination ecology of *Delphinium tricorne* (Ranunculaceae). *American Journal of Botany* 62 (10): 1009–1016.

MacRae, Ted C., and Gayle H. Nelson. 2003. Distribution and biological notes on Buprestidae (Coleoptera) in North and Central America and the West Indies, with validation of one species. *The Coleopterists Bulletin* 57 (1): 57–70.

Mahr, August C. 1955. Review of the botanical lore of the California Indians. *Ethnohistory* 2 (3): 284–286.

Malusa, James. 1992. Phylogeny and biogeography of the pinyon pines (*Pinus* subsect. *Cembroides*). *Systematic Botany* 17 (1): 42–66.

Manniche, Lise. 1989. *An Ancient Egyptian Herbal*. University of Texas Press, Austin.

Manning, Wayne E. 1978. The classification within the Juglandaceae. *Annals of the Missouri Botanical Garden* 65 (4): 1058–1087.

Marsh, Dick E. 1969. Two contemporary Papago recipes for indigenous plants and the American Southwest botanical implications. *The Kiva* 34 (4): 242–246.

Martin, Alexander C., Herbert S. Zim, and Arnold L. Nelson. 1951. *American Wildlife and Plants: A Guide to Wildlife Food Habits*. Dover Publications, Inc., New York.

Martin, Paul S., David Yetman, Mark Fishbein, Phillip Jenkins, Thomas R. van Devender, and Rebecca K. Wilson, eds. 1998. *Gentry's Río Mayo Plants: The Tropical Deciduous Forest and Environs of Northwest Mexico*. University of Arizona Press, Tucson.

Martin, William C., and Charles R. Hutchins. 1986. *Summer Wildflowers of New Mexico*. University of New Mexico Press, Albuquerque.

Martínez, Maximino. 1969. *Las Plantas Medicinales de Mexico*. Ediciones Botas, D. F., Mexico.

Martínez, Maximino. 1979. *Católogo de nombres vulgares y científicas de plantas mexicanas*. Fondo de Cultura Económica, D. F., Mexico.

Martínez-A., Miguel A., V. A. Oliva, M. Mendoza-Cruz, G. Morales-Garcia, G. Toledo-Olazcoaga, and A. Wong-León. 1995. *Catálogo de Plantas Útiles de la Sierra Norte de Puebla, México*. Cuadernos del Instituto de Biología 27, Universidad Nacional Autónoma de México, D. F.

Masamune, Satoru. 1964. Total syntheses of diterpenes and diterpene alkaloids. III. Kaurene. Garryine; V. Atisine. *Journal of the American Chemical Society* 86: 289–290.

Mathias, Mildred. 1994. Magic, myth, and medicine. *Economic Botany* 48: 3–7.

Mathias, Mildred, and Lincoln Constance. 1965. A revision of the genus *Bowlesia* Ruiz and Pav. (Umbelliferae-Hydrocotyloideae) and its relatives. *University of California Publications in Botany* 38: 1–73.

Mathiot, Madeleine. 1973. *A Dictionary of Papago Usage*. 2 vols. Indiana University Press, Bloomington.

Matlock, W. Gerald, and P. R. Davis. 1972. Groundwater in the Santa Cruz Valley, Arizona. *University of Arizona Agricultural Experiment Station Technical Bulletin* 194. University of Arizona Press, Tucson.

Mayes, Vernon O., and Barbara B. Lacy. 1989. *Nanisé: A Navajo Herbal*. Navajo Community College Press, Tsaile, AZ.

Mayes, Vernon O., and James M. Roeminger. 1994. *Navajoland Plant Catalog*. National Woodlands Publication Co., Lake Ann, MI.

McCool, Daniel. 1981. Federal Indian policy and the sacred mountains of the Papago Indians. *Journal of Ethnic Studies* 9 (3): 57–69.

McLaughlin, Steven P. 1982. A revision of the southwestern species of *Amsonia* (subgenera *Sphinctosiphon* and *Articularia*). *Annals of the Missouri Botanical Garden* 69 (2): 336–350.

McLaughlin, Steven P. 1986. A floristic analysis of the southwestern United States. *Great Basin Naturalist* 46: 46–65.

McLaughlin, Steven P. 1989. Natural floristic areas of the western United States. *Journal of Biogeography* 16 (3): 239–248.

McLaughlin, Steven P. 1992a. Vascular flora of Buenos Aires National Wildlife Refuge (including Arivaca Cienega), Pima County, Arizona. *Phytologia* 73 (5): 353–377.

McLaughlin, Steven P. 1992b. Are floristic areas hierarchically arranged? *Journal of Biogeography* 19: 21–32.

McLaughlin, Steven P. 1993. Additions to the flora of the Pinaleño Mountains, Arizona. *Journal of the Arizona-Nevada Academy of Sciences* 27: 5–32.

McLaughlin, Steven P. 1994. Apocynaceae. *Journal of the Arizona-Nevada Academy of Sciences* 27 (2): 164–168.

McLaughlin, Steven P., and Janice E. Bowers. 1999. Diversity and affinities of the flora of the Sonoran Floristic Province. Pp. 12–35. In: Robichaux, Robert H., ed. *Ecology of the Sonoran Desert Plants and Plant Communities*. University of Arizona Press, Tucson.

McLaughlin, Steven P., and Willard Van Asdall. 1980. Flora and vegetation of the Rosemont area. In: Davis, Russell, and Joan R. Callahan, eds. *An Environmental Inventory of the Rosemont Area in Southern Arizona*. 1: 64–98. University of Arizona Press, Tucson.

Mekeel, H. S. 1935. Subsistence plant foods and preparation. Pp. 48–57. In: Kniffe, Fred B., Gordon McGregor, Robert A. McKennan, H. Scudder Mekeel, and Maurice Mook, edited by Alfred L. Kroeber. *Hualapai Ethnography: Memoirs of the American Anthropological Association* No. 42. Menasha, WI.

Menges, Robert M. 1987. Allelopathic effects of Palmer amaranth (*Amaranthus palmeri*) and other plant residues in soil. *Weed Science* 35 (3): 339–347.

Merriam, C. Hart. 1979. *Indian Names for Plants and Animals among Californian and Other Western North American Tribes*. Heizer, Robert F., ed. Ballena Press, Socorro, NM.

Merrill, William L. 1977. An investigation of ethnographic and archaeological specimens of mescalbeans (*Sophora secundiflora*). *American Museum Technical Report* No. 6, Museum of Anthropology, University of Michigan, Ann Arbor.

Merrill, William L., and Christian F. Feest. 1975. An exchange of botanical information in the early contact situation: Wisakon of the southeastern Algonquians. *Economic Botany* 29: 171–184.

Michigan Botanical Club. 2001. Endangered plants of Michigan. http://www.michbotclub.org/plants_mich/endanger.htm.

Mickel, John T. 1979. *How to Know the Ferns and Fern Allies*. Wm. C. Brown Co., Dubuque, IA.

Miller, Wick R. 1996. *Guarijío: Gramática, Textos y Vocabulario*. Universidad Nacional Autónoma de México, Mexico City.

Millspaugh, Charles F. 1892. *American Medicinal Plants*. Reprinted 1974 by Dover Publications, Inc., New York.

Minnis, Paul E. 1992. Famine foods of the North American desert borderlands in historical context. *Journal of Ethnobiology* 11: 231–258.

Moerman, Daniel E. 1998. *Native American Ethnobotany*. Timber Press, Inc., Portland, OR.

Moir, William H. 1979. Soil-vegetation patterns in the Central Peloncillo Mountains, New Mexico. *American Midland Naturalist* 102 (2): 317–331.

Molina, Felipe S., Herminia Valenzuela, and David L. Shaul. 1999. *Hippocrene Standard Dictionary Yoeme-English, English-Yoeme*. Hippocrene Books, New York.

Monro, Alex K., and Peter J. Stafford. 1998. A synopsis of the genus *Echinopepon* (Cucurbitaceae: Sicyeae), including three new taxa. *Annals of the Missouri Botanical Garden* 85: 257–272.

Moore, Michael. 1979. *Medicinal Plants of the Mountain West*. Museum of New Mexico Press, Santa Fe.

Moore, Michael. 2003. *Medicinal Plants of the Mountain West*. Revised and Expanded Edition. Museum of New Mexico, Santa Fe.

Moore, Michael, ed. 2005. *Desert Plants and People* by Sam Hicks. The Southwest School of Botanical Medicine, http://www.swsbm.com.

Moreno, Nancy P., and Fred Essig. 1997. *Clematis*. Pp. 158–176. In: Editorial Committee. *Flora of North America North of Mexico*. Vol. 3. Oxford University Press, New York.

Morgan, David R., Douglas E. Soltis, and Kenneth R. Robertson. 1994. Systematics and evolutionary implications of rbcL sequence variation in Rosaceae. *American Journal of Botany* 81 (7): 890–903.

Mors, Walter B., Carlos T. Rizzini, Nuno A. Pereira, and Robert A. DeFilipps. 2000. *Medicinal Plants of Brazil*. Reference Publications, Inc., Algonac, MI.

Mulroy, Thomas W., and Philip W. Rundel. 1977. Annual plants: Adaptations to desert environments. *BioScience* 27 (2): 109–114.

Munro, Pamela. 1990. Stress and vowel length in Cupan absolute nouns. *International Journal of American Linguistics* 56 (2): 217–250.

Munz, Philip A., and David D. Keck. 1973. *A California Flora*. University of California Press, Berkeley.

Nabhan, Gary P. 1982. *The Desert Smells Like Rain: A Naturalist in Papago Indian Country*. North Point Press, San Francisco.

Nabhan, Gary P. 1983a. Guest editorial: A special issue on the desert tepary as a food source. *Desert Plants* 5 (2): 2–63.

Nabhan, Gary P. 1983b. Papago Fields: Arid Lands Ethnobotany and Agricultural Ecology. Ph.D. Dissertation. University of Arizona, Tucson.

Nabhan, Gary P. 1985. *Gathering the Desert*. University of Arizona Press, Tucson.

Nabhan, Gary P. 2002. *Coming Home to Eat: The Pleasures and Politics of Local Foods*. W. W. Norton and Company, Inc., New York.

Nabhan, Gary P., and Jan M. J. De Wet. 1984. *Panicum sonorum* in Sonoran Desert agriculture. *Economic Botany* 38: 65–82.

Nabhan, Gary P., and Richard S. Felger. 1978. Teparies in southwestern North America: A biographical and ethnohistorical study of *Phaseolus acutifolius*. *Economic Botany* 32: 2–19.

Nabhan, Gary P., and Amadeo M. Rea. 1987. Plant domestication and folk-biological change: The Upper Piman/Devil's Claw example. *American Anthropologist, New Series* 89 (1): 57–73.

Nabhan, Gary P., Richard C. Brusca, and Louella Holter., eds. 2004. *Conserving Migratory Pollinators and Nectar Corridors in Western North America*. Arizona-Sonora Desert Museum Studies in Natural History. Arizona-Sonora Desert Museum and University of Arizona Press, Tucson.

Nabhan, Gary P., Wendy Hodgson, and Francis Fellows. 1989. A meager living on lava and sand? Hia Ched O'odham food resources and habitat diversity in oral and documentary histories. *Journal of the Southwest* 31: 508–533.

Nabhan, Gary P., Amadeo M. Rea, Karen Reichardt, E. Eric Mellinck, and Charles Hutchinson. 1982. Papago influence on habitat and biotic diversity. Quitovac oasis ethnoecology. *Journal of Ethnobiology* 2 (2): 124–143.

Nahrstedt, A., M. Hungeling, and F. Petereita. 2006. Flavonoids from *Acalypha indica*. *Fitoterapia* 77 (6): 484–486.

Nellis, David W. 1997. *Poisonous Plants and Animals of Florida and the Caribbean*. Pineapple Press, Inc., Sarasota, FL.

Nesom, Guy L. 1991. Taxonomy of *Isocoma* (Compositae: Astereae). *Phytologia* 70: 69–114.

Nesom, Guy L. 2004. New distribution records for *Gamocheta* (Asteraceae: Gnaphalieae) in the United States. *Sida* 21 (2): 1175–1185.

Nesom, Guy L. 2006. *Ageratina*. Pp. 547–553. In: Editorial Committee, eds. *Flora of North America North of Mexico*. Vol. 21. Oxford University Press, New York.

Nixon, Kevin C., and Cornelius H. Muller. 1997. *Quercus* Linnaeus sect. *Quercus*. Pp. 471–506. In: Editorial Committee, eds. *Flora of North America North of Mexico*. Vol. 3. Oxford University Press, New York.

OED. 2007. *Oxford English Dictionary Online*. Oxford University Press, Oxford. http://www.oed.com.

O'Kane, Steve L., Jr., and Ihsan A. Al-Shehbaz. 2003. Phylogenetic position and generic limits of *Arabidopsis* (Brassicaceae) based on sequences of nuclear ribosomal DNA. *Annals of the Missouri Botanical Garden* 90: 603–612.

Ooststroom, Simon J. van. 1934. A monograph of the genus *Evolvulus*. *Mededeelingen van het Botanisch Museum en Herbarium van de Rijks Universiteit te Utrecht* 14: 1–267.

Ortiz, Alfonso, ed. 1983. *Southwest Handbook of North American Indians*. Vol. 10. Smithsonian Institution, Washington, DC.

Owen, Roger C. 1963. The use of plants and nonmagical techniques in curing illness among the Paipai, Santa Catarina, Baja California, México. *América Indígena* 23: 319–345.

Ownbey, Gerald B., Jeffrey W. Brasher, and Curtis Clark. 1998. Papaveraceae. In: A New Flora of Arizona. *Journal of the Arizona-Nevada Academy of Sciences* 30 (2): 120–132.

Palmer, Edward. 1871. Food products of the North American Indians. United States Department of Agriculture Report 1870: 404–428.

Palmer, Edward. 1878. Plants used by the Indians of the United States. *American Naturalist* 12: 593–606, 646–655.

Panter, K. E., and L. F. James. 1990. Natural plant toxicants in milk: A review. *Journal of Animal Science* 68: 892–904.

Paredes-A., Rafaela, Thomas R. Van Devender, and Richard S. Felger. 2000. *Cactáceas de Sonora, México: Su Diversidad, Uso y Conservación*. IMADES & Arizona-Sonora Desert Museum, Tucson.

Park, Marilyn M., and Dennis Festerling, Jr. 1997. *Thalictrum*. In: Editorial Committee, eds. *Flora of North America North of Mexico*. 3: 258–271. Oxford University Press, New York.

Parker, Kittie F. 1958. Arizona Ranch, Farm, and Garden Weeds. Circular 265, Agricultural Extension Service, University of Arizona, Tucson.

Parker, Kittie F. 1972. *An Illustrated Guide to Arizona Weeds*. University of Arizona Press, Tucson.

Parker, V. T., and C. H. Muller. 1982. Vegetational and environmental changes beneath isolated live oak trees (*Quercus agrifolia*) in a California annual grassland. *American Midland Naturalist* 107 (1): 69–81.

Peattie, Donald C. 1953. *A Natural History of Western Trees*. Reprinted 1980 by University of Nebraska Press, Lincoln.

Pellmyr, Olle, James Leebens-Mack, and John N. Thompson. 1998. Herbivores and molecular clocks as tools in plant biogeography. *Biological Journal of the Linnean Society* 63: 367–378.

Pennington, Campbell W. 1958. Tarahumar fish stupification plants. *Economic Botany* 12: 95–102.

Pennington, Campbell W. 1963. *The Tarahumara of Chihuahua: Their Environment and Material Culture*. University of Utah Press, Salt Lake City.

Pennington, Campbell W. 1969. *The Northern Tepehuan of Chihuahua: Their Material Culture*. University of Utah Press, Salt Lake City.

Pennington, Campbell W. 1973. Plantas medicinales utilizadas por el pima montanes de Chihuahua. *America Indigena* 33: 213–232.

Pennington, Campbell W. 1979. *The Pima Bajo of Central Sonora, Mexico*. Vol. II. *Vocabulario en la lengua Nevome*. University of Utah, Salt Lake City.

Pennington, Campbell W. 1980. *The Pima Bajo of Central Sonora, Mexico*. Vol. 1. University of Utah Press, Salt Lake City.

Pennington, Terrence D. 1990. Sapotaceae. In: *Flora Neotropica* 52: 1–771, New York Botanical Garden, Bronx.

Peoples, Alan D., Robert L. Lochmiller, David M. Leslie, Jr., Jon C. Boren, and David M. Engle. 1994. Essential amino acids in northern bobwhite foods. *Journal of Wildlife Management* 58 (1): 167–175.

Perkins, Kent D., and Willard W. Payne. 1978. *Guide to the Poisonous and Irritant Plants of Florida*. Institute of Food and Agricultural Sciences, Gainesville.

Peterson, Paul M., Robert J. Soreng, Gerrit Davidse, Tarciso S. Filgueiras, Fernando O. Zuloaga, and Emmet J. Judziewicz. 2001. Catalogue of New World Grasses (Poaceae): II. Subfamily Chloridoideae. *Contributions from the United States National Herbarium* 41: 1–255.

Pfefferkorn, Ignaz. 1794–1795. *Sonora. A Description of the Province*. T. E. Treulein (translator). Coronado Cuarto Centennial Publications, 1540–1940, Vol. 12. University of New Mexico Press, Albuquerque. Reprinted 1989 by University of Arizona Press, Tucson.

Phillips, Steven J., and Patricia W. Comus, eds. 2000. *A Natural History of the Sonoran Desert*. Arizona-Sonora Desert Museum Press, Tucson, and University of California Press, Berkeley.

Pickering, Jerry L., and David E. Fairbrothers. 1970. A serological comparison of Umbelliferae subfamilies. *American Journal of Botany* 57 (8): 988–992.

Pinkava, Donald J. 1995. Cactaceae, Part 1. The Cactus family. The Cereoid Cacti. In: A New Flora of Arizona. *Journal of the Arizona-Nevada Academy of Sciences* 29: 6–12.

Pinkava, Donald J. 1999. Cactaceae Part 3. *Cylindropuntia*. *Journal of the Arizona-Nevada Academy of Sciences* 32 (1): 32–47.

Pinkava, Donald J. 2003. http://www.lsvl.la.asu.edu/herbarium/pinkava.

Plunkett, Gregory M., Douglas E. Soltis, and Pamela S. Soltis. 1996. Higher level relationships of Apiales (Apiaceae and Araliaceae) based on phylogenetic analysis of rbcL sequences. *American Journal of Botany* 83 (4): 499–515.

Powell, A. Michael. 1988. *Trees and Shrubs of Trans-Pecos Texas*. Big Bend Natural History Association, Alpine.

Powell, A. Michael, and James F. Weedin. 2004. *Cacti of Trans-Pecos Texas and Adjacent Areas*. Texas Tech University Press, Lubbock.

Powers, Stephen. 1877. Tribes of California. *Contributions to North American Ethnology*. Vol. 3, Washington, DC.

Prasada Rao, K. E., Jan M. J. De Wet, D. E. Brink, and M. H. Mengesha. 1987. Intraspecific variation and systematics of cultivated *Setaria indica*, foxtail millet (Poaceae). *Economic Botany* 41: 108–116.

Price, Mary V., and Jamie W. Joyner. 1997. What resources are available to desert granivores: Seed rain or soil seed bank? *Ecology* 78 (3): 764–773.

Proctor, Michael, Peter Yeo, and Andrew Lack. 1996. *The Natural History of Pollination*. Timber Press, Inc., Portland, OR.

Puente, Raul, and Thomas F. Daniel. 2001. Garryaceae. In: A New Flora of Arizona. *Journal of the Arizona-Nevada Academy of Sciences* 33 (1): 31–34.

Puente, Raul, and Robert Faden. 2001. Commelinaceae. In: A New Flora of Arizona. *Journal of the Arizona-Nevada Academy of Sciences* 33 (1): 19–26.

Quattrocchi, Umberto. 1999. *CRC World Dictionary of Plant Names: Common Names, Scientific Names, Eponyms, Synonyms, and Etymology*. CRC Press LLC, Boca Raton, FL.

Quinn, Ronald D., and Sterling C. Keeley. 2006. *Introduction to California Chaparral*. University of California Press, Berkeley.

Ralph, Jolyon, and Ida Chau. 1993–2005. *Baboquivari Mountains*. http://www.mindat.org/loc-35387.html.

Rao, K. V., and Duvvuru Gunasekar. 1998. C-phenylated dihydroflavonol from *Rhynchosia densiflora*. *Phytochemistry* 48 (8): 1453–1455.

Rea, Amadeo M. 1997. *At the Desert's Green Edge: An Ethnobotany of the Gila River Pima*. University of Arizona Press, Tucson.

Reagan, Albert B. 1929. Plants used by the White Mountain Apache Indians of Arizona. *Wisconsin Archeologist* 8: 143–161.

Reed, Clyde F. 1971. *Common Weeds of the United States*. Dover Publications, New York.

Reed, Porter B., Jr., 1988. National List of Plant Species That Occur in Wetlands: National Summary. U.S. Fish and Wildlife Service. *Biological Report* 88 (24). 244 pp.

Reeder, John R. 1971. Notes on Mexican grasses IX. Miscellaneous chromosome numbers-3. *Brittonia* 23 (2): 105–117.

Reichman, Omer J. 1975. Relation of desert rodent diets to available resources. *Journal of Mammalogy* 56 (4): 731–751.

Reid, Jefferson, and Stephanie Whittlesey. 1997. *The Archaeology of Ancient Arizona*. University of Arizona Press, Tucson.

Reina-G., Ana Lilia. 1993. Contribución a la introducción de nuevos cultivos en Sonora: Las plantas medicinales de los Pimas Bajos del Municipios de Yécora. Undergraduate thesis, Universidad de Sonora, Hermosillo, Mexico.

Rendón, Beatriz, Robert Bye, and Juan Núñez-F. 2001. Ethnobotany of *Anoda cristata* (L.) Schl. (Malvaceae) in Central Mexico: Uses, management and population differentiation in the community of Santiago Mamalhuazuca, Ozumba, state of Mexico. *Economic Botany* 55 (4): 545–554.

Ribeiro, Antônia, Doria Piló-V., Alvaro J. Romanha, and Carlos L. Zani. 1997. Trypanocidal flavonoids from *Trixis vauthieri*. *Journal of Natural Products* 60 (8): 836–838.

Riley, Thomas J., Gregory R. Walz, Charles J. Bareis, Andrew C. Fortier, and Kathryn E. Parker. 1994. Accelerator mass spectrometer (AMS) dates confirm early *Zea mays* in the Mississippi River valley. *American Antiquity* 59: 490–498.

Robbins, G. Thomas. 1944. North American Species of *Androsace*. *American Midland Naturalist* 32 (1): 137–163.

Robbins, Wilfred W., John P. Harrington, and Barbara Freirre-Marreco. 1916. Ethnobotany of the Tewa Indians. *Bureau of American Ethnology Bulletin* 55, Smithsonian Institution, Washington, DC.

Rodríguez-J., Concepción. 1995. Distribución geográfica del género *Echiopepon* (Cucurbitaceae). *Anales del Instituto de Biología, Universidad Nacional Autónoma de México. Botánica* 66: 171–181.

Rollins, Reed C. 1993. *The Cruciferae of Continental North America*. Stanford University Press, Stanford, CA.

Romero, John B. 1954. *The Botanical Lore of the California Indians, with Side Lights on Historical Incidents in California*. Vantage Press, NY.

Rondeau, Renée, Thomas R. Van Devender, C. David Bertelsen, Philip Jenkins, Rebecca K. Wilson, and Mark A. Dimmitt. 1996. Annotated flora and vegetation of the Tucson Mountains, Pima County, Arizona. *Desert Plants* 12 (2): 3–46.

Rosengarten, Frederic, Jr. 1969. *The Book of Spices*. Livingston Publishing Co., Wynnewood, PA.

Roskruge, George J. 1888. *Survey of the Altar Valley of Pima County*. General Land Office, Pima County, Arizona. Copy at Buenos Aires National Wildlife Refuge, Sasabe, Arizona.

Russell, Frank. 1908. *The Pima Indians*. Twenty-sixth Annual Report, Bureau of American Ethnology, Smithsonian Institution, Washington, DC. Re-edition 1975, with introduction, citation sources, and bibliography by Bernard L. Fontana. University of Arizona Press, Tucson.

Salywon, Andrew. 1999. Vascular plants of Arizona: Sapindaceae Soapberry family. *Journal of the Arizona-Nevada Academy of Sciences* 32: 77–82.

Santamaría, Francisco J. 1959. *Diccionario de Mejicanismos*. Editorial Porrua, SA, Mexico City, Mexico.

Sapir, Edward. 1930. The Southern Paiute language. Reprinted 1990 by AMS Press, New York. From *Proceedings of the American Academy of Arts and Sciences* 65 (1–2): 1–536.

Sargent, Charles S. 1890–1902. *The Silva of North America*. Houghton Mifflin Co., Boston.

Saunders, Charles F. 1920. *Useful Wild Plants of the United States and Canada*. McBride, New York. Reprinted 1976 by Dover Publications Inc., New York.

Saxton, Dean, and Lucille Saxton. 1969. *Dictionary Papago and Pima to English (O'odham-Mil-gahn) English to Papago and Pima (Mil-gahn-O'odham)*. University of Arizona Press, Tucson.

Saxton, Dean, and Lucille Saxton. 1973. *O'odham Hoho'ok A'agitha: Legends and Lore of the Papago and Pima Indians*. University of Arizona Press, Tucson.

Saxton, Dean, Lucille Saxton, and Susie Enos. 1983. *Dictionary Papago/Pima–English O'otham–Mil-gahn English–Papago/Pima Mil-gahn–O'otham*, Second Edition/Revised and Expanded. University of Arizona Press, Tucson.

Sayre, Nathan F. 2002. *Ranching, Endangered Species, and Urbanization in the Southwest: Species of Capital*. University of Arizona Press, Tucson.

Schmutz, Ervin M., and Lucretia B. Hamilton. 1979. *Plants That Poison: An Illustrated Guide to Plants Poisonous to Man*. Northland Press, Flagstaff, AZ.

Schoen, Daniel J., Mark O. Johnston, Anne-Marie L'Heureux, and Joyce V. Marsolais. 1997. Evolutionary history of the mating system in *Amsinckia* (Boraginaceae). *Evolution* 51 (4): 1090–1099.

Schultes, Richard E., and Albert Hofmann. 1979. *Plants of the Gods*. McGraw-Hill Book Co., New York.

Seigler, David S. 1994. Phytochemistry and systematics of the Euphorbiaceae. *Annals of the Missouri Botanical Garden* 81 (2): 380–401.

SEINet. 2009. Southwestern Environmental Information Network. http://swbiodiversity.org/seinet/profile/newprofile.php.

Semple, John C. 1977. Cytotaxonomy of *Chrysopsis* and *Heterotheca* (Compositae-Astereae): A new interpretation of phylogeny. *Canadian Journal of Botany* 55: 2503–2513.

Semple, John C. 1981. A revision of the goldenaster genus *Chrysopsis* (Nutt.) Ell. nom. cons. (Compositae-Astereae). *Rhodora* 83: 323–384.

Semple, John C. 1996. A revision of *Heterotheca* sect. *Phyllotheca* (Nutt.) Harms (Compositae: Astereae): The prairie and montane goldenasters of North America. *University of Waterloo Biology Series* No. 37: 1–1164.

Semple, J. C., Vivian C. Blok, and Patricia Heiman. 1980. Morphological, anatomical, habit, and habitat differences among the goldenaster genera, *Chrysopsis*, *Heterotheca*, and *Pityopsis* (Compositae-Astereae). *Canadian Journal of Botany* 58: 147–163.

Shaul, David L. 1983. The position of Ópata and Eudeve in Uto-Aztecan. *Kansas Working Papers in Linguistics* 8 (2): 95–122.

Shaul, David L. 1994. A sketch of the structure of Oob No'ok (Mountain Pima). *Anthropological Linguistics* 36 (3): 277–365.

Shaul, David L. 2007. Dictionary to O'odham. Unpublished manuscript.

Shaul, David L., and Jane H. Hill. 1998. Tepimans, Yumans, and other Hohokam. *American Antiquity* 63: 75–396.

Sheridan, Thomas E. 1995. *Arizona*. University of Arizona Press, Tucson.

Sheridan, Thomas E. 2006. *Landscapes of Fraud: Mission Tumacácori, the Baca Float, and the Betrayal of the O'odham*. University of Arizona Press, Tucson.

Sheridan, Thomas E., and Nancy J. Parezo, eds. 1996. *Paths of Life: American Indians of the Southwest and Northern Mexico*. University of Arizona Press, Tucson.

Shinners, Lloyd H. 1965. Untypification for *Ipomoea nil* (L.) Roth. *Taxon* 14: 231–234.

Shreve, Forrest. 1942. The desert vegetation of North America. *Botanical Review* 8: 195–246.

Shreve, Forrest, and Ira L. Wiggins. 1964. *Vegetation and Flora of the Sonoran Desert*. Stanford University Press, Stanford, CA.

Shultz, Leila M. 2006. *Artemisia*. Pp. 503–534. In: Editorial Committee. *Flora of North America North of Mexico*. Vol. 19. Oxford University Press, New York.

Siméon, Rémi. 1885. *Diccionario de la lengua Náhuatl o Mexicana*. Reprinted 1981 by Siglo Veintiuno Editores, SA, Mexico City.

Simpson, Beryl B. 1989. Krameriaceae. *Flora Neotropica* 49. New York Botanical Garden, Bronx.

Simpson, Beryl B. 1991. The past and present uses of rhatany (*Krameria*, Krameriaceae). *Economic Botany* 45 (3): 397–409.

Simpson, Beryl B., and Molly Conner-Ogorzaly. 1995. *Economic Botany: Plants in Our World*. McGraw-Hill Book Co., New York.

Simpson, Beryl B., and Andrew Salywon. 1999. Krameriaceae. In: A New Flora of Arizona. *Journal of the Arizona-Nevada Academy of Sciences* 32 (1): 57–61.

Simpson, Beryl B., Andrea Weeks, D. Megan Helfgott, and Leah L. Larkin. 2004. Species relationships in *Krameria* (Krameriaceae) based on ITS sequences and morphology: Implications for character utility and biogeography. *Systematic Botany* 29: 97–108.

Singer, M. S., and J. O. Stireman III. 2001. How foraging tactics determine host-plant use by a polyphagous caterpillar. *Oecologia* 129: 98–105.

Small, Ernest. 1997. *Culinary Herbs*. Monograph Publishing, NRC Research Press, National Research Council of Canada, Ottawa, Ontario.

Smith, Alan R. 2006. *Zinnia*. Pp. 71–74. In: Editorial Committee, eds. *Flora of North America North of Mexico*. Vol. 21. Oxford University Press, New York.

Smith, Bruce D. 1992. *Rivers of change: Essays on early agriculture in eastern North America*. Smithsonian Institution Press, Washington, DC.

Smith, C. Earle, Jr. 1954. A century of botany in America. *Bartonia* 28: 1–30.

Smith, C. Earle, Jr. 1967. Plant remains. Pp. 220–255. In: *The Prehistory of the Tehuacán Valley*. Vol. 1. Byers, Douglas S., ed. University of Texas Press, Austin.

Smith, Charline G. 1973. *Selé*, a major vegetal component of the aboriginal Hualapai diet. *Plateau* 45 (3): 102–110.

Smith, Edwin B. 1989. A biosystematic study and revision of the genus *Coreocarpus* (Compositae). *Systematic Botany* 14 (4): 448–472.

Smith, Huron H. 1923. Ethnobotany of the Menomini Indians. *Bulletin of the Public Museum of the City of Milwaukee* 4 (1): 1–174.

Smith, Huron H. 1933. Ethnobotany of the forest Potawatomi Indians. *Bulletin of the Public Museum of the City of Milwaukee* 7: 1–230.

Smith, Janet H. 1972. Native pharmacopoeia of the eastern Great Basin: A report on work in progress. Pp. 73–86. In: Fowler, Don D., ed. *Great Basin Cultural Ecology: A Symposium*. Desert Research Institute Publications in the Social Sciences, No. 8. University of Nevada, Reno.

Smith, Nathan P., Scott A. Mori, Andrew Henderson, Dennis W. Stevenson, and Scott V. Heald. 2004. *Flowering Plants of the Neotropics*. New York Botanical Garden and Princeton University Press, Princeton, NJ.

Sobarzo, Horacio. 1991. *Vocabulario Sonorense*, Third Edition. Gobierno del Estado de Sonora, Instituto Sonorense de Culturas. Hermosillo, Sonora, Mexico.

Soholt, Lars F. 1973. Consumption of primary production by a population of Kangaroo Rats (*Dipodomys merriami*) in the Mojave Desert. *Ecological Monographs* 43 (3): 357–376.

Soreng, Robert J., Gerrit Davidse, Paul M. Peterson, Fernando O. Zuloaga, Emmet J. Judziewicz, Tarciso S. Filgueiras, and Osvaldo Morrone. 2007. *Catalogue of New World Grasses (Poaceae)*. http://mobot.mobot.org/W3T/Search/nwgc.html.

Southern Ute Tribe. 1979. *Ute Dictionary, Preliminary Edition. Núu-ˀApáĝa-Pi Pǫˀǫ́-Qwa-Tį.* Ute Press, Ignacio, CO.

Spellenberg, Richard W. 2004. Nyctaginaceae. Pp. 14–17. In: Editorial Committee. *Flora of North America North of Mexico*. Vol. 4. Oxford University Press, New York.

Spencer, Edwin R. 1957. *Just Weeds*. Charles Scribner's Sons, New York. Reprinted 1974 as *All About Weeds*, Dover Publications, New York.

Standley, Paul C. 1920–1926. *Trees and Shrubs of Mexico*. Reprinted 1971 by Smithsonian Press, Washington, DC.

Standley, Paul C., and Julian A. Steyermark. 1946. Flora of Guatemala, Part 4. Fieldiana: *Botany* 24 (4): 1–493.

Stebbins, G. Lenyard, Jr., J. I. Valencia, and R. Marie Valencia. 1946. Artificial and natural hybrids in the Gramineae, tribe Hordeae I. *Elymus*, *Sitanion*, and *Agropyron*. *American Journal of Botany* 33 (5): 338–351.

Steere, P. L. n.d. *How Did Brown Canyon Get Its Name?* Notes in Buenos Aires National Wildlife Collection from Special Collections, University of Arizona Library, Tucson.

Steggerda, M. 1941. Navajo foods and their preparation. *Journal of the American Dietetic Association* 17 (3): 217–225.

Stevens, O. A. 1920. The geographical distribution of North Dakota plants. *American Journal of Botany* 7 (6): 231–242.

Stevenson, Dennis W. 1993. Ephedraceae. Pp. 428–434. In: Editorial Committee. *Flora of North America North of Mexico*. Vol. 2. Oxford University Press, New York.

Stevenson, Mathilda C. 1915. *Ethnobotany of the Zuni Indians*. Smithsonian Institution–Bureau of American Ethnology Annual Report, Number 30.

Steward, Julian H. 1933. Ethnography of the Owens Valley Paiute. *University of California Publications in American Archaeology and Ethnology* 33: 233–350.

Stewart, Kenneth M. 1965. Mohave Indian gathering of wild plants. *The Kiva* 31 (1): 46–53.

Steyermark, Julian A., Paul A. Berry, and Bruce K. Holst, eds. 1998. *Flora of the Venezuelan Guayana*. Vol. 4. Caesalpinaceae–Ericaceae. Missouri Botanica Garden Press, St. Louis.

Stiles, Edmund W. 1980. Patterns of fruit presentation and seed dispersal in bird-disseminated woody plants in the eastern deciduous forest. *American Naturalist* 116 (5): 670–688.

Stiven, Gordon. 1952. Production of antibiotic substances by the roots of a grass (*Trachypogon plumosus* (H. B. K.) Nees) and of *Pentanisia variabilis* (E. Mey.) Harv. (Rubiaceae). *Nature* 170: 712–713.

Sundberg, Scott D., and David J. Bogler. 2006. *Baccharis*. Pp. 23–34. In: Editorial Committee. *Flora of North America North of Mexico*. Vol. 20. Oxford University Press, New York.

Tardío, Javier, Higinio Pascual, and Ramón Morales. 2005. Wild food plants traditionally used in the Province of Madrid, Central Spain. *Economic Botany* 59 (2): 122–136.

Thord-Gray, Ivor. 1955. *Tarahumara-English, English-Tarahumara Dictionary and an Introduction to Tarahumara Grammar*. University of Miami Press, Coral Gables, FL.

Timbrook, Jan. 1990. Ethnobotany of the Chumash Indians, California. Based on collections by John P. Harrington. *Economic Botany* 44: 236–253.

Timbrook, Jan. 2007. *Chumash Ethnobotany: Plant Knowledge among the Chumash People of Southern California*. Santa Barbara Museum of Natural History, Santa Barbara, and Heyday Books, Berkeley, CA.

Tomlinson, P. Barry. 1980. *The Biology of Trees Native to Tropical Florida*. Published by author, Allston, MA.

Toolin, Laurence J. 1980. Final Report on the Flora of Ramsey Canyon. Report to the Nature Conservancy, Arizona Natural Heritage Program, Tucson.

Toolin, Laurence J., Thomas R. Van Devender, and Jack M. Kaiser. 1979. The flora of Sycamore Canyon, Pajarito Mountains, Santa Cruz County, Arizona. *Journal of the Arizona-Nevada Academy of Sciences* 14: 66–74.

Train, Percy, James R. Henrichs, and Andrew W. Archer. 1957. *Medicinal Uses of Plants by Indian Tribes in Nevada*. Reprinted 1982 by Quartermaster Publications, Inc., Lawrence, MA.

TROPICOS. 2007. http://www.mobot.mobot.org/W3T/Search/vast.html.

TROPICOS. 2008. *Catalogue of New World Grasses*–July 6, 2008. http://www.mobot.mobot.org/W3T/Search/vast.html.

Tucker, Arthur O., and Robert F. C. Naczi. 2007. *Mentha*: An overview of its classification and relationships. Pp. 1–39. In: Lawrence, Brian M., ed. *Mint. The genus* Mentha. *Medicinal and Aromatic Plants—Industrial Profiles*. Vol. 44. CRC Press, Boca Raton, FL.

Tull, Delena. 1999. *Edible and Useful Plants of Texas and the Southwest: A Practical Guide*. University of Texas Press, Austin.

Turner, Billie L. 1996. Taxonomic study of the *Coreocarpus arizonicus–C. sonoranus* (Asteraceae, Heliantheae) complex. *Phytologia* 80: 133–139.

Turner, Nancy J., and Adam F. Szczawinski. 1991. *Common Poisonous Plants and Mushrooms of North America*. Timber Press, Inc., Portland, OR.

Turner, Raymond M., Janice E. Bowers, and Tony L. Burgess. 1995. *Sonoran Desert Plants: An Ecological Atlas*. University of Arizona Press, Tucson.

Turner, Raymond M., Robert H. Webb, Janice E. Bowers, and James R. Hastings. 2003. *The Changing Mile Revisited: An Ecological Study of Vegetation Change with Time in the Lower Mile of an Arid and Semiarid Region*. University of Arizona Press, Tucson.

Tyler, Jack D., Mike Hakylnie, Clark Bordner, and Michael Bay. 2000. Notes on winter habits of raccoons from western Oklahoma. *Proceedings of the Oklahoma Academy of Science* 80: 115–117.

Uphof, Johannes C. T. 1968. *Dictionary of Economic Plants*. J. Cramer, Lehre, Germany.

Urbatsch, Lowell E., Loran C. Anderson, Roland P. Roberts, and Kurt M. Neubig. 2006. *Ericameria*. Pp. 50–77. In: Editorial Committee. *Flora of North America North of Mexico*. Vol. 20. Oxford University Press, New York.

USFWS. 2003. Buenos Aires National Wildlife Refuge. Final Comprehensive Conservation Plan. U.S. Fish and Wildlife Service, Albuquerque, NM.

Vallès, Joan, Maria A. Bonet, and Antoni Agelet. 2004. Ethnobotany of *Sambucus nigra* L. in Catalonia (Iberian Peninsula): The integral exploitation of a natural resource in mountain regions. *Economic Botany* 58 (3): 456–469.

Van Devender, Thomas R., and Ana Lilia Reina. 2005. The forgotten flora of la Frontera. *USDA Forest Service Proceedings* RMRS-P-36: 158–161.

Van Devender, Thomas R., Karen Krebbs, Jean-Luc E. Cartron, Ana Lilia Reina-G., and William A. Calder. 2005. Hummingbird communities along an elevational gradient in the Sierra Madre Occidental of eastern Sonora, Mexico. Pp. 204–224. In Catron, Jean-Luc E., Gerardo Ceballos, and Richard S. Felger, eds. *Biodiversity, Ecosystems, and Conservation in Northern Mexico*. Oxford University Press, New York.

Vásquez-T., Mario, and Eloíse Jácome-C. 1997. *Flora medicinal de Veracruz. I. Inventario etnobotánico*. Universidad Veracruzana, Xalapa.

Verbylaité, R., B. Ford-Lloyd, and J. Newbury. 2006. The phylogeny of woody Maloideae (Rosaceae) using chloroplast trnl-trnF sequence data. *Biologija* 1: 60–63.

Vestal, Paul A. 1952. Ethnobotany of the Ramah Navaho. Reports of the Ramah project, Report No. 4, *Papers of the Peabody Museum of American Archaeology and Ethnology* 40 (4). Harvard University, Cambridge, MA.

Vestal, Paul A., and Richard E. Schultes. 1939. *The Economic Botany of the Kiowa Indians As It Relates to the History of the Tribe*. Botanical Museum Harvard University, Cambridge, MA.

Vickery, Roy. 1995. *A Dictionary of Plant-Lore*. Oxford University Press, Oxford.

Vieyra-O., Leticia, and Heike Vibrans. 2001. Weeds as crops: The value of maize field weeds in the Valley of Toluca, Mexico. *Economic Botany* 55: 426–443.

Vines, Robert A. 1977. *Trees of East Texas*. University of Texas Press, Austin.

Voegelin, Erminie Wheeler. 1938. Tübatulabal ethnography. *Anthropological Records* 2 (1): 1–90.

Voegelin, Erminie Wheeler. 1958. Working dictionary of Tübatulabal. *International Journal of American Linguistics* 24 (3): 221–228.

Von Reis-Altschul, Siri. 1973. *Drugs and Foods from Little-Known Plants*. Harvard University Press, Cambridge, MA.

Vora, Robin S. 1989. Seed germination characteristics of selected native plants of the lower Rio Grande Valley, Texas. *Journal of Range Management* 42 (1): 36–40.

Wagnon, H. Keith. 1952. A revision of the genus *Bromus*, section *Bromopsis*, of North America. *Brittonia* 7 (5): 415–480.

Walker, Michael, Jacque Nunez, Marion Walkingstick, and Sandra A. Bannack. 2004. Ethnobotanical investigation of the Acjachemen clapperstick from blue elderberry, *Sambucus mexicana* (Caprifoliaceae). *Economic Botany* 58 (1): 21–24.

Wallmo, O. C. 1954. Nesting of Mearns Quail in Southeastern Arizona. *The Condor* 56 (3): 125–128.

Warnock, Michael J. 1997. *Delphinium*. Pp. 196–240. In: Editorial Committee. *Flora of North America*. Vol. 3. Oxford University Press, New York.

Watahomigie, Lucille J., Malinda Powskey, Jorigine Bender, Elnora Matapais, and Francis H. Boone. 1982. *Hualapai Ethnobotany*.

Hualapai Bilingual Program, Peach Springs School Distict No. 8, Peach Springs, AZ.

Watanabe, Kuniaki, Robert M. King, Tetsukazu Yahara, Motomi Ito, Jun Yokoyama, Takeshi Suzuki, and Daniel J. Crawford. 1995. Chromosomal cytology and evolution in Eupatorieae (Asteraceae). *Annals of the Missouri Botanical Garden* 82 (4): 581–592.

Watson, Leslie, and Michael J. Dallwitz. 1992. *The Grass Genera of the World*. CAB International, London.

Weber, William A. 1990. *Colorado Flora: Eastern Slope*. University Press of Colorado, Niwot.

Webster, J. R. 1900. Cleistogamy in *Linaria canadensis*. *Rhodora* 2: 168.

Welsh, Stanley L., N. Duane Atwood, Sherel Goodrich, and Larry C. Higgins. 1987. *A Utah Flora*. Brigham Young University, Provo, UT.

Whiting, Alfred F. 1939. *Ethnobotany of the Hopi*. Museum of Northern Arizona Bulletin No. 15. Flagstaff.

Whitson, Tom D., ed. 1992. *Weeds of the West*. University of Wyoming, Jackson.

Wiersema, John W., and Blanca León. 1999. *World Economic Plants: A Standard Reference*. CRC Press LLC, Boca Raton, FL.

Wiggins, Ira L. 1980. *Flora of Baja California*. Stanford University Press, Stanford, CA.

Wiggins, Ira L., and Duncan M. Porter. 1971. *Flora of the Galapagos Islands*. Stanford University Press, Stanford, CA.

Wilbur-Cruce, Eva Antonia. 1987. *A Beautiful, Cruel Country*. University of Arizona Press, Tucson.

Wilcox, David R. 1981. The entry of Athapaskans into the American Southwest: The problem today. Pp. 213–256. In: Wilcox, David R., and W. Bruce Masse, eds. The Protohistoric Period in the North American Southwest, AD 1450–1700. *Anthropological Research Papers* No. 24, Arizona State University, Tempe.

Williams, Louis O. 1970. Jalap or Veracruz Jalap and its allies. *Economic Botany* 24: 399–401.

Williams, Louis O. 1981. The useful plants of Central America. *Ceiba* 24 (1–2): 343–381; *Ceiba* 24 (3–4). Escuela Agrìcola Panamericana, Tegucigalpa, Honduras.

Wilson, F. Douglas. 1963. Revision of *Sitanion* (Triticeae, Gramineae). *Brittonia* 15 (4): 303–323.

Windham, Michael D., and Eric W. Rabe. 1993. *Cheilanthes*, Pp. 152–169; *Pellaea*, Pp. 175–186; *Woodsia*, Pp. 270–280. In: Editorial Committee. *Flora of North America North of Mexico*. Vol. 3. Oxford University Press, New York.

Winter, Joseph C. 2001. *Tobacco Use by Native North Americans: Sacred Smoke and Silent Killer*. Civilization of the American Indian Series Vol. 236. University of Oklahoma Press, Norman.

Wright, Harold B. 1979. *Long Ago Told*. Appleton, New York.

Wunderlin, Richard P., and Bruce F. Hansen. 2003. *Guide to the Vascular Plants of Florida*, Second Edition. University Press of Florida, Gainesville.

Wyman, Leland C., and Stuart K. Harris. 1941. Navajo Indian Medical Ethnobotany. *University of New Mexico Bulletin* 366, *Anthropological Series* 3 (5): 1–75.

Wyman, Leland C., and Stuart K. Harris. 1951. The Ethnobotany of the Kayenta Navaho, an analysis of the John and Louisa Wetherhill ethnobotanical collection. *University of New Mexico Publications in Biology*, No. 5. University of New Mexico Press, Albuquerque.

Yanovsky, Elias. 1936. Food Plants of the North American Indians. *United States Department of Agriculture Miscellaneous Publications*, No. 237. U.S. Government Printing Office, Washington, DC.

Yetman, David, and Richard S. Felger. 2002. Ethnoflora of the Guarijíos. Pp. 174–230. In: Yetman, David. 2002. *The Guarijíos of the Sierra Madre: Hidden People of Northwestern Mexico*. University of New Mexico, Albuquerque.

Yetman, David, and Thomas R. Van Devender. 2001. *Mayo Ethnobotany: Land, History, and Traditional Knowledge in Northwest Mexico*. University of California Press, Berkeley.

Yoshihiro, Ohnishi, Nobuhiko Suzuki, Noboru Katayama, and Shin Teranishi. 2005. Seasonally different modes of seed dispersal in the prostrate annual, *Chamaesyce maculata* (L.) Small (Euphorbiaceae), with multiple overlapping generations. *Ecological Research* 20: 425–432.

Young, Robert W., and William Morgan, Sr. 1980. *The Navajo Language: A Grammar and Colloquial Dictionary*. University of New Mexico Press, Albuquerque.

Youngblood, Andrew. n.d. *Frangula* P. Mill. http://www.nsl.fs.fed.us/wpsm/Frangula.pdf. Accessed March 27, 2006.

Yuncker, Truman G. 1921. Revision of the North American and West Indian species of *Cuscuta*. *University of Illinois Biological Monographs*, 6 (2–3): 1–142 and 113 plates.

Zhou, Guang-Xiong, E. M. Kithsiri Wijeratne, Donna Bigelow, Leland S. Pierson, III, Hans D. Van Etten, and A. A. Leslie Gunatilaka. 2004. Aspochalasins I, J, and K: 7 from the rhizosphere of *Ericameria laricifolia* of the Sonoran Desert. *Journal of Natural Products* 67 (3): 328–332.

Zigmond, Maurice L. 1981. *Kawaiisu Ethnobotany*. University of Utah Press, Salt Lake City.

Zohary, Daniel, and Maria Hopf. 1993. *Domestication of Plants in the Old World*. Clarendon Press, Oxford.

Zuloaga, Fernando O., Osvaldo Morrone, Gerrit Davidse, Tarciso S. Filgueiras, Paul M. Peterson, Robert J. Soreng, and Emmet J. Judziewicz. 2003. Catalogue of New World Grasses (Poaceae): III. Subfamilies Panicoideae, Aristoideae, Arundinoideae, and Danthonioideae. *Contributions from the U.S. National Herbarium*, Washington, DC.

# INDEX TO NAMES

The binomials and author citations are those used on the International Plant Name Index (http://www.ipni.org/index.html) in comparison with the SEINet (http://seinet.asu.edu/collections/selection.jsp).

## ABOUT THE AUTHOR

Daniel F. Austin is a Research Associate at the Arizona-Sonora Desert Museum, Tucson, and of the Fairchild Tropical Garden, Miami, Florida. He is also Adjunct Professor, Department of Plant Sciences, University of Arizona, Tucson; Research Associate at the Department of Biological Sciences, Florida International University, Miami; Research Associate at the Desert Botanical Garden, Phoenix; and Emeritus Professor at Florida Atlantic University, Boca Raton. He attended graduate school at Washington University and Missouri Botanical Garden, St. Louis, Missouri. He received his M.A. in 1969 and his Ph.D. in 1970, and then spent 31 years teaching in the Florida State University System. He has received four awards from the Florida Native Plant Society, the Outstanding Member Award from the Florida Exotic Pest Plant Council, and was named Environmental Champion by NatureScaping at the Earth Day Celebration in 2001. He is technical advisor to several organizations, and is a Fellow of the Linnean Society of London. He has 324 publications, including eight books and 48 book chapters. His current research focuses on the ethnobotany of plant species shared between the Caribbean and northwestern Mexico and adjacent United States, and on the evolution of the Convolvulaceae.